PLANETARY RINGS

CONTENTS

COLLABORATING AUTHORS

P. Bodenheimer, *447*

A. Boischot, *73*

N. Borderies, *713*

A. Brahic, *3*

J. A. Burns, *200*

C. R. Chapman, *367*

S. A. Collins, *737*

J. N. Cuzzi, *73*

D. R. Davis, *367*

J. Diner, *737*

S. F. Dermott, *589*

R. H. Durisen, *416*

J. L. Elliot, *25*

L. W. Esposito, *73*

F. Franklin, *562*

G. W. Garneau, *737*

P. Goldreich, *713*

R. Greenberg, *3, 367*

E. Grün, *275*

A. W. Harris, *641*

J. B. Holberg, *73, 737*

A. L. Lane, *737*

M. Lecar, *562*

D. N. C. Lin, *447*

J. J. Lissauer, *73*

E. A. Marouf, *73*

D. A. Mendis, *275*

F. Mignard, *333*

E. D. Miner, *737*

G. E. Morfill, *200, 275*

P. D. Nicholson, *25*

M. R. Showalter, *200*

F. H. Shu, *513*

B. A. Smith, *704, 737*

G. R. Stewart, *447*

E. C. Stone, *687*

S. P. Synnott, *737*

R. J. Terrile, *737*

S. Tremaine, *713*

G. L. Tyler, *73, 737*

A. Van Helden, *12*

W. R. Ward, *660*

S. J. Weidenschilling, *367*

W. Wiesel, *562*

PREFACE

The human conception of planetary rings has undergone a revolution in recent years: Saturn's rings have changed from fuzzy ovals to finely detailed structures; the rings of Jupiter and Uranus have progressed from mere hypotheses to well-measured entities. Interpretation of the properties of rings is teaching us about the processes that shape them, processes that may have ramifications for other physical systems, such as galaxies or the nebulae from which planets form.

A major motivation and purpose for this book was to allow scientists, who have come from various disciplines to the study of planetary rings, to understand the work their colleagues from other backgrounds have accomplished. This objective required pedagogically clear chapters, leading up to explaining the very latest work on the subject. Chapters written in this way should have much broader use. Hence, we have edited this book to serve also as a textbook for graduate students and researchers in related fields. It is designed as an introduction for newcomers to the subject, leading them right up to the issues at the forefront of ring research. In that way, the book may play an important role in bringing about the resolution of outstanding problems and improving our understanding of what rings are, how they formed, and why they are the way they are.

As editors, we like to think of ourselves as representing the European-American alliance in the study of planetary rings. There is on the European side the long, continuing history of discovery and advances in the study of Saturn's rings, while American space exploration and analysis has clearly played a major role as well. More practically, our backgrounds help bridge an important interdisciplinary gap. Brahic came to the study of collisional mechanics and its effects on ring morphology from a background in galactic and stellar dynamics. Greenberg's past work involved solar system celestial mechanics and the origin and evolution of the solar system.

This book is part of the Space Science Series of the University of Arizona Press. T. Gehrels, who conceived the Series and established its style by editing several landmark volumes, contributed directly to this volume in his role as Series Consultant. M. S. Matthews, our editorial assistant, performed once again with the energy and skill that is by now legendary.

Like the other volumes in the Series, preparation of *Planetary Rings* was associated with an international meeting on the subject. That meeting was held in Toulouse, France, in September 1982. A separate book of proceedings of that conference, edited by Brahic, is being published by CNES, the French space agency. It represents a sort of "unretouched photograph" of the meeting, complete with all contributions and transcribed discussions. In contrast, the present volume consists of a selected set of review chapters based on reflection after the meeting, carefully reviewed and refined.

Financial support for preparation of this book came primarily from the Planetary Astronomy and Voyager Programs at NASA Headquarters, from the Galileo Project at the Jet Propulsion Laboratory, and from the University of Arizona Press. Additional support was provided by CNES, l'Observatoire de Paris, and the Planetary Science Institute of Science Applications, Inc. Research grants to our authors made their contributions possible; those grants are listed in the special *Acknowledgments* section at the end of the book.

We are honored to have had the opportunity to play a role in producing this book, but the credit for the high quality of its contents goes to the authors of the chapters. We simply served as shepherd satellites, keeping our herd in line and trying to prevent a phase lag. In the latter, we were not successful. In most cases, deadlines were missed, though not by much and for the best of reasons. We kept hearing the same excuse: "We have some new ideas that we want to develop for our own chapter." And in each case it was true. We were rewarded by receiving manuscript after manuscript with the latest ideas of the experts in this lively field. It is difficult to express adequately our gratitude and our affection to this delightful group of friends to whom we owe so much.

It is also a pleasure to acknowledge the contributions made by our scientific advisors: the members of the formal advisory committee for this book and the individuals who served as referees for the manuscripts. These people and others who helped in various ways are listed in the *Acknowledgments* section. Many of the referees were also authors. That was one way to keep the creative pot stirred.

It is our hope that this book will become out-of-date quickly, that new observations and theoretical connections will continue to revolutionize our knowledge of planetary rings. But we think the publishers can rest easy. We envision a more reflective period over the next several years in which the recent results are analyzed and digested, and we hope that this book will play a role.

Richard Greenberg
André Brahic

A Note on Notation. Every scientific field adopts its own set of notation, roughly optimized to its needs. In an interdisciplinary subject like planetary rings, it is inevitable that certain symbols will be firmly attached to more than one meaning. We have decided that imposition of an artificial, uniform notation throughout this book would only be confusing and would require an inordinate amount of uncreative work. Instead, we have tried to ensure that within each chapter the notation is internally consistent and well defined. In only a few cases have we encouraged limited redefinition of notation where we felt it could help clarify the text.

R.G. and A.B.

PART I

Introduction

PLANETARY RINGS: A DYNAMIC, EVOLVING SUBJECT

RICHARD GREENBERG
Planetary Science Institute

and

ANDRÉ BRAHIC
Observatoire de Paris

The study of planetary rings, like the rings themselves, is dynamic and evolving. Historically, understanding of the nature of rings has involved an interplay between tantalizing observations and conceptual model building. After the recent deluge of new data and images, theoretical studies based on a wide range of scientific disciplines are directed toward interpreting the observations and deriving their implications. In reviewing the motivations, historical advances, on-going methods, and goals of ring studies, this introduction provides the philosophical context for the chapters that follow.

Planetary rings have long captured human imagination. Their beauty makes Saturn's rings a prime objective for any observer with even a simple beginner's refracting telescope. Along with crescent moons and spiral galaxies, rings have served as a classical artistic motif, universally recognized as symbolic of the mystical wonders of the heavens and the brave future of science fiction. This book is about the mystery and the future. Thanks to the expertise and creativity of our authors, it is a compilation of practically all that is known about planetary rings. Yet, a pervasive theme is what we still do not know.

Despite the flood of new information on morphology and optical properties, we have very little direct evidence about what rings are, how they formed, or how they behave. We can only answer such questions by building theoretical models and comparing their implications with past and future observations. The interdisciplinary theories presented in this book represent a first generation of such models. This comprehensive, insightful work represents the best efforts of experts attracted from a variety of fields. But the most exciting aspect of this book, for us, is that these theories are nevertheless full of mutual contradictions and unanswered questions. This condition indicates a scientific inquiry in its most vital stage. By bringing present knowledge and ideas together and conveying them to a wider range of researchers and students, this book may play a significant role in the future resolution of the nature of planetary rings.

In this introduction, we review the history and the status of the subject in order to show the role this book is intended to play. We also describe the methods and policies we have adopted with that role in mind, and the rationale for topic selection and ordering of the chapters that follow.

Observations and Models: A Brief History

The interplay between observations and the complex model building required to interpret the data has pervaded the history of ring studies. The effort has always attracted outstanding scientific minds. From the beginning, observational resolution seemed to be just short of revealing the essential nature of rings. Galileo's first detection of something strange around Saturn was open to several interpretations: Was Saturn grossly oblate; were there two giant satellites; or did the planet have handles? Such alternative models failed to fit all the data. Even with all the observed apparent configurations catalogued by Gassendi, the true three-dimensional geometry was not obvious. Huygens' revelation of a disk-like ring structure was not the result of better observation, but rather of successful theoretical modeling (see Van Helden's chapter). But was the ring a solid chunk of material? Modeling was needed again. J. D. Cassini, whose discovery of a gap showed that the structure was not a monolith, suggested that the rings might consist of a myriad of small particles. Laplace and Maxwell showed that in fact a solid ring would be unstable. Only much later did observation confirm the theory, when Keeler spectrographically measured the Doppler shift, and hence the Keplerian orbital velocities, of the constituent particles. The rings must be a collection of tiny satellites.

Improved groundbased observations gradually revealed structural and optical details: the rings are flat but spread out; some light can pass through them; there is structural variation with radius; reflectivity depends on wavelength and viewing geometry. These observations were interpreted with models based on a swarm of small bodies, but results were ambiguous. The dynamical behavior of a colliding and self-gravitating swarm was not easy to model, nor were the optical properties, dependent as they are on shape, sizes, and material. Nevertheless, beginning with Poincaré, a general picture of collisional flattening and spreading emerged, with structure governed in part by resonances with the satellites. Dynamical theory was consistent with Earth-based observations of seemingly smooth, continuous rings. Optical and radio properties seemed consistent with a swarm of small, icy particles of various sizes. Theoretical models seemed adequately consistent with most observed properties.

Then came the deluge! In a golden half-decade our conception of rings underwent a revolution. We learned that rather than being smooth, continuous structures, rings are better characterized as sets of narrow ringlets with sharp edges, sometimes slightly elliptical or kinky in form. In March 1977, nine narrow, black rings were discovered around Uranus as they occulted a star observed from Earth. The pole-on aspect of Uranus allowed precise measurement of the rings during that and following occultations. In March and July 1979, very diffuse, thick rings were discovered around Jupiter by the Voyager spacecraft. Then three probes flew by the rings of Saturn (Pioneer 11 in September 1979, Voyager 1 in November 1980, and Voyager 2 in August 1981), revealing countless detailed features and structures that had never been imagined. During this same period, the Earth passed through Saturn's equatorial plane, allowing critical measurements of the rings' vertical and horizontal extent.

This book is about the observations of that golden period and the models they provoked, models that are necessary because, as usual, the observations once again seem to fall just short of revealing the essential nature of rings.

The Lure of the Rings

What is it about rings that has made them such an attractive scientific subject? We cannot deny the importance of aesthetic appeal: planetary rings are beautiful! They provide a sort of perceptual three-dimensionality that is missing from most celestial objects. The same strong appeal to human sensitivity that is recognized by commercial advertisers when they use pictures of planetary rings to sell merchandise also works on scientists.

Of course, there are strong, rational reasons for studying rings as well. They represent a fundamental class of planetary structure, and hence should be investigated just as are other properties of the solar system and of nature in general. As Burns et al. point out in their chapter, ''The amount of understanding to be gleaned from a ring system is not proportional to the system's mass.'' Burns also noted that with rings, as with perfume, even a small amount of substance can be very revealing.

We cannot ask, ''Why is there a solar system?'' without adding, ''Why are there rings around planets?'' Roughly speaking, the answer seems easy: close to a planet, tidal stresses are great. They might tend to break up any satellite or, alternatively, to prevent accretion of small particles into satellites in the first place. However, more quantitative consideration shows that real material may well be stable and even accrete within the known ring systems.

The correct answer is more subtle and still far from fully understood. This issue is an underlying theme throughout the book.

There are other issues related to the question of why rings exist. Only a few years ago, scientists were grappling with the question, "Why does only Saturn have rings?" That question is clearly obsolete now. Instead we ask, "Why are the ring systems of the planets so different from one another? Do they have anything in common? Why are there no rings around the terrestrial planets? Does Neptune have rings?" The last question at least is relatively straightforward: The negative results of the June 1983 occultation showed that Neptune cannot have much of a ring system; future observations may tell us more. The other questions are theoretical, and the ways to find the answers are not so clear. But in the mass of recent data, there must be clues.

Besides being fascinating subjects in their own right, ring systems provide an accessible laboratory for studying the sorts of dynamical processes that took place in the distant past, as planets (and satellites) were forming, and that occur in distant places, such as galaxies and accretion disks. The small planetesimals that coalesced into planets probably once formed an extended disk around the Sun, with its structure and evolution governed by mutual collisions and gravitational processes, in a way that may have some analogies to the way ring particles interact to govern ring structure. In this way, ring studies may be crucial to one of the fundamental objectives of planetary science: to understand the origin of the solar system. It seems inevitable that insights gained in the study of planetary rings will be applicable to any flat rotating disk in astronomy.

The Interdisciplinary Approach

In turn, the principles and techniques of several fields of astronomy are needed for the modeling of the structure, behavior, and evolution of planetary rings. For example, the scientific challenge presented by the data of the "golden years" has attracted the attention of a group of astronomers whose past work has generally not focused on the planets, but rather on the dynamics of galaxies and stellar accretion disks. The dynamical similarities between those systems and rings are extensive, and it was quite natural for much of the detailed work on ring-system dynamics to be based on the mathematical techniques of galactic dynamics.

A prescient connection between the study of rings and of galaxies was made by Maxwell in his seminal Adam's Prize essay of 1856:

> "I am not aware that any practical use has been made of Saturn's Rings, either in Astronomy or in Navigation.... But when we contemplate the Rings from a purely scientific point of view, they become the most remarkable bodies in the heavens, except, perhaps, those still less *useful* bodies—the spiral nebulae."

We now know that the connection between rings and galaxies is much stronger than simply a common degree of practical uselessness.

Although a few dynamicists are at home in both the galactic *and* planetary fields (several are authors in this book), much of this dynamical work has been foreign to members of the planetary community. The analyses are based on unfamiliar mathematical techniques; but, what is a more significant problem, they are based on a suite of implicit assumptions and approximations that have been incorporated into galactic dynamics. For example, it is not immediately clear how well the fluid mechanical treatment of galactic disks, which models particle interactions dominated by distant gravitational encounters, really can represent the dissipative interactions of ring particles colliding with one another and perturbing one another in the context of Keplerian motion. The problem of interdisciplinary communication is further compounded by the fact that many planetary workers have only recently begun to work on rings themselves, applying their expertise from relevant subjects such as celestial mechanics, the geology of small bodies, atmospheric optics, and instrumentation.

Researchers from these various fields need to understand what one another's disciplines tell us about rings. For example, just as the critical work in galactic dynamics is unfamiliar to planetary scientists, many galactic researchers are still uninformed about the physical properties of the planetary ring systems to which they are applying their analytical methods. As a case in point, a recent preprint describing a dynamical analysis began by describing Saturn's rings as comprising many small *rocky* particles. Legalistically, one might argue that ice is a kind of rock, but more realistically that slip reflects the uncoupled nature of past research. Dynamical properties (e.g., morphology, evolution) and physical properties (e.g., particle sizes, composition) have been treated largely independently.

That uncoupled approach was quite reasonable considering the freshness of the data and the need to simplify things for a first-cut interpretation. However, as illustrated repeatedly throughout this book, ring studies have now reached a level of maturity where we can no longer afford to treat the dynamics independently from the physical properties. The two are intimately related, and many of the unsolved problems depend on understanding those relationships. Throughout the book, we see the beginning of probes into the Terra Incognita of the interface.

A major objective of producing this book at this critical time is to present the results of theoretical dynamical studies in a manner that will be clear to all planetary scientists (not only dynamicists), and also to provide a sourcebook for relevant observational results on ring structure and physical properties as needed for meaningful dynamical interpretation. Such cross-fertilization of scientific ideas will certainly produce advances that might otherwise be impossible. Moreover, because each chapter is written in a pedagogical style designed to bring the reader from a state of relative ignorance in the particular subject up through the latest ideas, the book is naturally appropriate as a textbook for graduate students in astronomy and planetary science.

Arrangement of Contents

We have arranged the chapters in a way that we hope brings out the interdisciplinary relationships and exposes the remaining uncertainties. As a prologue, the historical chapter by Van Helden shows how the interplay between limited observational data and conceptual model building has been important from the beginning of the study of rings. He also makes it clear that awareness of the connection between the physics of rings and cosmogony is at least a couple of hundred years old.

The next two sections, the core of the book, are orthogonal to one another in subject matter and thus provide a sort of matrix coverage of the state of knowledge and understanding of rings after the "golden half-decade" of exploration. The section *Ring Systems* comprises three chapters, one about each of the known systems. These chapters provide definitive descriptions of each of the observed systems in the context of the broad range of dynamical processes that affect it. The next section comprises eight chapters on the various *Dynamical Processes*. Each of these chapters concentrates on a category of process with motivation from, and application to, all three ring systems.

Next we have two important chapters on *Origins*: one on the origin of ring systems, and the other treating the common processes in present rings and the past disk of planetesimals. Finally, we have three chapters that deal explicitly with the *Future* of ring studies: exploration by deep space probes to the planets, observation from the Earth or Earth orbit, and some critical unsolved theoretical problems.

Built into this matrix scheme with its orthogonal treatments are several layers of redundancy. The *Ring Systems* and *Origins* chapters deal with the relevant processes involved in shaping the observed structures. Thus, they cross the subjects of the *Dynamical Processes* chapters, which describe aspects of all the rings. Moreover, within each group of chapters there is considerable overlap. For example, the chapter on Jupiter's rings includes discussion of the similar ethereal rings of Saturn also discussed in the Saturn rings chapter. Resonances are discussed not only in the special chapter devoted to the subject, but naturally in the chapter on gravitational waves which are driven by resonances, as well as in nearly every other chapter. Likewise, the nature of interparticle collisions is described in the chapter on particle properties, the chapter on viscous instabilities, the chapter on the origin and evolution of the rings, and the chapter on collisional transport. These are only a few examples.

In a textbook designed to organize and teach a long well-understood subject, such redundancy would be inexcusable. Here it is essential. Our discipline is as dynamic and evolving as the rings themselves seem to be; any process, any interpretation is controversial. By having our authors present overlapping discussions from various points of view we emphasize the variety of interpretations and models under consideration. Our book would be misleading and incomplete if we presented only a single "party line." In fact, we would not know what party to join. There is an important message here for the serious reader. Read with a healthy skepticism and be alert for disagreements—that is where the action is.

The Roots of Conflict

Why is the nature of planetary rings so controversial, despite all the new data? One reason is that we have never seen a single ring particle. We have measured detailed optical properties and spatial structure of the ensembles of particles, and we know they must depend on the behavior of the individual particles themselves. But what is a particle really like?

The issue can be sidestepped to a certain degree by treating rings as continuous media, but implicit in all such models is some sort of idealization of the nature of the particles. For example, many influential theories (both optical and dynamical) have assumed that all particles are the same size. This approach led to scattering models for Earth-based observations of Saturn's rings that implied dominant particle size seemingly dependent on the wavelength of the observation. Thus, in the late 1970s, it was already realized that there must be some wide distribution of particle sizes. Voyager radio results have been interpreted in that context to give us some important constraints on the distributions. But it still remains to construct a full optical scattering model, consistent with Voyager results, to explain optical phenomena such as the opposition effect.

Like the earlier optical models, dynamical models also tend to assume simplified single-sized, smooth, round constituent particles. Though we have tried to ensure that all assumptions are explicit in this book, they are not always so in the literature. One sign of the single-size assumption is when a dynamical discussion uses the expression "optical thickness" (see Glossary) to mean "surface mass density." If all particles are the same and the optical thickness is moderate, that identification is reasonable. But a dynamically insignificant (i.e., low mass) amount of small particles may, and indeed probably does, dominate optical thickness. The reader must always bear in mind that we have never seen a ring particle and that the particles we do see as an optical ensemble probably contain a negligible fraction of the material.

A tendency to confuse the particles that dominate optical properties (the small ones, which have most of the areal cross section of the system) and the particles that dominate dynamics (the big ones, which contain most of the mass) has also muddled the monolayer issue. Photometric characteristics, notably the opposition effect, require Saturn's rings to be many particles thick, while collisions may well tend to produce a monolayer. These requirements are not inconsistent; most likely, the big particles form a monolayer, and the small ones form a swarm comparable in thickness to the scale of the big ones. Nevertheless, in most dynamical theories, ice particles in Saturn's rings are assumed to bounce off one another like billiard balls, in order to avoid a monolayer. Yet other collisional outcomes are possible: some workers envision gradual erosion of particles' surfaces, while others believe the particles must be ephemeral snow drifts. Do dynamical results on the system's structure depend strongly on such implicit assumptions? It is often hard to tell.

This latter problem illustrates a fundamental philosophical question about the process of conceptual model building. In order to perform a mathematical study that will reproduce an observed dynamical feature or optical property, one needs to make a model simple enough for its analysis to be tractable. Moreover, a simple elegant model can give much more insight into the relevant physics governing a problem than can a complicated one in which specific causes and effects are harder to sort out. The theorist is thus motivated to simplify his model, perhaps even with an awareness that some of the essential physics is neglected. In the latter case, the goal is simply to develop some insight into the problem rather than to reproduce precisely the mechanics of the real system. If the model does reproduce an observed property well, it suggests that the questionable assumptions may be correct after all, or at least that they are not critical. People tend to accept the model and hence the assumptions. This tendency is strong for the modeler himself who by now has made an investment of time and reputation in his work.

On the other hand, some researchers may have serious doubts about the assumptions. In an interdisciplinary field like planetary rings, what seem like insignificant assumptions to one model builder might loom very important (and very wrong) to specialists in a different discipline. Is it a legitimate scientific exercise to question such assumptions? One point of view (expressed to us in the course of preparing this book) is that the skeptic should withhold comment until he has constructed an alternative model based on the improved assumptions. We disagree. The skeptic has two unfair handicaps against producing an alternative model. First, it may be more difficult. Remember, the original assumptions were made in part to simplify the analysis. And second, the skeptic is likely to be an expert in a somewhat different discipline than the original model builder. He may know the assumptions related to his own field are wrong without knowing how to follow through to their implications in another field.

We think that reservations about proposed models must be aired. We have encouraged each of our authors to make his own assumptions explicit and to be critical of other models. Even if such criticism is not always immediately constructive in the sense of offering alternatives, it will discourage complacency and serve as a challenge to make things even better.

We conclude this introduction with a short list of outstanding problems that we subjectively identify as most pressing: Why are the rings of Jupiter, Saturn, Uranus, and Neptune so different from one another? What is their origin? Why are Saturn's A, B, and C Rings so different from one another? What is the time scale for radial spreading? Why have not Saturn's rings spread much farther than they have? What confines and shapes the various types of narrow structure? How can we explain differences in particle color and albedo among ring systems and within ring systems? Is the size distribution proposed for Saturn's rings consistent with photometric properties? Why is the surface density of Saturn's rings nearly constant no matter where or how measured? How important are nonlinear effects in governing wave mechanisms? Each of these questions with tentative answers arises repeatedly in this book, but the solutions are left as an exercise for the reader.

Despite Maxwell's concerns, quoted earlier, about practical applications, he went on to recognize other motivations for the work:

"When we have actually seen that great arch swung over the equator of the planet without any visible connexion, we cannot bring our minds to rest. We cannot simply admit that such is the case, and describe it as one of the observed facts in nature, not admitting or requiring explanation. We must either explain its motion on the principles of mechanics, or admit that, in the Saturnian realms, there can be motion regulated by laws which we are unable to explain."

Although in this introduction we have emphasized remaining uncertainties and controversies, the book provides definitive compilation of what we do know. We invite you to study these chapters and participate in the adventure of discovery that is still to come.

RINGS IN ASTRONOMY AND COSMOLOGY, 1600–1900

ALBERT VAN HELDEN
Rice University

Rings and disks became a part of astronomical thought only in the seventeenth century. In the 1640s Descartes put forward a cosmology in which the universe was filled with vortices of invisible matter, and in 1659 Huygens solved the problem of Saturn's telescopic appearances by postulating that the planet is surrounded by a ring. Since that time rings and disks have been a feature of astronomy and cosmology. As Cartesian vortices fell out of favor, in the middle of the eighteenth century, Kant and Lambert put forward cosmologies in which the galaxies were huge disks of rotating matter. For most of the nineteenth century Laplace's nebular hypothesis was the dominant cosmogony, and the increasing number of asteroids as well as the particle nature of Saturn's rings provided important supporting evidence for Laplace's theory. Although around 1900 the nebular hypothesis lost a great deal of its earlier appeal, rings of dispersed matter have continued to play an important role in theories accounting for the origin of the solar system.

Although the presence of ring and disk phenomena, such as planetary rings, asteroid belts, the solar system itself, and the Milky Way, may strike us as rather obvious, this has not always been the case. The astronomy of ancient Mesopotamia, as far as we know it, had no particular geometrical model at all, while the geometric theories of the Greeks and their intellectual heirs up to Copernicus were conceptually spherical. Not until several generations after the publication of Copernicus' *De Revolutionibus* (1543), with the elliptical astronomy of Kepler (1609), did astronomers begin to think of planetary paths as two-dimensional curves in three-dimensional space, and of the solar system as a disk of objects moving in roughly coplanar orbits. The geometry of the Milky Way had to wait for another century and a half. Rings consisting of rotating matter entered astronomy conceptually with the invisible vortices of Descartes (1644), and an actual material ring in the heavens was first postulated by Huygens twelve years later.

I. RINGS ENTER ASTRONOMY

In his ambitious attempt to create an entirely new system of natural philosophy, René Descartes (1596–1650) introduced a fundamental distinction between mind or thinking substance and body or extended substance. Science dealt only with the latter, and Descartes gave it only one property, extension. All other properties, including weight, were secondary and had to be explained. The main part of this explanation was the vortex, a disk of rotating matter, mostly invisible, that completely filled the space containing it. There was no void or vacuum. At the creation, God had put all extended substance in the universe in motion in vortices that preserved the total amount of motion. Our solar system was but one in an indefinite series of contiguous vortices with other suns as centers. The planets moving in the solar vortex, were themselves the centers of smaller vortices. Thus, the Earth had a vortex extending to and including the Moon, and Jupiter was the center of a vortex which accounted for the motions of the four Galilean satellites (Descartes 1644, pp.80–202).

The greatest strength of the vortex concept was that it could give a plausible, causal, mechanistic account of natural phenomena such as falling bodies and the nature and transmission of light. Its greatest weakness was that relationships derived directly from the phenomena, such as Kepler's Third Law, could not be mathematically deduced from it. It was this weakness that caused Cartesian physics ultimately to be eclipsed by Newtonian physics, in England by 1700, and in France half a century later. The vortex as a concept has, however, been invoked in science on a number of occasions since 1750.

At a very early age, Christiaan Huygens (1629–1695) became a follower of Descartes, who was a frequent guest of his father's. Huygens learned to think of the world in terms of Cartesian vortices, and when he discovered a satellite of Saturn in 1655 (Huygens 1656), the first since Galileo's discovery of Jupiter's satellites in 1610, he knew that Saturn's vortex had to extend from the planet at least to this satellite. For nearly half a century astronomers had been puzzled by the strange appearances of Saturn itself, and, partly because of his Cartesian convictions, Huygens was now able to solve this puzzle.

Since the seventeenth century it has been supposed by many that the problem of Saturn was simply a problem of telescopes, a supposition actively encouraged by Huygens himself. Because he discovered Saturn's satellite and the true nature of the planet's appearances, he argued that his telescopes must have been better than those of his contemporaries (Huygens 1659, p.270). In fact, Huygens was a neophyte at this game: the instrument in question was the first ambitious telescope made by him, and all indications are that among the research telescopes of that day it was by no means unusual in its magnifying power or optical quality (Van Helden 1970). Moreover, Huygens found his solution to the puzzle in the winter of 1655/6 when the ring was edge-on and therefore invisible (Van Helden 1974b, p.160). He obviously could not have solved the problem by simply *seeing* a ring. His telescope did have to be good enough to give him certain clues, but a number of his colleagues also had instruments of that quality (Van Helden 1970).

Huygens compared the vortices of the Earth and Saturn. Our Moon takes about 30 days to complete one circuit about the Earth, and the Earth itself turns on roughly the same axis in a day. He had measured the orbital period of Saturn's moon to be about 16 days, and he concluded, therefore, that the planet itself turns on the axis of this orbit in about half a day. Furthermore, all the matter between Saturn and its moon, including the ansae, had to rotate in Saturn's vortex with periods intermediate between 16 days and half a day. Although the ansae showed slow changes of shape over a period of years, they did not show any changes over a period of less than sixteen days. Only a body symmetrical about the axis of rotation of the vortex and planet could satisfy those conditions, and the solution to the problem was, therefore, that Saturn was surrounded by a ring (Huygens 1659, pp.294–296).

Three years after arriving at his solution, Huygens (1659) finally published his *Systema Saturnium*. What he put forward here was, however, not Saturn's ring as we know it. His ring was a relatively thick, solid structure (Fig. 1). Huygens had been one of the first to observe the shadow of the ring on the body (Huygens 1659, pp.239–247), but he interpreted it to be the dark exterior edge of a solid ring of perceptible thickness. He argued that this edge was dark either because it did not reflect light at all or because it was so smooth that it reflected the Sun's light to the Earth only at one point (Huygens 1659, pp.318–320).

What kept this huge, solid structure balanced around Saturn was its symmetrical placement and its rotation. For Huygens such a structure presented few problems since it was imbedded in, indeed part of, Saturn's vortex. For others, less committed to the Cartesian world view, the solid ring presented problems, as indeed any nonspherical structure in the heavens would. It had been an axiom since Greek antiquity that the proper shape for celestial bodies was spherical, and scientists had to become accustomed to another shape in the heavens. There were thus two questions: could Huygens' ring account for all the different appearances of Saturn seen since 1610, and could such a shape exist in the heavens?

Between 1610 when Galileo first directed a telescope at Saturn and 1659 when Huygens published *Systema Saturnium*, many conflicting and even mutually exclusive observations of the planet's strange shapes had found their way into print. It was, for instance, difficult to see how the little unattached globes seen by many observers to flank Saturn from time to time could grow

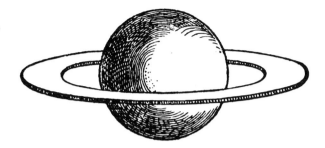

Fig. 1. Saturn surrounded by a relatively thick ring, as Huygens depicted it in *Systema Saturnium (Oeuvres Complètes de Christiaan Huygens*, XV :299).

into the large attached handles of various shapes seen by the same observers at different times (Van Helden 1974*a*). Huygens argued that since he had discovered the satellite, his telescopes were better than those of all others and this gave him the right to judge the merits of all observations since 1610 (Huygens 1659, p.270). He rejected many observations as due to optical illusions or inferior telescopes and explained those he retained by means of his ring theory (Huygens 1659, pp.270–284). Since others had, in fact, seen Titan before Huygens but had taken it to be a fixed star (Van Helden 1970, p.44), Huygens' claim was unfounded.

Needless to say, his approach offended some astronomers and telescope makers. One injured party was Eustachio Divini (1610–1685) in Rome, Europe's foremost telescope maker, mentioned several times by Huygens in *Systema Saturnium* (Huygens 1659, pp.278,310). Divini, an artisan without a university education, teamed up with the Jesuit Honoré Fabri (1607–1688), a member of the Inquisition and a very capable and versatile scientist. Fabri wished to attack Huygens for his overt Copernicanism and also to propose his own theory of Saturn's appearances. In 1660 a little book entitled *Brevis Annotatio in Systema Saturnium* appeared in Rome under Divini's name (Divini 1660); it was commonly known that Fabri was the instigator and main author (Guisony 1660, p.101).

Fabri supposed that besides the new moon recently discovered by Huygens, Saturn had four others. These satellites were much larger than Titan and their orbits and periods were very different. Two were light, two were dark, and all four revolved about points *behind* the plant (Fig 2) so that, with respect to the central Earth, they were always behind it. The various appearances that had puzzled astronomers for so long were caused by the different formations of these four satellites as they appeared, from the Earth, to flank the planet (Divini 1660, pp.422–437).

Huygens' *Systema Saturnium* had been dedicated to Prince Leopold de' Medici (1617–1675), brother of the Grand Duke of Tuscany, and founder of the Accademia del Cimento, a forerunner of modern scientific societies.

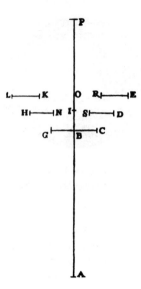

Fig. 2. Fabri's hypothesis put forward in *Brevis Annotatio*. *GC* represents Saturn; *HN* and *SD* are light absorbing satellites that move about point *I*; *LK* and *RE* are light reflecting satellites that move about point *O* (*Oeuvres Complètes de Christiaan Huygens*, XV:425).

Brevis Annotatio was written as an open letter to the prince, and it ended with a request that Leopold and his academy judge between the two theories (Divini 1660, p.436). Since it was commonly known that Fabri had written most of the book, Leopold was in the uncomfortable position of having to decide between an elegant theory put forward by a heretic and a Copernican, and a ridiculous theory put forward by a member of the Inquisition who insisted that the Earth was the center of the universe (Van Helden 1973, pp.242–243). In the intellectual climate in Italy following the trial of Galileo in 1633, this was no easy assignment.

The Prince solved the problem brilliantly. He had his academicians build models of both hypotheses and test them (Fig. 3). The models were placed in one of the long galleries of the Medici Palace, obliquely illuminated, and viewed from various distances with the naked eye and through low-powered spyglasses. There was even a control group of people untrained in astronomy who did not know what they were observing.

The results were devastating for Fabri's hypothesis. Huygens' ring was clearly superior in its ability to account for the various appearances, but it did not escape criticism. No matter how dark the academicians made the outside edge of the relatively thick ring, it never became entirely invisibile. They had to resort, therefore, to a very thin ring and even then it was difficult to make it disappear entirely in the edge-on position. Even the slightest unevenness or

Fig. 3. The model of Saturn and its ring used by the Accademia del Cimento to test Huygens'
hypothesis (*Oeuvres Complètes de Christiaan Huygens,* III:154–155).

roughness of the ring showed up as a bright spot when the eye was exactly in
the plane of the ring. This meant that, compared to its diameter, Saturn's ring
had to be extremely thin indeed for Huygens' theory to work (Van Helden
1973, pp.244–247).

This extreme thinness inevitably brought the ring's supposed solid construction into question. How could a solid structure, so vast and yet so thin,
persist? Although the science of mechanics was not equal to answering this
question formally, the members of the academy were uncomfortable with a
solid ring and concluded that it consisted either of a vapor or fluid, or a
collection of particles. One suggestion was that vapors rising from the planet
condensed into a ring of "very small stars of ice" (Van Helden 1973,
pp.250–253). Thus, from the very beginning, the notion of a solid ring was
questioned. And even before this report was written by the members of Prince
Leopold's academy, the French poet Jean Chapelain (1595–1674) had already
suggested to Huygens that the ring could consist of a large number of little
moons moving about Saturn (Chapelain 1660).

Although it is impossible to ascertain the opinions of all Huygens' scientific contemporaries (e.g., Newton) on the constitution of the ring, it appears that the most authoritative opinion was on the side of a satellite construction. Giovanni Domenico Cassini (1625–1712), his son Jacques Cassini (1677–1756), and Christopher Wren (1632–1723), are all on record as favoring this idea. Huygens himself, however, never abandoned his notion of a solid ring of substantial thickness (Huygens 1698, p.786).

II. SATURN'S RING AS A MODEL IN COSMOLOGY

Saturn's magnificent ring system was a wonderful celestial spectacle, a symbol of the young science of telescopic astronomy. For the time being, no formal dynamical problems were posed by either solid, fluid, or particle rings. That issue would only come to the fore towards the end of the eighteenth century. By the middle of the eighteenth century, however, Saturn's ring system had become important in cosmological and cosmogonical speculations. In 1750 Thomas Wright of Durham (1711–1786) published *An Original Theory or New Hypothesis of the Universe* (Wright 1750), in which he gave an explanation of the Milky Way. For theological reasons, Wright himself favored an arrangement of the fixed stars in a thin spherical shell which, he argued, was not inconsistent with the appearance of the Milky Way. He also mentioned, however, the notion of a flat disk of stars, revolving around a common center, like the rings of Saturn (Wright 1750, p.65):

> "Hence we may imagine some Creations of Stars may move in the Direction of perfect Spheres, all variously inclined, direct and retrograde; others again, as the Primary Planets do, in a general Zone or Zodiack, or more properly in the Manner of *Saturn's* Rings, nay, perhaps Ring within Ring, to a third or fourth order, ... nothing being more evident than that if all the Stars we see moved in one vast Ring, like those of *Saturn*, round any central Body, or Point, the general Phaenomena of our Stars would be solved by it; ... "

The young philosopher Immanuel Kant (1724–1804) took this ring or disk model from Wright and developed it to include the nebulae. Our Milky Way was, according to Kant, only one flat rotating system among many. These galaxies themselves might form a system which was, in turn, part of a super system, etc. Kant proposed a scientific cosmogony in which God had brought about the present condition of the universe by using natural forces. Matter that had been evenly distributed in the primitive chaos was concentrated in rotating nebulae by forces of attraction and repulsion, and as a result of this rotation, these nebulae flattened out into rotating disks which then separated into individual galaxies, stars, or planets (Kant 1755). Kant also gave an account of how Saturn's rings could have originated. When the planet had already formed and was cooling, vapors escaped from its surface and collected around it. Because they initially shared the planet's rotation, the vapor particles circulated around Saturn and concentrated near its equator as a ring (Kant 1755, II, Ch.7).

Just when the printing of Kant's *Universal Natural History,* in which these ideas were put forward, was complete, the publisher went bankrupt and the printed copies were impounded. Kant's cosmological ideas therefore remained unknown for years. Very similar notions were, however, independently put forward in 1761 by Johann Heinrich Lambert (1728–1777) in his very popular *Cosmological Letters* (Lambert 1761). It is, thus, fair to say that by the middle of the eighteenth century the ring or disk (which surely owed something to the Cartesian vortex) had become a model in cosmology and cosmogony.

The speculative cosmology of Kant and Lambert was supported by the painstaking work of William Herschel (1738–1822). With his powerful reflecting telescopes, Herschel mapped and "gauged" the Milky Way and even managed to locate the solar apex (Hoskin 1963); he also had definite and important ideas about the ring of Saturn. With him, the notion of a solid ring or rings, which had lived on in a number of books, took on new life. Early in his career, Herschel believed that the dark stripe discovered by Cassini in 1675 was a surface feature on the solid ring. When, after the edgewise appearance of 1789 he saw the stripe in the same position on the other face, he acknowledged it to be an actual division in the ring, and claimed that Saturn was surrounded by *two* solid rings (Herschel 1791). Herschel's prestigious opinion was one reason for the renewed popularity of solid rings in the first half of the nineteenth century; the other reason was the work of Laplace.

Dynamical studies of the ring system began in 1787, when Pierre Simon de Laplace (1749–1827) read a memoir on the stability of Saturn's rings before the French Academy. The conclusion of Laplace's mathematical argument was that the planet was surrounded by a large number of narrow solid rings whose centers of gravity did not coincide with their geometric centers (Laplace 1787). He repeated this argument in his tremendously influential *Exposition du Système du Monde* (Laplace 1796, Book IV, Ch.9) and *Mécanique Céleste* (Laplace 1799–1825, Book III, Ch.6).

Laplace was also the author of the so-called "nebular hypothesis" on the origin of the solar system. He argued that the solar system resulted from a spinning ball of nebular material, which, as it contracted, had thrown off successive rings of matter. These rings had aggregated and condensed into planets. Planets still in the nebular state had thrown off small rings of matter that gave rise to satellites. The system of Saturn presented the unique case in our solar system of nebular rings condensing into complete solid rings (Numbers 1977, pp.124–132;Jaki 1976).

III. ASTEROIDS, SATURN'S RING, AND THE NEBULAR HYPOTHESIS

Since Kepler, some astronomers had wondered about the large gap between Mars and Jupiter. In 1766 Johann Daniel Titius (1729–1796) put forward an *ad hoc* numerical series which yielded the heliocentric distances of the planets (Bonnet 1766, p.7):

$$A = 0.4 + 2^n \times 0.3 \tag{1}$$

where A is the heliocentric distance in astronomical units and $n = -\infty, 0, 1,$ .... This series served to formalize the vague idea that something was missing between Mars and Jupiter, for there was a planet at every term up to $n = 5$, except $n = 3$ ($A = 2.8$). When Herschel discovered a new planet (Uranus) beyond Saturn in 1781, its distance fit the term $n = 6$ fairly well, and this fact gave Titius' relationship almost the status of a law of nature. Astronomers now began to search for a body between Mars and Jupiter.

On 1 January 1801, Giuseppe Piazzi (1746–1801) discovered a small planet that turned out to have a period which put it in the gap between Mars and Jupiter. In fact, Ceres, as Piazzi named it, exactly fit the term for $n = 3$ in the Titius series. And, as it turned out, Ceres was only the beginning. In 1802 Wilhelm Olbers (1758–1840) discovered an asteroid (which he named Pallas) that had the same distance as Ceres. In 1804 K. L. Harding (1765–1834) found Juno and in 1807 Olbers found Vesta, both at the same distance from the Sun as the others (Clerke 1893, pp.87–94).

The hypothesis of an exploded planet, put forward by Olbers, seemed the best way to account for so many bodies moving at nearly the same distance from the Sun. Laplace, however, preferred to see these four bodies as an unfinished planet, evidence for his nebular hypothesis (Numbers 1977, pp.131–132). Olbers' explanation remained the more popular one for four decades, as no further asteroids were discovered. Starting in 1845, however, the number began to increase rapidly. By 1852, 20 asteroids had been discovered; by 1870, 110; and by 1900, 450 were known (Pannekoek 1961, pp.353–354). These bodies formed, as it were, a ring between Mars and Jupiter.

With so many bodies of different sizes scattered in a broad zone, Olbers' argument that they were the product of an exploded planet was deemed less and less likely. When, after 1850, calculations failed to show a common origin, the asteroid belt was considered an unfinished planet that could shed light on the origin of the solar system (Numbers 1977, pp.55–57).

In the meantime, the idea of solid rings around Saturn had been discredited. In 1848 the C Ring was discovered, and shortly afterwards it was found that the planet was visible through this ring. In the same year Édouard Roche (1820–1883) determined that within a certain limit from a planet, a fluid satellite (and presumably also a continuous ring) would be torn apart by tidal forces (see the chapter by Weidenschilling et al. in this book). Saturn's ring system lay within that limit (Darwin 1889). A series of dynamical studies followed, culminating in Maxwell's classic essay of 1857, after which no doubt remained that Saturn's rings could only consist of an indefinite number of very small particles (Alexander 1962, Ch.13). Moreover it was by no means the only instance of a ring made up of small particles. Richard A. Proctor (1837–1888) wrote (Proctor 1882, p. 135):

"An *a priori* argument in favour of such a supposition may be drawn from analogous instances in the solar system. In the zone of asteroids we have an undoubted instance of a flight of disconnected bodies travelling in a ring about a central attracting mass. The existence of zones of meteorites travelling around the sun has long been accepted as the only probable explanation of the periodicity of meteoric showers. Again, the singular phenomenon called the zodiacal light is, in all probability, caused by a ring of minute cosmical bodies surrounding the sun. In the Milky Way and in the ring-nebulae we have other illustrations of similar arrangements in nature, belonging, however, to orders immeasurably vaster than any within the solar system."

Just as the asteroid belt was now deemed to be an unfinished planet, Saturn's ring system was considered to be one or more unfinished satellites. Both were cited as evidence for the nebular hypothesis. Daniel Kirkwood (1814–1895) further strengthened the analogy by showing that in both systems the gaps occurred at distances associated with periods in resonance with the periods of larger bodies, e.g., Jupiter in the case of the asteroid belt and Mimas in the case of Saturn's rings (Kirkwood 1866,1872).

Thus, by the second half of the nineteenth century rings of disconnected particles had become important phenomena in astronomy and the young science of astrophysics, phenomena intimately related in the minds of scientists to the origin of the solar system. As the nebular hypothesis came under increasing fire, toward the end of the nineteenth century, capture and encounter hypotheses also became popular. In all cosmogonical theories entertained in the twentieth century, however, the asteroid belt and Saturn's ring system have occupied important positions. Any theory of the origin of the solar system has had to account not only for the planets and satellites, but also for the dispersed rings of matter surrounding the Sun, Saturn, and, lately, Jupiter and Uranus as well.

REFERENCES

Alexander, A. F. O'D. 1962. *The Planet Saturn: a History of Observation, Theory and Discovery* (London: Faber & Faber).

Bonnet, C. 1766. *Betrachtung über die Natur*, tr. J. D. Titius (Leipzig).

Chapelain, J. 1660. Letter to Huygens, 4 March 1660. In *Oeuvres Complètes de Christiaan Huygens*, vol. II (The Hague, 1890), pp.34–37.

Clerke, A. M. 1893. *A Popular History of Astronomy during the Nineteenth Century* (3d ed., London: Adam & Charles Black).

Copernicus, N. 1543. *De Revolutionibus Orbium Cœlestium Libri Sex* (Nuremberg).

Darwin, G. H. 1889. Saturn's Rings. *Harper's New Monthly Magazine* 79:66–76.

Descartes, R. 1644. Principia Philosophia. In *Oeuvres de Descartes*, eds. C. Adams and P. Tannery, vol. VIII (Paris, 1905), pp. 1–348. (The 1647 French edition in vol IX [Paris, 1904].)

Divini, E. 1660. Brevis Annotatio in Systema Saturnium Christiani Eugenii. In *Oeuvres Complètes de Christiaan Huygens,* vol. XV (The Hague, 1925), pp.403–437.

Guisony, P. 1660. Letter to C. Huygens, 1 August 1660. In *Oeuvres Complètes de Christiaan Huygens,* vol. III (The Hague, 1890), pp.101–104.

Herschel, W. 1791. On the ring of Saturn, and the rotation of the fifth satellite upon its axis. *Phil. Trans.* 82:1–22.

Hoskin, M. A. 1963. *William Herschel and the Construction of the Heavens* (London: Oldbourne).

Huygens, C. 1656. De Saturni Luna Observatio Nova. In *Oeuvres Complètes de Christiaan Huygens,* vol XV (The Hague, 1925), pp.172–177.

Huygens, C. 1659. Systema Saturnium. In *Oeuvres Complètes de Christiaan Huygens,* vol. XV (The Hague, 1925), pp.209–353.

Huygens, C. 1698. Kosmotheoros, sive de Terris Cœlestibus, earumque ornatu, Conjecturae. In *Oeuvres Complètes de Christiaan Huygens,* vol. XXI (The Hague, 1944), pp.677–821.

Jaki, S. J. 1976. The Five Forms of Laplace's Cosmogony. *Amer. J. Phys.* 44:4–11.

Kant, I. 1755. *Allgemeine Naturgeschichte und Theorie des Himmels* (Königberg and Leipzig). (The standard English translation is in *Kant's Cosmogony,* ed. tr. W. Hastie [Glasgow, 1900], pp.13–167.)

Kepler, J. 1609. *Astronomia Nova* (Heidelberg).

Kirkwood, D. 1866. On the Theory of Meteors. *Proc. AAAS 1866,* pp.8–14. (published 1867).

Kirkwood, D. 1872. On the formation and primitive structure of the solar system. *Proc. Amer. Phil. Soc.* 12:163–167.

Lambert, J. H. 1761. *Cosmologische Briefe über die Einrichtung des Weltbaues* (Augsburg); *Cosmological Letters on the Arrangement of the World Edifice,* ed. tr. S. L. Jaki (New York, 1976: Science History Publ.).

Laplace, P. S. de 1787. Mémoire sur la Théorie de l'Anneau de Saturne. *Mémoires de l'Académie Royale des Sciences de Paris 1787* (published 1789) 249 ff; *Oeuvres Complètes de Laplace,* 14 vols. (Paris, 1878–1912), XI:275–292.

Laplace, P. S. de 1796. *Exposition du Système du Monde* (Paris, 1796); *Oeuvres Complètes de Laplace,* 14 vols. (Paris, 1878–1912), VI.

Laplace, P. S. de 1799–1825. *Mécanique Céleste* (Paris, 1799–1825); *Oeuvres Complètes de Laplace,* 14 vols. (Paris, 1878–1912), I–V.

Numbers, R. L. 1977. *Creation by Natural Law: Laplace's Nebular Hypothesis in American Thought* (Seattle: Univ. of Washington Press).

Pannekoek, A. 1961. *A History of Astronomy* (New York: Barnes & Noble).

Proctor, R. A. 1882. *Saturn and its System* (2nd ed., London).

Van Helden, A. 1970. Eustachio Divini versus Christiaan Huygens: a reappraisal. *Physis* 12:38–50.

Van Helden, A. 1973. The Accademia del Cimento and Saturn's ring. *Physis* 15:237–259.

Van Helden, A. 1974a. Saturn and his anses. *J. Hist. Astron.* 5:105–121.

Van Helden, A. 1974b. 'Annulo Cingitur': the solution of the problem of Saturn. *J. Hist. Astron.* 5:155–174.

Wright, T. 1750. *An Original Theory or New Hypothesis of the Universe,* ed. M. A. Hoskin (London: MacDonald; New York: American Elsevier, 1971).

PART II

Ring Systems

THE RINGS OF URANUS

J. L. ELLIOT
Massachusetts Institute of Technology

and

P. D. NICHOLSON
Cornell University

Uranus is encircled by nine narrow rings, denoted 6, 5, 4, α, β, η, γ, δ and ϵ, in order of increasing distance from the planet. The rings have been observed at high spatial resolution (3.5 km) during 13 stellar occultations from 1977 to 1983. Sharp edges are characteristic of the ring structures, whose optical depths range from ~0.5 to values of ~1.5 and greater. The widths of most rings lie between 2 and 12 km, with two exceptions. The η Ring has a narrow core a few km wide adjacent to a broad component ($\tau \sim 0.1$) ~55 km wide. The width of the ϵ Ring varies, as a linear function of its radius, from 20 km at periapse to 96 km at apoapse. The widths of the α and β Rings also vary approximately linearly with their radii. A Keplerian model, including five orbital elements for each ring, has been successfully fitted to the occultation data. The semimajor axes are found to lie between 41877 km (Ring 6) and 51188 km (ϵ Ring). The largest eccentricity is 7.94×10^{-3} (ϵ Ring), while the γ and η Rings cannot be distinguished from circles ($e < 10^{-4}$). Seven of the nine rings are inclined to the planet's equatorial plane by a few hundredths of a degree. The orbit solution yields values for the Uranian gravitational harmonic coefficients $J_2 = (3.349 \pm 0.005) \times 10^{-3}$ and $J_4 = (-3.8 \pm 0.9) \times 10^{-5}$, which can be used to constrain interior models of Uranus. Infrared imaging observations confirm the derived precession rate for the ϵ Ring and place limits on the optical depth of inter-ring material. Spectra of the rings show their geometric albedo to lie between 0.02 and 0.03 in the wavelength range 0.89–3.9 μm, ruling out ice-covered particles. Several lines of evidence suggest that the "average" ring particles are larger than 1 μm. The main theoretical problems still posed by the Uranian rings concern their sharp edges and the internal structures exhibited by the α, η and ϵ Rings. New information about the rings will be provided by further groundbased observations, the Space Telescope and the Voyager encounter in January 1986.

In this chapter we survey current knowledge concerning the Uranian rings, and explain how this knowledge has been derived from the observations. For each ring system we have the ultimate goal of learning its origin and its evolution to its present state, as described by (i) the composition of the ring particles, (ii) the distribution function of particle sizes, and (iii) the dynamical processes that determine the particle orbits. Although we are far from this goal for the Uranian system, we do know a great deal, particularly about its kinematics. In fact, the discovery of the Uranian rings in 1977 marked the beginning of the current renaissance in studies of planetary rings, and the Uranian system has made some notable contributions to our studies of all ring systems. The phenomena first observed in the Uranian system are:

1. Narrow rings (inspiring confining satellite models);
2. Sharp edges;
3. Long-lived structure in the radial direction on kilometer scales (ϵ Ring);
4. Eccentric rings;
5. Inclined rings;
6. Uniform apsidal and nodal precession;
7. Adjacent broad and narrow ring components (η Ring);
8. Low albedo ring particles.

Most of the items in the above list were discovered through the high spatial resolution provided by stellar occultations, presently the principal method used to observed the Uranian rings. Observation of a stellar occultation provides a trace of the transmitted starlight as a function of position in the ring system. For groundbased observations of occultations at near-infrared wavelengths, the limit to the *spatial resolution* is ~ 3.5 km for the structural features in the rings and about 0.1 km in the *position* of a sharp feature, such as an edge. The 3.5-km resolution compares to a resolution of $\sim 10,000$ km achievable with groundbased imaging and is somewhat better than the resolution achieved by Voyager imaging and radio occultation observations of Saturn's rings. It is about one and a half orders of magnitude worse than the resolution of 0.1 km achieved by the Voyager photopolarimeter occultation observations at Saturn.

With these high-resolution occultation data, we have been able to construct a good kinematic model for the ring orbits and have begun to characterize the structure of individual rings. The picture that emerges from these studies (see Fig. 1) is of a system of nine, sharp-edged rings; they are relatively opaque, narrow ($\lesssim 2$ to 100 km) and separated by broad (300–2500 km) gaps. Reports of other material have been made, but not confirmed. Current upper limits on the optical depth of inter-ring material are ~ 0.05, based on occultation data, and ~ 0.003, based on images of the rings at 2.2 μm (see Sec. IV). At least six of the rings are measurably eccentric (i.e., $e \gtrsim 10^{-4}$), a

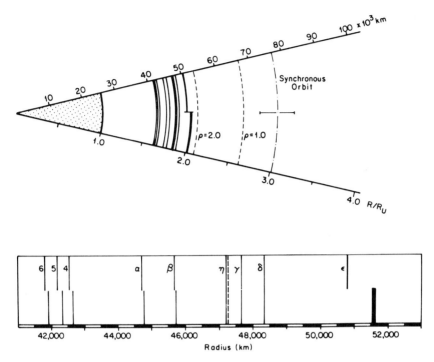

Fig. 1. Schematic diagram showing the locations of the nine confirmed rings of Uranus. The dashed arcs represent the Roche limits for satellite densities of 1.0 and 2.0 g cm^{-3}, while the dot-dashed arc corresponds to the synchronous orbit distance for a planetary rotation period of 15.6 ± 1.4 hr (see Sec. V). The lower portion of the diagram shows the radial region which spans the known rings, enlarged by a factor of 10. For each of the 6 definitely eccentric rings (see Sec. III) both the minimum (periapse) and maximum (apoapse) radii are indicated. The width of the ε Ring, which varies markedly from periapse to apoapse, is plotted to scale; the other rings all have widths of <10 km, below the resolution of this figure. The outer boundary of the η Ring's broad component (Sec. II) is indicated by a dashed line.

characteristic shared by several narrow ringlets in the Saturnian system (see chapter by Cuzzi et al.). Perhaps even more surprising is the fact that the nine rings do not lie exactly in the same plane. Their inclinations relative to Uranus' equatorial plane are of the same order as the eccentricities (with one notable exception), but are more difficult to measure because of the currently almost pole-on aspect of the planet.

The outermost ring, designated ε, is broad enough to reveal long-lived internal structure. The α Ring sometimes shows a "double-dip" in transmitted light during stellar occultation, perhaps indicating an internal division, and the η Ring has a broad component of low optical depth. The widths of the inner trio of rings, designated 6, 5, and 4, are below the diffraction limit of Earth-based observations (~3.5 km at a wavelength of 2.2 μm).

Near-infrared observations of the ring system indicate that the ring particles have extremely low geometric albedos (0.02–0.03) and an essentially flat
1–4 μm spectrum. Apart from their intrinsic interest, the rings serve both as
an ensemble of test particles for studies of Uranus' gravitational field and as a
precise reference system for studies of the shape and orientation of the planet.

Before proceeding to a detailed description of the structure and kinematics of the rings in Secs. II and III, we review briefly the techniques and
limitations of groundbased stellar occultation studies of ring systems in general, and of the Uranian system in particular. Previous reviews of the Uranian
ring system, from observational and theoretical viewpoints, have been given
by Brahic (1982), Elliot (1982), and Goldreich and Tremaine (1982).

I. STELLAR OCCULTATION OBSERVATIONS

The essential features of a stellar occultation observation are simply
described as follows. The occulting body (planet, satellite, or ring) casts a
shadow in the light from the star, through which the Earth passes in its orbital
motion. Since the distance of the star is effectively infinite, the shadow at the
Earth is a full-scale, two-dimensional projection of the occulting body. In the
case of the Uranian rings, the maximum dimension of the shadow is ~8 times
the diameter of the Earth and only a limited portion of the system can be
studied during any one occultation. Typically, the projected velocity of the
Earth across the shadow is 10–30 km s^{-1}, although lower velocities are
possible for occultations that occur near the "stationary" points in the
planet's geocentric motion.

When a stellar occultation is observed, the intensity of starlight is
recorded as a function of time as the Earth moves relative to the occulting
body. In other words, the intensity of starlight is measured along a single line
through the two-dimensional shadow pattern. This situation is illustrated
schematically in Fig. 2, for a central occultation by a ringed planet. Note that,
while the extinction of starlight by a ring can be related to the optical depth
(and width) of the ring, the decrease in intensity as the starlight penetrates the
planet's atmosphere is due to differential refraction, rather than to absorption
(see the review by Elliot [1979]).

The fundamental limit on the resolution achievable from a stellar occultation is set by diffraction. At typical Uranus-Earth distances, $D \sim 18$ AU, and
at the usual observational wavelength, $\lambda = 2.2$ μm, this limit is $\sim (2 \lambda D)^{\frac{1}{2}} =$
3.5 km, as measured by the full width at half maximum (FWHM) of an
occultation profile for a narrow ring.

A. Occultation Predictions

The first step in an observational program for occultations is to obtain
predictions for those occultations whose signal-to-noise ratio would be suffi-

SCHEMATIC LIGHT CURVE
(central occultation by a ringed planet)

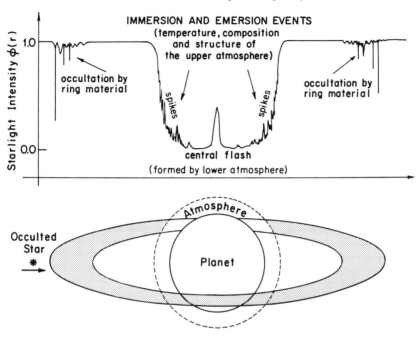

Fig. 2. Schematic lightcurve for a central occultation by a ringed planet. The upper portion of the figure shows the light intensity of the star that would be observed from Earth as a function of the position of the star behind the planet. The first attenuation of starlight is due to the extinction by the ring material. The occultation of the star by the atmosphere occurs through the process of differential refraction, with irregular variations (spikes) caused by atmospheric structure that deviates from being isothermal. The central flash is observed when the star is directly behind the center of the planet and can yield information about the extinction of the lower atmosphere (figure after Elliot 1979).

cient to produce useful information about the rings. One method to predict occultations is to compare the ephemeris of Uranus with stellar positions in the Smithsonian Astrophysical Observatory (SAO) catalog. This technique was used by Taylor (1973) to predict the occultation of SAO 158687 by Uranus on 10 March 1977, during which the rings were accidentally discovered (Elliot et al. 1977a; Elliot 1979). However, occultations of SAO catalog stars by Uranus are rare. Furthermore, the SAO catalog does not contain most of the stars whose occultations are observable at 2.2 μm. To identify additional events, one must obtain plates of the star fields ahead of Uranus, measure the positions of stars in Uranus' path, and then calculate the circumstances for the occultations observable from Earth. From these candidates, only the sufficiently bright stars are selected for the final list of events, which occur at a rate of about two per year. This method of predicting and evaluating occultations for the Uranian rings was developed by Klemola and Marsden (1977), Liller (1977) and Elliot (1977). A list of predictions through 1984 has been published by Klemola et al. (1981).

B. Observational Techniques

The goal of occultation photometry is to measure the intensity of the starlight as a function of time; these data are then readily converted to optical depth as a function of radius for the individual rings. Observations of the same occultation from different sites are valuable because they correspond to different tracks through the ring system and give information about the two-dimensional structure of the ring occultation pattern. Simultaneous observations at different wavelengths, from the same site, give information about the distribution of the ring particle sizes and allow discrimination against some forms of noise in the photometry discussed below. Primary concerns in carrying out the observation are (i) to obtain photometry with the greatest possible signal-to-noise ratio, and (ii) to be sure that the data are accurately synchronized with Universal Time (UT).

Let us now consider the factors involved in obtaining photometry of high signal-to-noise ratio. Since the rings are at most only 2 arcsec from the limb of Uranus, at least some, if not all, of the light from Uranus is included within the photometer aperture along with the starlight. For our purposes, the light from Uranus constitutes a background and is a source of noise. One method of minimizing this contribution is to observe in one of the methane bands, which become progressively deeper in the spectrum of Uranus at longer wavelengths (Fink and Larson 1979). Within wavelengths accessible to photomultipliers, the methane band at 8900Å is the deepest (Elliot et al. 1977c). In the near infrared, the 2.2. μm band is so deep that Uranus is even fainter than the rings themselves. This band nicely matches the passband of the K filter, which has become standard for observing ring occultations.

For the data obtained to date, a variety of noise sources have proven to be important. For each particular occultation, the significant noise sources depend on the prevailing conditions and observing set-up. The sources of noise that occur are the following:

1.   Photon (shot) noise from both Uranus and the occulted star;
2.   Scintillation noise from both planet and star;
3.   Light from both planet and star spilling out of the photometer aperture due to time variations in the seeing;
4.   Changes in transparency of the atmosphere due to clouds;
5.   Uncompensated sky background variations (for infrared observations);
6.   Variations in signal due to telescope guiding errors (if the beam profile of the photometer is not flat);

7. A host of infrequent, sometimes not understood, "glitches" which are probably due to momentary problems with the signal processing electronics or isolated, small clouds that are not otherwise noticed by the observer.

To obtain useful data for the analyses that will be discussed later in this chapter, the observer should perform several calibrations. Well before the event, the signals from both Uranus and the occulted star should be measured through the available filters in order to determine which would give the best signal-to-noise ratio (Elliot 1977). At this time it would also be wise to check the beam profile of the photometer and do whatever possible to make it flat. The timing system should be carefully traced and well understood. Timing based on hardware is usually better than software, since additional timing uncertainty is introduced by the computer-interrupt system. Just before the observations, these earlier calibrations should be verified and the size of the photometer aperture selected to be compatible with the seeing conditions. The aperture should not be so small that light spilling out during moments of the worst seeing is a source of noise. The step response of the entire system should be well measured. Also, it has proven prudent to carefully monitor the signal continuously, with a chart recorder, and to note any known causes for events seen in the signal as they occur. Frequent calibration of the background and star signal levels is desirable.

Synchronizing the observation with UT is accomplished through radio time signals. The accuracy achievable with this method is limited by the uncertainty in the propagation delay of the radio signals, which should be $\leqslant$ 0.01 s. The timing synchronization should be confirmed both immediately before and after the data recording for the event.

C. Limits to Spatial Resolution

The limiting spatial resolution for a given occultation depends on four main factors:

1. Diffraction;
2. Angular diameter of the star;
3. Signal-to-noise ratio of the data;
4. Impulse response of the observing equipment.

The effects of each of these factors can be best explained by tracing the path of the starlight through the rings, and into the telescope. Then we consider how the starlight is detected and how the resulting signal is processed and recorded. A plane wave from a distant point source encounters the rings, and the light not scattered or absorbed by the ring particles emerges on the other side of the rings and continues on its path through space. To learn its intensity pattern at a given distance from the rings, we must calculate the diffracted intensity for this distance. Complicated diffraction effects can arise, but for our present purpose, we can simply note that the spatial resolution is degraded by $(2\lambda D)^{\frac{1}{2}}$ where λ is the wavelength of the light and D is the distance from the rings to the observer. As discussed earlier in this section, this limiting resolution for $\lambda = 2.2\ \mu m$ is ~ 3.5 km in the sky plane, for typical distances between the Earth and Uranus. We can think of this limit as a circular ''beam profile'' in the sky plane, which is a plane passing through the center of Uranus, perpendicular to the line of sight. Projected into the plane of a ring, this beam profile becomes an ellipse that has a minor axis equal to the diameter of the circular beam profile in the sky plane. Along the minor axis of the ellipse, the resolution in the ring plane is the same as in the sky plane. However, the resolution is degraded by a factor csc B along the major axis of the ellipse, where B is the angle between the line of sight and the ring plane.

Presently, B is about 70° for the Uranian rings, so that the resolution limit in the ring plane essentially equals that in the sky plane. When B becomes significantly smaller, the csc B factor becomes appreciably > 1, as is always the case for Saturn. However, its effect on degrading the radial resolution in the ring plane depends on the orientation of the elliptical beam profile relative to the path of the star through the ring system. For example, if the occultation track were along the line joining the ansae of the rings, the *radial* resolution would not be degraded from the sky plane diffraction limit. However, the beam would be expanded in the *azimuthal* direction by the csc B factor.

Since the star has a finite angular diameter, the point-source diffraction pattern of a ring is convolved with the intensity distribution function across the stellar disk. For a bright star of later spectral type, the resulting degradation of spatial resolution can dominate diffraction. The relevant parameter is the length subtended by the angular diameter of the star over the distance between Uranus and the Earth (see Table I).

Relativistic bending of the starlight by the planet's gravitational field results in a nonuniform radial shrinkage of the shadow, but does not affect the spatial resolution. For light grazing the limb of Uranus, this shrinkage amounts to ~ 25 km at the distance of Earth.

Intrinsic to this light is the photon noise and whatever modulation is added on its path through the Earth's atmosphere. The critical parameter is the noise within the time that the star moves an element of spatial resolution in the ring plane, say, the diffraction limit. Hence, the noise in the time domain is scaled by the relative velocity (projected into the ring plane) of the Earth and Uranus. This velocity is typically 20 km s^{-1}, but can vary from a few to ~ 30 km s^{-1}.

After the light arrives at the telescope, it is detected and recorded. For observations at visible wavelengths, the detector is a photomultiplier, which has a negligibly short response time. The number of detected photons can be integrated over a short time and then recorded, with a negligible dead time between integrations. For infrared observations with star-sky chopping, the process is more complicated. For these data, we have a detector time constant to contend with, and the beam is chopped between star and sky so that the star is observed only half of the time. The chopped signal is filtered, integrated and recorded. All of these processes can add significant response time to the data. The combined effect of these processes can be determined through measuring the step response of the system. One method for doing this is to rapidly move a star out of the center of the photometer aperture, if the telescope can be accelerated quickly enough.

D. Summary of Observations

All known occultation observations of the Uranian rings, through the 1983 apparition, are summarized in Table I. The star designations KM and KME refer to the lists of occultation candidates published by Klemola and Marsden (1977) and Klemola et al. (1981). For each star, the V (0.56 μm) and K (2.2 μm) magnitudes are given, together with its estimated angular diameter. In the last two columns, the number of sites at which observations were obtained and references to published data are given. A typical set of recordings, showing all nine rings, is presented at low spatial resolution in Fig. 3.

Reports of possible other material in the ring system have been made by Millis and Wasserman (1978), Bouchet et al. (1980) and Sicardy et al. (1982). Also, Bhattacharyya and Bappu (1977) and Bhattacharyya et al. (1979) have asserted that an extended, Saturn-like ring system surrounds Uranus. However, only nine rings have been confirmed. In fact, the occultation data place a limit of ~0.05 on the optical depth of any inter-ring material that lies anywhere from the top of the Uranian atmosphere to well beyond the orbit of Miranda. The imaging data place more stringent limits on the integrated amount of inter-ring material, as we discuss in Sec. IV.

II. RING STRUCTURE

A. Observed Characteristics

Enough optical depth profiles have been accumulated for us to characterize the ring structures and to conclude that the radial structure is not, in general, the same from occultation to occultation. Hence the ring structure must vary with orbital longitude or time or both. Below we describe the structures that have been observed for all of the nine rings. These structures are illustrated in Figs. 4 through 7 and summarized in Table II. (See also Table III for figure parameters.)

Rings 6, 5 and 4. These three rings are unresolved in all occultation observations to date, which suggests intrinsic radial widths ≤ 2 km. Lower limits on their widths in the range 0.4–0.8 km follow from the fraction of

TABLE I

Observations of Occultations by the Uranian Rings

Date	Star[a]	Magnitude[b] V	K	V-K	Stellar Profile Angular Diameter[c] (milliarcsec)	FWHM at Uranus (km)	Sky Plane Velocity (km s^{-1})	Number of Sites	References[d]
10 Mar 1977	SAO 158687	9.0	5.2	3.8	0.55	6.5	17	6	1–8
23 Dec 1977	KM 2	10.4	7.0	3.4	0.24	2.9	27	1	9
4 Apr 1978	KM 4	13.4	11.9	1.5	0.024	0.3	18	1	10
10 Apr 1978	KM 5	11.6	10.1	1.5	0.055	0.6	20	1	10
10 Jun 1979	KM 9	13.7	11.0	2.7	0.037	0.4	19	1	11
20 Mar 1980	KM 11	13.0	10.1	2.9	0.057	0.7	10	1	12
15 Aug 1980	KM 12	12.3	8.7	3.6	0.11	1.3	8	3	13–16
26 Apr 1981	KME 13	10.0	7.4	2.6	0.19	2.2	20	2	17
22 Apr 1982	KME 14	11.6	5.2	6.4	0.54	6.1	18	5	18
1 May 1982	KME 15	11.4	7.4	4.0	0.20	2.3	21	1	18
4 Jun 1982	KME 16	13.0	9.5	3.5	0.075	0.8	22	1	18
3 Mar 1983	KME 17a	10.3	8.8	1.5	0.10	1.4	5	1	18
25 Mar 1983	KME 17b	10.3	8.8	1.5	0.10	1.3	5	2	18

[a] SAO refers to the Smithsonian Astrophysical Observatory Catalog; KM to the prediction list of Klemola and Marsden (1977); and KME to the prediction list of Klemola et al. (1981).

[b] Compiled by K. J. Meech from a variety of sources. The errors may be as large as a few tenths of a magnitude in some cases.

[c] Calculated by K. J. Meech, based on the data of Ridgway et al. (1980) and the magnitudes given here.

[d] References: 1. Elliot et al. (1977a,b); 2. Millis et al. (1977a,b); 3. Churms (1977); 4. Bhattacharyya and Kuppuswamy (1977); 5. Mahra and Gupta (1977); 6. Chen et al. (1978); 7. Elliot et al. (1978); 8. Hubbard and Zellner (1980); 9. Millis and Wasserman (1978); 10. Nicholson et al. (1978); 11. Nicholson et al. (1981); 12. Elliot et al. (1981b); 13. Elliot et al. (1981a); 14. Nicholson et al. (1982); 15. Sicardy et al. (1982); 16. Elliot et al. (1983); 17. French et al. (1982); 18. Unpublished.

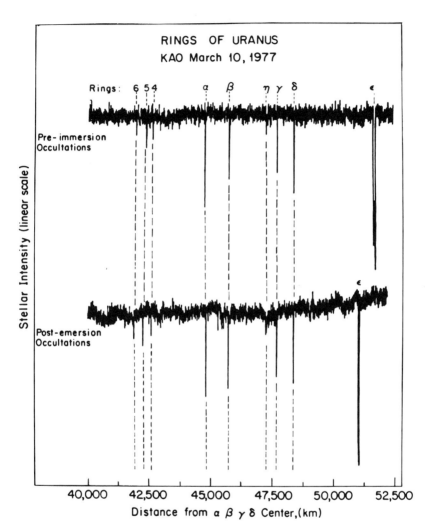

Fig. 3. Occultations by the rings of Uranus. The pre-immersion and post-emersion occulta-tions by the rings of Uranus observed with the Kuiper Airborne Observatory (Elliot et al. 1977c) have been plotted on the common scale of distance from the center of Uranus in the ring plane. Occultations corresponding to the nine confirmed rings are easily seen. Most (if not all) of the low-frequency variations in the lightcurves are due to a variable amount of scattered moonlight on the telescope mirror (figure after Elliot 1979).

TABLE II

Ring Structures

Ring	Radial Width[a] (km)	Normal Optical Depth	Confirmed Structural Features
6	0.4–2	$\geqslant 0.5$	—
5	0.8–2	$\geqslant 1.0$	—
4	0.7–2	$\geqslant 0.8$	—
α	$5 \rightarrow 10$	$\sim 1.4 \rightarrow 0.7$	"Double-dip" structure at widest part
β	$5 \rightarrow 11$	$\sim 1.5 \rightarrow 0.35$	—
η	$0.5–2^{b}$	$\geqslant 0.6^{b}$	Broad component terminated by a 2nd narrow ringlet ~ 55 km outside principal component
γ	~ 3	$\geqslant 1.5$	—
δ	2–3	$\geqslant 1.5$	2nd component ~ 12 km inside
ϵ	$20 \rightarrow 96$	$? \rightarrow 1.2^{c}$	Consistent internal structure, sharp edges

[a] $\rightarrow$ indicates established width variation, otherwise a range of possible widths is given.
[b] Narrow component.
[c] Optical depth of narrowest part of the ϵ Ring is too high to be reliably measured.

starlight that could be intercepted by a totally opaque ring. The integrated extinction of Ring 6 is somewhat less than that for Rings 5 or 4. There is little evidence for variations in extinction, although variations in width for all three rings might be expected by analogy with those established for the other eccentric rings (see below).

α *Ring*. This ring is almost always resolved, and varies in width by a factor of ~ 2. At its maximum width of ~ 10 km, on two occasions, it has shown a "double-dip" structure: two components separated by ~ 4 km, which is essentially at the limit of our spatial resolution. This could be indicative either of a structured ring or of an internal division that extends around the ring, but is resolved only where the ring is widest. At its widest part, the average normal optical depth of the α Ring is ~ 0.7 (Nicholson et al. 1982).

β *Ring*. This ring is also almost always resolved, and exhibits a range of widths similar to that of the α Ring. It shows, however, no evidence of internal structure except for a possible difference in the sharpness of the inner and outer edges. The average normal optical depth at the widest part of the β Ring is ~ 0.35, which suggests that it may contain about half as much material as the α Ring. This conclusion is supported below by an analysis of the width variations of both rings.

TABLE III

Parameters for Ring Profiles[a]

Ring	Figure in text	Zero-point Radius (km)	True Anomaly[b] (deg)	Radial Velocity (km s^{-1})
6	4	41872	276.7	8.29
5	4	42293	103.5	8.29
4	4	42634	239.8	8.30
α	5	44792	193.1	8.34
β	5	45713	132.1	8.36
η	4,6	47216	(142.4)	8.39
γ	5	47667	(262.0)	8.39
δ	5	48337	316.3	8.40
ϵ	7	51279	256.7	8.45

[a]These parameters refer to the Cerro Tololo Inter-American Observatory profiles and would be slightly different for the European Southern Observatory and Las Campanas Observatory profiles. (See Figs. 4–7.)
[b]Angle from periapse.

η *Ring*. This ring comprises both a narrow, unresolved component (see Fig. 4) and a broad ($\sim$55 km) component of low optical depth that is located primarily, and perhaps entirely, outside the narrow component. The optical depth of the broad component apparently varies considerably, with measured values ranging from $\sim$0.03 to $\sim$0.10, as shown in Fig. 6. A second unresolved sharp feature occurs at or near the outer edge of the broad component. The η Ring appears strikingly similar to the F Ring of Saturn (Lane et al. 1982), since they each have broad and narrow components of about the same widths and optical depths. One difference, however, is that the narrow component of the η Ring lies at the inner edge of the broad component while for the F Ring the situation is reversed. In this regard, it might prove significant that the radius of the η Ring is smaller than the corotation radius, while the radius of the F Ring is larger than the corotation radius.

γ *Ring*. This ring lies at the limit of our spatial resolution, and is probably $\sim$3 km in width and fairly opaque ($\tau \gtrsim 1.5$). It appears to have very sharp edges, judging from the frequently prominent diffraction fringes on the occultation profiles.

δ *Ring*. Although very similar in its occultation signature to the γ Ring, the δ Ring may be slightly narrower and/or less opaque. Its diffraction fringes are generally less prominent than those of the γ Ring. It now appears probable that this ring also has a secondary component, in this instance located $\sim$12 km inside, and with an integrated extinction $\sim$10% of the primary component.

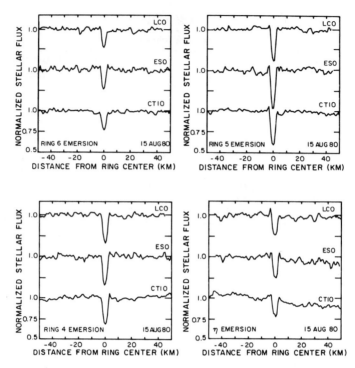

Fig. 4. Occultation profiles of Rings 6, 5, 4 and the narrow component of the η Ring. Here and in Fig. 5 the emersion data obtained from Las Campanas, the European Southern Observatory and Cerro Tololo on 15–16 August 1980 have been aligned according to their midradii. The width of each of these profiles is near the diffraction limit. See Table III for zero-point radius (figure after Elliot et al. 1983).

ϵ *Ring*. This ring, with a width varying from 20 to 96 km (see below), is always well resolved and shows considerable structural detail (see Fig. 7). This structure is largely consistent from one occultation to another, except possibly for the narrowest profiles, which are essentially featureless and correspond to large optical depths. (Six representative observations of the ϵ Ring, reproduced at a uniform radial scale, are presented in Fig. 8b.) The outer third of the ring consistently has the highest average optical depth, with a secondary maximum occurring at or near the inner edge. There are no resolved gaps within the ring, although the presence of numerous unresolved gaps seems quite likely. Both inner and outer edges are sharp, and frequently show low amplitude diffraction fringes. The average optical depth of the ϵ Ring is ~ 1.2 at its widest part, and varies approximately inversely with ring width (see Table IV).

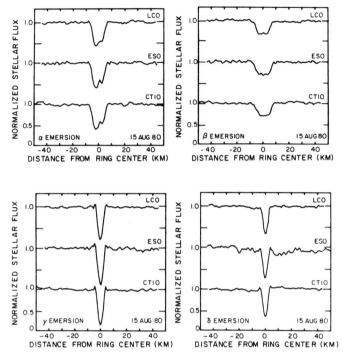

Fig. 5. Occultation profiles of Rings α, β, γ and δ (cf. Fig. 4). Note the "double-dip" structure of the α Ring, the lack of structure in the β Ring and the diffraction fringes on the γ Ring. See Table III for zero-point radius (figure after Elliot et al. 1983).

TABLE IV

Parameters for ϵ Ring Profiles[a]

Profile	Date I = Immersion E = Emersion	Site[b]	Zero-point Radius (km)	True Anomaly (deg.)	Radial Velocity (km s^{-1})	Radial Width[c] (km)
a	4/10/78 I	LCO	50789	9.2	18.65	20.5
b	4/10/78 E	LCO	51353	113.9	18.79	76.5
c*	4/22/82 I	LCO	51504	141.9	17.49	88.7
d	3/10/77 I	PO	51594	182.5	12.28	98.2
e	8/15/80 E	LCO	51278	256.8	8.44	68.0
f*	4/22/82 E	LCO	51086	284.3	17.63	46.5

[a]See Fig. 8 in Sec. II.C.

[b]LCO = Las Campanas Observatory, Chile;
PO = Perth Observatory, Australia.

[c]The widths plotted in Fig. 10 are based on a previous analysis and some may differ by up to 2 km from the values presented in this table. Those in the table are preferred.

*Since these data were not included in the orbit solution of Table VI, their zero-point radii, true anomalies and radial velocities may change, relative to the other data, when the fitted orbit model is updated to include these points.

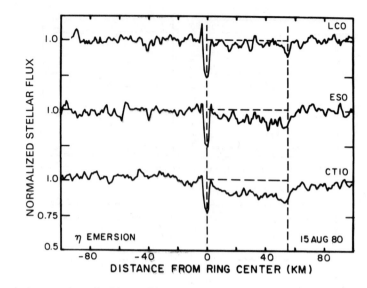

Fig. 6. A comparison of η Ring profiles from Las Campanas, the European Southern Observatory and Cerro Tololo obtained on 15–16 August 1980. Note the different occultation depths obtained at the three observatories. See Table III for zero-point radius (figure after Elliot et al. 1983.)

B. Model Profiles

From the above discussion, we see that the structures of most rings cannot be described by a simple model. However, deconvolution of the actual profiles from the data is not possible because we lack the phase of the detected light and have limited signal-to-noise ratio. While not the ideal solution to these problems, the fitting of simple models to the profiles does yield certain information. First, it provides a consistent method for obtaining the mid-times of ring occultations, which are needed to determine the ring orbits (see Sec. III). Second, model fitting is also a consistent method for defining the ring widths.

Two types of model have been fitted to the data. The first is a trapezoid model, which includes the effects of a finite star diameter and the time constant of the data recording system, but not diffraction (Elliot et al. 1981b). The second model treats the optical depth profile of the ring as a "square well" of uniform, nonzero transmission and includes the effects of diffraction, the passband of the filter, the stellar diameter and the time constant of the data-recording equipment (Nicholson et al. 1982). A typical model ring profile at various stages of the calculation is shown in Fig. 9. The above conclusions concerning intrinsic ring widths and optical depths are based on such modeling.

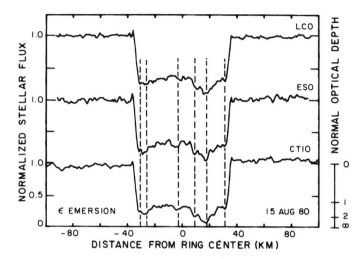

Fig. 7. A comparison of ε Ring profiles from Las Campanas, the European Southern Observatory and Cerro Tololo on 15–16 August 1980. Coincident features in the three profiles, most of which correspond to those pointed out by Nicholson et al. (1982), have been highlighted by dashed lines. See Table III for zero-point radius (figure after Elliot et al. 1983).

C. Width-Radius Relationships

The systematic variation in radial width of the ε Ring provided the first clue that at least one of the Uranian rings was eccentric. Nicholson et al. (1978) discovered a linear relation between the width and radius of different segments of this ring. Such a relation is expected for an eccentric ring whose inner and outer edges have slightly different eccentricities but the same longitude of periapse (see below). Subsequently, similar width-radius relations were established for the α and β Rings (Elliot et al. 1981b; Nicholson et al. 1982). Figs. 10–12 show these relations, revised to include the 1981 and 1982 observations.

Although this eccentric model is clearly in good agreement with the observations, especially for the ε Ring, differential apsidal precession between the inner and outer edges of each ring due to Uranus' oblate figure amounts to 180° in only a few hundred years. Rapid disruption and circularization of initially eccentric rings is thus inevitable, unless some mechanism exists to cancel the differential component of precession. Goldreich and Tremaine (1979b) have shown that this can be accomplished by the rings' self-gravity, although other possibilities also exist (Dermott and Murray 1980).

A uniformly precessing, eccentric ring can be described in terms of a set of nested, nonintersecting ring particle orbits, much like the streamlines of laminar fluid flow. Each orbit is a Keplerian ellipse with a certain semimajor axis a, eccentricity e, and longitude of periapse $\tilde{\omega}$. All orbits share the same

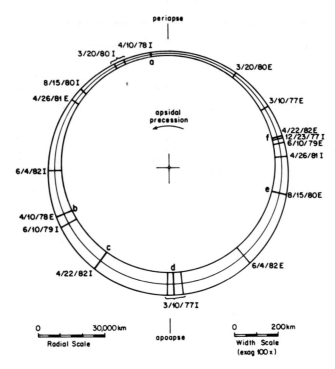

Fig. 8a. A diagram of the ε Ring, viewed from above Uranus' north pole, showing the locations of the various stellar occultation profiles. Note that the width of the ring is exaggerated by a factor of 100 for purposes of clarity, but that the center line of the ring is plotted with the correct eccentricity of 0.0079.

apsidal precession rate $\dot{\varpi}$, but there is a gradient (across the ring) of e, and possibly of ϖ as well. In terms of the total ranges in a, e and ϖ from the inner to the outer edge of the ring (δa, δe and $\delta\varpi$), the radial width of the ring as a function of azimuth θ, measured from the mean periapse, is

$$w \simeq (1 - q_e \cos\theta - q_\omega \sin\theta)\, \delta a$$
$$= (1 - Q \cos(\theta - \theta_0))\, \delta a \qquad (1)$$

where $q_e = a\delta e/\delta a$, $q_\omega = ae\delta\varpi/\delta a$, $Q = (q_e^2 + q\omega^2)^{\frac{1}{2}}$, and $\tan\theta_0 = q_\omega/q_e = e\delta\varpi/\delta e$. This expression is derived under the simplifying, but quite valid, assumptions that $\delta e \ll e \ll 1$, $\delta a \ll a$, $\delta\varpi \ll \pi/2$, and $\delta a/a \ll \delta e/e$. To $\mathcal{O}(e)$, the mean radius of the ring is given by $r = a(1 - e\cos\theta)$, where a and e are the mean semimajor axis and eccentricity, so that if $\delta\varpi = 0$ we have the linear relation:

$$w = \delta a + \frac{\delta e}{e}(r - a). \qquad (2)$$

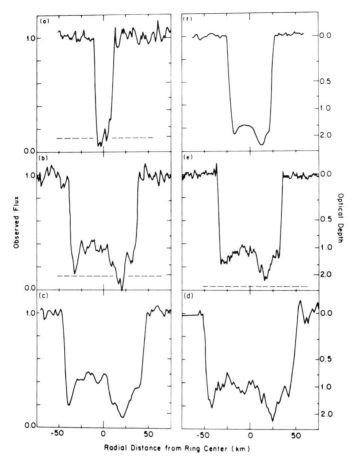

Fig. 8b. Six representative occultation profiles of the ε Ring, corresponding to observations (a-f) in Fig. 8a, plotted at a uniform radial scale. Details of the particular observations may be found in Table IV. The ordinate is observed flux, normalized to the average flux levels immediately preceding and following the event. Except for the March 1977 profile, no attempt has been made to subtract the small, but rather uncertain, contributions of Uranus and the rings to the total flux. Dashed lines indicate the upper limits to these contributions, where they are > 1% of the stellar flux. The unusually smooth appearance of the April 1982 profiles (c and f) is due, in part, to the rather large angular diameter of the occulted star (see Table I). The data is from Millis et al. (1977a), Nicholson et al. (1978, 1982), and K. Matthews (April 1982, unpublished).

From these relations, and from the linear fits to the observed widths in Figs. 10–12, values of δa and δe have been obtained for the α, β and ε Rings and are given in Table V. Mean values of a and e are taken from the kinematic models in Table VI. Also given are the inferred minimum and maximum

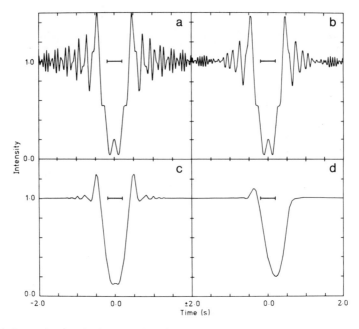

Fig. 9. Successive steps in the generation of a synthetic occultation profile for an opaque ring with a projected width of 3.0 km. The geometric shadow of the ring is indicated by a horizontal bar, and the velocity of the shadow relative to the observer is 7.5 km s^{-1}. (a) Monochromatic (λ = 2.2 μm) diffraction pattern for a point source; (b) averaged over the 2.0–2.4-μm passband; (c) convolved with the intensity distribution of a uniform circular source of angular diameter 1.3 $\times$ 10^{-4} arcsec; (d) convolved with the instrumental response function (figure after Nicholson et al. 1982).

widths, $\delta a(1-q_e)$ and $\delta a(1+q_e)$. Strictly speaking, these widths are FWHM occultation profile widths and are larger than the intrinsic ring widths by 0.5–1.0 km.

In the future, with an improved data base, it may be possible to determine $\delta\varpi$ from the distribution of ring widths with orbital longitude. We note that the effect of a small deviation in longitude of periapse between the inner and outer edges of an eccentric ring is to produce a much larger (by a factor of $e/\delta e$) offset in the longitude of minimum ring width from the mean longitude of periapse. Such an offset has been discussed by Dermott and Murray (1980) in connection with the development of the eccentricity gradient across the ϵ Ring. Measurement of a nonzero $\delta\varpi$ could provide information on dissipative processes operating within the rings.

The mechanism by which the α, β and ϵ Rings are maintained in a state of uniform precession is uncertain, but the simplest and most readily quantified model is that of Goldreich and Tremaine (1979a), who attribute the necessary small perturbations to the "free" precession rates (i.e., those in-

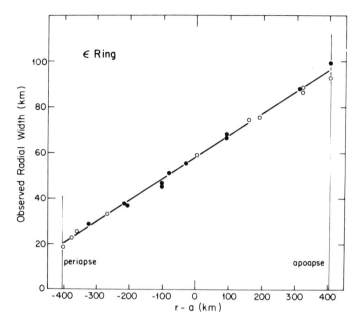

Fig. 10. Relation between radial width (occultation profile FWHM projected onto the radial direction in the ring plane) and the center-line radius exhibited by the ε Ring, including data from 8 stellar occultations (see Fig. 8a,b). The zero of the radius scale is the semimajor axis of the ring, as given in Table VI. Open and filled symbols indicate ring segments whose true anomalies (i.e., angles from periapse) lie in the respective ranges 0°–180° and 180°–360°. The line is a linear least-squares fit of Eq. (2) to the data, the resulting values of δa, δe and q_e being given in Table V. Deviations of individual points from the fitted relation are no more than a few km in radial width, comparable to the uncertainties in the measurements.

duced by Uranus) to the self-gravity of the rings. One prediction of this model is that the eccentricity gradient q_e must be positive, as observed. A second consequence is a relation between the total mass of each ring and the parameters δa and δe:

$$m_{\text{ring}}/M_u = \frac{21\pi}{8} f \, \frac{e}{\delta e} \left(\frac{\delta a}{a} \right)^3 J_2 \left(\frac{R}{a} \right)^2 \tag{3}$$

where J_2 is the usual coefficient in the zonal harmonic expansion of Uranus' gravitational field, R is the equatorial radius of Uranus, and f is a dimensionless factor which depends on the distribution of mass within the ring. Generally, $f \simeq 0.5$ (Goldreich and Tremaine 1979a,b). From this expression, and the value of J_2 derived from kinematic models of the rings in Sec. III, the ring masses and average surface densities σ given in Table V are derived. Note that the mass obtained for the α Ring is ~50% greater than that obtained for the β Ring, and that the inferred minimum surface densities, corresponding to

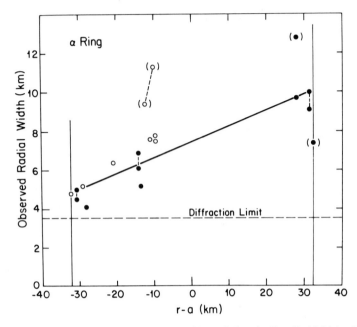

Fig. 11. Width-radius relation for the α Ring, with symbols as in Fig. 10. Multiple observations of the same event from nearby stations are connected by dashed lines. The horizontal dashed line represents the minimum FWHM, set by diffraction, of ~ 3.5 km at a wavelength of 2.2 μm. As in Fig. 10, the zero of the radius scale is the semimajor axis of the ring, given in Table VI. In constructing the linear least-squares fit of Eq. (2) to the data, the obviously discrepant points shown in parentheses have been discarded. (These discrepancies, which do not appear to be attributable to observational uncertainties or error, suggest that the structure of the α Ring is more complicated than the present simple model.) Values of δa, δe and q_e derived from the fit are given in Table V.

maximum widths, are in approximate proportion to the optical depths given above. In each case, the opacity, defined as τ/σ, is ~ 0.8 cm^2 g^{-1}. The implied mass of the ϵ Ring is much greater, and it seems likely that this ring contains $\sim 99\%$ of the total mass in the system. The opacity derived for the ϵ Ring, 0.08 cm^2 g^{-1}, is significantly lower than that obtained for α and β, as pointed out by Yoder (1982). This suggests a somewhat larger mean particle size for the ϵ Ring. In comparison, the opacity of Saturn's A and B Rings, as inferred from the study of density waves (see chapter by Cuzzi et al.), is ~ 0.01 cm^2 g^{-1} in most regions where it has been measured. Analyses of two narrow, eccentric ringlets in Saturn's C Ring yield opacities of 0.05 cm^2 g^{-1} (Esposito et al. 1983; Porco et al. 1983), comparable with that of the ϵ Ring.

Finally, we note that a similar argument about disruption by differential precession applies to inclined rings, so that for such rings the nodal line must regress uniformly. According to the self-gravity model, such uniform regression must be accompanied by gradients in inclination across the ring, with $\delta i/i$

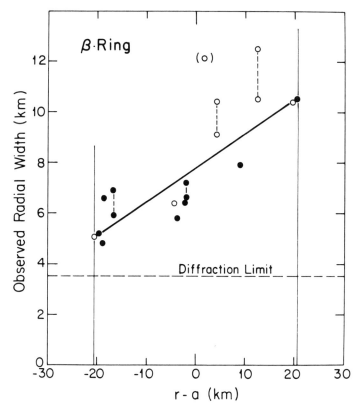

Fig. 12. Width-radius relation for the β Ring, with symbols as in Figs. 10 and 11. Note that the total range in radial width for the β Ring is comparable to that shown by the α Ring, although the eccentricity of β is significantly lower (see Table VI). Only one discrepant point, shown in parentheses, has been omitted from the least-squares fit of Eq. (2) to the data, the results of which are given in Table V.

$\simeq \delta e/e$ (Borderies et al. 1983). It may be possible to detect such gradients by their effects on the projected widths of the inclined rings (see Sec. III), although the present near pole-on aspect of the system reduces the predicted amplitude of the effect to $\lesssim 1$ km for any of the rings.

III. RING ORBITS

A remarkable feature of the Uranian rings is the high precision with which we can derive their orbits from multiple occultation observations. From each event we obtain a large number of points on the ring orbits whose relative positions are precisely known. The situation is illustrated in Fig. 13 for the discovery occultation on 10 March 1977. The basic reason for the high precision of the orbital parameters is the high precision of the occultation data; but

TABLE V

Width-Radius Relations and Ring Masses

Ring	α	β	ϵ
Directly Measured Parameters			
a (km)[a]	44758	45701	51188
δa (km)	7.5 ± 0.2	7.8 ± 0.3	58.0 ± 0.4
e [a]	7.8×10^{-4}	4.3×10^{-4}	7.94×10^{-3}
δe	5.8×10^{-5}	6.0×10^{-5}	7.4×10^{-4}
$q_e = a\delta e/\delta a$	0.35 ± 0.03	0.35 ± 0.05	0.65 ± 0.01
Minimum width (FWHM, km)	4.9	5.0	19.8
Maximum width (FWHM, km)	10.1	10.6	96.3
Normal optical depth at			
maximum width, τ_0	0.7^{b}	0.35^{b}	1.2^{c}
Parameters Inferred From			
Goldreich and Tremaine Model			
Total mass (g)	2.6×10^{16}	1.4×10^{16}	4.9×10^{18}
Average surface density σ (g cm^{-2})	1.2	0.65	26
σ at maximum width σ_0 (g cm^{-2})	0.9	0.5	16
Opacity $= \tau_0/\sigma_0$ (cm^2 g^{-1})	0.8	0.8	0.08

[a]From kinematic models. See Table VI.
[b]Nicholson et al. (1982). See Table II.
[c]Integrated optical depth/FWHM at maximum width.

another important factor is that the pole of Uranus is only 8° from its orbit plane. Hence, we see the rings from a continuously changing aspect that goes through a complete cycle during an orbit period of about 84 yr. In 1986 the ring system will appear almost pole-on to the Earth. A fundamental difference exists between determining satellite orbits and ring orbits from Earth-based observations: for satellite orbits we know the revolution period more precisely than the orbit dimensions; for ring orbits we know the dimensions precisely, but can measure only their precession periods—not their revolution periods.

The kinematic model has been developed through a series of papers (Elliot et al. 1978, 1981a,b; Nicholson et al. 1978, 1981, 1982; French et al. 1982) into its present state. The model treats the rings as ellipses, which are inclined to the equatorial plane of Uranus and precessing under the influence of the zonal harmonics of Uranus' gravitational field. Both apsidal and nodal precession rates are assumed to be uniform across each ring, for the reasons discussed in the previous section. Hence it is only necessary to fit the orbit of the center of mass of each ring (approximated by its center line), rather than the two edges separately. The uniform ring precession rates are taken to be the rates appropriate to an infinitely narrow ring located at the center line.

The origin of the apsidal and nodal precession can be understood most readily in terms of the three natural angular velocities of particle motion about

TABLE VI

Fitted Model Parameters: Orbital Elements[a]

Ring	Semimajor Axis a (km)	Eccentricity $e \times 10^3$	Azimuth of Periapse ϖ_0 (deg)[b]	Inclination i (deg)	Azimuth of the Ascending Node Ω_0 (deg)[b]
6	41877.3 ± 16.6	1.01 ± 0.10	243.6 ± 3.6	0.066 ± 0.012	15.7 ± 4.4
5	42275.2 ± 16.6	1.85 ± 0.08	170.1 ± 2.5	0.050 ± 0.010	288.5 ± 9.1
4	42609.6 ± 16.8	1.15 ± 0.04	125.3 ± 3.0	0.022 ± 0.005	117.2 ± 24.7
α	44758.3 ± 16.4	0.78 ± 0.02	329.4 ± 2.3	0.017 ± 0.003	65.4 ± 13.6
β	45701.0 ± 16.5	0.43 ± 0.02	223.9 ± 2.8	0.006 ± 0.002	296.3 ± 41.0
η	47214.9 ± 16.5	$(0.03 \pm 0.04)^d$	(129.0 ± 58.0)	(0.003 ± 0.004)	(263.0 ± 117.7)
γ	47666.3 ± 16.4	(0.04 ± 0.02)	(84.0 ± 22.3)	0.006 ± 0.002	100.2 ± 31.1
δ	48338.7 ± 16.5	0.06 ± 0.02	135.1 ± 15.8	0.012 ± 0.003	299.0 ± 11.5
ϵ	51188.1 ± 17.0	7.94 ± 0.02	214.9 ± 0.4	(0.003 ± 0.003)	(242.9 ± 29.5)

Harmonic Coefficients of the Gravity Potential[c]

$J_2 = (3.349 \pm 0.005) \times 10^{-3}$

$J_4 = (-3.8 \pm 0.9) \times 10^{-5}$

Pole of the Ring Plane

$\alpha_{1950.0} = 5^h\,06^m\,29^s.4 \pm 5^s.3$

$\delta_{1950.0} = +15°\,14'\,10'' \pm 1'31''$

[a] For $M_u = 8.669 \times 10^{28}$ gm and $G = 6.670 \times 10^{-8}$ dyn cm^2 gm^{-2}; this tabulation is for the model of French et al. (1982).

[b] At 20:00 UT on 10 March 1977.

[c] For a reference radius, $R = 26,200$ km.

[d] Those eccentricities, inclinations, and associated angles within parentheses cannot be distinguished from zero.

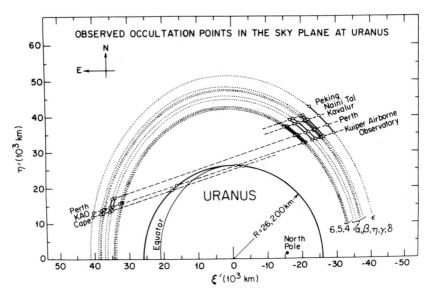

Fig. 13. Observed occultation points in the sky plane at Uranus for the 10 March 1977 occultation of SAO 158687. The observed points are indicated by open circles, pre-immersion points to the *right* and post-emersion points to the *left*. The dashed lines show the tracks of the observatories in the sky plane. The rings are indicated by dotted lines and the solid segments between the observed occultation points. The radius of Uranus corresponds to a number density level of $6 \times 10^{13}\,\mathrm{cm}^{-3}$ (figure after Elliot et al. 1978).

an oblate planet: the mean angular velocity* of the orbital motion n; the mean angular velocity of radial oscillations for a particle slightly displaced from a circular orbit κ; and the mean angular velocity* of vertical oscillations for a particle slightly displaced from the planet's equatorial plane ν. In terms of the mass and equatorial radius of the planet M and R and the usual zonal gravity harmonic coefficients J_2 and J_4, these three angular velocities are given by the following expressions, correct to first order in orbital eccentricity and inclination for terms involving J_2, J_2^2 and J_4:

$$n = \left(\frac{GM}{a^3}\right)^{\frac{1}{2}} \left[1 + \tfrac{3}{4} J_2 \left(\frac{R}{a}\right)^2 - \tfrac{9}{32} J_2^2 \left(\frac{R}{a}\right)^4 - \tfrac{15}{16} J_4 \left(\frac{R}{a}\right)^4\right] \tag{4}$$

$$\kappa = \left(\frac{GM}{a^3}\right)^{\frac{1}{2}} \left[1 + \tfrac{3}{4} J_2 \left(\frac{R}{a}\right)^2 - \tfrac{9}{32} J_2^2 \left(\frac{R}{a}\right)^4 - \tfrac{45}{16} J_4 \left(\frac{R}{a}\right)^4\right] \tag{5}$$

$$\nu = \left(\frac{GM}{a^3}\right)^{\frac{1}{2}} \left[1 + \tfrac{9}{4} J_2 \left(\frac{R}{a}\right)^2 - \tfrac{81}{32} J_2^2 \left(\frac{R}{a}\right)^4 - \tfrac{75}{16} J_4 \left(\frac{R}{a}\right)^4\right]. \tag{6}$$

---

*The units of "angular velocity" are radians/seconds.

Since $n > \kappa$, the longitude of periapse, measured from some inertially fixed direction, advances at a rate*

$$\dot{\tilde{\omega}} = n - \kappa = \left(\frac{GM}{a^3}\right)^{\frac{1}{2}} \left[\tfrac{3}{2} J_2 \left(\frac{R}{a}\right)^2 - \tfrac{15}{4} J_4 \left(\frac{R}{a}\right)^4\right] \ . \qquad (7)$$

Similarly, the longitude of the node regresses at a rate

$$\dot{\Omega} = n - \nu = -\left(\frac{GM}{a^3}\right)^{\frac{1}{2}} \left[\tfrac{3}{2} J_2 \left(\frac{R}{a}\right)^2 - \tfrac{9}{4} J_2^2 \left(\frac{R}{a}\right)^4 - \tfrac{15}{4} J_4 \left(\frac{R}{a}\right)^4\right] \ . \qquad (8)$$

The fixed reference direction is chosen as the ascending node of Uranus' equatorial plane on the Earth's equator of 10 March 1977, 20[h] UT.

In Eqs. (4) through (8), the quantity a is the *geometrical* semimajor axis of the ring, i.e., the semimajor axis of the best fitting ellipse traced by the ring particles. The deviation of the ring from an ellipse due to the harmonics of the Uranian potential would be order $ae^2J_2(R/a)^2$, only ~0.003 km for the ϵ Ring. Although a 2nd-order analysis is adequate for the instantaneous geometry of the rings, the secular effects of the precessions will require a 3rd-order treatment within the next few years. This will introduce terms in Eqs. (7) and (8) of order J_2^3, e^2J_2, J_2J_4, $i^2 J_2$, J_3 and J_6. The required expressions can be obtained by extending the analysis of Brouwer (1946) to the next order.

As discussed by Greenberg (1981), equivalent forms of Eqs. (7) and (8) have been written in terms of the osculating Keplerian elements (Brouwer 1959) and in terms of the observables (n and a) for satellite orbits (Kozai 1959; Brouwer 1946; Null et al. 1981). These three forms differ in the numerical coefficient of the J_2^2 term, since our a is not the osculating Keplerian semimajor axis and $n \neq (GM/a^3)^{\frac{1}{2}}$ (see Eq. 4).

These equations are the only dynamics included in the present kinematic model; the rest of the job just involves being careful with geometry. The method basically follows Smart's (1977) procedure, except that it becomes conceptually easier to project the occultation points into a three-dimensional coordinate system at Uranus instead of using Smart's fundamental plane that passes through the center of the Earth. Details of the method are discussed by Elliot et al. (1978, 1981b) and French et al. (1982).

We summarize the steps in the procedure as follows. First, the midtimes of the ring occultations are determined by either fitting a model profile or finding the midtime between the half-maximum signal points on the occultation profile. Next, the ephemerides of Uranus *and the star* are calculated, including the effects of precession and nutation but not stellar aberration, and these data are used to calculate coordinates and velocities (relative to the center of Uranus) for the star at the occultation times. This information is then available to a least-squares fitting program that minimizes the sum of the

---

*This rate is also denoted by $\dot{\tilde{\pi}}$ in the literature.

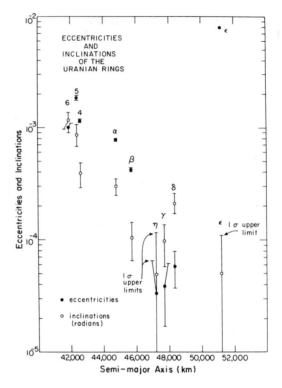

Fig. 14. Eccentricities (●) and inclinations (○) of the Uranian rings. The eccentricity and inclination (in radians) for each of the nine presently known rings are plotted against the semimajor axis. Except for the eccentricity of the ε Ring, the e's and i's show a decreasing trend with increasing semimajor axis. The eccentricities of the η and γ Rings, as well as the inclinations of the η and ε Rings, are not large enough to be statistically significant (figure after French et al. 1982).

squared residuals (in time) between the data and the orbit model. In accordance with our geometrical definition of the semimajor axis, the eccentricity e and inclination i (relative to Uranus' equatorial plane) are also defined in terms of the inclined ellipse traced by the ring particles. The free parameters in the final solution are:

1. a, e, and i for each ring;
2. The longitude of periapse (ϖ_o) and the longitude of the ascending node (Ω_o) for each ring at a specified epoch (10 March 1977 at 20^h UT);
3. The right ascension and declination (α_p and δ_p) for the rotation pole of Uranus;
4. J_2 and J_4 for Uranus;
5. Corrections in right ascension and declination ($\triangle\alpha$ and $\triangle\delta$) to the coordinates of each occulted star.

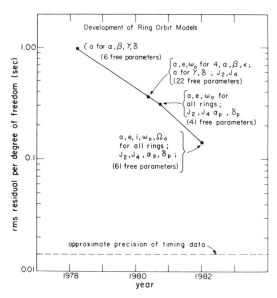

Fig. 15. Development of ring orbit models. As the model has been made more realistic by including more parameters, the rms error per degree of freedom has decreased. However, the rms error of the present model is still substantially larger than the timing errors in the data, indicating that the model is not yet complete.

The most recent model, incorporating data through April, 1981, fits 61 parameters to 140 data points and is given in Table VI, where the errors are the formal errors in the parameters from the least-squares fit. One interesting relation between the orbit parameters is shown in Fig. 14, where the eccentricity and inclination of each ring is plotted versus its semimajor axis. Except for the eccentricity of the ϵ ring, we see a generally decreasing trend of the eccentricities and inclinations as the semimajor axis increases. This effect may or may not prove significant, and no explanation has been proposed so far.

The precision of the orbit model has continually improved, as indicated by the rms error per degree of freedom in the graph in Fig. 15. Here we see that the rms error has steadily decreased with each major improvement to the model. However, we still have significant improvements to make before we reach the limit of the precision (~ 0.015 s) with which we can determine the midtime of a ring occultation profile in the presence of random noise in the data. Some of the effects that might be causing the large residuals are: (i) systematic errors in the times reported at different observing stations for the same occultation; (ii) velocity errors in the ephemeris of Uranus; (iii) the manner in which the midtimes are defined for the rings of irregular structure such as the α Ring; and (iv) unknown causes.

Other effects, which have been discussed by Freedman et al. (1983), are (i) the motion of Uranus relative to the barycenter of the system between the immersion and emersion occultations; (ii) the forced ring precession due to possible shepherd satellites between the rings; and (iii) perturbations of the orbits from ellipses, again due to the possible shepherds.

IV. IMAGING AND SPECTRA

While the systematic observation of stellar occultations has led to a detailed picture of the kinematics, widths, optical depths and radial structure of the Uranian rings (as discussed in the preceding sections), our knowledge concerning the nature of the ring particles themselves is much less complete. In particular, we know neither the composition of these particles nor their size distribution. The most direct way of approaching these questions is to study both the spectral and spatial distribution of sunlight reflected from the rings. Knowledge of the reflection spectrum will constrain, and perhaps uniquely specify, the surface composition of the ring particles, while information on the particle size distribution can be obtained from a study of the phase function of the rings at different wavelengths. Information on the particle size distribution can also be obtained from occultation observations at different wavelengths. A fourth potential source of valuable information concerning both composition and particle size is the spectrum of thermal flux emitted from the rings.

Unfortunately, the possibilities for making such observations with Earth-based telescopes are extremely limited for several reasons. Foremost of these is the extreme narrowness of the rings and thus their small projected area on the sky (1.3% of the area of Uranus' disk, under the most favorable conditions). This, combined with a maximum angular diameter of the ring system of only 8 arcsec, makes direct observations virtually impossible except at wavelengths corresponding to deep absorption bands in the spectrum of Uranus. Sufficiently deep bands, due largely to CH_4, occur only in the near infrared (0.8–4.0 μm), precluding useful visual photoelectric or photographic observations of the rings. Two further limitations on Earth-based observations stem from Uranus' great distance from the Sun: a maximum observable phase angle of only $3°.1$, and an equilibrium temperature for dark, isothermal ring particles of 65 K. At this temperature, and for a particle geometric albedo of 0.03 (see below), thermal emission from the rings exceeds reflected sunlight only for wavelengths longer than 11.5 μm, where the Earth's atmosphere is largely opaque. In the radio region, the anticipated thermal flux from the rings, $\sim$0.07 mJy at a wavelength of 3 cm, or $\sim$0.4% of the measured flux from Uranus, is almost certainly undetectable. Radar observations of the Uranian rings, such as have been successfully made of Saturn's rings, are rendered impractical by a combination of small target area, great distance, a two-way light-travel time of $\sim$5 hr, and, at present, Uranus' southerly declination (which puts it beyond the range of the Arecibo antenna).

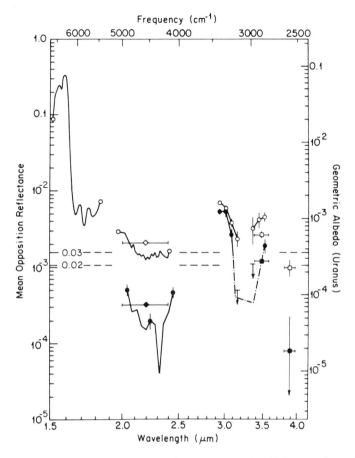

Fig. 16. Spectral reflectances of the Uranian system between 1.5 and 3.9 μm as observed with both large (open symbol) and small (filled symbol) circular apertures centered on Uranus. Solid lines and circles represent circular variable filter data (resolution 0.03 to 0.05 μm); squares represent narrowband filter measurements at 3.48 and 3.89 μm; diamonds represent broadband 2.2 μm measurements. Error bars are included for representative points, when larger than the diameter of the symbols. 1-σ upper limits are shown at 3.15 and 3.36 μm in the small aperture spectrum. Ordinate scale on right gives the geometric albedo of Uranus for an equatorial radius of 25,600 km, and ignores the contribution by the rings to the large aperture data. Horizontal dashed lines indicate the reflectance of the rings alone for ring geometric albedos of 0.02 and 0.03 (figure after Nicholson et al. 1983).

A. Infrared Spectrophotometry

Pending the encounter of Voyager 2 with Uranus in January 1986 and the advent of the Space Telescope, when many of these difficulties should be overcome, we are restricted to observations in selected regions of the near-infrared spectrum. Fig. 16 shows the spectral reflectance of Uranus and its rings in the wavelength range 1.5 to 4.0 μm (Nicholson et al. 1983). It is

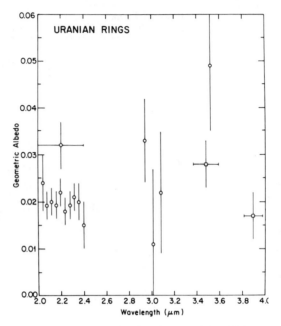

Fig. 17. The geometric albedo of the Uranian rings between 2.0 and 3.9 μm, as derived from the observations and an average integrated ring width of 85 km. Circles, squares, and diamonds have same meaning as in Fig. 16 (figure after Nicholson et al. 1983).

apparent that the rings should be most readily detected and studied in the 2.0–2.5 μm and 3.1–4.0 μm bands. The former is due to a combination of several CH_4 bands and the fundamental pressure-induced vibrational absorption of H_2, while the latter is due to the ν_3 fundamental of CH_4.

Spectrophotometric observations in the 2.0–2.5 μm region, with spectral resolutions of ~0.04 μm, have been published by Nicholson and Jones (1980) and Soifer et al. (1981). These observations have been repeated and extended to the 1.5–1.85 and 2.95–4.0 μm regions by Nicholson et al. (1983), whose derived albedo spectrum for the rings is shown in Fig. 17. This spectrum was obtained by subtracting the spectrum of Uranus, recorded with small circular apertures of 3.5 to 5.9 arcsec diameter, from the combined spectrum of Uranus plus rings recorded with 10 to 12 arcsec diameter apertures. This method can obviously provide reliable results only at wavelengths where the integrated flux from the rings is comparable to, or greater than, that from the planet itself. A significant source of uncertainty in the derived ring albedos lies in the seeing corrections that must be applied to the Uranus spectrum in order to account for planetary flux spilling out of the necessarily undersized small aperture. These corrections are estimated from multi-aperture observations of Uranus at 1.25 and 1.65 μm, where the contributions from the rings are insignificant (Matthews et al. 1982).

Within the uncertainties in the data, the spectrum of the rings appears to be essentially flat between 2.0 and 4.0 μm. Average geometric albedos, calculated for an integrated effective width for the 9 rings of 85 km, are 0.020 ± 0.003 between 2.08 and 2.40 μm, 0.024 ± 0.006 between 2.94 and 3.08 μm, 0.028 $\pm$ 0.005 at 3.5 μm, and 0.017 $\pm$ 0.005 at 3.9 μm. Broadband 2.0–2.4 μm photometry yields a slightly higher value of 0.032 $\pm$ 0.005 (Matthews et al. 1982). These albedos refer to the rings as a whole, and most particularly to the ϵ Ring, which contributes $\sim$70% of the surface area. The geometric albedos of the ring particles must be slightly higher because of the finite ($\tau \simeq 1$ to 2) optical depth of the rings.

At shorter wavelengths, only rough estimates of, or upper limits on, the ring albedo are available. At 0.89 μm, Thomsen et al. (1978) measured a geometric albedo of 0.02 $\pm$ 0.01, while Sinton (1977) and Smith (1977) set upper limits of 0.05 and 0.01, respectively, for integrated ring widths of 100 km. More recently, B. A. Smith and J. A. Westphal (personal communications) have each succeeded in detecting the rings at 1.0 μm using CCD imaging systems, but no useful albedo estimates are available. Nicholson et al. (1983) estimate an upper limit to the ring geometric albedo of 0.02 at 1.7 μm. It appears, therefore, that the rings are quite dark at all wavelengths at which they have been observed, covering, albeit very incompletely, the range 0.89 to 3.9 μm.

While no direct identification of the surface composition of the ring particles can be made from our present knowledge of the reflection spectrum, certain inferences can be drawn from the lack of prominent spectral features in the 2.0–4.0 μm region. First, the ring particles are not covered by substantial amounts of either H_2O or NH_3 frost. This follows from both the very low 2.0–2.4 μm albedo and the absence of a strong absorption feature at 3.0 μm exhibited by both of these materials (Larson and Fink 1977). In addition, H_2O frost absorbs at 2.0 and >2.5 μm quite strongly, while NH_3 frost absorbs at 2.0 and 2.2–2.3 μm (Kieffer and Smythe 1974; Larson and Fink 1977). This conclusion is significant in light of the identification of H_2O frost on the surfaces of four of the Uranian satellites (Cruikshank 1980; Cruikshank and Brown 1981; Soifer et al. 1981).

A second constraint from the spectral data is that the surface material of the ring particles does not contain appreciable quantities of bound or adsorbed H_2O. Hydrous materials invariably show a broad, strong absorption feature centered at 2.9–3.0 μm (Hovis 1965; Fink and Burk 1973; Pollack et al. 1978), which is not apparent in the ring spectrum. This feature persists even when opaque materials such as carbon are mixed with the hydrous material (Larson et al. 1979), although the weaker 1.4 and 1.9 μm H_2O features may be suppressed in such a mixture (Johnson and Fanale 1973).

A comparison of the spectrum, and more especially the albedo, of the Uranian rings with published spectra of other solar system objects does not reveal any likely matches. Among the meteorites, only carbonaceous chondrites, ureilites and "black chondrites" have 2.0–2.5 μm reflectances ($\simeq$ Bond albedos) $\leq$0.10 (Gaffey 1976). Of these, only the primitive C1 and C2 carbonaceous chondrites have reflectances $\leq$0.05, and Larson et al. (1979) have shown that these objects exhibit strong 3.0 μm absorptions due to H_2O. Several small satellites, notably Phobos, Deimos and Amalthea, have visual geometric albedos of 0.04 to 0.06 (Veverka 1977; Smith et al. 1979). Amalthea, however, has a considerably higher reflectance in the 2.0–2.5 μm region (Neugebauer et al. 1981). We still lack 3–4 μm observations of these objects. The leading (i.e., dark) side of Iapetus has a geometric albedo of $\sim$0.12 in the 2.0–2.5 μm region (Soifer et al. 1979), and a deep absorption feature at 3.0 μm which is evidently due to small amounts of H_2O frost (Lebofsky et al. 1982). A few data are available concerning the infrared spectra ($>$1.1 μm) of asteroids (Larson and Veeder 1979). C-type asteroids have relatively flat 0.6–2.2 μm spectra and geometric albedos $\leq$0.065, with some as low as 0.02–0.03 (Morrison and Lebofsky 1979). The best-studied asteroid of this type, 1 Ceres, exhibits a pronounced 3.0 μm absorption, interpreted in terms of bound H_2O in the surface material (Lebofsky 1978; Larson et al. 1979; Lebofsky et al. 1981), although not all C-type asteroids show this feature (Lebofsky 1980).

Is it possible that the rings are uniformly dark in backscattered light because the scattering cross section is dominated by particles with sizes comparable to, or smaller than, the wavelengths of observation, rather than because the particles themselves are intrinsically dark? A direct answer to this question should be provided by Voyager 2, which will observe the rings at a wide range of phase angles. At present, the particle size distribution in the rings is constrained by the mean surface densities derived from their observed eccentricity gradients using the Goldreich and Tremaine model (see Table V). For the ϵ Ring, a mean surface density of 26 g cm^{-2}, combined with an estimated mean optical depth of $\sim$2, suggests an effective particle diameter of 20 cm, for a particle density of 1 g cm^{-3}. It is difficult, although perhaps not impossible, to construct a size distribution which reconciles this effective, mass-weighted diameter with an effective diameter for light scattering of $\sim$1 μm. There also does not appear to be a strong dependence of the geometric albedo of the rings on wavelength, which provides a further argument against scattering by micron-sized particles. Finally, the occultation profiles of the rings have as yet shown no discernable differences for wavelengths of 0.62, 0.75, 0.85 and 2.2 μm, again arguing for a small population of micron-sized particles.

B. 2.2 μm Mapping

With the above-mentioned exception of CCD observations at either 0.89 or 1.0 μm, there are no presently available astronomical techniques capable of directly imaging the Uranian rings from the surface of the Earth. (This situation may, however, soon be remedied by the development of arrays of infrared detectors.) Because of the greatly reduced scattering of planetary flux, and the absence of atmospheric seeing, the Space Telescope should provide useful visual images after the Voyager 2 encounter. Even the Space Telescope, of course, will not be able to resolve the 10^{-4} to 10^{-2} arcsec widths of the rings, although it should permit studies of the photometric and spectroscopic properties of individual rings (see Chapter by Smith).

It is possible, however, to produce maps of the broad distribution of reflected light from the rings from scans of the system made with an infrared photometer in the 2.2 μm (K) band. Two such maps, with resolutions of 5 arcsec (or 65,000 km), are presented in Fig. 18, and again in Fig. 19 in the form of gray-scale images. These figures have been adapted from Matthews et al. (1982). The maps were actually produced from sets of simultaneous 1.6 and 2.2 μm scans. At 1.6 μm, only the planet appears, while at 2.2 μm the (integrated) brightness of the ring system is about three times that of Uranus. The 1.6 μm scans may, therefore, be used to subtract the planetary component from the 2.2 μm scans, leaving the contribution from the rings alone.

The principal significance of these maps is that they are sensitive to the possible presence of broad, optically thin, components of the ring system, such as have been reported by Bhattacharyya and Bappu (1977). Material with a normal optical depth ≤ 0.05, distributed smoothly, is extremely difficult to detect in stellar occultation recordings, although if spread over the 9000 km radial range of the 9 narrow rings, such material could account for up to 80% of the light reflected by the system. That this situation is, in fact, not the case is demonstrated by the approximately 2:1 azimuthal brightness variation exhibited by both maps. This variation is primarily due to the variable width of the ϵ Ring (see Figs. 8 and 10), which provides $\sim$70% of the total surface area of the narrow rings. The position angle of minimum ring brightness in each map coincides with the predicted position angle of the periapse, or narrowest part, of this ring based on the kinematic model described in Sec. III. In contrast to this observed azimuthal variation, any broad component of the ring system should appear axially symmetric, because of the rapidity with which any clumps of material are sheared out by the orbital motion of the particles (period $\sim$8 hr). A quantitative comparison of the ratio of observed brightnesses at the positions of the ϵ Ring apoapse and periapse with the predicted ratio based on computer-generated model maps leads to an upper limit of 0.003 on the optical depth of any *axisymmetric* ring component with a nominal width of 5000 km (Matthews et al. 1982). These model maps, shown also in Figs. 18 and 19, incorporate the known width variation and apsidal precession of the ϵ Ring, as well as a constant integrated width of 30 km for the other 8 rings (the width variations of the α and β Rings are comparatively unimportant, in this context). In addition, the above-mentioned comparison of brightness ratios takes into account a variation in optical depth around the ϵ Ring, in inverse proportion to the width variation. If the ϵ Ring's optical depth is, on the other hand, essentially constant (an unlikely possibility) then

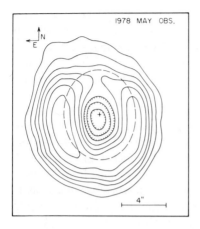
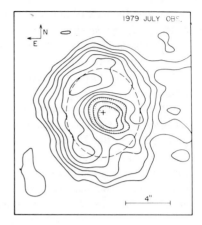
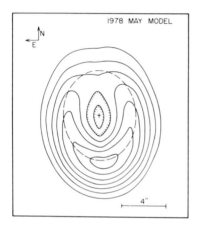
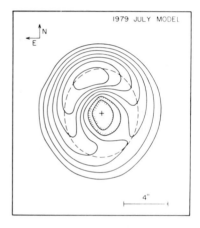

Fig. 18. Contour maps of reflected light from the rings of Uranus at a wavelength of 2.2 μm. Simultaneous observations at 1.5 or 1.65 μm were used to subtract the planetary component of flux at 2.2 μm, as described in the text. Also shown are model maps constructed for the dates on which the observations were made. The model consists of a single circular ring whose brightness is proportional to the projected integrated width of the nine known narrow rings, and whose orientation and precession rate are those determined for the ε Ring from occultation observations (see text). The location of the ε Ring is indicated by a dashed ellipse, while the cross represents the center of Uranus' disk, which has a diameter of ~4 arcsec. The maps cover an area of 16 × 16 arcsec. Contours represent flux levels on an arbitrary but linear scale, with the scale of the models adjusted to match that of the observations. Zero and negative contours are suppressed for clarity; the 1-σ noise level in the 1978 map is ±0.4 contour interval, and ±0.6 contour interval in the 1979 map. Resolution of all maps is ~5 arcsec, as determined by the convolution of the 4-arcsec scanning aperture with seeing of 1.5 to 2 arcsec. (a) Map constructed by averaging eight sets of scans made on 19 May 1978; (b) Map constructed by averaging four sets of scans on 6 and 7 July 1979; (c) Model map for 19 May 1978; (d) Model map for 6–7 July 1979 (figure after Matthews et al. 1982).

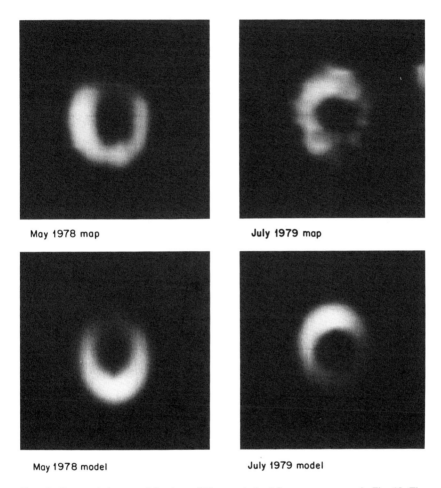

May 1978 map   July 1979 map

May 1978 model   July 1979 model

Fig. 19. Gray-scale images of the rings of Uranus, derived from contour maps in Fig. 18. The May 1978 map represents the average of 4 hr of observations on the Hale 5-m telescope on 19 May 1978, and the July 1979 map a total of 2 hr of observation on 6 and 7 July 1979. Resolution of the images is ~5 arcsec, as set by the convolution of the 4 arcsec diameter scanning aperture with atmospheric seeing. Each image covers an area of 20 × 20 arcsec, centered on Uranus. North up, east to the left. Below each observed map is a model map for that date, based on stellar occultation observations and a computer simulation of the scanning procedure used to obtain the real maps. The model incorporates the known width variation and apsidal precession rate of the ε Ring, as well as a constant integrated width of 30 km for the other 8 rings. The satellite Ariel ([2.2 μm] = 13.0) appears on the western edge and again, more faintly, on the eastern edge of the 6–7 July 1979 map. Miranda (estimated [2.2 μm] ~15) was too faint and/or fast moving to detect in these data (figure adapted from Matthews et al. 1982).

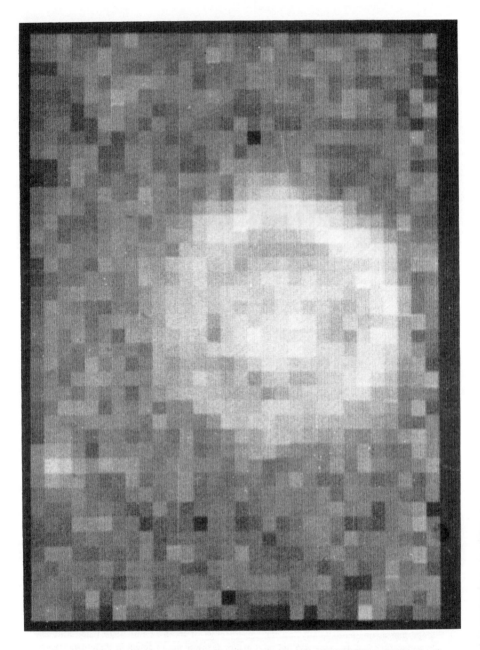

Fig. 20. Image of Uranus and its rings at 2.2 μm, taken in May 1982 by Allen (1983). East is to the left and north is upward on the page. Satellite Miranda appears near the left edge of the picture. (Copyright 1982, Anglo-Australian Telescope Board.) See Color Plate 1 for a more recent color image of the rings of Uranus by D. Allen.

the upper limit is raised to ~0.010. Either limit is considerably below the value of ~0.025 reported by Bhattacharyya et al. (1979), based on low amplitude, low-frequency photometric variations in an occultation recording from 10 March 1977. Implicit in the above discussion is the assumption that the 2.2 μm albedo of the particles in any extended, optically thin ring is equal to, or greater than, that of the particles in the narrow rings.

Although the principal features of the 2.2 μm maps may be understood in terms of the known widths and kinematics of the rings, the significant enhancement in brightness of the east side of the system relative to the west side observed in May 1978 is not. It has been shown (Matthews et al. 1982) that the effect cannot be explained in terms of a bright satellite or condensation within the rings, a background star, or differences between the brightness distributions across the disk of Uranus at 1.6 and 2.2 μm. Again, a broad, axisymmetric ring should not produce such an asymmetry in the brightness distribution. In May 1982, when the periapse of the ϵ Ring was located in the south rather than in the north as in May 1978, Allen (1983) obtained a third 2.2 μm map of the rings (Fig. 20) which shows an enhancement in brightness of the west side. This suggests that the anomalously bright region precesses with the apsidal line of the ϵ Ring, although this tentative conclusion should be tested by further observations. The location of the bright region in the quadrant following passage of the ring particles through periapse, where collisions are most probable (Dermott and Murray 1980), suggests further that the phenomenon may be due to scattering from collisional debris, which is subsequently reaccreted. However, no evidence of such debris has been found in occultation profiles of the ϵ Ring.

V. THE RINGS AS PROBES OF URANUS

Studies of the Uranian rings, and in particular the development of precise kinematic models for the ring orbits, have led to a considerable improvement in our knowledge of several important parameters of Uranus itself. First is the determination of the gravity harmonic coefficients J_2 and J_4 from the apsidal and nodal precession rates of the eccentric and/or inclined rings (see Table VI). The control on J_4, whose influence on precession rates drops off as $a^{-11/2}$, stems from the close proximity of the rings to the planet. From Table VII, which compares the gravity coefficients of Uranus with those of Jupiter and Saturn, obtained from spacecraft flybys, we see that the Voyager flyby of Uranus in 1986 is unlikely to improve on the precision obtainable from the ring orbits. Improved accuracy, however, should come from a more accurate determination of GM_u, which is a source of systematic error in J_2 that could well be greater than its current uncertainty (see Table VI).

Secondly, the rings serve as a precise reference system for locating occultation-derived atmospheric temperature and density profiles with respect to the center of mass of Uranus. This has permitted a determination of the

TABLE VII

Rotation Periods of the Jovian Planets[a]

	Jupiter	Saturn	Uranus	Neptune
Oblateness (ϵ)	0.0648 ± 0.0001[1]	0.088 ± 0.006[4]	0.024 ± 0.003[7]	0.021 ± 0.004[12]
Equatorial radius (R_e, km)	$71,541 \pm 4$[1]	$60,000 \pm 500$[4]	$26,145 \pm 30$[7]	$25,225 \pm 30$[12]
$GM \times 10^{-3}$ (km^3 s^{-2})	$126,686.9 \pm 0.5$[2]	$37,939 \pm 2$[5]	$5,784 \pm 4$[9]	$6,787 \pm 5$[13]
$J_2 \times 10^3$	14.733 ± 0.004[2]	16.479 ± 0.018[5]	3.349 ± 0.005[8]	3.5 ± 0.4[14]
$J_4 \times 10^6$	-587 ± 7[2]	-937 ± 38[5]	-38 ± 9[8]	?
Reference radius (R_r, km)	$71,398$	$60,000$	$26,200$	$25,225$
Calculated rotation period (Eq. [9] in text)	$9^h\ 54^m\ 52^s \pm 44^s$	$11^h\ 19^m \pm 37^m$	15.6 ± 1.4 hr	14.9 ± 1.6 hr
Measured rotation period	$9^h\ 55^m\ 29\overset{s}{.}71 \pm 0\overset{s}{.}04$[3]	$10^h\ 39^m\ 24^s \pm 7^s$[6]	16.2 ± 0.3 hr[10] (24 ± 3 hr[11])	17.7 ± 0.1 hr[15]
Difference (measured−calculated)	$38^s \pm 44^s$	$40^m \pm 37^m$	0.6 ± 1.4 hr	2.8 ± 1.6 hr

[a]Errors have been estimated when not explicitly given by the original source, and the error in the calculated period reflects that introduced by the error in the oblateness.

[b]References: 1. Lindal et al. (1981); 2. Null (1976); 3. Smoluchowski (1976); 4. Gehrels et al. (1980); 5. Null et al. (1981); 6. Desch and Kaiser (1981); 7. Elliot et al. (1981a); 8. Table VI in text; 9. Dunham (1971); 10. Goody (1982); 11. Hayes and Belton (1977); 12. Kovalesky and Link (1969); 13. Gill and Gault (1968); 14. Harris, personal communication, 1982; 15. Brown et al. (1981).

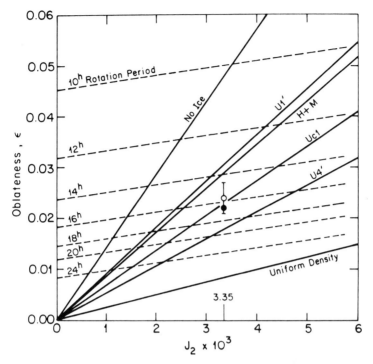

Fig. 21. Predicted values of oblateness, ϵ, and J_2 as a function of rotation period of several recently published interior models of Uranus. The open symbol represents the current best estimates of these two quantities based on stellar occultation observations of both planet and rings (see text and Table VI), while the filled symbol indicates Franklin et al.'s (1980) determination of $\epsilon = 0.022 \pm 0.001$, based on stratoscope images. Models U1' (Podolak and Reynolds 1981) and H + M (Hubbard and MacFarlane 1980) are essentially equivalent and consist of three compositionally distinct layers (a rocky core of mass $\sim 3\ M_{\oplus}$, an "ice" layer of $\sim 10\ M_{\oplus}$, and a solar-composition envelope of $\sim 1.5\ M_{\oplus}$) while models Uc1 and U4' (Podolak and Reynolds 1981) incorporate various degrees of enrichment of H_2O, NH_3 and CH_4 in the outer envelope. Also shown are extreme models, corresponding to (i) a uniform density planet, and (ii) a rocky core surrounded by a pure H_2-He envelope ("no ice"). Figure is adapted from Hubbard and MacFarlane (1980) and Podolak and Reynolds (1981).

planet's oblateness (ellipticity) ϵ which is independent of the recent redetermination of the optical oblateness by Franklin et al. (1980). The measured oblateness, at a molecular number density level of $\sim 8 \times 10^{13}\ cm^{-3}$, is 0.024 $\pm$ 0.003 (Elliot et al. 1981a), in good agreement with Franklin et al.'s 0.022 $\pm$ 0.001. The corresponding equatorial radius at this same level is 26,145 $\pm$ 30 km, corresponding to a cloud-top radius of $\sim 25,600 \pm 100$ km.

From a knowledge of both J_2 and ϵ, and the assumption of hydrostatic equilibrium, a third important parameter may be calculated, namely the rotation period of the planet. Uranus' rotation period has proved extremely resistant to determination by the usual photometric and spectroscopic techniques, which have produced ambiguous and/or discordant results (Goody 1982). The rotation period P is given by the relation (Brouwer and Clemence 1961):

$$P = 2\pi \left[\frac{R_e^3(1-\epsilon)(1+\frac{3}{2}J_2)}{2GM(\epsilon-\frac{3}{2}J_2-\frac{5}{8}J_4-\frac{9}{4}J_2^2)} \right]^{\frac{1}{2}}. \tag{9}$$

The quantity R_e refers to the equatorial radius for which the geometric oblateness is ϵ.* If the reference radius R_r for the harmonic expansion of the gravity potential does not equal R_e, then defining $\rho = R_r/R_e$ one must substitute $\rho^2 J_2$ for J_2 and $\rho^4 J_4$ for J_4 in Eq. (9).

From the occultation-derived values for J_2, J_4 and ϵ, a rotation period of $P = 15.6 \pm 1.4$ hr is obtained. If Franklin et al.'s value of ϵ is used, the result is $P = 16.6 \pm 0.5$ hr. Although the assumption of hydrostatic equilibrium may be questionable, the basic soundness of this procedure is shown by the comparison in Table VII of the periods derived for Jupiter and Saturn from their measured values of J_2, J_4 and ϵ with the directly determined (radio) periods.

Accurate values of J_2, J_4, ϵ and P are of great importance in constraining interior models of Uranus. In Fig. 21 the predicted values of J_2 and ϵ, as a function of rotation period, for several recently constructed interior models (Hubbard and MacFarlane 1980; Podolak and Reynolds 1981) are compared with the measured values of these quantities. Already the constraint is useful, although a more accurate determination of P, independent of ϵ, is clearly necessary.

The fourth parameter that has now been determined with improved precision from studies of the rings is the orientation of Uranus' rotation axis (see Table VI). This was previously obtained from astrometric studies of the orbits of Uranus' five satellites (Dunham 1971), but use of the ring-determined pole direction in a reanalysis of the satellite observations might lead to improvements in the other orbit parameters.

Finally, analysis of stellar occultations by the rings is resulting in an accumulation of data that can eventually be used to improve our knowledge of the orbit of Uranus in relation to the celestial coordinate system. The position of each occulted star, relative to the orbit of Uranus, is known to a precision of $\sim 0\overset{''}{.}01$ (see e.g., Elliot et al. 1981a).

VI. FUTURE OBSERVATIONS

We can expect to learn a great deal more about the rings of Uranus during the next few years through groundbased observations, the Space Telescope and the Voyager encounter during January 1986. From groundbased observations, we can improve the kinematic model for the orbits. The improved

---

*Geometric oblateness is defined by $\epsilon = 1 - R_p/R_e$, where R_e and R_p are equatorial and polar radii.

model may allow the detection of effects due to unseen satellites within the rings. These satellites would cause the shapes of the rings to deviate from ellipses and also cause them to precess faster than presently modeled (Freedman et al. 1983). Also, we should make significant progress in our understanding of the ring profiles. If a bright enough star is occulted, perhaps we can detect inter-ring material.

With the planetary camera on the Space Telescope, we hope to separate nearly all nine rings and determine their individual albedos. For occultation observations with the high-speed photometer, scintillation noise will be absent, which will allow a more extensive search for inter-ring material. Also, occultations observed in the far ultraviolet will have better spatial resolution than current 2μm data by a factor of three, because of the smaller Fresnel scale.

The Voyager encounter will provide much more information about the rings, particularly from the occultation, tracking and imaging data. According to A. L. Lane (personal communication), several stellar occultations will be visible. These should provide spatial resolution up to 100 times better than achievable from Earth-based occultations, which would revolutionize our knowledge of the ring structure. Two occultations of radio signals transmitted by the spacecraft will also occur (Stone 1982) providing information on the abundance of cm- and larger-sized ring particles.

The resolution of the imaging will not be as good as we have already achieved with groundbased occultations, but would provide phase functions of the rings at different wavelengths, hopefully shedding further light on particle size distribution. The imaging data should also provide a more sensitive search for inter-ring material and the postulated confining satellites (Goldreich and Tremaine 1979a). It is possible that an analysis of images of the rings with different viewing geometries could lead to a refinement in the orientation of Uranus' rotation axis, which would in turn lead to reduction of the current ±17 km uncertainties in the absolute semimajor axes of the rings.

The tracking data will yield values for GM_u and J_2. Although the Voyager value for J_2 will not be as precise as that already obtained from the occultation data (Stone 1982), it should prove useful as an independent check on the latter. Infrared observations would certainly be of interest, but the low temperature and small area of the rings may prevent their detection by the Voyager infrared radiation spectrometer (IRIS).

VII. DISCUSSION

Apparently the Uranian system has an intermediate amount of material in comparison with the low optical-depth Jovian ring and the impressive ring system that surrounds Saturn. Since the Saturnian system contains relatively few free gaps, much of the theory describing the Saturnian system deals with wave phenomena in a continuous distribution of particles. For Uranus, we have nine discrete rings, and the theoretical problems are mainly concerned with explaining the dynamics of narrow rings and their sharp edges.

Two types of mechanism of ring confinement, both involving unseen satellites, have been proposed and investigated (see chapter by Dermott). For the "shepherd" satellite mechanism of Goldreich and Tremaine (1979a), each of the narrow rings would have a pair of satellites keeping it from spreading. The inclinations and eccentricities of the rings would then be forced by the satellites, which would be in eccentric and inclined orbits (Goldreich and Tremaine 1981). The model of Dermott et al. (1979) invokes satellites within the rings, whose particles move on horseshoe orbits. A recent explanation for sharp edges in terms of satellite resonances (Borderies et al. 1982) has not yet been applied to the Uranian rings. In the shepherd picture, the uniform precession of the rings is explained by the self-gravity of the ring (Goldreich and Tremaine 1979b; Borderies et al. 1983), an explanation not universally accepted (Dermott and Murray 1980). An important potential achievement of the Voyager flyby would be the detection of, or placing a significant upper limit on, the presence of these postulated satellites.

Further observational constraints on the confinement mechanism would be placed through the detection of inter-ring material. Such material must exist, at some level, since the smallest particles would leak out of a narrow ring due to the Poynting-Robertson effect (see chapter by Mignard) or plasma drag (see chapter by Grün et al.), if Uranus possesses a significant magnetic field.

Other theoretical work is needed to produce models that explain the details of the ring structure for the α, η and ϵ Rings. Important information for this work would be the particle size distribution and velocity dispersion, which are not presently known.

Once the present dynamics of the rings are well understood, we can then extrapolate these processes back in time, with the hope of inferring the age and possibly the origin of the rings. In this context, another question that must be answered is why the surfaces of the ring particles are not icy, as are the surfaces of the Uranian satellites (Cruikshank and Brown 1981) and the particles in Saturn's rings.

For the next few years, at least, we expect significant progress towards answering these questions, both from comparative studies with Saturn's narrow ringlets and from new observations of the Uranian rings from the ground, the Space Telescope, and the Voyager 2 encounter in 1986.

Acknowledgments. We are grateful to referees S. Dermott and B. Sicardy for their constructive criticisms of this review. Helpful comments and information were also provided by D. Allen, L. Esposito, R. French, R. Greenberg, A. Harris, K. Matthews, K. Meech and S. Tremaine. Uranian ring research has been supported, in part, by grants from NASA and NSF.

REFERENCES

Allen, D. A. 1983. Infrared views of the giant planets. *Sky Telescope* 65:110–112.

Bhattacharyya, J. C., and Bappu, M. K. V. 1977. Saturn-like ring system around Uranus. *Nature* 270:503–506.

Bhattacharyya, J. C., and Kuppuswamy, K. 1977. A new satellite of Uranus. *Nature* 267:331–332.

Bhattacharyya, J. C., Bappu, M. K. V., Mohin, S., Mahra, H. S., and Gupta, S. K. 1979. Extended ring system of Uranus. *Moon Planets* 21:393–404.

Borderies, N., Goldreich, P., and Tremaine, S. 1982. Sharp edges of planetary rings. *Nature* 299:209–211.

Borderies, N., Goldreich, P., and Tremaine, S. 1983. Precession of inclined rings. *Astron. J.* 88:226–228.

Bouchet, P., Perrier, C., and Sicardy, B. 1980. Occultation by Uranus. *I.A.U. Circ.* No. 3503.

Brahic, A. 1982. Planetary rings. In *Formation of Planetary Systems,* ed. A. Brahic (Toulouse: Cepadues Editions), pp. 651–724.

Brouwer, D. 1946. The motion of a particle of negligible mass under the gravitational attraction of a spheroid. *Astron. J.* 51:223–231.

Brouwer, D. 1959. Solution to the problem of artificial satellite theory without drag. *Astron. J.* 64:378–397.

Brouwer, D., and Clemence, G. M. 1961. Orbits and masses of planets and satellites. In *The Solar System III,* eds. G. P. Kuiper and B. M. Middlehurst (Chicago: Univ. of Chicago Press), pp. 31–94.

Brown, R. H., Cruikshank, D. P., and Tokunaga, A. T. 1981. The rotation period of Neptune's upper atmosphere. *Icarus* 47:159–165.

Chen, D.-H., Yang, H.-Y., Wu, C.-H., Wu, Y.-C., Kiang, S.-Y., Huang, Y.-W., Yeh, C.-T., Chai, T.-S., Hsieh, C.-C., Cheng, C.-S., and Chang, C. 1978. Photoelectric observation of the occultation of SAO 158687 by Uranian ring and the detection of Uranian ring signals from the light curve. *Scientia Sinica* XXI:503–508.

Churms, J. 1977. Occultation of SAO 158687 by Uranian satellite belt. *I.A.U Circ.* No. 3051.

Cruikshank, D. P. 1980. Near-infrared studies of the satellites of Saturn and Uranus. *Icarus* 41:246–258.

Cruikshank, D. P., and Brown, R. H. 1981. The Uranian satellites: Water ice on Ariel and Umbriel. *Icarus* 45:607–611.

Dermott, S. F., and Murray, C. D. 1980. Origin of the eccentricity gradient and apse alignment of the ε ring of Uranus. *Icarus* 43:338–349.

Dermott, S. F., Gold, T. G., and Sinclair, A. T. 1979. The rings of Uranus: Nature and origin. *Astron. J.* 84:1225–1234.

Desch, M. D., and Kaiser, M. L. 1981. Voyager measurement of the rotation period of Saturn's magnetic field. *Geophys. Res. Letters* 8:253–256.

Dunham, D. W. 1971. Motions of the Satellites of Uranus. Ph.D. dissertation, Yale Univ.

Elliot, J. L. 1977. Signal-to-noise ratios for occultations by the rings of Uranus, 1977–1980. *Astron. J.* 82:1036–1038.

Elliot, J. L. 1979. Stellar occultation studies of the solar system. *Ann. Rev. Astron. Astrophys.* 17:445–475.

Elliot, J. L. 1982. Rings of Uranus: A review of occultation results. In *Uranus and the Outer Planets,* ed. G. A. Hunt (Cambridge: Cambridge Univ. Press), pp. 237–256.

Elliot, J. L., Dunham, E. W., and Millis, R. L. 1977a. Discovering the rings of Uranus. *Sky Telescope* 53:412–416, 430.

Elliot, J. L., Dunham, E. W., and Mink, D. J. 1977b. The rings of Uranus. *Nature* 267:328–330.

Elliot, J. L., Dunham, E. W., Wasserman, L. H., Millis, R. L., and Churms, J. 1978. The radii of Uranian rings α, β, γ, δ, ε, η, 4, 5 and 6 from their occultation of SAO 158687. *Astron. J.* 83:980–992.

Elliot, J. L., Elias, J. H., French, R. G., Frogel, J. A., Liller, W., Matthews, K., Meech, K. J., Mink, D. J., Nicholson, P. D., and Sicardy, B. 1983. A comparison of Uranian ring occultation profiles from three observatories. *Icarus.* In press.

Elliot, J. L., French, R. G., Frogel, J. A., Elias, J. H., Mink, D. J., and Liller, W. 1981a. Orbits of nine Uranian rings. *Astron. J.* 86:444–455.

Elliot, J. L., Frogel, J. A., Elias, J. H., Glass, I. S., French, R. G., Mink, D. J., and Liller, W. 1981b. The 20 March 1980 occultation by the Uranian rings. *Astron J.* 86:127–134.

Elliot, J. L., Veverka, J., and Millis, R. L. 1977c. Uranus occults SAO 158687. *Nature* 265:609–611.

Esposito, L. W., Borderies, N., Goldreich, P., Cuzzi, J. N., Holberg, J. B., Lane, A. L., Pomphrey, R. B., Terrile, R. J., Lissauer, J. J., Marouf, E. A., and Tyler, G. L. 1983. The eccentric ringlet in the Huygens gap at 1.45 Saturn radii: Multi-instrument Voyager observations. *Science.* Submitted.

Fink, U., and Burk, S. D. 1973. Reflection spectra, 2.5–7μ, of some solids of planetary interest. *Comm. Lunar Planetary Lab.* 10:8–20.

Fink, U., and Larson, H. P. 1979. The infrared spectra of Uranus, Neptune and Titan from 0.8 to 2.5 microns. *Astrophys. J.* 233:1021–1040.

Franklin, F. A., Avis, C. C., Colombo, G., and Shapiro, I. I. 1980. The geometric oblateness of Uranus. *Astrophys. J.* 236:1031–1034.

Freedman, A., Tremaine, S. D., and Elliot, J. L. 1983. Weak dynamical forcing of the Uranian ring system. *Astron. J.* 88:1053–1059.

French, R. G., Elliot, J. L., and Allen, D. A. 1982. Inclinations of the Uranian rings. *Nature* 298:827–829.

Gaffey, M. J. 1976. Spectral reflectance characteristics of the meteorite classes. *J. Geophys. Res.* 81:905–920.

Gehrels, T., Baker, L. R., Beshore, E., Blenman, C., Burke, J. J., Castillo, N. D., DaCosta, B., Degewij, J., Doose, L. R., Fountain, J. W., Gotobed, J., KenKnight, C. E., Kingston, R., McLaughlin, G., McMillan, R., Murphy, R., Smith, P. H., Stoll, C. P., Strickland, R. N., Tomasko, M. G., Wijesinghe, M. P., Coffeen, D. L., and Esposito, L. 1980. Imaging photopolarimeter on Pioneer Saturn. *Science* 207:434–439.

Gill, J. R., and Gault, B. L. 1968. A new determination of the orbit of Triton, pole of Neptune's equator, and mass of Neptune. *Astron. J.* 73:S95 (abstract).

Goldreich, P., and Tremaine, S. 1979a. Towards a theory for the Uranian rings. *Nature* 277:97–99.

Goldreich, P., and Tremaine, S. 1979b. Precession of the ϵ ring of Uranus. *Astron. J.* 84:1638–1641.

Goldreich, P., and Tremaine, S. 1981. The origin of the eccentricities of the rings of Uranus. *Astrophys. J.* 243:1062–1075.

Goldreich, P., and Tremaine, S. 1982. The dynamics of planetary rings. *Ann. Rev. Astron. Astrophys.* 20:249–283.

Goody, R. M. 1982. The rotation of Uranus. In *Uranus and the Outer Planets,* ed. G. A. Hunt (Cambridge: Cambridge Univ. Press), pp. 143–153.

Greenberg, R. 1981. Apsidal precession of orbits about an oblate planet. *Astron. J.* 86:912–914.

Hayes, S. A., and Belton, M. J. S. 1977. The rotational periods of Uranus and Neptune. *Icarus* 32:383–401.

Hovis, W. A. 1965. Infrared reflectivity of iron oxide minerals. *Icarus* 4:425–430.

Hubbard, W. B., and MacFarlane, J. J. 1980. Structure and evolution of Uranus and Neptune. *J. Geophys. Res.* 85:225–234.

Hubbard, W. B., and Zellner, B. H. 1980. Results from the 10 March 1977 occultation by the Uranian system. *Astron. J.* 85:1663–1669.

Johnson, T. V., and Fanale, F. P. 1973. Optical properties of carbonaceous chondrites and their relationship to asteroids. *J. Geophys. Res.* 78:8507–8518.

Kieffer, H. H., and Smythe, W. D. 1974. Frost spectra: Comparison with Jupiter's satellites. *Icarus* 21:506–512.

Klemola, A. R., and Marsden, B. G. 1977. Predicted occultations by the rings of Uranus: 1977–1980. *Astron. J.* 82:849–851.

Klemola, A. R., Mink, D. J., and Elliot, J. L. 1981. Predicted occultations by Uranus: 1981–1984. *Astron. J.* 86:138–140.

Kovalesvsky, J., and Link, F. 1969. Diamètre, aplatissement et propriétés optiques de la haute atmosphère de Neptune d'après l'occultation de l'étoile BD −17° 4388. *Astron. Astrophys.* 2:398–412.

Kozai, Y. 1959. The motion of a close earth satellite. *Astron. J.* 64:367–377.

Lane, A. L., Hord, C. W., West, R. A., Esposito, L. W., Coffeen, D. L., Sato, M., Simmons, K., Pomphrey, R. B., and Morris, R. B. 1982. Photopolarimetry from Voyager 2: Preliminary results on Saturn, Titan and the rings. *Science* 215:537–543.

Larson, H. P., Feierberg, M. A., Fink, U., and Smith, H. A. 1979. Remote spectroscopic identification of carbonaceous chondrite mineralogies: Applications to Ceres and Pallas. *Icarus* 39:257–271.

Larson, H. P., and Fink, U. 1977. The application of Fourier transform spectroscopy to the remote identification of solids in the solar system. *Appl. Spectrosc.* 31:386–402.

Larson, H. P., and Veeder, G. J. 1979. Infrared spectral reflectances of asteroid surfaces. In *Asteroids*, ed. T. Gehrels (Tucson: Univ. of Arizona Press), pp. 724–744.

Lebofsky, L. A. 1978. Asteroid 1 Ceres: Evidence for water of hydration. *Mon. Not. Roy. Astron. Soc.* 182:17–21.

Lebofsky, L. A. 1980. Infrared reflectance spectra of asteroids: A search for water of hydration. *Astron. J.* 85:573–585.

Lebofsky, L. A., Feierberg, M. A., Tokunaga, A. T., Larson, H. P., and Johnson, J. R. 1981. The 1.7–4.2 μm spectrum of asteroid 1 Ceres: Evidence for structural water in clay minerals. *Icarus* 48:453–459.

Lebofsky, L. A., Feierberg, M. A., and Tokunaga, A. T. 1982. Infrared observations of the dark side of Iapetus. *Icarus* 49:382–386.

Lindal, G. F., Wood, G. E., Levy, G. S., Anderson, J. D., Sweetham, D. N., Hotz, H. B., Bucklesa, B. J., Holmes, D. P., Doms, P. E., Eshleman, V. R., Tyler, G. L., and Croft, T. A. 1981. The atmosphere of Jupiter: An analysis of the Voyager radio occultation measurements. *J. Geophys. Res.* 86:8721–8727.

Liller, W. 1977. Colors and magnitudes of stars occulted by the rings of Uranus, 1977–1980. *Astron. J.* 82:929.

Mahra, H. S., and Gupta, S. K. 1977. Occultation of SAO 158687 by the Uranian rings. *I.A.U. Circ.* No. 3061.

Matthews, K., Neugebauer, G., and Nicholson, P. D. 1982. Maps of the rings of Uranus at a wavelength of 2.2 microns. *Icarus* 52:126–135.

Millis, R. L., and Wasserman, L. H. 1978. The occultation of BD-15° 3969 by the rings of Uranus. *Astron. J.* 83:993–998.

Millis, R. L., Wasserman, L. H., and Birch, P. 1977a. Detection of rings around Uranus. *Nature* 267:330–331.

Millis, R. L., Wasserman, L. H., Elliot, J. L., and Dunham, E. W. 1977b. The rings of Uranus: Their widths and optical thicknesses. *Bull. Amer. Astron. Soc.* 9:498 (abstract).

Morrison, D., and Lebofsky, L. A. 1979. Radiometry of asteroids. In *Asteroids*, ed. T. Gehrels (Tucson: Univ. of Arizona Press), pp. 184–205.

Neugebauer, G., Becklin, E.E., Jewitt, D. C., Terrile, R. J., and Danielson, G. E. 1981. Spectra of the Jovian ring and Amalthea. *Astron. J.* 86:607–610.

Nicholson, P. D., and Jones, T. J. 1980. Two-micron spectrophotometry of Uranus and its rings. *Icarus* 42:54–67.

Nicholson, P. D., Jones, T. J., and Gatley, I. 1983. The geometric albedo spectra of Uranus and its ring system, 1.5–4.0 μm. In preparation.

Nicholson, P. D., Matthews, K., and Goldreich, P. 1981. The Uranus occultation of 10 June 1979. I. The rings. *Astron. J.* 86:596–606.

Nicholson, P. D., Matthews, K., and Goldreich, P. 1982. Radial widths, optical depths and eccentricities of the Uranian rings. *Astron. J.* 87:433–447.

Nicholson, P. D., Persson, S. E., Matthews, K., Goldreich, P., and Neugebauer, G. 1978. The rings of Uranus: Results of the 10 April 1978 occultation. *Astron. J.* 83:1240–1248.

Null, G. W. 1976. Gravity field of Jupiter and its satellites from Pioneer 10 and Pioneer 11 tracking data. *Astron. J.* 81:1153–1161.

Null, G. W., Lau, E. L., Biller, E. D., and Anderson, J. D. 1981. Saturn gravity results obtained from Saturn Pioneer 11 tracking data and Earth-based Saturn satellite data. *Astron. J.* 86:456–468.

Podolak, M., and Reynolds, R. T. 1981. On the structure and composition of Uranus and Neptune. *Icarus* 46:40–50.

Pollack, J. B., Witteborn, F. C., Erickson, E. F., Strecker, D. W., Baldwin, B. J., and Bunch, T. E. 1978. Near-infrared spectra of the Galilean satellites: Observations and compositional implications. *Icarus* 36:271–303.

Porco, C., Borderies, N., Danielson, G. E., Goldreich, P., Holberg, J. B., Lane, A. L., and Nicholson, P. D. 1983. The eccentric ringlet at 1.29 R_s, In Proceedings of *I.A.U. Colloquium, Planetary Rings*, ed. A. Brahic, Toulouse, France, Aug. 1982.

Ridgway, S. T., Joyce, R. R., White, N. M., and Wing, R. F. 1980. Effective temperatures of late-type stars: The field giants from K0 to M6. *Astrophys. J.* 235:126–137.

Sicardy, B., Combes, M., Brahic, A., Bouchet, P., Perrier, C., and Courtin, R. 1982. The 15 August 1980 occultation by the Uranian system: Structure of the rings and temperature of the upper atmosphere. *Icarus* 52:454–472.

Sinton, W. M. 1977. Uranus: The rings are black. *Science* 198:503–504.

Smart, W. M. 1977. Occultations and eclipses. In *Textbook on Spherical Astronomy* (Cambridge: Cambridge Univ. Press), pp. 368–403.

Smith, B. A. 1977. Uranus rings: An optical search. *Nature* 268:32.

Smith, B. A., Soderblom, L. A., Beebe, R., Boyce, J., Briggs, G., Carr, M., Collins, S. A., Cook, A. F. II, Danielson, G. E., Davies, M. E., Hunt, G. E., Ingersoll, A., Johnson, T. V., Masursky, H., McCauley, J., Morrison, D., Owen, T., Sagan, C., Shoemaker, E. M., Strom, R., Suomi, V. E., and Veverka, J. 1979. The Galilean satellites and Jupiter: Voyager 2 imaging science results. *Science* 206:927–950.

Smoluchowski, R. 1976. Origin and structure of Jupiter and its satellites. In *Jupiter,* ed. T. Gehrels (Tucson: Univ. Arizona Press), pp. 3–21.

Soifer, B. T., Neugebauer, G., and Gatley, I. 1979. The near-infrared reflectivity of the dark and light faces of Iapetus. *Astron. J.* 84:1644–1646.

Soifer, B. T., Neugebauer, G., and Matthews, K. 1981. Near-infrared spectrophotometry of the satellites and rings of Uranus. *Icarus* 45:612–617.

Stone, E. C. 1982. The Voyager encounter with Uranus. In *Uranus and the Outer Planets,* ed. G. Hunt (Cambridge: Cambridge Univ. Press), pp. 275–291.

Taylor, G. E. 1973. An occultation by Uranus. *J. Brit. Astron. Assoc.* 83:352.

Thomsen, B., Baum, W. A., Wilkinson, D. T, and Loh, E. 1978. New results on the albedo of the rings around Uranus. *Bull. Amer. Astron. Soc.* 10:581–582 (abstract).

Veverka, J. 1977. Photometry of satellite surfaces. In *Planetary Satellites,* ed. J. A. Burns (Tucson: Univ. Arizona Press), pp. 171–209.

Yoder, C. F. 1982. The gravitational interaction between inclined, elliptical rings. Preprint.

SATURN'S RINGS: PROPERTIES AND PROCESSES

JEFFREY N. CUZZI, JACK J. LISSAUER
Ames Research Center

LARRY W. ESPOSITO
University of Colorado

JAY B. HOLBERG
University of Arizona

ESSAM A. MAROUF, G. LEONARD TYLER
Stanford University

and

ANDRÉ BOISCHOT
Observatoire de Paris

The structural and particle properties of Saturn's Rings are described, with particular emphasis on spacecraft observations. Properties are discussed generically rather than regionally, and we attempt in all cases to relate observed properties to favored causative processes. The ring particles are primarily icy, but there is evidence for albedo, and therefore possibly compositional, variation on both local and regional scales. Most of the particles are in the 1 cm to 5 m radius range, but there is reason to suspect the existence of some particles of all sizes up to 10 km in radius. Ring structure is, in general, dominated by gravitational and collisional dynamics. Orbital resonances with various satellites drive spiral density and bending waves, and define the sharp outer edges of the A and B Rings. From such features, the ring mass density and local vertical thickness may be determined. The structure of kinky ringlets in the F Ring and Encke Gap is probably a manifestation of gravitational perturbations by local shepherd-

[73]

ing moonlets. Electromagnetic processes are evident in the spokes, which are radially elongated regions enhanced in fine dust relative to their surroundings. However, the observed abundance of irregular, fine-scale structure is not well understood. It is also not yet clear what process creates and maintains the well-defined elliptical ringlets lying in several empty gaps. Several problems of a more global scale are also outstanding, including the morphology of the inner edges of the A and B Rings, and the short ring evolutionary time scale associated with the transfer of angular momentum in density waves.

I. HISTORICAL INTRODUCTION

The rings of Saturn were first observed in 1610 by Galileo as one of the first discoveries made with his newly-invented astronomical telescope. Poor image quality and varying ring inclination angle yielded, over the next few decades, many fanciful descriptions and hypotheses for the nature of the system (Alexander 1962). Huygens in 1655 discerned the true nature of the system to be a "thin, flat, ring, nowhere touching and inclined to the ecliptic" which varied in aspect with the Saturnian seasons (see chapter by Van Helden). In 1675, J. D. Cassini realized that the ring was more or less equally "divided" by a dark band; it was not until 1852 that this band, the Cassini Division, was found to be "empty" and to separate two actually distinct rings. The dynamical stability of such an unusual configuration was (and is) of great theoretical interest.

Cassini was one of the first to suggest, without proof, that the rings were composed of "swarms of small satellites." However, William Herschel, the foremost astronomer of the eighteenth century, strongly believed the rings to be solid. In 1785, Laplace showed that a solid ring, in uniform rotation, would break apart under "centrifugal" forces, and that the system would probably have to consist of a large number of independent "ringlets" (cf. Laplace 1802). Roche suggested in 1848 that the rings were debris resulting from tidal disruption of an unfortunate precursor object (see chapters by Harris and by Weidenschilling et al.), but his theory apparently went unnoticed for many years. It was nearly seventy-five years after Laplace's paper that Maxwell (1859) demonstrated mathematically that continued stability of the system required that it be composed of countless, independently orbiting satellites. In 1895, this hypothesis was confirmed by Keeler's and Campbell's spectroscopic observations of reflected solar Fraunhofer lines showing the rings to have a Keplerian radial velocity profile.

Not long after general acceptance of the idea that the rings consisted of swarms of independently orbiting satellites, dynamical perturbations were suggested as responsible for the observed structure. Kirkwood, who had suggested resonant perturbations by Jupiter as the cause of the radial gaps and clumps in the asteroid belt (1866), extended his theory one year later to the Cassini Division and in 1871 to "Encke's" Division, the location and even existence of which were highly uncertain at that time (cf. Osterbrock and

Cruikshank 1983). During the late 19th century, observers visually noted many elusive subdivisions in both the A and B Rings, and various numerological extensions of Kirkwood's theory of resonances were attempted, including the work of Goldsborough (1921) who extended it to the outer edge of the A Ring and the inner edge of the B Ring (see also Franklin and Colombo 1970; Franklin et al. 1971). Only very recently has this dynamical line of thought begun to reach fruition, as discussed in Sec. IV.B. Visual observers have also occasionally claimed to see transient, dark radial features and bright spots in the rings (Alexander 1962). These reports are especially intriguing in the light of Voyager discovery of spokes (Sec. IV.E); however, like many other visual reports, they are difficult to assess objectively.

Observations of the brightness attributable to the rings (Müller 1893) clearly demonstrated a strong brightening with decreasing phase angle, the so-called opposition effect, which was explained by Seeliger (1887, 1895) in the light of Maxwell's many-particle ring as due to variations in mutual shadowing by these many particles in a layer significantly thicker vertically than the size of a typical particle. This concept has come to be known as the classical, or many-particle-thick, ring model.

Important implications of the Maxwell-Seeliger many-particle-thick ring concept were pointed out by Jeffreys (1947a, b). Such a vertical distribution implies random vertical (and horizontal) velocities superposed on the systematic orbital motion. Jeffreys claimed that mutual collisions between the ring particles, with any reasonable inelasticity, would rapidly damp these random motions, causing the ring to flatten vertically. Due to the Keplerian radial variation of orbital velocity, continuing friction between the particles provides a viscosity which acts to transfer angular momentum outwards, causing the ring to spread radially until collisions asymptotically cease. He also realized that this spreading tendency would fill in gaps in the rings. Jeffreys' work has also been cited to rule out Seeliger-type vertical structure; however, it seems that Jeffreys (1947a) himself believed that an equilibrium vertical interparticle distance of a few diameters was allowed, with random motions "parasitic on the general motion," or maintained by the slight energy given up by the spreading ring. We further address the question of ring vertical structure in Sec. II.B.

The historical background summarized above is described in more detail in Van Helden's chapter and by Alexander (1962). The era of modern astronomy has been characterized by a great expansion of observational capability: in photometry and polarimetry, into previously inaccessible spectral regions (infrared, microwave, ultraviolet), and recently into local probing by spacecraft. We discuss the modern work in subsequent sections.

Simultaneously, great advances in our theoretical understanding have been made; many are fairly recent and are discussed in detail in other chapters of this book. Clearly, the history of thought on Saturn's ring system closely parallels, and in many cases has stimulated, progress in gravitational and fluid dynamics. Gravitational resonances in the rings have much in common with asteroidal Kirkwood gaps. Recent work on resonantly driven spiral waves in the ring context (see, e. g., chapters by Shu and by Borderies et al.) has profited from, and will undoubtedly have application to, work in galactic spiral structure. Collisional and viscous processes in Keplerian disks, pioneered by Jeffreys, appear to have more complex manifestations than previously expected, possibly being responsible for the observed "thousand-ringlet" structure (see Sec. IV.D and the chapter by Stewart et al.), and perhaps wider applications as well (chapter by Ward). More recently suggested forms of satellite-ring interactions (shepherding) rely on a complex interplay of gravitational and viscous processes. Evidence of their effects may be seen in several different planetary ring systems. Of course, answering the ultimate question of the origin of Saturn's fascinating ring system requires us to understand fully the ongoing processes (and possibly extinct ones as well) that have produced its current structure.

In this chapter we summarize and, to some extent, digest and analyze the wealth of knowledge about Saturn's ring system accumulated over three hundred years. We will devote special attention to the cascade of data obtained by the Pioneer and Voyager spacecraft. Our perspective in this chapter will be to emphasize generic aspects of Saturn's ring structure which are germane to planetary ring structure in general, and perhaps to other astrophysical disk applications as well. For reviews in greater detail of prespacecraft knowledge the reader is referred to Alexander (1962), Bobrov (1970a,b), Cook et al. (1973), Pollack (1975), Cuzzi (1978), and Ip (1980a,b). For a recent review dealing in greater detail with specific observed properties of Saturn's ring system, see Esposito et al. (1984). A valuable atlas of ring properties appears as an appendix in this book, and a more extensive version is being published separately by the Voyager project.

This chapter complements the review by Esposito et al. (1984) by focusing here more on aspects of Saturn's rings which have general application to planetary rings, whereas Esposito et al. (1984) focus on aspects which are specific to the Saturn system. However, both chapters are fairly self-contained. In addition, some of the material presented herein is derived from work in preparation by various coauthors. In that sense, it is both a review and a preview chapter.

In the next section (II), we give a general description of the overall structure of the ring system. Subsec. II.A describes the radial structure, including discussion of mass and optical depth, comparison of the main rings with the tenuous peripheral rings, and the relationship between rings and nearby moonlets. Subsec. II.B considers the vertical structure and introduces the concept of local thickness, global warping, monolayers vs. many-particle-thick layers, velocity dispersion, volume density, opposition effect, and *in situ* measurements of the E and G Rings. Subsec. II.C describes the ring atmosphere including observations, density, composition, sources, and remaining puzzles.

Section III covers the properties of the particles that constitute Saturn's rings. Subsection III.A. describes the reflection of sunlight. The first part of III.A discusses the particles' phase function, albedo, and polarization, including inferences for size, dust content (especially in the E and G Rings), and implications for dynamics. Sec. III.A.2 deals with spectral reflectivity: color, composition, and variation between regions.

In Subsec. III.B, we describe thermal emission and its implications regarding thermal balance, particle size, spin rates, thermal models, eclipse cooling, horizontal or vertical inhomogeneities, and infrared spectral variation.

Subsection III.C describes probing the rings with microwaves. First, we discuss groundbased radar and radio observations, including results on radar reflectivity, particle size and composition, and ring radio depth. In the second part, we cover the Voyager radio occultation experiment including wavelength variation of radio depth, size implications and vertical structure information. Knowledge of particle properties discussed in detail throughout Sec. III is summarized in III.D.

Section IV, which constitutes the bulk of this chapter is a detailed review of our understanding of the specific structural properties of the main rings and the F Ring. Subsection IV.A gives an overview including regional similarities and differences, correlations between structure and particle properties, classes of structure, and discussion of present vs. primordial structures.

In Subsec. IV.B, we review the essential dynamics of gravitational resonances and waves and describe their implications for defining ring edges and gaps, for determation of local mass density, viscosity, and ring thickness, and for angular momentum exchange with nearby satellites.

Subsection IV.C covers gaps and eccentric ringlets. Secs. IV.C.1 and 2 interpret, respectively, the edges of the gaps and narrow ringlets *à la* Uranus. The latter includes discussion of locations, optical depth, shape, precession, and mass density. Sec. IV.C.3 concentrates on the kinky ringlets, including their observed features, the theory of shepherding, properties of the F Ring and the Encke Division with its ringlets and wavy edges. Implications regarding the presence of local moonlets are emphasized.

Subsection IV.D describes the irregular structure of the main rings. The description includes locations, radial structures and length scales, optical depth, albedo, size, and phase function variations, and azimuthal variations, as well as various dynamical hypotheses to explain this structure.

Subsection IV.E covers electromagnetic effects and their manifestations including (1) the magnetospheric environment, cosmic ray effects, charges and forces on particles, particle lifetimes, and possible structural influences, (2) properties of spokes, and (3) Saturn's electrostatic discharges (SED).

Finally, Sec. V provides a summary of our knowledge with emphasis on unsolved problems and constraints on theories of the origin of the rings. An appendix to this chapter gives a tutorial review of radiative transfer theory, which presents ideas fundamental to much of the interpretative material of the chapter. The appendix covers absorption, scattering and extinction efficiencies, phase functions, Mie scattering, Babinet's paradox, diffraction, polarization, Voyager radio occultation implications, diffuse reflectivity of a particle layer, and properties of monolayer and of many-particle-thick systems.

II. GLOBAL STRUCTURE AND GENERAL DESCRIPTION

A. Radial Structure

In this section we introduce the ring system as conventionally subdivided, and describe its environment and related members of the Saturn system. In subsequent sections we elaborate in greater detail on the specific properties of the various regions.

The regions subtended by the classically known boundaries of the ring system are the A, B, and C Rings and Cassini Division (see Fig. 1, Color Plate 2 and Table I; also Ring Atlas Appendix of this book). They contain far more material than the peripheral D, E, F, and G Rings described below, as measured by their normal optical depth:

$$\tau(\lambda) = \int \int n(r,z) Q_e(r,\lambda) \pi r^2 \, dr \, dz,$$

where z is the direction perpendicular to the rings, $n(r,z)$ is the number density per radius increment (cm^{-4}) of particles of radius r, and $Q_e(r,\lambda)$ is the extinction efficiency of the particles at wavelength λ (see chapter appendix). Frequently in this chapter we will refer to the value of τ at radio wavelengths as the radio depth. For simplicity, we will often use the notation

$$n = \int_0^\infty n(r,z) \, dr \ (cm^{-3}).$$

The surface mass density or mass per unit area,

$$\sigma = \int \int n(r,z) \frac{4\pi\rho}{3} r^3 \, dr \, dz,$$

is dynamically more meaningful than the optical depth, but is not generally measured as directly. In certain regions of the rings, density wave measurements (Sec. IV.B) allow direct determinations of σ to be made. As τ is also known in these regions, the mass extinction coefficient

$$K = \tau/\sigma \ (cm^2 \, g^{-1})$$

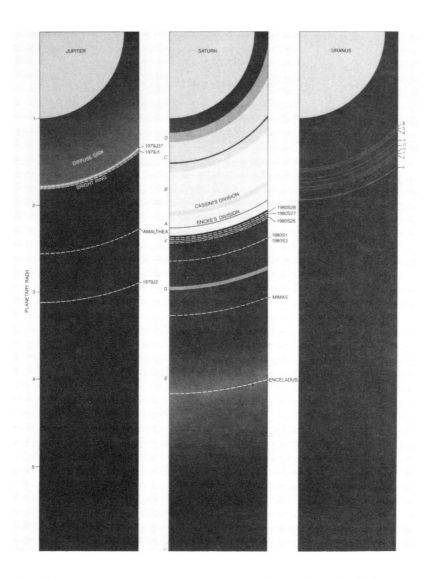

Fig. 1. Schematic of Saturn's ring structure, including the locations of the close-in ringmoons. The ring systems of Jupiter and Uranus are shown for comparison; all systems are scaled to the same planetary radius. The dimensions of Saturn's ring features are given in Table I. The A and B Rings contain the most material, with the Cassini Division and C Ring containing substantially less. The ring mass is actually dominated by meter-sized particles, whose distribution is better represented by the profile of optical depth at radio wavelengths (Sec. III.C). Only about a dozen "empty" gaps exist (see Sec. IV.C; Table III), and no gaps are seen at the A and B Ring inner edges. The structure of the C Ring shows a certain family resemblance to that of the Cassini Division, and the A and B Ring structures are also similar in many ways (Sec. IV.A). For a more detailed view of ring structure, see Ring Atlas Appendix. (Figure from Pollack and Cuzzi 1982.)

TABLE I

Ring Nomenclature and Dimensions[a]

R_S = Saturn's Radius = 60,330 km
M_S = Saturn's Mass = 5.685×10^{29} g
GM_S = 3.793×10^{22} cm^3 s^{-2}

Ring Region	Boundaries (R_S)	Boundaries (km)[b]	Orbital Frequency Ω $(10^{-4}$ s$^{-1})$	Mass
-----------	1.11	66,970	3.554	
D				?
—————	1.235	74,510	3.028	
C				$\sim 2 \times 10^{-9}$ M_S^c
—————	1.525	92,000	2.207	
B				$\sim 5 \times 10^{-8}$ M_S^c
—————	1.949[d]	117,580[d]	1.528	
Cassini Division				$\sim 1 \times 10^{-9}$ M_S^c
—————	2.025	122,170	1.442	
A				$\sim 1.1 \times 10^{-8}$ M_S^c
—————	2.267	136,780	1.218	
---Atlas--- (1980S28)	2.282	137,670	1.208	1.47×10^{-11} M_S^e
---1980S27---	2.310	139,350	1.186	1.03×10^{-9} M_S^e
——F——	2.324[d]	140,180[d]	1.176	
---1980S26---	2.349	141,700	1.157	6.38×10^{-10} M_S^e
Epimetheus (1980S3)	2.510	151,420	1.047	1.55×10^{-9} M_S^{eh}
Janus (1980S1)	2.511	151,470	1.047	6.48×10^{-9} M_S^{eh}
G	2.82	170,100	0.878	$(1-4) \times 10^{-17}$ M_S^f
---Mimas---	3.075	185,540	0.772	6.60×10^{-8} M_S $(8.00 \times 10^{-8}$ $M_S)^g$
E	3.0-8.0	181,000 483,000	0.8 0.2	?
---Enceladus---	3.946	238,040	0.531	1.48×10^{-7} M_S

[a]Sources: Smith et al. (1982); Esposito et al. (1984); Synnott et al. (1983).
[b]Distances rounded to the nearest 10 km, roughly the uncertainty of the observations.
[c]See Table III.
[d]Eccentric feature; semimajor axis is tabulated.
[e]These satellite masses assume mass density = 1.2 g cm^{-3}.
[f]Obtained from results of Van Allen (1983).
[g]Second mass is from Tyler et al. (1981).
[h]These satellites exchange orbits every 4 yr.

may be determined. Where known, K is roughly constant, so the value of σ may be estimated in the densest regions of the A and B Rings where it is not measured. The estimated value, $\sigma \sim 100-200$ g cm^{-2} in the densest regions, is in good agreement with the value obtained from cosmic-ray-produced neutron densities (see Sec. IV.E). Integration of σ over the main rings yields a total ring mass of $M_R \cong 5 \times 10^{-8} M_S$ or 3×10^{22} g, about the mass of Mimas (see Table III in Sec. IV.B). A previous upper limit from deflection of Pioneer 11 was $M_R < 5 \times 10^{24}$ g (Null et al. 1981).

Regional boundaries are defined by significant and fairly abrupt changes in optical depth, or quantity of material. For instance, the Cassini Division is a very real and well-defined division in quantity of material (Sec. IV.A) but is far from empty; in fact, it is comparable to the C Ring in optical depth (see Fig 10 and Ring Altas). Several of these regional boundaries are the result of gravitational resonances with various satellites, as classically suggested. These gravitation effects, and related ring structure, are discussed further in Sec. IV.B, and reviewed in detail in the chapter by Franklin et al. and Shu. Inner boundaries are, in general, less well-defined than outer boundaries and not understood dynamically (see Secs. IV.A and V). Not only is each of the main classical rings radially well-defined, with transitions on a radial scale ~ 10 km, but the particles in each region also show distinct differences in properties from those in adjacent ring regions. For instance, the brightness, color, and size distribution of the A and B Ring particles seem to differ from those of the C Ring and Cassini Division particles. Size variation has important implications for ring structure. At visible wavelengths, we see everything larger than on micron. The mass of the rings, however, is dominated by the largest common particles, whose radial distribution may be quite different, and is best determined at radio wavelengths (see Sec. III.C).

The peripheral rings are all far less opaque at visible wavelengths. The D Ring lies just inside of the C Ring (Smith et al. 1981, 1982). Its surface brightness is ~ 100 times less than that of the C Ring in general, and, if its particles have similar properties, it thus contains ~ 100 times less optical depth or cross-sectional area. Contrary to claims by groundbased observers, no division exists between the C and D Rings. In fact, the D Ring itself is invisible from Earth.

The F Ring, which lies about 3500 km outside the outer edge of the A Ring, is like the elephant in the Hindu tale of the five blind men; its apparent nature varies dramatically with the viewpoint of the observer. To the eye it is a narrow feature of azimuthally variable width and structure, sometimes having an unusual multi-stranded, kinked appearance. However, magnetospheric data provide a very different perspective on its structure, suggesting it is composed of a wider band of unseen objects (Sec. IV.C.3).

Also lying just outside of the main rings, but inward of Saturn's classical satellites, is a retinue of small ringmoons (Table I) which have intimate physical and structural relationships with the ring system. Tiny 15-km Atlas

(1980S28) barely skims the outer edge of the A Ring, while the larger 1980S26 and 1980S27 straddle the F Ring. 1980S26 and 1980S27 have been called the shepherds due to their supposed confinement of the F Ring flock of particles. Further out, at ~ 2.5 R$_S$, are located the coorbital satellites Janus (1980S1) and Epimetheus (1980S3). All of these ringmoons (except possibly Atlas) have important dynamical effects on the ring system, produced by gravitational resonances (Sec. IV.B). It may be that this retinue has many smaller and as yet unseen members, both exterior to, and embedded within, the main ring system (Sec. IV.C). The distribution of such mountain-sized moonlets in and around the rings has far-reaching cosmogonical implications, which are discussed in Sec. V and in the chapter by Harris.

Moving further outward, the G Ring at ~ 2.8 R$_S$ is similar in structure to Jupiter's ring. The optical depth of this ring is quite low ($\sim 10^{-4}$ to 10^{-5}); the visible ring is several thousand km wide with poorly defined radial boundaries (Cuzzi, unpublished). It also apparently shows a high degree of azimuthal symmetry, based on magnetospheric density observations (Simpson et al. 1980). Van Allen (1983) has shown from such observations that the G Ring seems to contain a narrow ($\leqslant 500$ km wide) core of particles which are between 10 and 10^3 μm in size. The tiny particles which have been inferred over a larger radial extent from photometry (Sec. III.A) or impacts on the Voyager spacecraft (Sec. II) could not produce the observed magnetospheric densities. Van Allen (1983) also concludes that not even a single object as large as 1 km can lie within the region.

The E Ring is a vast, diffuse sheet of material with very low optical depth, extending from <3 R$_S$ to beyond 8 R$_S$. It was first discovered from the ground at a time when the ring plane was edge-on to the Earth (Feibelman 1967). Studies during the 1980 ring-plane crossing have shown that the density of the E Ring peaks very near to the orbit of Enceladus. There is continuing disagreement between the various observing groups as to whether the core of the E Ring is exactly at, or slightly outside, the orbit of Enceladus (Baum et al. 1981; Lamy and Mauron 1981; Larson et al. 1981; Dollfus and Brunier 1982). However, because of the close radial association, and observed properties of E Ring particles (Sec. III.A), the E Ring is probably causally related to Enceladus rather than being an extended product or member of the main ring system.

All the rings lie within Saturn's magnetosphere, which extends from ~ 20 R$_S$ inward to the edge of the A Ring and contains radiation belts of trapped energetic electrons and ions. These magnetospheric species mirror rapidly across the magnetic equator, and drift both longitudinally and radially inward with rates that vary with species and radial distance (Van Allen et al. 1980a; Van Allen and Krimigis 1984; chapter by Grün et al.). They are thus absorbed into the surfaces of satellites and ring particles; because the magnetospheric species diffuse inward, and are effectively absorbed by the opaque main rings, the main ring radial region is the part of the solar system most nearly void of charged particles (see e.g., Simpson et al. 1980). However, at the A Ring's edge, and, more importantly, in the outlying rings, the erosive and radiation-damaging effects on the ring material of this bath of electrons and ions must be considered (see Cheng et al. 1982; also the chapter by Burns et al.). Spacecraft observations of fluctuations in the spatial density of these species have led to several discoveries of satellites and unique insights into properties of the region outside the main rings, as discussed further in Secs. IV.C and IV.E.

In summary, Saturn's ring system consists of a compact, opaque classical component and several extended, diffuse components. The various components are distinct in their structural and particle properties. Intermingled with these rings are a number of small, recently discovered satellites, or ringmoons, which interact gravitationally with the rings. The entire ring-ringmoon system is immersed in the trapped radiation belts of Saturn's magnetosphere.

B. Vertical Structure

In this section we summarize current knowledge of ring vertical structure, and the observational foundations for our beliefs. Subsequent sections address aspects of this problem in more detail.

The main rings lie, for practical purposes, in the plane of Saturn's equator and therefore exhibit a varying tilt angle as seen from the Earth (B) and from the Sun (B'), respectively, due to Saturn's obliquity of $26°7$ and its orbital seasonal cycle. When Saturn is near its equinox, the rings are viewed in their edge-on configuration. This geometry, occurring about every 15 yr (most recently in 1980), has allowed groundbased observers to establish the extreme flatness of the rings relative to their radial extent. Although the main rings may not be physically resolved at edge-on presentation, extrapolations of their photometric brightness to zero tilt angle have yielded a photometric effective thickness on the order of 1 km (Focas and Dollfus 1969; Lumme and Irvine 1979; Brahic and Sicardy 1981; Sicardy et al. 1982). This photometric thickness probably includes contributions that cause it to be far larger than the true or local thickness of the main rings. For instance, the symmetry plane of the rings is actually the invariable plane of the angular momentum of the total system; this plane is slightly warped like the brim of a hat, due primarily to long-term perturbations by Titan and Sun, with a vertical amplitude of ~ 400 m (Burns et al. 1979). Also, short-period vertical resonances with Mimas produce localized bending waves in the A Ring with full vertical extent of ~ 1 km (Shu et al. 1983; see also Sec. IV.B). Finally, the photometric contributions of the E, F, and G Rings in edge-on presentation are probably significant (Burns et al. 1979).

Recent spacecraft observations of the rings have shown directly that the edges of several ring features are locally far thinner than 1 km. A stellar occultation by the rings was observed with the Voyager photopolarimeter

(PPS) experiment with radial resolution on the rings of ~ 100 m. Even at this resolution the rings cut off the starlight nearly instantaneously, leading the PPS team to place an upper limit of 200 m on the local thickness at the outer edge of the A Ring (Lane et al. 1982). The Voyager spacecraft itself, and its microwave telemetry signal as seen from Earth, was also occulted by the rings, providing an independent upper limit of 150 to 200 m at several locations based on the observed knife-edge Fresnel diffraction patterns (Marouf and Tyler 1982; Tyler et al. 1983). Theoretical considerations suggest that the local input of energy consequential to maintaining a sharp edge would cause the edge region to be, if anything, of greater vertical thickness than regions of equal content away from the edges (Borderies et al. 1982). Other indirect measurements (Sec. IV.B) indicate a vertical thickness of ~ 10 to 100 m at locations away from the edges.

The vertical extent of the E and G Rings has been directly measured by several means, and is far greater than that of the main rings. During the 1980 ring-plane crossing, space age astronomical technology was first brought to bear on the ring thickness problem. Baum et al. (1981) used the space telescope planetary CCD camera, and Lamy and Mauron (1981) used a similar detector technology along with coronagraphic techniques. Both groups established that the E Ring has a vertical extent of several 1000 km. The data of Baum et al. (1981) also seem to show, at the limit of sensitivity, a thickness increasing away from Saturn, with possibly a slight depression near the orbit of Enceladus where the E Ring has its peak optical depth.

In situ measurements constraining the vertical distribution of material in the E and G Rings were made by the Pioneer and Voyager spacecraft, which apparently sensed numerous impacts by micron-sized dust particles when they crossed through the ring plane. From the time history of these impacts, an effective vertical thickness of ~ 2000 km has been determined for the E Ring (Humes et al. 1980) and of ~ 100 km for a region on the outer periphery of the G Ring at 2.88 R_S (Aubier et al. 1983; Gurnett et al. 1983). The tiny particles so detected exhibit equatorial symmetry consistent with a lack of strong influence from Saturn's magnetic field, which is essentially pole-aligned. In contrast, Jupiter's ring halo may show a distinct vertical offset from Jupiter's equator plane; if true, this is easily attributed to electromagnetic effects (see Sec. IV.E; also chapters by Grün et al. and Burns et al.).

We now discuss the meaning of ''local vertical thickness'' in greater detail, and then address more inferential means of constraining the vertical distribution of material in the main rings. The simplest vertical distribution of material is, of course, a monolayer in which the centers of all particles lie in the same plane. This situation could arise in the rings if all particles followed circular (zero eccentricity e), noninclined (zero inclination i), collisionless orbits.

A slight modification of the orbital plane, such as the large-scale warp demonstrated by Burns et al. (1979), or, on a more local scale, in the bending waves of Shu et al. (1983), allows the particles to have inclined orbits; however in these cases the particle orbital parameters vary slowly and systematically, i.e. coherently, with longitude and the ring could still be only a single particle thick *locally*, i.e., a wavy monolayer.

In order for the rings to be regarded as *locally* many particles thick in vertical extent, the particle orbits must not only have mean inclination $<i> \neq 0$, but also have randomly distributed nodal longitudes at any given location. Such a situation is generally described by an ensemble of particles with a range of orbital parameters having a probability distribution $\mathcal{P}(e,i)$ characterized by some sort of average expectation values $<e>$, $<i>$. As discussed in greater detail in the chapter by Stewart et al., Trulsen (1972) and Hämeen-Anttila (1978) showed that the form of $\mathcal{P}(e,i)$ is in general that of a Rayleigh distribution: $<e> \sim <i> \neq 0$. Effectively then, the particles exhibit locally random relative velocities $c\sqrt{3}$ where c is the one-dimensional dispersion velocity in the coordinate system rotating with the mean orbital motion Ω. From the likely distribution $\mathcal{P}(e,i) \sim e \cdot i \cdot \exp(-e^2/<e^2> - i^2/<i^2>)$, Trulsen (1972) and Hämeen-Anttila (1978) have shown that the vertical variation of particle density is

$$n(z) = n_0 e^{-(z^2/z_0^2)}$$

that is, a familiar Gaussian centered on the mean plane, with scale height $z_0 \sim c/\Omega$. Integration to infinite z shows that 95% of the particles lie between $\pm z_0$. Subsequently, we ignore vertical variation in number density. Of course, as was first realized by Jeffreys (see Sec. I) random relative velocities imply interparticle collisions. The likelihood of a randomly moving particle traversing the ring vertically without colliding with another similarly moving particle is $\sim e^{-\tau}$ where τ is the ring optical depth. The particles, moving in their slightly inclined orbits, traverse the ring twice in each orbital period; therefore the collision frequency is $\Omega\tau/\pi$, or several times per day for the A and B Rings. Any reasonable dissipation of energy per collision will easily damp any initial random velocities over the age of the solar system.

However, mechanisms exist which act to prevent the velocity dispersion from vanishing altogether, by transforming some of the energy of ordered motion into random velocities. For instance, orbital velocity V varies with distance R from Saturn as $V = R\Omega = (GM_S/R)^{\frac{1}{2}}$, and particles at a different semimajor axis a may collide due to their finite radius r even if in circular orbits. The relative velocity $V_{rel} \sim 2rR(d\Omega/dR)$ is then transformed to another direction, becoming a random component in the motion of one or both particles. Because of transfer of angular momentum during the encounter, the two particles separate by $\sim \Delta a$ in semimajor axis; a simple derivation assuming conservation of angular momentum during a collision between elastic particles of radius r in close circular orbits shows that $\Delta a \sim r$. The total energy in circular motion of the two particles in their newly separated orbits, is slightly less than before the collision by an amount $2E(a) - [E(a + \Delta a) + E(a - \Delta a)] \sim 2E(a)(r/a)^2$. For moderately elastic collisions, this excess energy primarily reappears as energy of random motion in the newly eccentric and/or inclined particle orbits with dispersion velocity $c \sim <i> \cdot a\Omega \sim <e> \cdot a\Omega$, where $<i> \sim <e> \sim r/a$. The ensuing true thickness is $z_0 \sim c/\Omega \sim r$, or a few particle radii. This replenishment of random velocities by the Keplerian shear motion led Jeffreys to conclude that identical particles would maintain a few (but not many) diameters separation or ring thickness as long as collisions and ring spreading persisted (see Bobrov 1970a,b). More recently, numerical studies (Brahic 1977) have confirmed these expectations. Furthermore, a distribution of particle sizes exists in the rings such that the largest particles provide the bulk of the energy input by both physical and gravitational encounters (Safronov 1972). There will probably ensue some equilibrium velocity dispersion, maintained by a balance between these inputs and inelastic damping, in which the larger particles may be disturbed from a monolayer state only slightly (Goldreich and Tremaine 1978a) but the smaller particles, having comparable dispersion velocities, could easily define a layer much thicker than their own sizes (Cuzzi et al. 1979a,b; Brahic and Sicardy 1981; chapter by Stewart et al.). Resonant forcing by satellites and subsequent dissipation of collective motions may also contribute to the velocity dispersions locally (see Secs. IV.B and III.A.1). A further byproduct of this line of thought is that, due to off-axis collisions, one would expect the particles to be spinning about randomly oriented axes with spin rate $\Omega_s \sim c/r$ (see Sec. III.B; also chapter by Weidenschilling et al.). The reader is referred to the chapters by Stewart et al. and Borderies et al. for a more complete discussion of velocity dispersion and ring thickness. The question is especially important because velocity dispersion is intimately related with the evolutionary time scale of the rings (chapter by Harris). We return to this discussion in Secs. IV.B and V.

We now discuss several observations that inferentially constrain ring vertical structure on a scale too small to be directly observed. The patriarch of these constraints, and possibly still the most powerful, is the opposition effect. This phenomenon refers to the strongly nonlinear brightening of the rings as they approach zero phase angle. Many solar system objects exhibit qualitatively similar opposition brightening, typified by the dramatic brightness of our own Moon when at full (zero) phase. The explanation is fairly well understood as being due to the fact that, as zero phase is approached, individual particles, or grains on the surface of a particle, rapidly begin to cover their own shadows (Hapke 1963, 1981; Irvine 1966). The usefulness of the effect as a diagnostic arises from the fact that the angular shape and magnitude of the brightness variation uniquely determine the volume fraction

$$D = n \, (4/3) \, \pi r^3$$

of shadowing particles (see, e.g., Irvine 1966; Veverka 1977). In recent years, careful observations (Franklin and Cook 1965) and modeling (Bobrov 1970a,b; Kawata and Irvine 1974; Esposito 1979; Lumme et al. 1983) have found the volume fraction of shadowers to be $D \lesssim 2 \times 10^{-2}$, a value far lower than found to characterize the volume fraction of the grainy surface of any single solar system object. It is especially difficult for bright surfaces such as those of the ring particles (see Sec. III.A) or icy satellites to show such a sharp opposition effect, because multiple scattering between and through translucent grains dilutes the shadows. Furthermore, it may be shown that the distinct shadow edges required of the hypothesis may only appear if the product of the mean ray path to the next scatterer

$$\ell \sim 1/(n\pi r^2) \sim r/D$$

and the diffraction angle $\lambda/2r$ for light of wavelength λ is much less than the typical particle size r. Thus, $r >> \lambda/D \sim 50\text{--}500$ μm. Such large particle sizes, with typical separations of approximately $r D^{-\frac{1}{3}} \sim (5-10)r$, would seem inconsistent with surface microstructure of particles continually bumping into each other, no matter how gently. Therefore, it seems that the observed opposition effect supports the many- (or at least several-) particle-thick ring in which individual ring particles, which themselves probably have some grainy surface structure, cast shadows on each other

On this basis we can obtain a full, main ring thickness. Using observed ring optical depth $\tau \sim n \pi r^2 2 z_0 \sim 1$, with $n \sim 3D/4\pi r^3$ and average particle radii of 10 to 10^2 cm (Sec. III.C), we obtain $2z_0 \sim 4r/3D \gtrsim 5-50$ m for $D \lesssim 2 \times 10^{-2}$ (Lumme et al. 1983). The above relations also imply that, as D decreases, larger particles are required in order to cast shadows on each other at all (B. Hapke, personal communication, 1982). For the very faint ($\tau \sim 10^{-7}$) and thick ($z_0 \sim 10^3$ km) E Ring, in which $D \sim 2r\tau/3z_0$, one sees that $r >> 1$ km radius particles are required in order for any shadowing-related opposition effect to occur (see Sec. III.A.1).

Several aspects of radar studies of the rings, both from the ground and by spacecraft, are also more readily understood in light of the classical, many-particle-thick model. The very existence of the substantial observed attenuation by the rings of the coherent spacecraft telemetry signal is very difficult to reconcile with a single-particle-thick ring (Marouf et al. 1983), although model studies suggest that the very largest particles may not be separated vertically by more than a few radii (H. A. Zebker and E. Marouf, personal communication, 1982). This result is fully consistent with theoretical expectations as mentioned earlier (chapters by Borderies et al. and by Weidenschilling et al.). Also, the variation of ring radar reflectivity with tilt angle is better fitted by many-particle-thick models than by monolayer models (Ostro et al. 1980a; Ostro and Pettengill 1983). We discuss these observations further in Sec. III.C. In addition, studies of the thermal infrared behavior of the rings may eventually help constrain vertical ring structure, by utilizing the relationship between a particle's spin and its dispersion velocity mentioned earlier (cf. Sec. III.B, and Esposito et al. 1984).

In summary, the main rings are extremely vertically thin locally, and only upper limits of ~ 150 to 200 m at several edges have been established from direct observations. Less direct evidence indicates that the actual thickness is probably somewhat smaller, although still considerably greater than the size of the most numerous particles. Typical dispersion velocities $c \sim \Omega z_0$ are probably on the order of 0.1 cm s^{-1}, or a few times larger, in the main rings. On much larger radial scales, the rings are warped to a vertical extent of ~ 1 km. On the other hand, the outlying, optically far thinner G and E Rings have a much greater vertical thickness, directly observed to be on the order of 10^2 and 10^3 km, respectively.

C. The Ring Atmosphere

The realization that the ring particles are predominantly composed of water ice naturally leads to the expected presence, at some level, of a ring atmosphere of H_2O and its dissociation products. Such an atmosphere would represent an equilibrium between those processes which produce or free molecules from ice surfaces in the rings and those mechanisms which lead to either the readsorption of gas molecules or their final loss from the rings due to ionization or neutral escape. Initial estimates (Dennefeld 1974; Blamont 1974) involving such sources as thermal sublimation, meteoritic impact, and solar and interstellar wind bombardment had led to low production rates and atmospheres so tenuous as to be undetectable.

The most sensitive method for detecting a ring atmosphere is by observations of solar Lyman-α emission scattered from neutral hydrogen. Such hydrogen, presumably formed by photodissociation of H_2O, OH and H_2, would form a bright Lyman-α cloud similar in some respects to those observed to develop around comets. The first detection of Lyman-α emission from Saturn's rings was due to a sounding rocket observation by Weiser et al. (1977) who measured 200 ± 100 Rayleighs (Ra) of Lyman-α emission from the general vicinity of Saturn's rings. This measurement was followed by observations of higher spatial resolution with the *Copernicus* (Barker et al. 1980), and IUE satellites (Clarke et al. 1981) both of which produced upper limits to the Lyman-α brightness of the rings less than that of Weiser et al. (1977). During the Pioneer 11 encounter, the long-wavelength ultraviolet photometer observed an enhanced emission while the spacecraft was beneath the B Ring. Judge et al. (1980) attributed this signal to Lyman-α emission which peaked in the vicinity of the B ring. A second weaker component, several R_S outside of the A Ring, was also suggested by the data.

The most extensive ultraviolet observations of the rings are those of the Voyager 1 and 2 ultraviolet spectrometers, which made numerous scans of the rings from widely differing observing geometries. The preliminary picture of ring-associated emission which emerged from these observations contains significant differences from those discussed previously. Although Lyman-α emission was detected from the rings (Broadfoot et al. 1981), it should be pointed out that, as viewed from Voyager in Lyman-α, the rings appear dark when seen against the Lyman-α background of the sky. This sky background is due to solar Lyman-α resonantly scattered from interstellar neutral hydrogen entering the solar system, and constitutes an important complicating factor in the interpretation of Lyman-α emission from the rings. A full analysis of the Voyager (and other) Lyman-α observations of the rings would necessarily include the following additional contributions to the apparent brightness:

1. A fraction of the sky background which is transmitted through the rings $e^{-\tau/\cos\theta}$, where τ is the normal optical depth of the rings and θ is the zenith angle of the line of sight (Fig. A3 in chapter appendix);
2. A foreground component, which for Voyager involved a contribution from the Titan torus (Sandel et al. 1982) and which, for near-Earth observations, also includes contributions from interplanetary and geo-coronal hydrogen columns;
3. A ring albedo component in which solar Lyman-α is diffusely scattered directly from the ring particles.

The Voyager observations had good spatial resolution and a wide range of viewing geometries, permitting a partial separation of the contribution from each of these components. For example, comparison of observations of the illuminated and unilluminated sides of the optically thick portion of the B Ring allows effective separation of components (1) and (3). A full analysis of the Voyager observations, modeling all the components of the apparent ring brightness, is not yet complete; however, preliminary analysis has yielded several conclusions:

1. There is a component of the Lyman-α ring brightness which can be explained by an atmosphere of neutral hydrogen associated with the rings. The magnitude of this component is ~ 360 Ra, and is the same on lit and unlit faces of the rings.
2. This ring-associated hydrogen does not appear to extend significantly beyond the outer edge of the A Ring.
3. The appearance of the rings in Lyman-α was remarkably similar during both Voyager encounters even though the solar elevation angle with respect to the rings increased from $3°6$ (Voyager 1) to $8°$ (Voyager 2).

The simplest, least contrived, explanation for these observations is that the bulk of the 360 Ra minimum observed on both sides of the B Ring is due to H associated with the rings. Since the Sun was at very low elevation angles during both Voyager encounters, any ring atmosphere having an appreciable scale height (>0.1 R_S) would display approximately equal illuminated columns on both sides of the rings. Beyond this and the previously mentioned observation that the H does not appear beyond the A Ring, little can currently be said concerning the extent of the ring atmosphere. Densities derived from the observed brightness are therefore model dependent. Assuming that *all* the 360 Ra of Lyman-α observed by Voyager 1 on the B Ring came from the ring atmosphere, Broadfoot et al. (1981) quoted an H column density of 10^{13} cm^{-2}. If this H were uniformly distributed in a spherical shell having the radial extent of the rings, the corresponding cloud would contain 5×10^{33} atoms with a density of 600 cm^{-3} and an implied production rate of 1×10^{28} s^{-1}. Confining the H more closely to the rings lowers the content of the cloud but increases its density, since the observed column does not depend critically on the assumed geometry. The reported Voyager 1 brightness assumed all of the signal was due to the ring atmosphere; however, more than 30% of the signal could be contributed by hydrogen from the Titan torus together with a sky background signal entering through the solar occultation port of the ultraviolet spectrometers.

The source mechanisms for the production of H in the rings considered by Dennenfeld (1974) and Blamont (1974) all fail by several orders of magnitude to account for the observed ring atmosphere. The principal reason for this is that the mean lifetime for readsorption of free H in the rings must be relatively short, $\sim 3 \times 10^5$ s (Carlson 1980). Following the Weiser et al. (1977) observation, which implied the existence of a substantial ring atmosphere, several potentially more potent H production mechanisms were proposed. Cheng and Lanzerotti (1978) suggested that H_2O was sputtered from the ring particles by energetic magnetospheric ions and Carlson (1980) considered a photosputtering mechanism in which H_2O is photodissociated on the surfaces of the ring particles by solar radiation. The latter mechanism is potentially effective when the rings are open with respect to the Sun, as they were at the time of the Weiser et al. observation. However, the fact that the Voyager observations, which occurred during low solar elevation angles, showed little if any change between the two encounters while the solar elevation more than doubled is not consistent with the expected behavior of the photosputtering mechanism. Cheng et al. (1982) and Ip (1983) reviewed proposed source mechanisms, including charged particle sputtering, and concluded that none is capable of supplying the observed amount of H. On the other hand, a recent reevaluation of both the meteoroid flux incident on the rings and the vapor phase yielded by the subsequent high-velocity impacts (Morfill et al. 1983), indicates that this mechanism may, after all, be a major source of the ring atmosphere.

One other way to account for the presence of a relatively substantial ring atmosphere is to consider possible sources of H external to the rings. Ip (1978) proposed that protons escaping from the midlatitudes of Saturn's exosphere could become neutralized in the rings and contribute to the H atmosphere.

This mechanism, however, appears to fail as a source by several orders of magnitude (Carlson 1980). A more promising source has recently been proposed by Shemansky and Smith (1982), in which dissociation of H_2 by electrons (Shemansky and Ajello 1983) in the exobase of Saturn's atmosphere would provide a source of hot H atoms with sufficient energy (<3 eV) to escape from the planet's exosphere. In this view, the rings would constitute a cold trap for a fraction of the escaping H. This source, which is estimated to produce a total of 10^{29} H atoms/second (Shemansky and Smith 1982) from Saturn's atmosphere, must also be considered as a potential or even major contributor to the observed H in the Titan torus at 8 to 20 R_S.

III. OVERVIEW OF PARTICLE PROPERTIES

The properties of individual ring particles, which bear directly on the question of ring origin and evolution processes, may or may not be correlated with ring structure. Properties of interest include particle size distribution, composition, and surface/subsurface structure. These basic properties are constrained by observations of the angular radiation scattering behavior or phase function, and the absorption coefficients, of the particles, both of which are functions of wavelength from ultraviolet to radio ($\lambda \sim 10^{-5}$ to 10 cm). Because the rings comprise an ensemble of particles acting collectively, radiative transfer provides an indirect link between the observations and the inferred particle properties which does not exist for, e.g., an individual satellite. The form of the link depends to some extent on whether the ring is vertically thick or a monolayer. Although the weight of current evidence favors the many-particle-thick slab of scatterers that is easily treated with many standard methods of radiative transfer, certain aspects of the observations require that we also consider monolayer behavior. The radiative transfer that relates remotely observed ring properties to intrinsic particle properties such as albedo and phase function is outlined in the chapter appendix, and is discussed in more detail by Esposito et al. (1984).

Below, we discuss what has been learned about the particles from reflected sunlight, thermal emission, and microwave probing. The wavelength variation of albedo through the visible (i.e., color) is, for instance, an important diagnostic on compositional differences in, at least, the surface layers of the particles. Spectral regions from 1 μm to 1 mm also contain direct information on composition from infrared absorption features. Far-infrared and millimeter observations provide constraints on the crucial, composition-dependent transition between primarily emitting (absorbing) behavior ($\lambda < 100$ μm) and primarily loss-free scattering behavior ($\lambda > 1$ cm). The infrared observations also, of course, provide thermal properties of the particles. Most size information comes from wavelengths close to the size of the typical particles, where the greatest sensitivity of scattering behavior to size exists; most ring particles are in the range of centimeters to meters in size, so that radar reflectivity, radio occultation, and radio brightness observations have been and will continue to be extremely valuable. In a similar way, observations at visible wavelengths have contributed some knowledge of the distribution of the component of particles of size comparable to visible wavelengths.

In Sec. III.D, we summarize the inferences as to particle properties obtained from remote observations. For a more detailed discussion, the reader is referred to Esposito et al. (1984).

A. The Rings in Reflected Sunlight

1. Broadband Observations of Particle Phase Function and Albedo. The brightness of the rings, as perceived under given conditions of illumination and observation geometry, depends both on the total reflecting power of the particles (albedo $\tilde{\omega}_o$) and upon the directional distribution of their reflected radiation (phase function PΘ, where Θ is the scattering angle [see chapter appendix, Figs. A2 and A3]). Groundbased observations are limited to the very small range of scattering phase angles $\alpha = -\Theta \leqslant 6°$ subtended by the Earth's orbit as seen from Saturn. Therefore, our knowledge of particle properties has been greatly advanced by spacecraft observations which allow the rings to be viewed at a large range of phase angles extending to about 160°, fairly close to the forward-scattering direction. Spacecraft trajectories, unfortunately, do not sample the various geometrical variables independently, and certain important geometries (e.g., very high phase angles) are undersampled, making modeling difficult. Also, photometric calibration of the Voyager imaging data is not yet fully satisfactory. For these reasons the particle phase function is still only approximately known.

It is, however, well-established that the main ring particles are rather strongly backscattering. Figure 2 shows the brightness of certain ring regions, averaged over several thousand kilometers radially, obtained from Voyager imaging observations (Smith et al. 1981). The points shown are in the clear filter of the wide-angle camera with an effective passband similar to the "B" filter of classical photometry (Smith et al. 1977). Clearly, the rings are 10 to 100 times brighter at low phase angles than at high phase angles. Model calculations are also shown for many-particle-thick model rings composed of scatterers with Lambertian surfaces, such as characterize moderately rough surfaces of particles far larger than the wavelength. Closer investigation and work in progress reveal that a better fit is obtained for somewhat more sharply-peaked backscattering phase functions, such as characterize "snowbanks" or the various icy satellites, i.e., realistic regoliths with some granular structure. Because a scattering particle's phase function becomes strongly forward-directed as its size approaches the size of the scattered wavelength, the implication of this lack of forward scattering is that, overall, the main rings contain at most only a small area fraction of ~ 10% of micron-and-smaller-sized particles. No stronger upper limits on this fraction than ~ 10%

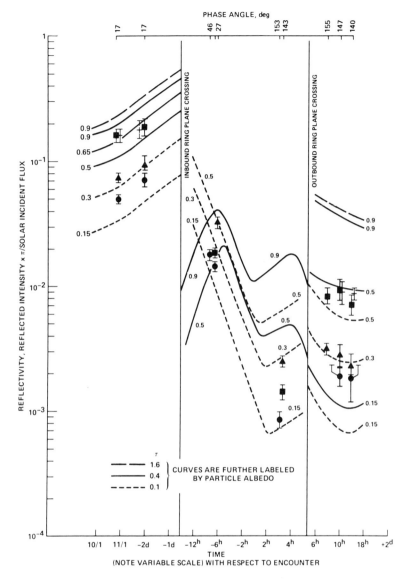

Fig. 2. Radially averaged reflectivity (I/F = geometric albedo) of several ring areas as observed with the Voyager wide-angle camera (clear filter) at a variety of phase angles (with varying emission angle; see chapter appendix for geometry). Mid-A Ring, 2.10–2.16 R_S (■); inner B Ring, 1.55–1.60 R_S (+); Mid-C Ring, 1.33–1.38 R_S (●); and mid-Cassini Division, 1.95–1.99 R_S (▲). The rings are more than ten times brighter at low phase angles than at high phase angles, showing the particles to be strong backscatterers. Model calculations are shown which assume particles with Lambertian surfaces, such as would approximate macroscopic, rough-surfaced objects. The model curves, given for several different ring optical depths, are also labeled by single-scattering albedo (adapted from Smith et al. 1981). The optical depths of the C Ring and Cassini region are approximately 0.1, and of the A and inner B regions are approximately 0.4 and 0.7, respectively (see Fig. 10 in Sec. III.C.2 and Ring Atlas).

may be established until the observations at the highest phase angles have been satisfactorily analyzed. The curves in Fig. 2 are further labeled by the particle albedo. It may be seen that the A and B Ring particles are fairly bright ($\varpi_0 \sim 0.6$) whereas the C Ring and Cassini Division particles are substantially darker ($\varpi_0 \sim 0.2-0.3$). These results are consistent with the albedos derived from groundbased observations (summarized by Esposito et al. 1984).

Observations of the polarization of the rings complement these photometric results. The polarization is quite weak and exhibits the characteristics of other bright, granular, macroscopic surfaces: it lies uniformly in the Sun-ring-observer plane of scattering, or "negative" at small phase angles (Kemp and Murphy 1973; Johnson et al. 1980; Esposito et al. 1980); at larger phase angles it is also quite weak ($<10\%$) although contamination by strongly-polarized Saturn-shine makes it difficult to determine even the sign of the polarization (Burke and KenKnight 1980; Esposito et al. 1980). Both very small particles, like Rayleigh scatterers, and very large, smooth particles, would be strongly positively polarized at large phase angles. Dollfus (1979a,b) has observed a small, time-variable component of polarization in the tangential direction, and has claimed this as evidence for some degree of orientation of the ring particles. This phenomenon has not been confirmed by other observers (Johnson et al. 1980).

There is, however, some evidence for small overall fractions ($<10\%$ by area), or higher localized concentrations, of micron-and-smaller-sized dust particles in the main rings. Certain areas are seen in spacecraft images to reverse contrast with their surroundings between low and high phase angles (see Fig. 3). They are thus often loosely said to brighten in forward scattering; strictly speaking, these areas are still darker at high than at low phase angles, but to a lesser degree than the average or surrounding areas. A likely explanation is that the average phase function for that area has, in fact, a small additional forward-directed, or even isotropic, component implying the presence of a small area fraction of wavelength-sized or smaller particles that are lacking in surrounding areas (see chapter appendix). Figure 4 demonstrates the phase variation of brightness for model rings composed of ensembles of Lambertian spheres with varying amounts of dust added. Dusty areas are darker than clean areas at low phase angles and brighter at high phase angles. This relative brightening is shown by several discrete, dynamically active zones in the outer A Ring (see Sec. IV.B and IV.C), and the entire outer A Ring itself in a more gradual way, as well as the spokes in the B Ring (Sec. IV.E). More precise statements about the size and optical depth of the dust will have to await further photometric analysis of Voyager images.

The photometric behavior of the optically thin E, F and G Rings is quite different from that of the main rings. All of these diffuse features are far brighter at high phase angles than at low phase angles, directly demonstrating the presence of a substantial component of forward-scattering, micron-sized dust particles; for the E and G Rings, this is consistent with *in situ* measure-

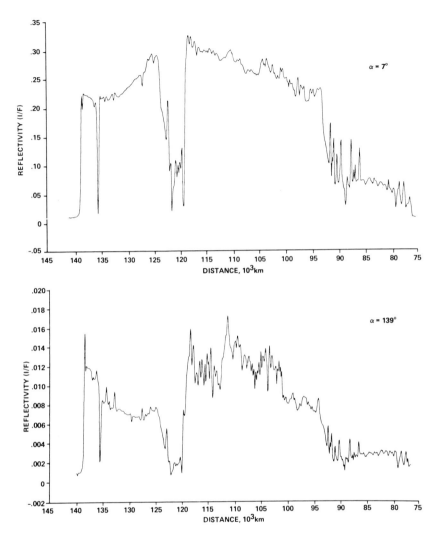

Fig. 3. Radial profiles of reflectivity of the rings at visual wavelengths, from Voyager observations with the wide-angle camera clear filter; upper: low phase angle ($\sim 7°$); lower: high phase angle ($\sim 139°$). The differences in profile indicate varying quantities of "small" forward- or isotropic-scattering particles. Of special note are the different reflectivity profiles of the A and B Rings at high and low phase angles. The "contrast" of B Ring features increases markedly in the outer 2/3, but not the inner 1/3, of the B Ring. Also, in the A Ring the entire sense of radial variation reverses between low and high phase, in a way that suggests an outwardly increasing fractional content of small particles within the A Ring (see Fig. 4). The several sharp peaks in the central half of the A Ring are locations of known resonantly driven density waves (Sec. IV.B), and their contrast reversals suggest enhanced dust content in these dynamically active regions. In a similar way, the general outwardly increasing dustiness may be ascribed to the increasing number density and strength of many other, unresolved, resonances in the region.

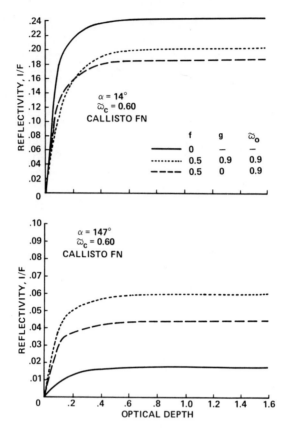

Fig. 4. Variation of reflectivity (I/F = geometric albedo) with normal optical depth for the geometry characterizing Voyager 1 pre- and post-encounter ($B' \sim 4°$). The particles are assumed to consist of (a) a large backscattering particle component with the phase function of Callisto (Pang et al. 1983a) and albedo $\widetilde{\omega}_c = 0.6$ and (b) a small dust component with optical depth fraction f of higher albedo $\widetilde{\omega}_o = 0.9$ and scattering anisotropy parameter g, where $g = 0$ is isotropic and $g = 0.9$ is highly forward directed scattering (see chapter appendix). Compared with clean areas ($f=0$), the addition of moderate amounts of dust simultaneously produces relative darkening at low phase angle ($\alpha = 14°$) and brightening at high phase angle, ($\alpha = 147°$) but with larger contrast at high phase angle. Note, however, that the scattering layer is in all cases darker in an absolute sense at high phase angles. This general behavior is characteristic of many ring areas which reverse contrast with varying phase angle (see Sec. III.A and Fig. 3).

ments of particles impacting the Pioneer and Voyager spacecraft (Humes et al. 1980; Gurnett et al. 1983; Aubier et al. 1983).

Groundbased observations of the E Ring, although extremely difficult, appear to show an opposition brightening even larger than that of the main rings (Larson 1983; Larson et al. 1981; Pang et al. 1983a; Dollfus and Brunier 1982). These observations are difficult to reconcile with a "shadowing" model (see Sec. II.B), but are consistent with the backscattering enhancement produced by transparent, micron-sized particles (Terrile and Tokunaga 1980; Pang et al. 1983; see, e.g., Esposito et al. 1984). The phase functions of such particles, of moderate optical size $X = 2\pi r/\lambda \sim 20$, are not strongly sensitive to the degree of nonsphericity (Zerull et al. 1980) so even somewhat irregular particles could be capable of producing the effect. Pollack (1981) has analyzed the F Ring phase variation and finds a substantial component of very small (~ 0.1 μm) particles, along with a possible macroscopic component. However, the important phase angles sensitive to 1 to 10 μm-sized particles were not sampled; also, the optical depth inferred for the 0.1 μm particles is far smaller than the total F Ring optical depth (Lane et al. 1982). Therefore, additional material of larger sizes must be present (see Sec. IV.C.3.b). Whether this is also true for the E and G Rings may be eventually inferred from their depletion of various magnetospheric species (Thomsen and Van Allen 1979; Grosskreutz 1982).

   2. Spectral Reflectivity: Color and Composition. Much can be learned from the full spectral variation of particle albedo; comparisons with laboratory mixtures of materials have been extremely useful for satellite surface observations (Johnson and Pilcher 1977). It must be remembered, however, that most spectral observations of the rings remain in the form of net ring reflectivity and have not been corrected for multiple scattering (see chapter appendix, Fig. A4). For this reason, only qualitative conclusions may be drawn. Figure 5 compares the spectral reflectivity of the A and B Rings with that of various icy satellites. The rings are distinctly reddish, apparently more so than, for instance, Europa. However, because of the steepening effect of multiple scattering, it is at least possible that the spectral albedos of individual particles could be fairly similar to that of Europa, except at ultraviolet wavelengths where the ring spectrum seems flatter. A quantitative analysis would be most useful, however. Color images of the rings obtained with Voyager imaging (Smith et al. 1981, 1982) clearly show the C Ring and Cassini Division to be less reddish than the A and B Rings (Fig. 6; see also Color Plate 3). Although the color differences appear between regions of comparable optical depth, it is conceivable that the combined albedo and optical depth variation between the different regions could explain (via increased multiple scattering) some of this apparent color difference, which is greatly exaggerated in the various Voyager color-enhanced images. Clearly, more quantitative analysis is needed in this area.

   The reflectivity of the A and B Rings has also been observed at higher spectral resolution at near-infrared wavelengths, where specific absorption features exist that contain direct compositional information; for instance, water-ice absorption features are seen at 1.6, 2.0, and 3.0 μm (Peutter and Russell 1977; Kuiper et al. 1970; Pilcher et al. 1970; Pollack et al. 1973), and there is a hint of an impurity band near 0.85 μm.

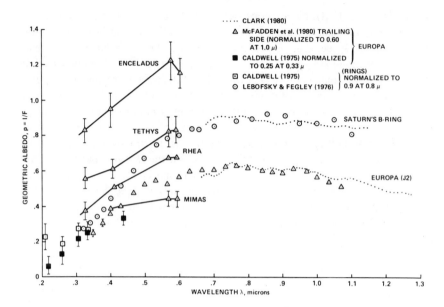

Fig. 5. Spectral reflectivity of the A and B Rings and several of Saturn's major satellites, as discussed in Sec. III.A.2. The rings have a distinctly reddish color, similar to that of most other nearby objects and the Galilean satellites. This reddish color is indicative of nonicy components in the rings, possibly with very small relative abundance. (Figure from Esposito et al. 1984.)

Recent observations of ring and laboratory frost reflectivity spectra in the near-infrared (0.84 to 4 μm) have been carefully analyzed by Clark (1980) and Clark and McCord (1980). The detailed shape of the 1.6 and 2.0 μm water-ice absorption bands are well-matched by laboratory spectra of medium-grained water frost at 100 K, contrasted even with the spectra of frost overlying solid ice such as characterizes Europa (Clark 1980; cf. also Pollack et al. 1973). It has long been realized, however, that the reddish color of the rings (Fig. 5) is not consistent with pure water ice and requires some impurity (Lebofsky et al. 1970; Lebofsky and Fegley 1976). Clark (1980) points out that the 1.6, 2.0, and 3.0 μm water-ice bands in the near infrared are so strongly absorbing that a small, uniformly mixed impurity would be easily masked; however, the ice becomes sufficiently transparent at shorter wavelengths that the presence of even quite small mass fractions of impurities could be easily seen. The impurity features at 0.6 and 0.85 μm, as well as the blue absorption and the high and constant reflectivity between 0.9 and 1.5 μm, are consistent with a variety of silicate materials.

The E Ring is a difficult subject for spectrally-resolved observations; however, at visible wavelengths, its spectrum is clearly bluish, different from that of the main rings or any other solar system object (Larson 1983; Esposito et al. 1984). Also, Terrile and Tokunaga (1980) have detected a spectral reflectivity variation in the near-infrared consistent with oscillations in backscattering efficiency characterizing wavelength-sized particles (Pang et al. 1983a). More detailed analysis is needed, however, to indicate whether these observations and the dramatic opposition effect are all quantitatively consistent.

B. Thermal Emission from Ring Particles

The thermal balance for ring particles is more complex than for isolated objects due to interparticle effects such as shadowing, diffuse scattering of the incident sunlight, and mutual heating by the particles themselves. The detailed form of the response varies slightly, depending on whether the ring is a monolayer or is many particles thick (multilayer). These differences are beginning to be significant diagnostics on ring vertical structure. Approximate solutions to the heating problem have been obtained by Aumann and Kieffer (1973), Kawata and Irvine (1975) and Kawata (1979, 1982) for the multilayer and by Froidevaux (1981) for the monolayer. In the chapter appendix we provide a simplified sketch of the solution; here we use calculated results to give useful constraints on ring structure.

The particle temperature is determined by a balance between absorbed and emitted energy. The total absorbed energy E_{abs} is due to three main sources: direct and reflected or scattered solar radiation, thermal radiation from other ring particles, and thermal radiation from the planet. Of course, the absorbed energy is strongly dependent on the particle bolometric Bond albedo A. The temperature T_p of the illuminated face of the particle is given by

$$f \epsilon \sigma T_p^4 = E_{abs}$$

where ϵ is infrared emissivity, σ (in this equation) is the Stefan-Bolzmann constant, and the term f characterizes whether the particle radiates over its whole surface area ($f \sim 4$) or over a hemisphere ($f \sim 2$). The former case would apply if the particles were either small or rapidly rotating, the latter, if large and slowly rotating. For thermal time constants in Saturn's rings, which are comparable to an orbital period, synchronous rotation is slow and corresponds to $f \sim 2$. However, $f = 2$ would not require that the particles' axes are really locked to Saturn; collision dynamics would anticipate that the spin angular velocity $\Omega_s \sim \Omega$ with random orientation in a several-particle-thick layer as shown below (see chapter appendix for more details).

The ring particles are known to be large enough that their 20 μm emission is very close to blackbody emission, as $\epsilon \sim 1$ for macroscopic particles of geologically abundant material at these wavelengths. Both groundbased (Rieke 1975) and Voyager infrared interferometer spectrometer (IRIS) observations (Hanel et al. 1982) show that the C Ring and Cassini Division particles

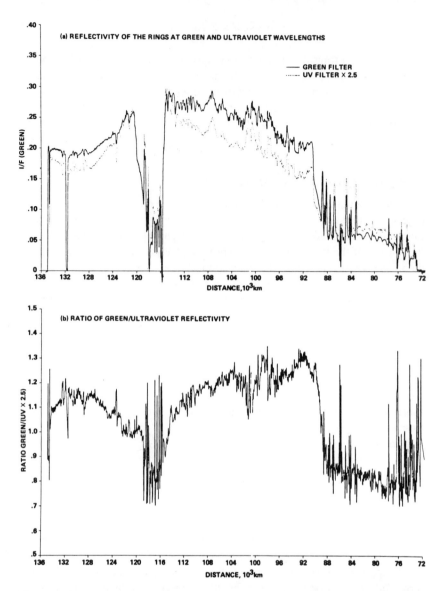

Fig. 6. (a) Reflectivity of the rings in the Voyager green (solid line) and ultraviolet filters (dotted line), compiled from simultaneous sets of high-resolution color composites (Smith et al. 1982). The radial scale has an absolute uncertainty of ~ 700 km. The ultraviolet scan has been scaled to match the green scan at the inner A and outer B Ring boundaries. A slight imperfect radial alignment remains in some regions due to camera distortion. (b) The relative brightness in green/ultraviolet is shown in a ratio of the scans of (a). Large noise spikes are seen at certain sharp edges due to imperfect alignment. However, the Cassini Division and C Ring are clearly less green than the A or B Ring. The B/C transition is quite abrupt, but the A/CD and B/CD transitions are less so. The B Ring and A Ring show gradual, systematic variations in color. Also, the very outer edge of the A Ring is quite blue (see panel a), due possibly to a particle size effect. Compare, for instance, its phase variation (Fig. 3) and see also Color Plate 3. (Figure from Cuzzi et al. 1984.)

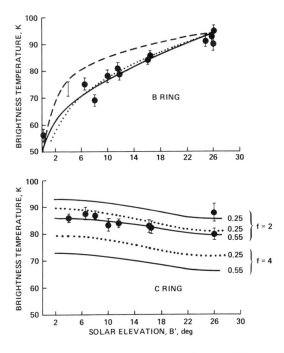

Fig. 7. Thermal brightness temperature (at 20 μm wavelength) of the B and C Rings as a function of solar elevation angle *B* '. The data points are from a variety of authors (for a tabulation see Esposito et al. 1984). The form of the A Ring variation is similar to that of the B Ring. Also shown are different thermal models of the rings; solid lines: monolayer model (Froidevaux 1981); dashed line: many-particle-thick, homogeneous model (Kawata and Irvine 1975; Kawata 1979, 1982); dotted lines: many-particle-thick inhomongeneous model assuming a central layer of dark particles surrounded by a haze of bright particles (Kawata 1982). In the case of the C Ring, the known particle albedo makes it possible to compare detailed predictions. In this case, the dotted lines indicate homogeneous many-particle-thick models (Kawata 1982). The theoretical curves are labeled by particle albedo (=0.25 or 0.55), and *f* is a measure of particle spin velocity (see text and chapter appendix for discussion).

have higher temperatures than the A and B Ring particles on the average, consistent with their lower albedos (Smith et al. 1981; Sec. III.A). The brightness temperature T_B of the rings varies with ring tilt angle, as shown in Fig. 7, due to a combination of effects (see chapter appendix). The form of the variation differs for the A/B Rings and the C Ring, and is most easily understood for the C Ring where the structure and particle properties are fairly uniform and interparticle effects are small.

In the C Ring, the particle single-scattering albedo at visual wavelengths is 0.2–0.3 without evidence of inhomogeneity (Smith et al. 1981; Esposito et al. 1984) and the flat spectrum obtained in Voyager IRIS observations of the unlit face suggests lack of thermal inhomogeneity, by comparison with Rhea's spectrum (Hanel et al. 1982). Both monolayer (Froidevaux 1981) and many-particle-thick (Kawata 1982) models can explain the increase in C Ring brightness temperature with decreasing ring tilt as a filling factor effect, but the monolayer models are somewhat too hot for $A \sim 0.2$–0.3 unless $f > 2$. The particles are known to be large from microwave (Sec. III.C) and thermal eclipse observations (Aumann and Kieffer 1973; Froidevaux et al. 1981). Thus, a value of $f > 2$ must be due to a large spin rate. However, for $f \sim 4$, $\Omega_s >> \Omega$ would be required. This could be maintained by off-axis interparticle collisions at dispersion velocity $c \sim r\Omega_s$ (Sec. II.B) but would then imply a ring much thicker than a single particle ($z_0 \sim c/r \sim r\Omega_s/\Omega$), inconsistent with the model. However, the multilayer models are in better agreement with the observations, even with $f \sim 2$ (Kawata 1982). Voyager 1 and 2 observations also contain evidence for $f \sim 2$. The C Ring particles show a larger eclipse cooling effect at low phase than high phase angles (Hanel et al. 1981), and the absolute particle temperature measured at phase angles larger than accessible from Earth (Voyager 1: 24°, Voyager 2: 31°, 46°) is lower than Earth-based values by an amount consistent with thermal beaming characterizing a slowly rotating object (J. Pearl, personal communication, 1983; Pettit 1961). There-fore, the albedos and temperatures of C Ring particles favor the many-particle-thick model. A complementary discussion is given in Esposito et al. (1984). Clearly, infrared data and theoretical analyses might be capable of providing useful constraints on ring vertical structure.

The A and B Ring cases are less clear, as greater inhomogeneity in particle properties and radial structure exists. Both monolayer and many-particle-thick models can provide reasonable fits to the data. Kawata (1982) has noted that the radiative-equilibrium, vertically homogeneous, many-particle-thick models of Kawata and Irvine (1975) exhibited a ring brightness temperature decreasing with tilt angle which, although in agreement with the data, was due to modeling errors. Currently existing many-particle-thick models require a bimodal, vertically inhomogeneous, albedo distribution con-sisting of dark particles surrounded by, or overlaid with, bright particles (Nolt et al. 1980; Kawata 1982). Again, the parameters are more reasonable for $f = 2$. It must be cautioned, however, that the existing multilayer models do not account for particle vertical motions which cause the particle heating to be time dependent. It may very well be that this effect largely removes the need for vertical albedo inhomogeneity. However, radial inhomogeneities certainly exist. Detailed inspection of IRIS spectra and imaging-derived reflectivities for the A and B Rings would provide valuable constraints on such in-homongeneity of particle properties.

The low unlit-face temperature (56 K) of the A and B Rings (Ingersoll et al. 1980; Tokunaga et al. 1980; Froidevaux and Ingersoll 1980) is consistent with either monolayer or many-particle-thick models, if the particles have the low thermal inertia implied by eclipse observations. A low thermal inertia implies a highly insulating, low conductivity surface capable of rapidly warm-

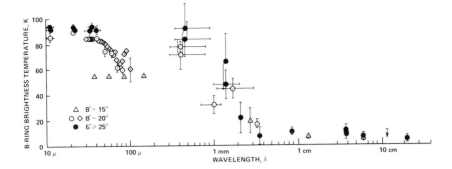

Fig. 8. Variation with wavelength of B Ring brightness temperature, as determined by groundbased observations. Observations are grouped according to their solar elevation angle B'. Throughout the millimeter-wavelength band, or possibly at even shorter wavelengths, the ring brightness temperature drops from a value close to the particle physical temperature (~ 90 K) to a very low value (~ 10 K) which primarily represents diffuse reflection of the planetary thermal emission. This is because the particles become essentially perfect reflectors as the wavelength approaches their circumference (see chapter appendix). There is a suggestion of variable brightness temperature in the region between 40-400 μm wavelength. (Figure from Esposito et al. 1984, where the data are tabulated.)

ing or cooling in response to variations in energy input. The fact that the edge-on rings do not cool to 50 K (Tokunaga et al. 1980) should not be held against the multilayer models, which have been recently improved and could probably have a still better treatment of heating by Saturn itself (see chapter appendix).

Somewhere between 100 μm and 1 mm wavelength, the ring brightness temperature T_B drops sharply below blackbody behavior (see Fig. 8). This important spectral range is discussed in more detail by Esposito et al. (1984). This decrease is primarily due to the increase of particle albedo, which is close to unity for $\lambda > 1$ cm due to a combination of particle size and composition effects (see chapter appendix). As particle albedo $\tilde{\omega}_0$ increases throughout the millimeter wavelength range, the thermal emission from the rings decreases as $(1-\tilde{\omega}_0)^{\frac{1}{2}}$ (Horak 1952; Cuzzi et al. 1980) and at centimeter wavelengths, the rings act essentially as an opaque mirror, diffusely reflecting Saturn's thermal emission and the cold of space.

In the transition range from 100 μm to 1 mm, the emission is especially sensitive to particle refractive indices. More observations in this spectral range would be quite useful. Because the microwave absorption coefficient of hexagonal phase water ice (*Ih*) is approximately 10^{-4} cm^{-1} (Mishima et al. 1983; Whalley and Labbé 1969), the penetration depth of microwaves is on the order of meters or longer. It seems that water-ice particles with a size distribution consistent with Voyager observations are marginally consistent with 3 mm-wavelength observations (Epstein et al. 1983; Muhleman and Berge 1983) although Epstein et al. (1980, 1983) favor a somewhat lower absorption coefficient than currently accepted for ice *Ih* at 85 K (Mishima et al. 1983). Ultimately, mm-wavelength observations may be our best means of constraining the bulk absorption coefficient of the ring material.

C. Probing the Rings with Microwaves

1. Groundbased Radar and Radio Observations. Our understanding of the sizes of the ring particles really only began in 1973, when the rings were observed to be strong reflectors of groundbased radar (Goldstein and Morris 1973). This implied immediately that typical particles were at least comparable to, and possibly much larger than, the radar wavelength (12.6 cm) in size because of the well-known transparency to radiation of particles much smaller than the wavelength (see chapter appendix, Fig. A1). However, this very transparency had long been advocated to explain the lack of observable emission at similar radio wavelengths (see, e.g., Berge and Read 1968; Berge and Muhleman 1973). Pollack et al (1973) interpreted this paradoxical behavior as characteristic of particles of radius comparable to a radar wavelength, i.e. centimeters to meters (reviewed by Pollack [1975] and Morrison [1977]). As the sensitivity and number of radar and groundbased radio observations increased (cf. Esposito et al. 1984 for a summary), further constraints could be put on particle composition and size distribution and their radial variation.

Existing groundbased radar observations of the rings contain both intensity and polarization information as functions of varying ring inclination angle. The radar reflectivity and polarization measurements of the rings are summarized in Table II. The total reflectivity (in both polarizations received) of the 12.6 cm wavelength data is plotted in Fig. 9 as a function of ring inclination angle B, along with theoretical calculations for various ring models. The reflectivity is given in terms of geometric albedo $p = I/F = S \div (4 \sin B)$ where S is the usual diffuse reflectivity summed over polarizations (see chapter appendix for definitions; also Cuzzi and Van Blerkom 1974). By comparison with the geometric albedos (in backscatter) of most solar system objects (e.g., Venus ~ 0.6, Moon ~ 0.2), the large value of the ring radar geometric albedo implies either a near monolayer distribution of largely backscattering objects with fairly high albedo (Goldstein et al. 1977) such as may characterize the Galilean satellites (Campbell et al. 1977; Ostro et al. 1980*b*), or a high optical depth, classical, many-particle-thick layer of less backscattering, nearly perfect scatterers (Pollack et al. 1973; Cuzzi and Pollack 1978). These different hypotheses produce very different inclination angle dependences, as shown in Fig. 9. The observations seem most consistent with the multiple-scattering, many-particle-thick layer composed of much smaller individual scatterers of high albedo. As the tilt angle decreases and the incident wave becomes more grazing, less energy is reflected back to the observer by the classical model due to the increasing fraction of single (for-

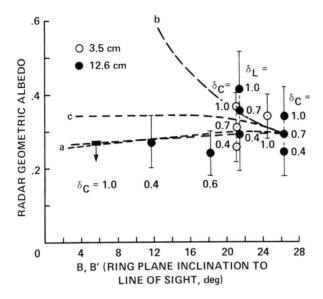

Fig. 9. Variation of A and B Ring radar geometric albedo I/F with ring tilt angle. For several values of tilt angle, the reflectivity was only measured in a single polarization (see, e.g. Table II); therefore the total albedo depends on assumed values of the linear L or circular C depolarization ratio δ, as seen in the ranges shown at $B \sim 21°$ and 26°. Model predictions are also shown. Curves (a) are for a many-particle-thick ring of ice particles with a particle size distribution $n(r) \sim r^{-3}\, dr$. Curve (b) is for a monolayer (neglecting shadowing) of large, singly and preferentially highly backscattering objects such as the Galilean satellites. Curve (c) is an estimate of the reflectivity of such a monolayer including shadowing following Froidevaux (1981).

ward) scattering compared to the backscattering monolayer model where the reflectivity *increases* due to the decreasing cross section of void areas (see chapter appendix). However, Froidevaux (1981) has noted that shadowing effects may diminish the increase in monolayer reflectivity with decreasing B (see Fig. 9). Also, consideration of the finite size and shadowing effects of the particles in the exact backscatter direction ($\alpha = 0°$) intrinsic to radar observations introduces a small multiplicative correction $e^{+\tau_s/\sin B}$ to the singly-scattered contribution to the reflected brightnesses calculated in the usual manner (Hapke 1963, 1981; Lumme and Irvine 1976b), where τ_s is the normal optical depth of particles large enough to cast shadows. Expressions in Sec. II.B. indicate that particles with $r >$ several meters will produce significant shadowing at 10 cm wavelength if $\tau_s > 1$. Because the normal optical depth τ_s of such large particles is probably, in fact, less than unity and possibly much less (Sec. III.C; Marouf et al. 1983; Cuzzi et al. 1980), the reflectivity correction is not large unless $\sin B \ll 1$, in which case it could become quite important.

TABLE II

Summary of Saturn's Rings Radar Results to Date[a]

B	λ(cm)	$\rho = I/F$	δ_C	σ_{OC}	σ_{SC}	δ_L	σ_{SL}	Reference
26°4	12.6			0.68 ± 0.17				Goldstein and Morris (1973)
24°4	3.5	0.34 ± 0.06	1.00 ± 0.25	0.68 ± 0.13	0.68 ± 0.13			Goldstein et al. (1977)
	12.6					1.0 ± 0.3		
21°4	12.6			0.75 ± 0.08			0.83 ± 0.21	Ostro et al. (1982)
	3.5							
18°2	12.6	0.24 ± 0.06	0.57 ± 0.1	0.61 ± 0.15	0.35 ± 0.09			Ostro et al. (1980a)
11°7	12.6	0.27 ± 0.07	0.40 ± 0.05	0.76 ± 0.19	0.30 ± 0.08			"
5°6	12.6	≤0.27		≤0.52	≤0.55			"
6°3	3.5			0.54 ± 0.15				Goldstein and Jurgens (1982)

[a] Measured values of geometric albedo I/F, normalized radar cross section σ, circular polarization ratio, δ_C, and linear polarization ratio δ_L, measured at ring tilt angle B, and wavelength λ, are listed in chronological order of the radar observations. Receiver polarization is designated as SC (same sense of circular polarization as transmitted), OC (sense of circular polarization orthogonal to that transmitted), or SL (same sense of linear polarization as transmitted). Table adapted from Ostro and Pettengill (1983); see also Esposito et al. (1984).

The reflected radar signals are all highly depolarized, whether the sense of transmitted radiation is linear or circular (Ostro et al. 1980a; see Table II). The observed reflectivities and depolarization ratios are well within the realm of possibility for realistic, moderately irregular particles (Cuzzi and Pollack 1978; Ostro et al. 1980). Detailed calculations of average ring reflectivity led Cuzzi and Pollack (1978) to suggest that a broad distribution of particle sizes, such as a power law distribution of the number of particles $n(r)$ with radii between r and $r+dr$, $n(r) \propto r^{-3} dr$, provided the most natural explanation for the rough equality of the reflectivities at 3.5 and 12.6 cm wavelengths. This is because the scattering behavior of a narrower size distribution varies too rapidly with wavelength over the wavelength range of the observations. In addition, Cuzzi and Pollack (1978) showed that primarily silicate materials are inconsistent with the radar observations, as their microwave albedos are simply too low. Either fairly pure ice or metal are necessary as bulk constituents. Metal may now be ruled out on grounds of mass density (Secs. III.C.2 and IV.B). However, the exact interpretation of the variation of both total intensity and depolarization with tilt angle is unclear, due to the lack of observed depolarization ratios at $B = 21°4$ and $26°4$ (cf. Esposito et al. 1984). Careful study of the polarization data and cross sections in Table II and Fig. 9 indicates that either the single-scattering depolarization ratio or the fraction of single to total scattering, and the total cross section, must change slightly between 3.5 cm and 12.6 cm wavelengths (Ostro et al. 1980a; Ostro and Pettengill 1983). It would be quite reasonable behavior for moderately irregular wavelength-sized particles to look rougher or less bright at the shorter wavelength (see, e.g., Zerull et al. 1980; Pollack and Cuzzi 1980).

The quantity directly measured in radar observations is the power spectrum of the energy reflected by the rings. These spectra peak at Doppler-shifted frequencies produced primarily by particles orbiting within the A and B Rings; these rings are, in fact, the location of most of the radar reflectors (Goldstein and Morris 1973; Goldstein et al. 1977). Ostro et al. (1982) have further deconvolved the ring radar reflectivity into radial segments. The observed reflectivities of the A and B Rings do not differ by more than 20%, even though their microwave optical depths differ by nearly a factor of two (Sec. III.C.2 and Fig. 10). This is a natural result of either multiple scattering in layers of different optical depth composed of identical particles or of the monolayer variation in $(1-e^{-\tau})/\sin B$ for the two ring elements (see chapter appendix). The very low relative reflectivity of the C Ring is also of interest. Further analysis is necessary to determine whether it may be explained by the relatively lower optical depth of the region, or whether bulk composition or size differences leading to lower particle radar albedos are implied. There has been no reappearance, in these low tilt-angle data (Table II), of the low-Doppler excess power, comprising 18% of the total, which was observed at 24° ring tilt (Goldstein et al. 1977).

Passive groundbased observations of the rings at microwave frequencies have also helped in the unraveling of this story. Radio interferometry allows a spatial resolution to be attained which is on the order of several seconds of arc. Thus, not only can the ring brightness be measured but also the occultation of the planet by the rings, giving a direct measurement of the optical depth at radio wavelengths, the radio depth. It was quickly realized that although the rings were extremely cold, their radio depth was comparable to that at visual wavelengths (Briggs 1974; Cuzzi and Dent 1975; Schloerb et al. 1979a,b, 1980). This implied immediately that the low observed ring brightness temperature and large radar reflectivity must be due to a particle albedo effect, not a transparency effect with a sharply peaked backscatter phase function.

The observed ring brightness at wavelengths >1 cm is so low ($T_B \sim 10°$; see Fig. 8) that it may be essentially accounted for merely by scattering of Saturn's thermal microwave emission by particles with albedo $\tilde{\omega}_o > 0.95$ (Cuzzi and Dent 1975; Schloerb et al. 1979a,b, 1980). Pollack et al. (1973) and Pollack (1975) had shown that such unfamiliarly high albedoes could be attained by particles of commonly occurring material, if they were in the centimeter-to-meter size range. If the particles are indeed in this size range, their scattering is strongly forward directed and so is the net scattering function of an ensemble comprising the ring layer (Cuzzi and Van Blerkom 1974; Cuzzi et al. 1980; see chapter appendix). Thus, the ring brightness in the zone where the rings occult the planet may be considerably higher than that at the ansae, making it difficult to obtain accurate estimates of optical depth (cf. dePater and Dickel 1983). Recent Voyager results, discussed below, provide more accurate, and less model-dependent, determinations of maximum particle size and total optical depth.

Groundbased observations of the ring brightness at shorter microwave wavelengths ($\lambda < 1$ cm) do display a thermal emission component which increases toward shorter wavelengths, and is sensitive to both particle size and material properties. These observations are shown in Fig. 8 and discussed in Sec. III.B. Because the size distribution is fairly well constrained by Voyager radar observations, as discussed below, especially accurate estimates of material properties may eventually be made.

2. Voyager Radar Observations. Voyager observations of radio occultations and bistatic scattering by the rings at 3.6 and 13 cm wavelength have greatly advanced our knowledge of the particle size distribution and its variation with distance from Saturn. These experiments apply many of the techniques of atmospheric occultation, and owe their strength to the excellent match between wavelength and typical particle size, as well as to the coherent nature of the illumination which is provided by the spacecraft communication system. The details of this experiment are given by Eshleman et al. (1977) and by Marouf et al. (1982), and are discussed in the chapter appendix. The radial

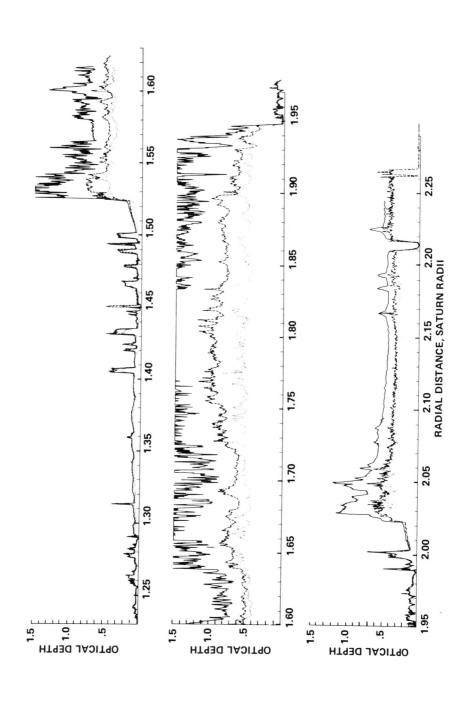

Fig. 10. Normal optical depth of the rings at visual wavelengths (solid line), and at 3.6 cm
(dashed) and 13 cm (dotted) wavelengths (radio depth) as a function of distance from Saturn.
The radial resolution and sensitivity vary with optical depth. Specifically, between 1.72 and
1.85 R_S the radio depth values shown are probably lower limits (Tyler et al. 1984). The
difference between the visual and 3.6 cm data is a measure of the area fraction of particles
with radii between about a micron and a centimeter in size. The difference between the 3.6
cm and the 13 cm data is a measure of the fraction with radii between about 1 and 4 cm. The
radio depths have been scaled down by a factor of 2, relative to those at visual wavelengths,
to account for coherency effects which differ between the two methods of measurement (see
chapter appendix). However, this factor may not be accurate in closely-packed, optically-
thick regions. For example, the good agreement between the optical and scaled radio depth
in the C Ring and Cassini region, where close-packing effects are negligible, demonstrates
that few sub-centimeter-radius particles exist in these regions. However, the large
wavelength variation seen in the optically thicker regions is probably due to a combination
of the presence of sub-centimeter-sized particles and close-packing effects, as discussed in
Sec. III.C. A detailed analysis is presented by Tyler et al. (1984).

resolution of the experiment is limited by diffraction to the size of a Fresnel
zone on the rings (15 km radially at 3.6 cm λ). Subject to the limitation
of signal-to-noise ratio, inversions and deconvolutions are able to improve
this resolution to better than 1 km radially (Tyler et al. 1983). The occultation
experiment consists of measuring the optical depth and forward-scattered
energy at both 3.6 and 13 cm as a function of position in the ring plane.
It must be recalled that the coherence of the transmitted signal introduces
an additional factor of two in these radio depths relative to the optical
depth of the same ensemble of particles. That is, if all the particles were
larger than both radio wavelengths, we would expect the radio depths so
measured to be twice as large as the optical depths measured at visual wavelengths (see chapter appendix).

Radio depth measurements by Voyager are limited by sensitivity to the
approximate dynamic range $0.05 < \tau/\sin \theta < 15$, where θ is the ring opening
at the time of occultation. Thus, for the Voyager 1 encounter, the small
ring opening ($\theta = 5°9$) caused the relatively opaque regions of the B Ring
to nearly block the radio signal, requiring substantial radial averaging for
signal detection (Tyler et al. 1983). The small opening angle, on the other
hand, provided better sensitivity to the optically thin regions of the C Ring
and Cassini Division.

If not all of the particles are larger than both wavelengths, transparency
effects will cause the two wavelengths to be differentially transmitted (see
chapter appendix). The difference in transmission between 3.6 and 13 cm
wavelength is generally (but not completely) insensitive to particle sizes
greater than 10 cm but is sensitive to particles of radius between 1 and 4 cm
(see chapter appendix, Fig. A1). In Fig. 10 we show the radio depth of
Saturn's rings at 3.6 and 13 cm wavelengths. From the relative differences in
these curves, it may be seen that the fractional abundance of 1 to 4 cm radius
particles does vary with location; for instance, it seems to increase outward in

the outer A Ring and to vanish entirely in the Cassini Division. Further implications may be obtained by comparison of these radio depths with optical depths measured at visible wavelengths (also shown in Fig. 10). The main impression one gets from Fig. 10 is that, after correction for the coherency factor of 2, the radio and optical depths are in very good agreement in the C Ring and Cassini Division, but the radio depth is substantially less than the optical depth in the A and B Rings. This differential transparency might imply the presence, only in the A and B Rings, of a large number of sub-centimeter-sized particles. However, if the ring thickness is really only a few times the size of the largest particles, as discussed below, two different nonideal effects come into play which cause the normal radio depth inferred in the optically thicker regions to be systematically underestimated. These effects are (1) the location of some particles in the near-field zone of others, which generally causes a decrease in the particle extinction efficiency below the value of 2 assumed in Fig. 10, and (2) the possibility that the large ring particles act more like a monolayer than a many-particle-thick one. For a ring volume density in the range of 10^{-1} to 10^{-2}, crude estimates (by JNC) indicate a near-field-related effect of ~ 20–30% (cf. appendix of Cuzzi et al. 1980), and an additional effect of comparable or larger amplitude due to the use of a many-particle-thick expression, rather than a monolayer expression for attenuation (see chapter appendix). The Voyager radio occultation experiment is especially sensitive to the latter effect because of its high grazing angle of incidence ($\mu \sim 0.1$).

Therefore, it appears that the central A Ring and inner B Ring radio/optical depth differences shown in Fig. 10 could possibly be accounted for by systematic nonideal effects in their inference from observed intensities. However, it seems that the even larger differences characteristic of the outer 2/3 of the B Ring and certain other areas may require, in addition, some component of sub-centimeter particles which are lacking in the C Ring and Cassini regions, at least. This would be consistent with observations of the ring phase behavior (Sec. III.A and Fig. 3). Clearly, more detailed analysis of these data is needed.

A measurement of the phase of the occulted Voyager coherent radio signal also contains information on the abundance of smaller particles, ($r < 1$ cm), as discussed in the chapter appendix and in Marouf et al. (1982). The abundance of such small ($r < 1$ cm) particles seems to increase at certain localities in the C Ring. However, the general lack of large phase shifts, and the close agreement between visible and radio optical depths, rules out the existence of large quantities of particles of size < 1 cm in these regions. Because the signal amplitude and phase depend on the particle size, number density, and refractive indices in different ways (Marouf et al. 1982), it may be possible in the near future to determine the particle absorption coefficient directly by using the signal phase and amplitude at 3.6 and 13 cm wavelengths, thereby constraining bulk composition.

Determinations of particle size in the several-meter-radius range of the size distribution may be derived from the incoherent signal or forward-scattered intensity (Tyler et al. 1981, 1983; Marouf et al. 1982, 1983; see chapter appendix). A characteristic size may be obtained from the occultation intensity either by combining the optical depth at a given point with the total forward-scattered intensity, or by actually observing the forward diffraction lobe of a Doppler-resolved region moving through the radar illumination pattern on the rings. The latter technique applies only if the scattering lobe of local particles is *narrower* than the illumination lobe of the antenna, which is true for particles larger than the antenna (1.8 m radius). In both of these approaches, the particle radius inferred is essentially that of the *largest* typical particle r_{max}. This is because the quantity that is measured is forward-scattered *intensity*, or scattered flux ($Q_s \pi r^2$) divided by the solid angle into which it is essentially diffracted ($\sim \lambda^2/r^2$). The effective size r_{eff} determined by this intensity weighting is thus the ratio of the observed forward-intensity to the observed τ:

$$r_{eff}^2 \sim \int_{r_{min}}^{r_{max}} n(r) r^4 \, dr / \int_{r_{min}}^{r_{max}} n(r) r^2 \, dr;$$

where, for $n(r) = n_o r^{-3}$, $r_{eff}^2 = r_{max}^2 / 2 \ln(r_{max}/r_{min})$.

Values of r_{max} derived in this way are approximately 1 and 5 m radius in the C Ring and outer Cassini Division, respectively (Tyler et al. 1981; Marouf et al. 1983).

A more detailed treatment is also possible using variation of the diffusely scattered signal over a limited range of scattering angles attained during the Voyager 1 radio occultation experiment. In this experiment, the diffuse scattering was measured over a range of $\Theta < 0°.7$, corresponding to the diffraction lobe half widths $\lambda/_{2r}$ of $r \gtrsim 1$ m particles. An inversion of these data provides a direct determination of the particle size distribution in this size range (Marouf et al. 1983). Results for several ring regions are shown in Fig. 11. A distinct break in the size distribution is seen at several meters radius, consistent with the estimates from the forward-scattered intensities mentioned above and the groundbased results. In most cases, the slope is so steep above the cutoff radius as to be indistinguishable from a complete absence of particles, within the limits of the inversion technique.

However, there are several caveats to consider. The inversions make certain assumptions as to the importance of multiple scattering, which acts to broaden the forward lobe. If the largest particles were confined to a near monolayer instead of being vertically dispersed by many times their own radii, as is usually assumed in radiative transfer treatments, *their* multiple scattering would decrease, and a slightly smaller "largest particle" would be inferred from the data. It is, in fact, precisely this tendency that is implied by comparison of the inversion results (e.g. Fig. 11) and the total optical depth

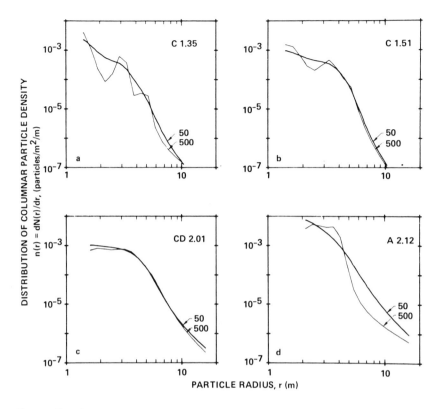

Fig. 11. Size distribution for particles with radius > 1 m, as determined by direct inversion of Voyager radio occultation data. Comparison with simulations (Marouf et al. 1983) establishes that the cutoff at roughly several meter radius is real, and the observed distribution for larger sizes is generally consistent with a complete absence of particles. However, model uncertainties such as the vertical thickness of the layer may cause the numerical value shown of the cutoff radius to be slightly too large (see text, Sec. III.C). Also, these observations do not preclude the existence of smaller numbers of much larger particles (Sec. III.D).

measurements (Fig. 10). That is, the optical depth in the $r > 1$ m component alone, inferred from the *inversions,* is as large as the total optical depth observed in the occultation experiment. However, as $\tau(3.6) > \tau(13.6)$ we know that there must be significant numbers of sub-meter-sized particles. The problem would be resolved if the largest ($r >$ several meter) particles were restricted to a layer of thickness only a few times their own radius. This would somewhat reduce the maximum particle size and the optical depth of the $r > 1$ m component (H. Zebker, personal communication, 1982). The slightly lower net number densities of the resulting $r > 1$ m distribution allow a smoother joining to the number densities of $r < 1$ m radius particles inferred from the differential opacity measurements, which are not strongly model dependent. The smooth joining is consistent with a power-law size distribution between 1 cm and the cutoff radius of several meters: $n(r) \sim r^{-q}$, with $2.8 < q < 3.4$ depending on the feature considered. The nonexistence of strong forward scattering from larger particles implies that $q > 5$ for $r >$ several meters.

The intriguing possibility of constraining ring vertical structure using the radio occultation results requires further investigation. It also cautions us that the value of the particle radius at the upper-size cutoff is somewhat model dependent. The existence of a cutoff of the relative number density in the several-meter-radius range is, however, unquestioned. This conclusion does not preclude the existence of larger objects in the rings, as long as their numbers are far less than obtained by an upward extrapolation of the $r^{-2.8}$ or $r^{-3.4}$ size distribution which characterizes the centimeter-to-several-meter-sized particles, as further discussed below.

Finally, the direct measurement of the particle size distribution from radio occultation observations may be combined with the direct measurement of surface mass density from gravitational (density and bending) waves (Sec. IV.B) to determine particle mass density. Not knowing the exact value of the upper cutoff radius induces some uncertainty, but the particle mass density so determined is on the order of 1 g cm^{-3}. This is consistent with a primarily icy composition but not a primarily metallic or rocky bulk particle composition (Marouf et al. 1983). More precise determination of particle density, specifically any indication of an unusually porous or underdense particle, is simply not possible at present.

D. Particle Properties: A Summary

Most of the material contributing to the observed opacity and mass of the rings of Saturn is found in the form of centimeter-to-several-meter-sized particles, which are distributed following a differential power law of the approximate form $n(r) = n_0 r^{-q}$, $2.8 < q < 3.4$. Power laws of this sort are common in geophysical processes (Hartmann 1969). Lower-size cutoffs are regionally variable, sometimes on the order of a centimeter, sometimes (perhaps) less (Fig. 10). Distinct, and possibly regionally variable, upper-size cutoffs are seen at several meters radius. The Voyager inversions strongly suggest that relatively very few particles exist with $r > 10$ m (Sec. III.C.2). Supporting this direct observation are the various occultation measurements of a thickness < 150 m of several ring edges. Clearly this could not be true if *many* particles with radii much larger than 100 m presented a significant cross section (Sec. II.B). However, small numbers of mountain-sized objects are inferred in certain regions (see Sec. IV.C). It is instructive to consider whether a smooth distribution of particle sizes exists through intermediate size regimes. The fact of a sharp "knee" in the size distribution at $r \sim 5$ m is well-established. The particle size distribution must be as steep or steeper than r^{-5} for larger sizes. Suppose the distribution is given by

$$n(r) = \begin{cases} n_0 r^{-3} \text{ for } r_0 = 1 \text{ cm} < r < 5 \text{ m} = r_1 \\[2mm] n_1 r^{-q} \text{ for } r_1 = 5 \text{ m} < r \, 5 \text{ km}. \end{cases}$$

We may then determine the value of q which connects the observed number of $r < 5$ m particles (Sec. III.C) smoothly with several particles of $r \sim 5$ km such as inferred to lie within, at least, the Encke Division (Sec. IV.C.3). The total number of particles in the rings between distance R_0 and R_1 from Saturn and of radius r within radius increment $dr \simeq r$, may be written (for particles larger than $r = 5$ m $= r_1$:

$$N \simeq z_0 \pi (R_0^2 - R_1^2) n_0 r_1^{q-3} r^{1-q}.$$

Using $\tau \simeq z_0 \pi n_0 \ln (r_1/r_0)$, and $\tau \sim 1$, we find that $q = 6$ yields one object of ~ 5 km radius and $q = 5$ yields several hundred. Without making too much of the specific numbers, it seems quite possible that a smooth and continuous distribution could exist of particles with radii ranging from 5 m to 10 km. The total ring mass would still be dominated by the several-meter-radius objects, due to the r^{3-q} weighting of the mass integral. However, the larger objects could be influential for global structure.

Small, regionally variable, quantities of dust-sized ($\sim 0.1 - 10 \ \mu$m radius) particles are seen with fractional optical depth never greater than ~ 0.1. Possible sources for such dust are discussed by Smoluchowski (1983) and by Morfill et al. (1983). In contrast, certain regions such as the E, F, and G Rings contain quite substantial fractions of micron-sized dust, possibly 100% for the E Ring. Particle size estimates of $10-100 \ \mu$m derived from near-infrared reflectance bands and far infrared spectral variations may be characteristic of granulation on the surfaces of the ring particles (Secs. III.A and III.B; also Esposito et al. 1984).

Direct observations of water-ice spectral features and the generally high particle albedo (Sec. III.A), as well as the need for an extremely low-loss dielectric to great depth in the particle subsurfaces (Sec. III.C) suggest that the particles are composed primarily of icy material. Because various clathrate hydrates are spectrally indistinguishable from pure water ice out to $\sim 100 \ \mu$m wavelength (Smythe 1975; Bertie et al. 1973), the possibility that the icy material is clathrate-rich must be considered. Also, the somewhat low thermal inertia and microwave absorption coefficient indicated, relative to pure hexagonal phase water ice Ih (Sec. III.B), suggest that other phases of water ice might also be considered, such as amorphous ice (Smoluchowski 1978). Significant quantities of metal, a conceivable (but cosmogonically unlikely) constituent on grounds of radar reflectivity, may be ruled out by a combination of particle size and optical depth measurements (Secs. III.A and III.C) and measurements of mass density from density wave properties (see Sec. IV.B). The inferred particle bulk density is close to unity, an independent support of a primarily icy composition.

The ring particles must contain a nonicy component, however. The visible-wavelength albedos in the C Ring and Cassini Division ($\varpi_0 \sim 0.2-0.3$) are too low for impurity-free ice; also, there is a large overall decrease in reflectivity shortwards of 0.6 micron wavelength. Finally, there is a hint of a 0.85 μm absorption band in the ring spectrum. All of these properties may be satisfied by a wide range of silicate materials. The quantity of impurity absorber required to produce these effects may be quite small, ($<< 10^{-1}$) and is probably well mixed with the ice at least in particle surfaces (Sec. III.A). The distribution of this component is distinctly nonuniform, with (at least) gradual variations on a radial scale of $\sim 10^4$ km. The nature of the nonicy component, and the reason for its regional variation, are exciting questions. Whether there is vertical segregation or size segregation by composition remains an open question (Sec. III.B).

IV. SPECIFIC STRUCTURAL PROPERTIES: MAIN RINGS AND F RING

A. Overview

In this section (IV) we discuss specific aspects of ring structure in more detail. As with particle properties such as albedo and size, discussed in Sec. III, structural properties are correlated with the classical ring regions to a large extent. Many of the particle properties have been discussed in Sec. III. For example, the A and B Rings may contain a significantly larger population of centimeter- and sub-centimeter-sized particles than the C Ring and Cassini region (Fig. 10), and the A and B Ring particles are, on the average, significantly brighter than the C Ring and Cassini particles (Fig. 2). Color differences in the rings are also regionally correlated (Fig. 6 and Color Plates 3 to 6), although the radial scale of color variation is much less abrupt (except at the B-C boundary) than that of optical depth variation. More detailed study is needed to determine the scale of particle size, albedo and color variation. It is possible that the observed color variations (Fig.6) are due to multiple-scattering effects arising from a slowly varying particle albedo (see chapter appendix). It is also possible that some (but probably not all) of the radio-depth variation with wavelength in the B and A Rings may be due to near-field effects in the rings.

In addition to the particle property similarities between the A and B Rings, and between the C Ring and Cassini Division, there are analogous similarities in the structure of these regional pairs. For instance, the structure of the dark, optically thin pair (C Ring and Cassini Division) has a similar banded look to it, in that both regions are characterized by a series of optical depth plateaus which convey the impression of regularity in spacing and length scale. These plateaus are often, but not always, separated from their surroundings by clear gaps. Generally these features are several hundred km wide, with little internal structure, and show a transition at their edges through a factor of 5 to 10 or so in optical depth within a scale of tens of km. These plateau edge transitions are thus less abrupt, in general, than those seen at the edges of the Encke, outer A, outer B, or Maxwell ringlet edges, and more similar in sharpness to the edges of the Maxwell Gap. However, some plateau edges are fairly sharp, especially if they are bordered inwards by an empty gap. These specific edges are discussed further in Sec. IV.C and by Marouf and Tyler (1984a). Tyler et al. (1983) have investigated several of the C Ring plateau features in detail, using Voyager 1 occultation amplitude and phase at both 3.6 and 13 cm wavelengths (see chapter appendix). In both features studied, it seems that a sort of W-shape may be seen, with the smaller (1–4 cm radius) particles concentrated in the optically thicker edge regions (at the "feet" of the W), as seen by the qualitatively different 3.6 cm and 13 cm optical depth profiles. Signal phase variations, due (at 3.6 cm wavelength) to particles of radius < 1 cm, are even more sharply confined to these edge regions. Simultaneously, very little difference is seen between 3.6 cm and visible wavelength optical depths, once the factor of two is accounted for (see chapter appendix, and Fig. 10). Therefore, we infer that the size distribution in these regions cuts off at around a centimeter radius, and that centimeter-sized particles are concentrated in narrow regions near the edges of the features shown. The very largest particles, however, are concentrated in the center of these features, as seen by comparing the 3.6 cm and the 13 cm optical depth profiles.

In another striking similarity, the outer edge region of both the C Ring and the Cassini Division consists of a broad, nearly featureless, band with optical depth and particle albedo increasing slowly, and nearly linearly, outwards. These broad bands (Figs. 10 and Color Plate 4) end abruptly at the inner edge of their optically much thicker neighbor, the B or A Ring, respectively. The pair similarity continues with increasing distance from Saturn. Inspection of Fig. 10 shows the optical depth of the inner region of the B and A Rings to be sharply peaked, increasing inward toward the ring boundary; at both the inner B and inner A Ring edges, there is a fairly abrupt transition in optical depth, over a radial scale of tens of km, and showing no sign of an empty gap (Lane et al. 1982; Esposito et al. 1983c). Also, the inner third of the A Ring and the entire B Ring exhibit a qualitatively similar irregular, fine-scale radial structure, as discussed in more detail in Sec. IV.D.

A certain morphological similarity may be seen by comparing the inner B/inner A edge structure (Fig. 10) with high-resolution, diffraction-corrected observations of four of the five bands in the inner Cassini Division; these structures show sharp, peaked inner edges and gradual, somewhat rounded outer edges (Marouf and Tyler 1984a). A similar characteristic inner edge/ outer edge structural pattern is produced by numerical models of erosion processes (see, e.g., chapter by Durisen; Ip 1983). Of course, both the theory and the observational data are in an extremely preliminary state. However, it

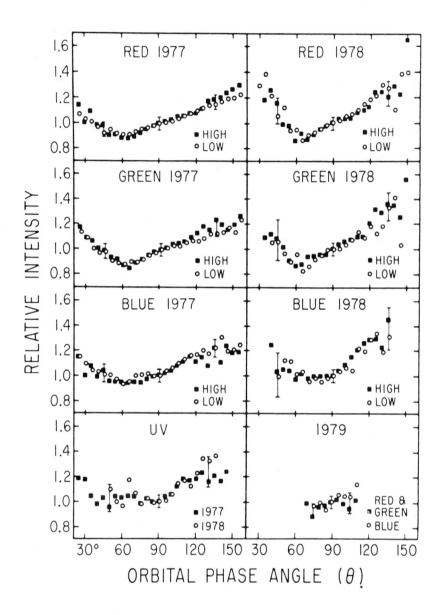

Fig. 12. Azimuthal asymmetry in the backscattered reflectivity of the A Ring. The variation with orbital longitude, shown only for one ansa, is nonsinusoidal. Similar effects have been seen in Voyager data (Smith et al. 1981). The shape of the variation in the groundbased data seems to change with tilt angle and color, possibly due to the varying importance of multiple-scattering effects. (Figure from Thompson et al. 1981.)

is intriguing to consider that a destructive process may play a role in sharpening some ring edges, and to consider ways in which other regional similarities may be so explained. It also cautions us that any inferences of primordial distribution merely from existing morphology and locally active processes may be extremely misleading.

For all their similarities, the A/B Ring and C/CD Region twins are fraternal, not identical. For example, the C Ring optical depth exhibits a smooth, axisymmetric wavelike pattern in the region between 79,000 and 84,500 km (see Fig. 6a, and Color Plate 4) which is unique in the ring system, and is not understood.

Significant differences also exist between the A and B Ring structure. For instance, the A Ring exhibits an azimuthally asymmetric reflectivity, as shown in Fig. 12 (Lumme and Irvine 1976a; Reitsema et al. 1976; Lumme et al. 1977) in its inner regions for which no analogue exists elsewhere in the rings. The effect is believed from groundbased observations to be isolated to the inner half of the A Ring between 125,000 and 132,000 km (Thompson et al. 1981) and has been observed by Pioneer 11 (Esposito et al. 1980) and Voyager on both the lit and unlit faces of the rings (Smith et al. 1981, 1982). The effect is characterized by enhanced (decreased) brightness of ring quadrants preceding (following) orbital conjunction with the observer. The brightness variation has an amplitude of 15% and is distinctly nonsinusoidal (see Fig. 12; Thompson et al. 1981). Because theoretical considerations (Peale 1976; chapter by Weidenschilling et al.) render untenable any explanation involving particles in synchronous, planet-aligned rotation, dynamical processes invoking trailing density wakes excited by locally large particles are regarded as most likely (Colombo et al. 1976; Franklin and Colombo 1978; see also Julian and Toomre 1966). The causative particles could be as large as 100 m in size, larger than the typical largest particle ($>$ 5 m radius) and small enough to avoid shepherding a clear gap for themselves ($<$ 1 km radius) as discussed in Sec. IV.C. The presence of a sprinkling of 100 m-sized particles in the A Ring is not consistent with the radio occultation determination of a largest typical radius of 5 m in a distribution following $n(r) \sim r^{-3}$ (see Sec. III.D). The lack of such an effect in the B Ring could be due to the lack of discernible differential brightness which may be produced by a similar fractional change in the higher overall B Ring optical depth (Colombo et al. 1976; cf. also Fig. 4).

In the following parts of Sec. IV, we discuss in more detail several classes of structure which are, in general, characteristic of different ring regions and (probably) different structural influences. In some cases, the causative process has been identified. In other cases, it is unknown or controversial. For instance, practically all of the radial structure in the outer two-thirds of the A Ring has been identified with isolated trains of gravitational waves driven at satellite resonances (Sec. IV.B). However, the Encke and Keeler Gaps in the A Ring are notable exceptions, and driven waves have been observed at resonances in the B Ring and the Cassini Division. The positive identification of compressional density waves and vertical corrugation waves in Saturn's rings has been a strong confirmation of spiral gravitational wave theory in general (chapter by Shu), and the observed waves provide extremely useful diagnostics as to local ring mass density and viscosity (velocity dispersion). The ultimate influence of driven spiral waves on momentum and energy transport in ring-ringmoon systems is certain to be important (see, e.g., Lissauer et al. 1984), although the details are far from well understood at present (chapter by Borderies et al.).

Clear gaps with or without embedded ringlets are a specific class of structure (Sec. IV.C) found in the A, C, and Cassini regions but not in the B Ring. Several, but not all, of the gaps in the optically thin C Ring and Cassini region are associated with strong gravitational resonances. In the C Ring and Cassini region, the embedded or adjacent ringlets are distinct and opaque, with a well-defined eccentricity which increases outward similarly to the behavior of the narrow Uranian rings. In the Encke Division, the embedded ringlets are irregular and optically thin, and similar in general appearance and particle properties to the F Ring. In the case of the F Ring and the Encke Division, the presence of shepherding moonlets (cf. also chapter by Dermott) is either known or strongly inferred. Extension of this association to the remaining nonresonant clear gaps with or without embedded ringlets is tempting, but as yet unsupported.

The B Ring structure is best described as irregular with no obvious association with either strong known gravitational resonances or local perturbing moonlets. The individual features are not in the nature of isolated ringlets but are more or less smooth fluctuations in optical depth, with only a few possible (and very narrow) clear gaps seen in the photopolarimeter (PPS) data (Evans et al. 1982), but not confirmed by other occultation data. The inner A Ring also exhibits similar optical depth fluctuations on a variety of length scales. The association of the irregular structure with the generally optically thickest ring regions, as well as its lack of regular radial spacing or length scales, has led to suggestions that it arises from unforced, collective instabilities within the disk arising, perhaps, from viscous effects (see, e.g., Lin and Bodenheimer 1981; Ward 1981, Hämeen-Anttila 1981; Lukkari 1981; see also chapter by Stewart et al.). However, this inference is still somewhat tentative. For instance, the existence in the rings of many objects with radii between 100 m and ~ 1 km, with a distribution sufficiently steep ($n(r) \sim r^{-6}$ or r^{-5}) as to render them invisible to the radio occultation, is still quite possible (see Sec. III.D). Most objects of these sizes would be incapable of clearing empty gaps in the B Ring or inner A Ring, but could still have some structural influence (Lissauer et al. 1981; Hénon 1981, 1983; Goldreich and Tremaine 1982). Therefore, the cause of the irregular structure is probably best left open. In Sec. IV.D we describe various aspects of the irregular structure.

Certain aspects of ring behavior are known to be related to electromagnetic phenomena, as discussed in Sec. IV.E. The B Ring spokes, for instance, have a well-defined periodicity and longitude preference matching the properties of Saturn's magnetic field. Also, one would expect that the tiny E, F, and G Ring particles, embedded in Saturn's magnetosphere, would exhibit some electromagnetic influence. A full understanding of the F Ring and its complement of moonlets also depends on a good understanding of the ring-magnetosphere relationship. Finally, the question of sporadic Saturn electrostatic discharges (SED) arising in the Saturn system, and possibly in the rings, has generated considerable interest. In Sec. IV.E we review and discuss the relevant observations.

In reading the remainder of Sec.IV, remember that the observed regional differences in structural appearance may only reflect the response of regions of different optical depth (and therefore different collective properties) to the same forcing. For instance, it seems that optically thick ($\tau \gtrsim 0.5$) regions respond to strong satellite resonances by launching spiral propagating waves (Sec. IV.B), whereas optically thin ($\tau \lesssim 0.1$) regions respond by opening gaps and/or coalescing ringlets (Sec. IV.C). However, optically thin regions may launch wave trains in response to weak resonances, as apparently seen in the Cassini Division. Similarly, it may be that the nonresonant irregular structure in the B Ring may merely reflect a different response to embedded moonlets than the clear gaps theoretically expected (cf. chapter by Borderies et al.). Clearly, our understanding of even the most favored processes is in a very rudimentary state.

B. Gravitational Resonances and Waves

Much of the structure observed in Saturn's rings results from resonant forcing of ring particles by known satellites of Saturn. Forcing due to resonances between the ring particles and a hypothetical nonaxisymmetric component of Saturn's gravitational field may also be important (Franklin et al. 1982; Stevenson 1982). However, as nothing observed in Saturn's rings is currently identified as a planet-induced feature (Holberg et al. 1982), and the asymmetry in Saturn's gravitational field necessary to produce such resonances has never been detected, we shall restrict our discussion to satellite-caused resonant effects.

Bodies in orbit about Saturn at radius R move with mean angular frequency $\Omega(R)$. If they are in slightly eccentric orbits, their epicyclic (radial) frequency is K (R), and if their orbits are inclined to Saturn's equatorial plane they oscillate about it with (vertical) frequency μ (R). If Saturn were a point mass or perfectly spherical, these three frequencies would all be equal and orbits would be closed. However, the oblateness of Saturn causes the orbital node to regress and the line of apsides to advance (see, e.g., Burns 1976). Therefore, the various frequencies split such that $\mu > \Omega > K$. Resonances occur where the natural horizontal (vertical) frequency of the ring particles is equal to that of a component of the satellite's horizontal (vertical) forcing, as sensed in the rotating frame of the particle. We can view the situation as the resonating particle always being at the same phase in its radial or vertical oscillation when it experiences a particular phase of the satellite's forcing. This situation enables continued coherent "kicks" from the satellite to build up the particle's radial or vertical motion, and significant forced oscillations may thus result.

The locations and strengths of resonances can be determined by decomposing the gravitational potential of the satellite into Fourier components (see chapter by Shu for details). The disturbance frequency ω, can be written as the sum of integer multiples of the satellite's circular, vertical and radial frequencies:

$$\omega = m\Omega_M \pm n\mu_M \pm pK_M \tag{1}$$

where m, n and p are nonnegative integers, with n being even for horizontal forcing and odd for vertical forcing. Horizontal forcing, which can excite density waves and possibly open gaps by angular momentum transport, occurs at the inner Lindblad resonance R_L where

$$K(R_L) = m\Omega (R_L) - \omega . \tag{2a}$$

Vertical forcing occurs at the inner vertical resonance R_V where

$$\mu (R_V) = m\Omega (R_V) - \omega . \tag{2b}$$

From Eqs. (1) and (2), approximating with $\Omega \sim \mu \sim K$, one obtains

$$\frac{\Omega (R_L)}{\Omega_M} = \frac{m + n + p}{m-1} = l/(m-1) . \tag{3}$$

The $l/(m-1)$ or $l: (m-1)$ notation is commonly used to identify a given resonance. Because the strength of the forcing by the satellite depends, to lowest order, on the satellite's eccentricity e and inclination i as $e^p (\sin i)^n$, the strongest horizontal resonances have $n=p=0$, and are of the form $m:(m-1)$.

The radial locations and relative strengths of such orbital resonances are easily calculated from known satellite masses and orbital parameters and the gravitational moments of Saturn (Lissauer and Cuzzi 1982; chapters by Franklin et al. and by Shu). By far the lion's share of the strong resonances lie within the outer A Ring (Fig. 13).

Several types of features are observed to occur at resonance locations within Saturn's rings. One class of feature includes sharp, noncircular outer ring edges. The outer edges of the B and A Rings are located quite close to the Mimas 2:1 and Janus 7:6 resonances, respectively. In the case of the B Ring edge, the shape of the edge is that of an oval centered on Saturn. That is, there

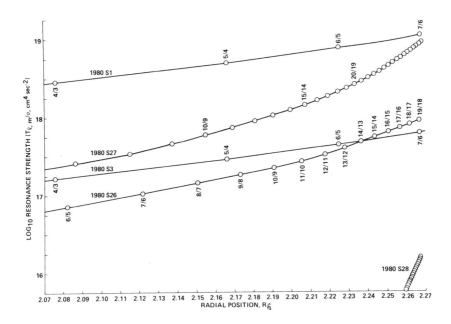

Fig. 13. Locations and strength (torque) of the strongest gravitational resonances in the A Ring. $T\ell_{,m}/\sigma$ is the torque per unit mass density due to the (ℓ/m) Lindblad resonance at R_L with a satellite M, where Ω (R_L) $\simeq$ (ℓ/m) Ω_M. Less than ten resonances of strength $> 10^{17}$ cm^4 s^{-2} exist elsewhere in the rings. The closer satellites have more closely spaced resonances with strength increasing outward more rapidly than for more remote ones. (Figure from Lissauer and Cuzzi 1982.)

are two radial minima and two maxima along the edge circumference (see, e.g., Fig. 24 in Sec. IV.D). The variation in radius is > 140 km, and the axis of minimum radius is closely aligned with the direction to Mimas (Smith et al. 1983). The entire pattern rotates at the orbital rate of Mimas. This morphology is in good agreement with theoretical expectations for particle orbit streamlines which would result from forcing at an $m:(m-1)$ resonance with $m=2$ (Borderies et al. 1982). The particle orbits are, in fact, very nearly normal Keplerian ellipses. However, the resonance constrains particles to cross periapse at each conjunction, occuring every m^{th} orbit. For example, in the time taken by a particle to move from periapse to apoapse, the edge pattern for a 2:1 resonance rotates by 90°. The amplitude of the forced radial oscillation of a particle near resonance is easy to calculate in the collisionless, zero-gravity limit (Goldreich and Tremaine 1982; Lissauer and Cuzzi 1982), and increases as the inverse of the distance from resonance. The "natural" amplitude, or resonance width, is often cited to be R_L $(M/M_S)^{\frac{1}{3}}$, where M_S is Saturn's mass and M is the mass of the satellite. This is the distance from resonance at which perturbed streamlines cross the resonance and intersect

TABLE III

Distribution of Mass in the Rings, as Scaled from Optical Depth τ/σ Determined in Density-Wave Regions

Region	Boundaries (R_S)	Mean Optical Depth (PPS)	Mass (10^{-8} M_S) PPS[a]	UVS[b]
Inner C	1.23–1.39	0.08	0.05	0.2
Outer C	1.39–1.53	0.15	0.09	
Inner B	1.53–1.66	1.21	0.80	
Middle B	1.66–1.72	1.76	0.60	5.0
Outer B	1.75–1.95	1.84	2.0	
Cassini Division	1.95–2.02	0.12	0.06	0.1
Inner A	2.02–2.16	0.70	0.70	
Outer A	2.16–2.27	0.57	0.40	1.1
Total Ring Mass[c]			5.0	6.0

[a]Esposito et al. 1983b.

[b]Holberg et al. 1982.

[c]Total ring mass obtained from Voyager radio occultation particle size distribution, assuming solid water ice (Marouf et al. 1983) $\sim 3 \times 10^{-8}$ M_S.

Average A and B Ring mass density from cosmic ray albedo neutrons (Cooper et al. 1982) $\sim 100–200$ g cm^{-2} ($\sim 5–10 \times 10^{-8}$ M_S).

(cf. chapter by Franklin et al.). For the Mimas 2:1 resonance, this amplitude is 40 km; however, the observed radial variation of the B Ring edge is > 140 km. Borderies et al. (1982) show that the larger observed amplitude may result from the local disk self-gravity, which alters the precession rate of particle orbits and maintains a lower, and more stable, eccentricity gradient further into the resonance, thereby yielding a larger edge eccentricity and radial variation. The observed amplitude is consistent with B Ring surface density on the order of 100 g cm^{-2}, in agreement with independent constraints (Table III).

If the outer edge of the A Ring is under the control of the Janus 7:6 resonance, a similar situation would be anticipated at this boundary, with the pattern differing by having seven maxima and seven minima. The amplitude of the effect would be smaller due to the smaller "natural" width of the Janus 7:6 resonance and the lower mass density of the outer A Ring. Some observational support for this situation is given by Porco (1983).

Another class of features includes empty gaps with embedded, opaque ringlets. Several of these features have been observed at known resonance locations in optically thin regions of the rings (Holberg et al. 1982) and are probably caused by a resonance-related process; however, no explanation for the embedded ringlets currently exists. Several empty gaps with embedded ringlets occur at nonresonant locations as well (see Sec. IV.C).

A third class of resonantly produced features, which are generally seen in regions of moderate-to-large optical depth, are spiral density waves and spiral bending waves. There are several dozen examples of these waves in Saturn's rings (Fig. 14), so they warrant discussion in some detail. Density waves are (horizontal) oscillations in local surface mass density of the rings, and are excited at horizontal (Lindblad) resonances with Saturn's satellites, whereas bending waves are vertical warps (corrugations) excited at vertical resonances (see Fig. 2 in the chapter by Shu). Both types of waves propagate by virtue of the self-gravity of the ring disk. The theory of spiral density waves was developed by Lin and Shu (1964) to explain the spiral structure observed in most disk galaxies. Goldreich and Tremaine (1978b) were the first to suggest the presence of density waves in Saturn's rings, speculating that these waves, resonantly excited by external satellites, may have been responsible for clearing the Cassini Division and other gaps within the rings. Density waves have since been observed at many locations within the rings (Cuzzi et al. 1981; Lane et al. 1982; Holberg et al. 1982; Holberg 1982; Esposito et al. 1983b). Bending waves were first discussed in connection with warps in disk galaxies (Hunter and Toomre 1969; Bertin and Mark 1980), and have been identified at several locations within Saturn's rings (Shu et al. 1983). The theory underlying both types of spiral waves has been exhaustively reviewed in the chapter by Shu (see also Shu et al. 1983 and chapter by Borderies et al.), therefore we shall not discuss these in detail here.

Analysis of spiral waves (see, e.g., Cuzzi et al. 1981; Shu et al. 1983) has yielded the most accurate measurements of local ring surface mass density, which, in turn, have been used to estimate the mass of the entire ring system (Holberg et al. 1982; Esposito et al. 1983b). Measurements of the damping of bending waves and density waves also yield values of the viscosity of the rings, which can be used to estimate the local thickness (scale height) of the rings (Lissauer et al. 1983). We shall now summarize the theory behind these measurements following Lissauer (1982) and Goldreich and Tremaine (1979a) and summarize the results thus far obtained. Subsequently, we will discuss the major unresolved questions of wave phenomena in the rings: nonlinearity and the transport of angular momentum and energy.

The analysis below is for density waves. However, the procedure for bending waves is essentially identical to that for density waves except for (a) the substitution of $\mu(R)$ (the vertical frequency) for $K(R)$ (the radial frequency), and (b) certain changes in sign due to the fact that bending waves propagate inward from inner vertical resonance whereas density waves propagate outward from inner Lindblad resonance (see Shu et al. [1983] for details). Crudely speaking, the opposite directions of propagation result from the opposite effects of self-gravity on horizontal oscillation frequencies (which are diminished) and on vertical oscillation frequencies (which are augmented). Therefore, each type of wave may only propagate in the direction in which the "beat" frequency at which the forcing is experienced becomes

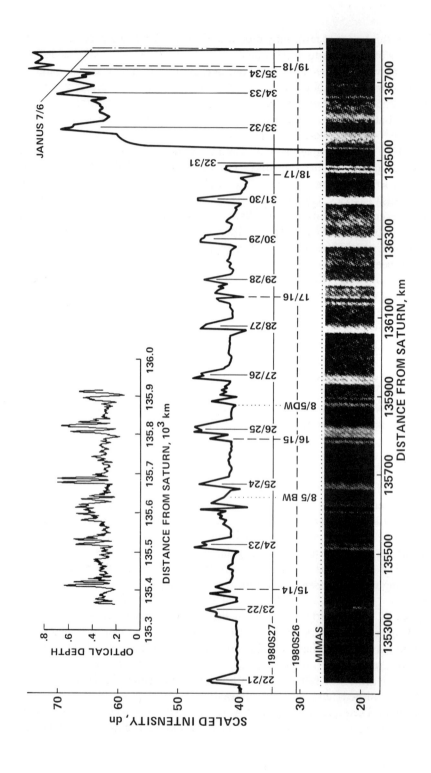

Fig. 14. Gravitational resonances and their driven density waves in the outer A Ring, as discussed in Sec. IV.B. The lower panel shows part of the highest resolution Voyager 2 image of the A Ring (~3 km/pixel). The middle panel is a profile of reflectivity obtained from this image. The Keeler Gap is the dark band at right; the ring region between the Keeler Gap and the A Ring edge is anomalously bright, probably due to enhanced forward scattering (see, e.g. Fig. 3). The regularly spaced bright features are also regions of enhanced forward scattering, because they are darker than their surroundings at low phase angles. By aligning the edge of the ring in this image with the absolute radius determined by the stellar occultation experiment (Holberg et al. 1982), it is seen that all but two of the significant features correspond to $m:(m-1)$ resonances with 1980S26 and 1980S27 (shown in Fig. 13). The other two features, which have a longer wavelength and are actually resolved in the image data (lower panel), are a density wave/bending wave pair, excited at the Mimas 8:5 resonance (see, e.g., Shu et al. 1983). Measurements of the optical depth of the region at even higher (~100 m) resolution, from the PPS stellar occultation experiment (upper panel), reveal that each feature is actually a short sequence of fairly intense fluctuations in optical depth. These are short trains of density waves, one excited at each resonance and propagating outward by virtue of the self-gravity of the disk. More recognizable density wavetrains are shown in Fig. 15, and discussed in the text.

more positive (outward) and more negative (inward), respectively (cf. chapter by Borderies et al).

Consider free, linear, long spiral density waves, of arbitary frequency ω and azimuthal periodicity $2\pi/m$. Ignoring small "pressure" effects due to the velocity dispersion of ring particles, the wavelength and frequency of these waves are related by the dispersion relation (Lin and Shu 1964; chapter by Shu):

$$(\omega - m\Omega)^2 = K^2 - 2\pi G\sigma|k| \qquad (4)$$

where σ (except where otherwise stated in this chapter) is the local surface mass density and k is the radial wave number. Equation (4) can be rearranged to give

$$|k| = D/2\pi G\sigma \qquad (5)$$

where the frequency difference

$$D \equiv K^2 - (\omega - m\Omega)^2 \qquad (6)$$

is a measure of distance from the exact Lindblad resonance (see Eq. 2). Near the resonance, D approaches zero and thus k approaches zero because the Doppler-shifted, or locally experienced, forcing frequency, $\omega - m\Omega$, is close to the particles' natural frequency K (see Eq. 1). Waves can be excited most easily near inner Lindblad resonance because their long wavelength, $\lambda = 2\pi/|k|$, in this region enables them to couple to the forcing potential of the satellite, which varies very slowly in space (chapter by Shu).

Equation (5) contains the information necessary for determining surface density from observations of a density wavetrain excited by a known surface. A Taylor series expansion of D about $R = R_L$ for driven waves yields:

$$D = \mathcal{D} x \qquad (7)$$

where

$$\mathcal{D} \equiv \left(R \frac{dD}{dR} \right)_{R_L} = \left(2KR \frac{d}{dR} (K - m\Omega) \right)_{R_L} \qquad (8)$$

and $x \equiv (R - R_L)/R_L$ is the fractional distance from inner Lindblad resonance. We thus see from Eqs. (5)–(7) that the wavelength between successive crests should be inversely proportional to the nondimensional distance from resonance x:

$$\lambda = 4\pi^2 G\sigma/\mathcal{D} x . \qquad (9)$$

Thus,

$$\sigma = \lambda \mathcal{D} x/4\pi^2 G . \qquad (10)$$

In order to use Eq. (10) for mass density determination, we must evaluate $\mathcal{D}$. We use expansions of Ω and K given by Lissauer and Cuzzi (1982), truncated after the second term for simplicity, to obtain

$$\mathcal{D} = K \left(\frac{GM_S}{R^3} \right)^{\frac{1}{2}} \left[3(m-1) + \frac{21}{4}(m+1) J_2 \left(\frac{R_S}{R} \right)^2 \right] \qquad (11)$$

where J_2 is Saturn's quadrupole moment. If we use the Keplerian approximation, $\Omega = (GM_S/R^3)^{\frac{1}{2}} \approx K$, for the term outside the parentheses, and also ignore the second term in the parentheses for all resonances except apsidal ($m = 1$) ones, the following simple equations result:

$$\mathcal{D} = \frac{21}{2} J_2 \left(\frac{R_S}{R} \right)^2 \Omega^2(R_L) \quad (m = 1) \qquad (12a)$$

$$\mathcal{D} = 3(m-1) \Omega^2 (R_L) \qquad (m > 1) . \qquad (12b)$$

When Eqs. (10) and (12) are combined, and numerical values appropriate to Saturn are inserted, the following formulae are derived:

$$\sigma = 0.0185 \, \lambda d \left(\frac{R_S}{R} \right)^6 \qquad (m = 1) \qquad\qquad (13a)$$

$$\sigma = 0.325 \, (m - 1) \, \lambda d \left(\frac{R_S}{R} \right)^4 \qquad (m > 1) \qquad\qquad (13b)$$

where σ is in units of g cm^{-2} and where λ and $d \equiv R - R_L$ are each measured in km.

In actual applications, the location of the resonance relative to the observed wavetrain is somewhat uncertain. If small perturbations to a uniform surface density are assumed, the measured wavelengths can be fitted to a function of the form

$$\lambda = \frac{c_1}{d' - c_2} \qquad\qquad (14)$$

where d' is the distance from the assured resonance location and where c_1 and c_2 are varied to minimize residuals. When a best fit is obtained, one sets $d = d' - c_2$. Then $c_1 = \lambda d$ can be substituted into Eq. (13) to determine σ. When the background average σ varies in the region of wave propagation, a good fit cannot be obtained in this manner, and other methods must be used. Holberg et al. (1982) assumed that the mass extinction coefficient, $K = \tau / \sigma$ was constant in each wavetrain, even if the (wavelength-averaged) optical depth varied through the region of observed propagation. Equation (13b) was replaced by

$$\frac{1}{K} = \frac{\sigma}{\tau} = 0.325 \, (m-1) \, \frac{\lambda d}{\tau} \left(\frac{R_S}{R} \right)^4 \quad (m > 1). \qquad (15)$$

Equation (15) was then solved by a similar least residuals fit using observed $\lambda \, (R)$ and $\tau \, (R)$ as input parameters. In general, the ultraviolet spectrometer date of Holberg et al. (1982) show a far better match to the constant K assumption than to the constant σ assumption. However, the photopolarimeter spectrometer data of Esposito et al. (1983b) do not show such a clear preference.

Several authors have obtained surface densities using the methods mentioned above at over 20 resonance locations; Esposito et al. (1984) give a complete listing of results obtained prior to the writing of this chapter. In Table IV we show several characteristic mass density determinations obtained from such waves. Regions of higher average optical depth generally have higher measured surface mass density; this correlation is especially strong if we only include measurements with low-quoted uncertainties, and those in which the resonance position derived using Eq. (14) agrees closely with the theoretically predicted location (Table IV). The mass extinction coefficient K

TABLE IV

Surface Mass Densities and Mass Extinction Coefficient as Determined from Density and Bending Waves

Resonance[a]	Lindblad Radius (R_S)	σ_{in}[b] (g cm^{-2})	τ_{in}[b]	$K_{in} \times 10^2$[b] (cm^2 g^{-1})	τ_{out}[c]	σ_{out}[c] (g cm^{-2})	References[d]
Janus 2:1	1.5953	60,70 ± 10,70	0.73	1.2,1.0,1.0	~0.70	70 ± 10	Lane et al. (1982); HFL; EOW
Iapetus 1:0	2.0087	10 ± 3	0.12	1.2*	—	—	This work; Cuzzi et al. (1981)
(5)	2.027–2.087	69 ± 35	0.68	1.0	0.70	69 ± 35	EOW
(3)	2.115–2.155	44 ± 13	0.47	1.1	0.45	42 ± 13	EOW; H
Janus 5:4	2.1664	100,57	0.8,0.48	0.8*, 0.8	0.45	50 ± 12, 51	HFL; EOW
(2)	2.173–2.185	51 ± 2	0.45	0.9	0.45	51 ± 3	EOW; H
Mimas 5:3 (BW)	2.1863	45 ± 4*	0.60	1.3	0.45	35 ± 4	SCL
Mimas 5:3	2.1929	80,40	0.85	1.0*, 2.1	0.45	42 ± 8, 21	HFL; EOW
(3)	2.200–2.220	46 ± 13	0.50	0.9	0.50	46 ± 13	EOW; H
Janus 6:5	2.2255	110,96	1.10,0.59	1.0*, 0.6	0.50	50 ± 6, 80	HFL; EOW
(3)	2.227–2.234	57 ± 14	0.52	0.9	0.50	57 ± 14	H
Mimas 8:5 (BW)	2.2483	42 ± 13*	0.65	1.4	0.50	32 ± 10	SCL
Mimas 8:5	2.2522	20 ± 3	0.50	2.4	0.50	20 ± 3	This work; (PPS data)
1980S27 27:26	2.2533	28 ± 15	0.60	1.7	0.50	23 ± 15	This work; (PPS data)

[a] The resonance is given from which the mass density was determined. Where only a number in parentheses is cited, that number of published results (with large errors or without quoted errors) have been averaged together. The error given is determined from the internal scatter. Also, all results are from density waves except where denoted (BW) for bending waves.

[b] The results denoted by subscript "in" are characteristic of the wave zone itself. As discussed in the text (cf. Eq. 15), either surface mass density σ or mass extinction coefficient K may be the primary parameter determined; in cases where a difference has been found, the primary parameter is indicted by an asterisk (*).

[c] A more representative value of σ is probably given *outside* the wave zone. These columns scale typical local ring optical depths out by observed mass extinction coefficient K.

[d] References: HFL = Holberg et al. (1982); EOW = Esposito et al. (1983b); H = Holberg (1982); SCL = Shu et al. (1983).

seems to maintain an approximately constant value of 10^{-2} cm^2 g^{-1} for the regions in the A and B Rings previously measured. In preparing this review, we have analyzed several waves in the outer portions of the A Ring (Fig. 14), and find significantly larger mass extinction coefficients than in other areas of the rings. This result indicates a larger portion of small particles, which block more light per unit mass. An increase in the fraction of small particles in this region is also suggested by other measurements; the outwardly increasing difference between radio depths at 3.6 and 13 cm wavelengths indicates an increasing fraction of centimeter-sized particles (Fig. 10; also Sec. III.C), and the general outwardly increase in brightness of the outer A Ring in forward scattering seen in Voyager images (Fig. 3; also Sec. III.A) implies an outwardly increasing fraction of micron-sized particles in the region.

Holberg et al. (1982) and Esposito et al. (1983b) have extrapolated the surface mass densities from values derived at resonances to estimate the mass of the rings as a whole, under the assumption of constant mass extinction coefficient $K = \tau/\sigma$. Holberg et al. used the optical depth profile of the ultraviolet spectrometer stellar occultation and assumed a mass extinction coefficient $K = 0.01$ cm^2 g^{-1} to derive a ring mass of $6 \pm 2 \times 10^{-8}$ M$_S$; Esposito et al. used the photopolarimeter optical depth profile and $K = 0.013$ cm^2 g^{-1} to yield a ring mass of $5 \pm 3 \times 10^{-8}$ M$_S$. A more detailed breakdown of their results, by regions of the rings, is given in Table III. Note that if τ/σ increases in the outer part of the A Ring, as we suspect from our new measurements (Table IV), the A Ring mass is probably slightly less than listed in Table III. Other estimates of ring mass and their sources are also shown in Table III.

While the surface mass density of the region in which density waves and bending waves are propagating can be determined by wavelength measurements alone, viscosity estimates require, in addition, information on the amplitude of the waves. If these gravitational waves propagate without damping in a uniform region of the rings, they conserve their angular momentum luminosity F_J (azimuthally integrated angular momentum flux). Because their wavelength is linearly decreasing, the amplitude of the density variations increases linearly with distance from inner Lindblad resonance (Goldreich and Tremaine 1978b):

$$\frac{\Delta\sigma}{\sigma} = \frac{x}{x_{NL}} , \qquad (16)$$

where $\Delta\sigma$ is the increase (decrease) in surface density at the peaks (troughs) of the waves. Then, we define x_{NL} as the scaled distance at which point the fractional perturbation is equal to the unperturbed density, and the waves become nonlinear:

$$x_{NL} = \frac{2\pi^2}{\mathcal{D}} \left(\frac{R\sigma}{F_J}\right)^{\frac{1}{2}} (G\sigma)^{\frac{3}{2}} . \qquad (17)$$

In the linear regime, wavetrains display smooth, nearly sinusoidal, variations; in the nonlinear regime, the crests are sharply peaked and the gradients extremely steep, although the wavelength is not significantly affected for moderate nonlinearity (Fig. 15). This behavior is not unlike that of more familiar ocean waves. Beyond x_{NL}, nonlinear effects and related damping processes (e.g. shocks) become important, as discussed further below.

Even for $x \ll x_{NL}$ the presence of interparticle collisions causes small amplitude density waves to suffer viscous damping by a factor of:

$$\exp\left[-\int_o^x k_I(x')R_L \, dx' \right] \tag{18}$$

where $k_I = (\frac{7}{3}\nu + \zeta)(K\mathcal{D}^2/2\pi G \sigma^3)x^2$ (Goldreich and Tremaine 1978b; Cuzzi et al. 1981; chapter by Shu). In Eq. (18) ν is the kinematic shear viscosity and ζ is the bulk viscosity. Taking into account both amplitude variation factors (Eqs. 16 and 18), we obtain the wave amplitude profile near the position of the inner Lindblad resonance:

$$\frac{\Delta\sigma}{\sigma} = \frac{x}{x_{NL}} \exp\left[-(x/x_D)^3 \right] \tag{19}$$

where

$$x_D \equiv \left(\frac{9}{7} \right)^{\frac{1}{3}} \frac{2\pi G \sigma}{[K(\nu+\frac{3}{7}\zeta)^2 R]^{\frac{1}{3}}} \tag{20}$$

is the characteristic dimensionless damping length. Equation (20) can be inverted to give the viscosity (actually a combination of the shear and bulk viscosities) as a function of x_D:

$$(\nu + \tfrac{3}{7}\zeta) = \frac{9}{7}\left(\frac{2\pi G \sigma}{x_D} \right)^3 \frac{1}{K\mathcal{D}^2 R} \, . \tag{21}$$

As with Eqs. (13a,b), Eqs. (20)–(21) assume a uniform surface density in the region of wave propagation. Variable surface density induces severe complications in any attempt to determine ring viscosity from wave damping. The fractional amplitude of undamped waves no longer simply increases linearly with distance from resonance, but it also varies as σ^{-2} (see Eq. 17), and integration of Eq. (18) becomes more complex. Additionally, the assumption of constant viscosity, also required for Eq. (21), is likely to be violated in regions where the surface density changes.

The analysis of the Iapetus wavetrain in the region between 120,000 and 122,000 km from Saturn in the outer Cassini Division (Fig. 15a) by Cuzzi et

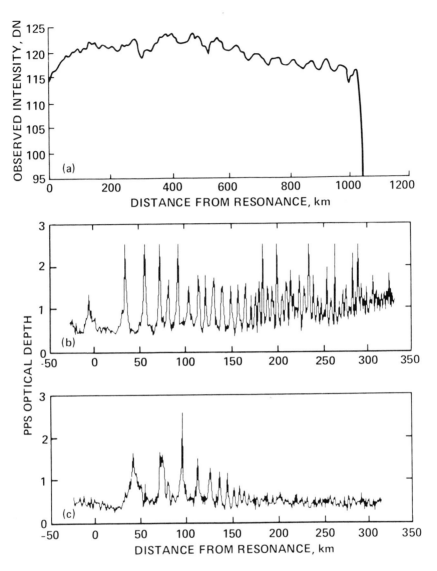

Fig. 15. Comparison of several characteristic density wavetrains. (a) The Iapetus apsidal density wavetrain (Cuzzi et al. 1981), here shown in a brightness profile obtained by the Voyager 1 narrow-angle camera, is quite weak and highly linear in the sense that $\Delta\tau/\tau \ll 1$. The shape of the waves is smooth and nearly sinusoidal (see text for further discussion). (b) The Janus 2:1 density wavetrain (Lane et al. 1982) becomes nonlinear ($\Delta\tau/\tau \sim 1$) fairly close to its causative resonance (Janus 2:1), and its "crests" are quite sharply peaked. Nonetheless, it propagates for nearly 100 waves before being damped out. (c) The Mimas 5:3 density wavetrain (Esposito et al. 1982) is of comparable strength, in the sense of torque, and is equally nonlinear. However, it propagates less than half as far before being damped. The damping of density and bending waves has been used to determine the local viscosity and thereby velocity dispersion, or ring thickness (see Sec. IV.B).

al. (1981) suffers from neglect of these effects. More careful analysis of this wavetrain using the recently determined actual optical depth profile of the region (see, e.g., Fig. 10; also Ring Atlas in the Appendix) and assuming mass density to increase proportionately outwards, yields two results. First, the wavetrain source is seen to lie at ~ 200 km into the region, ~ 170 km or 3 to 4 standard deviations inward (Holberg et al. 1982) of the theoretical loca- tion of the Iapetus apsidal resonance (Lissauer and Cuzzi 1982) believed by Cuzzi et al. (1981) to be responsible for the wavetrain. A revised mass density is given in Table IV. In addition, the initial analysis assumed that surface mass density variations were simply proportional to the observed brightness varia- tions. In reality, a combination of varying particle albedo and varying depen- dence of transmitted brightness on optical depth through the region invali- dated this assumption and caused erroneous estimates of the wave amplitude, and consequently damping and viscosity, to be made. A careful radiative transfer reanalysis (Cuzzi, unpublished) reveals that uncertainties in the radia- tive transfer model (possibly even as to whether monolayer or many-particle- thick expressions should be used to relate observed intensity to optical depth), make it extremely difficult to understand the amplitudes of these waves and to determine their damping. Thus, the first (and most linear) density wave may be the hardest to interpret.

Bending waves also suffer viscous damping but as these waves are not compressional in nature, only shear viscosity is important (Shu et al. 1983). The variation of the maximum slope $\mathcal{S}_{max} = 2\pi h/\lambda$, where h is the vertical amplitude of a linear bending wave, has the same form as that of the amplitude of a linear density wave:

$$\tan(\mathcal{S}_{max}) = \frac{x}{x_{NL}} \exp\left[-\left(\frac{x}{x_D}\right)^3\right]. \tag{22}$$

However, in the case of bending waves,

$$x_D \equiv 2\pi G\sigma \left(\frac{3\mathcal{D}^2 R}{\mu\nu}\right)^{\frac{1}{4}} \tag{23}$$

(Shu et al. 1983; chapter by Shu). As before, the viscosity is determined by inverting the equation for the damping length:

$$\nu = (2\pi G\sigma)^3 \frac{3\mathcal{D}^2 R}{\mu}. \tag{24}$$

As with Eq. (21), Eq. (24) assumes constant surface density and constant viscosity over the region in which x_D has been determined.

Lissauer et al. (1983) have used this method and a value of $x_D = 167$ km measured for the Mimas 5:3 bending waves to deduce a value of $\nu = 260^{+150}_{-100}$ cm^2 s^{-1} for the viscosity in the mid-A Ring. Although these waves are the strongest and most prominent bending waves in Saturn's rings, their slope never exceeds $\sim 1/3$; thus, nonlinear affects should not be significant. From the approximate relationship between viscosity ν, particle dispersion velocity c, and optical depth τ:

$$\nu \sim \frac{c^2\tau}{2\Omega(1+\tau^2)} \tag{25}$$

(Goldreich and Tremaine 1978a), a dispersion velocity of $c \sim 0.4$ cm s^{-1} and a half ring thickness $z_0 \sim c/\Omega \sim 30$m is inferred. Lane et al. (1982) used Eq. (21) to estimate the viscosity in the inner portion of the B Ring from the damping of density waves excited at Janus' 2:1 resonance (Fig. 15b). Their result of $\nu = 20$ cm^2 s^{-1} suggests a ring thickness (scale height) of roughly 5 m. However, the B Ring density waves have distinctly nonsinusodial profiles, in agreement with the theoretical expectation that they should become nonlinear within a distance from their resonant source of less than a single wavelength. The angular momentum carried by highly nonlinear waves is not a simple function of the squared peak amplitudes as expressed in linear theory by Eqs. (16) and (17). The linear theory used to derived Eq. (21) may be further invalidated by the fact that the form of the damping changes, because nonlinear effects (possibly shocks) become important (Lissauer et al. 1983). Thus, the value of viscosity derived by Lane et al. may be best regarded as an upper limit on viscosity in this region. However, comparison between these waves and the Mimas 5:3 density waves, another nonlinear wavetrain (Fig. 15c) located at 2.195 R_s in the A Ring, strongly suggests that the inner B Ring indeed does have lower viscosity than the A Ring. That is, the Janus 2:1 wavetrain propagates for more than 400 km or nearly 90 wavelengths, while the Mimas 5:3 wavetrain, with similar wavelength and forcing amplitude (Lissauer and Cuzzi 1982) propagates less than half as far and only ~ 10 wavelengths. It is of interest that the 5:3 density wavetrain (which is more strongly forced) propagates almost exactly the same distance as the 5:3 bending wavetrain, arguing for similar, probably viscous, damping of both wavetrains. Esposito et al. (1983b) have determined viscosity and ring thickness in the outer A Ring from about a dozen wavetrains, and find them to increase outward in a regular fashion. The larger dispersion velocity inferred in the outer A Ring may be maintained by the more numerous strong resonances in the A Ring region (Lissauer et al. 1983); an outwardly increasing dynamical activity is also indicated by the outwardly increasing number of small particles inferred from Voyager images (Fig. 3; Sec. III.A). This trend is also supported by mass extinction coefficient measurements, which generally increase outward

(Table IV). However, Esposito et al. (1983b) find no regular trend in their A Ring mass extinction coefficient.

Resonantly excited spiral waves and resonantly produced gaps and edges are probably the best understood forms of structure observed in Saturn's rings; however, certain first-order problems remain. Chief among these are the large torques predicted at resonances, which should cause rapid outward orbital evolution for Saturn's inner satellites and rapid inward motion for the A Ring where most of the strong resonances occur. Calculations based on a linear theory of density-wave excitation (cf. Eq. 17) suggest that the ringmoons Janus, Epimetheus, 1980S26, 1980S27 and Atlas are evolving so rapidly that, at their current rate, they should have been at the outer edge of the A Ring between 3 million and 300 million years ago. Because the combined mass of these ringmoons is comparable to that of the A Ring, the time scale for half of the A Ring to be pushed into the B Ring is also on the order of 300 million years (Goldreich and Tremaine 1982; chapter by Borderies et al.; Lissauer et al. 1984).

Density waves excited at the strongest (largest torque) resonances are observed to be highly nonlinear, very much as predicted by theory. Several nonlinear effects will probably reduce the actual torques significantly at many of the strongest resonances. It becomes difficult for a ringmoon to transfer more momentum to a wave once the perturbed density becomes comparable to the background density in the region of coupling. However, the minimum angular momentum deposition which is required in order to cause observed waves to become nonlinear (Eq. 17) as rapidly as they do is only a few times smaller than predicted by the linear theory. Because distance to nonlinearity is proportional to the square root of the torque, it seems that, in most cases, the actual torques are no less than a tenth or so of those calculated using linear theory (Lissauer and Cuzzi 1982). Thus the time scales for orbital evolution remain much less than the age of the solar system (see the chapter by Borderies et al.). If some of the ringmoons are or were locked in resonances with much more massive outer satellites, then the ringmoons could transfer the angular momentum they receive from the rings to the more massive bodies, and the ringmoons would undergo little orbital evolution (Lissauer et al. 1984). The angular momentum lost by the rings, especially the outer A Ring, would have to be supplied by outward viscous torques, and ultimately by whatever is maintaining the inner edge of the A Ring against inward diffusion. The problem of maintaining the moderately sharp inner A and B Ring edges is a difficult one, and is not understood at the present time. We return to this problem in Sec. V.

To summarize, it is known that several dozen examples of spiral gravitational waves exist in Saturn's rings. The observed density wavetrains are primarily excited at orbital resonances with the major, close-in ringmoons: Janus, Epimetheus, 1980S26, and 1980S27, as well as Mimas and Iapetus. Due to its inclined orbit, Mimas also drives several bending or "corrugation"

TABLE V

Known "Empty" Gaps and their Ringlets

Ring Region	Gap Name[a]	Location (R$_S$)[a]	Width (km)[a]	Ringlet?	Ringlet Eccentricity (10^{-4})	Ringlet Width Δa (km)	Causative Process	Reference
A Ring	Keeler	2.263	~35	?	—	—	?	Holberg et al. (1982).
	Encke	2.214	325	several, kinky	?	?	moonlets	Cuzzi and Scargle (1984).
Cassini Division[b]	TBD	1.988	246	opaque	?	37	?	Tyler et al. (1983); Lissauer et al. (1981); outer rift.
	TBD	1.972	42	?	—	—	?	Marouf and Tyler (1984a); Lissauer et al. (1981).
	TBD	1.966	40	?	—	—	?	
	TBD	1.960	28	?	—	—	?	
	TBD	1.959	38	?	—	—	?	
	Huygens	1.953	285–440	opaque, eccentric	4.0 ± 1.7	43	Mimas 2:1	Porco (1983); Lissauer et al. (1981).
B Ring[c]	TBD	(1.81)	<1	?	?	?	?	Evans et al. (1982).
C Ring	TBD	1.495	20	opaque, on border	?	60	Mimas 3:1	Holberg et al. (1982); Lane et al. (1983).
	TBD	1.470	?	opaque, on border	?	20	Mimas (3:1 vertical); 1980S27 (2:1)	

TABLE V (continued)

Known "Empty" Gaps and their Ringlets

Ring Region	Gap Name[a]	Location (R_S)[a]	Width (km)[a]	Ringlet ?	Ringlet Eccentricity (10^{-4})	Ringlet Width Δa (km)	Causative Process	Reference
C Ring	Maxwell	1.450	270	opaque, eccentric	3.5 ± 0.4	64	?	Esposito et al. (1983a); Lane et al. (1983); Porco (1983).
	TBD	1.290	184	opaque, eccentric, + other feature?	2.6 ± 0.2	25	Titan (1:0)	Holberg et al. (1982); Lane et al. (1983); Porco (1983).
	TBD	1.241	~70?	several features	?	?	Mimas (4:1)	Holberg et al. (1982); Lane et al. (1983).
(F Ring)	—	2.32	100–500	several, kinky, complex	26 ± 6	3–30	moonlets?	Synnott et al. (1983).

[a]Most radial distances, widths, etc. have uncertainty ~10 km; widths are measured between half-amplitude points in ring optical depth. Some gap names are to be determined (TBD).

[b]Several narrower gaps (~1 km wide) may exist in the Cassini Division. The structure of this division is discussed in more detail by Marouf and Tyler (1984a).

[c]Detection of a narrow (~150 m wide) gap in the B Ring at 1.81 R_S (Evans et al. 1982). However, this result is unconfirmed by other occultation measurements.

wavetrains which may be largely responsible for the edge-on appearance of the rings as seen from Earth. The properties of these wavetrains are well enough understood that the wavetrains have become diagnostic tools for measuring the local ring surface mass density ($20-100$ g cm^{-2}) and, to some extent, vertical scale height or local thickness ($5-30$ m) at a discrete set of locations. The apparent rate of momentum transfer seen in these wavetrains leads to a disturbingly short evolutionary time scale for the orbits of ring particles and ringmoons alike.

C. Gap and Eccentric Ringlet Structure

1. Gaps and their Edges. In spite of the often-heard remark that Saturn has thousands of rings or ringlets, the number of truly isolated ringlets, and of truly empty ($\tau < 10^{-2}$) gaps, is actually quite small (Table V). Overall, structure consisting of essentially empty gaps with or without narrow, opaque ringlets is most characteristic of the C Ring and Cassini Division, optically thin regions. The Encke Division and Keeler Gap are also members of this class, although the Encke Ringlets have more properties in common with the F Ring than with the opaque, well-defined eccentric ringlets. We have included the F Ring as a narrow, eccentric ringlet in this section.

Several of the gaps and/or ringlets are close to the locations of specific resonances or clusters of resonance (Sec. IV.B.), such as the Titan 1:0 (apsidal) resonance at 1.29 R_S and the Mimas 3:1 and 1980S26 2:1 resonances at 1.495 R_S resonances in the C Ring (Lane et al. 1983). The narrow ringlet and the Huygens Gap just outside the B Ring outer edge may be affected by the Mimas 2:1 resonance which causes the edge itself (Porco 1983). The 2:1 resonance with 1980S27 and the Mimas 3:1 vertical (1.47 R_S) and 4:1 Lindblad (1.24 R_S) resonances also show features; when studied at even higher resolution, these features show intriguing fine structure on the radial scale of $\leqslant 1$ km, as shown in Fig. 16 (Lane et al. 1983). Due to the high order of eccentricity involved (chapter by Shu), the Mimas 4:1 feature is, in a torque sense, the second weakest identified feature in the rings, the weakest being the Iapetus apsidal resonance (Sec. IV.B). Several other resonances of comparable strength (Lissauer and Cuzzi 1982) should lie in regions of comparably low optical depth, such as the 1980S26 9:7 (1.9897 R_S) and the 1980S27 10:8 (1.9936 R_S). No careful search has, as yet, been made for features with such correspondences. On the other hand, the prominent Maxwell Gap (1.45 R_S) and its eccentric ringlet are not associated with any known strong satellite resonance, neither are the Cassini Division outer rift (1.99 R_S) and its ringlet, the Encke Division, nor the Keeler Gap. Several minor gaps within Cassini's Division are also without currently accepted resonance assignments (Lissauer et al. 1981; Marouf et al. 1984).

Several of these gaps are so narrow (< 10 km wide) that their true nature has only recently been established (e.g., Tyler et al. 1983). It may be that a similar but more serious lack of resolution has prevented us from seeing any

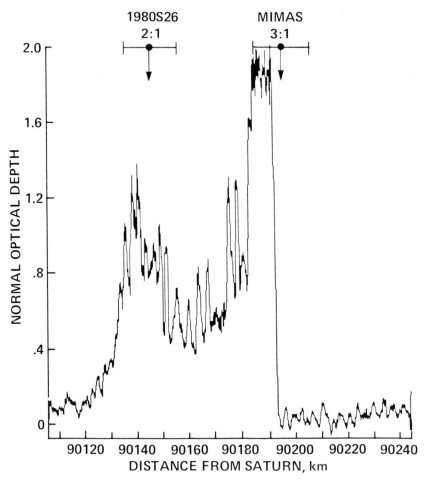

Fig. 16. Fine structure in the 1.495 R_S ringlet just interior to the Mimas (3:1) resonance (from Lane et al. 1983). This feature shows a very sharp outer edge, bordering on a narrow gap. The outer edge of the feature agrees well with the theoretical resonance location. The weaker 1980S26 (2:1) resonance at 90145 km may also cause a second sharp edge. There is a good deal of fine-scale structure within this feature, as well as within other similar features for which no ready explanation exists. A misprint in the radial scale of the previously published version has been corrected (A. Graps, personal communication, 1983).

gaps at the inner edges of the A and B Rings, and at the edges of the C Ring plateaus. The boundaries of these features have a common property of steep gradients in optical depth with no hint of any empty gap at highest photo-polarimeter resolution. Of course, gaps narrower than the highest resolution of the photopolarimeter (~ 100 m) could only exist if the local ring thickness were also $<< 100$ m. It may be, of course, that such discontinuities may exist without a clear gap (see Sec. IV.D).

Aside from their various embedded ringlets, the gaps listed in Table V are mostly empty to within the measurement capabilities of the Voyager experiments. From the ultraviolet spectrometer and photopolarimeter occultations, an upper limit to average τ of 10^{-2} may be established. From imaging results at high and low phase and tilt angles, and in regions where the planet's shadow, crossing the area, provides a sensitive tool for the detection of faint material, similar limits may conservatively be set. Due to the highly grazing angle of incidence and large dynamic range of the radio occultation experiment, an upper limit may be placed on τ of 5×10^{-3} in these gaps for particles > 1 cm. However, in certain locations within the Maxwell, Encke, and other gaps, features of marginal statistical significance ($\tau \sim$ several $\times 10^{-2}$ or ~ 3 sigma) have been detected by the photopolarimeter experiment at the few-km radial scale (Lane et al. 1983; Lane et al. 1982).

The morphology of gap edges varies significantly. Some edges, such as the A Ring outer edge and Encke Division edges, are extremely radially sharp (Lane et al. 1982; Marouf and Tyler 1982). The implication of these studies is that the A Ring is truly < 150 m thick at its outer edge, as discussed in Sec. II. The inner and outer edges of the Keeler Gap and Encke Division are similarly sharp, as seen from their Fresnel diffraction of the Voyager 1 spacecraft radio signal, and do not display significant structure (i.e. other gaps) on a radial scale of a Fresnel zone (~ 15 km). However, the photopolarimeter stellar occultation, obtained at a different time and ring longitude, shows considerable structure near the inner edge of the Encke Division, and in fact a narrow, opaque, 2-km-wide ringlet separated from the actual (?) edge by a clear gap of similar width (Lane et al. 1982). Although extremely sharp edges are also seen in these data, this structure is not consistent with the well-defined, persistent Fresnel diffraction fringes seen in the radio occultation result (Marouf, unpublished). We infer, therefore, that the radial structure of the Encke Division edge is time, or azimuthally, variable. Azimuthal variation is directly seen in the appearance of both inner and outer edges in Voyager images, as discussed further in Sec. IV.C.3.c.

In contrast, the structure of the Maxwell Gap edges (~ 1.45 R$_S$) is much more smoothly varying, on a scale more like 10–20 km (see Fig. 17). A similar scale is seen at radio and visible wavelengths, implying that the relevant physics is particle-size independent.

2. Opaque, Sharp-edged Ringlets. These somewhat different types of gap also differ noticeably in the nature of their embedded ringlets. Gaps in low optical-depth material seem to contain or be bordered by sharp-edged, well-defined, often eccentric, ringlets of substantial optical depth and radial widths of ~ 20 to 60 km (see Table V). These ringlets are similar in many, but not all, ways to the rings of Uranus (chapter by Elliot and Nicholson).

The very edges of these ringlets are indeed quite sharp; in several cases, the optical depth goes from zero to several tenths on a scale of ≤ 1 km

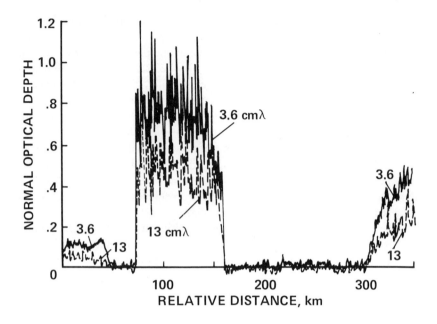

Fig. 17. The Maxwell Gap (1.45 R_S) and its ringlet, as seen in radio occultation data which have had diffraction effects removed (Marouf and Tyler 1984a,b). At the displayed resolution of 0.6 km, the optical depth in the center of the ringlet itself is completely in the noise; however, the sharpness of the edges is faithfully reproduced. The ringlet edges are notably sharper than the gap edges. The sharpness difference is not a function of particle size, as seen by comparing scans at both radio wavelengths of 3.6 and 13 cm.

(Esposito et al. 1983a; Porco 1983; Lane et al. 1983). It is interesting that the Maxwell Ringlet edges are far sharper than the edges of the gap in which it is found (Fig. 17). This is true at both optical and radio wavelengths. This same behavior is apparently seen in the Cassini Division outer rift (1.99 R_S) gap; the ringlet edges there also seem sharper than the edges of the outer rift in which it is embedded, at least at radio wavelengths (Tyler et al. 1983). It may be that this is telling us something about the viscosity in regions of different optical depth. The periapse longitudes $\tilde{\omega}$ of these opaque ringlets all precess freely at rates determined primarily by Saturn's oblateness

$$\dot{\tilde{\omega}} \approx \dot{\tilde{\omega}}_f = \Omega(R)\left[\frac{3}{2} J_2\left(\frac{R_S}{R}\right)^2 - \frac{15}{4}J_4\left(\frac{R_S}{R}\right)^4 + \frac{105}{16}J_6\left(\frac{R_S}{R}\right)^6\right] \qquad (26)$$

where $J_2 = (16{,}299.1 \pm 18) \times 10^{-6}$, $J_4 = (-916.7 \pm 38) \times 10^{-6}$, and $J_6 = 81.3 \times 10^{-6}$ (corrected to $R_S = 60{,}330$ km from Null et al. 1981). However, in cases where the gaps are associated with satellite resonances, resonant perturbations also contribute to $\dot{\tilde{\omega}}$.

The Maxwell Gap (1.45 R_S) ringlet, for instance, precesses at a rate in very good agreement with the rate determined by planetary oblateness alone (14°68/day) (Esposito et al. 1983a; Porco 1983). This is consistent with our current lack of a resonant cause for the gap and ringlet. The Huygens Gap (1.95 R_S) ringlet is at present poorly understood. Its precession rate is consistent with the value expected from planetary oblateness but there is some evidence that it is being perturbed by the nearby Mimas 2:1 resonance and the mass in the outer B Ring (Sec. IV.B). The Titan Gap ringlet (1.29 R_S) is a somewhat degenerate case; as it is near an apsidal resonance, the free precession rate is equal to Titan's orbital rate by definition. The observed precession rate is virtually identical to this value. The observed close alignment of the ringlet apoapse with Titan is the expected orientation, if the ringlet is controlled by and also lies somewhat outside the Titan (1:0) resonance (Porco 1983). The ringlet is, in fact, 80 km outside of the predicted apsidal resonance (Holberg et al. 1982).

Because $\tilde{\omega}_f$ varies as a function of radius R (Eq. 26), the fact that inner and outer edges are seen to precess at the same rate, as for the Uranian rings, must be explained. Goldreich and Tremaine (1979b) suggested for the Uranian ring that the ringlet mass itself could contribute a component $\tilde{\omega}_r$ such that $\tilde{\omega}_r (R) + \tilde{\omega}_f (R)$ is constant, and the ringlet can precess as a unit. Using this criterion for the Maxwell Ringlet, Esposito et al. (1983a) obtain a ringlet surface mass density of $\sigma \sim 20$ g cm^{-2}, similar to that of the Uranian rings. Porco (1983) finds a similar value of σ for the Titan ringlet. The self-gravity hypothesis also leads naturally to another observed trait of the Uranian and narrow Saturnian rings, that their eccentricity increases outward. This means that the ringlet width is greater at apoapse than at periapse. The "precessional pinch" of Dermott and Murray (1980) also preserves this property. Other hypotheses for maintaining the ringlet eccentricity do not predict a preferred sign for the effect (Dermott et al. 1979; cf. also chapter by Dermott). There is at present no indication that Saturnian eccentric, opaque ringlets are inclined in addition to being eccentric, as are the rings of Uranus (French et al. 1982; Porco 1983).

The increase in eccentricity of the Maxwell Ringlet with distance is such that, at apoapse, the ringlet comes extremely close to the outer edge of the gap. The fits obtained by Esposito et al. (1983a) are consistent with the ringlet touching the outer edge of the gap at apoapse, within their $\pm$ 20 km error in apoapse distance. Porco (1983) finds that the outer edge of the ringlet misses the outer gap edge by 21 $\pm$ 5 km.

Comparing the mass density determined for the Maxwell Ringlet (20 g cm^{-2}) with its average optical depth ($\tau \sim 0.9$) gives an estimate of the ringlet mass extinction coefficient $\tau/\sigma \sim .045$ cm^2 g^{-1}. This is significantly larger than the mass extinction coefficient inferred from density-wave measurements in the A and B Rings, ($\tau/\sigma \sim 0.01$) and is most simply interpreted in terms of a correspondingly larger proportion of smaller particles in the ringlet. Small particles, with a higher surface area/mass ratio, are more effective per unit mass at blocking light. Comparison of τ (radio) with τ (optical) in the ringlet supports this conclusion, although particle close-packing and near-field effects have not been accounted for (see Sec. III.C). However, the albedo of the ringlet particles is fairly similar to that of the surrounding C Ring material (Esposito et al. 1983a).

The analogy between these narrow, isolated ringlets and the Uranian rings is not a complete one, however. The α and ϵ Rings of Uranus show a clear increase of optical depth toward their inner and outer edges, although the optical depth does not necessarily peak right at the edges. On a similar scale of resolution, the Maxwell Ringlet (at least) is far more optically thick toward its center than near its edges, although a hint of a double-peaked structure is seen in the very core (central 10 km) of the ringlet (Esposito et al. 1983a). The narrow ringlet at 1.495 R_S shows a more Uranian double-peaked structure at low resolution, but that ringlet is not clearly detached and may be the result of several independent, nearly overlapping resonances (Holberg et al. 1982; Lane et al. 1983; see Fig. 16).

3. Kinky Ringlets and Moonlet Gaps. Another type of ringlet is the optically thin, azimuthally irregular, kinky variety represented by the F Ring and the Encke Division ringlets. We discuss both features together here because of their similarities in morphology and particle properties. In addition, there is strong indirect evidence in both cases for an intimate relationship with moonlets (ring particles of radius greater than several kilometers) which, although few in number, are capable of strongly influencing ring dynamics in their vicinity. Below, we briefly outline the theoretically expected influence of local moonlets and expected observational aspects. These topics are also discussed in the chapters by Dermott and Borderies et al. and by Goldreich and Tremaine (1979c, 1980, 1981, 1982), Lissauer et al. (1981), and Hénon (1981). In Secs. IV.C.3.b and IV.C.3.c (immediately following this section), we discuss the F Ring and the Encke Division, respectively, in detail.

(a) Observable Effects of Shepherding by Local Moonlets. The concept of gravitational shepherding was originally proposed in the ring context by Goldreich and Tremaine (1979c) in order to explain the confinement of the rings of Uranus. Useful basic work along these lines had previously been done in different contexts by Julian and Toomre (1966) and Lin and Papaloizou (1979). Below we illustrate how the concept of gravitational torque, or shepherding, may be derived from the local response of a ring edge to the close passage of a perturbing moonlet. Along the way, we will also note how several characteristics of the process seem to be observed in both the F Ring and the Encke Division.

Lin and Papaloizou simplify the interaction between a moonlet and a nearby, differentially-rotating stream of particles into an impulse which acts only at closest approach, once every synodic period. The duration of the encounter is, indeed, much shorter than the synodic period, and is comparable to the orbital period. More detailed treatments must account for acceleration due to the rotating frame as well (see, e.g., Goldreich and Tremaine 1982; Lin and Papaloizou 1979). In the impulse approximation, the net effect of the encounter is to deflect the particle orbit by adding a small velocity $V_\perp$ in the radial direction, where $V_\perp$ is the product of the radial acceleration by the moonlet and the effective duration of the encounter. Consider a moonlet of mass M and a stream of ring particles in initially circular orbits of mean semimajor axis a, but separated by distance $s \ll a$. Then the relative velocity is $V_{\rm rel} \sim (3/2)\,\Omega s$ where Ω is the local orbital frequency and the effective encounter time is $t_{\rm enc} \sim 2s/V_{\rm rel} \sim 4/(3\Omega)$, which is indeed small compared to the synodic period $P_{\rm syn} \sim 4\pi a/3\Omega s$. In the "impulse" approximation, the particle acquires $V_\perp \sim (GM/s^2)t_{\rm enc} \sim 4\,GM/(3\Omega s^2)$ during the encounter, which gives the perturbed particle an eccentricity $e \sim V_\perp/(\Omega a) \sim 4GM/(3as^2\Omega^2)$. A more detailed calculation by Julian and Toomre (1966) yields a different numerical coefficient (see Goldreich and Tremaine 1982). Including the numerical factors, we may rewrite the expression for the perturbed eccentricity as

$$e = 2.24 \left(\frac{M}{M_{\rm S}}\right)\left(\frac{a}{s}\right)^2, \qquad (27)$$

where $M_{\rm S}$ is Saturn's mass. Once the moonlet moves past, there is no further perturbation for a time $P_{\rm syn}$ during which the particles maintain their new, eccentric, Keplerian orbit. One orbital period later, the ring particles have executed a complete epicycle and are again at the same orbital phase: near apocenter (for particles exterior to the moonlet) or near pericenter (for particles interior to the moonlet). Because of the relative motion of the moonlet, successive cycles of exterior (interior) particles lag behind (move ahead) of the moonlet by a distance $\lambda \sim V_{\rm rel}\,2\pi/\Omega \sim 3\pi s$ during each orbital period, yielding a train of waves of radial amplitude ae and wavelength $3\pi s$ (see Fig. 18; Julian and Toomre 1966, their Fig. 12). The characteristic wavelength of the F Ring kinks was first attributed in this way to the effect of shepherding satellites by Dermott (1981). However, good agreement was only attained (Showalter and Burns 1982) after an initial error in the wavelength was corrected (Smith et al. 1982). The actual wavelength is $\sim 10^4$ km, which is in good agreement with the typical separation of the F Ring and its shepherds (1980S26 and 1980S27) of $s \sim 10^3$ km. However, this existence of a characteristic wavelength should not obscure the fact that the kinks and braids differ somewhat in appearance and spacing from one to another.

The exact form of the perturbed streamlines depends on the relative eccentricity of the ring and perturbing moonlet. In the circular case, the full treatment (see, e.g., Goldreich and Tremaine 1980, 1982) yields the streamline equations (in the coordinate system fixed to the moonlet):

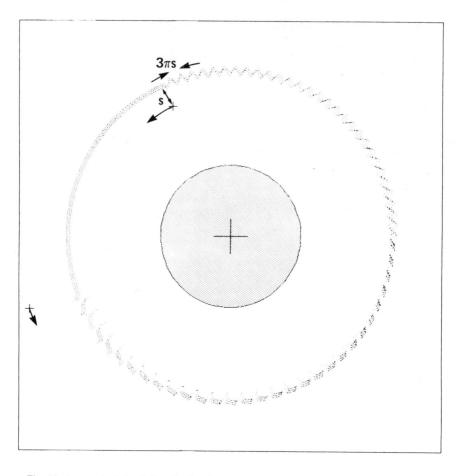

Fig. 18. A numerical simulation of epicyclic radial perturbations, of wavelength $3\pi s$, induced in ring particle streamlines by shepherding satellites (shown by +) at distance s from a ring (adapted from Showalter and Burns 1982; the radial scale is magnified for visibility). The features trail interior satellites and lead exterior satellites, due to differential Keplerian motion; but in each case they are fixed in the frame rotating with the satellite. Three characteristic forms are shown: (top): nearly sinusoidal epicycles are induced when both particle and satellite orbits are circular; (bottom left): kinky features are observed when either ring particles or satellite, or both, are on eccentric orbits; (bottom right): kinky features evolve into clumpy features of the same wavelength. (See Fig. 19; cf. also Fig. 41 of Smith et al. 1982).

$$x = s + ea \cos(2y/3s + \psi), \qquad (28)$$

where x is the outward radial coordinate, y the tangential one, e is from Eq. (27), and ψ is a phase shift. In the noncircular case, the strength of the perturbation depends on where in its orbit the moonlet encounters the ring material. This is because both s and the relative velocity (and thus effective encounter time) will vary as the moonlet moves from apocenter to pericenter. Showalter and Burns (1982) have shown numerically that highly nonsinusoidal streamlines may result in cases with eccentric perturbers, because angular motion is affected in addition to radial motion (see Fig. 18). Notice also in Fig. 18 that perturbations by eccentric moonlets may lead in places to an azimuthally clumpy, but not notably radially kinked, ringlet. The clumps, however, also maintain their characteristic wavelength $3\pi s$.

Because λ is a function of separation s, the streamlines of different orbits in the perturbed ring will eventually intersect, and collisional damping will ensue if the local optical depth is nonnegligible. A rough upper limit on the number of waves permitted before streamline crossing comes from the criterion that the streamlines over a region of radial width $\sim ae$ may shift by no more than $\sim \lambda/2$; this shift occurs after N waves where $N \sim (\lambda/2)/(ae\ d\lambda/ds) \sim \lambda/6\pi ae$. A rough lower limit may easily be obtained by determining the first intersection point from the streamline equations by setting $dx(y,s)/ds = 0$, giving $N \sim \lambda/6\pi^2 ae$. These two limits provide useful bounds on the azimuthal wave damping length.

The damping of these perturbations, and recircularization of the particle orbits, is one way of allowing the process of gravitational shepherding or ringlet confinement to occur (cf. Greenberg 1983; see also chapter by Dermott). That is, secular transfer of angular momentum, or torque, between the moonlet and ring will occur if e is significantly reduced between successive moonlet encounters. We demonstrate this using an argument by Goldreich (personal communication, 1979). During any planar three-body encounter with perturber M on a circular orbit, Jacobi's constant $E_J = E - \Omega_M J$ is conserved, where E and J are the ring particle orbital energy and angular momentum per unit mass and Ω_M is the moonlet's angular velocity. Noting that $J^2 = GM_S a(1-e^2)$ and $E = -\frac{1}{2}[(GM_S)/J]^2 (1-e^2)$, we see that

$$\Omega_M \Delta J = \Delta E = \frac{(GM_S)^2 \Delta J}{J^3} + \frac{(GM_s)^2}{2J^2} \Delta(e^2)$$

which may be written as

$$\Delta J (\Omega - \Omega_M) \simeq -(\Omega a)^2 \Delta(e^2). \tag{29}$$

If the particle eccentricity prior to encounter is much less than the perturbed eccentricity e, $\Delta(e^2) \sim e^2$ and it may be seen that angular momentum always flows outward. That is, relative to the moonlet, interior particles lose momentum and move inward, while exterior ones gain it and move outward. The moonlet, being so much heavier, moves relatively little. There are other ways for the shepherding process to occur; in general, the perturbed particles must only "lose their memory" of the prior moonlet encounter before undergoing the next one. For instance, a large number of purely elastic collisions would also suffice. If all of the prior "memory" is lost (specifically, if e is fully damped to zero), the shepherding process is fully efficient and ΔJ will be transferred over a time $\Delta t \sim P_{syn}$. The averaged angular momentum transfer rate or torque per unit mass exerted on the ring edge is determined by substituting for e and Δt: $\Delta J/\Delta t = T \simeq \pm (\Omega^2 a^6/s^4)(M/M_S)^2$, essentially the result reached by Goldreich and Tremaine (1979c) and Lin and Papaloizou (1979) (see also Goldreich and Tremaine 1982; chapter by Borderies et al.).

Thus, we see that the collisional damping of particle streamline perturbations is of prime importance in the shepherding process. The original shepherding theory visualized pairs of moonlets concentrating a ringlet between themselves by this process (Goldreich and Tremaine 1979c); moonlets embedded within a ring disk will clear gaps around themselves by the same physics (Lissauer et al. 1981; Hénon 1981, 1983). Dermott et al. (1979), in an alternative hypothesis, have also claimed that a single moonlet may gather material around itself. Below we discuss the observations and show that certain of these varieties of behavior are probably active in various gap and ringlet features in the rings.

(b) F Ring: The Classic Prototype. The discovery of the F Ring by Pioneer 11 (Gehrels et al. 1980), and of its complex structure and the two ringmoons that straddle it (1980S26 and 1980S27) by Voyager, have generated considerable scientific and popular interest (see Fig. 19). Several well-separated strands were seen in Voyager 1 images, each of radial width ~ 20 to 30 km, and spanning a range of ~ 120 km. In some photographs, covering ~ 50–$60°$ longitude, two of the strands appear to intersect and to be radially or vertically kinked. Synnott et al. (1983) note that the longitude of greatest kinkiness is close to the longitude where 1980S26 and 1980S27 last underwent conjunction with each other. Where the kinks are observed, the azimuthal scale length of $\sim 10^4$ km (Smith et al. 1982) is in fairly good agreement with that predicted from Eq. (28) for 1980S26 (~ 7800 km) and 1980S27 (~ 14300 km), respectively. Also, the observed radial perturbations (10–30 km) are comparable to the predictions of Eqs. (27), and (29), and numerical experiments (Showalter and Burns 1982). At other longitudes, with less total coverage, the three strands appear straight and parallel, with about the same radial span. Simultaneously, at still another longitude, the Voyager 1 radio occultation experiment observed only a single strand (Tyler et al. 1983). During Voyager 2 encounter, only one image out of a total of several dozen covering 15% of the F Ring in longitude shows any structure other than a single dominant strand (Smith et al. 1982; Terrile 1982). Also, the Voyager 2 stellar occultation experiment demonstrated that the F Ring consisted (at the longitude of occultation) of a 60-km-wide, optically-thin, "halo" and a single, optically-thick, few-km-wide "core" which has internal structure at the sub-kilometer scale (Lane et al. 1982). An even broader (~ 500-km-wide) halo of

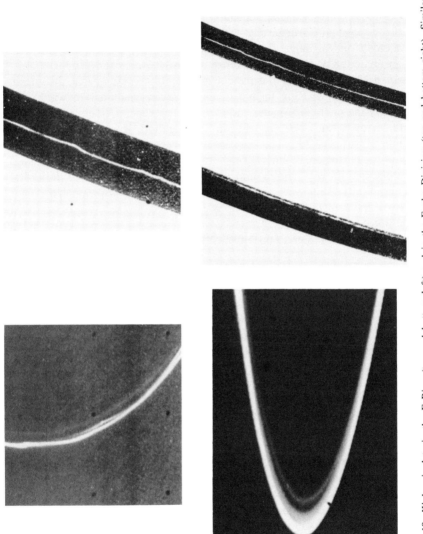

Fig. 19. Kinky ringlets in the F Ring (top and bottom left) and in the Encke Division (top and bottom right). Similar morphology is seen in both instances (cf. also Fig. 18). The F Ring has a broad halo of material extending several hundred km in width, but of a much lower optical depth than its narrow core region. These narrow, kinky ringlets also share the common property of an unusually large fractional component of micron-sized particles, as revealed by their variations in brightness with phase angle (Sec. III.A).

presumably lower optical depth is also seen in images taken by the more sensitive Voyager 2 cameras (Smith et al. 1982), but would not have been visible in the less sensitive Voyager 1 narrow-angle images, even if it were present (Fig. 19).

The optical depth of the F Ring core (corrected for the coherency factor of 2 as discussed in the chapter appendix) is much smaller at 3.6 cm λ than at visible wavelengths, suggesting that a substantial fraction of the visible-wavelength optical depth is in the form of sub-centimeter-sized particles (Tyler et al. 1983). However, no broad halo of low optical depth material is observed at 3.6 cm λ by the radio experiment, suggesting that the entire extended, diffuse F Ring feature of one-to-several-hundred-km width observed by the photopolarimeter and imaging experiments (Lane et al. 1982; Smith et al. 1981, 1982), if uniform in longitude, is comprised of sub-centimeter particles. Although significant numbers of cm-sized particles in the core of the F Ring are required by the 3.6 cm radio depth, the F Ring was not even detected in the 13-cm radio occultation data, giving an upper limit of τ (13 cm) <0.01. Thus, very few particles > 4 cm radius exist even in the core of the F Ring, at least at the longitude observed by the radio occultation experiment.

Issues that continue to pose difficulties in a full understanding of the F Ring are:

1. Details of the ring structure such as its multi-stranded nature, the significance of the large abundance of micron-sized ring particles (Sec. III.A), and the relative lack of particles larger than a few centimeters radius;
2. The difference between the location and shape of the observed kinks and numerical predictions (Showalter and Burns 1982);
3. The lack of significant relationship between longitudes of the kinky features and the simultaneous positions of 1980S26 and 1980S27 (Showalter 1982; Synnott et al. 1983);
4. The nonequilibrium location of the ring (it is closer to the more massive moonlet, implying imbalance of torques).

Some of these phenomena could be explained if the visible F Ring were actually only part of a little "asteroid belt" of much larger objects (0.1–10 km) extending across a substantial part of the entire region between 1980S26 and 1980S27. Some support for this speculation comes from interpretation of Pioneer 11 observations of magnetospheric particle fluxes, as discussed below.

Van Allen et al. (1980b) distinguish several distinct effects of embedded objects upon the phase-space density of magnetospheric charged particles (see also Sec. IV.E). Certain forms of time-stationary radial density distributions (macrosignatures) result from the long-term effect of circular rings (which are azimuthally uniform) and satellites (averaged over all longitudes). Very different radial scales are exhibited by longitudinally localized absorption features (microsignatures), which are attached to their responsible moon. The microsignatures may trail or lead the moon, depending upon the charge and energy of the magnetospheric species being sampled (Van Allen et al. 1980b). Specific particle absorption features seen in the Pioneer 11 data are shown in Fig. 20. Prior to Voyager, it had been noted that the inner and outer features, separated by $\sim$ 1200 km, were fairly symmetric about the planet, and could be the usual macrosignature of two narrow rings of material, possibly variable in longitude, while the central feature was clumpy and might be due to a single object (see, e.g., Simpson et al. 1980; Ip 1980c; Dermott et al. 1980). The central feature (β) is obviously azimuthally variable, and this has led to the conclusion that the (visible) F Ring, which is usually assumed responsible for the central feature, is azimuthally clumpy. In fact, Terrile (1982) has noted several point-like objects ($<$5 km diameter) which lie within the visible F Ring. No optical counterpart of the inner and outer features (α and γ) was detected by Voyager, which led many to believe that these features were the macrosignatures of 1980S26 and 1980S27. However, after a careful analysis of the data from his instrument (equivalent to that shown in Fig. 20) on the shapes of the charged particle absorption features α, β, and γ, Van Allen (1982) has concluded that all of the observed signatures are microsignatures and that none of them are due to 1980S26 or 1980S27. The immediate implication would be that Pioneer sampled a different set of moonlets or longitudinal clumps on each pass, and that they are distributed in a band of width $\sim$ 1200 km which has fairly abrupt edges. Isolated mountain-sized objects smaller than $\sim$ 10 km; would be too small to have been detected in Voyager images (Synnott, personal communication, 1983). Further study is necessary to determine whether the width, depth, and number of the observed microsignatures are consistent with the imaging observations. Overall however, the characteristic wavelength and appearance of the F Ring kinks is consistent with shepherding effects due to 1980S26, 1980S27, or other local moonlets with similar semimajor axes. The clumpy, nonaxisymmetric behavior of the F Ring is quite different from that of the G Ring, which appears smooth and axisymmetric in similar data (Simpson et al. 1980; Van Allen 1983).

(c) Encke's Division: Moonlets within the Rings? From the example of the F Ring, even though as yet incompletely understood, we obtain a sense of confidence as to the interaction between ring particles and moonlets. We do not completely understand the origin of the particles that delineate the F Ring in all of its complex detail, but the general appearance of the kinks (and even strands) is related to close encounters as predicted by conventional theory (Julian and Toomre 1966; Dermott 1981; Showalter and Burns 1982). The related byproduct of these perturbations, in the presence of damping, appears to be an effective repulsion by moonlets of particulate material, much as suggested by Goldreich and Tremaine (1979c) and Lin and Papaloizou (1974).

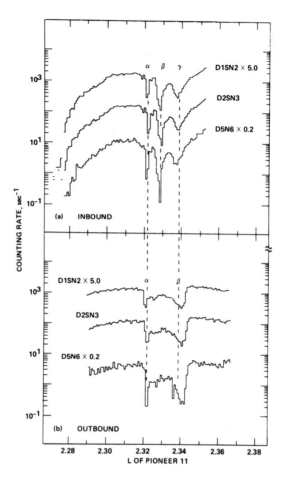

Fig. 20. Absorption of magnetospheric species by material in the vicinity of the F Ring, as observed in several energy ranges by Pioneer 11 (radially corrected data from Simpson et al. [1980]; see also Van Allen [1982], in which (a) is "S5 and Ring F" and (b) is "S6 and Ring F"). The radial variable L is equivalent to distance measured in Saturn radii. Several of the dips are extremely narrow (~ 130 km), and none of the features may be azimuthally symmetric. The radial range of the features (~ 1400 km) is significantly greater than that of the visible F Ring (~ 730 km). Simpson et al. (1980) have differentially shifted their observations to line them up radially; compare their figures 21 and 22.

However, the fact that large objects may be seen *within* the F Ring strands (Terrile 1982) may allow such optically thin strands to contain or even be maintained by individual massive objects (cf. Dermott et al. 1979).

In studying Encke's Division, we see other, complimetary pieces of the shepherding jigsaw puzzle. Several kinky ringlets are seen with properties essentially identical, qualitatively, to those of the F Ring strands (Fig. 19).

There are at least two, lying in the center and inner third of the division (Smith et al. 1981, 1982). The ringlets have kinks and clumps very similar to the predictions of gravitational perturbation theory for nearby moonlets on eccentric orbits (see Fig. 18). As in the case of the F Ring, the photopolarimeter occultation reveals a wealth of fine ringlet detail not seen by the cameras or radio occultation (Lane et al. 1982). The known kinky ringlets contain, as with the F Ring particles, a substantial quantity of forward-scattering dust which is short-lived and suggests locally vigorous dynamics. These ringlets are also azimuthally discontinuous (Smith et al. 1982). The edges of the division are quite abrupt (Marouf and Tyler 1982), and require some forces to counteract the inexorable tendency of a ring to diffuse by viscous momentum tranfer. No gravitational resonance of significant strength lies in its vicinity, although identifiable effects of known resonances are abundant elsewhere in the region.

In the Voyager 1 and 2 images of Encke's Division, an azimuthal wave pattern is sometimes discernible on either or both inner and outer edges (see Fig. 21). Careful Fourier spatial frequency analysis of many images reveals that edge waves are present on one or both edges in nearly all Encke Division images with resolution better than 20 km/line pair (Cuzzi and Scargle 1984). The wavelength λ of the features may be interpreted in terms of the distance $s = \lambda/(3\pi)$ of a hypothetical perturber from the edge; in all cases, s is $1/3-1/2$ of the width of the division. In several cases, edge waves on inner (i) and outer (o) edges are seen in pairs at different longitudes separated by 10 to 30 degrees and respectively leading and trailing their average longitude. If these pairs are indeed caused by the same object, the sum of their associated distances $s_i + s_o = \lambda_i/(3\pi) + \lambda_o/(3\pi)$ should equal the width of the Encke Division; in each case it does so to within measurement error.

The largest amplitude seen for any edge wave is $ae \sim 2$ km (Cuzzi and Scargle 1984). Using the value of s implied by the wavelength $\lambda = 1800$ km, the mass required of a perturbing moonlet on circular orbit is $M \sim 10^{-11}$ M_S; for a density of $1-3$ g cm^{-3}, such a moonlet would have a radius of $r \sim 5-10$ km. If in eccentric orbit, the moonlet could be smaller (Showalter and Burns 1982).

The longitudinal sampling at required resolution is quite sparse ($\sim 10\%$), but given expected wave damping lengths (see Sec. IV.C.3.a) it seems necessary to have at least 3 or 4 independent moonlets to explain all of the observed wave patterns. The presence of several moonlets is, certainly, in good agreement with the presence of several kinky ringlets. The combined mass of several 10-km moonlets is sufficient to clear and maintain the 325-km-wide Encke Division by gravitational shepherding (see, e.g., Hénon 1983; Lissauer et al. 1981). Perhaps the Encke ringlets and the F Ring strands are both transient features, resulting from only the most recent collision between unseen objects.

The photopolarimeter occultation also reveals an apparently detached,

Fig. 21. Voyager 2 image of the Encke Division, reprojected and horizontally magnified. Running along the right-hand edge are seen roughly sinusoidal "edge waves," or radial oscillations, which are not noticeable on the left edge: These edge waves may be the visible trace of particle streamlines which have been perturbed by the recent passage of a moonlet embedded within the Encke Division, as indicated schematically in Fig. 18. Such waves have been observed on both inner and outer edges of the Encke Division, at different longitudes. One of several kinky ringlets may be seen in the center of the division. (Figure from Cuzzi and Scargle 1984.)

opaque ringlet of width and separation from the inner edge of several km (Lane et al. 1982). However, the existence of such a structure is not consistent with the clean, knife-edge diffraction pattern seen in the Voyager 1 radio occultation at a different ring longitude (Marouf and Tyler 1982; Marouf, unpublished). Therefore, this feature is also azimuthally asymmetric. Future studies of edge phenomena of the Encke Division must reconcile these phenomena with the edge waves and kinky ringlets. It is already clear that the smoothly tapered radial profile calculated for shepherded edges in viscous balance by Lissauer et al. (1981) is not applicable to the observations, which show extremely sharp edges much like the resonantly caused A and B Ring outer edges. The reconciliation is probably in terms of reversals of the angular momentum flux caused by interactions between particles on distorted, non-

Keplerian, streamlines (Borderies et al. 1982, 1983; also, their chapter). On-going analysis of the relationships between the positions of kinky ringlets and edge phenomena will be most instructive.

As of this writing, there has been no direct detection of isolated moonlets anywhere within Saturn's main rings. Voyager sequencing conflicts prevented an adequate search from being made in any but the (then believed) most promising clear gaps which were in the Cassini Division. Careful searches of this division have failed to show the presence of moonlets as large as ~ 6 km diameter in either of the two large Cassini Division rifts at 1.95 R_S (Huygens) and 1.99 R_S (Table V; Smith et al. 1982) while a moonlet-clearing hypothesis had estimated the required sizes to be 15 to 30 km diameter (Lissauer et al. 1981). During this search, fortuitous coverage of the Maxwell Gap was also obtained, which remains to be analyzed. However, the Encke Division was not covered sufficiently to provide good detection statistics. That is to say, it is not likely that we will see any of these moonlets directly until the next Saturn mission. Ongoing work is exploring the edges of other gaps for edge waves in order to provide another means of establishing the existence of moonlets in other clear gaps in the rings.

The combination of the edge waves and kinky ringlets in a remarkably wide gap with no known resonant cause makes a circumstantial, but compelling, case for the existence of several unseen moonlets embedded within the Encke Division. If this is indeed the case, the excitation and damping of the edge waves provides an illuminating insight into different phases of the shepherding process discussed in Sec. IV.C.3.a: (a) excitation of eccentricity (see, e.g., Goldreich and Tremaine 1982); (b) damping which allows secular transfer of angular momentum to circular orbits of different energy (see, e.g., Greenberg 1983); and (c) interaction between particles with systematically non-Keplerian velocity gradients, which may help maintain the extremely sharp edges. Increasingly accurate occultation data may eventually allow the inference of edge waves in the rings of Uranus (see, e.g., Freedman et al. 1983).

From the standpoint of cosmogony, the existence of 10-km-sized objects within the rings may be of crucial importance in constraining hypotheses for the origin of the rings, discussed further in Sec. V.

To summarize, the clear gaps in Saturn's rings almost always are associated with embedded or adjacent narrow ringlets. Some of these gaps are associated with satellite resonances. Properties of the F Ring and of the Encke Division strongly suggest the presence of the local moonlets of ≤ 10 km radius as shepherding agents. The evidence includes the presence of optically thin, kinky ringlets, magnetospheric particle absorption in the F Ring, and wavy edges of the Encke Division. In still other gaps, no evidence of an association with either resonances or moonlets has yet been found. The unexplained gap-ringlet cases include the most Uranian-type examples seen in the ring system, such as the Maxwell Gap and its eccentric ringlet at 1.45 R_S.

D. Irregular Structure

The irregular structure, which represents by far the bulk of the ring structure by any measure (area covered, number of features, etc.), is also the least well characterized. Although it is perhaps natural to attack the easy problems first, other hindrances have been the low quality and quantity of observations of the regions involved. The irregular structure is seen primarily in the B and inner A Rings. These regions, with a few exceptions, have optical depths greater than unity, and are essentially opaque over much of their area to the Voyager stellar and radio occultation experiments. Also, for this reason Voyager imaging observations at high resolution (~ 5 km/pixel) exist mostly (although not only) on the lit face for the B Ring. In the outer B Ring, the system of irregular features appears to be azimuthally variable on radial scales of < 100 km, which seriously complicates the problem in light of the sparse data. Finally, the existing images are obtained at a variety of illumination and viewing angles, over which range many recognizable ring features (e.g. density and bending waves) exhibit contrast reversals. In this section we will present and discuss certain observed properties of the irregular structure. We will then briefly sketch existing hypotheses for the structure. It is likely that progress in observational and theoretical understanding will be steady in this area as increasingly refined techniques of analysis are applied to the data.

The irregular structure is so called because it appears to show no regular (or slowly varying) spacing or occurrence rate for features of similar appearance as is, e.g., strikingly obvious in the resonance sequence seen in nearly all images of the outer A Ring, or within individual density or bending wave-trains. Admittedly, a few limited regions do seem to show some degree of short-term order (several adjacent similar features with roughly equal or slowly varying spacing). The significance of this occasionally regular appearance remains to be seen, even for these few features.

It is not completely obvious what property determines the visibility of the irregular features. Many features are seen as fluctuations in ring optical depth (see, e.g., Fig 22). These fluctuations, in the B Ring at least, occur on different spatial scales at different locations. In the outer B Ring, the ultraviolet spectrometer occultation recorded substantial fractional optical depth fluctuations ($\Delta\tau/\tau \sim 1/2$) on radial scales nearly as small as their instrumental resolution of 3.3 km (Fig. 22a). However, in the inner B Ring, occultation and imaging observations (with a resolution of 5 km/pixel) resolve nearly all the structure which locally does not show $\Delta\tau/\tau \sim 1/2$ on radial scales < 50–60 km (Fig. 22b).

The optical depth fluctuations in the inner B Ring are of sufficiently broad radial scale (50–60 km half width) that they may be resolvable, although marginally detectable, in the 3.6 cm wavelength occultation data. In this way, the fluctuation amplitude $\Delta\tau/\tau$ may be compared at radio and visible wavelengths, thereby addressing the question of differential diffusion ability

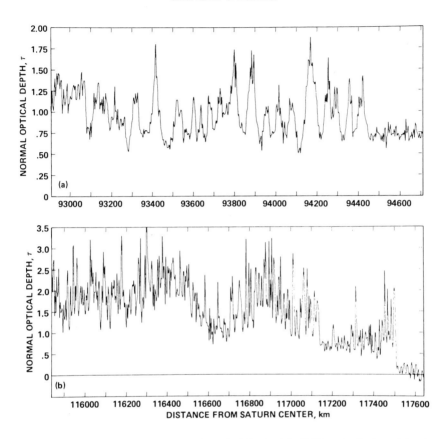

Fig. 22. Irregular radial fluctuations in the optical depth of the B Ring near its inner (a) and outer (b) edges, as seen in the Voyager ultraviolet spectrometer stellar occultation and radio occultation data. There is a noticeable regional variation in the radial scale of the fluctuations. In the outer B Ring (b) fluctuations occur on all scales down to the ultraviolet spectrometer resolution limit (of ∼ 3 km). However, fluctuations with significant amplitude are not seen on radial scales less than ∼ 50 km in the inner B Ring (a).

for particles of sub-centimeter and super-centimeter sizes. As yet, the resolution and sensitivity of the radio data is insufficient for the needs of this comparison. Further data filtering and manipulation may improve the situation. Otherwise, we are forced to await another spacecraft mission at a time when the ring tilt angle is at least 15°.

Imaging observations cover the lit face of the B Ring at a resolution of 5 km/pixel. This is probably the resolution limit for the detection of radially narrow features; however, the amplitude of the brightness fluctuations of such narrow features will be reduced by the instrumental function of the camera.

Throughout the B Ring, brightness fluctuations as large as 20% are seen, on a variety of length scales not as yet quantified, but certainly from several tens to several hundred km in width (see Figs. 23 and 4a,b). These features do not appear to be as sharply bounded as the C Ring fluctuations, possibly due to their larger optical depth.

It must be recalled that the brightness of these features is a function of the geometry of observation as well as the albedo and phase function of the average local scatterer. It is also, of course, a function of the optical depth of the layer of particles (see chapter appendix). However, at the time of the Voyager 1 and 2 encounters, the solar elevation angle B' was only $\sim 4°$ and $8°$, respectively. Thus, sunlight propagates only to a normal optical depth τ of a few tenths ($\tau_{\text{eff}} \sim 1 \sim \tau/\sin B'$). Material at greater depths, therefore, has no influence on the scattering properties of the layer. This is shown quantitatively in Fig. 4: the reflectivity I/F of layers of normal optical depth >0.6 is essentially independent of optical depth.

All of the central half of the B Ring has $\tau > 0.6$. Therefore, variations in I/F can only be caused by variations in particle albedo or phase function. For instance, Fig. 4 shows the variation in I/F (for the geometry of the observations), which results from changing either the particle albedo or phase function in slight ways consistent with the overall ring scattering behavior (cf. Fig. 2). The addition of an isotropic, $g=0$ (see chapter appendix) or forward-scattering, $g=0.9$ dust haze with a fractional optical depth of a few tens of percent could produce the brightness variations at the particular geometry shown, as could variations of $\sim 50\%$ in particle albedo.

These alternatives may be distinguished by comparison of the ring reflectivity at different phase angles. As shown in Fig. 4, the phase function variations needed to produce the I/F variations observed at a *given* phase angle produce even larger variations over a *range* of phase angles; furthermore, these variations are also essentially independent of optical depth. The differential phase variation of part of the B Ring has been determined by using two Voyager 2 images obtained at phase angles of 45° and 70°, respectively. The images were digitally reprojected into a rectangular grid with one axis in the radial direction. They could then be radially scaled to account for resolution differences, superposed, and ratioed. The original brightness map at 70°, and the ratioed result, are shown in Fig. 23a and b, respectively. A spoke crosses the upper left of the ratio frame, reminding us that areas of relatively enhanced forward-scattering are brighter in the ratio plot.

In spite of the large and continuous range of I/F values seen in the scan of Fig. 3 for the region shown, the ratio plot in Fig. 23b is essentially bimodal, showing only dark grey and light grey regions, as seen in the scan across the frame. Furthermore, several areas of significantly different total brightness have an identical phase ratio. That is to say, in the context of the above discussion, the brightness variation between these several areas and their surroundings must be due primarily or only to variations in albedo, not phase

ab

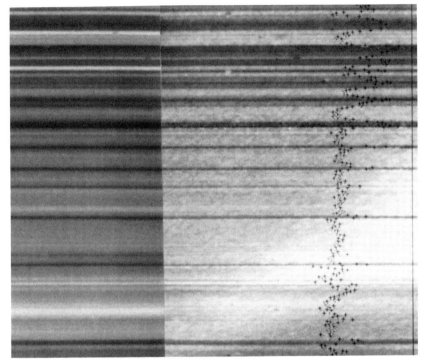

Fig. 23. Evidence for small-scale variations of particle albedo within the B Ring. Left panel
(a) shows Voyager 2 image of the B Ring at ~90° phase angle, reprojected to a rectangular
grid. Another image at lower phase angle (~30°), with comparable resolution, was similarly
reprojected and divided into (a); the ratio image is shown in (b). Right panel (b): in the ratio
"image" brighter areas are more forward scattering, as seen in the spoke which lies in the
left center of (b). Several areas of different absolute brightness in (a) show no difference in
their phase ratio (b), implying similar phase functions. As may be seen in the superposed scan,
the ratio image seems to be bimodal, having one (dark-grey) component which is more
strongly backscattering than the other (light-grey) component. This is in contrast to the
fluctuating total brightness of the B Ring (e.g., Fig. 3). These facts lead one to suspect
that at least some of the brightness fluctuations are due only to particle albedo variations
(Sec. IV.D).

function or optical depth, of the particles. The range in I/F for these albedo-
related features, of length scale from 50 to 300 km, indicates an albedo
variation of 0.4 to 0.6 for a "typical" particle phase function of Callisto (or
Lambert-surfaced spheres).

The fact that the ratio image shows essentially two levels of grey indicates that there are essentially two families of phase function. Although all
areas are darker in forward scatter in the absolute sense (Fig. 2), the light-grey
regions in the ratio plot are less so than the dark-grey regions. This could

result if the light-grey areas contained, in addition to a primarily backscattering macroscopic particle component, some fraction of less backscattering dust. Such a situation is consistent with another generally qualitative observation, that the B Ring contrast increases in forward scattering (Fig. 3). Because the most abundant particles are strong backscatterers but fairly bright, a little bit more multiple scattering has a large fractional contribution to the brightness at high phase angles.

To summarize the above results, some of the irregular features seen in Voyager imaging observations of the lit face vary in brightness due primarily to variations in particle albedo. Some show definite regional variations in phase function. Some features seem to indicate both forms of variation. There is not, as yet, enough data to constrain the fractional dust optical depth in greater detail than the above order of magnitude ($\leqslant 10\%$ by optical depth). One current frustration is that with the exception of the inner B Ring, it has not as yet been determined whether there is any correlation between the features seen in reflected light, which are thought to be primarily defined by variations in particle properties, and the optical depth fluctuations seen in the occultation experiments.

Another property of the irregular structure, at least in the outer B Ring, is that, on length scales < 50 km, much of it seems to be azimuthally asymmetric. Most of the 50-km- or larger-scale features are azimuthally symmetric, as testified by the fairly good registration and systematic ratio behavior seen in Fig. 23. However, most of the very fine-scale features seen in higher resolution observations (e.g. Fig. 24), are seemingly unrecognizable from one ring longitude to another. Some nonalignment is to be expected due to the eccentricity of the B Ring edge (Sec. IV.B) but should be systematic, and also negligible at a distance from the edge farther than the left center of the image. One must be a little careful about detailed comparison of features, as the photographs used in Fig. 24 were obtained at a variety of phase and solar hour angles. For example, several of the most noticeable variable features (especially in the third panel from the top) are due to the Mimas 4:2 bending wave which has vertical structure of a spiral nature, and would exhibit nonaxisymmetric brightness due to shadowing (Shu et al. 1983).

Clearly, the irregular structure of the B Ring and inner A Ring is an enormously complex problem, and is in need of more observational constraints. Below, we address alternative hypotheses in a fittingly cursory fashion.

If the structure results from the semishepherding effects of numerous moonlets of radii < 1 km (too small to clear an empty gap; Sec. IV.C.3.a), a preferred scale, as in the inner B Ring, could reflect overall torque balance and the moonlet size distribution. If the irregular structure is the result of viscous or thermal instability, it could be that the detailed form of the relation between restitution coefficient and impact velocity, or the detailed form of the size distribution, produces a preference for certain radial scales and optical

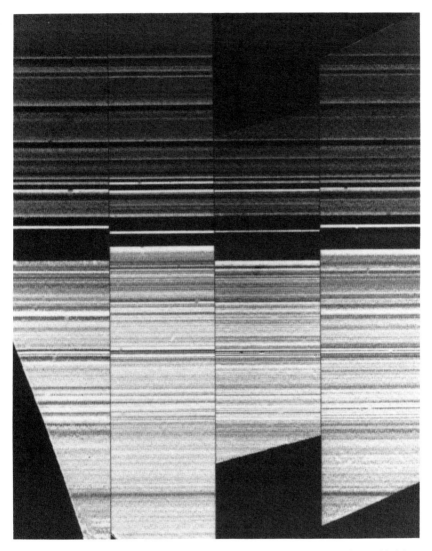

Fig. 24. B Ring outer edge, Huygens gap, and inner Cassini Division, at four orbital longitudes. Saturn is towards the bottom. This image has been processed to emphasize fine-
scale variations. The B Ring outer edge is seen to be doubly periodic by comparison with an
adjacent narrow ringlet which is singly periodic. The double periodicity is a result of forcing
by the Mimas 2:1 resonance (Sec. IV.B). Also visible in the B Ring is the irregular structure.
The fact that it is very difficult to follow the narrower features between panels has led to the
belief that the narrow features are nonaxisymmetric. Although this may be true, detailed
analysis is needed. For instance, the left and right panels were taken at a solar hour angle
differing by ~ 180° from the middle two panels. The markedly contrasting feature in bottom
center is the Mimas 4:2 spiral bending wave (Sec. IV.B), which shows a brightness profile
varying with both ring longitude and hour angle. (Photograph courtesy of A. Collins and
G. Garneau, Jet Propulsion Laboratory.)

depths. These and other related effects are discussed in the chapters by
Stewart et al. and Borderies et al.

Several aspects of the irregular structure seem in conflict with the behavior expected of a moonlet hypothesis. In the inner B Ring, the scale length
of the fluctuations is > 50 km, which is far longer than the effective range of a
moonlet too small ($r < 1$ km) to even clear a gap around itself. As far as can be
determined, these features are azimuthally symmetric and invariant over the
Voyager 1–2 time frame (< 1 yr).

If it represents a real material distribution, the azimuthal asymmetry of
the irregular structure in the outer B Ring would argue against a moonlet
hypothesis there; the viscous diffusion time for a ringlet of radial width ΔR is

$$t_{\text{diff}} \sim \Delta R^2 / \nu.$$

For $\Delta R \sim 20$ km and $\nu \sim 100$–260 cm^2 s^{-1} (Sec. IV.B), $t_{\text{diff}} \sim 500$–1000 yr.
This is far longer than the synodic period P_{syn} of the ring particles and their
hypothetical moonlet:

$$P_{\text{syn}} \sim \frac{2\pi}{\Delta\Omega} \sim \left(\frac{2R}{3\Delta R}\right) P_{\text{orb}}$$

or, for $P_{\text{orb}} \sim 1/2$ day, P_{syn} (20 km) ~ 5 yr. Therefore, it would seem that
moonlets would maintain axisymmetric features by a semi-shepherding effect;
if they were the cause of the irregular structure, variations with ring longitude
would be hard to understand. Finally, it is also difficult to accept that such
putative moonlets are so numerous in the optically thicker regions, and practically nonexistent in regions of only slightly lower optical depth.

Similar objections might seem to apply to the viscous instability
hypothesis. The formation time of the instability is roughly t_{diff}; because this
is much longer than P_{syn} for radial scales larger than a few hundred meters,
one would expect full communication of the instability around 360° longitude,
whatever the initial perturbation. However, a detailed theoretical analysis of
the nonaxisymmetric case has yet to be done. Of course, the inner B Ring
irregular features (and, for the most part, all the B Ring features of radial width
$\geqslant 100$ km) show azimuthal symmetry. It remains to be seen whether this
radial scale results in a natural way from viscous or thermal instability theory
(see, e.g., chapter by Stewart et al.). It may be worthwhile to point out that
the plateau structure of the C Ring and Cassini Division (which is azimuthally
symmetric) also seems to have a characteristic radial scale. However, the
inner B Ring features are nearly sinusoidal in radial profile (Fig. 22a) while
the C Ring and Cassini features have fairly abrupt edges. Furthermore, the
viscous instability is generally thought of as a large-optical-depth process
(chapter by Stewart et al.) and regions of intermediate optical depth ($\tau \sim 0.5$–
0.7 at visible wavelengths), occurring in the A and inner B regions (1.57–
1.58 R_S), do not show irregular structure.

In summary, the B Ring and inner third of the A Ring display characteristically irregular fluctuations in optical depth and/or particle properties. In general (but not invariably) no preferred radial length scale or periodicity is seen in this irregular structure. It seems that it may be azimuthally asymmetric on radial scales <50 km or so. In the various hypotheses discussed for the irregular structure, none makes an especially compelling case for itself as yet and there are apparent difficulties with all. Perhaps no single process is responsible for all the forms of irregular structure. Further observational and theoretical study is sorely needed.

E. Electromagnetic Effects

1. Magnetospheric Environment and Forces on Small Particles. Most of the ring features may probably be explained by gravitational forces alone. However, there are exceptions. In this section we present observational evidence for, and discussion of, ring-related electromagnetic phenomena. The physics of electromagnetic phenomena, along with potential applications to other ring structure, are covered in more detail in the chapters by Grün et al. and Burns et al. and by Hill and Mendis (1981*a*) and Mendis et al. (1982).

Pioneer 11 explored Saturn's inner magnetosphere as far inward as 1.325 R_S. In addition to providing magnetospheric properties of intrinsic interest, this deep probe of the ring environment has determined several unique and unrelated ring properties (Secs. II and IV.C.3; cf. also Van Allen 1984). Saturn's magnetic pole is very nearly centered on Saturn and antiparallel to the rotation axis, as with Earth. Its field has surface strength ~ 0.2 G at the equator.

In situ measurements show that the magnetosphere contains a large corotating inner plasma torus which extends from 8 R_S inward. It contains primarily oxygen ions (O^{2+} and O^{3+}), electrons, and some protons, all with energies between several eV and several keV. The electron density, with a value of ~ 10 cm^{-3} at 6 R_S, varies approximately as L^{-3}, where L represents the equatorial distance of a magnetic field line measured in units of R_S (Bridge et al. 1981). The extent of the torus perpendicular to Saturn's equatorial plane is large (several R_S) except in the inner part, <4.5 R_S, where it is more localized near this plane. Inward of ~ 4 R_S, the plasma density likely increases still further, but was not detected due to a rapid drop of the temperature from 5×10^6 K to 2×10^5 K at that distance. The torus could actually extend to the limit of the A Ring (2.267 R_S). The inner torus is relatively dense near the orbits of Dione and Tethys (6.26 R_S and 4.88 R_S, respectively) where ion densities up to 50 cm^{-3} have been measured (Frank et al. 1980). This indicates that the ions and electrons probably come from ionization of molecules and atoms formed by sputtering of the ice-covered surfaces of these satellites. Such erosion effects are discussed by Cheng et al. (1982).

It appears that the Jovian-terrestrial-type, inwardly-diffusing, magnetosphere is effectively absorbed by the rings at ~ 2.3 R_S; the region within has the lowest charged particle density ever observed by Pioneer 11 (Simpson et al. 1980; Van Allen et al. 1980a). Even within the central region (1.3–2.3 R_S), however, there are detectable levels of protons and electrons which apparently derive from neutrons splashed off the ring and satellite material by cosmic rays (Chenette et al. 1980; Fillius and McIlwain 1980). From the known strength of the cosmic ray source, an average mass density of 100 to 200 g cm^{-2} is obtained (Fillius and McIlwain 1980; Cooper et al. 1982; also Cooper and Simpson 1980). This is in good general agreement with values obtained in other ways (Secs. IV.B and III.C; Table III) and therefore with an overall icy composition.

Particles in such an environment may acquire electric charges by a variety of mechanisms. This problem has been studied in relation with the study of comet tails (Hill and Mendis 1981b; Mendis et al. 1981), interplanetary and interstellar dust (Spitzer 1978), and the charging of space vehicles moving in a plasma (Prokopenko and Laframboise 1980). The sign, as well as the amount, of charge depends on the charging process, on which there is no consensus at the present time. The main processes are the emission of photoelectrons, which leave a positive charge on the grain, and interaction with a plasma in contact with the grain. The simplest theory predicts that grains embedded in a plasma will become negatively charged, because they will capture more rapidly-moving electrons than slowly-moving ions. Under certain conditions (not common in the outer solar system) the charge can be positive (Mukai 1981). Also, if the plasma is bimodal (one cold and one hot component) the charge equation can have two stable solutions so that grains can acquire positive or negative charges depending on their previous history (Meyer-Vernet 1982). In any event, the charge on a grain will fluctuate sporadically. Regardless of the sign of the charge, a limit to its magnitude seems to be reached due either to field emission (Burns et al. 1980) or self-fracture (Öpik 1956; Fechtig et al. 1979), when the grain potential ϕ attains a value in the range of 10 to 100 volts. An indication of the relative importance of electromagnetic to gravitational forces may then be obtained in a simple fashion. A particle of radius r, density ρ, and mass m, carrying a charge of $q = \phi r$ and moving with velocity V_{rel} perpendicular to magnetic field B (see, e.g., Mendis and Axford 1974; Hill and Mendis 1979; chapter by Grün et al.), experiences an acceleration

$$\ddot{R}_{\text{EM}} \sim \frac{3qV_{\text{rel}}B}{4\pi\rho\, r^3 c} \sim \frac{3\phi\, V_{\text{rel}}B}{4\pi\rho\, r^2 c} \tag{30}$$

where c is the speed of light. The ratio of $\ddot{R}_{\text{EM}}$ to the gravitational acceleration $\ddot{R}_{\text{G}}$ on the particle orbiting Saturn at radius R is

$$\frac{\ddot{R}_{\text{EM}}}{\ddot{R}_{\text{G}}} \sim (1.3 \times 10^{-3}) \left[\frac{\phi\ (\text{volts})}{r^3\ (\text{microns})} \right] \left[\frac{1.36 - (R/R_S)^{\frac{1}{2}}}{1.36\ (R/R_S)^{\frac{1}{2}}} \right] \tag{31}$$

where R_S is the radius of Saturn and a dipolar magnetic field of surface strength 0.2 G has been assumed. At $R/R_S = 1.36^2 = 1.86$, electromagnetic forces vanish. This is the corotation radius, where the orbital period is equal to the period of Saturn's magnetic field ($P_S = 10^h 39^m4$) and consequently there is no velocity difference to induce a Lorentz force. Away from this radius, electromagnetic effects increase in magnitude. For $R \sim 1.5$–2.0 R_S and $\phi \sim 10$–100 volts, we obtain $r < 0.1$ μm for electromagnetic forces to be important. A more stringent criterion, for a particle to be tied to the corotating magnetic field, is that the particle's gyrofrequency (qB/mc) must be much greater than its orbital frequency, yielding $r \ll 0.07$ μm. It seems that electromagnetic control may be important for some of the Jupiter ring halo particles (Hill and Mendis 1979, 1981a; Burns et al. 1980; chapter by Burns et al.; Morfill et al. 1980a,b,c,d; Grün et al. 1980); however, the majority of the visible Saturn ring particles are in the size range of centimeters to meters in radius (Secs. III.C and III.D) and their dynamics is overwhelmingly dominated by gravity and collisions.

However, evidence does exist for electromagnetic processes in the rings superposed on the overall gravitational dynamics. For instance, the E and G Rings are relatively optically thin and do not deplete the charging plasma. Also, they are regions populated by significant numbers of micron-sized particles. This problem has been discussed by Hill and Mendis (1981a, 1982a,b). The F Ring, by contrast, seems to be less influenced by electromagnetic forces (chapter by Grün et al.). In addition, the B Ring spokes are believed to contain concentrations of small particles from their light-scattering properties, and show independent evidence of electromagnetic influence (Sec. IV.E.2). The B Ring overall may have a small fractional population of micron-sized particles, whose motion is a complex product of a blend of electromagnetic and gravitational forces with $\ddot{R}_G \sim \ddot{R}_{EM}$ (Hill and Mendis 1979, 1981a; Mendis et al. 1982; chapter by Grün et al.).

For instance, Northrop and Hill (1982) propose that the boundary between the inner and outer B Ring at around 1.63 R_S (corresponding to a noticeable inward decrease of the optical depth) is due to a stability limit for the orbit of particles of high negative charge to mass ratio. Inward of this limit, particles slightly perturbed normal to the ring plane will move away to higher magnetic latitude and be lost to the planet's atmosphere; outside this limit, they oscillate back and forth across the equatorial plane. Northrop and Hill (1983) have also suggested that the inner edge of the B Ring, at ~ 1.52 R_S, is just located at the stability limit of charged particles with an extremely large negative charge to mass ratio (essentially molecules), launched in the ring plane at the Keplerian velocity. Eshleman et al. (1983) also suggest that electromagnetic forces could increase the stability of certain kinds of entwined orbits. Other electromagnetic effects may be found in the rings. Although not fully understood, the spokes appear to be the best-established example of such forces at work.

2. *Spokes.* *(a) Description.* While still 1/3 AU from Saturn, Voyager 1 observed dark, radially-extended features which were immediately called spokes after their shape and appearance (Collins et al. 1980). Spokes have a characteristic length and width of 10,000 km and 2000 km, respectively (Smith et al. 1981). They are observed only in the range from 103,900 km to 117,000 km (1.72 to 1.94 R_S) from Saturn's center (Grün et al. 1983). Interestingly, this radial range includes the most optically thick regions in the entire ring system (Fig. 10). Their inner radial limit is particularly sharp and the outer limit is within that of the B Ring. Their observed radial range covers both sides of the synchronous orbit (Eq. 31).

The spokes are $\sim 10\%$ darker than surrounding regions when viewed in backscattered light; however, their contrast reverses when they are viewed in forward-scattered light, and they become 10–15% brighter than surrounding areas (Smith et al. 1981, 1982; Cuzzi, unpublished). It should be recalled that both spoke and nonspoke areas are darker in an absolute sense in forward scatter than in backscatter (see Fig. 2). Contrast between spokes and the underlying ring decreases at intermediate phase angles of 30–70° (Grün et al. 1983). The visibility of the spokes, specifically their larger ratio of forward-to-backscattering than that of the background ring particles, implies that the spoke particle size is very small, of the order of a few microns or less (see Sec. III.A and chapter appendix). These scattering particles do not need to be physically removed or elevated above the main ring layer, but may be uniformly mixed throughout it (Sec. III.A and Fig. 4).

A few spokes have also been seen on the unilluminated face of the ring at intermediate phase angles ($\sim 90°$), and appear bright against the very dark opaque B Ring background. Their brightness in this geometry has been interpreted as scattering of light from Saturn's disk, and demonstrates the presence of spoke material on the unilluminated face of the rings with abundance comparable to that on the illuminated face (Smith et al. 1982).

Very few clear examples of spoke formation seem to exist in the Voyager 1 and Voyager 2 images. In the few cases where spokes in the process of formation have been observed, they were near-radial. Grün et al. (1983) describe a spoke which appears suddenly across its full length of 6000 km between two successive frames taken only 5 minutes apart. This gives an upper limit on the time scale for the onset of spoke activity. On the first frame where it appears, the contrast of a newly formed spoke is low, but it increases rapidly in the next 20 min or so. The subsequent motion of spokes has been studied by comparing several successive images of the ring taken over 30-min intervals. In general, the velocity corresponds clearly to a Keplerian or near-Keplerian motion, and not to the velocity of the magnetic corotation motion. Some spokes have been followed for longer than a rotation period, but it was not clear whether they were physically the same spokes or if new spokes were merely reprinted from time to time in the same region of the B Ring (Smith et al. 1982). This last possibility is more in accord with the fact

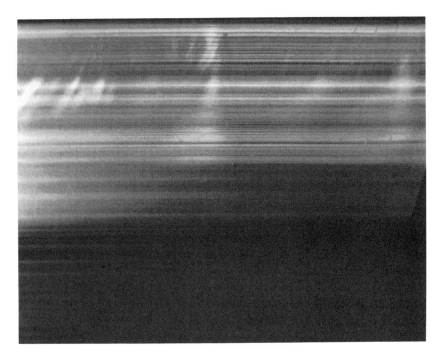

Fig. 25. The characteristic spoke "wedge-shaped" morphology (Sec. IV.E) is seen in this
  Voyager 1 image at ∼ 130° phase angle, in which geometry the spokes are brighter than their
  surroundings. The image has been reprojected to a rectangular grid with radius the vertical
  axis; Saturn is toward the bottom. The corotation radius is near the narrowest part of the
  wedge; the leading inward edge is characteristically tilted, the trailing inward edge is nearly
  radial. The opposite behavior is observed outward of corotation. Close inspection indicates
  that the spoke enhancement is greatest in the already brighter (relatively more forward-
  scattering) features. (Photography courtesy of A. Collins and G. Garneau, Jet Propulsion
  Laboratory.)

that spokes with very large tilt relative to the radial direction are uncommon
(after one rotation a spoke would make an angle of 75° with the radial direction)
(Grün et al. 1983).

    Actually, the ages of the spokes can be inferred from their tilt. This has
been done by Grün et al. (1983) who found that the ages of spokes observed in
the morning ansa range from less than 1000 s to 14,000 s, which is less than a
quarter rotation of Saturn. It is then likely that the spokes disappear by losing
their contrast with the ring background in much less than a rotation. An
obvious explanation is that the lifetime is limited by the free lifetime of the
dust particles, which collide with other ring particles several times per orbit. If
the spoke particles stick upon encounter, a lifetime $(1/3–1/4) P_S t \sim 3–4$ hr
would result, consistent with observed spoke tilts (Grün et al. 1983).

    It was realized by Smith et al. (1981) that the spokes frequently display a wedge-shaped morphology, with the narrow apex of the wedge lying near to the ring radius at which the Keplerian orbit period is equal to the period of Saturn's magnetic field: the so-called corotation radius (Eq. 31). Often, the spokes broaden both inwards and outwards of this location (Fig. 25). The actual shapes of spokes are not always so simple (Grün et al. 1983). However, the wedge shape suggested that the feature might be formed in a system fixed to the magnetic field. If formation continued over some extended active period, the relative velocity, which increases away from the corotation radius, would broaden the spoke increasingly with radial distance inward and outward of corotation.

More detailed studies of the narrowest and least trailing spokes by Grün et al. (1983) and Eplee and Smith (1983) have tended to support this idea. The velocities of a few young spokes, and perhaps especially their trailing edges (inwards of corotation) show substantial deviations from Keplerian, in the direction of corotational velocities. That is, they may be still fixed to the magnetic field. However, the more commonly observed broader spokes show nearly Keplerian velocities in both leading and trailing edges. This means that, even though the magnetic field may play a dominant role in spoke formation, the subsequent motion of the particles is governed by gravitation after the termination of some limited period of activity (Smith et al. 1981; Terrile et al. 1981).

(b) Relationship of Spokes with the Magnetic Field and the Solar Hour Angle. The spokes are not permanent features of the rings, even if some spokes are nearly always present on a ring picture. Smith et al. (1981) first noted that, on the average, they are more numerous and show greater contrast in the morning ansa than on the evening ansa. The first firm connection between the spokes and Saturn's magnetic field was noted by Porco and Danielson (1982).

These authors performed Fourier spectral analysis of the spoke activity level, and found a pronounced peak near 10 hr 21 min ± 22 min. This value is consistent with the period of rotation of Saturn's magnetic field (10 hr 39.4 min) but also with other physically reasonable periods within Saturn's system as, for example, the range of Keplerian periods in the B Ring and the period of the SED (Warwick et al. 1981; Evans et al. 1981). However, the correlation of spoke activitiy with Saturn's magnetic field is further strengthened by the correlation of the level of spoke activity with Saturnicentric magnetic longitude at the ansa, in the Saturn longitude system (SLS; Desch and Kaiser 1981). Porco and Danielson (1982) showed that the activity level is highest when the ansa is aligned with the magnetic field quadrant $\lambda_{SLS} \sim 115°$, the magnetic sector most often aligned with the Sun during periods of Saturn kilometric radiation (SKR) (Kaiser et al. 1981) and also associated with intense ultraviolet (auroral) emission (Sandel and Broadfoot 1981). Subsequently, Grün et al. (1983) showed that spokes are more nearly radial when they lie at this magnetic longitude. The level of activity shows a strong longitudinal effect even over the period between Voyager 1 and Voyager 2. This indicates that the actual periodicity is probably even closer to the field period than indicated by the Fourier analysis of the Voyager 1 data alone, described above. More extensive analysis of Voyager 2 data (Porco 1983) gives a very sharp peak in the Fourier spectrum at 10 hr 40.6 min ± 3.5 min. However, the interpretation of this spoke-field relationship is not straightforward, because the range of B Ring radii where the spokes are observed contains only magnetic lines of force which have their feet at latitudes much lower (39°–45°) than those of the SKR and auroral source (Kaiser et al. 1981; Sandel and Broadfoot 1981).

In summary, the spokes can form anywhere on the ring, but appear in the morning ansa with greater frequency and higher contrast. The spoke activity is significantly larger when the SKR active sector is facing the morning ansa. Spokes form in a very short time and simultaneously over their full radial extent. Underlying radial structure is not affected. Their lifetime is probably shorter than one rotation, the contrast decreasing from the morning ansa onward, but often other new spokes form on the fading ones. The angular velocity of very narrow, nearly radial spokes may be close to uniform, at the corotation velocity of the magnetic field. The angular velocity of inactive spokes, with both leading and following edges trailing, is at or very close to the Keplerian velocity. All of this suggests that the formation of spokes is controlled by Saturn's magnetic field while the subsequent dynamics of the spoke particles corresponds to Keplerian motion.

(c) Spoke Formation Hypotheses. Two types of mechanisms have been proposed for the formation of spokes: theories using purely electric phenomena and theories which postulate an electric charging of small grains and their subsequent motion in the electric and magnetic fields of the planet. Each type must explain both the mechanism by which the spokes become visible, and the general causative process which explains the occurrence, periodicity and shape of the spokes.

As an example of the first type of theory, it has been suggested that micron-sized ring particles are polarized and aligned in a preferential direction by a radial electric field. However, theories using alignment of dust particles predict that the scattering properties of the spokes will be highly anisotropic around the azimuthal direction in the ring, i.e. that the spoke characteristics, especially the contrast, will be doubly periodic with longitude. This is contrary to the observations; whether the spokes are dark or bright, they are so all around the ring. Also, it has been shown by Weinheimer and Few (1982) that the calculated alignment torque is probably too small for the process to work efficiently, especially upon particles large enough $(r \gg \lambda)$ for the alignment of their physical cross section to affect the reflected light. Therefore, grain alignment scenarios may be discarded.

However, the existence of strong radial electric fields may still be of importance in the overall picture, especially the connection noted by Carbary et al. (1982) between the B Ring radial region and the latitudes of possible ionospheric jet streams which are connected along magnetic field lines. The feet of the magnetic lines of force which go through the B Ring are in a latitude region of high winds ($V \sim 100$ km s^{-1}). It is possible that such winds, moving eastward relative to the magnetic field, carry along part of Saturn's ionosphere, and create an electric field $\mathbf{E} = \mathbf{V} \times \mathbf{B}$ (where $\mathbf{B}$ is the magnetic field) and hence an electric current in the north-south direction. Since the high wind is localized in a small range of latitudes, it can be shown that the current will leave the ionosphere at the boundary of the wind region to flow along the magnetic field lines and close the circuit between northern and southern hemispheres. This creates, at the level of the B Ring, a radially varying electrical potential. From the wind velocity and the characteristics of Saturn's ionosphere (Tyler et al. 1981; Atreya and Waite 1981), Carbary et al. (1982) estimate this potential to be 16 kV, giving a radial field of several V km^{-1} in the rings.

The second type of theory makes use of some process which may locally enhance the fractional optical depth of micron-sized particles. Their presence would cause the ring scattering in spoke regions to be less strongly backscattering, i.e. relatively dark in backscatter and relatively bright in forward scatter (Fig. 4). This is because the dominant Lambertian behavior of the majority of the (macroscopic) ring particles is augmented by the dominant forward- (or even isotropic-) scattering behavior of wavelength-and-smaller-sized particles (chapter appendix).

The nature of this enhancement process is uncertain at present, even given the prior existence of the dust particles in the rings (Smoluchowski 1983; Morfill et al. 1983). It has been suggested by Smith et al. (1981, 1982) and Hill and Mendis (1982a) that the micron-sized particles could be initially lying on the surfaces of larger ring parent bodies (Burns et al. 1980) and then jettisoned off these surfaces by electrostatic repulsion, if the surfaces of the parent bodies are sporadically charged to high electrostatic potential.

Another possibility is that a large number of micron-sized particles independently preexist deep within the rings, but are in general hidden by the large ring optical depth and contribute very little to the total brightness of the ring, dominated by the quasi-Lambertian scattering of larger particles. Their charging (which does not change their scattering characteristics by itself) causes them to levitate to some height *above* the ring and become visible (Grün et al. 1983). Such a physical separation of the spoke grains above the normal ring layer is, however, not at all necessary to explain any of the photometric observations (see Sec. III.B).

The magnetospheric plasma density is greatly decreased in the ring vicinity relative to values in less cluttered regions of the magnetosphere outward of the main rings (see, e.g., Simpson et al. 1980). The charging might, however, arise internally from plasma instabilities in current loops running between the rings and Saturn's magnetosphere, as has been suggested informally by many (Alfvén, Gold, Mendis, and others) or by impact-induced clouds of plasma (Goertz and Morfill 1983; Morfill and Goertz 1983).

No explanations have been proposed so far for the facts that the charging mechanism giving rise to the spoke occurs in localized regions (radially and in longitude) and suddenly over a large range of radial distances. Also, although various authors have claimed evidence from the observations that the grain charge is primarily negative (Hill and Mendis 1982a; Thomsen et al. 1982), more recent data and interpretation (see, e.g., Grün et al. 1983) seem to require that this question be left open for the present.

3. Saturn Electrostatic Discharges. During a few days around Voyager Saturn encounters the planetary radio astronomy (PRA) instruments detected a novel kind of nonthermal radio emission. It consisted of impulsive bursts ($<$ 30 ms to 400 ms duration), appearing randomly within the PRA frequency spectrum (1.2 kHz to 40 MHz); it was interpreted as a succession of broadband emissions which are only detected by the channels being sampled during the duration of an event. The bursts appear in separate episodes, with a periodicity of roughly 10 hr 10 min, close to Saturn's atmospheric and magnetic field rotation periods (10 hr 10 min and 10 hr 39.4 min, respectively). The occurrence rate and intensity of the bursts vary roughly as the inverse square of the distance of the spacecraft from the planet. Warwick et al. (1981) concluded that the bursts were coming from the close environment of Saturn and, due to their similarity to the radio emission properties of terrestrial lightning strokes, called them Saturn electrostatic discharge (SED).

SED intensities cover a large range of values, from the threshold of sensitivity of the PRA instrument up to $10^3 - 10^4$ times this threshold when observed close to the planet. The peaks of the largest SED's correspond to instantaneous radiated powers between 10^8 and 10^{10} W (Zarka and Pedersen 1983). The average power radiated by all the SED, taking account of the occurrence rate of the events (1.4% for Voyager 1, 0.4% for Voyager 2), is of order 10^6 to 10^8 W.

The observed characteristics of the SED, in particular the periodicity of the episodes, suggest two possible locations for the source: in the planet or in the ring at a distance of 1.81 R_S from Saturn's center where the Keplerian period is equal to the SED periodicity.

The atmospheric solution was initially rejected, and the ring chosen as a source, by the PRA team (Warwick et al. 1981, 1982; Evans et al. 1981, 1982, 1983) on the basis of the rotation period, which differs from that of the magnetic field, and also because of the fact that frequencies as low as 20 kHz, where SED were thought to be observed, are unable to propagate through ionospheric densities as high as those measured by Pioneer 11 and Voyager 1 and 2 (Kliore et al. 1980; Tyler et al. 1981, 1982). Another argument leading to this choice was that emissions similar to SED were not detected close to Jupiter by the Voyager PRA experiment, in spite of the detection of lightning flashes by the Voyager cameras (Cook et al. 1979) and the indirect evidence of the existence of lightning provided by the observation of whistler emission by the Voyager plasma wave science (PWS) experiment (Gurnett et al. 1979). On the other hand, these experiments did not find any such evidence for lightning in Saturn's atmosphere (Smith et al. 1981; Gurnett et al. 1981).

Burns et al. (1982, 1983) showed that the arguments against an atmospheric source location are, however, not fatal. The first argument—concerning the SED periodicity—is not a strong one, as it is known that the atmosphere does not rotate with the same periodicity as the magnetic field. Indeed, equatorial cloud velocities have been obtained from groundbased observations and by Voyager which indicate a period of 10 hr 14 min, within the uncertainty of the SED value. Concerning the putative observations of SED down to 20 kHz, Burns et al. (1982, 1983) pointed out that the characteristics of Saturn's ionosphere, especially in the region perpetually shadowed by the rings, are not sufficiently well known to predict accurately the ionospheric electron density, and thus the low-frequency cutoff of radio propagation (Atreya and Waite 1981).

The atmospheric origin was strongly supported by Kaiser et al. (1983) who scrutinized the complete PRA data set and concluded that: (1) no SED are observed at frequencies $\lesssim 4$ MHz during the pre-encounter episodes, while they appear routinely at frequencies as low as a few hundred kHz after encounter. This difference (see, e.g., Zarka and Pedersen 1983) cannot be explained by a difference in antenna directivity due to the change in the direction of observation.

The low-frequency cutoff at 4 MHz which is seen before, but not after, encounter is very difficult to explain for a source in the ring, but easily understood in terms of a low-frequency propagation cutoff for radiation originating in the atmosphere and passing through the dayside Saturn ionosphere. The implied dayside electron density n_e ($\gtrsim 2 \times 10^5$ electrons cm^{-3}), is consistent with that measured (at the terminator) by Pioneer 11 and the Voyagers (2.3×10^4 electrons cm^{-3}). After encounter the storm system is observed through the nighttime ionosphere, the density of which could be low enough to allow SED observations to a frequency as low as 100 kHz ($n_e \lesssim 100$ electrons cm^{-3}). The absence of low-frequency SED before encounter might explain the nondetection of SED, even around close encounter, at frequencies below 56.2 kHz by the PWS experiment (Kurth et al. 1983). In addition to the above, Kaiser et al. (1983) also noted (2) that no SED events are detected between the episodes; (3) the episodes begin and end abruptly and in much better agreement with a radial location on the planet's surface than at 1.81 R$_S$ in the rings; and (4) the onset and termination of SED episodes are independent of frequency except for the onset of the episode centered on Voyager 1 closest approach (Evans et al. 1981), in which the SED appear first at high frequencies, then gradually at lower frequencies, until the entire frequency range is covered. Kaiser et al. (1983) point out that such a gradual frequency dependence of the onset and decay of episodes could be due to refraction effects in the ionosphere, but would be detectable only at the onset of the Voyager 1 encounter episode because at that time the effective rotation rate of the planet, as seen by the PRA instrument, is diminished due to the motion of the spacecraft along its trajectory. No such ready explanation for this behavior exists for a ring source. For reasons given above, it is now generally believed that the SED are not a ring phenomenon at all, but are generated (by unknown means) in the atmosphere or ionosphere of the planet. Consequently, we do not discuss the effect further, and refer the reader to a review chapter by Kaiser et al. (1984) and to the original papers.

In summary, there is strong evidence for the effects of electromagnetic forces in the play of spokes on the face of the B Ring. There are also several hypotheses which might relate the global structure of the rings to small electromagnetic effects acting over time, such as the correspondence of the stability limits of extremely small charged particles with certain jumps in optical depth in the B Ring. In this regard, it may not be coincidental that the radial region of highest optical depth, which is also the location of much of the spoke activity, is the radial region nearest to corotation (Eq. 31) where tiny grains are least affected by electromagnetic forces and therefore most stable. On the other hand, it seems that the SED bursts are not a ring-related phenomenon at all, but arise in the equatorial atmosphere of the planet.

V. SUMMARY OF KNOWNS AND UNKNOWNS

In this section, we present a snapshot of current understanding of the properties of the rings, and discuss some of the major open issues. The reader is also referred to a review chapter by Esposito et al. (1984).

The rings are, of course, a vast number of individual, marble-to-house-sized, probably irregularly shaped, orbiting lumps of water ice containing trace amounts of rocky material or possibly other geological ices (Sec. III). Most of the cross-sectional area of the rings is presented by particles of ~ 1 cm to ~ 5 m radius. Between these limits, the differential number density $n(r)$ $dr = dN(r)/dr$ behaves very much like a power law of the form $n(r) = n_0 r^{-q}$, with $q \cong 3$ (Sec. III.C and III.D). Such a form is characteristic of objects undergoing countless mutual collisions (Hartmann 1969). For radii $\lesssim 1$ cm, the number density must cut off to some extent; less than 10% of the visible area can consist of sub-centimeter particles in many regions, although more sub-centimeter particles *may* exist in the B Ring (Sec. III.C). About 10% fine dust is seen in various specific localized areas of the rings, like resonance zones (Sec. IV.B), spokes (Sec. IV.E) and possibly throughout large parts of the B and A Rings. The tenuous, outlying E, F, and G Rings contain larger fractions of fine dust (Sec. III.A). For radii larger than $r \sim 5$ m, the number of particles drops extremely rapidly. The total mass of the rings ($\sim 5 \times 10^{-8}$ M_S) is dominated by the several-meter-radius particles (Table III).

However, several objects of radius ~ 10 km probably lie within the main rings (Secs. III.D and IV.C). Continuing analysis may soon extend the locations and numbers of such objects. The radial distribution of these moonlets will surely be of interest. Specifically, the real nature of the enigmatic F Ring, lying in a possible zone of transition between rings and ringmoons, is still not well understood. Generally, moonlets are structurally important only in a local sense. However, their mere presence, as well as that of the more massive ringmoons (Janus et al.), may have important implications for ring origins, as discussed subsequently.

The question of particle composition is still far from fully answered. Even though we are currently quite certain that the bulk of the particle material is icy (Secs. III.B and III.C), the nature of the remaining material and the cause of its apparent radial variation as manifested in varying particle albedo (and color), are of great interest. Particle albedos vary dramatically within the rings, spanning the range from icy to rocky (Secs. III.A and IV.A). The darker particles are primarily found in the C Ring and Cassini region, but significant, 50–100 km scale, radial variations are also seen in the B Ring. It has been pointed out (Epstein et al. 1983) that rocky material could be more abundant than inferred from studies of thermal microwave emission if it were segregated into separate chunks instead of being uniformly mixed, as is usually assumed. Even for the icy material, no firm identification exists for any material but water ice, and even the crystalline phase of the ice, or whether clathrates with other icy materials are present, is uncertain (Secs. III.A and III.B).

Space Telescope observations of spectral reflectivity will be of great value in answering these questions as will further groundbased and spacecraft studies of radio occulation and radar reflectivity, and 0.1 to 10 mm wavelength studies of the thermal emission of the rings.

In the area of ring structure, we have actually begun to understand some of the dynamical processes that structure the rings on a relatively small scale. The smallest of these scales is the vertical thickness of the rings. We have learned that the vertical scale height z_0 of the rings lies in the range of several to several tens of meters in various internal locations (Sec. IV.B), and that the ring full thickness $2z_0$ is < 150–200 m at the sharp outer edge of the A Ring and at both inner and outer edges of the Encke Division (Sec. II). There are currently hints that the larger ring particles lie in a thin layer, a few radii thick, near the ring symmetry plane, and the smaller particles are spread throughout a layer of comparable thickness which is many times their own size, consistent with both dynamical expectations and remote-sensing experience (Secs. II.B and III.C).

From the standpoint of small-scale radial structure, the discovery of Saturn's close-in ringmoons (Janus et al.) and the high-quality Voyager and Pioneer data have allowed the subject of gravitational dynamics to come into its own. Gravitational resonances, a subject of interest for over a century in the ring context, are finally seen to play an important role in constraining major ring edges (Sec. IV.B). Also, these satellites are seen to excite propagating spiral waves of density or vertical corrugation at many resonances within the rings. The physics of spiral density waves, now clearly demonstrated for, perhaps, the first time, allows us to measure directly local ring properties such as surface mass density and viscosity (Sec. IV.B). However, many of these waves are strongly nonlinear and their properties are not well understood in full detail. Inward transport of material by the momentum flux carried by these waves is less dramatic than had been anticipated. The absence of gaps in wave regions is understandable if inward wave transport is offset by outward viscous spreading processes (cf. chapter by Borderies et al.). This is an area of ongoing active research.

Shepherding effects by both external and embedded moonlets have apparently been observed in the F Ring and the Encke Division (Sec. IV.C); the details of the observed behavior, not yet fully analyzed, will tell us a great deal about the ability of moonlets to repel or concentrate ring material. However, we still do not know what, if indeed anything, constrains the narrow, opaque, smoothly eccentric, Uranian-type ringlets found in most otherwise empty gaps. In several locations, careful searches have found no shepherding objects of the sizes expected.

We have learned that the spokes are seen when the rings become locally more dusty, nearly instantaneously over a vast radial extent (Sec. IV.E). The initiation of this behavior is somehow influenced by a sporadic process relating a special magnetic longitude on Saturn, which is also responsible for aurorae and nonthermal planetary radio emission, to the diurnal cycle of night and day. However, the specific nature of the process is completely unknown at present.

Another form of sporadic electromagnetic process is responsible for the Saturn electrostatic discharges (Sec. IV.E). It seems that the weight of current evidence indicates that these discharges originate somewhere in the equatorial atmosphere of the planet, and not in the rings at all.

Although we are starting to acquire a familiarity with some of the "trees," we are still strangers in general to the "forest." For instance, we have little understanding of the large-scale structure of the rings. Why are there A, B, and C Rings at all? Why are the C Ring and Cassini Division so similar and so unlike the A and B pair in both structure and particle properties? To some extent, the abrupt outer edges of the A and B Ring are explained by resonant angular momentum transfer to satellites. However, nearby resonances of only slightly lesser strength produce mere waves in the disk, with no hint of a gap. A possibly related problem of central interest is the cause of the fairly abrupt inner edges of the A and B Rings. Dynamical relationships with the (external) known satellites are ineffective at preventing material from evolving inward. Perhaps erosion processes, operating on the very largest scales, act to preferentially erode already optically thin areas and sharpen inner edges (chapter by Durisen). Perhaps, also, other processes, long since extinct, have provided the rings with some aspects of their underlying large-scale structure. An example of this could be the initial creation of the Cassini Division by an extremely strong, possibly primordial, density wave driven at the Mimas 2:1 resonance (Goldreich and Tremaine 1978b; Sec. IV.B).

Also, the very fine-scale irregular structure, widespread in radial extent, is not well understood even as to whether it is primarily a variation in quantity or in properties of material, or both (Sec. IV.D). Another spacecraft mission will be needed eventually in order to fundamentally understand the structure and the cause of the large optical depth of these regions. The C Ring structure (abrupt optical depth transitions without intervening gaps) also remains one of the major currently unattacked structural problems. An improved understanding of viscous spreading and instability effects could possibly aid us with several of the above mysteries (cf. chapter by Stewart et al.).

The tenuous, peripheral rings continue to puzzle us by their very existence. The lifetimes of the tiny E, F and G Ring particles to magnetospheric and meteoroidal destruction mechanisms (chapter by Burns et al.) are extremely short, only $10^2 - 10^4$ yr. Although many are comfortable with ascribing replenishment of the tiny F Ring particles to small local source satellites, and of the E Ring to Enceladus in some unspecified way, the nature and ultimate source of the G Ring material is a mystery (Secs. III.A and IV.C). These features should not be neglected; information content is not necessarily proportional to optical depth (chapter by Burns et al.).

Perhaps the most global scale puzzle is that of the short overall time scale for significant orbital evolution of the ring-ringmoon system. The amount of angular momentum which is observed transferred between the rings and the shepherding satellites alone (Sec. IV.B) is apparently sufficient to collapse the A Ring and fully evolve the ringmoons from the edge of the current A Ring to their current locations in a time of $10^7 - 10^8$ yr (see chapter by Borderies et al.; Lissauer et al. 1984). Also, current estimates of the erosion lifetime of the main rings are disturbingly short (see, e.g., chapter by Durisen). These are, naturally, both areas of intense current interest.

From the standpoint of the origin of the rings (cf. chapter by Harris; Pollack and Consolmagno 1984), it seems as a first impression that the combination of compositional (albedo) inhomogeneities and embedded 10-km-sized moonlets, which are isolated from their surroundings, seems to suggest that the visible ring material (currently dominated in mass by 1 to 10-meter size particles) is a *derivative* rather than a *primordial* population. Pollack (1975) and Shoemaker (1983) have discussed ways in which extra-Saturnian meteoroid erosion can grind to fragments an initially small number of large parent objects which might have been of heterogeneous composition. An erosion hypothesis is also favored for Jupiter's ring (chapter by Burns et al.). If this is in fact the case, it may be necessary to reexamine current thinking on models of the evolution of the proto-Saturnian nebula, which are currently of the thermal-equilibrium type and make specific predictions of gradual temporal and radial gradients in temperature, and therefore in composition, of condensates in the ring region. The rapid orbital decay of condensed objects due to gas drag in such a proto-Saturnian nebula has led Pollack et al. (1976) to the conclusion that no record of the primarily early condensation of silicates will be observed in the ring regions of the nebula.

However, a population of precursor objects would have had to come from somewhere. Perhaps heterogeneous condensation occurs in different nebular regions simultaneously, with significant radial mixing and more rapid dissipation of the nebula than currently believed. Alternatively, precursors of different composition, formed at different stages of nebular evolution may, by chance, grow large enough to survive orbital decay by gas drag; they would remain in the ring region, surviving the subsequent thermal evolution and dissipation of the nebula but not the external bombardment (from whatever source) to follow. Eventually, we may attempt the difficult task of relating the primordial distribution of material to the observed distribution which, very possibly, has been smeared in various ways by global transport processes, many of which are currently poorly understood and some of which may even be long extinct.

However, before we can confidently begin such extrapolations, we must first better understand the details of particle size evolution in a collisional disk. Of course, this understanding will be directly relevant to the question of the formation of the terrestrial planets (at least) in the protoplanetary disk of planetesimals. Specifically, it is not known for certain that the currently suspected 10-km-sized objects must be primordial. Processes of accretion may proceed to some point even within the Roche limit (see, e.g., chapter by Weidenschilling et al.). Already we know that relatively large ring particles may have effects which range through enhanced stirring (Safronov 1972; Cuzzi et al. 1979a,b; chapter by Stewart et al.) to repulsion or shepherding, of the local material. Perhaps the accretion process is self-limiting in this way; however, Goldreich (1982) has suggested that the largest objects may be able to migrate radially through the disk under their own gravitational torques and collect in gaps, where close encounters and collisions such as those we may be seeing today in the kinky ringlet areas could continue the accretion process.

In providing us with a giant analogue laboratory in which to observe the complex interactions between a disk and its most dominant masses, interactions which surely predated the formation of the Earth and terrestrial planets, the rings of Saturn are functional as well as beautiful. For hundreds of years, astronomers and physicists have devoted many hours at the eyepiece of a telescope, or standing before a blackboard, to the study of Saturn's fascinating ring system. We have been truly fortunate in our time to receive a great harvest of knowledge, made possible by enormous technological advances and the dedicated efforts of the hundreds of women and men of the Pioneer and Voyager projects. It is clear, however, that our main questions about the rings, as to how they were originally created and what gives them their elegant form, remain largely unanswered. Our recent work has only explored the tip of the iceberg of knowledge we must eventually possess. However, true synthesis of this knowledge has only begun, and a deeper understanding will surely follow.

Acknowledgments. We would like to thank our colleagues for many helpful conversations and critical reviews of various preliminary versions of this manuscript; especially T. Ackerman, J. Burns, N. Borderies, R. Clark, A. Collins, P. Goldreich, R. Greenberg, B. Hapke, W. Irvine, L. Lane, D. Lin, S. Ostro, J. Pearl, J. Pollack, C. Porco, R. Reynolds, R. Samuelson, J. Scargle, M. Showalter, F. Shu, G. Stewart, S. Synnott, J. Van Allen, and H. Zebker. We are indebted to K. Bilski, M. Legg, K. Fischer and G. Garneau for their help with data analysis, and we are very grateful to M. Gomes for her help with the manuscript preparation. Figure 1 was reproduced from *Scientific American* by permission. J.J.L. is a NAS-NRC Resident Research Associate at NASA Ames Research Center. Most of the work described here has been supported by the Planetary Astronomy Discipline Office, the Voyager Project, and the Voyager Saturn Data Analysis Program of NASA.

APPENDIX

In this appendix, we briefly sketch the concepts and approach of radiative transfer theory as applied to the remote sensing of planetary rings. The approach most often applied, which assumes a many-particle-thick slab of scattering and/or absorbing particles, generally follows that of Chandrasekhar (1960). Many practical aspects of such solutions are reviewed by Hansen and Travis (1974) and Irvine (1975). It must be kept in mind, however, that certain aspects of the ring behavior may require monolayer-like treatment. One example is the scattering of radio waves by 10-meter-diameter particles in a layer of thickness only several times larger (Sec. III.C).

Individual particles of radius r have a total interaction with radiation of wavelength λ described by their cross sections for pure (lossless) scattering $Q_s \pi r^2$ and for pure absorption $Q_a \pi r^2$, where Q_s and Q_a are efficiency factors. The total extinction of energy from an incident beam is proportional to $Q_e \pi r^2$ where $Q_e = Q_a + Q_s$. The particle's albedo for single scattering $\tilde{\omega}_0 = Q_s / Q_e$. In general, the efficiencies Q_a and Q_s are not strongly dependent on particle shape (Pollack and Cuzzi 1980). The angular distribution of the scattered radiation is described by the phase function $P(\Theta)$, where the scattering angle Θ is generally measured from the direction of the incident beam, and $P(\Theta)$ is normalized over solid angle such that $\frac{1}{2} \int_0^\pi P(\Theta) \sin \Theta \, d\Theta = 1$. Particle efficiencies and phase functions depend primarily on the ratio of particle size to wavelength, usually expressed by the nondimensional size parameter $x = 2\pi r/\lambda$, but also on the complex refractive index of their constituent material (see, e.g., Kerker 1969; van de Hulst 1957; Hansen and Travis 1974). All efficiencies are low for $x \ll 1$ (as seen, e.g., in Fig. A1); i.e., the particles are essentially transparent to the wave. The scattering ability of a particle increases rapidly as x increases towards unity, and in the Mie scattering regime $1 < x < 100$, Q_s may be much larger than Q_a, so that, for low-to-moderately-absorbing materials, $Q_s/Q_e = \tilde{\omega}_0 \sim 1$, while $Q_e \gtrsim 1$. It is possible to find $Q_e > 2$ in this range of x (see, e.g., Hansen and Travis 1974). This domain characterizes the observed scattering behavior of the ring particles in the wavelength range of roughly 1 to 10 cm, and explains why the thermal emission from the particles, proportional to $(1 - \tilde{\omega}_0)^{\frac{1}{2}}$ (see, e.g., Horak 1952), is so small at these wavelengths while their combined optical depth is substantial ($Q_e \sim Q_s \sim 1$).

For $x \gg 1$, the total extinction efficiency approaches the physical optics limit: rigorously, $Q_e = 2$; i.e., a particle removes exactly twice its cross-sectional area from the incident beam. This is known as Babinet's paradox (or principle) and arises because the particle *reflects* or *absorbs* an energy of πr^2 times the incident flux, and *diffracts* an equal amount of energy. This behavior is basic to the understanding of certain observations, as seen below.

Figure A2 demonstrates typical variation of particle phase functions with x. For $x \ll 1$ the particle is a Rayleigh scatterer $P(\Theta) \sim (1 + \cos^2\Theta)$ and scatters equal amounts of energy into the forward and backward directions. As x increases, the scattering becomes more forward-directed and concentrated in an increasingly narrow lobe about the forward direction. This is in essence a diffraction effect, and the width of the forward lobe ($\sim \lambda/2r$) is practically independent of shape and composition for randomly oriented particles (Pollack and Cuzzi 1980, and references therein). For extremely large particles ($x \gg 1$ or $r \gg \lambda$), the forward lobe becomes so narrow that it is indistinguishable from the incident beam under normal conditions. By Babinet's principle, the forward diffraction lobe contains the energy equivalent of $Q_s \sim 1$. If this energy is no longer perceived as removed from the beam, the familiar geometric or ray optics limit ($Q_e \sim 1$) is attained. However, in the special case of the Voyager radio occultation experiment, the phase-coherent nature of the directly transmitted signal distinguishes it unmistakably from even infinitesimally displaced forward-scattered radiation which, arriving from a different region of the rings, has a slight, but measureable, Doppler frequency shift relative to the directly transmitted beam. Thus, for this experiment, $Q_e \sim 2$ for $x \gg 1$. This difference must be borne in mind when comparing radio occultation results with "geometric optics" results at visible wavelengths obtained without coherent detection.

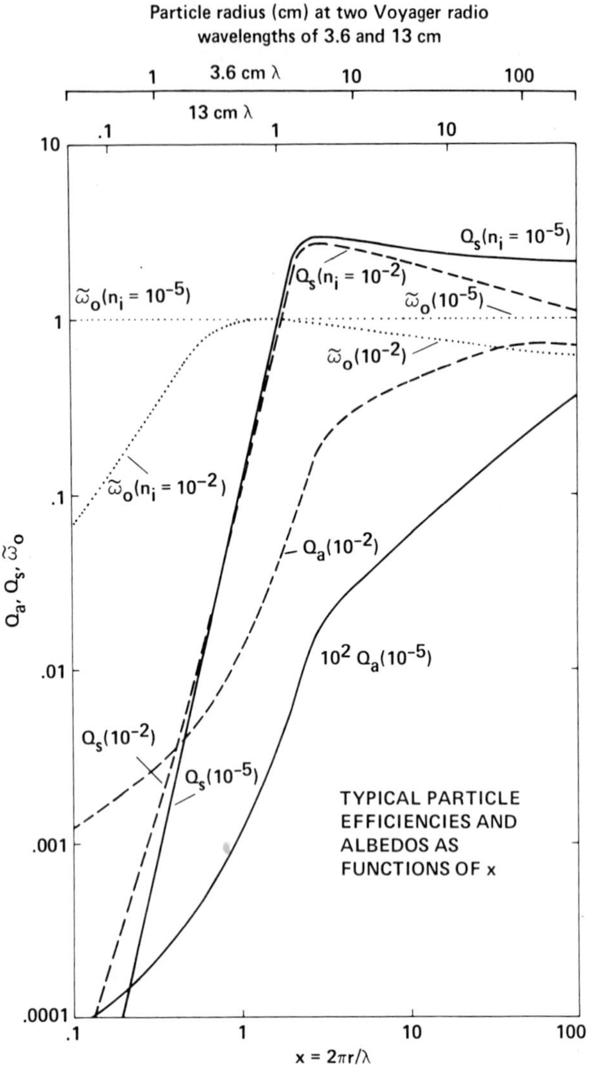

Fig. A1. Efficiency for scattering Q_s, and for absorption, Q_a, as a function of size parameter $x = 2\pi r/\lambda$. A Hansen-Hovenier (unimodal) size distribution was used with fractional width of 0.3 (Hansen and Travis 1974). Values are shown for two different imaginary refractive indices, 10^{-5} (typical of ice particles at radio wavelengths) and 10^{-2} (more absorptive particles, typical of ice-rock mixtures). Also shown are the particle albedos $\tilde{\omega}_0$ (dotted lines). Particle radii corresponding to the given values of x at the two wavelengths of the Voyager radio occultation are shown along the top.

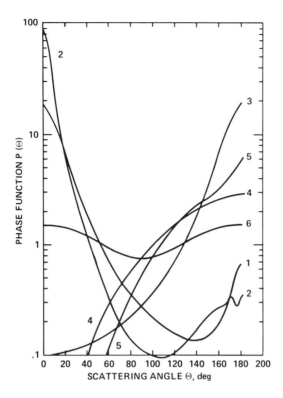

Fig. A2. Typical particle phase functions $P(\Theta)$, where Θ is the scattering angle from the incident direction. Phase functions (1) and (2), with $g = 0.7-0.9$, are representative of wavelength-sized, forward-scattering particles. Phase function (1) is a head-to-tail Henyey-Greenstein approximation, and (2) is an exact Mie-scattering calculation. Function (3) is a Henyey-Greenstein ($g = -0.3$) approximation to (4), the Lambert sphere phase function. Function (5) is the Calliso phase function, which is somewhat more sharply peaked towards backscatter than that of Lambert spheres (Pang et al. 1983a; see also Fig. 4).

In the geometric optics limit, only radiation reflected off the surface of the particle is perceived as scattered. A simple model for the phase function of macroscopic particles is obtained by assuming that their surfaces are Lambertian (i.e. perfectly diffusing [Harris 1961]). This produces a primarily backscattering phase function (see Fig. A2). Realistic surfaces such as that of the Moon are even more highly backscattering with phase functions that exhibit a narrow, intense peak at direct backscatter due to high-porosity surface structure (see Sec. II). Ring particles exhibit all of the above varieties of scattering behavior over the range of wavelengths observed. The degree of forward scattering of a particle phase function is generally expressed by its anisotropy parameter g:

$$g = \int_0^\pi P(\Theta) \cos\Theta\, d(\sin\Theta) / \int_0^\pi P(\Theta)\, d(\sin\Theta)$$

where $-1 \le g \le 1$; $g = 1$ corresponds to pure (undeviated) forward-scattering, $g = 0$ to isotropic scattering and $g = -1$ to pure backscattering. As a meaningful approximation, the Henyey-Greenstein phase function, parameterized by g, is often used (cf., Irvine 1975):

$$P_{H-G}(\Theta) = (1-g^2)/(1 + g^2 - 2g\cos\Theta)^{\frac{3}{2}}.$$

The phase function P (Θ, λ, r) is in general a phase matrix which describes the entire polarization state of the scattered radiation (see, e.g., Hansen and Travis 1974; Chandrasekhar 1960, p. 40). Very small (Rayleigh scattering) particles $(x \ll 1)$, may produce a large ($\sim 100\%$) polarization at intermediate phase angles because they scatter as dipoles. For larger particles which are smooth and spherical, each scattering produces a smaller ($\sim 10-60\%$) polarization in any direction. Because multiple scatterings further randomize scattering angles, multiply-scattered, even initially polarized, radiation is essentially depolarized after multiple scattering by spheres; this is certainly even more true of randomly oriented, irregular particles which depolarize more on single scattering (Cuzzi and Pollack 1978; Liou and Schotland 1971). However, the polarization remnant from the single scattering can contain useful information on particle shape and surface structure on the scale of a wavelength or larger.

The scattering, emissive, and absorptive properties at wavelength λ of a homogeneous layer of vertical thickness $2z_0$ containing particles of size distribution $n(r)$ are determined by the size-averaged layer optical depth $\tau(\lambda)$, phase function $P(\Theta, \lambda)$, and particle albedo $\tilde{\omega}_0$ (λ), where for example (Hansen and Travis 1974):

$$\tau(\lambda) = 2z_0 \int n(r)\, \pi r^2 Q_e(\lambda, r)dr \quad . \tag{A1}$$

In transversing a layer of optical depth τ at angle θ to the layer normal, "direct" radiation is extinguished, or reduced in intensity, by a factor $e^{-\tau/\cos\theta}$, where τ contains, when $x \gg 1$ for example, $Q_e = 1$ or 2 for incoherent or coherent radiation, respectively. In contrast, for a monolayer the extinction is proportional to $(e^{-\tau})\, C_1/\cos\theta$ where C_1 (τ, θ) is a particle blockage function (Froidevaux 1981).

Coherent, monochromatic incident radiation, as in the Voyager radar occultation and bistatic scattering experiment, allows two other measurements to be made (Marouf et al. 1982; Eshleman et al. 1977). For $x \ll 1$, individual particles are sensed primarily as an homogeneous dielectric layer of some effective refractive index which is a simple function of the number density and refractive index of the particles (see, e.g., van de Hulst, 1957, Ch. 6.). Passage of a beam of coherent radiation through such a refractive medium produces both a decrease in amplitude and a phase shift of the coherent signal. The decrease in amplitude is very small, but the phase shift "bends" the ray so that it appears to be coming from a slightly different part of the rings. Because each part of the rings has different relative velocity with respect to the spacecraft, the bent ray, although still unscattered and coherent, arrives with a different Doppler shift, or frequency offset, from the nominal signal. This difference may easily be detected with narrowband telemetry receivers, making the phase of the coherent signal a very sensitive tool for measuring $n(r)$ for $r \ll \lambda$. Large particles $(r > \lambda)$ mainly diminish the intensity of the coherent signal by scattering and absorption, and do not affect the phase significantly. Secondly, the power scattered by the particles is randomized in phase and in direction, and comes from a range of angles surrounding the incident direction. It is therefore seen Doppler-shifted into a band of frequencies because of the differences between the orbital velocities of adjacent ring regions and the velocity of the spacecraft. This "incoherent signal" is analogous to the scattered intensity of traditional (scalar) radiative transfer discussed earlier with one important exception; namely, the scattered energy is measureably Doppler shifted and this provides information on its source in the ring plane. The Voyager 1 radio occultation trajectory was designed specifically to orient Doppler contours with lines of constant orbital radius, to optimize radial resolution (Marouf et al. 1982). The incoherent, or diffusely scattered, radiation may be computed by a variety of techniques for the many-particle-thick layer (Chandrasekhar 1960; Hansen and Travis 1974; Irvine 1975). For the Voyager experiment, the scattering geometry is highly forward, allowing certain special methods to be used. In radiative transfer approaches, it is generally assumed that the particle spacing is random and sufficiently large compared to the wavelength that interference or near-field effects are negligible. This assumption may be called into question for certain microwave observations. Although coherence effects are not likely to be major for the rings, particle close-packing effects will be significant (Sec. III.C) and certainly deserve further investigation.

In the typical radiative transfer approach, the scattering layer is characterized by a diffuse reflectivity function $S[\tau, P(\Theta), \tilde{\omega}_0, \theta', \phi', \theta, \phi]$ and a diffuse transmissivity function $T[\tau, P(\Theta), \tilde{\omega}_0, \theta', \phi', \theta, \phi]$ where the parameters are defined in Fig. A3. These functions are, in general, matrices which determine the complete polarization state of the scattered radiation. We will treat them as scalars and speak only of total intensity. Then, for incident radiation of flux πF (erg cm^{-2} s^{-1}) in the plane perpendicular to the beam from direction (θ', ϕ'), the intensity (erg cm^{-2} s^{-1} str^{-1}) viewed from direction (θ, ϕ) is

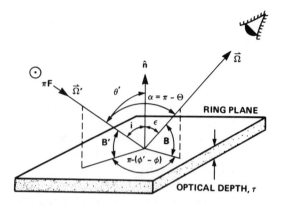

Fig. A3. Geometric parameters: πF = incident flux (W cm^{-2}), $i = \theta'$ = incidence angle, $\epsilon = \theta$ = emission angle, $\Theta \geqq$ scattering angle, α = phase angle = $\pi - \Theta$; B and B' are the usual viewer and solar elevation angles used to describe groundbased observations. The surface normal is $\hat{n}$. The azimuthal difference $\phi - \phi'$ is also shown.

$$I_{\text{scatt}}(\Theta < \pi/2) = \frac{F}{4 \cos \Theta} S'[\ldots]$$

$$\tag{A2}$$

$$I_{\text{trans}}(\Theta > \pi/2) = \frac{F}{4 \cos \Theta} T[\ldots].$$

(Below we use the standard notation $\mu = \cos \theta$, $\mu' = \cos \theta'$.) For comparison, the classical geometric albedo p of such a layer is $I/F = p = S/4\mu$, and the standard radar cross section of such a slab (the fractional area filled with perfect, isotropic reflectors; see, e.g., Muhleman [1966] and Pettengill [1978], is equal to S/μ (Cuzzi and Van Blerkom 1974). Of course, because transmitted radar power is of a single polarization and reflected power is always at least partially depolarized, the radar cross section must be measured in both orthogonal polarizations, and added to give a true measure of S (see, e.g., Ostro et al. 1980a). For incident circular polarization, smooth spheres backscatter the "orthogonal" sense O. For incident linear polarization, they reflect the "same" sense S. A combination of particle irregularity and multiple scattering causes radiation to appear in the orthogonal polarization. The linear depolarization ratio is defined as $\delta_L = p_{OL}/p_{SL}$, and the circular depolarization rate is $\delta_C = p_{SC}/p_{OC}$. The radar depolarization ratio δ is useful for constraining particle surface structure on a macroscopic scale.

For an optically thin region, or for low-albedo particles, single scattering dominates (Chandrasekhar 1960, p 172):

$$S_1(\tau, P(\Theta), \tilde{\omega}_0, \mu', \phi', \mu, \phi) = \left(\frac{\mu\mu'}{\mu + \mu'}\right) P(\mu', \phi', \mu, \phi) \times$$

$$\left[1 - \exp\left(\frac{-\tau(\mu' + \mu)}{\mu\mu'}\right)\right] \tag{A3}$$

$$T_1(\tau, P(\Theta), \tilde{\omega}_0, \mu', \phi', \mu, \phi) = \left(\frac{\mu\mu'}{\mu' + \mu}\right) P(\mu', \phi', \mu, \phi) \left[e^{-\tau/\mu'} - e^{-\tau/\mu}\right].$$

The multiple scattering contributions $(S - S_1)$ and $(T - T_1)$ vary with particle albedo and phase function, layer optical depth, and illumination and viewing geometry. Figure A4 shows the variation of $S - S_1$ with various parameters. For moderate optical depths $\tau \gtrsim 1$ and high particle albedos $\tilde{\omega}_0 \gtrsim$ 0.7, they are substantial unless the ring layer is illuminated and viewed near grazing incidence. Because $I_1 = S_1/4\mu$ is essentially independent of μ for $\tau \gtrsim$ 1 and the range of μ relevant to the rings' situation (cf. Eq. A3), it is obvious that multiple scattering in a many-particle-thick ring of moderate optical depth provides a natural explanation for the observed increase in ring brightness with increasing ring tilt angle, as in the case of the tilt effect at visible wavelengths (Secs. II and III.A). Similar arguments will eventually become useful concerning the radar tilt effect, when it becomes better characterized.

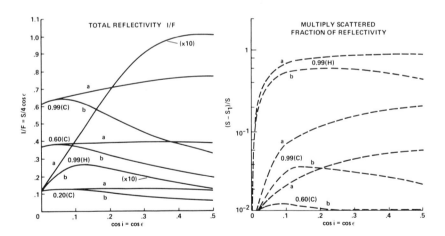

Fig. A4. Total reflectivity I/F, and its multiply-scattered fraction $(S - S_1)/S$, as a function of optical depth and elevation angle for several particle albedos and phase functions. The curves are in pairs with (a) for $\tau = 1.5$ and (b) for $\tau = 0.2$. Each pair is labeled by the particle single-scattering albedo (0.99, 0.60, 0.20) and whether the phase function is that of Callisto (C) or a forward-scattering Henyey-Greenstein (H) with $g = 0.7$. The calculations are for phase angle $\alpha = 6°$.

For a monolayer, there is less particle surface area available for high-order scatterings and they are, thus, less important. Any geometrical variation in brightness must arise either from variations in fractional area obscured, including finite-particle shadowing effects, or from surface microstructure (Hämeen-Anttila and Vaaraniemi 1975; Froidevaux 1981). These approaches have not been successful at visible wavelengths. Hämeen-Anttila and Vaaraniemi have found that the observed tilt effect may only be produced if the individual particle surfaces have zero reflectivity, like a Lambert surface, as (B', B) approach 0. This is quite unlike the behavior of any real solar system object. The Moon, for instance, has constant reflectivity as (B', B) approach zero (Harris 1961). Also, it would be impossible to produce the observed opposition effect in the surfaces of such particles. Jantunen (1982) has extended this approach to a slightly thickened inhomogeneous layer which is less dense at its upper and lower limits. This removes the Lambert surface constraint from the tilt effect, but implies a tilt-angle-dependent opposition effect which is not observed (Esposito et al. 1979; Thompson et al. 1981; Lumme et al. 1983). Therefore, it seems that the visible-wavelength observations really do require a many-particle-thick layer. This model is theoretically justified for short-wavelength observations, because the bulk of particle surface area (τ) is presented by the numerous small particles $1 \text{ cm} < r < 1 \text{ m}$, which are probably distributed in exactly such a layer (Sec. II). However, the monolayer approaches may find use in the analysis of longer-wavelength radio and radar observations which are most sensitive to the largest ring particles which may, in fact, lie in a layer of thickness only a few times their own size (see Sec. III.C; chapter by Stewart et al.).

Thermal emission from the ring particles is also an important diagnostic, as discussed in Sec. III.B. The solution to this problem consists of two parts: determining the particle physical temperature from a thermal balance, and interpreting the observed flux, or brightness temperature, which arises from summing the thermal radiation from each particle along the line of sight. An important difference between the models is that particles at optical depth τ' in a multilayer of total optical depth τ are illuminated by sunlight of reduced intensity whereas all the monolayer particles not physically blocking each other are in full sunlight. Thus, the thermal balance may be written for the monolayer:

$$\pi r^2 [(1-A) F_\odot C_1 (\tau, \mu') + \sigma T_S^4 V_S C_2 (\tau)/\pi + (f/4) \left(\frac{\epsilon \sigma T_p^4}{\pi} \right) \Omega_R]$$
$$= f \pi r^2 \epsilon \sigma T_p^4 \qquad (A4)$$

and for the multilayer:

$$\pi r^2 [(1-A) F_\odot e^{-\tau'/\mu'} + \frac{\sigma T_S^4 \Omega_S}{\pi} + H(\tau' \tau)] = f \pi r^2 \epsilon \sigma T_p^4 (\tau') \qquad (A5)$$

where r is particle radius, A is bolometric Bond albedo, T_S and Ω_S are Saturn's effective temperature and subtended solid angle, T_p and T_p (τ') are particle temperature, Ω_R is the effective solid angle subtended by ring particles, ϵ is infrared emissivity, and σ is the Stefan-Boltzmann constant. Also, f describes whether the particles radiate their energy primarily over one hemisphere ($f=2$) or over their whole surface ($f=4$). For isothermal particles, either fairly small or rapidly rotating on the scale of a thermal relaxation time, f=4. Also, C_1 (τ, μ') and C_2 (τ) are physical blockage functions (Froidevaux 1981) and $H(\tau', \tau)$ is an integral over thermal emission from particles in the layer, of the form

$$H(\tau',\tau) = 2\pi \int_0^1 \int_0^{\tau'} \frac{\tau' \, \sigma \, T_p^4(\xi)}{\pi} \, \exp \frac{-(\tau'-\xi)}{\mu} \, \frac{d\xi}{\mu} \, d\mu \, +$$
$$+ \, 2\pi \int_{-1}^0 \int_{\tau'}^\tau \frac{\sigma T_p^4(\xi)}{\pi} \, \exp \left(\frac{-(\xi-\tau')}{\mu} \right) \, \frac{d\xi}{\mu} \, d\mu \qquad \text{(A6)}$$

which may be rewritten in the form

$$H(\tau', \tau) = 2\pi \left[\int_0^{\tau'} \frac{\sigma T_p^4(\xi)}{\pi} E_1 (\tau' - \xi) d\xi + \int_{\tau'}^\tau \frac{\sigma T_p^4(\xi)}{\pi} E_1 (\xi - \tau') d\xi \right] \qquad \text{(A7)}$$

where E_1 (τ', ξ) is the first exponential integral (Kawata 1979; Mihalas 1970). The form of $H(\tau', \tau)$ also depends on whether f=2 or f=4 (Kawata 1982).

Saturn's contribution to the energetics is given by the second term on the left-hand side of Eq. A4 or A5. It is this term which establishes the lower limit to particle temperature, when the rings are edge-on. An initial estimate of this was given by Cuzzi and Van Blerkom (1974) and used by Kawata and Irvine (1975) and Kawata (1979, 1982). More careful estimates (see, e.g., Froidevaux 1981) are certainly possible. We have not included terms explicitly representing the diffusely reflected solar radiation, because the contribution is small (Aumann and Kieffer 1973); however, any exact treatment should include it (Kawata and Irvine 1975). It must also be noted that Eqs. A4 and A5 neglect the time-dependent nature of the problem; particles move vertically throughout the layer and its radiation field on a time scale comparable to the thermal radiative relaxation time. This must eventually be considered in thermal modeling of the many-particle-thick layer.

The brightness temperature T_B is the temperature producing the intensity observed at a given wavelength:

$$B(T_B, \mu) = \epsilon \int_0^\tau B\left(T_p(\tau')\right) e^{-\tau'/\mu} \, d\tau' /\mu,$$

where B is the Planck function. Assuming T_p (τ') is constant $= \bar{T}_p$, we get the familiar multilayer result

$$B(T_B, \mu) = \epsilon\, B(\bar{T}_p)\, [1 - e^{-\tau/\mu}]$$

and the somewhat similar monolayer result

$$B(T_B, \mu) = \epsilon B(\bar{T}_p)\, C_1\, (\tau, \mu)\, \frac{(1-e^{-\tau})}{\mu}.$$

The low optical depth of the C Ring (~ 0.1) explains the low *brightness* temperature for the C Ring and its increase with decreasing ring tilt (Sec. III.B), in spite of the high *physical* temperature of the C Ring particles which results from both their low albedo and their larger energy input from Saturn's thermal emission.

At far-infrared and millimeter wavelengths, the particle albedo increases, and scattering may not be neglected. The thermal emission problem then becomes much more complex (Horak 1952; Cuzzi et al. 1980). At longer microwave wavelengths ($\lambda > 1$ cm), the particle albedo is so high that emission may be neglected. The ring brightness is then determined by essentially conservative ($\tilde{\omega}_0 \sim 1$) scattering of the planetary thermal microwave emission, and the solution may be obtained using the S and T functions (Cuzzi and Van Blerkom 1974; Cuzzi et al. 1980).

REFERENCES

Alexander, A. F. O'D. 1962. *The Planet Saturn* (New York: McMillan Co.).

Atreya, S. K., and Waite, Jr., J. H. 1981. Saturn's ionosphere: Theoretical interpretation. *Nature* 292:682–683.

Aubier, M.G., Meyer-Vernet, N., and Pedersen, B. M. 1983. Shot noise from grain and particle impacts in Saturn's ring plane. *Geophys. Res. Letters* 10:5–8.

Aumann, H. H., and Kieffer, H. H. 1973. Determination of particle sizes in Saturn's rings from their eclipse cooling and heating curves. *Astrophys. J.* 186:305–311.

Barker, E., Cazes, S., Emerich, C., Vidal-Madjar, A. and Owen, T. 1980. Lyman-alpha observations in the vicinity of Saturn with Copernicus. *Astrophys. J.* 242:383–394.

Baum, W. A., Kreidl, T., Westphal, J., Danielson, G. E., Seidelmann, P. K., Pascu, D., and Currie, D. G. 1981. Saturn's E Ring. I: CCD observations of March 1980, *Icarus* 47:84–96.

Berge, G. L., and Muhleman, D. O. 1973. High-angular-resolution observations of Saturn at 21.1 cm wavelength. *Astrophys. J.* 185:373–381.

Berge, G. L., and Read, R. B. 1968. The microwave emission of Saturn. *Astrophys. J.* 152:755–764.

Bertie, J. E., Othen, D. A., and Solinas, M. 1973. The infrared spectra of ethylene oxide hydrate and hexamethylenetetramine at 100°K. In *Physics and Chemistry of Ice*, eds. E. Whalley, S. J. Jones, and L. W. Gold (Ottawa: Royal Soc. of Canada), pp. 61–65.

Bertin, G., and Mark, J.W-K. 1980. On the excitation of warps in galaxy disks. *Astron. Astrophys.* 88:289–297.

Blamont, J. 1974. The "atmosphere" of the rings of Saturn. In *The Rings of Saturn*, (Washington, D. C.: NASA SP-343), pp. 125–130.

Bobrov, M. S. 1970a. *The Rings of Saturn* (translation), NASA TTF-701.

Bobrov, M. S. 1970b. Physical properties of Saturn's rings. In *Surfaces and Interiors of Planets and Satellites*, ed. A. Dollfus (New York: Academic Press), pp. 377–458.

Borderies, N., Goldreich, P., and Tremaine, S. 1982. Sharp edges of planetary rings. *Nature* 299:209–211.

Borderies, N., Goldreich, P., and Tremaine, S. 1983. Perturbed particle disks. *Icarus* 55:124–132.

Brahic, A. 1977. Systems of colliding bodies in a gravitational field: Numerical simulation of the standard model. *Astron. Astrophys.* 54:895–907.

Brahic, A., and Sicardy, B. 1981. Apparent thickness of Saturn's rings. *Nature* 289:447–450.

Bridge, H. S., Belcher, J. W., Lazarus, A. J., Olbert, S., Sullivan, J. D., Bagenal, F., Gazis, P. R., Hartle, R. E., Ogilvie, K. W., Scudder, J. D., Sittler, E. C., Eviatar, A., Siscoe, G. L., Goertz, C. K., and Vasilyunas, V. M. 1981. Plasma observations near Saturn: Initial results from Voyager 1. *Science* 212:217–225.

Briggs, F. H. 1974. The microwave properties of Saturn's rings, *Astrophys. J.* 189:367–377.

Broadfoot, A. L., Sandel, B. R., Shemansky, D. E., Holberg, J. B., Smith, G. R., Strobel, D. F., McConnell, J. C., Kumar, S., Hunten, D. M., Atreya, S. K., Donahue, T. M., Moos, H. W., Bertaux, J. L., Blamont, J. E., Pomphrey, R. B. and Linick, S. 1981. Extreme ultraviolet observations from Voyager I encounter with Saturn, *Science* 212:206–211.

Burke, J. J., and KenKnight, C. E. 1980. An extraordinary view of Saturn's rings. *J. Geophys. Res.* 85:5925–5928.

Burns, J. A. 1976. An elementary derivation of the perturbation equations of celestial mechanics. *Amer. J. Phys.* 44:944–949.

Burns, J. A., Hamill, P., Cuzzi, J. N., and Durisen, R. H. 1979. On the "thickness" of Saturn's rings caused by satellite and solar perturbations and by planetary precession. *Astron. J.* 84:1783–1801.

Burns, J. A., Showalter, M. R.; Cuzzi, J. N., and Durisen, R. H. 1982. Saturn's electrostatic discharges (SED): Exotic ring phenomenon or just lightning? *EOS* 63:156.

Burns, J. A., Showalter, M. R.; Cuzzi, J. N.; and Durisen, R. H. 1983. Saturn's electrostatic discharges: Could lightning be the cause? *Icarus* 54:280–295.

Burns, J. A., Showalter, M. R., Cuzzi, J. N. and Pollack, J. B. 1980. Physical processes in Jupiter's ring—clues to its origin by Jove. *Icarus* 44:339–360.

Campbell, D. B., Chandler, J. F., Pettengill, G. H., and Shapiro, I. I. 1977. The Galilean satellites: 12.6 cm radar observations. *Science* 196:650–653.

Carbary, J. F., Bythrow, P. F., and Mitchell, D. G. 1982. The spokes in Saturn's rings: A new approach. *Geophys. Res. Letters* 9:420–422.

Carlson, R. W. 1980. Photo-sputtering of ice and hydrogen around Saturn's rings. *Nature* 283:461.

Chandrasekhar, S. 1960. *Radiative Transfer* (New York: Dover Publications).

Chenette, D. L., Cooper, J. F., Eraker, J. H., Pyle, K. E., and Simpson, J. A. 1980. High-energy trapped radiation penetrating the rings of Saturn. *J. Geophys. Res.* 85:5785–5792.

Cheng, A. F., and Lanzerotti, L. J. 1978. Ice sputtering by radiation belt protons and the rings of Saturn and Uranus. *J. Geophys. Res.* 83:2597–2602.

Cheng, A. F., Lanzerotti, L. J., Pirronello, V. 1982. Charged particle sputtering of ice surfaces in Saturn's magnetosphere. *J. Geophys. Res.* 87:4567–4570.

Clark, R. N., 1980. Ganymede, Europa, Callisto, and Saturn's rings—compositional analysis from reflectance spectroscopy. *Icarus* 44:388–409.

Clark, R. N., and McCord, T. B. 1980. The rings of Saturn—new near-infrared reflectance measurements and a 0.326–4.08 micron summary. *Icarus* 43:161–168.

Clarke, J. T., Moos, H. W., Atreya, S. K. and Lane, A. L. 1981. IUE detection of burst of H-Lyman alpha emission from Saturn. *Nature* 290:226–267.

Collins, S. A., Cook, II, A. F., Cuzzi, J. N., Danielson, G. A., Hunt, G. E., Johnson, T. V., Morrison, D., Owen, T., Pollack, J. B., Smith, B. A., and Terrile, R. J. 1980. First Voyager view of the rings of Saturn. *Nature* 288:439–442.

Colombo, G., Goldreich, P., and Harris, A. 1976. Spiral structure as an explanation for the asymmetric brightness of Saturn's A Ring. *Nature* 264:344–345.

Cook, A. F. II, Duxbury, T. C., and Hunt, G. E. 1979. First results on Jovian lightning. *Nature* 280:794.

Cook, A. F., Franklin, F. A. and Palluconi, F. D. 1973. Saturn's Rings—A Survey. *Icarus* 18:317–337.

Cooper, J. F., Eraker, J. H., and Simpson, J. A. 1982. Origin of the high energy protons in Saturn's inner magnetosphere: The secondary nuclear radiation from cosmic ray interactions in the main rings, Paper presented at "Saturn" Conference (abstract p. 39) Tucson, Arizona, May 1982.

Cooper, J. F., and Simpson, J. 1980. Sources of high-energy protons in Saturn's magnetosphere. *J. Geophys. Res.* 85:5793–5802.

Cuzzi, J. N. 1978. Rings of Saturn: State of current knowledge and some suggestions for future studies. In *The Saturn System* (NASA Conf. Publ. 2068).

Cuzzi, J. 1983. Physical properties of Saturn's rings. Proceedings of *I.A.U. Colloquium 75 Planetary Rings*, ed. A. Brahic, Toulouse, France, Aug. 1982.

Cuzzi, J. N., Bilski, K., Bunker, A., Collins, S. A., Danielson, G. E., Johnson, T. V., and Pollack, J. B. 1984. Photometry and spectrophotometry of Saturn's rings from Voyager. In preparation.

Cuzzi, J. N., Burns, J. A., Durisen, R. H., and Hamill, P. 1979b. The vertical structure and thickness of Saturn's rings. *Nature* 281:202–204.

Cuzzi, J. N., and Dent, W. A. 1975. Saturn's rings: The determination of their brightness temperature and opacity at centimeter wavelengths. *Astrophys. J.* 198:223–227.

Cuzzi, J. N., Durisen, R. H., Burns, J. A., and Hamill, P. 1979a. The vertical structure and thickness of Saturn's rings. *Icarus* 38:54–68.

Cuzzi, J. N., Lissauer, J. J., and Shu, F. H. 1981. Density waves in Saturn's rings. *Nature* 292:703–707.

Cuzzi, J. N., and Pollack, J. B. 1978. Saturn's rings: Particle composition and size distribution as constrained by microwave observations. I: Radar observations. *Icarus* 33:233–263.

Cuzzi, J. N., Pollack, J. B., and Summers, A. L. 1980. Saturn's rings: Particle composition and size distribution as constrained by microwave observations. II. Radio interferometric observations. *Icarus* 44:683–705.

Cuzzi, J., and Scargle, J. 1984. Encke's division has wavy edges. In preparation.

Cuzzi, J. N., and Van Blerkom, D. 1974. Microwave brightness of Saturn's rings. *Icarus* 22:149–158.

Dennefeld, M. 1974. Theoretical studies of an atmosphere around Saturn's rings. In *IAU Symp. No. 65, Exploration of the Planetary System*, eds A. Woszczyk and C. Iwaniszewski (Dordrecht Holland: D. Reidel Publ. Co.), p. 471.

dePater, I., and Dickel, J. R. 1983. New information on Saturn and its rings from VLA multifrequency data. Proceedings of *I.A.U. Colloquium 75 Planetary Rings*, ed. A. Brahic, Toulouse, France, Aug. 1982.

Dermott, S. F. 1981. The braided F ring of Saturn. *Nature* 290:454–457.

Dermott, S. F., Gold, T., and Sinclair, A. T. 1979. The rings of Uranus—nature and origin. *Astron. J.* 84:1225–1232.

Dermott, S. F., and Murray, C. D. 1980. Origin of the eccentricity gradient and the apse alignment of the epsilon ring of Uranus. *Icarus* 43:338–349.

Dermott, S. F., Murray, C. D., and Sinclair, A. T. 1980. The narrow rings of Jupiter, Saturn, and Uranus. *Nature* 284:309–313.

Desch, M. D., and Kaiser, M. L. 1981. Voyager measurements of the rotation period of Saturn's magnetic field, *Geophys. Res. Letters* 8:253–256.

Dollfus, A. 1979a. Optical reflectance polarimetry of Saturn's globe and rings. *Icarus* 37:404–419.

Dollfus, A. 1979b. Optical reflectance polarimetry of Saturn's globe and rings. *Icarus* 40:171–179.

Dollfus, A., and Brunier, S. 1982. Observation and photometry of an outer ring of Saturn. *Icarus* 49:194–204.

Eplee, E., and Smith, B. A. 1983. Reflective measurements and radial growth of an anomalous spoke in Saturn's rings. *Bull. Amer. Astron. Soc.* 15:814 (abstract).

Epstein, E. E., Janssen, M. A., and Cuzzi, J. N. 1984. Saturn's rings: 3 mm low-inclination observations and derived properties. *Icarus.* In press.

Epstein, E. E., Janssen, M. A., Cuzzi, J. N., Fogarty, W. G., and Mottmann, J. 1980. Saturn's rings—3 mm observations and derived properties, *Icarus* 41:103–118.

Eshleman, V. R., Breakwell, J. V., Tyler, G. L., and Marouf, E. A. 1983. W-shaped occultation signatures: Inference of entwined particle orbits in charged planetary ringlets. *Icarus* 54:212–226.

Eshleman, V. R., Tyler, G. L., Anderson, J. D., Fjeldbo, G., Levy, G. S., Wood, G. E., and Croft, T. A. 1977. Radio science investigations with Voyager. *Sp. Sci. Revs.* 21:207–232.

Esposito, L. W. 1979. Extensions to the classical calculation of the effect of mutual shadowing in diffuse reflection—applied to Saturn's rings. *Icarus* 39:69–80.

Esposito, L. W., Borderies, N., Cuzzi, J. N., Goldreich, P., Holberg, J. B., Lane, A. L., Lissauer, J. J., Marouf, E. A., Pomphrey, R. B., Terrile, R. J., and Tyler, G. L. 1983a. Voyager observations of an eccentric ringlet in Saturn's C ring *Science* 222:57–60.

Esposito, L. W., Cuzzi, J. N., Holberg, J. H., Marouf, E. A., Tyler, G. L., and Porco, C. C. 1984. Saturn's rings: Structure, dynamics, and particle properties. In *Saturn*, eds. T. Gehrels and M. S. Matthews (Tucson: University of Arizona Press).

Esposito, L. W., Dilley, J. P., and Fountain, J. W. 1980. Photometry and polarimetry of Saturn's rings from Pioneer 11. *J. Geophys. Res.* 85:5948–5956.

Esposito, L. W., Lumme, K., Benton, W. D., Martin, L. J., Ferguson, H. M., Thompson, D. T., and Jones, S. E. 1979. International planetary patrol observations of Saturn's rings. *Astron. J.* 84:1408–1415.

Esposito, L. W., O'Callahan, M., and West, R. A. 1982. Spiral density waves as a probe of the ring structure. Paper presented at "Saturn" Conference, Tucson, Arizona, May 1982.

Esposito, L., O'Callahan, M., and West, R. A. 1983b. The structure of Saturn's rings: Implications from the Voyager stellar occultation. *Icarus*. In press.

Evans, D. R., Romig, J. H., Hord, C. W., Simmons, K. Warwick, J. W., and Lane, A. L. 1982. The source of Saturn electrostatic discharges. *Nature* 299:236–237.

Evans, D. R., Romig, J. H., and Warwick, J. W. 1983. Saturn's electrostatic discharges: Properties and theoretical considerations. *Icarus* 53:267–279.

Evans, D. R., Warwick, J. W., Pearce, J. B., Carr, T. D., and Schauble, J. J. 1981. Impulsive radio discharges near Saturn. *Nature* 292:716–718.

Fechtig, H., Grün, E., and Morfill, G. 1979. Micrometeoroids within 10 Earth radii. *Planet. Sp. Sci.* 27:511–531.

Feibelman, W. A. 1967. Concerning the "D" ring of Saturn. *Nature* 214:793–794.

Fillius, W., and McIlwain, C. 1980. Very energetic protons in Saturn's radiation belt. *J. Geophys. Res.* 85:5803–5811.

Focas, J. H., and Dollfus, A. 1969. Propriétés optiques et épaisseur des anneaux de Saturne observés par la tranche en 1966. *Astronomy Astrophys.* 2:251–265.

Frank, L. A., Burek, B. G., Ackerson, K. L., Wolfe, J. H., and Mihalov, J. D. 1980. Plasmas in Saturn's magnetosphere. *J. Geophys. Res.* 85:5695–5708.

Franklin, F. A., and Colombo, G. 1970. A dynamical model for the radial structure of Saturn's rings. *Icarus* 12:338–347.

Franklin, F. A., and Colombo, G. 1978. On the azimuthal brightness variations of Saturn's rings. *Icarus* 33:279–287.

Franklin, F. A., Colombo, G., and Cook, A. F. 1971. A dynamical model for the radial structure of Saturn's rings. *Icarus* 15:80–92.

Franklin, F. A., Colombo, G., and Cook, A. F. II. 1982. A possible link between the rotation of Saturn and its ring structure. *Nature* 245:128–130.

Franklin, F. A., and Cook, A. F. 1965. Optical properties of Saturn's rings. II: Two-color phase curves of the two bright rings. *Astron. J.* 70:704–720.

Freedman, A. P., Tremaine, S., and Elliot, J. L. 1983. Weak dynamical effects in the Uranian ring system, *Astron. J.* 88:1053–1059.

French, R. G., Elliot, J. L., and Allen, D. A. 1982. Inclinations of the Uranian rings. *Nature* 298:827–829.

Froidevaux, L. 1981. Saturn's rings—infrared brightness variation with solar elevation. *Icarus* 46:4–17.

Froidevaux, L., and Ingersoll, A. P. 1980. Temperatures and optical depths of Saturn's rings and a brightness temperature for Titan. *J. Geophys. Res.* 85:5929–5936.

Froidevaux, L., Matthews, K., and Neugebauer, G. 1981 Thermal response of Saturn's ring particles during and after eclipse. *Icarus* 46:18–26.

Gehrels, T., Baker, L. R., Beshore, E., Blenman, C., Burke, J. J., Castillo, N. D., DaCosta, B., Degewij, J., Doose, L. R., Fountain, J. W., Gotobed, J., KenKnight, C. E., Kingston, R., McLaughlin, G., McMillan, R., Murphy, R., Smith, P. H., Stoll, C. P., Strickland, R. N., Tomasko, M. G., Wijesinghe, M. P., Coffeen, D. L., and Esposito, L. 1980. Imaging photopolarimeter on Pioneer Saturn. *Science* 207:434–439.

Goertz, C., and Morfill, G. 1983. A model for the formation of spokes in Saturn's ring. *Icarus* 53:219–229.

Goldreich, P. 1982. Solved and unsolved problems in ring dynamics. Proceedings of *IAU Colloquium 75 Planetary Rings*, ed. A. Brahic, Toulouse, France, Aug. 1982.

Goldreich, P., and Tremaine, S. 1978a. The velocity dispersion in Saturn's rings. *Icarus* 34:227–239.

Goldreich, P., and Tremaine, S. 1978b. The formation of the Cassini Division in Saturn's rings. *Icarus* 34:240–253.

Goldreich, P., and Tremaine, S. 1979a. The excitation of density waves at the Lindblad and corotation resonances by an external potential. *Astrophys. J.* 233:857–871.

Goldreich, P., and Tremaine, S. 1979b. Precession of the epsilon ring of Uranus. *Astron. J.* 84:1638–1641.

Goldreich, P., and Tremaine, 1979c. Towards a theory for the Uranian rings. *Nature* 277:97–99.

Goldreich, P., and Tremaine, S. 1980. Disk-satellite interactions. *Astrophys. J.* 241:425–441.

Goldreich, P., and Tremaine, S. 1981. The origin of the eccentricities of the rings of Uranus, *Astrophys. J.* 243:1062–1075.

Goldreich, P., and Tremaine, S. 1982. The dynamics of planetary rings. *Ann. Rev. Astron. Astrophys.* 20:249–283.

Goldsborough, G. R. 1921. *Phil. Trans. Roy. Soc. London* 222:101.

Goldstein, R. M., Green, R. R., Pettengill, G. H., and Campbell, D. B. 1977. The rings of Saturn: Two-frequency radar observations. *Icarus* 30:104–110.

Goldstein, R. M., and Jurgens, R. 1982. Radar observations of the rings of Saturn. Paper presented at "Saturn" Conference, Tucson, Arizona, May 1982.

Goldstein, R. M., and Morris, G. A. 1973. Radar observations of the rings of Saturn. *Icarus* 20:260–262.

Greenberg, R. 1983. The role of dissipation in shepherding of ring particles. *Icarus* 53:207–218.

Grosskreutz, C. 1982. Distribution of energetic electrons (0.040 < E_e < 21 Mev) in Saturn's inner magnetosphere. Master's thesis, Dept. of Physics and Astronomy, University of Iowa.

Grün, E., Morfill, G., Schwehm, G., and Johnson, T. V. 1980. A model for the origin of the Jovian ring, *Icarus* 44:326–338.

Grün, E., Morfill, G. E., Terrile, R. J., Johnson, T. V., Schwehm, G. 1983. The evolution of spokes in Saturn's B ring, *Icarus* 54:227–252.

Gurnett, D. A., Grün, E., Gallagher, D., Kurth, W. S., and Scarf, F. L. 1983. Micron-sized particles detected near Saturn by the Voyager plasma-wave experiment. *Icarus* 53:236–254.

Gurnett, D, A., Kurth, W. S., and Scarf, F. L. 1981. Plasma waves near Saturn: Initial results from Voyager 1. *Science* 212:235–239.

Gurnett, D. A., Shaw, R. R., Anderson, R. R., Kurth, W. S., and Scarf, F. L. 1979. Whistlers observed by Voyager 1: Detection of lightning on Jupiter. *Geophys. Res. Letters* 6:511–514.

Hämeen-Anttila, K. A. 1978. An improved and generalized theory for the collisional evolution of Keplerian systems. *Astrophys. Sp. Sci.* 58:477–520.

Hämeen-Anttila, K. A. 1981. Quasi-equilibrium in collisional systems. *Moon Planets* 25:477–506.

Hämeen-Anttila, K. A., and Vaaraniemi, P. 1975. A theoretical photometric function of Saturn's rings. *Icarus* 25:470–478.

Hanel, R., Conrath, B., Flasar, F. M., Kunde, V., Maguire, W., Pearl, J., Pirraglia, J., Samuelson, R., Cruikshank, D., Gautier, D., Gierasch, P., Horn, L.,and Ponnamperuma. C. 1982. Infrared observations of the Saturnian system from Voyager 2. *Science*

215:544–548.

Hanel, R., Conrath, B., Flasar, F. M., Kunde, V., Maguire, W., Pearl, J., Pirraglia, J., Samuelson, R., Herath, L., Allison, M., Cruikshank, D., Gautier, D., Gierasch, P., Horn, L., Koppany, R., and Ponnamperuma, C. 1981. Infrared observations of the Saturnian system from Voyager 1 *Science* 212:192–200.

Hansen, J. E., and Travis, L. D. 1974. Light scattering in planetary atmospheres, *Space Sci. Rev.* 16:527–610.

Hapke, B. 1963. A theoretical photometric function for the lunar surface. *J. Geophys. Res.* 68:4571–4586.

Hapke, B. 1981. Bidirectional reflectance spectroscopy. I. Theory. *J. Geophys. Res.* 86-B4:3039–3054.

Harris, D. 1961. Photometry of planets and satellites. In *The Solar System, Vol. III*, eds. G. Kuiper and B. Middlehurst (Chicago: University of Chicago Press).

Hartmann, W. K. 1969. Terrestrial, lunar, and interplanetary rock fragmentation. *Icarus*. 10:201–213.

Hénon, M. 1981. A simple model of Saturn's rings. *Nature* 293:33–35.

Hénon, M. 1983. A simple model of Saturn's rings, revisited. Proceedings of *IAU Colloquium 75 Planetary Rings*, ed. A. Brahic, Toulouse, France, Aug. 1982.

Hill, J. R., and Mendis, D. A. 1979. Charged dust in the outer planetary magnetospheres. Physical and dynamical processes. *Moon Planets* 21:3–16.

Hill, J. R., and Mendis, D. A. 1981*a*. Charged dust in the outer planetary magnetospheres. 2. Trajectories and spatial distribution. *Moon Planets* 23:53–71.

Hill, J. R., and Mendis, D. A. 1981*b*. On the origin of striae in cometary dust tails. *Astrophys. J.* 242:395–401.

Hill, J. R., and Mendis, D. A. 1982*a*. The dynamical evolution of the Saturnian ring spokes. *J. Geophys. Res.* 87:7413–7420.

Hill, J. R., and Mendis, D. A. 1982*b*. On the dust ring current of Saturn's F ring. *Geophys. Res. Letters* 9:1069–1071.

Holberg, J. B. 1982. Identification of 1980S27 and 1980S26 resonances in Saturn's A Ring, *Astron. J.* 87:1416–1422.

Holberg, J. B., Forester, W., and Lissauer, J. J. 1982. Identification of resonance features within the rings of Saturn. *Nature* 297:115–120.

Horak, H. G. 1952. The transfer of radiation by an emitting atmosphere. *Astrophys. J.* 116:477–490.

Humes, D. H., Oneal, R. L., Kinard, W. H., and Alvarez, J. M. 1980. Impact of Saturn ring particles on Pioneer II. *Science* 207:443–444.

Hunter, C., and Toomre, A. 1969. Dynamics of the bending of the galaxy. *Astrophys. J.* 155:747–776.

Ingersoll, A. P., Neugebauer, G., Orton, G. S., Munch, G., and Chase, S. C. 1980. Pioneer Saturn infrared radiometer-preliminary results. *Science* 207:439–443.

Ip, W-H. 1978. On the Lyman-alpha emission from the vicinity of Saturn's rings. *Astron. Astrophys.* 7:435–437.

Ip, W-H. 1980*a*. Physical studies of the planetary rings. *Space Sci. Rev.* 26:36–96.

Ip, W-H. 1980*b*. New progress in the physical studies of the planetary rings. *Space Sci. Rev.* 26:97–109.

Ip, W-H. 1980*c*. Discussion of the Pioneer 11 observations of the F ring of Saturn. *Nature* 287:126–128.

Ip, W-H. 1983. On planetary rings as sources and sinks of magnetospheric plasma. Proceedings of *I.A.U. Colloquium 75 Planetary Rings*, ed. A. Brahic, Toulouse, France, Aug. 1982.

Irvine, W. M. 1966. The shadowing effect in diffuse reflection. *J. Geophys. Res.* 71:2931–2937.

Irvine, W. M. 1975. Multiple scattering in planetary atmospheres. *Icarus* 25:175–204.

Jantunen, H. 1982. The photometric function for Saturn's rings, *Moon Planets* 26:383–387.

Jeffreys, H. 1947*a*. The effects of collsions on Saturn's rings. *Mon. Not. Roy. Astron. Soc.* 107:263–267.

Jeffreys, H. 1947*b*. The relation of cohesion to Roche's limit. *Mon. Not. Roy. Astron. Soc.* 107:260–262.

Johnson, P. E., Kemp, J. C., King, R., Parker, T. E., and Barbour, M. S. 1980. New results from optical polarimetry of Saturn's rings, *Nature* 283:146–149.

Johnson, T. V., and Pilcher, C. P. 1977. Satellite spectrophotometry and surface compositions. In *Planetary Satellites,* ed. J. Burns (Tucson: University of Arizona Press), pp. 232–268.

Judge, D. L., Wu, F. M., and Carlson, R. W. 1980. Ultraviolet photometer observations of the Saturnian system, *Science* 207:431–434.

Julian, W., and Toomre, A. 1966. Non-axisymmetric responses of differentially rotating disks of stars, *Astrophys. J.* 146:810–830.

Kaiser, M. L., Connerney, J. E. P., and Desch, M. D. 1983. Atmospheric storm explanation of Saturnian electrostatic discharges. *Nature* 303:50–53.

Kaiser, M. L., Desch, M. D., and Lecacheux, A. 1981. Saturnian kilometric radiation: Statistical properties and beam geometry. *Nature* 292:731–733.

Kaiser, M. L., Desch, M. D., Kurth, W. S., Lecacheux, A., Genova, F., and Petersen, B. M. 1984. Saturn as a radio source. In *Saturn,* eds. T. Gehrels, and M. S. Matthews (Tucson: University of Arizona Press).

Kawata, Y. 1979. The infrared brightness temperature of Saturn's rings based on the multiparticle layer assumption. In *Proc. 12th Lunar and Planetary Symposium,* Inst. of Space and Aeronautical Sci., Tokyo, Japan, July 1979.

Kawata, Y. 1982. Thermal energy balance of Saturn's rings. *Proc. 15th ISAS Lunar and Planetary Symposium,* pp. 8–18. Inst. of Space and Astronautical Sci., Tokyo, Japan, July 1982. Also submitted to *Icarus.*

Kawata, Y., and Irvine, W. M. 1974. Models of Saturn's rings which satisfy the optical observations. In *Exploration of the Planetary System,* eds. A. Woszczyk and C. Iwaniszewska (Dordrecht Holland: D. Reidel Publ. Co.), pp. 441–464.

Kawata, Y., and Irvine, W. M. 1975. Thermal emission from a multiply scattering model of Saturn's rings. *Icarus* 24:472–482.

Kemp, J. C., and Murphy, R. E. 1973. The linear polarization and transparency of Saturn's rings, *Astrophys. J.* 186:679–686.

Kerker, M. 1969. *The Scattering of Light and Other Electromagnetic Radiation* (New York: Academic Press).

Kliore, A., Patel, I. R., Lindal, G. F., Sweetnam, D. N., Hotz, H. B., Waite, Jr., J. H., and McDonough, T. R. 1980. Structure of the ionosphere and atmosphere of Saturn from Pioneer 11 Saturn radio occultation. *J. Geophys. Res.* 85:5857–5870.

Kuiper, G. P., Cruikshank, D. P., and Fink, U. 1970. The composition of Saturn's rings. *Sky Telescope* 39:14.

Kurth, W. S., Gurnett, D. A., and Scarf, F. L. 1983. A search for Saturn electrostatic discharges in the Voyager plasma wave data. *Icarus* 53:255–261.

Lamy, P., and Mauron, N. 1981. Observations of Saturn's outer ring and new satellites during the 1980 edge-on presentation. *Icarus* 46:181–186.

Lane, A. L., Graps, A. L., and Simmons, K. E. 1983. The C ring of Saturn: A high-resolution view of some of its structure. Proceedings of *I.A.U. Colloquium 75 Planetary Rings,* ed. A. Brahic, Toulouse, France, Aug. 1982.

Lane, A. L., Hord, C. W., West, R. A., Esposito, L. W., Coffeen, D. L., Sato, M., Simmons, K. E., Pomphrey, R. B., and Morris, R. B. 1982. Photopolarimetry from Voyager 2: Preliminary results on Saturn, Titan, and the rings. *Science* 215:537–543.

Laplace, P. S., Marquis de 1802. *Mechanique Celeste,* translated by N. Bowditch (Boston, 1832: Hilliard, Gray, Little, and Wilkins), Vol. III, Sec. 46, p. 512.

Larson, S. M. 1983. Observations of Saturn's E ring. Proceedings of *I.A.U. Colloquium 75 Planetary Rings,* ed. A. Brahic, Toulouse, France, Aug. 1982.

Larson, S. M., Fountain, J. W., Smith, B. A., and Reitsema, H. J. 1981. Observations of the Saturn E ring and a new satellite. *Icarus* 47:288–290.

Lebofsky, L., and Fegley, M. B. Jr., 1976. Chemical composition of icy satellites and Saturn's rings. *Icarus* 28:379–388.

Lebofsky, L. A., Johnson, T. V., and McCord, T. B. 1970. Saturn's Rings: Spectral reflectivity and compositional implications. *Icarus* 13:226–230.

Lin, D. N. C., and Bodenheimer, P. 1981. On the stability of Saturn's rings. *Astrophys. J.* 248:L83–L86.

Lin, D. N. C., and Papaloizou, J. 1979. Tidal torques on accretion disks in binary systems with extreme mass ratios. *Mon. Not. Roy. Astron. Soc.* 186:799–812.

Lin, C. C., and Shu, F. H. 1964. On the spiral structure of disk galaxies. *Astrophys. J.* 140:646–655.

Liou, K., and Schotland, R. M. 1971. Multiple backscattering and depolarization from water clouds for a pulsed lidar system. *J. Atmos. Sci.* 28:772–784.

Lissauer, J. 1982. Dynamics of Saturn's rings, unpublished thesis, Dept. of Mathematics and Astronomy, University of California, Berkeley.

Lissauer, J. J., and Cuzzi, J. N. 1982. Resonances in Saturn's rings, *Astron. J.* 87:1051–1058.

Lissauer, J. J., Peale, S. J., and Cuzzi, J. N. 1983a. Ring torque on Janus and the tidal heating of Enceladus. *Icarus.* In press.

Lissauer, J. J., Shu, F.H., and Cuzzi, J. N. 1981. Moonlets in Saturn's rings? *Nature* 292:707–711.

Lissauer, J. J., Shu, F. H., and Cuzzi, J. N. 1983b. Viscosity in Saturn's rings. Proceedings of *I.A.U. Colloquium 75 Planetary Rings,* ed. A. Brahic, Toulouse, France, Aug. 1982.

Lukkari, J. 1981. Computer simulations of self-focusing particle streams. *Astrophys. Sp. Sci.* 61:111–120.

Lumme, K., Esposito, L. W., Irvine, W. M., and Baum, W. A. 1977. Azimuthal brightness variations of Saturn's rings. II. Observations at an intermediate tilt angle. *Astrophys. J.* 216:L123–L126.

Lumme, K., and Irvine, W. M. 1976a. Azimuthal brightness variations of Saturn's rings. *Astrophys. J.* 204:L55–L57.

Lumme, K., and Irvine, W. M. 1976b. Photometry of Saturn's rings. *Astron. J.* 81:865–893.

Lumme, K., and Irvine, W. M. 1979. Low tilt angle photometry and the thickness of Saturn's rings. *Astron. Astrophys.* 71:123–130.

Lumme, K. A., Irvine, W. M., and Esposito, L. W. 1983. Theoretical interpretation of the ground-based photometry of Saturn's B ring. *Icarus* 53:174–184.

Marouf, E. A., Holberg, J. B., and Cuzzi, J. N. 1984. Irregular structure in Saturn's B ring: Particle size variation. In preparation.

Marouf, E. A., and Tyler, G. L. 1982. Microwave edge diffraction by features in Saturn's rings observations with Voyager 1. *Science* 217:243–245.

Marouf, E. A., and Tyler, G. L. 1984a. Radial structure of Saturn's rings from diffraction-removed Voyager 1 radio occultation observations. In preparation.

Marouf, E. A., and Tyler, G. L. 1984b. Removal of diffraction effects from Voyager 1 radio occultation data. In preparation.

Marouf, E. A., Tyler, G. L., and Eshleman, V. R. 1982. Theory of radio occultation by Saturn's rings. *Icarus* 49:161–193.

Marouf, E. A., Tyler, G. L., Zebker, H. A., and Eshleman, V. R. 1983. Particle size distributions in Saturn's rings from Voyager 1 radio occultation. *Icarus* 54:189–211.

Maxwell, J. C. 1859. *On the Stability of the Motions of Saturn's Rings* Cambridge and London: MacMillan and Co.), Also reprinted in *Scientific Papers of James Clerk Maxwell* (Cambridge Univ. Press 1890), Vol. 1.

Mendis, D. A., and Axford, W. I. 1974. Satellites and magnetospheres of the outer planets. *Ann. Rev. Earth Planet. Sci.* 2:419–474.

Mendis, D. A., Hill, J. R., Houpis, H. L. F., and Whipple, Jr., C. E. 1981. On the electrostatic charging of the cometary nucleus, *Astrophys, J.* 249:787–797.

Mendis, D. A., Houpis, L. F. and Hill, J. R. 1982. The gravito-dynamics of charged dust in planetary magnetospheres. *J. Geophys. Res.* 87:3449–3455.

Meyer-Vernet, N. 1982. "Flip-flop" of electric potential of dust grains in space, *Astron. Astrophys.* 105:98–106.

Mihalas, D. 1970. *Stellar Atmospheres.* (San Francisco: W. H. Freeman and Co.), pp. 21–23.

Mishima, O., Klug, D. D., and Whalley, E. 1983. The far infrared spectrum of ice *Ih* in the range 8-25 cm^{-1}; Sound waves and difference bands, with application to Saturn's rings. *J. Chem. Phys.* 78:6399–6404.

Morfill, G. E., Fechtig, H., Grün, E., and Goertz, C. K. 1983. Some consequences of meteoroid impacts on Saturn's rings. *Icarus* 55:439–447.

Morfill, G., and Goertz, C. 1983. Meteoroid production of plasma clouds in Saturn's rings. *Icarus* 55:111–123.

Morfill, G. E., Grün, E., and Johnson, T. V. 1980a. Dust in Jupiter's magnetosphere: Physical processes. *Planet Space Sci.* 28:1087–1100.

Morfill, G. E., Grün, E., and Johnson, T. V. 1980b. Dust in Jupiter's magnetosphere: Origin of the ring. *Planet Space Sci.* 28:1101–1110.

Morfill, G. E., Grün, E., and Johnson, T. V. 1980c. Dust in Jupiter's magnetosphere: Time variations. *Planet Space Sci.* 28:1111–1114.

Morfill, G. E., Grün, E., and Johnson, T. V. 1980d. Dust in Jupiter's magnetosphere: Effect of magnetospheric electrons and ions. *Planet Space Sci.* 28:1115–1123.

Morrison, D., 1977. Radiometry of satellites and the rings of Saturn. In *Planetary Satellites*, ed. J. Burns, (Tucson: University of Arizona Press).

Muhleman, D. O. 1966. Planetary characteristics from radar observations. *Space Sci. Rev.* 6:341–364.

Muhleman, D., and Berge, G. 1983. Microwave emission from Saturn's rings. Proceedings of *I.A.U. Colloquium 75 Planetary Rings*, ed. A. Brahic, Toulouse, France, Aug. 1982.

Mukai, T. 1981. On the charge distribution of interplanetary grains. *Astron. Astrophys.* 99:1–6.

Müller, G. 1893. Helligkeitsbestimmungen der grossen Planeten und einiger Asteroiden, *Pub. d. Astrophys. Obs. zu Potsdam* 8:193–389.

Nolt, I. G., Caldwell, J., Radostitz, J. V., Tokunaga, A. T., Barrett, E. W., Gillett, F. C., and Murphy, R. E. 1980. IR brightness and eclipse cooling of Saturn's rings. *Nature* 283:842–843.

Northrop, T. B., and Hill, J. R. 1982. Stability of negatively charged dust grains in Saturn's ring plane. *J. Geophys. Res.* 87:6045—6051.

Northrop, T. G., and Hill, J. R., 1983. The adiabatic motion of charged dust grains on rotating magnetospheres. *J. Geophys. Res.* 88:1–11.

Null, G. W., Lau, E. L., Biller, E. D., and Anderson, J. D. 1981. Saturn gravity results obtained from Pioneer 11 tracking data and Earth-based Saturn satellite data. *Astron. J.* 86:456–468.

Öpik, E. J. 1956. Interplanetary dust and terrestrial accretion of meteoritic matter. *Irish Astron. J.* 4:84–135.

Osterbrock, D. E., and Cruikshank, D. P. 1983. J. E. Keeler's discovery of a gap in the outer part of the A ring, *Icarus* 53:165–173.

Ostro, S. J., Campbell, D. B.; Pettengill, G. H., and Shapiro, I. I. 1980b. Radar observations of the icy Galilean satellites. *Icarus* 44:431–440.

Ostro, S., and Pettengill, G. 1983. A review of radar observations of Saturn's rings. Proceeding of *I.A.U. Colloquium 75 Planetary Rings*, ed. A. Brahic, Toulouse, France, Aug. 1982.

Ostro, S. J., Pettengill, G. H., and Campbell, D. B. 1980a. Radar observations of Saturn's rings at intermediate tilt angles. *Icarus* 41:381–388.

Ostro, S. J., Pettengill, G. H., Campbell, D. B., and Goldstein, R. M. 1982. Delay-Doppler radar observations of Saturn's rings. *Icarus* 49:367–381.

Pang, K. D., Lumme, K., Bowell, E., and Ajello, J. M. 1983b. Interpretation of integrated-disk photometry of Calliso and Ganymede. *J. Geophys. Res.* 88:A569–A576.

Pang, K., Voge, C., and Rhoads, J. 1983a. Macrostructure and microphysics of Saturn's E ring; Proceedings of *I.A.U. Colloquium 75 Planetary Rings*, ed. A. Brahic, Toulouse, France, Aug. 1982.

Peale, S. 1976. Orbital resonances in the solar system. *Ann. Rev. Astron. Astrophys.* 14:215–246.

Pettengill, G. H., 1978. Physical properties of the planets and satellites from radar observations, *Ann. Rev. Astron. Astrophys.* 16:265–292.

Pettit, E. 1961. Planetary temperature measurements. In *The Solar System, Vol. III*, eds. G. Kuiper and B. Middlehurst (Chicago: University of Chicago Press).

Peutter, R. C., and Russell, R. W. 1977. The 2-4 micron spectrum of Saturn's rings. *Icarus* 32:37–40.

Pilcher, C. B., Chapman, C. R., Lebofsky, L. A., and Kieffer, H. H. 1970. Saturn's rings: Identification of water frost. *Science* 167:1372–1373.

Pollack, J. B. 1975. The rings of Saturn, *Space Sci. Rev.* 18:3–93.

Pollack, J. B. 1981. Phase curve and particle properties of Saturn's F ring. *Bull. Amer. Astron. Soc.* 13:727 (abstract).

Pollack, J. B., and Consolmagno, G. 1984. Origin and evolution of the Saturn system. In *Saturn*, eds. T. Gehrels, and M. S. Matthews (Tucson: University of Arizona Press).

Pollack, J. B., and Cuzzi, J. N. 1980. Scattering by nonspherical particles of size comparable to a wavelength: A new semi-empirical theory and its applications to tropospheric aerosols. *J. Atmos. Sci.* 37:868–881.

Pollack, J. B., Grossman, A. S., Moore, R., and Graboske, H. C. Jr. 1976. The formation of Saturn's satellites and rings as influenced by Saturn's contraction history. *Icarus* 29:35–48.

Pollack, J. B., Summers, A. L. and Baldwin, B. 1973. Estimates of the size of the particles in the rings of Saturn and their cosmogonic implications. *Icarus* 20:263–278.

Porco, C. 1983. Voyager Observations of Saturn's Rings. 1. The eccentric rings at 1.29, 1.45, 1.95 and 2.27 R_S. 2. The periodic variation of spokes. Unpublished thesis, Calif. Inst. of Technology, Pasadena (Part 1 submitted to *Icarus*).

Porco, C. C., and Danielson, G. E. 1982. The periodic variation of spokes in Saturn's rings. *Astron. J.* 87:826–833.

Prokopenko, S. M. L., and Laframboise, J. G. 1980. High voltage differential charging of geostationary spacecraft, *J. Geophys. Res.* 85:4125–4131.

Reitsema, H. J., Beebe, R. F., and Smith, B. A. 1976. Azimuthal brightness variations in Saturn's rings. *Astron. J.* 81:209–215.

Rieke, G. H. 1975. The thermal radiation of Saturn and its rings. *Icarus* 26:37–44.

Safronov, V. S. 1972. *Evolution of the Proplanetary Cloud and Formation of the Earth and Planets.* (Translated from Russian), NASA TTF-677.

Sandel, B. R., and Broadfoot, A. L. 1981. Morphology of Saturn's aurora. *Nature* 292:679–682.

Sandel, B. R., Shemansky, D. E., Broadfoot, A. L., Holberg, J. B. Smith, G. R., McConnell, J. C., Strobel, D. F., Atreya, S. K. Donahue, T. M., Moos, H. W., Hunten, D. M., Pomphrey, R. B., and Linick, S. 1982. Extreme ultraviolet observations from the Voyager 2 encounter with Saturn. *Science* 215:548–553.

Schloerb, F. P., Muhleman, D. O., and Berge, G. L. 1979. Interferometric observations of Saturn and its rings at a wavelength of 3.71 cm. *Icarus* 39:214–231.

Schloerb, F. P., Muhleman, D. O., and Berge, G. L. 1979. An aperture synthesis study of Saturn and its rings at 3.71 cm wavelength. *Icarus* 39:232–250.

Schloerb, F. P., Muhleman, D. O., and Berge, G. L. 1980. Interferometry of Saturn and its rings at 1.30 cm wavelength. *Icarus* 42:125–135.

Seeliger, H. von. 1887. Zur Theorie der Beleuchtung der grossen Planeten, insbesondere des Saturn. *Abhandl. Bayer. Akad. Wiss.* Kl. II 16:405–516.

Seeliger, H. von. 1895. Theorie der Beleuchtung staubformigen kosmischen Masses insbesondere des Saturn-ringes. *Abhandl. Bayer. Akad. Wiss.* Kl. II 18:1–72.

Shemansky, D. E., and Ajello, J. M. 1983. The Saturn spectrum in the EUV: Electron-excited hydrogen. *J. Geophys. Res.* 88:459–464.

Shemansky, D. E., and Smith, G. R. 1982. Whence comes the "Titan" hydrogen torus? *EOS* 63:1019.

Shoemaker, E. 1984. The Oort cloud of comets and bombardment of the planets and satellites. To be submitted to *Protostars and Planets. II.* eds. D. C. Black and M. S. Matthews (Tucson: Univ. of Arizona Press).

Showalter, M. R. 1982. Effects of shepherd conjunctions on the F ring of Saturn. Proceedings of *I.A.U. Colloquium 75 Planetary Rings* ed. A. Brahic, Toulouse, France, Aug. 1982.

Showalter, M. R., and Burns, J. A. 1982. A numerical study of Saturn's F ring, *Icarus* 52:526–544.

Shu, F. H., Cuzzi, J. N., and Lissauer, J. J. 1983. Bending waves in Saturn's rings. *Icarus* 53:185–206.

Sicardy, B., Lecacheux, J., Laques, P., Despiau, R., and Auge, A. 1982. Apparent thickness and scattering properties of Saturn's rings from March 1980 observations. *Astron. Astrophys.* 108:296–305.

Simpson, J. A., Bastian, T. S., Chenette, D. L., McKibben, R. B., and Pyle, K. R. 1980. The trapped radiations of Saturn and their absorption by satellites and rings. *J. Geophys. Res.* 85:5731–5762.

Smith, B. A., Briggs, G. A., Danielson, G. E., Cook, A. F., Davies, M. E., Hunt, G. E.,

Masursky, H., Soderblom, L., Owen, T., Sagan, C., and Suomi, V. E. 1977. Voyager imaging experiment. *Space Sci. Rev.* 21:103–128.

Smith, B. A., Soderblom, L. A., Batson, R., Bridges, P., Inge, J., Masursky, H., Shoemaker, E., Beebe, R., Boyce, J., Briggs, G., Bunker, A., Collins, S. A., Hansen, C., Johnson, T., Mitchell, J., Terrile, R., Cook, A., Cuzzi, J., Pollack, J., Danielson, E., Ingersoll, A., Davies, M., Hunt, G. E., Morrison, D., Owen, T., Sagan, C., Veverka, J., Strom, R., Suomi, V. 1982. A new look at the Saturn system: The Voyager 2 images. *Science* 215:504–537.

Smith, B. A., Soderblom, L., Beebe, R., Boyce, J., Briggs, G., Bunker, A., Collins, S. A., Hansen, C., Johnson, T., Mitchell, J., Terrile, R., Carr, M., Cook, A., Cuzzi, J., Pollack, J., Danielson, G. E., Ingersoll, A., Davies, M., Hunt, G., Masursky, H., Shoemaker, E., Morrison, D., Owen, T., Sagan, C., Veverka, J., Strom, R., Suomi, V. 1981. Encounter with Saturn: Voyager 1 imaging results. *Science* 212:163–191.

Smoluchowski, R. 1978. Amorphous ice on Saturnian rings and on icy satellites. *Science* 201:809–811.

Smoluchowski, R. 1983. Formation of fine dust on Saturn's rings as suggested by the presence of spokes. *Icarus* 54:263–266.

Smythe, W. D. 1975 Spectra of hydrate frosts: Their application to the outer solar system. *Icarus* 24:421–427.

Spitzer, L. Jr. 1978. *Physical processes in the interstellar medium.* (New York: Wiley, Interscience).

Stevenson, D. 1982. Are Saturn's rings a seismograph for planetary internal oscillations? *EOS* 63:1020.

Synnott, S P., Terrile, R. J., Jacobson, R. A., and Smith, B. A. 1983. Orbits of Saturn's F-ring and its shepherding satellites. *Icarus* 53:156–158.

Terrile, R. J. 1982. Unusual imaging observations of Saturn's rings. Proceedings of *I.A.U. Colloquium 75 Planetary Rings*, ed. A. Brahic, Toulouse, France, Aug. 1982.

Terrile, R., and Tokunaga, A. 1980. Infrared photometry of Saturn's E ring. *Bull. Amer. Astron. Soc.* 12:701.

Terrile, R. J., Yagi, G., Cook, A. F., Porco, C. C. 1981. A morphological model for spoke formation in Saturn's rings (abstract). *Bull. Amer. Astron. Soc.* 13:728.

Thompson, W. T., Irvine, W. M., Baum, W. A., Lumme, K., and Esposito, L. W. 1981. Saturn's rings: Azimuthal variations, phase curves, and radial profiles in four colors. *Icarus* 46:187–200.

Thomsen, M. F., Goertz, C. K., Northrop, T. G., and Hill, J. R. 1982. On the nature of particles in Saturn's spokes. *Geophys. Res. Letters* 9:423–426.

Thomsen, M. F., and Van Allen, J. A. 1979. On the inference of properties of Saturn's E ring from energetic charged particle observations. *Geophys. Rev. Letters* 6:893–896.

Tokunaga, A. T., Caldwell, J., and Nolt, I. G., 1980. The 20-micron brightness temperature of the unilluminated side of Saturn's rings. *Nature* 287:212–214.

Trulsen, J. 1972. On the rings of Saturn. *Astrophys. Sp. Sci.* 17:330–337.

Tyler, G. L., Cuzzi, J. N., Zebker, H. A., and Marouf, E. A. 1984. Saturn's rings: Radial variation of optical depth at visual and radio wavelengths, and implications for particle size variations. In preparation.

Tyler, G. L., Eshleman, V. R., Anderson, J. D., Levy, G. S., Lindal, G. F., Wood, G. E., and Croft, T. A. 1981. Radio science investigations of the Saturn system with Voyager I: Preliminary results. *Science* 212:201–206.

Tyler, G. L., Eshleman, V. R., Anderson, J. D., Levy, G. S., Lindal, G. F., Wood, G. E., and Croft, T. A. 1982. Radio science with Voyager 2 at Saturn: Atmosphere and ionosphere and the masses of Mimas, Tethys, and Iapetus. *Science* 215:553–558.

Tyler, G. L., Marouf, E. A., Simpson, R. A., Zebker, H. A., and Eshleman, V. R., 1983. The microwave opacity of Saturn's rings at wavelengths of 3.6 and 13 cm from Voyager 1 radio occultation. *Icarus* 54:160–188.

Van Allen, J. A. 1982. Findings on rings and inner satellites of Saturn by Pioneer 11. *Icarus* 51:509–527.

Van Allen, J. A. 1983. Absorption of energetic protons by Saturn's G ring. *J. Geophys. Res.* 88:6911–6918.

Van Allen, J. A. 1984. The inner magnetosphere of Saturn (L ≤ 10. In *Saturn,* eds. T. Gehrels and M. S. Matthews (Tucson: University of Arizona Press).

Van Allen, J. A., Randall, B. A., and Thomsen, M. F. 1980*a*. Sources and sinks of energetic electrons and protons in Saturn's magnetosphere. *J. Geophys. Res.* 85:5679–5694.

Van Allen, J. A., Thomsen, M. F., and Randall, B. A. 1980*b*. The energetic charged particle signature of Mimas. *J. Geophys. Res.* 85:5709–5718.

van de Hulst, H. C. 1957. *Light Scattering by Small Particles* (New York: John Wiley and Sons).

Veverka, J. 1977. Photometry of satellite surfaces, in *Planetary Satellites*, ed. J. Burns, (Tucson: University of Arizona Press).

Ward, W. R. 1981. On the radial structure of Saturn's rings. *Geophys. Res. Letters* 8:641–643.

Warwick, J. W., Evans, D. R., Romig, J. H., Alexander, J. K., Desch, M. D., Kaiser, M. L., Aubier, M., LeBlanc, Y., Lechcheux, A., and Pedersen, B. M. 1982. Planetary radio astronomy observations from Voyager 2 near Saturn. *Science* 215:582–587.

Warwick, J. W., Pearce, J. B., Evans, D. R., Carr, T. D., Schauble, J. J., Alexander, J. K., Kaiser, M. L., Desch, M. D., Pedersen, M., and Lecacheux, A. 1981. Planetary radio astronomy observations from Voyager 1 near Saturn. *Science* 212:239–243.

Weinhermer, A. J., and Few, A. A. 1982. The spokes in Saturn's rings: A critical evaluation of possible electrical processes. *Geophys. Res. Letters* 9:1139–1142.

Weiser, H., Vitz, R. C., and Moos, H. W. 1977. Detection of Lyman-alpha emission from the Saturnian disk and from the ring system. *Science* 197:755–757.

Whalley, E., and Labbé, H. J. 1969. Optical spectra of orientationally disordered crystals. III. Infrared spectra of the sound waves. *J. Chem. Phys.* 51:3120–3127.

Zarka, P., and Pedersen, B. M. 1983. Statistical study of Saturn electrostatic discharges. *J. Geophys. Res.* 88:9007–9018.

Zerull, R. H., Giese, R. H., Schwill, S., and Weiss, K. 1980. Scattering by particles of non-spherical shape. In *Light Scattering by Irregularly Shaped Particles*, ed. D. W. Schuerman (New York: Plenum Press).

THE ETHEREAL RINGS OF JUPITER AND SATURN

JOSEPH A. BURNS, MARK R. SHOWALTER
Cornell University

and

GREGOR E. MORFILL
Max-Planck-Institut für Extraterrestrische Physik

The Jovian ring is tenuous, with little known substructure. Its brighter main band lies in Jupiter's equatorial plane at about 1.7–1.8 R_J, well within the Roche limit. Micron-sized grains account for most of the visible ring's brightness, but particles with sizes of centimeters and larger must also be present to absorb charged particles. Since dynamical evolution times and survival lifetimes for micron grains are quite short ($\lesssim 10^{2-3}$ yr), the visible Jovian ring must be continually replenished; probably most small grains are generated by micrometeoroids colliding into unseen parent bodies that reside in the main band. The halo is composed of yet smaller particles, is radially localized (~ 1.3 $R_J - 1.7$ R_J) and has significant vertical extent ($\sim 10^4$ km), probably due to electromagnetic forces. The E, F, and G Rings of Saturn share many properties with Jupiter's ring: they have low optical depths, are immersed in a magnetospheric plasma, and contain a significant complement of micron-sized particles. Their natures are not as clearly understood as that of Jupiter's ring. All of these rings are distinguished in two fundamental ways from the rings of Uranus and the major Saturnian rings: single-particle dynamics, rather than collective effects, most likely govern their form, and the majority of ring particles have quite limited lifetimes.

Oftentimes trivial amounts of matter can provide enticing clues as to possible past or future events. Isotopic anomalies in meteorites illustrate the point, as does perfume in a more personal realm. Jupiter's ring and the faint outlying Saturnian rings may be other examples, because their insubstantiality and relatively simple form allow fundamental processes to be highlighted. In other ring systems these same phenomena, though active, may be obscured by collective effects.

The three outermost rings of Saturn—its E, G, and F Rings—and Jupiter's ring each contain a significant component of micron-sized grains. All of these rings reside in fierce magnetospheric environments. Accordingly, the local thermal plasma/trapped particle population plays an important role for the typical particle; it may govern the particle's lifetime either through orbital evolution by plasma drag or through the particle's elimination by sputtering. In addition, the local plasma influences the particle's dynamics by determining the particle's electric potential and thereby the strength of the electromagnetic force perturbation. Moreover, all these rings (except the G Ring) are known to have nearby or embedded satellites, and all (except certain parts of the F Ring) have very low optical depths.

In this chapter we first summarize the observations of the Jovian ring (Sec. I); emphasis is placed on the findings of Showalter et al. (1983*a,b*,1984), since this as yet unpublished work refines several previous interpretations of Jupiter's ring. Section II then outlines the available observations of Saturn's tenuous rings. In Sec. III we discuss processes that might be important in the workings of ethereal rings (cf. chapter by Grün et al.). Finally we show how these processes may explain the observed ring structures of Jupiter (Sec. IV) and Saturn (Sec. V); however, we largely defer to the chapters by Cuzzi et al. and Dermott insofar as the F Ring's dynamics is concerned.

Properties of Jupiter's ring have been compiled previously by Burns et al. (1980), Jewitt and Danielson (1981), and Jewitt (1982). General reviews of planetary ring systems, including Jupiter's, have been made by Ip (1980*b*), Goldreich and Tremaine (1982), and Cuzzi (1983). The chapters by Cuzzi et al. and Grün et al. also contain information on the diffuse rings.

I. DESCRIPTION OF JUPITER'S RING

Jupiter's is the only ring system to have been discovered by spacecraft. When energetic particle fluxes were measured by Pioneer 10 to be unexpectedly low inwards of Amalthea's orbit (see Fig. 1), Acuña and Ness (1976; see also Fillius 1976) proposed that an undiscovered ring or a satellite might be the cause. Since other explanations for the reduced flux levels were available, little heed was paid to the suggestion. The only image of the ring obtained on Voyager 1's passage through Jupiter's equatorial plane (see Fig. 2; Smith et al. 1979*a*) demonstrated undeniably that a ring was the cause of the dips in the charged particle data. Jupiter's ring system was much more clearly viewed by Voyager 2 (Smith et al. 1979*b*; Owen et al. 1979) from a few degrees out of the ring plane in both forward-scattered (Figs. 3 and 4) and backscattered (Fig. 5) light. These figures are not only representative of Voyager images but

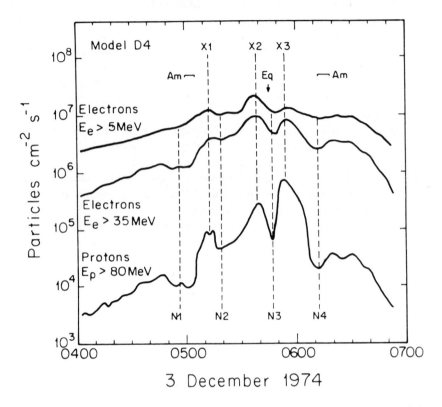

Fig. 1. Intensities of trapped radiation measured by Pioneer 11 as it approached and receded from Jupiter. Particle absorptions are due to Amalthea ($N1$ and $N4$) and the then-undiscovered Jovian ring ($N2$ and $N3$). (Figure from Fillius et al. 1975.)

they are also about the only representatives; as tabulated by Owen et al. (1979), Voyager took six wide-angle and eighteen narrow-angle images, of which we show ten.

A. Morphology

Due to its diaphanous nature (normal optical depth $\tau \sim 10^{-5}$ typically) and the limited information available, even the most general structure of Jupiter's ring remains in dispute. It has been described by Jewitt and Danielson (1981; cf. Owen et al. 1979; Smith et al. 1979b; Ip 1980b) as having three compo-nents: *bright band, faint disk, halo* (see Fig. 6a). Significant smearing in the images, due to spacecraft motion during the long exposures required by the

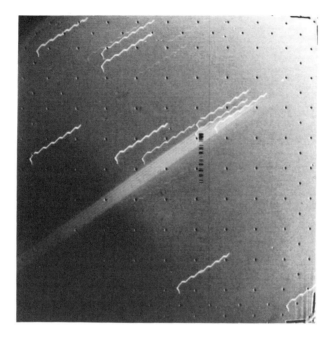

Fig. 2. The first view of the Jovian ring actually showing six separate images (the parallel straight lines extending from lower left to upper right), each obtained as the spacecraft paused during its nodding limit-cycle motion. The jagged hairpin lines are star trails smeared by the same spacecraft motion. Each ring image has an effective exposure of about one minute. The ring is seen here in backscattered light as Voyager 1 passed near the ring plane (narrow-angle frame FDS 16368.19). (Figure from Smith et al. 1979a.)

Fig. 3. A mosaic of wide-angle frames (FDS 20691.27–20691.47, 20693.02) taken at high phase angles while Voyager 2 was in Jupiter's shadow looking back toward the Sun. Jupiter's limb is highlighted as a result of forward-scattering by the hazes comprising the planet's upper atmosphere. (Figure from Jewitt 1982.)

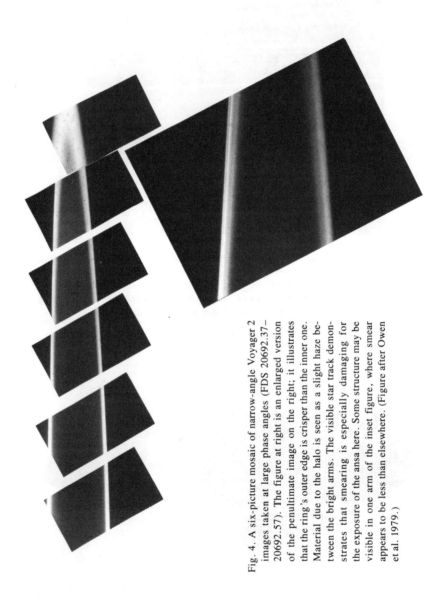

Fig. 4. A six-picture mosaic of narrow-angle Voyager 2 images taken at large phase angles (FDS 20692.37–20692.57). The figure at right is an enlarged version of the penultimate image on the right; it illustrates that the ring's outer edge is crisper than the inner one. Material due to the halo is seen as a slight haze between the bright arms. The visible star track demonstrates that smearing is especially damaging for the exposure of the ansa here. Some structure may be visible in one arm of the inset figure, where smear appears to be less than elsewhere. (Figure after Owen et al. 1979.)

Fig. 5. Backscatter image of Jovian ring. Substantial contrast enhancement allows the near
and far arms of the ring to be seen in backscattered light. These are Voyager 2 frames FDS
20612.27 and 20612.31, taken from about 2°.5
above the ring plane at a range of 2×10^6 km.
(Figure from Owen et al. 1979.)

faintness of the rings, limits the resolution to ~600 km (~0.01 R_J). To this
accuracy the *bright band*, as seen in forward-scattered light, starts abruptly at
1.81 R_J and ends more gradually at about 1.72 R_J, where it has been considered to blend into a *faint disk* which Owen et al. (1979; cf. Jewitt 1982) state
to be continuous down to the planet's cloudtops (see next paragraph, however). The *halo* in this model enshrouds both the band and the disk, and is
lens-shaped, being thickest ($h \sim 10^4$ km) near the planet but considerably
reduced over the main band. By contrast the full vertical thickness of the main
band as seen edge-on in the discovery image is $h \leqslant 30$ km. The entire ring
lies easily within the planet's Roche limit.

  Showalter et al. (1983*a,b*, 1984) have reanalyzed the Voyager 2 images
with a modern video display system. They were surprised by the information

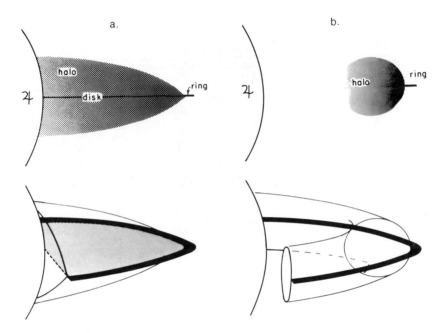

Fig. 6. (a) The model of Jewitt and Danielson (1981) showing main band, halo and faint disk. (b) A sketch of the main band and a radially confined torus according to the Jovian ring model of Showalter et al. (1983a, 1984).

contained in these fuzzy images and have used it to improve the understanding of ring morphology and particle properties. Isophotes of the observed brightness levels are given in Color Plate 9 for the ansa and in Color Plate 10 (see Color Section) for the planetary shadow region. Showalter et al. (1983a, 1984) point out that an abrupt dropoff in brightness would be seen across the intersection of Jupiter's shadow with the ring plane if an equatorial disk of material were present. Since no drop is apparent (compare Color Plate 10 with Fig. 6), they argue that the faint disk does not exist at all; rather it is merely a manifestation of the halo.

Two features at the shadow boundary (Plate 10) indicate that the halo itself must be radially confined, with most of its material concentrated just inside the bright band. First, the isophotes near the point where the planet's shadow cuts the main ring are parallel to Jupiter's limb and brightness levels drop swiftly. Second, a series of slight bumps in the isophotes above the far arm mimic the shape of the planet's limb; this suggests radially localized material extending well above the ring plane. The halo must have substantial vertical extent and an intensity that gradually diminishes with height. This is additionally demonstrated by the shapes of the isophotes at the ansa which, as Jewitt and Danielson (1981) also contend, indicate out-of-plane material rather than a radial extension to the ring. The putative faint disk, first described by Owen et al. (1979) and later by Jewitt and Danielson (1981), can now be understood as due to two overlapping halo sheets, one associated with the near arm of the ring and the other with the far arm. The Jovian ring, as sketched in Fig. 6b (Showalter et al. 1983a,1984), has a main band (with the properties already described) which circumscribes a toroidal halo, whose full vertical extent is $\sim 2 \times 10^4$ km and whose radial extent is perhaps $\sim 3 \times 10^4$ km, depending upon what density level one wishes to consider as the boundary. It therefore stretches from the main ring's inner edge at $\sim 1.7\,R_J$ to $\sim 1.3\,R_J$.

Beyond the gross breakdown into these components, little other structure is noticeable in the images. A brightness intensification of $\sim 10\%$ extends over $\sim 0.01\,R_J$ centered around $\sim 1.79\,R_J$ (see Jewitt's Fig. 3, 1982); in backscattered light this enhancement seems heightened to $\sim 50\%$. The Voyager 2 narrow-angle images may hint at additional radial structure,[1] but they were smeared too much during the long exposures to be unambiguous. Haemmerle et al. (1982) have attempted to desmear these images, using visible star trails as guides. They find several low contrast features in the bright band, but in view of how difficult it is to deconvolve a noisy, smeared image, these periodic ringlets may merely be artifacts of the Fourier Transform processing.

Attempted occultations of the Voyager 2 radio signals (Tyler et al. 1981) at wavelengths of 3.6 cm and 13 cm identified no structures larger than 40 km (120 km) with $\tau > 10^{-4}$ (2×10^{-4}). A stellar occultation (Dunham et al. 1982), which was capable of detecting any rings of $\tau > 8 \times 10^{-3}$ broader than 13 km, discerned nothing. The charged particle absorption data (Fillius 1976; Pyle et al. 1983) can be interpreted as caused by a ribbon of boulders (see Sec. I.D. below) somewhat narrower than the visible bright band (Fillius, personal communication, 1980).

A few other points complete the information available on the Jovian ring's morphology. The main ring is not appreciably eccentric ($e \leqslant 0.003$). The halo may be slightly asymmetric about the ring plane; Jewitt and Danielson (1981) reported it to be shifted about 300 km south, a small but significant inclination of $0°1$. The precise nature of the halo's asymmetry is not yet certain in Showalter et al.'s revised morphology, because it is nontrivial to separate ring morphology cleanly from the photometric properties of the halo grains. The only published groundbased picture (see Fig. 7; Jewitt et al. 1981) places the edge at 1.81 (± 0.01) R_J, but at 1.78 (± 0.01) R_J after correction for smear and seeing-broadening. Although this value seems contradicted by the Voyager measurements, it has not received comment.

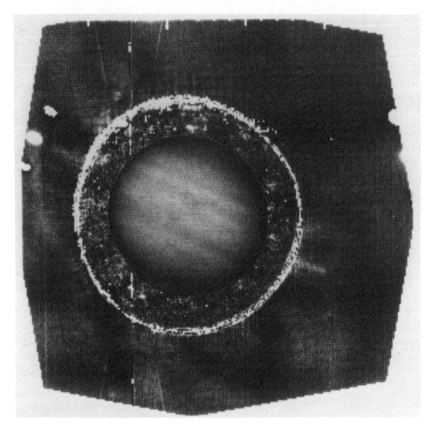

Fig. 7. A picture of Jupiter and its ring taken at 0.9 μm with a CCD attached to the 5-m Hale telescope. Amalthea is visible at the extreme left edge. (Figure from Jewitt et al. 1981.)

Limits can be placed on the presence of any additional rings. As noted above, occultations of the spacecraft and a star rule out rings that are narrow and relatively dense (although still ethereal by most standards); the stellar occultation did not, however, probe inside ~1.75 R$_J$, and so narrow rings could exist close to the planet. In addition, Jewitt et al. (1981) would have noticed any other rings with more than one-tenth the integrated backscattered brightness of the known ring at 0.9 μm, and they did not. Showalter et al. (1983a; Burns et al. 1983) have detected a very faint ring that extends beyond the main band to 180 (± 6) × 10^3 km; Amalthea at 181,300 km likely defines its outer edge. With a brightness of about 1% of the main ring, the optical depth of this gossamer ring would be like that of Saturn's E Ring. However, owing to the difficulty of properly subtracting the variable background off the Voyager images, this "discovery" must remain tentative.

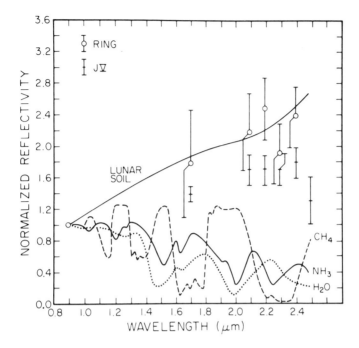

Fig. 8. Reflection spectra of the Jovian ring and Amalthea between 0.9 and 2.5μm. (Figure from Neugebauer et al. 1981.)

B. Spectrophotometry

Difficult Earth-based measurements of the Jovian ring reflectance spectrum have been performed by Smith and Reitsema (1980) as well as by Neugebauer et al. (1981) to determine ring particle composition. The former measurements were made in the optical band (0.55 to 1 μm) while the latter were taken at five infrared wavelengths between 1.7 and 2.4 μm. As displayed in Fig. 8, they indicate that the spectrum is flat to reddish, quite like that of the Moon or Amalthea. From the absence of absorption features, water, methane, and ammonia ice may all be ruled out as major ring constituents. A silicate or carbonaceous composition has therefore been inferred. Such compositions are consistent with possible solutions for the particle's refractive index deduced by Grün et al. (1980) and by Tyler et al. (1981) from photometry.

One should be cautioned, however, that the measured spectra by no means require carbonaceous or silicaeous grains. More exotic compositions, such as metallic grains coated by Io-derived sulfur, could also match satisfactorily. Gradie et al. (1980) contended with similar uncertainties when they

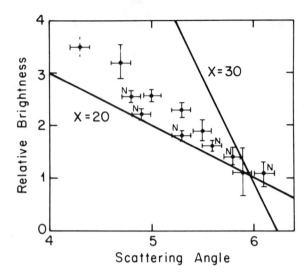

Fig. 9. Phase diagram for Jupiter's ring in forward-scattered light. Jewitt and Danielson (1981) measured the values shown by crosses; "N" designates near-arm photometry. They are compared against single particle-sized phase functions for $x = 20$ ($r \approx 1.5\ \mu$m) and $x = 30$ ($r \approx 2.2\ \mu$m).

studied the surface of Amalthea, for which much better data exist. Carbonaceous grains are perhaps most likely on cosmogonic grounds, but this reasonable inference cannot be verified. This issue is of more than passing interest because composition implies electrical and mechanical properties which enter models of origin and evolution.

C. Photometry

The Voyager 2 cameras revealed the ring to be roughly twenty times brighter in forward-scattered light than in backscattered light (Owen et al. 1979); the stronger forward-scattered signal is apparent in a quick comparison of Fig. 5 (backscatter) with Figs. 3 and 4 (forward-scatter). These relative brightnesses imply that diffraction is dominant and thus that particle circumferences are on the order of the wavelength of the light. Jewitt and Danielson (1981) measured the ring particles' phase function (see Fig. 9) for a limited range of scattering angles (4° to 6°) from Voyager 2 images taken through a clear filter (effective wavelength $\sim 0.47\ \mu$m). They concluded that the equivalent Mie scattering spheres would have a mean radius of about 2.5 μm, correct to within a factor of two.

More sophisticated photometric models by Tyler et al. (1981) and Grün et al. (1980) have attempted to specify simultaneously both size distribution and optical properties. Such attempts are instructive, though doomed to provide nonunique identifications since more parameters are introduced than are needed to represent the measurements. Both groups model the size spectrum as a power law

$$n(r)\, dr = Cr^{-p}\, dr, \tag{1}$$

where $n(r)dr$ is the number of particles in the size range from r to $r + dr$ and where C and p are constants; for such a power law, equal cross-sectional areas are present in equal increments for $p = 2$, but in equal log increments for $p = 3$. Collisional ejecta have $p \simeq 3.4$ (Grün et al. 1980) so then small particles account for most of the area, but large ones contain most of the mass. Dohnanyi (1978) demonstrates that $2.5 < p < 4.5$ for interplanetary debris. Grün et al. (1980) attempt to best duplicate the phase function in the visible while Tyler et al. (1981) also include upper limits on optical depths at longer wavelengths (see below) to guide their solution. The latter authors find that the data fit a steep spectrum ($3 \leq p \leq 4$) with a lower cutoff at $r = 1$ to 2 μm. On the other hand, Grün et al. (1980) choose a very flat spectrum ($p = 0.4$) and fix the lower cutoff ($r \geq 0.2$ μm) for arguable dynamical reasons, and then find that the photometry is best matched with an upper cutoff at $r \approx 2.5$ μm. Both of these models contain a preponderance of 2 μm particles, and so in a sense they are not very different from Jewitt and Danielson's (1981) monodispersed description.

In this connection, we point out that forward-scattered light tends to highlight those particles in the population which are comparable in size to λ, the wavelength of the observation. This occurs because diffraction dominates the forward-scattered signal and the half-angle of the diffraction lobe $\sim \pi/x$ with $x \equiv 2\pi r/\lambda$. Hence measurements taken at scattering angle θ emphasize particles of radius $r \sim \lambda/2\theta$. For the existing Jovian ring images, this radius is ~ 2.5 μm and so it is perhaps not surprising that the presence of micron-sized particles has been inferred. Indeed, by the above argument, this general size range was predetermined before any measurements were taken. That is to say, while the photometry indicates that micron-sized grains must be present in Jupiter's ring, it should not be construed as *prima facie* evidence against particles much larger or smaller than a micron.

As a specific example of this principle, in the neighborhood of $\theta = 5°$, phase functions of single spheres with $x = 10$ to 20 are quite similar to those for continuous power law size distributions with $p = 1$–4. Fig. 10 shows that at small scattering angles the phase function slopes for $p = 1$–4 overlap those for $x = 10$ to 50 (cf. Showalter et al. 1983b, 1984). However, for any of these untruncated power laws, the concept of a "typical particle size" makes no sense, even though some particle size might duplicate the phase behavior near some θ. This simple exercise calls into question the oft-observed fact that the particles of Jupiter's and Saturn's diffuse rings (when detected at small scat-

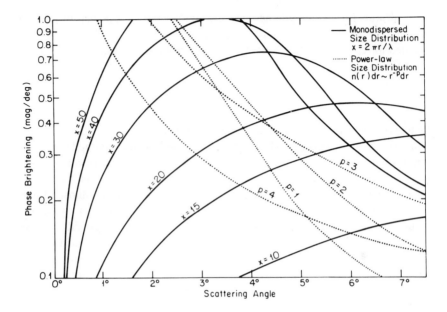

Fig. 10. Phase function slopes for scattering by single particles ($x = 2\pi r/\lambda = 20, \ldots$) and untruncated power law size distributions (see Eq. 1, $p = 1$–4).

tering angles) always turn out to be of order 1 μm, comparable in size to the Voyager wavelengths. On the basis of our simulations (see Fig. 10) we expect that if the measurements had been taken at other small scattering angles, other x values (but still of order 10 to 10^2) would dominate the scattering for the same suite of power laws. Hence, to well constrain a particle size distribution, better phase coverage is necessary.

Showalter et al. (1983b, 1984) have incorporated the pairs of wide-angle violet ($\lambda \approx 0.41$ μm) and orange ($\lambda \approx 0.62$ μm) images into their reexamination of the forward-scattering photometry. They find comparable phase brightening of the main band at both wavelengths, which is incompatible with a single particle size. Instead, a distribution of particle sizes must be present, with perhaps a broad peak in the 2 μm range. They observe less phase brightening in the halo than in the bright band, though it is quite difficult to deconvolve phase effects from the ill-determined halo geometry; nevertheless this suggests that halo particles are smaller. Furthermore, they find evidence that the combined populations of both band and halo obey a broad power law distribution, with index $p = 2$ to 3. This distribution extends from tenths to tens of microns, with the larger particles in the bright band and smaller ones in the halo.

Once the particle size distribution is specified, the observed 20:1 ratio of backscattered to forward-scattered intensity can be used to restrict the material's refractive index to a narrow range. In this way Grün et al. (1980) maintain that the ring particles are silicates and not water ice. In a somewhat more complete survey, but one with a similar philosophy, Tyler et al. (1981) consider a variety of relatively steep power laws ($2 < p < 4$) with x ranging between a lower limit of 1 to 20 and an upper limit of 40 to 75 in order to try to match the data in the visible. They find that $p = 3.5 \pm 0.5$, $x_{min} \simeq 19$ ($r_{min} \simeq 1.5 \ \mu m$) and the real refractive index $m \simeq 1.62 \pm 0.02$ while the imaginary part is less than 0.005. These refractive indices are fairly consistent with silicates but clearly exclude water ice. One should be warned, however, that any conclusions drawn by combining forward-scattered and backscattered photometry are extremely model-dependent. The forward-scattered images highlight micron-sized particles, while the backscattered images emphasize only the largest ones; in effect the data apply to mutually exclusive portions of the size distribution. Also, for the largest grains, the effects of nonsphericity are both important and unpredictable (see, e.g., Pollack and Cuzzi 1980).

The presence of a halo is also intimated by the measurements shown in Fig. 9; the far arm of the ring always appears brighter than the near arm (Jewitt and Danielson 1981). Slight differences occur in the slant path, accounting for the near arm being about 10% fainter, but this effect is insufficient to explain the entire discrepancy. The difference is most likely attributable to a small nonsymmetric contribution from halo-scattered light, but specifically what this indicates of halo geometry is still uncertain.

D. Ring Opacity

Compared to most other planetary rings, the Jovian system is quite diffuse; the usually quoted optical depths τ in visible light being 3×10^{-5} for the main band, 7×10^{-6} for the disk and $< 5 \times 10^{-6}$ for the halo (Jewitt 1982); recall that visible light is most affected by particles larger than $\sim 0.1 \ \mu m$. These values for τ are all scaled to the first, which ratios the measured brightness of the backscattered radiation off the ring relative to that from the Jovian disk; since these τ's assume a ring particle's geometric albedo $k = 0.04$, the absolute values of τ will differ as they scale with the actual albedo which probably lies in the range $0.02 < k < 0.20$. Indeed, Tyler et al. (1981) maintain that all should be raised 50% because of an improper definition of k by Jewitt and Danielson (1981). In addition, Tyler et al. (1981) propose a different expression to obtain the optical depth from the backscattered intensity. Their alternate equation depends on the optical properties of the ring material which they claim are no more poorly known than the albedo; assuming refractive indices of silicate particles, they derive that τ for the bright band should be 4×10^{-4}, more than an order of magnitude larger than the nominal value. However, this result is questionable; it relies on the backscattering photometry which is interpreted in terms of a Mie phase function for absorbing spherical particles. Yet nonspherical grains are known to backscatter radiation quite differently than do spheres, even though particle shape is unimportant for forward-scattering intensities.

A triad of groundbased detections of the ring allow estimates of τ at different wavelengths, under the same limitation of proportionality to k as with the spacecraft measurements. Becklin and Wynn-Williams (1979) find that to reproduce the Jovian ring's signal at 2.2 μm (sensitive to particles with radii larger than ~ 0.4 μm), the ring should contain 10^{-5} the number in Saturn's rings if both systems were of the same composition. Likely they are not; nevertheless, if they were, $k\tau_{2.2\mu m} \sim 10^{-5}$ and τ would be about 2×10^{-4} for dark particles. Jewitt et al.'s 0.9 μm image (see Fig. 7; cf. Neugebauer et al. 1981) corresponds to 18.2 mag (linear arcsec)$^{-1}$, or $\tau_{0.9\ \mu m} \sim 2 \times 10^{-4}$; Smith and Reitsema (1980) conclude that the ring's brightness at 0.7 μm is also ~ 18 mag (linear arcsec)$^{-1}$.

Thermal emission from the ring was not detected by the Voyager IRIS instrument operating at three infrared wavelengths (Hanel et al. 1979) nor by the Hawaii 2.2-m telescope at 20 μm (Becklin and Wynn-Williams 1979). Of the two, the Voyager experiment was much more sensitive; assuming particles emitting at 125 K (the equilibrium temperature at Jupiter for a rotating 0.60 albedo sphere), the upper limits placed on optical depth are $< 2 \times 10^{-4}$ (40 μm), 5×10^{-4} (25 μm), and 3×10^{-3} (16 μm); a more reasonable choice for the albedo would lower these values even further.

If successful, occultations would yield τ directly. However, neither a stellar occultation at ~ 0.5 μm (sensitive to ≥ 0.1 μm particles) nor spacecraft occultations at 3.6 cm (sensitive to ≥ 0.6 cm particles) and 13 cm (sensitive to ≥ 2 cm particles) saw any attenuation; the corresponding limits on τ are $< 8 \times 10^{-3}$ over 13 km (Dunham et al. 1982), $< 5 \times 10^{-4}$ with 40 km resolution, and $< 2 \times 10^{-4}$ with 120 km resolution (Tyler et al. 1981), respectively.

One can estimate the cross-sectional area of the material responsible for the diminution of charged particle fluxes but this requires substantially more modeling than in the case of the photometry. Several complications arise. First, a charged particle, as it bounces along and drifts across the planetary magnetic field, takes some time to diffuse past an obstacle. During this period, it can be totally absorbed by a large obstacle or can have its energy progressively lowered while penetrating several thinner targets; simultaneously an electron's energy will also be gradually drained away by synchrotron emission. Second, a ring and a satellite affect magnetospheric density in different ways (Fillius, personal communication, 1980; cf. Van Allen 1982). Third, the magnetic field, which for Jupiter is tilted and has appreciable high-order terms, should be correctly represented. And finally, a good expression is needed for the diffusion time (which depends on the nature of the magnetic field and the pitch angle distribution of the energetic particles). With a crude absorption model, Ip (1980a) estimated an absorption optical depth of 6×10^{-6} while a preliminary attempt at improved modeling by Fillius (personal communication, 1980) leads to τ of order, but somewhat larger than, 10^{-6}. The electron data, which require only $\sim 10^{-1}$ the absorbing area indicated by the proton data, show how incomplete absorption models are; this discrepancy might be due to synchrotron losses, which affect electrons alone, but the energy dependence of the electron data contradicts this possibility (Fillius, personal communication, 1980). Since the diffusion times are believed to be brief, particles are lost mainly by total absorption rather than by a gradual energy degradation. Hence, these optical depths refer to particles larger than about 5 cm, the absorption length of 80 MeV protons (Ip 1980a).

The data in Fig. 1 are the best observational evidence for the presence in the ring of larger parent bodies; as we will describe below, these bodies are required to replenish continually the supply of smaller, visible grains. For comparison, two satellites associated with the ring have about one-half the corresponding cross-sectional area (or $\tau \sim 4 \times 10^{-7}$). Even though these correspond to substantial fractions of the ring area, the approximate longitudinal constancy of the absorption profiles in Fig. 1 argues that the dips are produced in part by ring material and not entirely by satellites; in Van Allen's terminology (1982), these are "macrosignatures" and not "microsignatures." Furthermore, the fact that the signature of Thebe (larger than either Adrastea or Metis) is not discernible in the data also implies that more than the ringmoons is needed. Pyle et al. (1983) have identified the ring's absorption signature for 0.5 to 8.7 MeV protons, for >3.4 MeV electrons and for medium Z nuclei with energies >70 MeV/nucleon. An energy-dependent absorption model has not yet been attempted although its computation should be a valuable exercise. Additional constraints on the nature of the absorbing material may come after modeling the Very Large Array (VLA) observations of synchrotron emission from electrons spiraling in Jupiter's magnetic field (de Pater and Jaffe 1984).

E. Nearby Satellites

A pair of satellites (which one could perhaps consider to be merely the largest of the ring size distribution) have been discovered by Voyager spacecraft in the vicinity of the ring. The satellite Adrastea (JXV, or 1979J1) has a radius ~ 10 km and is at semimajor axis 1.8064 R_J, i.e., very close to, if not coincident with, the ring's outer edge (Jewitt et al. 1979; cf. Jewitt 1982; Synnott 1983); no orbital eccentricity or inclination was detectable. Metis (JXVI, or 1979J3), a somewhat larger object ($R \sim 20$ km), was discovered through the shadow it cast on the Jovian atmosphere and has been seen only in transit across Jupiter's disk (Synnott 1980, 1983); assuming no orbital eccentricity or inclination, it is located at 1.7922 R_J. This satellite may be near the brightness enhancement in the main band but, with current observational uncertainties, one cannot know for sure. Both objects are dark (albedos of $\sim 5\%$) like Amalthea.

TABLE I

Nominal Characteristics of Jupiter's Ring

	Main Band	Halo
Radial location	$1.72\,R_J - 1.81\,R_J\ (\pm 0.01\,R_J)$	$\sim 1.3\,R_J - 1.7\,R_J$
Thickness	$\leqslant 30$ km	$\sim 2 \times 10^4$ km
Normal optical depth τ	$\sim 5 \times 10^{-5}$	$\leqslant 7 \times 10^{-6}$
Particle size	broad distribution about micron; power law ($p = 2-3$) for halo plus band.	submicron (?)
Parent bodies	$\tau \sim 10^{-6}$ for $r \gtrsim 5$ cm	none (?)
Associated satellites	JXV Adrastea (1979J1) $R \sim 10$ km, $a = 1.8064\,R_J$ JXVI Metis (1979J3) $R \sim 20$ km, $a = 1.7922\,R_J$	— —

F. Summary of Jovian Ring Properties

The observations summarized above show Jupiter's ring to have a relatively simple form and to be located near the planet. It has a low optical depth and a substantial complement of micron-sized grains, although particles at other sizes, up to at least centimeters across, are present as well. The ring is bathed in an intense radiation environment.

A listing of the Jovian ring properties is provided in Table I for quick reference. Naturally, all of these numbers are surrounded by many caveats and the reader is always advised to refer back to the basic literature.

II. DESCRIPTION OF SATURN'S OUTLYING ETHEREAL RINGS

Saturn's E, G, and F Rings share many properties with Jupiter's tenuous ring system. Constituents of all these rings are affected by similar physical processes, to be described in the next section. In order to allow comparison of these rings, we now outline the little that is known about the ethereal rings of Saturn; Table II summarizes these properties. Additional descriptions of these rings may be found in the chapters by Cuzzi et al. and Grün et al.

Burns et al. (1983) have noticed a broad diffuse band inward of the nominal F Ring. Graps et al. (1983) interpret Voyager occultation data to indicate that a faint ring lies between the outer edge of the A Ring and Atlas' (SXV, or 1980S28) orbit. The radio occultation experiment may also see some of this debris (H. A. Zebker, personal communication, 1983). These gossamer rings have only just been discovered and will not be discussed in detail.

A. Saturn's G Ring

Of all the Saturnian rings given a letter designation (except for the obscure D Ring), the least is known about the G Ring. Its presence was

TABLE II
Nominal Characteristics of Saturn's Tenuous Rings

	E Ring	G Ring	F Ring
Radial location eccentric (0.0026)	$\sim (3-8)\ R_S$	$2.82\ R_S$	$2.324\ R_S$ (eccentricity $e = 0.0026$)
Width	3×10^5 km	$\sim 10^3$ km	30–500 km
Thickness	10^3 km	10^2–10^3(?) km	little
Normal optical depth	$\sim 10^{-6} - 10^{-7}$	$\sim 10^{-4} - 10^{-5}$	$\sim 10^{-1} - 10^{-2}$ (over ~ 100 km width) ~ 1 (2 km core)
Typical grain radius	micron	0.035 cm (model dependent)	submicron (optical) $\leqslant$ several cm (radio)
Character	diffuse, peaks at/near Enceladus	diffuse	kinky, clumped, braided core, diffuse outer reaches
Nearby satellites	Mimas (3.09 R_S) Enceladus (3.95 R_S) Tethys (4.91 R_S) Dione (6.29 R_S) Rhea (8.78 R_S)	none	shepherds: 1980S26 (2.349 R_S, $e = 0.0043$) 1980S27 (2.310 R_S, $e = 0.0027$)

detected by Pioneer 11 through its modest absorption of several species of charged particles (Van Allen 1983); the feature labeled "Janus?" in Fig. 11 of Simpson et al. (1980) is now known to be caused by the G Ring, which also accounts for the dip denoted 1979S3 in Fig. 4 of Van Allen et al. (1980). Decreases in counting rates were similar inbound and outbound, and were several tens of percent for 7- to 17-MeV electrons and for 0.5- to 1.8-MeV protons (Simpson et al. 1980), as well as for energetic (>80 MeV) protons (Van Allen et al. 1980). The signature extends $\sim \pm 0.1 R_S$, is centered at 2.81 R_S (Simpson et al. 1980) or at 2.82 R_S (Van Allen 1983), and seems to be caused by an azimuthally symmetric ring.

Voyagers 1 (Fig. 31c of Smith et al. 1981) and 2 (see Fig. 11; Smith et al. 1982) each sighted the ring in forward-scattered light. Since images were faint and badly smeared, no structure was discernible and, in fact, the actual dimensions of the ring are in question. Cuzzi (personal communication, 1983; cf. Smith et al. 1982) feels that, once smear is accounted for, the ring has a width $\Delta r \sim 2000$ km with poorly determined edges and a mean optical depth of 10^{-4} to 10^{-5}, depending upon particle albedo and phase function.

No satellites are known to be in the vicinity of the G Ring and the charged particle features (Simpson et al. 1980; Van Allen et al. 1980) appear to be macrosignatures, indicative of a ring rather than a satellite. Indeed, according to Van Allen's (1983) interpretation of the Pioneer charged particle results (cf. Voyager 2 data; Vogt et al. 1982), no satellites having radii on the order of a kilometer or larger are associated with the ring while objects with $r \gtrsim 10$ cm contribute less than 10^{-3} of the ring's opacity. Van Allen's analysis also leads to an effective particle radius of $\gtrsim 0.035$ cm.

G Ring particles may have impacted the Pioneer 11 and Voyager 2 spacecraft. The Pioneer penetration experiment was sensitive to grains with $r \gtrsim 10$ μm; its first two impacts in the neighborhood of Saturn occurred at 3.1 R_S, just outside the G Ring, when the spacecraft was 900 km above the ring plane. Two Voyager 2 instruments received intense noise near ring-plane passage at ~ 2.87 R_S, a little beyond the G Ring; the signals are thought to be produced by impact ionization of particles striking the spacecraft. Aubier et al. (1983) interpret the noise detected by the Voyager 2 radio astronomy experiment as being caused by micron-sized grains mainly spread over ± 70 km but measured as high as 700 km off the ring plane. The plasma wave experiment's results (Gurnett et al. 1983) can be explained by micron-sized grains (0.3 μm $< r < 3$ μm) with a m^{-3} mass distribution; such a distribution would have $\tau \approx 4 \times 10^{-5}$. The effective ring thickness is ~ 100 km, but some bursts were detected ~ 1000 km above the ring plane. These *in situ* measurements are each consistent with the G Ring being symmetric about Saturn's equatorial plane, a situation probably not true for the Jovian halo (see Sec. I.A); likely, the near-alignment of the Saturnian magnetic field with the planet's rotation axis allows the observed symmetry.

Fig. 11. A Voyager 2 long exposure image of the G Ring, F Ring and outer edge of the A Ring at a low scattering angle (19°3). Parts of the image are badly smeared as indicated by the visible star trails lying beyond the G Ring. (Figure from Smith et al. 1982.)

B. Saturn's E Ring

Telescopes on Earth have provided the bulk of the information available on Saturn's E Ring, which occupies more area than all other known planetary rings combined. Groundbased studies (summarized by Larson 1983; Pang et al. 1983b) have yielded results on ring structure as well as particle size and composition. Pioneer and Voyager data on charged particle absorptions and impacts, as well as very limited Voyager imaging, have helped elucidate the makeup of the E Ring. The overall structure of the ring is listed in Table II.

Marginal evidence for the E Ring's existence came during the 1966 ring-plane passage (Feibelman 1967; Feibelman and Klinglesmith 1980), when the slant optical depth through this extremely rarified ring became large enough to be barely perceivable; since the turn of the century observers had

WEST   EAST

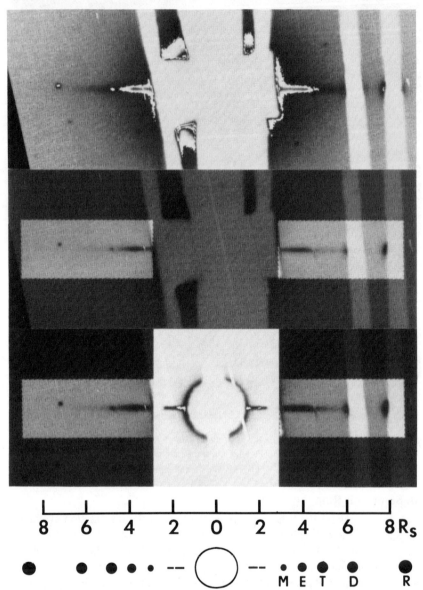

Fig. 12. A CCD frame taken by the USNO 1.5-m Flagstaff telescope near ring-plane passage showing the E Ring extending on both sides of Saturn. The planet's scattered light has been subtracted from the two exterior rectangles, the central rectangle is the area masked in the original E Ring images but here shown with a superimposed image of the A and B Rings. The scale at the bottom of the image shows the orbital distances of the classical inner Saturnian satellites. (Figure from Baum et al. 1981.)

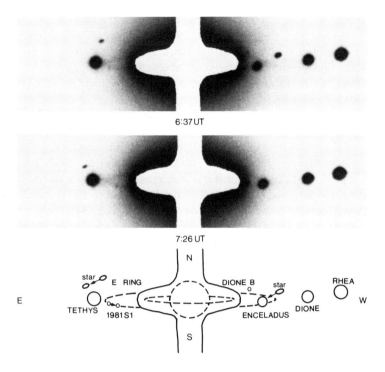

Fig. 13. Coronographic photographs of the E Ring taken by the 1.54-m Catalina Observatory reflector when the ring plane was inclined by 5°.4. The mask is the white central shape. The E Ring is narrow (indistinguishable from a line source which in these images has a FWHM of 8000 km) and is located at 4.07 ($\pm$0.07) R_S. Various objects are identified in the bottom sketch. (Figure from Larson et al. 1981.)

been suggesting the ring's presence but had been unable to prove their case. A good reference for our knowledge of the E Ring prior to the last ring-plane crossing is Smith (1978).

From observations made during the 1979–1980 ring-plane crossing, the E Ring is known to commence abruptly at about 3 R_S, to peak near Enceladus' orbit at 4 R_S (see Figs. 12, 13), and to disappear gradually at, or slightly beyond, 8 R_S (Baum et al. 1981; Dollfus and Brunier 1982; Lamy and Mauron 1981; Larson et al. 1981). Baum et al. (1983) convert the observed edge-on profile into a radial brightness plot; with the Rhea value ($\sim$ 8.8 R_S) taken as 1, the brightness is 13 at Dione (6.3 R_S), 125 at Tethys (4.9 R_S) and 10^3 at Enceladus (4.0 R_S); an alternative optical depth plot is presented by Hood (Fig. 1 of Hood 1983). The peak near the latter satellite has a full-width-at-half-maximum (FWHM) in the equatorial profile estimated to be 0.54 R_S according to the edge-on observations in Fig. 12 (Baum et al. 1980); it was unresolvable ($\leqslant$0.13 R_S) when seen at 5°.4 by Larson et al. (1981) in Fig. 13.

Estimates of the ring's maximum normal optical depth are 4×10^{-7} (Baum et al. 1980) and 10^{-6} (Reitsema et al. 1980; Larson et al. 1981). Voyager 2 imaging (Smith et al. 1982) confirms the E Ring's outer terminus and indicates that rings of similarly low τ are absent from the region 8–20 R_S; a groundbased CCD search for the tenuous rings between 10 and 36 R_S (Baron and Elliot 1983) has also been fruitless.

Some disagreement exists as to whether the peak is precisely at (Baum et al. 1981, 1983), just inside (Reitsema et al. 1980), or barely outside (Larson et al. 1981; Larson 1983) Enceladus' orbit; those placements derived from edge-on observations (Baum et al. 1981; Reitsema et al. 1980) are quite model-dependent. Further controversy concerns the E Ring's spatial (Lamy and Mauron 1981) and temporal (Dollfus and Brunier 1982) variabilities which have not been noticed by all observers (Baum et al. 1981; Larson 1983). Some slight radial brightness enhancements/depressions may be visible (Baum et al. 1981; Larson et al. 1981; Larson 1983), and charged particle microsignatures have been interpreted as indicating some clumpiness (Carbary et al. 1983).

The ring may be thicker in its outskirts than near its core (Baum et al. 1981); Baum and Kreidl (1982) estimate the ring's FWHM thickness at 7500 km (~ 0.12 R_S) near Enceladus' orbit but four times this value at its outer periphery. Larson (1983) qualitatively confirms this observation but cautions that signal-to-noise, especially bad for the faint outer reaches, can produce a false impression of broadening. Micrometeoroid impacts recorded by Pioneer and Voyager suggest that the maximum vertical extent of the ring is $\sim 10^3$ km at ~ 3 R_S, as described in the preceding section on the G Ring. Interestingly, evidence for grain impacts was not obtained by Voyager 1 during its only ring-plane crossing through the E Ring (~ 6 R_S); at this point intense electrostatic noise bursts were received within ± 2400 km of the ring plane but these seem to be shot noise due only to electrons and ions (Aubier et al. 1983). Information on the ring's radial and vertical profile also comes from the Voyager 2 Canopus star tracker (Pang et al. 1983a,b). The inversion of these data is not yet complete, but preliminary results on the radial profile (Pang et al. 1983b) are consistent with previous models. Pang et al. (1983a) claim to see vertical layering which is ascribed to electrostatic levitation; without a more complete presentation of the data, we are skeptical of this conjecture.

At visible wavelengths the E Ring exhibits a surprisingly strong opposition effect, in which its brightness doubles between phase angles of 6° and 0° (Dollfus and Brunier 1982; Larson 1983). Because of the low optical depth, this surge cannot be readily explained by mutual shadowing or multiple scattering, the classical interpretations for the B Ring's somewhat weaker opposition effect (see Cuzzi et al.'s chapter). Instead, Dollfus and Brunier (1982) interpret it as the backscattering enhancement of transparent particles several microns in radius; particles of moderate x (~ 20), even when somewhat nonspherical, exhibit this property (Cuzzi et al.'s chapter).

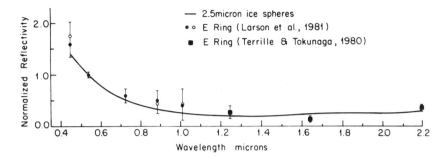

Fig. 14. The spectral reflectivity of the E Ring, normalized to unity at 0.54 μm. The solid line is a synthetic spectrum generated by Mie-scattering calculations for a size distribution of pure water-ice spheres with an effective radius of 1.25 μm and an effective variance of 0.1. (Figure after Pang et al. 1983*b*.)

A number of other observations suggest a significant component of micron-sized particles in the E Ring. First, Voyagers 1 and 2 could only image the ring during forward-scattered geometry (specifically at phase angles of 160° and 166°; Smith et al. 1981, 1982); the discussion of the Jovian ring photometry may, however, be pertinent. Second, micron grains have been called upon to produce the emissions detected by the particle wave science (Gurnett et al. 1983) and planetary radio astronomy (Aubier et al. 1983) experiments, although these signals were received at ~ 2.87 R$_S$, somewhat inward of the nominal E Ring. Lastly, a predominance of small particles has been invoked (Terrile and Tokunga 1980; Dollfus and Brunier 1982; Larson 1983) to account for the unusual spectrum of the E Ring (Fig. 14), which shows it to be the outer solar system's only blue object.

Pang et al. (1983*a,b*) have estimated the back-to-side scattering ratio (brightness at 0° to 6° phase angles vs. at 87° phase angle) as 10 to 20%. By comparing a large variety of phase functions for particles of different shapes, sizes, and complex refractive indices, they conclude that only dielectric spheres greater than a micron in radius match this phase behavior. They further argue that the backscattering strength fixes the real refractive index to lie outside the $\sqrt{2}\rightarrow 2$ range, and that the shape of the E Ring's backscatter peak indicates a very low imaginary refractive index ($\leq 10^{-3}$). Finally Pang et al. (1983*b*) contend that all optical and infrared observations of the E Ring are compatible with Mie scattering by a narrow size distribution of nearly pure water-ice spheres 1 to 1.25 μm in radius. Nevertheless, we caution once again that this model is merely consistent with the limited photometric data; other models may satisfy the measurements equally well.

Charged particle absorption signatures give additional clues as to the nature of the E Ring (McDonald et al. 1980; Van Allen et al. 1980; Sittler et al. 1981); unfortunately, these data have not yet yielded the information they purportedly contain (see Thomsen and Van Allen 1979; Hood 1983). Sittler et al. (1981; cf. Haff et al. 1983) employ the electron velocity distribution encountered by Voyager 1 to locate the E Ring in roughly the same place as do other techniques. However, as measured in this way, the ring is not radially symmetric about Saturn, but extends to nearly 9.0 R_S in the noon hemisphere and only to 6.5 R_S in the midnight hemisphere; similar values were observed by Pioneer 11. Such an asymmetry is not unheard of in magnetospheric physics but would be startling, if true, for a dust ring.

Particle sizes and optical depths may be suggested by variations in the spectra of energetic electrons and protons. McDonald et al. (1980) attribute a roll-over seen at low energies to ~ 2 mg cm^{-1} of light Z (atomic number) material (i.e., Si or O); assuming a 10-day lifetime for their measured species in the E Ring neighborhood, they estimate grain radii to be $\sim$ a few μm and optical depths $\sim 10^{-3}$ but much more material could be present. Van Allen et al. (1980) also ascribe features in the distribution of low-energy electrons (0.040 to 0.56 MeV) to a tenuous ring of fine particulate matter; Frank et al. (1980) point out that gas could produce similar effects.

C. Saturn's F Ring

The F Ring's structure (see Fig. 20 in Cuzzi et al.'s chapter; Figs. 1–3 in Dermott's chapter) has been the focus of attention since 1979 when the ring was discovered by Pioneer 11 (Gehrels et al. 1980) and characterized as clumpy. Detached from the A Ring by about 4000 km (~ 0.07 R_S), the F Ring has two attending satellites that are thought to confine it (Goldreich and Tremaine 1982). As noted in Table II, the ring and the satellites lie on eccentric orbits. The orbital shapes are such that the inner satellite's apoapse almost overlaps the ring's pericenter (Synnott et al. 1983; Fig. 14 in Dermott's chapter) while the outer satellite's orbit gets only as close as ~ 500 km; since the orbits differentially precess in Saturn's oblate gravity field, ring particles should pass within 70 km of the satellite's surface every eighteen years, having last done so eight years before Voyager's arrival. The resulting complex interaction has been explored by Borderies et al. (1983) who note that, including the ring's forced eccentricity, the inner satellite actually penetrates at least part of the bright ring. (See also the next paragraph.) Borderies and co-workers propose that these extraordinary dynamical events may explain the peculiar morphology of the ring. To close the circle, we remind the reader that the ring's eccentricity is the outcome of competing processes (Goldreich and Tremaine 1982; see also Showalter and Burns 1982)—satellite perturbations and internal dissipation.

The orbital inclinations of the ring and its shepherds are all $0°0$ ($\pm0°1$). Pollack (personal communication, 1983) interprets ring photometry as indicating that the ring is flattened rather than toroidal. Marouf and Tyler (1983) infer thickness of < 30 m from an occultation of the Voyager radio signal by the F Ring's core.

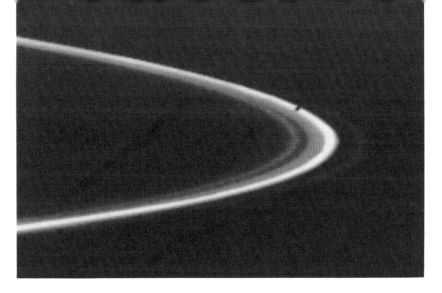

Fig. 15. Voyager 2 image (FDS 44006.49) of the F Ring. At least four ringlets surround a bright core in this image, showing 25° in longitude near the F Ring ansa. The faint strands appear smooth and may blend together; in contrast the edges of the bright section are irregular. (Figure from Smith et al. 1982.)

As with so many other aspects of the F Ring, one must be very precise when describing the ring's radial extent for it differs when measured in different ways and/or at different epochs/places. Three components, each about 20 to 30 km wide, were visible to Voyager 1 cameras and were separated *in toto* by about 100 km; the innermost component was much fainter and less kinked than the other two. The latter pair occasionally merged and varied in brightness (see below) but at other spots appeared straight and parallel. Voyager 2's more sensitive cameras identified faint material spread over ~ 500 km (Smith et al. 1982). This structure was broken into three or four ringlets by relatively clear lanes and also contained an irregular bright ring, much like the F Ring spied by Voyager 1, located toward the outer edge of the overall feature (Fig. 15). The faint inner ringlets are reminiscent of Jupiter's rings, although of substantially higher τ and much narrower. Their apparent azimuthal smoothness is hard to understand since the inner shepherd should pass through this region. These rings deserve the scrutiny that Jupiter's ring is now receiving (Showalter et al. 1984; Burns et al. 1983).

Using the same image processing system that allowed us to produce Color Plates 9 and 10 of the Jovian ring, we have begun study of the F Ring. In the Voyager 1 image shown as Fig. 19 in Cuzzi et al.'s chapter, we find several faint rings both inside and outside the three components already mentioned. The total ring width is several hundred kilometers and the faint components exhibit periodic brightness variations. With the same equipment even fainter material can be seen inward of the F Ring, all the way to the A Ring's perimeter (Burns et al. 1983); this material is noticed in images crossing the shadow boundary as well as highly enhanced versions of Fig. 15. The radio and photopolarimeter occultation profiles (Graps et al. 1983; H. A. Zebker, personal communication, 1983) suggest a tenuous band between Atlas and the A Ring as well as clumps in the vicinity of the F Ring itself.

A stellar occultation tracked by Voyager 2's photopolarimeter (PPS) found a 50 km wide ring with structure on scales $\leqslant 0.5$ km. The mean optical depth $\bar{\tau}$ was ~ 0.1 but a dense 2 km wide ringlet near the structure's outer edge had $\bar{\tau}$ about 1. Voyager 1's 3.6 cm radio signal was affected by the F Ring, which from the occultation appears to be a single strand 2.4 km wide with $\bar{\tau} = 0.115$ (Tyler et al. 1983); at 70 m resolution, a 300 m inner core with $\tau = 0.5$ is seen to be surrounded by a 2.4 km pedestal (Marouf and Tyler 1983). It is tempting to associate the core identified by the Voyager 2 PPS scan with the entire Voyager 1 feature. Since the occultation was not detected at 13 cm ($\bar{\tau} < 0.005$), particles larger than 4 cm occupy less than 4% of the total absorbing area. Furthermore, even the 3.6 cm τ is much smaller than the value derived from the stellar occultation, indicating that, if the same structure is being viewed, it contains mainly particles of microns or smaller. Since the radio science experiment did not see the faint halo, it too must be composed of small grains. The failure of the PPS occultation to identify broad flanks on the F Ring is not disturbing since it is sensitive only to $\tau \geqslant 0.01$ and clumps of this optical depth may be present (Graps et al. 1983) in the vicinity of the F Ring.

Charged particle data obtained by Pioneer 11 (Van Allen et al. 1980; Simpson et al. 1980; see Cuzzi et al.'s Fig. 20) clearly show absorbing material located between 2.32 R_S and 2.36 R_S. On the inbound pass, three obstacles appear to be present, while outbound only two absorptions are noticeable. Although one might interpret the data as the ring's macrosignature added to microsignatures of the satellites (or a microsignature of an eccentric or displaced ring; Ip 1980a), the most complete published study claims that the ring and the satellites are not properly placed to account for the absorption (Van Allen 1982; cf. Simpson et al. 1980). Van Allen (1982) therefore argues that the data testify to as yet unseen clumpy material scattered throughout the F Ring locale, a view also subscribed to by Cuzzi et al. (their chapter). Even though this interpretation of the data has most interesting dynamical implications for producing the F Ring's odd form, we do not feel that a full understanding of the data has yet been achieved. Indeed unpublished work (1983) by Cuzzi and Burns seems to indicate that the ring is responsible for two of the five signatures.

The F Ring increases in brightness at phase angles from 140° to 155° (Pollack 1981; Smith et al. 1981), indicating the presence of particles with sizes of a few tenths of microns ($\bar{\tau} \leqslant 0.1$) throughout the central core. This brightening happens especially in knotted or clumped regions suggesting that particle sizes differ in these regions versus the rest of the ring. However, recall from the section on Jovian ring photometry (Sec. I.C) that such photometric observations by their very nature do not exclude ring members of other sizes.

Indeed, additional ring photometry at phase angles of 20° to 30° provides evidence for a larger particle component ($\bar{\tau} \leqslant 0.01$). Photometry was not carried out at the small scattering angles that would highlight particles in the 1 to 10 micron range; presumably these objects are present too and account for much of the ring's optical depth as measured in the PPS occultation (Lane et al. 1982), since the latter values are almost an order of magnitude larger than Pollack's (1981).

Techniques do not exist for discriminating ring particles with radii much larger than meters; nevertheless circumstantial evidence hints that some very large (1 km $< r <$ 10 km; i.e., satellite-sized) objects are embedded in the bright ring. Smith et al. (1982, see their Fig. 40) have identified clumps of this scale in the ring. The edges of several of these bright spots are sharply defined, suggesting individual solid bodies in the ring; at other places one side is diffuse, raising the question of whether all bright spots could merely be due to enhanced grain densities (see above). Weidenschilling et al. (Fig. 8 in their chapter) numerically simulate the development of bright knots in a many-particle ring, which contains at least one large member.

The knots, kinks, and braids visible in the Voyager images have characteristic wavelengths of $\sim 10^4$ km (Smith et al. 1982); originally, the scale was thought to be an order of magnitude smaller because image foreshortening had been overlooked. Even though statements about wavelengths imply a regular phenomenon, the reader should not be misled; spacings vary from 5×10^3 km to 13×10^3 km, often the ring appears smooth (perhaps especially during the Voyager 2 encounter), and features falling into the same category (kinks, braids, etc.) are not carbon copies of one another.

A variety of observations clearly tell that the ring is either time-variable or spatially variable (see above). Nevertheless, Voyager 2, while following clumps over periods of at least 15 orbits, discerned no apparent variation in their longitudinal distribution. We suspect that additional information on periodic brightness variations is contained in the Voyager data set; a Fourier analysis should be performed on the entire data set to identify specific wavelengths and/or to indicate correlations with satellite orbits.

III. PHYSICAL PROCESSES

This section of the chapter will describe a variety of processes that are especially important for rings of low optical depth which contain many small grains and reside in a magnetosphere. In particular, we will discuss the consequences of the expected electric charges on ring particles, grain destruction mechanisms and lifetimes, orbital evolution processes for small particles, the production of tiny grains, thickening and the vertical structure of the ring, the effects of embedded satellites, and factors that influence the particle size distribution. Owing to the similarity of the Jovian ring in many regards to Saturn's E, G, and F Rings, we will consider them simultaneously.

Even though these rings all probably contain a spectrum of particle sizes, we will generally consider processes to act on 1 μm grains both to be specific and because particles near such sizes are the best-observed constituents of ethereal rings.

A. Electric Charge and Some of its Dynamical Effects

The question of the electric potential developed by a surface in space is one of general astrophysical interest (see, e.g., Lafon et al. 1981; Whipple 1981). Electrical effects are more important for the ethereal rings under discussion than for optically thicker rings because the small sizes of the constituent particles imply relatively large charge-to-mass ratios and because the charge on an individual particle becomes larger as grains are further separated from one another. Isolated grains in a plasma develop a potential such that the net electrical current to them averages to zero; the most important processes in defining this potential in usual planetary situations are collisions with electrons/ions and photoionization by solar ultraviolet photons (see chapter by Grün et al.). Because the solar insolation in the outer solar system is small and the electron fluxes of the magnetospheric plasmas surrounding Jupiter and Saturn exceed the proton fluxes, grains maintain a negative potential Φ in order to ward off the faster moving electrons. Once an isolated grain's potential is known, its charge q is determined from $\Phi = q/r$.

Since photoelectric charging can be ignored, day/night variations are not important for the Jovian and Saturnian cases even though they are for circumterrestrial artificial satellites. A particle's potential is therefore fairly constant. Nevertheless, some variations occur due to the stochastic nature of the collisions with plasma components (see Morfill et al. 1980a) and due to some systematic effects (see chapter by Grün et al.). A reasonable approximation is that the potential developed is of the same order as the kinetic energy corresponding to the electron temperature. We select a nominal value of -10 V for the potential of dust in the Jovian and Saturnian systems; this is in fair agreement with more complete treatments for the electric potentials on isolated circumplanetary grains (see Grün et al.'s chapter).

Collective effects and absorption of the plasma by the ring, however, will lower *potentials* in regions of even modest optical depth such as the F Ring. In addition, collective effects will substantially reduce the *charge* that resides on individual grains in rings once the particle spacing d is less than the Debye length λ_D (the distance over which a charge in a plasma is effectively shielded). We estimate d from

$$d^{-3} \sim N = \tau/(\pi r^2 h) \quad , \tag{2}$$

where N is the spatial density of grains. Using the values in Table I, for Jupiter's main ring $d_R \lesssim 20$ cm while for the halo $d_H \sim 10^2$ cm; from Table II, in Saturn's diffuse rings $d_E \sim 10^2$ cm, $d_G \sim 10$ cm and $d_F \sim 1$ cm. As Grün et al. show, in planetary magnetospheres $\lambda_D \approx 10^3 / n_e^{\frac{1}{2}}$ cm with n_e the electron density. Thus $\lambda_D \approx 10^2$ cm for the main Jovian ring. Since measurements of the plasma have not been made close to Jupiter, λ_D is not well specified for the halo, although it is likely to also be of order 10^2 cm. Throughout the Saturn system $\lambda_D \sim 10^2$ cm. Hence in terms of the charge that resides on a grain, our selection of the charge corresponding to a -10 V potential is likely to be correct for particles in the Jovian halo and Saturn's E Ring, may be an acceptable approximation for the grains in Saturn's G Ring and the Jovian main band, but clearly is a gross overestimate of the charge residing on the particles of Saturn's F Ring (cf. Table II in Grün et al.'s chapter).

Electrical effects can influence particle dynamics and perhaps even cause electrostatic breakup as described below (see also the chapter by Grün et al.). Particle dynamics depend upon Lorentz forces which are proportional to electric charge q, so that the electromagnetic perturbation relative to gravity will vary as q/m. For a uniform surface charge distribution on a sphere, $q = \Phi r$; since in most circumstances of interest (see chapter by Grün et al.), Φ is constant and fixed by the kinetic energy of the plasma electrons, $q/m \sim r^{-2}$ and so small grains are most affected. However, as still smaller sizes are considered, surface electric fields (which otherwise vary as r^{-1}) ultimately exceed the material's work function, and excess charges are expelled so as to maintain the surface *field* constant ($E = qr^{-2}$). (See Grün et al.'s Fig. 7 which plots the perturbation as this field emission limit is approached.) In this case, $\Phi \sim r$ and, accordingly, $q/m \sim r^{-1}$. Thus, at very small particle sizes (<0.1 μm for the nominal -10 V potential)—just when one might expect electromagnetic stresses to be large and Lorentz forces to overwhelm gravity totally—field emission weakens both effects.

A good measure of the relative importance of electromagnetic forces in influencing dynamics is the strength of the Lorentz force (due to the particle's motion relative to the planet's magnetic field) compared to the planet's gravitational attraction. The Lorentz force experienced by a charged grain moving at orbital velocity ΩR through the local B field (here taken to be a dipole normal to the orbit and of magnitude $B_0(R_p/R)^3$, where B_0 is the magnetic field at the planet's surface $R = R_p$) is

$$F_m = qB \frac{(\Omega - \omega)R}{c} \tag{3a}$$

with Ω the orbital angular velocity, ω the planet's spin rate and c the speed of light. This expression shows that only the particle's velocity with respect to the moving field is considered; that is to say, the Lorentz force $q[(\mathbf{v}/c) \times \mathbf{B} + \mathbf{E}]$ is measured in a reference frame rotating with the field (and with the plasma) because in such a frame $\mathbf{E} = 0$.

The planet's gravitational attraction can be written from the centrifugal acceleration it produces:

$$F_g = \frac{-GmM_p}{R^2} = m\Omega^2 R. \qquad (3b)$$

The relative strength of the electromagnetic force then is

$$\frac{F_M}{F_g} = \frac{3\Phi B_o c (R_p/R)^3}{4\pi\rho r^2 \Omega} \; (1 - \omega/\Omega) \qquad (4a)$$

for spherical particles charged to a potential Φ. We rearrange Eq. (4a) and define the so-called critical radius, the particle size at which the Lorentz force equals gravity, as

$$r_c \approx \frac{1}{4} \left[\frac{c\Phi B_o (R_p/R)^3}{\rho\Omega} \right]^{\frac{1}{4}} , \qquad (4b)$$

where we have approximated the second parenthesis in Eq. (4a) by 0.3. If $\Phi = -10$ V and $\rho = 2$ g cm^{-3}, then $r_c \approx 0.3$ μm in the Jovian ring but is about an order of magnitude smaller for Saturn's outlying rings.

Several dynamical regimes can now be described using Eqs. (4) as a guide. For those particles which are so large (say, $r \gtrsim 10$ μm) that the electromagnetic force is negligible compared to gravity, the local Keplerian velocity in the Jovian ring is 31 km s^{-1}; the opposite limit, in which the particle is so tiny (say, $r \lesssim 0.03$ μm) as to essentially corotate with the field, corresponds to an orbital velocity of 23 km s^{-1}. There is a smooth transition between these two cases as the particle size varies (see Fig. 1 of Burns et al. 1980).

For particles in the micron-size range the added perturbation is small and accordingly the orbital velocity which satisfies centrifugal balance deviates only slightly from that for gravity alone. The relative velocity Δv which a slightly perturbed, charged grain drifts with respect to a neutral (or, equivalently, a massive) grain can be found from

$$F_g + F_m = m(\Omega R + \Delta v)^2/R. \qquad (5)$$

Substituting Eqs. (3) into Eq. (5), we have

$$\Delta v \approx \frac{q}{2mc} \frac{B_o R_p^3}{R^2} (1 - \omega/\Omega). \qquad (6)$$

For circumjovian particles,

$$\Delta v \approx 300(\Phi/10 \text{ V}) (1 \text{ }\mu\text{m}/r)^2 \text{ m s}^{-1}, \qquad (7)$$

so that the typical micron-sized grain strikes other objects at ~ 300 m s^{-1}, somewhat less than the speed of a bullet. Because of the weaker Saturnian field, Δv in the outlying rings is about an order of magnitude smaller.

We have just described a relative drift motion that would be present between a pair of grains, one charged and the other neutral, when both are on perfectly circular orbits. Such a configuration would be difficult to achieve since the typical ring particle probably originates on a larger parent body, is abruptly ejected from that body by a collision, and then swiftly acquires its charge. In this case, it is more meaningful to ask how large is the deviation of the now-charged grain's orbit from its uncharged one. Writing the circular velocity as v_c, this excursion can be estimated as

$$\Delta a \sim (\Delta v/v_c)\, a \qquad (8)$$

which from Eq. (6) is of the order of 10^3 km for a 1 μm grain in Jupiter's ring (cf. Fig. 18 of Consolmagno 1983); the corresponding orbital eccentricity then is $e \sim \Delta v/v_c$ or ~ 0.01.

The role of electromagnetic forces in producing orbital evolution and in thickening the ring will be appraised in later sections (see Secs. III.C and III.F).

B. Grain Destruction and Particle Lifetimes

Many of the particles making up the visible Jovian ring are quite tiny, as mentioned above. Such motes cannot survive very long because sputtering by magnetospheric ions gradually eats them away and because impacts by micrometeoroids and other dust can abruptly cause their demise. Elimination through sublimation is not competitive with these processes on the nonicy materials suggested for the Jovian ring, but could be if instead they were water ice. Near Jupiter a black water-ice grain would vaporize in $40(r/1\ \mu m)$ yr; a more realistic 60% albedo grain would last $10^7(r/1\ \mu m)$ yr (Watson et al. 1963; Harrison and Schoen 1967; Pilcher 1979). For the colder temperatures in the Saturnian system, even a totally absorbing water-ice grain would survive for the age of the solar system; however other cosmically significant ices can have quite short lifetimes (e.g., black ammonia ice sublimates at 12 μm yr^{-1}; Watson et al. 1963).

Sputtering by energetic ions and electrons is surprisingly effective in destroying ring particles because the ejection process is efficient, and because large fluxes of energetic particles are found in planetary magnetospheres. This process has rates with substantial variations and uncertainties, owing to our incomplete knowledge of the impinging particle fluxes and to large differences in the sputtering yield depending upon target properties and impacting particle energy; occasional surprises have shown up in recent laboratory measurements (Brown et al. 1982). The sputtering lifetime T_s of a grain with mass M can be estimated from the mass loss rate $\dot{M}$ as

$$T_s(r) \sim \frac{M}{\dot{M}} \sim \frac{Nr}{\sum_i F_i \xi_i} \qquad (9)$$

where N is the number density of molecules in the grain and a sum is taken for all ions of the product F_i (the flux of the i-th ion) times ξ_i (the sputtering yield per impact by the i-th ion).

Using Eq. (9) with spacecraft data on the makeup of the charged particle flux and laboratory measurements on yields, Burns et al. (1980) conclude that in the Jovian ring $T_s \approx 30$ ($r/1$ μm) yr (water ice) and $\approx 10^3$ ($r/1$ μm) yr (silicates) to within an order of magnitude (cf. Morfill et al. 1980a; Johnson et al. 1981; chapter by Grün et al.). Since the flux of damaging ions depends sharply on distance from Jupiter, grain lifetimes will too, being briefest at the ring's outer edge and rising within the ring. According to the flux observations (Fig. 1), sputtering proceeds more rapidly once again near the planet; however, some caution should be applied in directly interpreting the flux observations since the spacecraft did not equally sample all equatorial radial locations (Fillius, personal communication, 1980).

The water-ice erosion rate in Saturn's E Ring due to protons above 50 kev is $\sim 10^{-4}$ μm yr^{-1} (Cheng et al. 1982; cf. Morfill et al. 1983b; chapter by Grün et al.); while several orders of magnitude slower than the Jovian case, this is still fast enough to require that the visible E Ring particles be recently introduced. Particle fluxes are such that sputtering is highest between ~ 4.5 R_S and ~ 8 R_S, where most of the E Ring is located. Photosputtering has a comparable rate (Harrison and Schoen 1967). Sputtering by heavy ions, notably oxygen, which is tied to the corotating field lines, results in an erosion rate of $\sim 10^{-2}$ to 10^{-4} μm yr^{-1} (Morfill et al. 1983b; Haff et al. 1983) for water-ice targets. Nearer to Saturn, where particle fluxes may be reduced, lifetimes may lengthen.

Collisions with micrometeoroids (Burns et al. 1980) or with Io-derived volcanic dust (Morfill et al. 1980b) also destroy Jovian ring particles. Catastrophic fragmentation, in which a grain is shattered by a single impact, dominates over progressive erosion (Burns et al. 1980; Grün et al. 1980); this implies that number fluxes (and not *mass* fluxes) of projectiles determine grain lifetimes and, assuming a lower cutoff to the projectile size, that small particles survive longer than large ones. By modeling the impact process, Burns et al. (1980; see below) estimate that the collision lifetime $T_c \approx 10^5$ (1 μm/r)^2 yr to within an order of magnitude. The number of interplanetary micrometeoroids in Saturn's vicinity is similar to that in Jupiter's neighborhood (Humes 1980) and, accordingly, catastrophic disruption rates ought to be comparable in the two systems once the effects of gravitational focusing are removed.

The projectiles that cause sputtering and chipping also produce orbital evolution, which will be discussed in the next section.

Electrostatic bursting has been considered as another possible means for destroying ring particles. Stresses develop in any electrically-charged material as the charges try to repel one another. This electrostatic stress is $\sigma_E \sim \Phi^2/r^2$, or $\sim 10^4$ dyne cm^{-2} for the nominal -10 V potential on a micron grain (cf. Fig. 4 in the chapter by Grün et al.). When σ_E exceeds the material's strength Σ, the grain will rupture into smaller pieces. This criterion states that breakup will occur once

$$\left(\frac{\Phi}{10 \text{ V}}\right)^2 \left(\frac{1 \text{ }\mu\text{m}}{r}\right)^2 \gtrsim 10^{-4} \left(\frac{\Sigma}{\text{dyne cm}^{-2}}\right). \tag{10}$$

Given likely strengths (see Grün et al.'s chapter; Burns et al. 1980), fracture is improbable. Nevertheless, it may act to remove tiny surface asperities, since r is effectively less at points with small local curvatures and so the local field (and hence stress) can be substantially enhanced. In this way, a grain might be gradually smoothed and eroded away.

The lifetimes estimated above are summarized in Table III. None of these estimates is well determined because the physics of sputtering and catastrophic fragmentation are poorly understood, and because the pertinent parameters are imperfectly known. Nevertheless, the basic finding for Jupiter's ring and Saturn's diffuse outer rings is inescapable: particle lifetimes are brief. This is crucial for understanding the nature of these structures, since short lifetimes require that the visible particles be continually replenished.

C. Orbital Evolution and Grain Removal Mechanisms

The particle lifetimes calculated above for erosion and destruction should be compared against characteristic times for several orbital evolution mechanisms: do particles disappear entirely from the system by their destruction or do they simply drift elsewhere? Drag forces will especially dominate the evolution of small grains since such forces are proportional to area whereas gravity is proportional to volume; the ratio of drag to gravity therefore varies inversely with particle size and so is largest for small particles.

Poynting-Robertson drag drains orbital angular momentum in the process by which particles absorb and reemit radiation (see Burns et al. 1979b; chapter by Mignard). Although radiation that is reflected and emitted from the planet can be a substantial fraction of direct solar radiation for ring particles near the giant planets, it is adequate for our approximate purposes to consider only the solar component. Burns et al. (1979b) have shown that the orbital collapse age under Poynting-Robertson drag is approximately the time it takes for a particle (whether in circumplanetary or circumsolar orbit) to absorb the equivalent of its own mass in radiation:

$$T_{\text{PR}}(r) \sim 10^3 \, (R_p/\text{AU})^2 \, (\rho/\text{g cm}^{-3}) \, (r/\mu\text{m}) \, Q_{\text{pr}}^{-1} \text{yr} \quad , \tag{11}$$

TABLE III
Characteristic Times for Micron Grains in Tenuous Rings[a]

| | Jovian Ring | | Saturn's Rings | | |
	Main Band	Halo	E	G	F
Loss					
Evaporation (H$_2$O ice with albedo 0.6)	10^7	10^7	10^{17}	10^{17}	10^{17}
Sputtering (Eq. 9) (silicate) (ice)	$10^{3\pm1}$ $3 \times 10^{1\pm1}$	$10^{3\pm1}$ (??) 10^2(??)	$10^{3\pm1}$	$10^{4\pm1}$	$10^{4\pm1}$
Meteoroid erosion	$10^{5\pm1}$	$10^{5\pm1}$	$3 \times 10^{6\pm1}$		$10^{6\pm1}$
Evolution					
Poynting-Robertson drag (Eq. 11)	$\geqslant 10^5$	$\geqslant 10^5$	$\leqslant 10^6$	$\leqslant 10^6$	$\leqslant 10^6$
Plasma drag (Eq. 12)	$2 \times 10^{2\pm1}$	10^{3+2}_{-1}(??)	$2 \times 10^{2\pm1}$	$2 \times 10^{3\pm1}$	$2 \times 10^{3\pm1}$(??)
Ring flattening (Eq. 15)	<10	(75)	10^{3-4}	10^{2-3}	10^{-2}
Generation					
Micrometeoroid impact into moons (Eq. 13)	$<2 \times 10^{3-4}$ (hard regolith) 50–500 (soft regolith)	source unknown	??	??	??
Recollision into satellite (Eq. 21)	30		50		

[a]Times are in years. The reader is cautioned to use this table only as a quick guide. Some caveats on these numbers are given in the body of this chapter or that by Grün et al.; better yet, see the referenced papers.

where R_p is the planet's orbital radius, ρ and r are the particle density and radius, and Q_{pr} is the radiation pressure coefficient which is ~ 1. According to Eq. (11), individual Jovian ring grains will evolve to the planet's surface in $\gtrsim 10^5$ yr, while at Saturn this approaches 10^6 yr. However, since such grains should collide with larger ring members that are virtually unaffected by Poynting-Robertson drag, these times are distinctly lower limits. Over the age of the solar system, Poynting-Robertson drag can eliminate cm-sized pebbles from Jovian orbit, and mm-sized ones from Saturnian orbit.

In addition to the gradual spiral induced by Poynting-Robertson drag, the solar radiation through its direct force causes the eccentricity of circumplanetary orbits to oscillate with a period like the planet's orbital period about the Sun. Accordingly, during a single cycle of this eccentricity oscillation, a particle's orbit is on average effectively smeared radially. However, as Mignard mentions in his chapter (cf. Burns et al. 1980), the precession of a ring orbit due to the planetary oblateness severely limits the effectiveness of this process. For small perturbations, the typical eccentricity developed by solar radiation is $(3\pi/2)\beta(v_\odot/v)(P_{prec}/P_\odot)$, where β is the ratio of the radiation pressure force to solar gravitational attraction, $v_\odot$ and $P_\odot$ are the planet's orbital velocity and period, v is the particle's velocity about the planet, and P_{prec} is the orbital precession period. For a perfectly absorbing particle in the Jovian ring, this forced eccentricity is $\sim 2 \times 10^{-3} (\mu m/r)$. The corresponding radial excursion is about 100 km for the visible Jovian grains and, as such, would not have been observable. For particles about Saturn, the eccentricity due to solar radiation is also fairly small, reaching at most $10^{-1} (\mu m/r)$ for the outer reaches of the E Ring.

Plasma drag (see Grün et al.'s chapter) is caused by the momentum transmitted as the particles making up the thermal plasma flow past a ring grain; since the grains are charged, momentum transfer occurs both through physical collision and through long-range charged particle interactions. Both can be included by using the particle's physical cross section or its Coulomb cross section, whichever is larger (see Grün et al.'s chapter). The ions and electrons in the plasma are tied to the planet's rotating magnetic field, and hence move more slowly than grains traveling within synchronous orbit, but more rapidly beyond that distance. Therefore, plasma drag causes orbital *collapse* for Jovian ring particles (since there $R_{ring} < R_{syn} = 2.29$ R_J) but orbital *expansion* for the particles in the tenuous outer Saturnian rings (where $R_{ring} > R_{syn} = 1.82$ R_S). Charged particles more energetic than the thermal plasma drift with respect to the rotating magnetic field. For both Jupiter's and Saturn's magnetic field directions, ions drift in the direction of the planet's rotation. Since these particles carry much more momentum than the oppositely-drifting electrons, the drag due to energetic charges switches signs at a position closer to the planet than synchronous orbit.

The orbital evolution time scale for plasma drag (Morfill et al. 1980a; Burns et al. 1980; Consolmagno 1983; cf. Dermott's chapter) is

$$T_{PD} \approx 2\rho r \xi / (3\rho_p v) \quad , \tag{12}$$

where ρ_p is the mass density of the thermal plasma, v is the grain's velocity relative to the plasma; ξ (the reduction applied when the particle's charge has an influence) is 1 for effectively uncharged grains (or in the case when the flow of the plasma past the grain is highly supersonic) and is relatively constant $\sim 10^{-2}$ (Eq. 49 in chapter by Grün et al.) once the flow is subsonic and the Coulomb cross section must be included.

The thermal plasma in Jupiter's inner magnetosphere (Belcher 1983; cf. Bagenal and Sullivan 1981) seems to be dominated by heavy ions (S^+, O^+, S^{2+}, O^{2+}, Na^+) in supersonic flow (thermal velocities are ~ 1 to 2 km s^{-1} compared to relative velocities of about 10 km s^{-1}) rather than by protons in subsonic flow, as Pioneer experimenters had thought. In terms of drag rates, the two cases do not differ much (because ξ's effect is counteracted by the lower proton mass). Under the assumption that heavy ions are most important, we find $T_{PD} \approx 2 \times 10^2$ yr but we caution that the plasma density varies rapidly inside Io's orbit and is not well constrained by the available measurements. Hence T_{PD} is likely a lower bound and, in view of the short lifetimes already mentioned, we cannot be certain whether plasma drag is the principal process to remove material from the ring.

The plasma drag time scale for micron grains in Saturn's E Ring has been estimated by Morfill et al. (1983b) to be 20 yr at 3 to 7.5 R_S; the Voyager 2 observations (Bridge et al. 1982) have found similar plasma parameters in as far as 2.7 R_S. Throughout this region heavy ions (principally O^+ with $n \sim 10^2$ cm^{-3}) determine the mass density of the plasma but here, in contrast to the Jovian magnetosphere, thermal speeds seem to be ~ 10 km s^{-1} so that the Coulomb correction must be applied; an alternate interpretation of the data (Bridge et al. 1982) yields lower thermal speeds (~ 2 km s^{-1}), and therefore supersonic flow with the plasma drag time scale lengthened by two orders of magnitude. Closer to Saturn, nearer the F Ring, plasma densities are not well specified and may be lower, meaning that the drag is reduced.

An orbital evolution also arises from the random impacts of micro-meteoroids, the same projectiles that we earlier considered to erode and destroy ring grains. Our approach to this time scale will be general. The characteristic evolution times for both plasma drag and Poynting-Robertson drag can be thought of as approximately the interval it takes for a grain to modify its original momentum by an amount approximately equal to that momentum. If the material bringing in the new momentum has no mean momentum of its own, then the evolution time is approximately the time for the particle to interact with its own mass. In this framework the evolution time for an uncharged grain of mass M and of cross section $A \sim M/(A \rho v)$ due to plasma drag, ρv being the plasma flux density with ρ the plasma density and v the relative velocity. Here we ignore that the plasma carries a mean momentum due to being tied to the rotating magnetic field; the associated error is a factor of several. With Poynting-Robertson drag the impacting density can be obtained from $E = mc^2$ (see Burns et al. 1979b, pp. 31–33). An expression similar to $t \sim M/A\rho v$ should hold roughly for micrometeoroid impacts (cf. Eq. 12 of Harris and Kaula 1975) where ρv is now the meteoroid flux density at the ring; this material presumably approaches the planet randomly and so contains no mean momentum. Before considering further the orbital evolution time scale due to micrometeoroids, we note that the time before disappearance through erosion is $\sim M/\dot{M} = M/(A\rho vY)$, where Y is the impact's yield (discussed in Sec. III.G with Fig. 17); for catastrophic disruption, the lifetime is even less. Since $Y > 1$ (and often $>> 1$) for all these processes except sputtering, particles can orbitally evolve little before they disappear. Hence the precise evolution time scale is not important. One exception may concern material leaving Saturn's F Ring; since it is so near the main ring system, orbits would only have to collapse a few percent in order to collide with the A Ring.

The orbits of charged grains diffuse by two separate processes, one due to stochastic variations in the particle's charge and the other caused by fluctuations in the magnetic field. In the first of these, changes of the electric charge abruptly produce corresponding alterations of the particle's gyroradius; these continually modify the field line about which the particle gyrates and thereby generate a random walk (see Morfill et al. 1980a,d; Grün et al. chapter). Morfill et al. (1980a) conclude that such a diffusion process could transport Io-derived volcanic dust grains into the neighborhood of the Jovian ring where they might serve as projectiles to excavate the parent bodies of the visible grains; Northrop and Hill (1983) caution that this diffusion will be counterbalanced by a radial drift due to systematic charge fluctuations at the particle's gyrofrequency and that this will drive particles toward synchronous orbit. However, the latter effect relies on charge fluctuations of less than one electron, even though particles are considered to be charged up to a surface potential of a few hundred volts. Other processes—for instance, systematic radial drifts imposed by the plasma density (and energy) gradients expected near rings—are more significant by far, as are the stochastic surface charge variations of Morfill et al.; these may completely mask the Northrop and Hill effect.

As a second diffusion process, Consolmagno (1980; cf. Morfill and Grün, 1979a,b) proposed electromagnetic scattering as a way to thicken and spread the ring. This case was treated using an orbit diffusion theory into which Consolmagno inserted higher-order (i.e., nondipole) components of the Jovian field as the perturber of particle orbits. Unfortunately diffusion is not the consequence of the periodic forces due to offset and tilt of the field since a true diffusion results only if the process is strictly Markovian. Orbit scattering is resonant in the sense that electromagnetic waves, with wave periods similar to the orbit period, are most effective. For a period ~ 10 hr, random magnetic field fluctuations are very small in the inner Jovian magnetosphere. Therefore electromagnetic scattering is negligible for dust grains (cf. Morfill et al.

1980*a*). Of course, this does not imply that deterministic orbit variations due to the known tilt of the field are absent. We discuss these in Section III.F below, which includes other out-of-plane forces. Table III lists the characteristic times for the above evolutionary processes.

D. Grain Generation Mechanisms

Clearly, if one accepts the brevity of particle lifetimes, today's visible Jovian ring must have been created not long ago. Presuming that the ring is a long-lived feature of the solar system, a continual replenishment of small particles is required. Virtually all who have considered the problem believe that the visible grains are ejecta from collisions into essentially unseen parent bodies, which Burns et al. (1980) dub "mooms" (this terminology, hardly universally accepted, comes from mating "moon" to "Mom," a parent body).

While the continual regeneration scenario may at first glance seem implausible (how can something not even visible produce a ring as lovely as that shown in Fig. 3?), its success comes from the observed ring containing so little mass. For example, a 6000-km wide band at 1.8 R_J with an optical depth 10^{-4}, consisting only of grains with 2 μm radii, contains material with a volume of 10^{-3} km^3 (i.e., a sphere 60 m in radius). Hence the supply of this miniscule amount of matter to the ring during 10^2 to 10^3 yr becomes believable. Over the solar system's age, a total volume of 5×10^3 to 5×10^4 km^3 ($R = 20$ to 40 km) would be consumed. A broad particle size distribution would have more or less mass than this depending upon the precise form of the distribution.

Evidence for the mooms is three-fold. First, without a sizable contribution to the ring cross section from large objects, charged particles would not be sufficiently attenuated to explain the Pioneer 10 observations (Ip 1980*b*; Burns et al. 1980). (Note that the nearby satellites do not have quite enough area to do this job and produce the wrong sort of signature; on the other hand, micron-sized grains are easily penetrated and therefore do not provide the required opacity.) Similar arguments have been given by Simpson et al. (1980) and Van Allen (1982) to infer characteristics of Saturn's F and G Rings, and by Lazarus et al. (1983) to suggest a hitherto unknown ring between Rhea and Titan; the latter ring has been sought but not found in a CCD imager search by Baron and Elliot (1983). The second hint that parent bodies may lie hidden within the Jovian ring is that the main band, especially the slightly brighter annulus at 1.79 R_J, backscatters more radiation than 2 μm particles alone should (Jewitt 1982). However, this enhanced backscatter can also be accomplished by, for example, a power-law particle size distribution between 0.2 μm and 2.5 μm (Grün et al. 1980); on the other hand, nonspherical grains backscatter less than spherical ones and make the problem more severe. Finally, the ring emits more infrared radiation than should be expected for micron-sized particles alone (Neugebauer et al. 1981); this conclusion is model dependent but again does hint at the presence of some large bodies.

These parent bodies originate within the Roche limit in presumably the same manner as do the components of the other ring systems, but that origin is not well understood (see Sec. IV. E; cf. Harris' chapter). Perhaps the distinctive character of the faint Jovian ring might result if primordially it contained less mass than did Saturn's system, and fewer large objects than did the Uranian system.

Given that at least some large objects are present in the main band, we now ask whether they provide sufficient target area that their erosion by micrometeoroids, for example, can generate the visible ring. To state this another way: How long does it take for today's micrometeoroid flux to produce the observed ring mass in the micron-size range by excavating parent bodies? This time must be comparable to the lifetime of a typical grain if the process is to be a continuing one.

The current ring mass $M_{\mu \text{ ring}}$ of micron-sized particles will be produced by a meteoroid mass flux density $\dot{m}_m$ in a time T_c according to

$$M_{\mu \text{ ring}} = Y \dot{m}_m A_p T_c f \tag{13}$$

where Y is the yield from a typical hypervelocity impact (i.e., mass excavated/impacting mass), A_p is the cross-sectional area of the parent bodies, and f is the mass fraction of ejecta that are micron-sized.

Laboratory experiments (see Fig. 17 in Sec. III.G) suggest that $Y \approx 5 [v/ (\text{km s}^{-1})]^2$ for impacts into a hard target (basalt) at speed v (Gault et al. 1963; cf. Grün et al. 1980), or $Y \sim 10^4$ for collisions that occur at roughly the Jovian escape velocity in the ring. Controlled collisions into a soft material (sand) suggest a similar functional dependence for Y but that the coefficient is 200 (Stöffler et al. 1975). Greenberg et al. (1978; cf. Dobrovolskis and Burns 1983) introduce a hypothetical material with $Y \approx 50 [v/ (\text{km s}^{-1})]^2$, intermediate between the hard and soft cases, as possibly representative of asteroid regoliths. It is unclear which of these expressions is the best choice for the Jovian ring situation. In fact, it is quite likely that at least two apply. For the small mooms, since they are unable to retain a soft regolith, the basalt model is probably appropriate. On the other hand, the satellites probably can sustain a regolith (recall the close-up Viking images of Deimos; Veverka and Thomas 1979) in which case the yield is substantially higher. It may be that, for the Saturnian rings, impact experiments into ice (see Kawakami et al. 1983) should be used.

For A_p in Eq. (13), the most conservative estimate suggested by the charged particle absorption data described earlier is $A_p \gtrsim 5 \times 10^3 \text{ km}^2$ (Fillius, personal communication, 1980). However, this area is not well constrained. We suspect that Fillius' value should be raised by a factor of several to include particles large enough to be targets for the micrometeoroids but not big enough to absorb the MeV protons. (See the discussion in Burns et al. 1980, who believed that a factor of two will account for the increase but who underestimated it because they did not recognize that impacts will totally shatter particles of 10 to 100 μm, producing a higher yield and an especially large proportion of μm-sized detritus.) The fraction f can be estimated by adopting a power law size distribution for the ejecta, like that given in Eq. (1).

The mass flux density of micrometeoroids arriving at the Jovian ring specifies the coefficient C in Eq. (1) but, of course, is poorly known. For meteoroid mass density $\bar{\rho}$, the mass flux density $\dot{m}_m$ that penetrates the ring with velocity v is

$$\dot{m}_m = Cv \int_{r_\ell}^{r_u} \frac{4\pi}{3} \bar{\rho} \, r^{3-p} \, dr \ . \tag{14}$$

We here assume that the power law size distribution holds between r_ℓ and r_u. The lower limit is fixed at 0.1 μm, since smaller interplantary particles are likely to be excluded from the inner Jovian magnetosphere by electromagnetic effects (cf. Hill and Mendis 1980; Grün et al.'s chapter) and since the strength of many materials increases substantially for small samples because tiny grains contain few, if any, weakening flaws. The upper limit r_u is chosen to be $\sim 10^2$ μm because the near-Earth micrometeoroid spectrum appears to steepen at about that size (cf. Stanley et al. 1979). With these choices, the fraction $f \sim 10^{-1}$ to 10^{-2} throughout most of the above range of p. We select C so as to match Pilcher's (1979) intermediate estimate for the erosive flux at Io which is based on the Pioneer micrometeoroid penetration experiment; however, when we apply Eq. (14) at the ring, the coefficient is raised by an order of magnitude to include the effect of Jupiter's gravitational focusing from Io to the ring. Finally, then, we have $\dot{m}_m \approx 3 \times 10^{-16}$ g cm^{-2} s^{-1}.

This estimate of the mass flux could be low for several reasons. According to Eq. (14), the mass flux through the ring is insensitive to r_ℓ but does increase if the power law actually extends to larger sizes. Even if it does not, some additional mass is obviously contained in meteoroids larger than 10^2 μm. Indeed, if the projectiles are themselves collisionally derived, then the largest objects would provide the bulk of the mass. Indications are that 10^2 μm objects may carry most of the mass in interplanetary micrometeoroids at 1 AU and Morfill et al. (1983a) contend the same may be true in the outer solar system.

If mm-sized meteoroids determine the mass flux of interplanetary material in the outer solar system, then the erosion rate for larger bodies will be considerably faster than the value given in Table III, which applies to small visible grains. Morfill et al. (1983a) estimate that millimeter projectiles erode Saturn's major rings at an average rate of 0.1 to 1 μm yr^{-1}. Assuming that these particles are derived from cometary debris, the flux density of material should not be very different at 5.2 AU from that at 9.5 AU. Accordingly, the erosion rate of solid surfaces will be even faster at Jupiter's ring than at Saturn's, since collisional velocities are higher in Jupiter's deeper gravitational well. Although this increased rate will not apply to the Jovian ring *grains* and will not thereby shorten the lives of the visible particles appreciably, it will mean that any unseen parent bodies will erode more rapidly. Indeed the rate proposed by Morfill et al. (1983*a*) is so high that only parent bodies with original sizes of ~ 1 km will survive over the solar system's age. In any event, a faster erosion rate for parent bodies means that such objects can more readily supply the visible grains.

If the estimated mass flux is low, two implications arise: first, as it becomes larger for any of these reasons, T_c will shorten and the putative generation of the ring by erosion with micrometeoroids becomes even more convincing. Second, if a few large objects dominate the mass of the projectiles, as would be true if the meteoroid population were collisionally evolved, then the brightness of Jupiter's ring could be episodic. It would vary substantially over time with puffs appearing temporarily after major impacts, whereas the usual brightness of the ring would be sustained by the micrometeoroid population. Even though it would be reasonable to have excavated today's entire ring mass with a single blast, that is not likely to be its origin for two reasons. First, the typical time for such a large impact into the mooms is much longer than the time to eliminate the ring altogether, so that the probability of finding the ring in such a state is low (cf. McKinnon 1983). Secondly, the ring width and three-dimensional form should mirror the ejecta's momentum distribution to some extent, whereas the ring is observed to be roughly uniform in brightness and quite thin.

Substituting the stated values of the parameters into Eq. (13), we find that debris from micrometeoroid impacts into hard targets will produce the observed Jovian ring mass in 2×10^3 or 2×10^4 yr; these drop to 50 to 500 yr for impacts into soft targets. Even though we have chosen a conservative value for the cross-sectional area of parent bodies, any of these production times is short relative to a particle's lifetime due to fragmentation. They are similar to, although for strong targets a bit longer than, the sputtering lifetime for silicate grains. The computed T_c is also much briefer than the Poynting-Robertson orbital evolution time. However, for the hard material, the plasma drag time seems less by an order of magnitude. While this may mean that targets are soft or that a source of projectiles in addition to micrometeoroids is called for, we caution that the plasma drag time is a poorly determined lower bound, especially at any place other than the ring's outer perimeter. Indeed, based on these numbers, meteoroid impacts alone are probably capable of maintaining Jupiter's visible ring.

Any other mechanism that can help supply the visible ring will make it even more certain that the ring is derived from unseen parent bodies. The only well-developed idea other than micrometeoroid erosion (Burns et al. 1980) is the suggestion (Morfill et al. 1980*b,c*; Grün et al. 1980) that Io-derived dust

(Johnson et al. 1980) might instead be the principal agent to erode the parent bodies, as described at the end of Sec. IV. D.

Ring material may be additionally supplied at some level by jostling collisions amongst the inhabitants of the bright band. The ring thickness h $\lesssim 30$ km (see Sec. III.F) limits the typical random (collision) velocity between parent bodies to $\delta v \sim v_c h/a$, where v_c is the local Keplerian velocity. Hence $\delta v \lesssim 10$ m s^{-1} in the ring, and collisions of one parent body with another may occur at speeds which are comparable to the escape speeds off the largest objects in the ring—the known satellites. It might seem then that gentle bumping among parent bodies could provide the visible material but, because very little ejecta leaves an impact site with speeds even close to the collision velocity (see Sec. III.G), this is unlikely.

A more promising provider of small particles might be impacts into the parent bodies by the tiniest ring grains since the latter drift appreciably relative to the mooms. While, according to Eq. (7), a collision between the nominal grain of 1 μm and a moom takes place at ~ 300 m s^{-1}, even this does not achieve the hypervelocities essential for excavating large amounts of mass. On the other hand, slightly smaller projectiles will impact at speeds in excess of 1 km s^{-1} and these can produce a cascade process in which first-generation ejecta begets further debris; unfortunately, this debris seems too small to account for much of the visible ring. In addition, these projectiles have the capacity to destroy catastrophically larger grains, perhaps thereby generating some fine material for the halo (Fig. 2 of Burns et al. 1980). Not knowing the source(s) of the grains in Saturn's diffuse rings, we do not list times for their generation in Table III.

E. Theoretical Considerations of the Particle Size Distribution

Shortly after the discovery of Jupiter's ring, there was general agreement that its particles were restricted in size to a micron or two. However, as discussed in Sec. I.C, power law size distributions that terminate near two microns or so (such that these sizes dominate the ring's cross-sectional area) also satisfy the limited photometry extant, at the scattering angles of 4° to 6° through Voyager's clear filter (~ 0.47 μm; Jewitt and Danielson 1981). Some additional data are available on unreduced Voyager 2 images and soon the ring's brightness should be known at scattering angles of 3° to 5° in both orange (~ 0.41 μm) and violet (~ 0.62 μm) (Showalter et al. 1983b, 1984). A preliminary assessment of these new measurements suggests a broad distribution of sizes about the micron range. However, we remind the reader again that the available forward-scatter observations emphasize just these particles. The population of large particles, say millimeters and more, in such a diffuse ring simply cannot be specified on the basis of the available observations.

Notwithstanding these caveats and partly for historical reasons, we list several explanations that have been proffered for the putative prominence of micron particles in the Jovian ring. Presumably some of these processes will

be as active in the Saturnian case. Burns et al. (1980) as well as Jewitt and
Danielson (1981) have pointed out that interplanetary dust and meteorites are
constructed of elements with typical sizes of 1 to 10 microns. Morfill et al.
(1980b) and Grün et al. (1980) advocate the selective generation of micron-
sized grains through impacts of 0.1 μm dust from Io. Electrostatic bursting
has been considered as a possible mechanism for fragmenting small grains but
is ineffective unless the particles are very weak (see references given in the
chapter by Grün et al.). Burns et al. (1980) mention that ring debris of just
under a micron will be preferentially destroyed, owing to mutual collision of
drifting dust grains. In addition, grains less than tenths of microns can be
swept out of the main Jovian ring by electromagnetic forces (see Eq. 4).

F. Random Orbital Velocities and Ring Thickness

An indication of the velocity fluctuations about some mean orbital speed
is given by a ring's thickness, since the latter is a consequence of any out-of-
plane component of velocity. The mean thickness of any ring represents a
natural end state in which out-of-plane perturbations are balanced by damping
during collisions (Burns et al. 1979a; Cuzzi et al. 1979a,b; Goldreich and
Tremaine 1982). Vertical impulses might be imparted by direct interparticle
collisions and by gravitational stirring (both brought about through Kepler
shear) as well as by electromagnetic forces (which can be most important of
all for ethereal rings). Thus thickness, with its straightforward geometric
meaning, has important dynamical implications.

In general, a distribution of vertical velocities is maintained in a ring by
sporadic interparticle collisions; the ring thickness can then be understood in
the sense of a kinetic temperature. Unfortunately, this concept may be ambiguous for the rings under consideration because the ring thickness will vary
drastically at small particle sizes, owing to the heightened role then played by
electromagnetic forces (see, e.g., Eq. 21 below). In this connection we note
here that the measured upper limit of 30 km thickness (corresponding to a
random velocity of ~ 10 m s^{-1}) for Jupiter's ring is based on an image taken in
backscattered light, which accentuates the largest objects present.

Any random component of velocity, whether due to perturbations or
merely reflecting the initial state at which a grain is injected into the ring, will
be damped by collisions which take place on a time scale

$$T_D \sim P/2\tau. \tag{15}$$

Damping of the tiny visible grains into the ring plane can be accomplished by
several collisions with other grains of similar size, or by just one impact into a
parent body. Grains collide with parent bodies somewhat more frequently than
Eq. (15) suggests because their electromagnetically-induced drift velocity
(Eq. 7) is larger than typical random velocities. Nevertheless Eq. (15) provides a rough estimate of the damping time; however then τ should only

include the area of parent bodies. In Eq. (15) $P \sim 10^{-3}$ yr is a typical particle's orbital period for Jupiter's ring, while $P \sim 10^{-3}$–10^{-2} yr for Saturn's tenuous rings. With the optical depths given by Jewitt (1982), including Tyler et al.'s (1981) correction, T_D is about 10 yr for Jupiter's main band and more than 75 yr for the halo; these would be shortened by an order of magnitude if we used the alternative τ expression proposed by Tyler et al. (1981). The damping times for Saturn's diffuse rings span the Jovian values.

The damping times of Jupiter's ring and Saturn's F Ring are quite short relative to the period in which the rings seem to be generated (or eliminated); see Table III but compare against the discussion in Sec. IV.C. Therefore, many collisions will ensue following a particle's injection into such rings, so that its initial velocity will no longer be identifiable. In the absence of continuing large perturbations, the rings should be quite thin, as observed; however Saturn's G and E Rings should be less collapsed. Any recently created particles, with as yet undamped motions, should form a very tenuous "atmosphere" of great vertical extent ($h \sim v_{ej} P/2 \sim 10^{2}$–^{3} km for injection at 100 m s^{-1}) about the primary Jovian ring. However, since a typical bright band particle spends such a small fraction ($\sim T_D$/lifetime, or $\sim 10^{-3}$–10^{-2}) of its lifetime before settling into the equatorial plane, this "atmosphere" would be invisible in the few available Voyager images. The thinness of Jupiter's ring could also be explained in part by a special family of small impacting projectiles, an idea entertained by Grün et al. (1980).

The thickness (and its variation with radial position) thus may suggest the flattening time relative to T, the time for an orbit to evolve across the ring. For example, if $T_D << T$, the ring should be thin but if $T_D >> T$, the ring should everywhere have approximately its injection thickness. If $T_D \sim T$ and the ring is principally generated near its outer edge, then the thickness should decrease roughly as

$$h(a) = h(a_o)e^{-(1-a_o/a)T_D/T} \qquad (16)$$

with $h(a_o) = (v_{ej}/v_c)a_o$. Since the Jovian main band is known to be flattened, T_D must be much less than T; since this contradicts the values summarized in Table III, it seems probable that we have overestimated plasma drag, or that indeed the measured thickness only pertains to that of the parent bodies.

Damping to a mean plane requires that physical collisions actually transpire. Electrically charged grains having the same sign will not touch one another until their relative velocities are large enough to overcome their mutual repulsion. Hence for damping collisions

$$\Delta v > \frac{\Phi}{r}\left(\frac{3}{2\pi\rho}\right)^{\frac{1}{2}} \qquad (17)$$

or, for the nominal Jovian ring grain, $\Delta v \gtrsim 0.8$ m s^{-1}, corresponding to a thickness of 5 km. Since this is not very far from the upper limit of 30 km, the Jovian ring thickness might be partially self-regulated such that perturbations inflate the ring up to the point at which dissipative collisions transpire. Of course, bigger particles will not be nearly so affected.

Electromagnetic forces also can have nonequatorial components, and so they should account for some thickening. For Jupiter's ring, out-of-plane motion can be produced through the tilted magnetic field which introduces a vertical component to the Lorentz force (Jewitt and Danielson 1981; Jewitt 1982). If we assume that collisions may be ignored, that the Jovian magnetic field is a centered dipole which is slightly tilted ($\sim 10°$), and that vertical displacements are small compared to the radial distance, Newton's equation normal to the equatorial plane is

$$m\ddot{z} \simeq -q \, \frac{v_\Theta}{c} \, B_r(t) - \frac{Gm \, M_J}{R^2} \, \frac{z}{R} , \qquad (18)$$

considering only first order forces. The second term on the right is the z-component of the Jovian gravity field. The lead term on the right is the Lorentz force and employs $v_\Theta = (\Omega - \omega)R$, the particle's velocity relative to the rotating field, as well as the magnitude of the radial component of the magnetic field felt by the particle: $B_r = B_{r_0} \cos(\Omega - \omega)t$. Substituting into Eq. (18) and rearranging it, we have

$$\ddot{z} + \Omega^2 z = - \frac{qB_{r_0}}{mc} (\Omega - \omega) \cos (\Omega - \omega)t. \qquad (19)$$

The asymptotic amplitude of this driven harmonic oscillator is

$$z_{\max} = \frac{gB_{r_0}}{mc} R(\Omega - \omega) \, \frac{1}{\Omega^2 - (\Omega - \omega)^2} \qquad (20)$$

or

$$\frac{z_{\max}}{R} = \frac{B_{r_0}}{B} \left(\frac{r_c}{r}\right)^2 \left[\frac{\Omega(\Omega - \omega)}{\omega(2\Omega - \omega)}\right] \qquad (21)$$

where $r_c \sim 0.3$ μm is the particle radius at which $F_g = F_m$ (see Eq. 4b). For the Jovian halo $z/R \sim 0.1$ and $B_{r_0}/B \sim 0.1$, while the ratio in brackets of the angular velocities varies between ~ 0.2 and 0.5, so that $r \sim r_c$ in order to produce the observed thickness in this simple model. Of course, at such small sizes the model breaks down and should not be taken too literally. We note that the solution (Eq. 21) blows up at $\omega = 0$ and at $2\Omega = \omega$, both of which correspond to the particle seeing a magnetic field that varies with frequency Ω. In each case the particle is then forced vertically at precisely the frequency it orbits the planet. The first situation ($\omega = 0$) is nonphysical and the second is located at ~ 3 R$_J$, well outside the Jovian ring. The Saturnian magnetic field seems to be closely aligned with the planet's rotation axis, to within observational error, so that there the above model is inappropriate.

Numerical calculations by Consolmagno (1983) have explored the consequences of perturbations due to differences of the Jovian magnetic field from a perfect, centered, aligned dipole. Earlier integrations (Consolmagno 1980) in part carried out a similar analysis with a simpler magnetic field but these contained an error of roughly a factor of three. Slightly altered paths are produced depending upon the magnetic coordinates at which particles first enter the field. Consolmagno (1983) followed the paths of 2.5 μm grains ($\rho = 2$ g cm^{-3}, $\Phi = -10$ V) through a model Jovian field, containing dipole and quadrupole terms which are derived from Pioneer data. Following their injection in the equatorial plane at random longitudes, the grains undergo systematic out-of-plane orbit variations, such that after a few orbits they spread ± 200 km off the mean ring plane (see Fig. 16). Our simple model (Eq. 21), including only the dipole's inclination, yields about half as much.

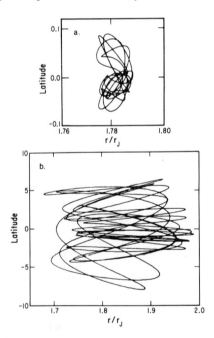

Fig. 16. (a) Path of a 2.5 μm grain, charged to -10 V, through a model Jovian magnetic field during 100 hr after starting at 1.88 R$_J$. (b) The same for a 0.3 μm particle. (Figure from Consolmagno 1983; cf. Eq. 21 in the text.)

To reverse the problem, if the measured $\leqslant 30$ km thickness were to be due to electromagnetic forces alone, then the grains would have to be $\geqslant 7$ μm in radius, and this seems incompatible with available photometry. Once again, either potentials are overestimated, perhaps due to collective self-shielding, or the backscattered thickness is not caused by the micron grains. Stochastic effects, as described by Consolmagno (1980), will not be pertinent to thickening questions because large amplitudes only develop over long intervals and, in the meantime, collisional damping will always bring the material back to the mean plane.

The solar radiation pressure force has a component normal to the ring plane because, in general, the planet's equatorial plane is not its orbital plane about the Sun. For the Jovian ring, this component is trivial in comparison to Jupiter's gravity, both because Jupiter's obliquity is small and because its ring is close-in. For the edges of Saturn's E Ring, we estimate that the solar radiation force is at most 10^{-3} to 10^{-4} that of Saturn's gravity and so cannot generate much thickness (see Mignard's chapter).

G. The Role of Embedded Satellites

It is as yet unclear what part, if any, satellites play in the structure of ethereal rings. Upon the discovery of Adrastea and Metis in the vicinity of Jupiter's ring, some (Ip 1980b; Jewitt and Danielson 1981; Jewitt 1982) supposed that they must bound the ring via the Goldreich Tremaine shepherding mechanism (1982). However, the small gravitational perturbations of these satellites versus the expected size of electromagnetically induced drift motions for visible grains and the infrequency of collisions (due to low optical depth) preclude a strong gravitational coupling between any tenuous ring and small satellite.

The motion of material near either Jovian satellite is suggested in numerical models of Dobrovolskis and Burns (1980; cf. Davis et al. 1981) and Burns et al. (1980, their Fig. 4). Although the escape speed from an isolated sphere of 2 g cm^{-3} density and 12.5 km radius is 13 m s^{-1}, Burns et al. (1980) find that escape off a triaxial satellite ($10 \times 12.5 \times 15$ km) orbiting at 1.79 R_J occurs at speeds ranging from 3 to 12 m s^{-1}, depending upon launch angle and location (cf. Papadakos and Williams 1983). Since all the speeds are positive, material cannot leak off such a satellite's surface without a push, even though it resides within Jupiter's Roche limit, this, of course, is not inconsistent with the concept of Roche's limit, which only rigorously applies to a deformable fluid object (see chapter by Weidenschilling et al.).

Satellites can be either sources or sinks for rings, depending upon how much ejecta is created in an impact and upon what fraction of that ejecta has sufficient speed to escape (see next paragraph). Therefore the question of source vs. sink cannot be resolved without knowing the precise nature of the satellite's regolith, which determines the velocity distribution of the ejecta. Laboratory experiments (Gault et al. 1963; Stöffler et al. 1975) indicate why

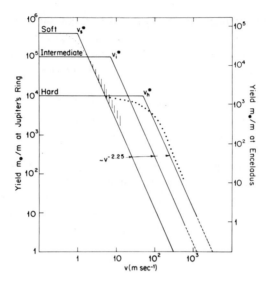

Fig. 17. The mass m^* of debris ejected at speeds greater than v by an impact of mass m at a speed of 44.7 km s^{-1}, representative of collisions between Jovian ring material and interplanetary particles (left ordinate). The right ordinate pertains to the Saturnian case with a typical impact velocity of 18 km s^{-1}. The dotted curve roughly indicates data from hypervelocity laboratory impacts into solid basalt targets (Gault et al. 1963), scaled to higher speeds; the superimposed shelf is the analytical approximation used by Dobrovolskis and Burns (1983). The vertical dashes similarly provide the data for hypervelocity impacts into loose sand (Stöffler et al. 1975) alongside the corresponding approximation. The final curve is for a hypothetical material intermediate between sand and rock, and supposedly pertinent to satellite regoliths.

regolith properties are so crucial in this matter. Studies (see Fig. 17; cf. Greenberg et al. 1978; Fig. 1 of Dobrovolskis and Burns 1983) show that all ejecta attain velocities greater than some cutoff value v* which varies with surface type. They also indicate that the proportion of mass exiting a crater in excess of a given velocity is a strong inverse function of that velocity ($m \sim v^{-k}$; $k \approx 2.25$, the sloping solid lines in Fig. 17) for both hard and soft materials. In addition, the yield from a hypervelocity impact varies with the kinetic energy delivered (i.e., with impact speed squared). The total excavated mass depends upon the nature of the impacted surface and differs by a factor of 40, as listed following Eq. (13).

Based on these data, a satellite could bound a ring even without a shepherding mechanism being active. Two possibilities exist. Given a soft regolith, a satellite may absorb other ring members that gently bump it and thereby impede their diffusional spreading. More likely, a satellite could be the primary fount from which ring material springs. If the vast majority of ejecta initially stays near the satellite, and if the debris subsequently evolves inward (outward), the satellite would seem to bound the ring's outer (inner) edge.

To illustrate this apparent bounding, consider the case of Adrastea and imagine that escape off that satellite occurs at 10 m s^{-1}. Since the brightness of Jupiter's ring is observed to develop within a single resolution element (600 km), we ask what fraction of the escaping mass leaves Adrastea at high enough speeds to move farther away than 600 km. To step that far, ejecta must depart with more than $\Delta v > v_c$ ($\Delta a/a$) ≥ 150 m s^{-1}, according to Eq. (8). If the satellite's regolith is characterized as soft or intermediate (Greenberg et al. [1978] think that such a substance might represent asteroid regoliths), only $(1/15)^{2.25} = 0.002$ of the escaped matter reaches more than 600 km away. The situation is not so extreme for hard surfaces like basalt, for which v* ≈ 50 m s^{-1}; hence, of the ejecta that escapes from basaltic targets, as much as $(50/150)^{2.25} \approx 0.08$ moves faster than 150 m s^{-1}. However, in view of the surfaces seen in the close-up photos of Phobos and Deimos (Veverka and Thomas 1979), a soft or intermediate characterization seems likely to be more appropriate for Metis and Adrastea.

H. Dust Rings from Satellites

If Adrastea and/or Metis are accepted as important sources of Jupiter's ring, should not debris rings be seen about all satellites, in particular nearby and bigger Amalthea (cf. Soter 1971; Davis et al. 1981; Burns and Showalter 1983)? Not necessarily. While Amalthea's cross-sectional area is much greater than that of Adrastea, it also has a higher escape velocity so that, following the arguments of the preceding paragraph, a smaller fraction of material will be released into independent orbit. Since an isolated satellite's escape velocity $v_e \sim r$, f_e (the fraction of ejecta above v_e; that is, the fraction that escapes) $\sim r^{-2.25}$ (as long as $v_e > $ v*). Hence the efficacy of a satellite in supplying ring material varies as $f_e r^2$, or as $r^{-0.25}$: *small satellites can provide more total ejecta to a ring than large.*

To compare the yield of Adrastea with Amalthea, we first consider them both to have the escape velocities of isolated spheres, in which case Adrastea produces about twice as much ring debris as Amalthea. If instead we say that Adrastea's escape velocity is lowered to 5 m s^{-1} on average due to its presence within Jupiter's Roche limit (see above), the smaller satellite yields 17 times what its larger neighbor does. Recall that a hint of a very faint ring may have been noticed outside Jupiter's known system; while it is tempting to assert that this could be produced by debris from Amalthea, plasma drag should cause its material to evolve *outward.*

A small object produces more than a large one only until it becomes so tiny that all ejecta escape. This happens once $v_e < $ v*, in which case $f_e = 1$ so that the production of ring material $\sim r^2$. The transition to "small-is-less-efficient" occurs at v* $= v_e$, which for soft regoliths happens at 1 m s^{-1}, corresponding to isolated satellites smaller than $r \approx 1$ km. With intermediate regoliths, all ejecta depart satellites smaller than about 7 km radius while, for hard surfaces, satellites must be greater than 50 km to retain some ejecta. These sizes are those that deliver the most debris per source. Might it be meaningful that Adrastea's size is close to the best for a satellite with the intermediate properties that we favor?

At what radius does a satellite stop being a ring source and instead become an absorber of interplanetary meteoroids? We determine this by setting f_e times the cratering yield Y equal to 1. This is equivalent to finding the velocity above which $m*/m = 1$ on Fig. 17 (note that this does not mean no ring—since part of the meteoroid mass does stay in orbit—but just no multiplicative factor). Jovian satellites with soft regoliths, the most absorbent case, must be larger than about 300 km in order to not yield extra ring material. This number is comparable to the situation for hard asteroids (Dobrovolskis and Burns 1983; cf. Housen et al. 1979) and is about ten times the radius of an accreting soft asteroid; the difference in these cases is due to the high yield caused by the swift impact velocities encountered deep in Jupiter's gravitational well. A soft regolith on a Saturnian satellite at 4 R_S will absorb once the satellite is larger than about 100 km while a satellite with a hard regolith would need to be larger than about 1000 km. Enceladus, if it had a regolith of intermediate properties, would be on the border line between a source and a sink, but would be a substantial source if its surface were hard.

The important issue of whether or not collisions between satellites and ring members generate material cannot be resolved for several reasons. Interparticle collision velocities are not well determined and, even if they were, lie outside the hypervelocity range for which impact experiments have been performed. In addition, the majority of ejecta velocities are likely to fall in the velocity regime where trajectories off the satellite are convoluted and highly dependent upon initial surface location (see Dobrovolskis and Burns 1980); hence, escape yields are unsure.

Considering the earlier arguments of yield alone, we observe that the classical satellites of Jupiter, Saturn and Uranus should not give rise to *dust belts*. Possible exceptions to this statement are Mimas and Enceladus which are relatively small and which suffer collisions at fairly high speeds. Hyperion is of similar size but impact velocities are substantially lower.

The smaller satellites close to Jupiter (Metis, Adrastea, Amalthea, and Thebe) and Saturn (Atlas, the F Ring shepherds, and the coorbitals) are all probable sources of material. Of these, Metis and Adrastea are far and away the best, followed by Thebe (JXIV, or 1979J2) or Atlas. In this regard we recommend that a search be made just beyond the outer edge of the A Ring for Atlas's predicted ring, which should be outside its orbit if plasma drag overwhelms Poynting-Robertson drag.

The material in the gap between the A Ring and Atlas' orbit found by Burns et al. (1983) and Graps et al. (1983) could be debris spalled off the A Ring itself. During half an orbit a 1-μm grain at the A Ring's edge will evolve about 50 m through plasma drag. Hence ejecta from a 50 m-wide strip at the ring's perimeter will not recollide with (and be reabsorbed by) the ring. Elsewhere in the ring, recollision is inevitable; plasma drag time scales are also much longer and so any gaps (e.g., Keeler Gap or Encke Division) can be quite free of debris. The total area of this strip is several hundred times Atlas' cross section and accordingly should produce a ring 10^2 to 10^3 times that of Atlas; $\tau \approx 10^{-2}$ to 10^{-3} and it is this sort of optical depth that the photopolarimeter is sensitive to, with significant averaging. As this debris evolves outward, it may strike Atlas and settle on it. It should be possible to see this material and perhaps we do in the gap between the A Ring and the F Ring in Fig. 13. Other Voyager images, particularly of the shadow boundary in forward scatter, should be examined.

Whether the ejecta ever becomes dense enough to be visible and identifiable as a ring, depends not only on the source strength but also on how long grains stay in adjacent orbits. Table III suggests that orbital evolution and grain survival times do not differ too much in the vicinity of the satellites of interest. Another possible loss process is recollision of a grain with the satellite from which it originated.

The expected collision lifetime of an orbiting particle with a satellite has been considered by Öpik (1951) and may be written as

$$T_c = \pi \left(\frac{a}{r} \right)^2 \frac{U_x}{U} P_0 \sin i \quad , \tag{22a}$$

where r is the satellite's radius and a its orbital semimajor axis, U is the encounter velocity in units of the mean orbital velocity of the satellite and U_x is the component of U directed along the radius from the planet during the encounter; P_0 and i are the grain's orbital period and the inclination of its orbit with respect to that of the satellite. Soter (1971) has averaged Eq. (22a) over all ejection angles, assuming that escape is equally likely in any direction (cf. Dobrovolskis and Burns 1980), and has found that the typical recollision time is

$$T_c = \frac{1}{3\pi} P_0^2 \, v_{ej} \frac{a}{r^2} \quad , \tag{22b}$$

where v_{ej} is the ejecta velocity after leaving the satellite's gravity field. Averaged over the ejecta velocity distribution we have chosen ($m \sim v^{-2.25}$); v_{ej} is close to the escape velocity, which itself is proportional to r. Hence $T_c \sim P_0^2 a/r$, assuming that orbital evolution does not take the material away; i.e., it applies to large grains. We estimate that ejecta from Adrastea or Metis will reimpact the parent satellite in 30 yr or 15 yr, respectively. Times for Amalthea, Thebe, and Io are also of order 10 yr. For Saturn's satellites, recollision times are slightly longer, e.g., about 50 yr for both Enceladus and Atlas.

IV. THE NATURE AND ORIGIN OF JUPITER'S RING

Let us now consider how the various processes described in the preceding section might influence the development of the Jovian main band and halo, and then similarly discuss Saturn's tenuous rings.

A. The Main Band

Three aspects of the main band's radial structure are of interest: the overall width and approximate uniformity of the main ring; the different appearances of the ring's inner and outer edges; and the $\sim 10\%$ brightness enhancement located near the embedded satellite Metis.

Overall Width. We argue that the width of the ring is principally governed by the distribution of the underlying sources rather than being due to ejecta broadcast away from a radially localized source. Using Eq. (8), one might maintain that the overall width of the belt indicates ejecta leaving a single site at $\Delta v \lesssim 1$ km s^{-1}, what appears to be a relatively low speed for collisions that transpire at tens of km s^{-1}. However, the laboratory experiments (Fig. 17) cited above show that the proportion of mass exiting a crater in excess of a given velocity is a strong inverse function of velocity; hence most escaping mass barely makes it off the satellite (cf. Dobrovolskis and Burns 1983) and should be located in a nearby orbit.

Accordingly, the ring's width cannot be ascribed to ejecta velocities. Neither can the ring be noticeably broadened by radiation pressure-induced oscillations of the dust's orbital eccentricities, as once proposed, because this mechanism only generates a width of ~ 100 km, as already mentioned. Direct electromagnetic forces and even long-term diffusion due to electromagnetic scattering cannot be effective in radially spreading material from a narrow source region (Consolmango 1980, 1983; cf. Eq. 21 and the ensuing discussion) because then the ring would also be much thicker than observed. Lastly, since orbits seem to evolve in times comparable to (or even shorter than) particle lifetimes (see Table III), some of the ring's breadth may be produced by particles that originate at sites near the ring's outer edge which then drift inward. If such is the case, the ring brightness or the sizes of particles may vary with radial position; unhappily, this might not be discernible with our current resolution on the ring.

A clear expectation of any model that generates visible grains by impacts into parent bodies is that, to some degree, the visible ring intensity should reflect the distribution of underlying sources. The brightness near Metis and the association of Adrastea with the ring edge may bear this out. We maintain that the width and smoothness of the ring indicate the approximate location of parent bodies and their rough spatial uniformity; charged particle absorption data agree with this interpretation to first order although they suggest a somewhat narrower band (Fillius, personal communication, 1980).

The overall width of the bright band today is consistent with parent bodies that have radially diffused over the age of the solar system from a more confined region. These parents could have originated from the collisional breakup of a single satellite (see Harris' chapter). To illustrate this diffusion, consider an orbiting disk composed of identical spheres of radius R. A typical particle collides with another during an interval T'_D by Kepler shear (see Eq. 15) and alters its path by $\sim R$. Since diffusion is a random walk process (here with step size R), the disk will spread through Δa in a time

$$T_K \sim T'_D \, (\Delta a / R)^2 \tag{23}$$

(cf. Brahic 1977). This expression is also approximately true for a power law size distribution with R now the maximum radius (see Cuzzi et al. 1979a). If τ of the parent bodies were $\sim 10^{-6}$ and the largest object in the size distribution for the Jovian ring were 10 km (the satellites?), the ring would diffuse $\Delta a \sim 2 \times 10^4$ km over 4.5 byr; for $R = 1$ km, $\Delta a \sim 2 \times 10^3$ km. The breadth of the ring is 6×10^3 km and the inner edge's brightness decays in $\sim 1.5 \times 10^3$ km, so diffusional spreading of parent bodies could (entirely or in part) account for the gross radial structure of the ring. An alternative mechanism to spread the mooms, Alfvén drag, has been proposed by Anselmo and Farinella (1983), but this requires rather special conducting properties for the mooms.

Different Appearance of the Ring Edges. How can the outer edge be sharp if diffusion due to Kepler shear has smeared radially the entire band of parent bodies? As previously described, Adrastea, either as a source or as a sink, may be crucial in developing the apparent crispness of the outer perimeter. Orbital evolution under plasma drag and, to a lesser extent, under Poynting-Robertson drag may help account for the distinction between the inner and outer edges. Since the visible particles drift inward across the ring in short times, any ejecta generated at the outer edge of the moom's disk rapidly leave the area; thus the brightness of the outer edge is reduced. Closer to the planet, the ring is brighter since it contains some locally derived particles in addition to material produced elsewhere which is now drifting toward the planet. The longer-surviving grains continue to evolve past the inner boundary of the mooms (see discussion of time scales below). The inner boundary's diffuse nature thus both reflects the varying lifetimes of its constituents and the ragged edge of a diffusing disk.

The flux density distribution of high-energy magnetospheric particles may also partially explain the different appearance of the inner and outer edges. Adrastea may partly shield the ring from sputtering by energetic magnetospheric particles since its cross section is an appreciable fraction of that of the mooms. The flux of magnetospheric particles, which are diffusing inward by pitch-angle scattering, is about 10 times greater on the outer border of the rings than at the inner boundary (Fillius 1976) due to their absorption by satellites/mooms in the intervening space. Near the outer fringe, sputtering lifetimes of grains are concomitantly shorter, perhaps being reduced to a decade or even less. These grains themselves may ultimately be projectiles, so nonlinear variations in the number of grains, and therefore the ring brightness, may accrue. As seen in the Pioneer data (Fig. 1), the flux increases again, once inward of the main band. The complex interplay between the distribution, plus precise character, of the parent bodies and the ring sculpturing by magnetospheric particles deserves further exploration.

Metis' Influence. The brightening of the ring near the orbit of Metis may intimate that this satellite furnishes material to that region; however, the absolute distance scale of the ring is so poorly known that we cannot be certain any association is real. If Metis does supply the ring, most ejecta leave it at $\leqslant 100$ m s^{-1} (cf. discussion of Adrastea's influence Sec. III.G). Note that the satellite's cross section accounts for $\sim 20\%$ of the expected target area of parent bodies (using Fillius's conservative A_p) and so it would not be surprising if the satellite noticeably perturbed the ring density in its vicinity. Furthermore, backscattered measurements suggest that additional parent bodies may be located in this same region so that their detritus might instead account for the enhanced brightness; these bodies, of course, may ultimately have been derived from Metis.

B. The Halo

The halo is probably composed of material which is evolving inward from the bright band, due to the several evolutionary mechanisms described earlier (see Sec. III.C). Over the time of orbital evolution, the size of the typical particle shrinks appreciably since lifetimes are similar to evolution times. This explains why the visible halo grains are simultaneously closer to the planet and tinier than those in the bright band. The smaller grains are more susceptible to electromagnetic forces (Burns et al. 1980; Grün et al. 1980; Jewitt and Danielson 1981; Jewitt 1982; Consolmagno 1983; compare Figs. 16a and b) which contain nonequatorial components. These are produced by the 10° tilt of Jupiter's magnetic field, by a possible poleward shift of the dipole as a whole, by terms of higher order than dipolar, and by the stochastic nature of the field. Furthermore, since no parent bodies populate this region, grains are not as readily damped back to the ring plane.

Typical particle sizes in the halo must be tenths of microns or less (see Sec. I.C and Eq. 4). Otherwise, as for the larger grains in the main band, electromagnetic forces would be insignificant compared to gravity, and vertical motions would be confined to less than injection velocities, giving a relatively flat ring. On the other hand, if halo particles were too tiny, say hundredths of microns or less, electromagnetic forces would overwhelm gravity. In this circumstance a particle's motion would be entirely determined by the field's structure. Accordingly, the halo would be symmetric about the magnetic equator and flap about the planetary equator as Jupiter rotated. Since symmetry about the magnetic equator is not observed, halo particles are not mere motes. However the seeming asymmetry of the halo structure, if real, probably indicates that the visible grains are intermediate between these two extremes. This hypothesis can be tested by a more complete photometric analysis for the halo; as yet, we merely know that the halo grains are small, in the sense that they do not exhibit much phase brightening, according to the ongoing study of Showalter et al. (1983b, 1984).

The halo's inner boundary can be interpreted as the point at which the gradually evolving and gradually eroding small grains finally disappear, or are too far dispersed to be seen. The vertical extent of the halo is likely determined by the limited cross-sectional area possessed by the few tiny grains that achieve such altitudes.

C. A Self-Consistent Discussion of Time Scales and a Model of the Jovian Ring

What are the main processes in the Jovian ring and how might its form be best understood? Table III indicates that loss is principally due to sputtering [time scale of $T_s \approx 10^{3\pm1}$ $(r/1\ \mu m)$ yr for silicates], while orbital evolution is caused mainly by plasma drag [time scale of $T_{PD} \approx 2 \times 10^{2\pm1}$ $(r/1\ \mu m)$ yr but just one-tenth that to cross the ring; recall however that self-shielding could lengthen this substantially; e.g., Consolmagno (1983) interprets ring thickness to suggest that a particle's potential could be only one-tenth that used, raising T_{PD} by a factor of ten]. The ring's form imposes a restriction on the relative length of the evolution time scale to that of destruction: for the ring to extend radially over only about 5% (i.e., $\sim 0.1\ R_J/1.8\ R_J$) of its semimajor axis, the orbital evolution time scale *must* be much longer than the survival time. For example, we could choose $T_{PD} = 2 \times 10^3$ yr while $T_s = 10^2$ yr, and still satisfy the experimental bounds on these numbers. (Recall that sputtering is much more rapid in the region outside the main band.) The ring would then need to be generated in $T_g \sim 10^2$ yr, which would require either high impact yields (soft regoliths) or a large target area (e.g., the value of Ip [1980a] rather than that of Fillius [personal communication]). Obviously with the large uncertainties in all these numbers, other combinations having the same general ratios are just as acceptable. For instance, one might choose to raise T_{PD} to 10^4 yr, arguing that the plasma conditions are poorly known in the ring's locale; in this case, the nominal generation and sputtering lifetimes could satisfy the model.

Since the main evolution and destruction processes are proportional to radius r, particles of all sizes will evolve equally as far before their demise. Hence, if the ring had only a single source, say Adrastea, one might anticipate that such a ring would have a square box profile. But, since particle sizes shrink as orbits decay, the brightness of such a ring will in fact lessen as the planet is approached. The observed profile—nearly constant brightness across most of the main band followed by a gradual decay on the inner edge (see Fig. 4)—can be explained if the major fount for the ring is near the ring's outer edge (Adrastea?). Closer in lie additional sources, Metis included. Mooms, over and above the satellites, are needed in order to flatten the ring, to provide additional sources, and to absorb charged particles. The ejecta coming off embedded sources replaces to an appreciable degree the reduced surface area of the eroding debris generated farther out in the ring. Once the band of mooms is no longer present, the ring's brightness slowly drops as the particles shrink, scatter into the halo, and disappear.

Toward the inner edge of the ring where parent bodies become dispersed (or are even absent), sputtering increases as the flux of high energy magnetospheric particles grows since they are no longer being absorbed by the mooms. The relatively short sputtering lifetimes in this region mean that all particles are quickly reduced to the size where electromagnetic forces become appreciable. Effectively, submicron particles are injected instantaneously at about 1.7 R_J. At that point they undergo the sorts of motions simulated by Consolmagno (1983; see Fig. 16). These small specks have only quite limited remaining lifetimes, however, because they are still subject to the intense magnetospheric environment. If they are a few tenths of microns in size, they last perhaps $10^{2 \pm 1}$ yr. While this size, they oscillate $\pm 5°$ in latitude as well as in and out by ~ 0.1 R_J (see Fig. 16b). The volume of space which they traverse in these Lorentz-driven motions is the region encompassed by the visible halo. Before wholly disappearing, they spend a very brief period as grains of ~ 0.01 μm and this allows them to be pushed to great elevations and to undergo large radial excursions. In this model, every ring grain ultimately becomes a member of the halo. Additional material is supplied as the small detritus from impacts. The halo's faintness arises for three reasons: individual particles are small, survive for short periods, and are dispersed through a large volume.

D. Grain Generation by Io Dust: Its Comparison to the Micrometeoroid Source

In this section we lay out an alternative model for the generation of Jupiter's ring through impacts into parent bodies by Io dust (Johnson et al. 1980; Morfill et al. 1980b; Grün et al. 1980) rather than by interplanetary micrometeoroids (Burns et al. 1980) as usually considered (see above). We first summarize the evidence for particulate matter leaving Io (Strom and Schneider 1982) and then point out the weaknesses of the two contenders, micrometeoroids and Io dust, as the causative agents of the small particle population in Jupiter's ring.

Photometry of Io's volcanic plumes indicates that they contain an inner core consisting of larger particles (1 to 1000 μm radius) surrounded by an outer component of extremely fine particles (0.01 to 0.1 μm radius). The volcanic plumes of Io range in height from ~ 60 to >300 km. However, particles are not believed to escape directly by ballistic means since implied vent velocities (~ 0.5 to 1.0 km s^{-1}) are well below the ~ 2.5 km s^{-1} escape velocity. Furthermore, the eruptions appear to be explosive volcanism driven by S_2 or SO_2 gas expanding up a vent; with such a mechanism, particles are entrained in a gas column so that their velocities are fairly uniform without any high-velocity fraction that might achieve escape. Some of this particulate matter may only develop at altitude as solid SO_2 snows out during the plume's expansion. Nevertheless, even though particles are not directly injected into the Jovian magnetosphere, fine volcanic material may attain heights at which it is exposed to the Jovian plasma to be electrically charged and thereby escape (see Johnson et al. 1979, 1980). This volcanic dust can be transported from 6 R$_J$ through the Jovian magnetosphere to the ring 1.8 R$_J$ by the electromagnetic diffusion processes (Morfill et al. 1980a) that we have already mentioned. There appears to be a narrow range of particle sizes such that the particles are small enough to be pulled off the satellite and also to evolve rapidly inward toward the Jovian ring but large enough to survive the journey.

The Io dust source will now be critically compared against the interplanetary micrometeoroid source. We start by emphasizing that both models agree on the fundamental point that continuous replenishment of the visible ring necessitates erosive collisions between small projectiles and unseen mooms. This avoids the unpalatable alternative that today we just happen to be lucky enough to observe an extremely evanescent feature of our solar system. The models differ in their production mechanism and in that Io dust is considered by Morfill et al. (1980b,c) as the primary agent eroding the visible ring grains, whereas Burns et al. (1980) contend that destruction occurs mainly through sputtering by magnetospheric species.

Burns et al. (1980) utilize the known population of interplanetary micrometeoroids and derive ring properties on that basis. Several possible difficulties (see Grün et al. 1980) with this model arise:

1. One may question whether the interplanetary meteoroids can penetrate as deeply as Jupiter's ring. Numerical calculations by Hill and Mendis (1980) of the trajectories of charged grains find that the plasma polarization electric field can prevent them from approaching the planet. However, this only occurs at unrealistically high potentials (hundreds of volts). And, in fact, micrometeoroid penetration detectors on board Pioneers 10 and 11 registered eleven impacts ($r \gtrsim 5 \mu$m) and two impacts ($r \gtrsim 10 \mu$m), respectively, in Jupiter's vicinity (Humes 1976); this corresponds to a flux increase of 2 to 3 orders of magnitude over the interplanetary value. While the enhanced flux could be caused by a dust population orbiting Jupiter, it is generally believed that the impacts are due to interplanetary dust that was gravitationally focused by Jupiter.

2. Most of the mass flux for interplanetary meteoroids in circumterrestrial space comes from particles at sizes centered around $\sim 10^2 \mu$m. If this is

also true near Jupiter, some micron-sized ejecta from hypervelocity colli-sions may have speeds approaching 1 km s^{-1}, depending upon the nature of the regolith (see Fig. 17). Even including the expected relatively rapid damping to the ring plane (see Eq. 15), the observed ring thickness should be in excess of $\sim$ 100 km, whereas the Voyager 1 ring-plane crossing gives an upper limit of 30 km.

3. The ejecta population, if interplanetary meteoroids are the projectiles, should range up to at least millimeters. This is not compatible with some Mie-scattering solutions that match the preliminary photometry (cf. Fig. 1 of Grün et al. 1980). However, as we remarked earlier, such calculations cannot uniquely unfold particle size distributions; moreover, the photo-metry of Showalter et al. (1983b, 1984) which contains greater phase and color coverage should be included before firm statements are made. Nevertheless, if one adopts the Burns et al. model, the material in the micron-size range is only a fraction of the total ejecta mass, implying the existence of a significant ring component which has eluded observation.

Arguable parts of the alternative Io dust model (Morfill et al. 1980b) include:

1. No observational evidence of the postulated source of projectiles is cur-rently available, even though its presence is theoretically plausible.
2. Survival of the volcanic dust against erosion by sputtering during its pas-sage from Io to the ring is problematic, especially if the gains are just SO$_2$ snow.
3. The importance attached by Morfill et al. (1980b) to Io dust was in part motivated by the observed small upper limit on the ring thickness in backscatter which, without a flattening mechanism, argues for small ejecta velocities and hence low mass projectiles. It was also driven by the claim that the ring photometry restricts the solution to be one that emphasizes micron-sized grains which then forces the projectiles to be no larger than sub-micron in size. However, as indicated previously, the Grün et al. specification of a narrow-size range for ring members is actually a nonunique match. In fact, the Tyler et al. (1981) solution to the photometry has none of Grün et al.'s tiny particles.

In reality, both Io dust and micrometeoroids may play roles in develop-ing the Jovian ring.

E. Theoretical Radial Ring Profiles

Once the dominant evolutionary processes affecting the life of a grain are identified, the expected brightness from a point source can be constructed. Conversely, ring profiles might be invertable to infer the evolutionary proc-esses that are operating, or to specify the parent body distribution necessary to account for the observed profile.

Morfill et al. (1980*b*) have carried this process out for a ring in which the grains are generated by Io dust impacting mooms and are destroyed during single impacts by the same flux of dust; in the meantime the grain orbits have evolved due to Poynting-Robertson drag. Hence the particle number density decays like

$$dn = -n \frac{dt}{T_L} , \tag{24}$$

where $1/T_L$ is the rate of destruction, taken to be constant for micron-sized, grains, although in fact it varies like r^{-2}. The orbit collapses according to

$$\frac{da}{a} = - \frac{dt}{T_{PR_0}(r/r_o)} \tag{25}$$

where the characteristic time T_{PR}, a linear function of particle radius r, is given by Eq. (11). Hence

$$\frac{dn}{n} = \frac{da}{a} \frac{T_{PR}}{T_L} \quad \text{or} \quad n = n_o \left(\frac{a}{a_o} \right)^{T_{PR}/T_L} \tag{26}$$

where n_o is the particle number density at the source location $a = a_o$. Since $T_{PR} >> T_L$ (otherwise the ring's brightness would not decay rapidly), n decreases as a high power of a. Even at that, profiles are relatively flat; for example, between the outer and inner edge of the Jovian ring $(a_i/a_o) \approx 0.95$ so that, with $T_{PR}/T_L = 20$, $[n(a_o)]/[n(a_i)] \approx 3$. The reader is referred to Figs. 2 and 3 of Morfill et al. (1980*b*) which show moderately successful preliminary attempts at matching the observed Jovian ring profile.

In the alternate model of Burns et al. (1980), a particle is born as ejecta from micrometeoroid impact, orbitally evolves through plasma drag, and sputters away. Hence, the birth rate is independent of the death rate, unlike the Morfill et al. (1980*b*) case. On the other hand, the typical lifetime is proportional to the initial particle radius due to the assumption that grains erode away: from Eq. (9), $T_s(r) = T_{s_0}(r/r_o)$. Orbits evolving by plasma drag (see Eq. 12) have

$$\frac{da}{dt} \approx \dot{a}_o (r_o/r) \tag{27}$$

if the plasma properties are taken as constant across the ring. Hence all particles, regardless of size, disappear at precisely the same distance from their point of origin:

$$a_i = a_o + \dot{a}t$$

$$= a_o (1 - T_{s_0}/T_{PD_0}).$$ (28)

If the observed Jovian ring brightness drops abruptly solely for this reason, T_{PD_0} must be at least twenty times T_{s_0}.

The distribution function $n(r,a,t)$ representing differential particle density (see Eq. 1) in general evolves according to

$$\frac{\partial n}{\partial t} + \frac{1}{a^2} \frac{\partial}{\partial a} (a^2 \dot{a} n) + \frac{\partial}{\partial r} (\dot{r} n) = S(a,r,t)$$ (29)

where S is a source function representing the injection of particles at point a at time t. This equation's steady state solution can be found by separation of variables for a region with no parent bodies (no sources):

$$n(r,a) = n_o (a_o/a)^2 e^{-k(T_{PD_0}/T_{s_0})(1 - a/a_o)} \left(\frac{r}{r_o} \right)^k$$ (30)

where k is the (as yet undetermined) separation constant, to be specified by the boundary conditions. Particles leave the source point ($a = a_o$) according to

$$\dot{a}_o (r_o/r)^n \Big|_{a = a_0} = N,$$ (31)

in which N is the flux of particles in a size interval that is created as a_0. Following the discussion below Eq. (1), we write $N = N_0(r_0/r)^p$ with $p = 3.4$ for impact detritus. Hence $k = 1 - p$ in the general solution and, for collisional ejecta, $k = -2.4$. Thus everywhere in the ring the particle size distribution $\sim r^{-2.4}$; the shallower dependence with r than for stationary collisional ejecta arises because smaller particles evolve swiftly away from the source region. Finally, the solution is

$$n(r,a) = n_o (a_o/a)^2 e^{-2.4 \left(\frac{a_o - a}{a_o - a_i} \right)} \left(\frac{r}{r_o} \right)^{-2.4}.$$ (32)

The exponential determines the decaying number density as a drops from a_o to a_i; for example, $n(a_o)/ n(a_i) \approx 10$ for $a_i/a_o = 0.95$. The number decreases inward because at any a fewer particles of a given size exist than further out in the ring. One could use this solution as a Green's function to be convolved with a hypothetical source distribution, but the assumptions have been so extreme that such an exercise would have little meaning.

F. Other Ideas on Jovian Ring Origin

Shortly after Jupiter's ring was found by Voyager but before its properties were fully appreciated, several brief papers that proposed ring origins were written. Owen et al. (1979; see also Smith et al. 1979b), in the report that gave the first details about the ring, advanced that the ring was a steady state configuration. Its particles were thought to come from various sources outside the Roche limit which then halted for some unknown reason, perhaps an orbital resonance, during their inward evolution under various drags (see also Ip 1980b).

Arguing from trajectory calculations, Hill and Mendis (1980) maintained that the ring was made of captured interplanetary meteoroid fragments. They numerically traced the equatorial paths of disrupted, highly charged grains after their entry into the Jovian magnetosphere at 35 R_J and found that many particles tended to pass as close as a few planetary radii from the planet. We find their arguments unconvincing for, as they admit, the grains in their simulations never penetrate as far as the Jovian ring. Furthermore, there is no tendency for concentration in a narrow band nor in the equatorial plane; collisional damping cannot be called upon to localize grains because, with highly elliptic paths, collisions will cause fragmentation and perhaps even vaporization.

Material that leaks off, or has been nudged off, the surface of a satellite has been investigated by Dermott et al. (1980) and Smoluchowski (1979). Both papers contend that surface debris will be readily lost because the satellite's orbit, being within Jupiter's synchronous orbit has tidally evolved inward (see Burns 1977), past Jupiter's Roche limit. Several possible definitions of the Roche limit in terms of satellite shape and strength are considered.

In this last case, as well as in the similar moom idea of Burns et al. (1980) and Morfill et al. (1980b), there is the question as to where the parents came from. In most scenarios (Pollack and Fanale 1982; Prentice and ter Haar 1979) they condense directly out of a protoplanetary nebula rather than being tidally captured or being the remnants of a tidal fragmentation as originally envisioned by Roche himself. A combination of these mechanisms would permit gas drag capture by the nebula of Jupiter just before its disappearance. For further discussion of such issues regarding the origins of ring systems, see the chapters by Harris and by Ward.

G. Possible Galileo Studies of Jupiter's Ring

Future groundbased observations, except perhaps for a successful stellar occultation, are unlikely to change our ideas of the Jovian ring in any fundamental way (see chapter by Smith). In view of the possibility of a faint ring exterior to the known one, it might be worthwhile to search at the edges of Jupiter's ring during ring-plane passage (every five years) to see whether very tenuous rings might exist. A ring at 2.5 R_J − 3.0 R_J might be detectable outside the glare of Jupiter much as Saturn's E Ring can be seen at 4 R_S; the Space Telescope, due to its low scattered light contamination, might be especially suitable for this search.

The next major advance in our understanding of Jupiter's ring should take place with the Galileo Orbiter spacecraft in the late 1980s (see chapter by Stone). Galileo's solid state imaging system will substantially improve resolution on the ring. This should allow any internal ring structure to be noted, and additional moonlets to be discovered. It should also accurately position the known satellites within the ring, which is vital in elucidating the connection between the satellites and the ring. Coupled with the improved resolution, the opportunity to see the ring from various aspects should yield a better idea of the ring's morphology. The ring/planetary shadow region, as we have already seen (see Fig. 6 and Color Plate 10), should be especially useful in this regard.

Improved phase coverage of the ring should permit the particle size distribution, and any variation in it, to be ascertained across the ring. The photopolarimeter/radiometer, near-infrared mapping spectrometer and the ultraviolet spectrometer may all help in this regard, as will the radio science experiments.

Unfortunately, the inclination of the Orbiter's planned trajectory is kept uniformly low in order to aid in multiple satellite flybys. (Ignoring the busy inbound approach to the planet, the inclination of the Orbiter will rarely reach a few degrees.) This will make it more difficult to determine the ring's three-dimensional structure, since it will often present a nearly edge-on view. Furthermore, the spacecraft will spend most of its time relatively far out in the magnetosphere so as to alleviate spacecraft radiation damage and to explore that region. This means that resolution on the ring is often less than ideal and again lowers the various perspectives from which the ring can be viewed. However, the Orbiter does reach Europa (9.5 R_J) on many orbital passes. Thus the Galileo survey, while not optimum, will be as good or better than the available Voyager observations ($2°.5$, ~ 20 R_J) in terms of perspective and resolution.

During Galileo's extensive mission, we should be able to understand the apparent asymmetry of the halo. If the structure is tied to the planet's magnetic field, Galileo with its time coverage should be capable of detecting the halo's movement with the field.

Other equipment aboard the Orbiter will indirectly influence Jovian ring studies. Galileo's full complement of magnetospheric survey instruments should better delineate the magnetosphere's makeup, providing some modest help in tying down time scales for ring sputtering and plasma drag. However, because of the high pericenter of the spacecraft orbit, Galileo will not measure these important parameters in the ring's vicinity but only much farther out in the magnetosphere. The dust detector is a vast improvement over the Pioneer penetration experiment and will give the mass, velocity and charge of grains between 10^{-2} and 10^2 μm. This information, even though principally available beyond Europa's orbit, should indicate the nature of the micrometeoroid flux striking the ring and should thereby favor either the erosion model of Burns et al. (1980) or that of Morfill et al. (1980b).

V. IDEAS ON SATURN'S ETHEREAL RINGS

Like the Jovian ring, the outlying Saturnian rings contain numerous micron-sized particles whose lives are relatively brief. Since the visible rings are presumably in steady state, as small grains evolve out of the system or are destroyed by sputtering, they must be continually replaced. The similarity to the Jovian ring may end at this point; in the Jovian case we have argued that collisions into unseen parent bodies generate the ring, but this mechanism seems to face difficulties for each of the Saturnian rings. For the *G Ring*, too few parent bodies may be present. The close physical association of Enceladus with the *E Ring* has lead many modelers to feel that the satellite is involved in the ring's genesis. For the *F Ring*, most thought has been devoted to dealing with the ring's baffling dynamics rather than with why the ring is there.

An important distinction between the tenuous rings of Saturn and Jupiter is that, with the possible exception of the F Ring (cf. Dermott et al. 1980; Smoluchowski 1979), Saturn's lie outside the Roche limit; conventional wisdom would thus allow objects to accrete in these rings (cf. chapter by Weidenschilling et al.). Yet they do not, perhaps because collision velocities are too high, or densities too low.

In the following sections, we address such issues for each of the various tenuous rings of Saturn.

A. Saturn's G Ring

Because of the scant information available on the G Ring, little effort has been spent on understanding it. Puzzles include what causes its possibly abrupt edges, why the ring does not accrete, and where the small particles that may comprise it (Van Allen 1983) originate. In addition, one wonders whether the little mass in this region given by Van Allen's solution is enough to reduce the energetic electron and proton flux (Simpson et al. 1980).

B. Saturn's E Ring

Attempts at an overview of the E Ring have been hindered because two key observations are ambiguous.

1. Is the E Ring brightest precisely at Enceladus (Baum et al. 1981) or just nearby (Larson et al. 1981)? In the former case, the ice-covered satellite almost certainly is the source of the ring; however, if the ring merely peaks in the vicinity of Enceladus, the connection is more problematic. Even if ring particles originate at this satellite of Saturn, one must question how they evolve both inward and outward. The ring is so tenuous that, even at

its core, a specific particle collides only every 10^3 yr, according to Eq. (15). Hence the ring cannot diffuse apart, and plasma drag (see Eq. 12) should drive grains outward. Injection velocities would need to be too high (see Eq. 8) to account for the ring's breadth as an initial condition.

2. Are the ring grains indeed spherical and narrowly confined in size (Pang et al. 1983*a,b*)? If so, a liquid or gas origin is clearly suggested. Straightforward arguments have been made to associate Enceladus' virgin surface with outgassing or volcanism (Squyres et al. 1983), although the energy source driving these processes is uncertain (see, however, Stevenson 1982, and Lissauer et al. 1983).

We now list a few scenarios that have been proffered to account for the E Ring. Cook and Terrile (1980) believe that tidal heating supplies particles by intermittent venting of water vapor through a thin ice crust on Enceladus. Pang et al. (1983*a,b*) also argue for episodic volcanic injections to replenish the ring. Haff et al. (1983) consider grain production by geysering and by impact, and point out that impact fluxes fail to sustain the ring by an order of magnitude. However, recall from our discussion of the generation of the Jovian ring (Sec. IV) and the development of dust rings from satellites (Sec. III.H), that fluxes are uncertain, and that regolith properties can dramatically alter yields. McKinnon (1983), accepting that ring particles are spherical but not recognizing a plausible way to allow tidal heating to produce volcanism, calls upon the melt and vapor generated in a single large impact to develop small spherical ring grains; he admits that such collisions are rare and calculates that the probability of observing such a ring is only ∼1%. Hill and Mendis (1982*b*) maintain that the ring is simply a concentration of interplanetary dust, but that seems improbable for the same reasons that a similar proposal for the Jovian ring was questioned.

The two most developed papers on the physics of the E Ring both reach impasses. Haff et al. (1983) identify various paradoxes that exist in the interaction between the E Ring and its plasma environment, while Morfill et al. (1983*b*) consider transport processes in the E Ring and find no reason for its outer terminus.

C. Saturn's F Ring

Since chapters by Cuzzi et al., and Dermott describe the dynamics of the F Ring in some detail, we will content ourselves with listing references for basic ideas and with addressing a few specific points.

The shepherding satellites on either side of the F Ring are generally acknowledged to narrow the ring (Goldreich and Tremaine 1982; Dermott chapter). Usually the nonaxisymmetric structures in the F Ring are ascribed to the gravitational perturbations of the satellites. The larger inner satellite appears to be—as it should be—most influential. Over a single orbit, ring particles drift $3\pi\Delta a \approx 7800$ km (see Eq. 28 in the chapter by Cuzzi et al.)

with respect to 1980S27; as demonstrated numerically by Showalter and Burns (1982), the impulses at the satellite's closest approach produce periodic clumping of the same wavelength (Fig. 19 in Cuzzi et al.'s chapter), and this is approximately what is observed (Smith et al. 1982). However such models have not been able to produce braids; in addition perturbations due to the satellites should develop only a single wavelength rather than the variety seen. Synnott et al. (1983) find that the braids, etc., are especially formed near the most recent satellite conjunction; however, according to Showalter (1983), such conjunctions have no special dynamical effect on the ring, so any observed association, given the observational errors, may be coincidental. The long-time evolution of an elliptical ring perturbed by a satellite on an elliptical path has been addressed by Borderies et al. (1983), who note that the apoapse of 1980S27's orbit might periodically penetrate the F Ring with presumably disastrous consequences. Indeed this disruption should be continually occurring for the recently found faint ringlet inward of the nominal F Ring (Burns et al. 1983).

Perturbations by satellites on circular orbits were once suggested to braid the ring across resonances (Dermott 1981); because the fine spacing between resonances would confuse trajectories and because the actual elliptical orbits introduce qualitatively different dynamics, this is no longer accepted. Goldreich (see Smith et al. 1982) has suggested that the motions of large particles within the F Ring could induce the observed twisted and clumped morphology (see Marginal Figs. in Weidenschilling et al.'s chapter). Several schemes involving charged grains have also been put forward to explain the clumpy and braided nature of the F Ring. Suggestions have included electromagnetically disturbed orbits (Hill and Mendis 1981; Smoluchowski 1981), a current instability (Hill and Mendis 1982a) and coherent orbit perturbations (Morfill et al. 1983b). All of these are less viable now that Grün et al. (see their chapter) have inferred that charges will be substantially reduced in any relatively dusty ring such as the F Ring. An additional difficulty faced by any electromagnetic origin for the F Ring structure is that charge-to-mass ratios need be quite uniform in addition to being unusually large; but, according to Burns and Showalter (1981), stochastically varying potentials and a distribution of masses make this improbable.

VI. SUMMARY

We have appraised a class of planetary rings—the ethereal rings—that have low optical depths, generally contain a large complement of miniscule particles, and interact significantly with other magnetospheric residents. These same rings are likely sculpted by adjacent and/or embedded satellites. The dominant physical processes for diffuse rings are often different from those that apply to the more substantial rings of Saturn and Uranus. In particular, small particles last for limited times and so the specks seen today must be continually replenished if these rings are to be longstanding members of the solar system. Grains also undergo appreciable orbital evolution and are subject to electromagnetic forces. On the other hand, collisions and collective effects are probably not so important for the faint rings as they are for the denser and more massive rings.

A reanalysis (Showalter et al. 1983a, b, 1984) of Voyager data on the Jovian ring is providing improved understanding of its morphology and composition. Jupiter's ring is made up of a relatively bright band plus a toroidal (i.e. radially-confined and vertically-extended) halo. The band is flat, thin and approximately uniform radially. A distribution of particles with sizes centered near a few microns constitutes much of the band while smaller particles are present in the halo.

Many of the Jovian ring's observed properties can be explained by, and indeed apparently require, the presence of innumerable macroscopic (but as yet unseen) parent bodies within the ring. The visible ring grains are derived from these mooms by meteoroid impacts, although dust from Io could also be a factor. These tiny grains are destroyed in short times by sputtering; micrometeoroid fragmentation is a minor contributor to their demise. The primary cause of orbital collapse is plasma drag, which moves grains significantly over their lifetimes. Collisions amongst particles and with parent bodies damp out-of-plane motions in the bright ring to determine its thickness. The ring's radial structure is produced by the radial distribution of the parent bodies (including the two known satellites), by the ability of the large mooms to serve as sources or sinks for visible grains, and finally by radial variations in rates of orbital evolution and destruction. The halo particles are likely derived from grains that evolve inward from the main belt. Once sizes become small enough, perturbations from the tilted Jovian magnetic field push particles out of the ring plane; since few (or no) parent bodies lie in the halo region, these grains are not damped back to the plane.

The Saturnian ethereal rings are not as well understood as their Jovian counterpart although they seem to share many of its characteristics. As with Jupiter's ring, the inferred particles are micron-sized, have orbits which evolve rapidly, and perish in times short compared to the solar system's age. Once again, plasma drag and sputtering are the dominant evolution processes. The sources to regenerate the small particles in these rings are uncertain. For the F Ring, hidden parent bodies are a likely source; micrometeoroid crosion and jostling collisions, perhaps driven in part by the shepherding satellites, may be instrumental in providing material. In the case of the E Ring, Enceladus is a likely source although the precise mechanism to inject material remains elusive. So little is known about the G Ring that its origin must be considered a total enigma.

The class of ring structures discussed in this chapter has much to teach about processes important both today and at the time of the solar system's origin. One should not expect that the amount of understanding to be gleaned from a ring system be proportional to its mass. Nor, as we have well demonstrated, should a ring's mass bear a simple relationship to the amount that can be written about it.

Acknowledgments. We are grateful for discussions and for improvements in this research suggested by J. Gradie, J. Lissauer, F. Mignard, J. Pollack, and S. Soter, and especially by G. Consolmagno, J. N. Cuzzi, and D. Jewitt. We thank B. Boettcher, M. Talman, and especially L. Vrooman for help in preparing the manuscript. We also appreciate the editors' forbearance in giving us additional time to pursue our reinvestigation of Jupiter's ring. J.A.B. and M.R.S. were partially supported by grants from the National Aeronautics and Space Administration. Some of this work was done while J.A.B. and M.R.S. were temporarily at the Space Science Division of the NASA Ames Research Center and while J.A.B. was temporarily at the Astronomy Department of the University of California at Berkeley.

REFERENCES

Acuña, M. H., and Ness, N. F. 1976. The complex main magnetic field of Jupiter. *J. Geophys. Res.* 81:2917–2922.

Anselmo, L., and Farinella, P. 1983. Alfvén drag for satellites orbiting in planetary plasmaspheres. Paper at IAU Colloquium 77, "Natural Satellites," Cornell University.

Aubier, M. G., Meyer-Vernet, N., and Pederson, B. M. 1983. Shot noise from grain and particle impacts in Saturn's ring plane. *Geophys. Res. Letters* 10:5–8.

Bagenal, F., and Sullivan, J. D. 1981. Direct plasma measurements in the Io torus and the inner magnetosphere of Jupiter. *J. Geophys. Res.* 86:8447–8466.

Baron, R. L., and Elliot, J. L. 1983. A CCD imager search for possible material orbiting Saturn between 10 and 36 Saturnian radii. *Astron. J.* 88:562–564.

Baum, W. A., and Kreidl, T. J. 1982. Implications of the three-dimensional distribution of material in Saturn's E ring. *Bul. Amer. Astron. Soc.* 14:749 (abstract).

Baum, W. A., Kreidl, T., Westphal, J. A., Danielson, G. E., Seidelman, P. K., Pascu, D., and Currie, D. G. 1980. Profile of Saturn's E ring. *Bul. Amer. Astron. Soc.* 12:700–701 (abstract).

Baum, W. A., Kreidl, T., Westphal, J. A., Danielson, G. E., Seidelman, P. K., Pascu, D., and Currie, D. G. 1981. Saturn's E ring. 1. CCD Observations of March 1980. *Icarus* 47:84–96.

Baum, W. A., Kreidl, T. J., and Wasserman, L. H. 1983. Saturn's E ring. In Proceedings of *IAU Colloquium 75 Planetary Rings*, ed. A. Brahic, Toulouse, France, Aug. 1982.

Becklin, E. E., and Wynn-Williams, C. G. 1979. Detection of Jupiter's ring at 2.2 μm. *Nature* 279:400–401.

Belcher, J. W. 1983. The low-energy plasma in the Jovian magnetosphere. In *Physics of the Jovian Magnetosphere*, ed. A. J. Dessler, (Cambridge: Cambridge University Press), pp. 68–105.

Borderies, N., Goldreich, P., and Tremaine, S. 1983. The variations in eccentricity and apse precession rate of a narrow ring perturbed by a close satellite. *Icarus* 53:84–89.

Brahic, A. 1977. Systems of colliding bodies in a gravitational field. I. Numerical simulation of the standard model. *Astron. Astrophys.* 54:895–907.

Bridge, H. S., Bagenal, F., Belcher, J. W., Lazarus, A. J., McNutt, R. L., Sullivan, J. D., Gazis, P. R., Hartle, R. E., Ogilvie, K. W., Scudder, J. D., Sittler, E. C., Eviatar, A., Siscoe, G. L., Goertz, C. K., and Vasyliunas, V. M. 1982. Plasma observations near Saturn: Initial results from Voyager 2. *Science* 215:563–570.

Brown, W. L., Lanzerotti, L. J., and Johnson, R. E. 1982. Fast ion bombardment of ices and its astrophysical importance. *Science* 218:525–532.

Burns, J. A. 1977. Orbital evolution. In *Planetary Satellites,* ed. J. A. Burns (Tucson: University of Arizona Press), pp. 113–156.

Burns, J. A., and Showalter, M. R. 1981. The puzzling dynamics of Saturn's F ring. *Proc. U.S./Brazil Conference on Motions of Planets and Satellites,* ed. S. Ferraz-Mello (Saõ Paulo: U. Saõ Paulo Press), pp, 201–213.

Burns, J. A., and Showalter, M. R. 1983. Dust rings from satellites. *Bul. Amer. Astron. Soc.* 15:814–815 (abstract).

Burns, J. A., Hamill, P., Durisen, R. H., and Cuzzi, J. N. 1979a. On the "thickness" of Saturn's rings caused by satellite and solar perturbations and by planetary precession. *Astron. J.* 84:1783–1801.

Burns, J. A., Lamy, P. L., and Soter, S. 1979b. Radiation forces on small particles in the solar system. *Icarus* 40:1–48.

Burns, J. A., Showalter, M. R., Cuzzi, J. N., and Pollack, J. B. 1980. Physical processes in Jupiter's ring: Clues to its origin by Jove! *Icarus* 44:339–360.

Burns, J. A., Cuzzi, J. N., and Showalter, M. R. 1983. Discovery of gossamer rings. *Bul. Amer. Astron. Soc.* 15:1013–1014.

Carbary, J. F., Krimigis, S. M., and Ip, W.-H. 1983. Energetic particle signatures of Saturn's satellites. *J. Geophys. Res.* 88:8947–8958.

Cheng, A. F., Lanzerotti, L. J., and Pirronello, V. 1982. Charged particle sputtering of ice surfaces in Saturn's magnetosphere. *J. Geophys. Res.* 87:4567–4570.

Consolmagno, G. J. 1980. Electromagnetic scattering lifetimes for dust in Jupiter's ring. *Nature* 285:557–558.

Consolmagno, G. J. 1983. Lorentz forces on the dust in Jupiter's ring. *J. Geophys. Res.* 88:5607–5612.

Cuzzi, J. 1983. Planetary ring systems. *Rev. Geophys. Space Phys.* 21:173–186.

Cuzzi, J. N., Burns, J. A., Durisen, R. H., and Hamill, P. M. 1979a. Vertical structure and thickness of Saturn's rings. *Nature* 281:202–204.

Cuzzi, J. N., Durisen, R. H., Burns, J. A., and Hamill, P. 1979b. The vertical structure and thickness of Saturn's rings. *Icarus* 38:54–68.

Davis, D. R., Housen, K. R., and Greenberg, R. 1981. The unusual dynamical environment of Phobos and Deimos. *Icarus* 47:220–233.

dePater, I., and Jaffe, W. J. 1984. VLA observations of Jupiter's non-thermal radiation. Submitted to *Astrophys. J. Suppl.*

Dermott, S. F. 1981. The braided F ring of Saturn. *Nature* 290:454–457.

Dermott, S. F., Murray, C. D., and Sinclair, A. T. 1980. The narrow rings of Jupiter, Saturn and Uranus. *Nature* 284:309–313.

Dobrovolskis, A. R., and Burns, J. A. 1980. Life near the Roche limit: Behavior of ejecta from satellites close to planets. *Icarus* 42:422–441.

Dobrovolskis, A. R., and Burns, J. A. 1983. Angular momentum drain: A mechanism for despinning asteroids. *Icarus.* In press.

Dohnanyi, J. S. 1978. Particle dynamics. In *Cosmic Dust,* ed. J. A. M. McDonnell (New York: John Wiley and Sons), pp. 527–605.

Dollfus, A., and Brunier, S. 1982. Observation and photometry of an outer ring of Saturn. *Icarus* 49:194–204.

Dunham, E., Elliot, J. E., Mink, D., and Klemola, A. R. 1982. Limit on possible narrow rings around Jupiter. *Astron. J.* 87:1423–1427.

Feibelman, W. A. 1967. Concerning the "D" ring of Saturn. *Nature* 214:793–794.

Feibelman, W. A., and Klinglesmith, D. A. 1980. Saturn's E ring revisited. *Science* 209:277–279.

Fillius, W. 1976. The trapped radiation belts of Jupiter. In *Jupiter,* ed. T. Gehrels (Tucson: University of Arizona Press), pp. 896–927.

Fillius, W., McIlwain, G. E., and Mogro-Campero, A. 1975. Radiation belts of Jupiter: A second look. *Science* 188:465–467.

Frank, L. A., Burek, B. G., Ackerson, K. L., Wolfe, J. H., and Mihalov, J. D. 1980. Plasmas in Saturn's magnetosphere. *J. Geophys. Res.* 85:5695–5708.

Gault, D. E., Shoemaker, E. M., and Moore, H. J. 1963. Spray ejected from the lunar surface by meteoroid impact. NASA Tech. Note D-1767.

Gehrels, T., Baker, L. R., Beshore, E., Blenman, C., Burke, J. J., Castillo, N. D., Da Costa, B., Degewij, J., Doose, L. R., Fountain, J. W., Gotobed, J., KenKnight, C. E., Kingston, R., McLaughlin, G., McMillan, R., Murphy, R., Smith, P. H., Stoll, C. P., Strickland, R. N., Tomasko, M. G., Wijesinghe, M. P., Coffeen, D. L., and Esposito, L. W. 1980. Imaging photopolarimeter on Pioneer Saturn. *Science* 207:434–439.

Goldreich, P., and Tremaine, S. 1982. The dynamics of planetary rings. *Ann. Rev. Astron. Astrophys.* 20:249–283.

Gradie, J., Thomas, P., and Veverka, J. 1980. The surface composition of Amalthea. *Icarus* 44:373–387.

Graps, A. L., Lane, A. L., Horn, L. J., and Simmons, K. E. 1983. Evidence for material between Saturn's A and F ring from the Voyager 2 photopolarimeter. *Bul. Amer. Astron. Soc.* 15:814 (abstract).

Greenberg, R., Wacker, J. F., Hartmann, W. K., and Chapman, C. R. 1978. Planetesimals to planets: Numerical simulation of collisional evolution. *Icarus* 35:1–26.

Grün, E., Morfill, G., Schwehm, G., and Johnson, T. V. 1980. A model of the origin of the Jovian ring. *Icarus* 44:326–338.

Gurnett, D. A., Grün, E., Gallagher, D., Kurth, W. S., and Scarf, F. L. 1983. Micron-sized particles detected near Saturn by the Voyager plasma wave experiment. *Icarus* 53:236–254.

Haemmerle, V. R., Danielson, G. E., Currie, D. G., Ravine, M., and Hatzes, A. 1982. Structure in the Jovian ring? *Bul. Amer. Astron. Soc.* 14:747 (abstract).

Haff, P. K., Eviatar, A., and Siscoe, G. L. 1983. Ring and plasma: The enigmae of Enceladus. *Icarus* 56:426–436.

Hanel, R. A., Conrath, B., Flasar, M., Herath, L., Kunde, V., Lowman, P., Maquire, W., Pirraglia, J., Samuelson, R., Gautier, D., Gierasch, P., Horn, L., Kumar, S., and Ponnamperuma, C. 1979. Infrared observations of the Jovian system from Voyager 2. *Science* 206:952–956.

Harris, A. W., and Kaula, W. M. 1975. A co-accretional model of satellite formation. *Icarus* 24:516–524.

Harrison, M., and Schoen, R. 1967. Evaporation of ice in space: Saturn's rings. *Science* 157:1175–1176.

Hill, J. R., and Mendis, D. A. 1980. Charged dust in the outer planetary magnetospheres. II-Trajectories and spatial distribution. *Moon Planets* 23:53–71.

Hill, J. R., and Mendis, D. A. 1981. On the braids and spokes in Saturn's ring system. *Moon Planets* 24:431–436.

Hill, J. R., and Mendis, D. A. 1982a. On the dust ring current of Saturn's F ring. *Geophys. Res. Letters* 9:1069–1071.

Hill, J. R., and Mendis, D. A. 1982b. The origin of the E-ring of Saturn. *EOS* 63:1010.

Hood, L. L. 1983. Radial diffusion in Saturn's radiation belts: A modelling analysis assuming satellite and ring E absorption. *J. Geophys. Res.* 88:808–818.

Housen, K. R., Wilkening, L. L., Chapman, C. R., and Greenberg, R. 1979. Asteroidal regoliths. *Icarus* 39:317–351.

Humes, D. H. 1976. The Jovian meteoroid environment. In *Jupiter*, ed. T. Gehrels, (Tucson: University of Arizona Press), pp. 1052–1067.

Humes, D. H. 1980. Results of Pioneer 10 and 11 meteoroid experiments: Interplanetary and near-Saturn. *J. Geophys. Res.* 85:5841–5852.

Ip, W.-H. 1980a. Discussion of the Pioneer 11 observations of the F ring of Saturn. *Nature* 287:126–128.

Ip, W.-H. 1980b. New progress in the physical studies of the planetary rings. *Space Sci. Rev.* 26:97–109.

Jewitt, D. C. 1982. The rings of Jupiter. In *Satellites of Jupiter*, ed. D. Morrison (Tucson: University of Arizona Press), pp. 44–64.

Jewitt, D. C., and Danielson, G. E. 1981. The Jovian ring. *J. Geophys. Res.* 86:8691–8697.

Jewitt, D. C., Danielson, G. E., and Synnott, S. P. 1979. Discovery of a new Jupiter satellite. *Science* 206:951.

Jewitt, D. C., Danielson, G. E., and Terrile, R. J. 1981. Groundbased observations of the Jovian ring and inner satellites. *Icarus* 48:536–539.

Johnson, R. E., Lanzerotti, L. J., Brown, W. L., and Armstrong, T. P. 1981. Erosion of Galilean satellite surfaces by Jovian magnetospheric particles. *Science* 212:1027–1030.

Johnson, T. V., Cook, A. F., Sagan, C., and Soderblom, L. A. 1979. Volcanic resurfacing rates and implications for volatiles on Io. *Nature* 280:746–750.

Johnson, T. V., Morfill, G. E., and Grün, E. 1980. Dust in Jupiter's magnetosphere: An Io source? *Geophys. Res. Letters* 7:305–308.

Kawakami, S., Mizutani, H., Takagi, Y., Kato, M., and Kumazawa, M. 1983. Impact experiments on ice. *J. Geophys. Res.* 88:5806–5814.

Lafon, J-P. J., Lamy, P. L., and Millet, J. 1981. On the electrostatic potential and charge of cosmic grains. 1. Theoretical background and preliminary results. *Astron. Astrophys.* 95:295–303.

Lamy, P. L., and Mauron, N. 1981. Observations of Saturn's outer ring and new satellites during the 1980 edge-on presentation. *Icarus* 46:181–186.

Lane, A. L., Hord, C. W., West, R. A., Esposito, L. W., Coffeen, D. L., Sato, M., Simmons, K. E., Pomphrey, R. B., and Morris, R. B. 1982. Photopolarimetry from Voyager 2: Preliminary results on Saturn, Titan, and the rings. *Science* 215:537–543.

Larson, S. 1983. Summary of optical ground-based E ring observations at the University of Arizona. In Proceedings of *IAU Colloquium 75, Planetary Rings*, ed. A. Brahic, Toulouse, France, Aug. 1982.

Larson, S. M., Fountain, J. W., Smith, B. A., and Reitsema, H. J. 1981. Observations of the Saturn E ring and a new satellite. *Icarus* 47:288–290.

Lazarus, A. L., Hasegawa, T., and Bagenal, F. 1983. Long-lived particulate or gaseous structure in Saturn's outer magnetosphere? *Nature* 302:230–232.

Lissauer, J. J., Peale, S. J., and Cuzzi, J. N. 1983. Torque on the co-orbital satellites due to Saturn's rings and the melting of Enceladus. *Icarus*. In press.

Marouf, E. A., and Tyler, G. L. 1983. Radio occultation of Saturn's rings: Is less than 75 meters radial resolution achievable? *Bul. Amer. Astron. Soc.* 15:817 (abstract).

McDonald, F. B., Schardt, A. W., and Trainor, J. H. 1980. If you've seen one magnetosphere, you haven't seen them all: Energetic particle observations in the Saturn magnetosphere. *J. Geophys. Res.* 85:5813–5830.

McKinnon, W. B. 1983. Origin of the E ring: Condensation of impact vapor . . . or boiling of impact melt? *Lunar Planet. Sci.* 14:487–488 (abstract).

Morfill, G. E., and Grün, E. 1979a. The motion of charged dust particles in interplanetary space. I. The zodiacal cloud. *Planet. Space Sci.* 27:1269–1282.

Morfill, G. E., and Grün, E. 1979b. The motion of charged dust particles in interplanetary space. II. Interstellar grains. *Planet. Space Sci.* 27:1283–1292.

Morfill, G. E., Grün, E., and Johnson, T. V. 1980a. Dust in Jupiter's magnetosphere: Physical processes. *Planet. Space Sci.* 28:1087–1100.

Morfill, G. E., Grün, E., and Johnson, T. V. 1980b. Dust in Jupiter's magnetosphere: Origin of the ring. *Planet. Space Sci.* 28:1101–1110.

Morfill, G. E., Grün, E., and Johnson, T. V. 1980c. Dust in Jupiter's magnetosphere: Time variations. *Planet. Space Sci.* 28:1111–1114.

Morfill, G. E., Grün, E., and Johnson, T. V. 1980d. Dust in Jupiter's magnetosphere: Effect on magnetospheric electrons and ions. *Planet. Space Sci.* 28:1115–1123.

Morfill, G. E., Fechtig, H., Grün, E., and Goertz, C. K. 1983a. Some consequences of meteoroid impacts on Saturn's rings. *Icarus* 55:439–447.

Morfill, G. E., Grün, E., and Johnson, T. V. 1983b. Saturn's E, G, and F rings: Modulated by the plasma sheet. *J. Geophys. Res.* 88:5573–5579.

Neugebauer, G., Becklin, E. E., Jewitt, D., Terrile, R., and Danielson, G. E. 1981. Spectra of the Jovian ring and Amalthea. *Astron. J.* 86:607–610.

Northrop, T. G., and Hill, J. R. 1983. The adiabatic motion of charged dust grains in rotating magnetospheres. *J. Geophys. Res.* 88:1–11.

Öpik, E. J. 1951. Collision probabilities with the planets and the distribution of interplanetary matter. *Proc. Roy. Irish Acad.* 54A:165–199.

Owen, T., Danielson, G. E., Cook, A. F., Hansen, C., Hall, V. L., and Duxbury, T. C. 1979. Jupiter's rings. *Nature* 281:442–446.

Pang, K. V., Voge, C. C., and Rhoads, J. W. 1983*a*. Macrostructure and microphysics of Saturn's E ring. In Proceedings of *IAU Colloquium 75, Planetary Rings,* ed. A. Brahic, Toulouse, France, Aug. 1982.

Pang, K.V., Voge, C. C., Rhoads, J. W., and Ajello, J. M. 1983*b*. The E-ring of Saturn and its satellite Enceladus. Submitted to *J. Geophys. Res.*

Papadakos, D. N., and Williams, I. P. 1983. Escape of ejecta from cratered solar system satellites. *Mon. Not. Roy. Astron. Soc.* 204:635–645.

Pilcher, C. B. 1979. The stability of water on Io. *Icarus* 37:559–574.

Pollack, J. B., 1981. Phase curve and particle properties of Saturn's F ring. *Bul. Amer. Astron. Soc.* 13:727 (abstract).

Pollack, J. B., and Cuzzi, J. N. 1980. Scattering by non-spherical particles of size comparable to a wavelength: A new semi-empirical theory and its application to tropospheric aerosols. *J. Atmos. Sci.* 37:868–881.

Pollack, J. B., and Fanale, F. 1982. Origin and evolution of the Jupiter satellite system. In *Satellites of Jupiter,* ed. D. Morrison (Tucson: University of Arizona Press), pp. 872–910.

Prentice, A. J. R., and terHaar, D. 1979. Origin of the Jovian ring and the Galilean satellites. *Nature* 280:300–302.

Pyle, K. R., McKibben, R. B., and Simpson, J. A. 1983. Pioneer 11 observations of trapped particle absorption by the Jovian ring and the satellites 1979, J1, J2, and J3. *J. Geophys. Res.* 88:45–48.

Reitsema, H. J., Smith, B. A., and Larson, S. M. 1980. The E Ring of Saturn. *Bull. Amer. Astron. Soc.* 12:1807 (abstract).

Showalter, M. R. 1983. Effects of shepherd conjunctions on Saturn's F-ring. In Proceedings of *IAU Colloquium 75, Planetary Rings,* ed. A. Brahic, Toulouse, France, Aug. 1982.

Showalter, M. R., and Burns, J. A. 1982. A numerical study of Saturn's F-ring. *Icarus* 52:526–544.

Showalter, M. R., Burns, J. A., Cuzzi, J. N., and Pollack, J. B. 1983*a*. Re-examining Jupiter's ring: Morphology. *Bul. Amer. Astron. Soc.* 15:815 (abstract).

Showalter, M. R., Burns, J. A., Cuzzi, J. N., and Pollack, J. B. 1983*b*. Re-examining Jupiter's ring: Photometry. *Bul. Amer. Astron. Soc.* 15:815 (abstract).

Showalter, M. R., Burns, J. A., Cuzzi, J. N. and Pollack, J. B. 1984. An alternate model of the Jovian ring. In preparation.

Simpson, J. A., Bastian, T. S., Chenette, D. L., McKibben, R. B., and Pyle, K. R. 1980. The trapped radiations of Saturn and their absorption by satellites and rings. *J. Geophys. Res.* 85:5731–5762.

Sittler, E. C., Scudder, J. D., and Bridge, H. S. 1981. Distribution of neutral gas and dust near Saturn. *Nature* 292:711–714.

Smith, B. A. 1978. The D and E rings of Saturn. In *The Saturn System* (NASA Conference Publication 2068), pp. 105–111.

Smith, B. A., and Reitsema, H. J. 1980. CCD observations of Jupiter's ring and Amalthea. Presented at IAU Colloquium 57, Kona, Hawaii.

Smith, B. A., Soderblom, L. A., Beebe, R. F., Boyce, J., Briggs, G. A., Carr, M. H., Collins, S. A., Cook, A. F. II, Danielson, G. E., Davies, M. E., Hunt, G. E., Ingersoll, A. P., Johnson, T. V., Masursky, H., McCauley, J. F., Morrison, D., Owen, T., Sagan, C., Shoemaker, E. M., Strom, R. G., Suomi, V. E., and Veverka, J. 1979*a*. The Galilean satellites and Jupiter: Voyager 2 imaging science results. *Science* 206:927–950.

Smith, B. A., Soderblom, L. A., Johnson, T. V., Ingersoll, A. P., Collins, S. A., Shoemaker, E. M., Hunt, G. E., Masursky, H., Carr, M. H., Davies, M. E., Cook, A. F. II, Boyce, J., Danielson, G. E., Owen, T., Sagan, C., Beebe, R. F., Veverka, J., Strom, R. G., McCauley, J. F., Morrison, D., Briggs, G. A., and Suomi, V. E. 1979*b*. The Jupiter system through the eyes of Voyager 1. *Science* 204:951–971.

Smith, B. A., Soderblom, L. A., Beebe, R. F., Boyce, J., Briggs, G. A., Bunker, A., Collins, S. A., Hansen, C. J., Johnson, T. V., Mitchell, J. L., Terrile, R. J., Carr, M. H., Cook, A. F. II, Cuzzi, J. N., Pollack, J. B., Danielson, G. E., Ingersoll, A. P., Davies,

M. E., Hunt, G. E., Masursky, H., Shoemaker, E. M., Morrison, D., Owen, T., Sagan, C., Veverka, J., Strom, R. G., and Suomi, V. E. 1981. Encounter with Saturn: Voyager 1 imaging science results. *Science* 212:163–191.

Smith, B. A., Soderblom, L. A., Batson, R., Bridges, P., Inge, J., Masursky, H., Shoemaker, E. M., Beebe, R. F., Boyce, J., Briggs, G. A., Bunker, A., Collins, S. A., Hansen, C. J., Johnson, T. V., Mitchell, J. L., Terrile, R. J., Cook, A. F. II, Cuzzi, J. N., Pollack, J. B., Danielson, G. E., Ingersoll, A. P., Davies, M. E., Hunt, G. E., Morrison, D., Owen, T., Sagan, C., Veverka, J., Strom, R. G., and Suomi, V. E. 1982. A new look at the Saturn system: The Voyager 2 images. *Science* 215:504–537.

Smoluchowski, R. 1979. The ring systems of Jupiter, Saturn and Uranus. *Nature* 280:377–378.

Smoluchowski, R. 1981. The F-ring of Saturn. *Geophys. Res. Letters* 8:623–624; erratum 8:946.

Soter, S. 1971. The dust belts of Mars. *Cornell Center Radiophys. Space Phys. Rept. 472.*

Squyres, S. W., Reynolds, R. T., Cassen, P. M., and Peale, S. J. 1983. The evolution of Enceladus. *Icarus* 53:319–331.

Stanley, J. E., Singer, S. F., and Alvarez, J. M. 1979. Interplanetary dust between 1 and 5 AU. *Icarus* 37:457–466.

Stevenson, D. J. 1982. Volcanism and igneous processes in small icy satellites. *Nature* 298:142–144.

Stöffler, D., Gault, D. E., Wedekind, J., and Polkowski, G. 1975. Experimental hypervelocity impact into quartz sand: Distribution and shock metamorphism of ejecta. *J. Geophys. Res.* 80:4062–4077.

Strom, R. G., and Schneider, N. M. 1982. Volcanic eruption plumes on Io. In *Satellites of Jupiter*, ed. D. Morrison (Tucson: University of Arizona Press), pp. 598–633.

Synnott, S. P. 1980. 1979J2: The discovery of a previously unknown Jovian satellite. *Science* 210:786–788.

Synnott, S. P. 1983. Orbits of the small satellites of Jupiter. Paper at IAU Colloquium 77, "Natural Satellites," Cornell University. Submitted to *Icarus*.

Synnott, S. P., Terrile, R. J., Jacobson, R. A., and Smith, B. A. 1983. Orbits of Saturn's F-ring and its shepherding satellites. *Icarus* 53:156–158.

Terrile, R. J., and Tokunaga, A. 1980. Infrared photometry of Saturn's E ring. *Bul. Amer. Astron. Soc.* 12:701 (abstract).

Thomsen, M. F., and Van Allen, J. A. 1979. On the inference of properties of Saturn's ring E from energetic charged particle observations. *Geophys. Res. Letters* 6:893–896.

Tyler, G. L., Marouf, E. A., and Wood, G. E. 1981. Radio occultation of Jupiter's ring: Bounds on optical depth and particle size, and a comparison with infrared and optical results. *J. Geophys. Res.* 86:8699–8703.

Tyler, G. L., Marouf, E. A., Simpson, R. A., Zebker, H. A., and Eshleman, V. R. 1983. The microwave opacity of Saturn's rings at wavelengths of 3.6 and 13 cm from Voyager 1 radio occultation. *Icarus* 54:160–188.

Van Allen, J. A. 1982. Findings on rings and inner satellites of Saturn by Pioneer 11. *Icarus* 51:509–527.

Van Allen, J. A. 1983. Absorption of energetic protons by Saturn's ring G. *J. Geophys. Res.* 88:6911–6918.

Van Allen, J. A., Randell, B. A., and Thomsen, M. F. 1980. Sources and sinks of energetic electrons and protons in Saturn's magnetosphere. *J. Geophys. Res.* 85:5679–5694.

Veverka, J., and Thomas, P. 1979. Phobos and Deimos: A preview of what asteroids are like? In *Asteroids*, ed. T. Gehrels (Tucson: University of Arizona Press), 628–651.

Vogt, R. E., Chenette, D. L., Cummings, A. C., Garrard, T. L., Stone, E. C., Schardt, A. W., Trainor, J. H., Lal, N., and McDonald, F. B. 1982. Energetic charged particles in Saturn's magnetosphere: Voyager 2 results. *Science* 215:577–582.

Watson, K., Murray, B. C., and Brown, H. 1963. The stability of volatiles in the solar system. *Icarus* 1:317–327.

Whipple, E. C. 1981. Potentials of surfaces in space. *Rep Progr. Phys.* 44:1197–1250.

PART III

Dynamical Processes

DUST-MAGNETOSPHERE INTERACTIONS

EBERHARD GRÜN
Max-Planck-Institut für Kernphysik, Heidelberg

GREGOR E. MORFILL
Max-Planck-Institut für Extraterrestrische Physik, Garching

and

D. ASOKA MENDIS
University of California, San Diego

Micron-sized dust grains were identified by their light scattering characteristics in most rings of the outer planets. A multitude of interactions between the magnetospheric particles, fields, and dust grains has been proposed. We review the major effects and indicate the pertinent observations. Energetic particle absorption signatures observed by Pioneer 11 and Voyager 1 and 2 trace the mass concentrations of particulates in the magnetospheres of Jupiter and Saturn. Particulates immersed in the magnetospheric plasma and exposed to solar ultraviolet radiation will charge up to a surface potential which depends on the density and electron energy E_e of the plasma as well as on the concentration n_d of the dust particles. An isolated dust grain ($n_d < \lambda_D^{-3}$; $\lambda_D = Debye$ length, typically $10^2 - 10^4 cm$) becomes negatively charged if the plasma electron flux exceeds the photoelectron flux ($\sim 2.5 \times 10^{10} r^{-2} cm^{-2} s^{-1}$, at distance r in AU from the Sun) from its surface. Its surface potential will reach $V_0 \sim E_e/e$ ($e = electronic charge$). At high dust concentrations ($n_d > \lambda_D^{-3}$) the charge on the dust grains will be significantly reduced. Kinetic effects of charged dust particles arise from the interaction with the planetary magnetic field. Radial drift of dust particles is induced by systematic and stochastic charge variation and by the plasma drag. Sputtering and mutual collisions affect the sizes of grains. Electromagnetic effects are discussed that lead to the halo of Jupiter's ring, the dust distribution in Saturn's E Ring and to levitated dust in the B Ring (spokes) as well as on the Moon.

Recent space probe measurements detected high dust concentrations in the magnetospheres of Jupiter and Saturn. Voyager 1 discovered the Jovian ring (Smith et al. 1979; see also chapter by Burns et al.) after it has been suspected by Pioneer 11 data analysis (Fillius 1976; Acuna and Ness 1976). At Saturn, Pioneer 11 and Voyager 1 and 2 disclosed a number of dust phenomena (cf. chapter by Cuzzi et al.). Some of them are obviously correlated with magnetosphere characteristics; for example, the spoke phenomena in Saturn's B Ring showed a periodicity with Saturn's rotation period (Porco and Danielson 1982) which indicates a coupling of spoke activity to the planetary magnetic field.

Most dust observations were made in the inner magnetospheres where the magnetic fields and also the plasma densities are relatively high. Absorption effects of magnetospheric charged particle populations by material concentrations in Saturn's ring system were immediately evident in the observations (Simpson et al. 1980; Vogt et al. 1981). Effects of the magnetospheric particles and fields on the dust grains are more subtle and generally not easy to observe directly.

Dust-magnetosphere interactions have been studied since the first measurements of particulates in the Earth's vicinity became available. Several investigators studied the magnetospheric effects on dust concentrations in the Earth's neighborhood (Öpik 1956; Belton 1966; Shapiro et al. 1966). These effects could not be verified by measurements because of the low dust concentration near the Earth. Lunar observations both by remote sensing and *in situ* experiments showed effects of electrostatically levitated dust (Rennilson and Criswell 1974; Severny et al. 1974; Berg et al. 1976). Recent satellite measurements indicate that electrostatic disruption of large fluffy meteoroids occurs in the auroral zones of the Earth's magnetosphere (Fechtig et al. 1979). In the Jovian system Mendis and Axford (1974) proposed that dust-magnetosphere interactions are responsible for the albedo variations of the Galilean satellites. From all these studies it became apparent that only for micron-sized grains do electromagnetic forces become important compared to gravitational force.

Indeed, largely from observation of the relative strengths of forward- and back-scattered light, it has become clear that such small (micron- and submicron-sized) grains dominate the populations of certain regions of the known planetary ring systems. These regions include the extended E Ring and the thin F and G Rings of Saturn, as well as the extended ethereal ring of Jupiter (see chapters by Cuzzi et al. and by Burns et al.). Also the near radial spokes seen to rotate across the dense B Ring of Saturn, which seem to be elevated above the ring plane, are apparently composed of such small grains. Since these discoveries, the number of scientists interested in dust-magnetosphere interactions is rapidly increasing.

A variety of effects has to be considered to study dust-magnetosphere interactions. Plasma and energetic particles are absorbed by the particulates.

Neutral atoms, molecules, ions, and electrons are emitted upon the impact of energetic ions and by mutual collisions among the dust particles. Simultaneously the dust grains are eroded. Dust particles are charged by electron and ion capture from the plasma as well as by the photoelectric effect from the solar ultraviolet radiation. Electrostatic disruption of individual grains as well as mutual repulsion and levitation may occur. Momentum exchange with the plasma exerts a drag on the grains which causes a radial drift towards or away from the planet. Interaction of charged dust particles with the planetary gravitational and magnetic fields (gravito-electrodynamics) becomes an important factor in the dynamics of highly charged small dust particles. For the description of radiation pressure effects see the chapter by Mignard. Stochastic fluctuations of the charge state of grains causes a diffusion of small dust particles throughout the magnetosphere.

This chapter gives an overview of the field of dust-magnetosphere interactions and shows the areas of current research. In Sec. I we describe the environment to which the dust grains are exposed: magnetospheric particles and fields. In the following sections we discuss the physical processes (Sec. II) and kinetic effects (Sec. III) of dust-magnetosphere interactions. Finally in Sec. IV we discuss relevant observations in the magnetospheres of the Earth, Jupiter, and Saturn and indicate the status of their interpretations.

I. CHARACTERISTICS OF JUPITER'S AND SATURN'S MAGNETOSPHERES

Within a magnetosphere the magnetic field is controlled by the field inherent in the planet. Close to the planet the magnetic fields of Jupiter and Saturn can be approximated by a centered dipole (this neglects higher multipole moments which are generally small [cf. Connerney et al. 1982]). The equatorial magnetic field strength is then expressed by

$$B = B_0 \cdot L^{-3} \tag{1}$$

where L is the magnetic shell parameter which corresponds (in the magnetic equatorial plane) to the distance from the center of the planet in units of the planetary radius R (with index J for Jupiter and S for Saturn). The equatorial radii are $R_J = 7.14 \times 10^9$ cm and $R_S = 6.03 \times 10^9$ cm. Space probe measurements yielded values of $B_{0J} = 4.2$ Gauss (Smith et al. 1974; Acuna and Ness 1976) and $B_{0S} = 0.20$ Gauss (Smith et al. 1980; Acuna and Ness 1980). The same measurements showed that Jupiter's magnetic dipole axis is tilted by 10° with respect to the planetary rotation axis, with longitude of the magnetic pole in the northern hemisphere 230°. Saturn's dipole axis is within 1° of being parallel to the rotation axis. Both dipole fields have their magnetic north pole in the northern hemispheres of the planets, hence their polarities are opposite to that of the Earth's magnetic field.

The inner part of the magnetosphere out to about 10 planetary radii (the exact distance varies from planet to planet), where the magnetic field is still dipolar, is called the plasmasphere. It contains plasma which rigidly rotates with the planet at a speed

$$u(r) = \Omega r \qquad (2)$$

where r is the distance from the rotation axis and Ω is the rotational angular velocity. The values are $\Omega_J = 1.7585 \times 10^{-4}$ radians s^{-1} and $\Omega_S = 1.6378 \times 10^{-4}$ radians s^{-1}. The distance at which the angular velocity of the planet equals the circular Keplerian orbital motion (synchronous orbit)

$$\Omega = \omega_{Kep} = (\mu/r^3)^{\frac{1}{2}} \qquad (3)$$

where $\mu = GM$, with G = gravitational constant and M = mass of the planet ($\mu_J = 1.25 \times 10^{23}$ cm^3 s^{-2} and $\mu_S = 3.80 \times 10^{22}$ cm^3 s^{-2}), is commonly referred to as the corotation distance r_{co}. This distance $r_{co\,J} = 1.59 \times 10^{10}$ cm (2.22 R$_J$) and $r_{co\,S} = 1.12 \times 10^{10}$ cm (1.86 R$_S$).

Inside the plasmasphere the plasma density n increases towards the planet from about 1 electron cm^{-3} at $L = 10$ to about 10^2 cm^{-3} at $L = 3$, and the electron energy E_e decreases from about 100 eV to 10 eV (Frank et al. 1976; Bridge et al. 1979, 1981). Beside this general trend there are some important local deviations (see Fig. 1). Particularly high plasma density $n \sim 3000$ cm^{-3} is found in the Io plasma torus (Bagenal and Sullivan 1981; Scudder et al. 1981). Especially low values ($n \approx 10^{-2}$ cm^{-3}) have been estimated by Goertz and Morfill (1983) for the region above and below Saturn's A and B Rings. The composition of the ions in Jupiter's plasmasphere is mainly oxygen and sulfur (0^+ and S^{2+}) or some combination of both ions (Frank et al. 1976; Bridge et al. 1979) and in Saturn's plasmasphere 0^+ has been identified (Bridge et al. 1981, 1982).

Outside the plasmasphere the magnetic dipole field is distorted by ring currents carried by the plasma. The densities are low, 10^{-2} cm$^{-3} \leq n \leq 1$ cm^{-3} but the energies may be > 1 keV (Bridge et al. 1979, 1981).

Besides this low-energy plasma, a second but distinctly different charged particle population also occupies the inner part of the magnetosphere. This is the high-energy (MeV) particle population of the radiation belts. The fluxes may reach 10^7 cm^{-2} s^{-1} in the regions of highest intensities (Fillius 1976; Fillius et al. 1980).

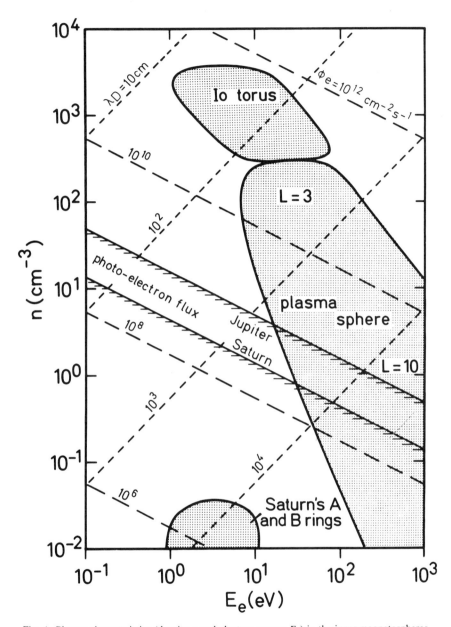

Fig. 1. Plasma characteristics (density n and electron energy E_e) in the inner magnetospheres of Jupiter and Saturn. The plasma in the plasmasphere shows a systematic trend from high densities/low energies close to the planets (small L) to low densities/high energies farther away. Especially high densities are found in the Io torus, and especially low plasma densities have been predicted above Saturn's A and B Rings. The dashed lines show the corresponding Debye length λ_D and the plasma electron fluxes, respectively. For comparison the photoelectron flux from a metal surface is shown, at the distances of Jupiter and Saturn.

From the plasma density and the electron energy E_e other important parameters like the Debye length λ_D (shielding length in the plasma)

$$\lambda_D = \left(\frac{E_e}{4\pi e^2 n_e} \right)^{\frac{1}{2}} \tag{4}$$

or

$$\lambda_D = 740 \left(\frac{E_e \, (\mathrm{eV})}{n_e (\mathrm{cm}^{-3})} \right)^{\frac{1}{2}} \; (\mathrm{cm}). \tag{4a}$$

The electron flux

$$\phi_e = n_e \left(\frac{2E_e}{m_e} \right)^{\frac{1}{2}} \tag{5}$$

can be derived, where n_e is the electron density, e is the electronic charge, and m_e is the electron mass. These quantities are also shown in Fig. 1. The Debye length varies from 10^2 to 10^4 cm in the inner magnetosphere and the electron flux ranges from 10^6 to 10^{11} cm^{-2} s^{-1}.

For comparison we also show the flux of photoelectrons released from a metal surface illuminated by the Sun at the distance of Jupiter (5.2 AU) and Saturn (9.5 AU). The flux of photoelectrons from such a surface at the Earth's distance (1 AU) is 2.5×10^{10} cm^{-2} s^{-1} (Wyatt 1969). This number is based on the solar ultraviolet flux measurement by Hinteregger (1965) and on measurements of the wavelength dependent photoyield. Feuerbacher and Fitton (1972) extended this work to other materials including nonconducting materials like silica and indium oxide, and carbonaceous materials like Aquadag, vitreous carbon, and graphite. Graphite showed the lowest yield, approximately one order of magnitude lower than the photoyield of metals. The yields of indium oxide and silica were intermediate. Therefore the photoelectronic flux from natural dielectrics like ices and stones may be somewhat lower than that from metals. The comparison of the fluxes shows that in the inner plasmaspheres of Jupiter and Saturn the plasmaelectron flux exceeds the photoelectron flux by far, except in the regions of the A and B Rings of Saturn.

II. PHYSICAL PROCESSES

A. Energetic Particle and Plasma Absorption

Cold (eV) plasma and energetic (MeV) particles coexist in inner magnetospheres. The energetic particle population, which consists mainly of electrons and protons, is trapped by the planetary magnetic field. There is a hierarchy of motion represented by the time scales involved for a trapped

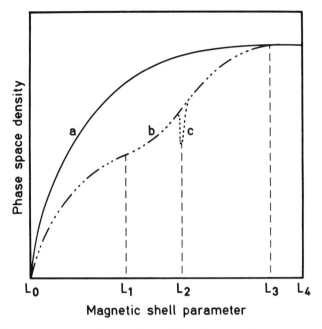

Fig. 2. Phase space density of energetic particles. For curve (a) no sources or sinks are assumed between L_0 and L_4. For curve (b) a partially absorbing ring of particulates has been assumed between L_1 and L_3. The same macro-signature of the time averaged phase space density would be obtained from a satellite orbiting in the same distance interval. A time-dependent micro-signature (curve c) would be observed close to a satellite or a clump of ring material.

particle in a magnetic dipole field (the times given below are valid for 1 MeV protons in Jupiter's and Saturn's magnetospheres, respectively, at $L = 3$):

1. Gyration about its guiding field line (2×10^{-3}s, 5×10^{-2}s);
2. Bounce motion (33 s, 28 s) due to mirroring in the stronger magnetic field regions closer to the poles of the planet; and
3. Longitudinal drift motion (2×10^{6}s, 6×10^{4}s).

Both spatial and temporal inhomogeneities in the magnetic and the related electric fields which are comparable to the gyroradius or gyroperiod will introduce nonadiabatic particle orbits, which in consequence will cause a radial diffusion of the trapped particles (cf. Schulz and Lanzerotti 1974). Figure 2, curve a, shows the distribution of the phase space density of energetic particles as a function of the magnetic shell parameter L. The phase space has been specified at an outer boundary (L_4) and is assumed to be zero at an inner boundary (L_0). Between these boundaries no local sources or sinks have been assumed. The phase space density declines monotonically as L decreases from L_4 inward and exhibits no maxima or minima.

If there is a partially absorbing ring of particulates in that region of space, then the number of charged particles is reduced in the tubes of magnetic lines of force passing through the ring. The local phase space density of energetic particles is reduced in the range of the ring between L_1 and L_3 (Fig. 2, curve b). Since there is no source assumed in that region, there is no maximum. The same macro-signature (curve b) is seen for a satellite orbiting in the radius interval between L_1 and L_3 if one measures the time-average phase space density. Close to a satellite, however, total depletion of energetic particles can be observed, which is due to the absorption of all particles from a tube of magnetic field lines intersecting the satellite. The particle shadow of a satellite for a particular class of particles either precedes or follows, depending on the relative drift motion of the satellite in its orbital motion, and is gradually filled in (as a function longitude ahead of or behind the satellite) by radial diffusion. Such a time-dependent signature localized in longitude is called micro-signature (Fig. 2, curve c). The time-averaged or longitudinally-averaged effect on the radial distribution of trapped particles is the above described macro-signature of a longitudinally uniform distribution of dispersed particulate matter in the form of a ring. A longitudinally localized distribution of particulates can masquerade as a satellite and would be difficult to distinguish from a satellite by the particle absorption technique.

The power of this technique for the identification of concentrations of particulates has been demonstrated by the Pioneer 11 and Voyager 1 and 2 measurements at Jupiter's and Saturn's rings (Fillius 1976; Goertz and Thomsen 1979; Van Allen et al. 1980a, b, c; Van Allen 1982). Energetic particle absorption signatures near Saturn's G Ring reported by Van Allen et al. (1980a) were interpreted as a mass density per unit area normal to the ring plane of $\geq 7 \times 10^{-9}$ g cm^{-2} (Van Allen, personal communication, reported by Gurnett et al. 1983). In the region of the F Ring Van Allen et al. (1980c) reported three individual micro-signatures, interpreted by Van Allen (1982) as three nearby satellites or longitudinally-localized distributions of dispersed particulate matter at radial distances 141,179 $\pm$ 100 km, 140,630 $\pm$ 80 km, and 140,150 $\pm$ 80 km. For comparison the (elliptical) F Ring has been optically identified at a mean distance of 140,300 km from Saturn's center.

Morfill et al. (1980d) suggest that in the Jovian system also, outside the optically visible ring, a concentration of small dust particles leads to significant losses throughout the inner magnetosphere that are in agreement with the available energetic electron measurements (Van Allen 1977).

Also, thermal plasma is affected by the absorption by a tenuous ring. Voyager 1 and 2 observations of the electron distribution showed a depletion of higher energy electrons (> 700 eV) by Saturn's E Ring (Sittler et al. 1981). The lower energy electrons are not affected because their bounce frequency and collision frequency, which are proportional to the particle energy, are lower and hence their absorption rate is reduced. Sittler et al. (1981) showed that the E Ring extends out to about 9 R_S and is not symmetrical about Saturn.

They also found some absorption signatures between 15 and 17 R_S and outside 20 R_S.

B. Charging of Dust Particles

A dust particle immersed in a plasma will be hit by electrons and ions. These electrons and ions will stick to the dust grain and change its charge state until the flux of positive charges onto the surface equals the flux of negative charges. If there are other currents of charged particles to and from the surface, such as photoelectron and secondary particle emission, they must also be taken into account. Charge equilibrium is reached when the sum of all currents is zero. Two cases can be distinguished: (1) there is only one isolated particle within a sphere of the radius of the Debye length λ_D, and (2) there are many particles within a Debye sphere and the electric fields of neighboring particles overlap. We shall begin with the first case.

1. Isolated Grains. We want to calculate the particle's surface charge and its variation with time when immersed in a plasma and irradiated by solar ultraviolet radiation. The plasma has density $n=n_e=n_i$, thermal electron energy E_e, and ion energy E_i. We shall assume that the thermal electron and ion velocities are much greater than the dust grain velocity so that we may consider the particle at rest with respect to the plasma. In a Maxwellian plasma with electron temperature T_e and ion temperature T_i (or thermal energies $E_e = kT_e$ and $E_i = kT_i$, respectively, where k is the Boltzmann constant), the rate of incidence of electrons on a grain with radius s and with (positive) charge $N e$ (e = electronic charge) results in a charging rate

$$\left. \frac{dN}{dt} \right|_e = -\pi s^2 \alpha_e \, n_e \, c_e \exp \left(Ve/kT_e \right) \tag{6}$$

where α_e is the sticking efficiency of electrons (~ 1) to the grain and $V = Ne/s$ is the potential of a (spherical) particle which has surface charge $q = Ne$ (number of electrons $N = 700 \, V \, s_\mu$, with V in volts and grain radius s_μ in μm). The rate at which positive ions are incident leads to a gain in (positive) charge at a rate

$$\left. \frac{dN}{dt} \right|_i = \pi s^2 \alpha_i \, n_i \, c_i \exp \left(-eV/kT_i \right). \tag{7}$$

α_i is the sticking efficiency of ions (≈ 1) and c_e and c_i are the thermal velocities of electrons and ions, respectively, in a Maxwellian plasma;

$$c_e = \left(\frac{2kT_e}{m_e} \right)^{\frac{1}{2}} \quad \text{and} \quad c_i = \left(\frac{2kT_i}{m_i} \right)^{\frac{1}{2}}. \tag{8}$$

The rate of increase of (positive) charge on the grain due to photoemission is (Wyatt 1969)

$$\frac{dN}{dt}\bigg|_{pe} = \pi s^2 K \, f(V) \qquad (9)$$

where we can represent f(V) approximately by

$$f(V) = \begin{cases} 1, & \text{if } V \leq 0 \\ e^{-V/V*}, & \text{if } V > 0. \end{cases} \qquad (10)$$

Wyatt (1969) estimates a yield $K = 2.5 \times 10^{10} \, (R_0/r)^2$ photoelectrons cm^{-2} s^{-1} in the solar ultraviolet spectrum at $r = R_0 = 1$ AU for a metal target. This number may be reduced by a factor of 0.1 for dielectrics (Feuerbacher and Fitton 1972). The constant $V*$ depends on the mean kinetic escape energy of the photoelectrons, and is $\sim 3\,V$.

The equilibrium potential V_0 is then found by setting

$$\frac{dN}{dt}\bigg|_e + \frac{dN}{dt}\bigg|_i + \frac{dN}{dt}\bigg|_{pe} = 0. \qquad (11)$$

If the photoelectron flux dominates (cf. Fig. 1), the equilibrium charge will be positive, and the exp $(-V/V*)$ factor automatically ensures small potentials $V_0 \sim V*$. The charge on the dust particle is given by

$$q = \frac{sV_0}{300} \qquad (12)$$

where q is in e.s.u., s in cm and V_0 in volts.

This situation applies to interplanetary space (Rhee 1967; Wyatt 1969) and to regions where the plasma flux is low (e.g. above Saturn's A and B Rings). In the dense plasma regions of the magnetospheres, the thermal electron flux exceeds all other fluxes on a grain at potential $V = 0$. In the case where plasma electron flux and photoelectron flux are the main fluxes, the equilibrium potential V_0 is given approximately by

$$eV_0 \approx -E_e \ln \frac{n_e c_e}{K} . \qquad (13)$$

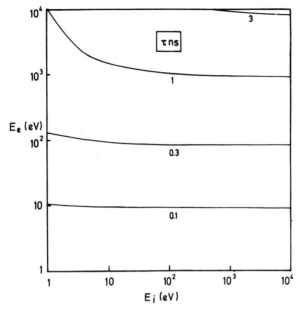

Fig. 3. Curves of constant τ ns as a function of ion (E_i) and electron (E_e) thermal energies. For plasma density n (in cm^{-3}) and grain radius s (in cm), τ is the time scale for charging up, in seconds (from Johnson et al. 1980).

If photoelectron flux is negligible compared to the plasma electron flux (i.e. $n_e c_e \gg K$) then balance is achieved by Eqs. (6) and (7) and we get

$$eV_0 \approx -E_e \frac{1}{2} \ln \left(\frac{E_e m_i}{E_i m_e} \right) \bigg/ \left(1 + \frac{E_e}{E_i} \right) . \qquad (14)$$

In other words, if $n_e c_e < K$ (by a factor $\geqslant 2$), the equilibrium potential is $V_0 \sim E_e/e$ and is negative.

The charging time constant τ can be obtained from Fig. 3 (after Johnson et al. 1980) for the case where plasma fluxes dominate. For given plasma energies E_e and E_i, the product τns is obtained; e.g., if $E_e = E_i = 100$ eV, $\tau ns = 0.3$, and for plasma density $n = 10$ cm^{-3} and $s = 10^{-4}$ cm a charging time constant $\tau = 300$ s is calculated. The surface potential V of the dust grain varies with time as

$$V = V_0 \left[1 - g \exp \left(-t/\tau \right) \right] \qquad (15)$$

where the initial charge is $V(t = 0) = V_0(1-g)$. These short time scales show that the potential on the grain surface follows variations of the plasma parameters (n_e, E_e) which have comparable or longer time scales. Therefore the charge state of a dust grain is determined by the local plasma conditions.

The negative equilibrium potential may be altered toward less negative values by several effects. If the dust particle is in motion relative to the plasma with a speed $w = v - u$ (v is the orbital speed of the dust particle and u is the corotational speed of the plasma) which becomes comparable with the thermal speed of the ions, or even with that of the electrons, then the currents onto the dust grain are enhanced. Typical relative speeds range from $w = 0$ at the corotation distance to several 100 km s^{-1} in the outer plasmasphere. The thermal speed of 100 eV electrons is 6×10^3 km s^{-1}, whereas the thermal speed of oxygen ions of the same energy is 35 km s^{-1}. With respect to the electrons the dust particles move with subsonic speed in all regions of the magnetosphere, but with respect to the ions the motion is supersonic in the outer parts of the plasmasphere. Therefore the ion flux onto the dust grain is enhanced and hence the negative equilibrium potential V_0 is reduced. A full treatment of this case is given by Wyatt (1969) and Mendis (1981).

At plasma energies > 10 eV secondary emission from solid particles becomes important and causes a reduction of the negative equilibrium potential. At plasma energies > 100 eV the yield of secondary electrons may become > 1 for some materials, and the charging effect of the impacting electrons reverses because more electrons are released than picked up. The equilibrium potential will then become ~ 10 V positive, since the energies of the secondary electrons are typically 10 eV. This effect is discussed in more detail by Meyer-Vernet (1982).

2. *Collective Effects.* An ensemble of dust particles embedded in a plasma is, of course, not necessarily described by the single, isolated particle approach used above. The question is, when are collective effects important? By collective, we mean not the mass, momentum, and energy interchanges that occur in a mass loaded plasma, but the interactions between neighboring particles and the resultant influences on the plasma, and in turn the backreaction on the dust particles. Collective effects are important when the plasma relaxation time scales and natural length scales become larger than the determining time and length scales of the dust distribution.

One important process is the redistribution of the plasma in the presence of a foreign point charge. The associated length scale is the Debye length λ_D, and the time scale is given approximately by λ_D/c_e, where c_e is the electron thermal velocity. The redistribution of the plasma implies that the foreign point charge is shielded from the rest of the system outside a distance of a few λ_D. If the mean separation between dust particles $d = n_d^{-\frac{1}{3}}$ (where n_d is the spatial density of the particles) is smaller than λ_D, then neighboring dust particles are not shielded and isolated from each other; they begin to act like a solid dielectric. For a flat extended ring, this implies that the ring particle surface potential must be calculated as if it were a flat plate. The influence of the neighboring particles then simply leads to a reduction in the total ring surface charge. As shown by Morfill (1983a, b) and Goertz and Morfill (1983) for the case of Saturn, the ring potential can be calculated from

$$V_R = V^* \left\{ \ln \frac{K}{2K_I} \left[1 - \exp(-\tau_R) \right] \right\} \tag{16}$$

with the work function for ice $V^* = 3\ V$, the photoelectron yield for ice at Saturn $K = 2.5 \times 10^7\ cm^{-2}\ s^{-1}$, and the photoelectron flux from the ionosphere $K_I = 2.5 \times 10^5\ cm^{-2}\ s^{-1}$. This yields values for the ring potential $V_R \simeq +10$ volts for optical depth $\tau_R = 1$ and $V_R \simeq +4.7$ volts for optical depth $\tau_R = 0.1$. The corresponding surface charge density is

$$\sigma = V_R/(4\pi\lambda_D) . \tag{17}$$

Again for Saturn's rings, this yields $\sigma \simeq 500$ electrons cm^{-2} ($\tau_R \simeq 1$). Ring particles with a systematic charge $>+e$ (electron charge) must have radii $\gtrsim 0.05$ cm. Smaller particles have only statistically fluctuating charges. We will see that only micron- or submicron-sized particles are at all likely to be electromagnetically affected. If these particles carry essentially no charge, as seems indicated by these considerations, then we must conclude that, on average, electromagnetic processes are unimportant in sufficiently dense planetary rings. They may be of interest for ring halos, if these consist of dust grains sufficiently small and not too abundant. If the halo were for some reason too dense, it would shield itself electrically, the perturbing electromagnetic forces would be turned off, and the dynamical evolution would be governed by dust-dust collisions finally leading to a flattened ring, unless it is continually replenished.

It may also be of interest that small practically neutral ring particles fluctuate among -1, 0, and $+1$ e over time scales given by the solar ultraviolet flux and its photoionization efficiency. At Saturn, this time scale is

$$\tau_{pe} = 1/(\pi s^2 K) \sim 1.3/s_\mu^2 \quad \text{seconds} \tag{18}$$

where s_μ is the particle radius in μm. The material is assumed to be ice. Further, if the electron energies are $\sim$ few eV (and thus $c_e \approx 10^8\ cm\ s^{-1}$), any plasma density $\gtrsim 0.25\ cm^{-3}$ suffices to charge isolated dust particles in Saturn's magnetosphere negatively. The Debye length (in cm) is then

$$\lambda_D \cong 10^3\ n_e^{-\frac{1}{2}} \tag{19}$$

and the time scale for plasma readjustment is

$$\tau_d \equiv \lambda_D/c_e \cong 10^{-5}\ n_e^{-\frac{1}{2}} \quad \text{seconds.} \tag{20}$$

This time scale is much shorter than any time scale involving dust grains (such as the time scales for charge fluctuation, translational motion, collisions, momentum exchange, etc.). Hence no higher order corrections are ever required in the situations discussed here.

Let us consider quite generally a dust ring with normal optical depth τ, mean particle size s, and scale height h. The dust particle number density is then

$$n = \tau/(\pi s^2 h) \tag{21}$$

and the mean separation between dust particles is

$$d = n^{-\frac{1}{3}}. \tag{22}$$

From Eqs. (19) and (22) we define the parameter η:

$$\eta \equiv \lambda_D/d \approx 3.2 \times 10^5 n_e^{-\frac{1}{2}} (\tau/h)^{\frac{1}{3}} \tag{23}$$

where h is the scale height in cm, and s is set equal to 1 μm. In Table I we list estimates of η for some specific rings (note that we have always used $c_e = 10^8$ cm s^{-1}; some plasma environments may be somewhat hotter). As can be seen from this table, the only clearcut cases where dust particles are isolated are the Jovian ring halo and the E Ring of Saturn. All other systems (at least when fully evolved, e. g. in the case of the spokes) must be regarded as collective, in the sense defined above. This implies that electromagnetic perturbations are unlikely to be important for most ring phenomena, except for Saturn's E Ring and Jupiter's ring halo, and perhaps for Saturn's G Ring and any disk component of Jupiter's ring system.

TABLE I

Electrical Conditions in Planetary Rings

Ring	Plasma Density (cm^{-3})	Normal Optical Depth	Scale Height (km)	η
Jupiter ring	< 100	~ 10^{-5}	< 30	> 5
Jupiter halo	~ 100	~ 10^{-6}	10^4	0.5
E Ring	≳ 10	10^{-6}	10^4–10^5	0.5–1
G Ring	~ 10	10^{-6}	10^3	2
F Ring	< 10	~ 0.1	< 1	> 10^3
A and B Ring	~ 10^{-2}	1	1	~ 10^5
Spokes region	~ 100	0.1	~ 30	~ 100

C. Electrostatic Disruption and Particle Levitation

A possible consequence of the electrical charging of dust particles is bursting due to the electrostatic stresses and the mutual repulsion of homogeneously charged grains. If one particle is much larger than the other and if the repulsion is counteracted by the gravitational attraction, then this situation is commonly referred to as electrostatic levitation (see below).

1. Electrostatic Disruption. This mechanism was first described for interplanetary meteoroids by Öpik (1956). Electrostatic disruption of charged dust particles occurs if the tensile strength F_t of the particle in dynes cm^{-2} is exceeded by the electrostatic repulsive force acting on a sphere of radius s (in cm) at a surface potential V_0 (V):

$$F_t < 8.85 \times 10^{-7} V_0^2/s^2. \tag{24}$$

The relevant tensile strength (and the corresponding maximum surface field strength V_0/s) is $\sim 10^4$ dyn cm^{-2} (10^5 V cm^{-1}) for fluffy aggregates, 10^6 to 10^8 dyn cm^{-2} (10^6 to 10^7 V cm^{-1}) for ice, 10^7 to 10^9 dyn cm^{-2} (3×10^6 to 3×10^7 V cm^{-1}) for silicates, $\sim 7 \times 10^9$ dyn cm^{-2} (9×10^7 V cm^{-1}) for glass, and $\sim 2 \times 10^{10}$ dyn cm^{-2} (1.5×10^8 V cm^{-1}) for metals (Öpik 1956; Rhee 1976; Pollack et al. 1979; Burns et al. 1980). Figure 4 shows lines of constant tensile strength (i.e. field strength) as a function of particle radius and surface potential.

Micron- and submicron-sized particles have increased strength because they may consist of individual crystals. For these particles the maximum field strength attainable at the surface is limited by ion field emission in the case of positive grain charge, or by electron field emission in the case of negative grain charge. The maximum field strength is then $\sim 5 \times 10^8$ and $\sim 10^7$ V cm^{-1} for ion and electron field emission, respectively (Müller 1956). Charging of spherical dust particles made of carbon, glass and metals in the laboratory resulted in surface field strengths within a factor of 5 of the ion field emission limit, without destroying the dust particles (Vedder 1963; Friichtenicht 1964; Fechtig et al. 1978). However, dust particles in planetary rings may be more friable than generally undamaged laboratory specimens, due to collisions among ring members (Burns et al. 1980; Hörz et al. 1975) and radiation damage (Mukai 1980; Smoluchowski 1980). Both effects enhance flaw densities and thereby weaken grains.

While Öpik (1956) calculated the electrostatic pressure for spherical grains, a more general derivation for prolate and oblate spheroids is given by Hill and Mendis (1981b). Another more general description of electrostatic bursting is obtained for particles with rough surfaces, if the dimension s in Eq. (24) is that of a typical asperity, not that of a whole grain. Since any roughness is usually much smaller than the grain itself, the stresses acting on

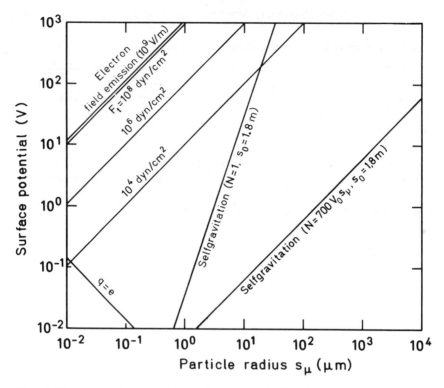

Fig. 4. Maximum surface potentials for the electrostatic disruption of different materials, characterized by their tensile strengths F_t. Electron field emission occurs at a surface field strength $\sim 10^7$ V cm^{-1}. The line $q = e$ indicates dust particles carrying only one electronic charge. Also shown are the limits of self-gravitation between a small particle with radius s_μ and a large particle of radius $s_o = 1.8$ m on which it sits, with the indicated surface potential for two charge states (number of electrons N) of the small particle: $N = 1$ and $N = 700\, V_o s_\mu$ (free space charge).

surface elements are considerably higher than given by Eq. (24). Such surface bumps will first be eroded away, and the grain will become more spherical. Eventually, if the tensile strength of the grain is sufficiently large, the end chipping process will cease; otherwise, chipping will continue until the grain becomes spherical, at which time it will explode in toto.

2. Electrostatic Levitation and Blow-off of Small Particles. In this subsection we consider the combined effect of electrostatic repulsion and gravitational attraction of spherical particles with radii s_0 and s, where $s_0 \gg s$. The small particle may be one which is electrostatically chipped off from the larger one or which inelastically collided with the larger one, and got stuck due to gravitational attraction when there was less electrostatic repulsion between the particles. The effects of spinning particles and of particles within the Roche limit will be discussed at the end of this subsection. In a plasma with Debye length λ_D the large particle will be charged to the surface potential V_o which depends on the plasma conditions and the solar ultraviolet flux. The electric potential U at a point r ($r > s_0$) from the center of the large particle is given by

$$U = \frac{V_0 s_0}{r} e^{-[(r-s_0)/\lambda_D]} \tag{25}$$

and the electric field strength $\mathbf{E}$ is

$$|\mathbf{E}| = V_0 \frac{s_0(\lambda_D + r)}{r^2 \lambda_D} e^{-[(r-s_0)/\lambda_D]} \quad . \tag{26}$$

The small dust grain lying on the surface would acquire a charge q (we assume $s << \lambda_D$). This charge is proportional to its projected surface area, as was first pointed out by Singer and Walker (1962):

$$q = \pi s^2 \sigma_0 = \pi s^2 \cdot \frac{E_0}{4\pi} \tag{27}$$

where σ_0 is the surface charge density on the large particles and E_0 is the surface field strength. Setting $r = s_0$ and substituting Eq. (26) in Eq. (27) yields

$$q = \frac{1}{4} s^2 V_0 \frac{\lambda_D + s_0}{\lambda_D \cdot s_0} \quad . \tag{28}$$

Note that this charge is much smaller than the charge ($q = sV_0$) that would be acquired by the grain if it were at a potential V_0 in free space. The number of electronic charges N on such a grain is given by

$$N = \frac{q}{e} = 1.74 \times 10^{-4} s_\mu^2 V_0 \frac{\lambda_D + s_0}{\lambda_D \cdot s_0} \tag{29}$$

with s_μ in μm, V_0 in volts and λ_D and s_0 in meters. The value N for a micron-sized grain lying on a particle with $s_0 = 1$ m at $V_0 = 10$ V and $\lambda_D = 10$ m is 1.9×10^{-3}. Obviously a grain cannot have a fractional electronic charge, and the proper interpretation of this number is that the net charge of N^{-1} small grains lying on the surface of the bigger one is unit charge, i.e., some of them are positively charged due to the loss of a photoelectron, some are negatively charged by the capture of a plasma electron, and many are uncharged (we assumed zero conductivity through the large particle).

As soon as the small grain leaves the larger particle it may acquire its full free space charge $N = 700\,V_o s_\mu$. Therefore we will discuss in the following sections both extreme cases, $N = 1$ and $N = 700\,V_o s_\mu$. The condition that the small grain escapes the gravitational field of the larger particle is that the energy E_{el} gained from the electric field of the larger particle exceeds the gravitational energy E_{gr}, or

$$Ne\,V_o > G\,M_o\,m/s_o \tag{30}$$

where G is the gravitational constant and M_o is the mass of the large particle. If both particles are spherical and have densities of $1\;g\;cm^{-3}$, then the critical grain radius $s_{\mu c}$ (in μm) is given by

$$s_{\mu c} = 5.1 \left(\frac{V_o}{s_o^2}\right)^{\frac{1}{3}} \text{ for } N = 1 \tag{31a}$$

and

$$s_{\mu c} = 310\;\frac{V_o}{s_o} \quad \text{for } N = 700\,V_o s_\mu \tag{31b}$$

with V_o in V and s_o in m. A particle of radius $s_\mu < s_{\mu c}$ will escape the gravitational attraction of the larger particle. This dependence is also shown in Fig. 4 for $V_o = 10\;V$, $s_o = 1.8\;m$ (a typical particle size in Saturn's B Ring) and for both extreme charge states. Figure 5 shows the dependence of the critical radius on the radius of the large particle at $V_o = 10\;V$.

If the large particle is much larger than the Debye length λ_D, then the small grain may be levitated without being able to escape from the gravitational field of the large particle (e.g. a satellite). This effect is described by comparing the forces acting on the small grain. The electric force of repulsion F_{el} on a grain carrying the charge Ne at the surface of the larger particle is given by

$$F_{el} = Ne\,V_o\,\frac{\lambda_D + s_o}{\lambda_D \cdot s_o} \approx 1.60 \times 10^{-14} N \cdot V_o\,\frac{\lambda_D + s_o}{\lambda_D \cdot s_o} \tag{32}$$

with the same units as in Eq. (29), and F_{el} in dynes.

The gravitational force F_{gr} on a grain of radius s lying on a larger particle of radius s_o is given by

$$F_{gr} = \frac{4}{3}\,\pi s^3 \rho \cdot \frac{4}{3}\,\pi s_o^3 \rho_o\,\frac{G}{(s + s_o)^2} \tag{33}$$

where ρ and ρ_o are the densities of the small and the large particle, respectively. With $\rho = \rho_o = 1\;g\;cm^{-3}$ and $s_o \gg s$ we get

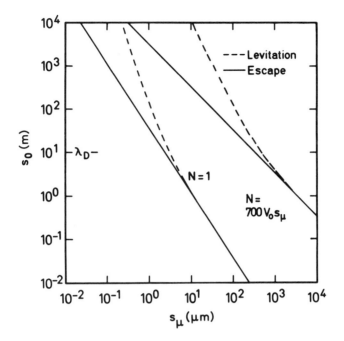

Fig. 5. Levitation and blow-off of small particles (radius s_μ) from a large particle (radius s_o). A surface potential $V_o = 10$ V has been assumed, and two charge states for the small particle $N = 1$ and $N = 700\, V_o s_\mu$. The Debye length was taken as $\lambda_D = 10$ m. All small particles with radii to the left of the solid line will escape from the large particle. Particles with radii between the solid and the broken lines will levitate, and those to the right of the broken line will stick to the surface of the large particles.

$$F_{\mathrm{gr}} \approx 1.17 \times 10^{-16}\, s_\mu^3 \cdot s_o \tag{34}$$

in dynes. The condition for such a grain to leave the surface is $F_{\mathrm{el}} > F_{\mathrm{gr}}$. This gives a critical grain radius

$$s'_{\mu c} = 5.1 \left(V_o\, \frac{\lambda_D + s_o}{\lambda_D \cdot s_o} \right)^{\frac{1}{3}} \text{ for } N = 1 \tag{35a}$$

and

$$s'_{\mu c} = 310\, \frac{V_o}{s_o} \left(\frac{\lambda_D + s_o}{\lambda_D} \right)^{\frac{1}{2}} \text{ for } N = 700\, V_o s_\mu. \tag{35b}$$

Both relations are displayed in Fig. 5 for $V_o = 10$ V and typical Debye length $\lambda_D = 10$ m. If $s_o > \lambda_D$, small particles may be levitated although they are not able to escape from the gravitational attraction of the large particle. This effect will lead to a halo of small particles around the bigger one at a height $\sim \lambda_D$.

A spin of the large particle with period T leads to a modification of the attractive force (Eq. 33). The effective gravitational force F'_{gr} on a small grain lying at latitude λ (i.e. angle λ from the rotational equator) on a large particle is given by

$$F'_{gr} = F_{gr} \left(1 - \frac{3\pi \cos^2 \lambda}{G \rho_0 T^2} \right) . \tag{36}$$

At the critical spin period

$$T_c = \left(\frac{3\pi \cos^2 \lambda'}{G \rho_0} \right)^{\frac{1}{2}} \tag{37}$$

small grains are only gravitationally bound to the surface of the large particle at latitudes $\lambda > \lambda'$. For density $\rho_0 = 1$ g cm^{-3} small particles will not stick (by gravitation) to the equatorial regions of the large particle if its rotation period is shorter than T_c ($\lambda' = 0$) = 1.19×10^4 s.

Another modification of the attractive forces occurs because both particles move in the gravitational field of the planet. The gravitational field of a particle of radius s_0 is altered by the tidal forces exerted by the planet with radius R. Qualitatively, the gravitational attraction at the sub- and anti-planet points of the spherical particle's surface is weakened, whereas the attraction is increased at the poles (with respect to the orbit plane) of the particles. At the leading and trailing edges the gravitational force of the particle remains approximately unchanged. The Roche limit is given by

$$r_L = \alpha(\rho/\rho_0)^{\frac{1}{3}} R \tag{38}$$

where ρ is the density of the planet ($\rho_J = 1.33$ g cm^{-3}; $\rho_S = 0.66$ g cm^{-3}), ρ_0 is the density of the particle, and α is a factor describing the shape of the particle ($\alpha = 1.44$ for a spherical particle and $\alpha = 2.46$ for a particle relaxed to hydrostatic equilibrium; see Dermott et al. [1979] and chapter by Weidenschilling et al.).

Most of the rings of Jupiter, Saturn and Uranus are located close to or even inside the Roche limit (Dermott et al. 1980). Inside the Roche limit for spherical bodies only particles with finite tensile strength can exist. Therefore loose dust grains lying on the surface of a large particle will be lost from the caps that are closest to and farthest from the planet. Only in the polar regions a dust coverage (regolith) may exist even inside the Roche limit. For more detailed discussion of effects occurring near and inside the Roche limit see Dobrovolskis and Burns (1980), Davis et al. (1981), and the chapter by

Weidenschilling et al. The additional effect of electric charging, i.e. electrostatic repulsion, will clear larger areas around the sub- and anti-planet points from the dust. Therefore, close to and inside the Roche limit, an increase of the surface potential of particles due to changes in the plasma parameters and the ultraviolet illumination will lead to the levitation and escape of dust grains which otherwise would stick to the surface of larger particles.

D. Sputtering

The bombardment of solid particles by the intense flux of magnetospheric ions at energies above several tens of eV releases atoms, molecules, and ions from the target. The sputtering yield S (the number of secondary particles per incident ion) depends strongly on the target material and on the energy and atomic number of the incident ions (for a review see, e.g., Carter and Colligon [1968], or Oechsner [1975]). Sputtering yields S are determined by laboratory simulation and have been reported by Wehner et al. (1963) for astrophysically important systems such as hydrogen and helium ions onto metal and stone targets in the energy range 1 to 20 KeV. For protons typical values of S_p are 0.01 to 0.04, and for α particles values of $S_\alpha = 0.1$ to 0.4 have been found. For higher primary ion masses S passes through a maximum ~ 10 when the mass of the primary ion is similar to that of the substrate atom (Wechsung 1977). At higher energies the sputtering yield increases. Recently Brown et al. (1978 and 1980) reported the sputtering yield of ice bombarded by hydrogen, helium, carbon, and oxygen ions at energies of 1.5 MeV. Hydrogen showed $S_p = 0.2$ to 0.4, helium $S_{He^+} = 10$, and carbon and oxygen $S_{C^+, \, O^+} \sim 500$. Sputtering is a source for magnetospheric atoms and ions; this effect is now discussed as an important source for the heavy ions in the Io torus (see Matson et al. 1974) and in Saturn's plasma sheet (cf. Cheng et al. 1982).

On the other hand, sputtering erodes particulate matter. The sputter erosion of lunar rocks by solar wind ions has been determined both theoretically and experimentally (for a review see, e.g., Ashworth 1978). The best value for the sputter rate on the lunar surface is 1.6×10^{-17} cm s^{-1}. This value refers to a flux of solar wind ions (roughly 95% protons and 5% α particles) of $\phi_{sw} \sim 2 \times 10^8$ cm^{-2} s^{-1} at a speed of ~ 400 km s^{-1}.

Lanzerotti et al. (1978) studied the sputter erosion of the icy surfaces of the Galilean satellites Europa, Ganymede, and Callisto by high-energy protons ($E > 100$ keV) of the Jovian radiation belt. Their estimated sputter rates of water ice are presented in Table II. Matson et al. (1974) estimated the sputter rate at Io from the observed sodium cloud to be 1.5×10^{-15} to 1.5×10^{-14} cm s^{-1}. An upper limit of the sputter rate at Io of 8×10^{-14} cm s^{-1} has been determined by Haff et al. (1979), who considered the heavy ion impact on ice at velocity 50 km s^{-1}, which corresponds to an energy 300 eV for ions of atomic mass 20. This sputter mechanism is most effective in the densest parts of Io's plasma torus.

TABLE II

Erosion Rates Due to Ion Sputtering and Life Times of 1 μm Ice Particles

Satellite	Distance (in planetary radii)	Erosion Rate (cm s^{-1})	τ_s for 1 μm (yr)	References
a) Jovian System				
Io	5.90	$1.5 \times 10^{-15} - 1.5 \times 10^{-14}$	$2 \times 10^{3} - 2 \times 10^{2}$	Matson et al. (1974)
		$< 8 \times 10^{-14}$	> 40	Haff et al. (1979)
Europa	9.40	$10^{-14} - 3 \times 10^{-12}$	$3 \times 10^{2} - 1$	Lanzerotti et al. (1978)
Ganymede	14.99	$10^{-15} - 3 \times 10^{-13}$	$3 \times 10^{3} - 10$	Lanzerotti et al. (1978)
Callisto	26.33	$6 \times 10^{-18} - 2 \times 10^{-15}$	$5 \times 10^{5} - 2 \times 10^{3}$	Lanzerotti et al. (1978)
b) Saturnian System				
Mimas	3.12	3×10^{-16}	10^{4}	Cheng et al. (1982)
Enceladus	3.98	3×10^{-16}	10^{4}	Cheng et al. (1982)
		8×10^{-16}	6×10^{3}	Morfill et al. (1983b)
Tethys	4.92	2×10^{-14}	1.5×10^{2}	Morfill et al. (1983b)
Dione	6.28	3×10^{-14}	1.5×10^{2}	Morfill et al. (1983b)
Rhea	8.75	2×10^{-15}	2×10^{3}	Morfill et al. (1983b)

The shortest lifetimes τ_s with respect to sputtering (~ 10 yr) for 1 μm ice particles are found in the Io torus and between Io and Europa. Silicate particles, however, would be expected to have ~ 10 times longer lifetimes than ice particles.

In the inner (4.5 to 8 R_S) Saturnian system Cheng et al. (1982) calculate that high-energy sputtering by protons > 50 keV at a flux $\sim 10^7$ cm^{-2} s^{-1} yield a water ice erosion rate 3×10^{-16} cm s^{-1}. This rate may also apply to the inner satellites Mimas (3.1 R_S) and Enceladus (4.0 R_S), since at their locations the observed high-energy proton flux is comparable to that farther out (Krimigis et al. 1981, 1982). Morfill et al. (1983b) consider the sputtering by low-energy heavy plasma ions. Corotational energies of oxygen ions in the E Ring region (3.5 to 9 R_S) range from 50 to 500 eV. Sputtering yields from low-energy heavy ion impact are $S = 1$ to 10 secondary particles per ion. This effect becomes dominant over the high-energy ion sputtering in the inner Saturnian magnetosphere because the high-energy particle flux (> 100 keV) is much reduced compared to that in the Jovian system. Ice particle lifetimes range from 100 to 10^4 yr in the region of the E Ring.

E. Plasma Drag

Dust grains traveling through a plasma will exchange momentum with the plasma particles, which exerts a drag on the grain. This drag force is a function of the relative speed of the dust grain with respect to the plasma. At the corotation distance r_{co} no drag force is exerted because the grain is at rest with respect to the plasma frame of reference. Inside the corotation distance the plasma drag decelerates the orbital speed of the grain and hence the orbit decays towards the planet. Outside the corotation distance, however, the plasma drag accelerates the grain's orbital speed and so the orbit expands and the particles are pushed outward from the planet. This effect competes with the drag force exerted by the radiation pressure, which causes the grains to lose orbital energy (cf. chapter by Mignard).

1. Direct Particle Drag. In this case we assume that the bulk energy of the plasma ions is much greater than the potential of the surface charge. The grain moves with a velocity $\mathbf{w} = \mathbf{v} - \mathbf{u}_p$ with respect to the plasma, where $\mathbf{v}$ is the orbital speed of the dust particle, $\mathbf{u}_p = r\Omega\hat{\phi}$ is the corotational bulk speed of the plasma at distance r from the rotation axis, Ω is angular velocity, and $\hat{\phi}$ is an azimuthal unit vector. Then the force acting on the dust particle is

$$F_D = \int_{-\infty}^{\infty} n(u) \, \pi s^2 \, |u - w| \, m_i \, (u - w) \, \mathrm{d}u \qquad (39)$$

where u is the thermal ion speed and m_i is the mass of a plasma ion. The direction of the drag force is antiparallel to $\mathbf{w}$. We assume that the distribution function for the plasma is Maxwellian, i.e.,

$$n(u) = n_i \frac{1}{\sqrt{\pi} c_i} \exp\left(u^2/c_i^2\right) . \tag{40}$$

Substituting Eq. (40) into Eq. (39) yields, for direct collisions between plasma ions and dust particles

$$F_D = - n_i \pi s^2 m_i c_i^2 \left[\frac{w}{c_i} \frac{1}{\sqrt{\pi}} \exp -(w^2/c_i^2) + \left(\frac{w^2}{c_i^2} + \frac{1}{2} \right) \mathrm{erf}\left(\frac{w}{c_i} \right) \right] \tag{41}$$

where

$$\mathrm{erf}(x) = \frac{1}{\sqrt{\pi}} \int_0^x e^{-t^2} dt . \tag{42}$$

In the strongly subsonic case, $w/c_i \ll 1$, we get

$$F_D = - 2\sqrt{\pi} \, n_i s^2 m_i c_i w . \tag{43}$$

In the strongly supersonic case, $w/c_i \gg 1$, we get

$$F_D = - \pi n_i s^2 m_i w^2 . \tag{44}$$

2. *Distant Coulomb Drag.* This is calculated using standard collision theory with minimum impact parmeter $= s$ and maximum impact parameter $= \lambda_D$, the Debye length. In addition, $m \gg m_i$ and the plasma is again assumed to have Maxwellian velocity distribution. The solution is

$$F_C = 2\sqrt{\pi} n_i m_i \alpha^2 \frac{1}{c_i} \int_{-\infty}^{\infty} du \frac{u-w}{|u-w|^3} \exp -(u^2/c_i^2) \cdot \ln \frac{\alpha^2 + \lambda_D^2(u-w)^4}{\alpha^2 + s^2(u-w)^4} \tag{45}$$

where $\alpha = qe/m_i$.

Figure 6 shows the relative strength of direct particle drag and distant Coulomb drag as a function of velocity w in units of the thermal speed c_i, for conditions corresponding to those in the inner plasmaspheres of Jupiter and Saturn. As can be seen, when the dust grain moves at subsonic speed with respect to the rest frame of the plasma, distant Coulomb collisions dominate by far, and the drag forces for both interactions are proportional to w. At supersonic speed the drag force due to direct collisions increases in proportion to w^{+2}, whereas drag force due to distant Coulomb collisions decreases rapidly.

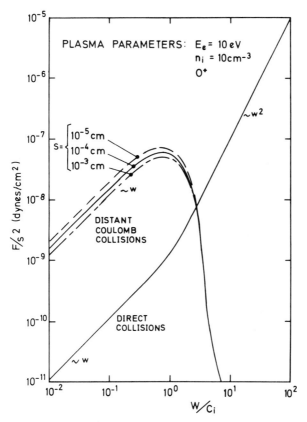

Fig. 6. Pressure exerted on a dust grain by magnetospheric plasma. The effect both of direct collisions of ions (oxygen$^+$) and of Coulomb collisions as a function of relative velocity (in units of the speed of sound c_i) between plasma and grain. The electrostatic potential of the dust particle was taken to be $V_0 = 10$ V.

It has been shown by Morfill et al. (1980a) that in the strongly subsonic case, if one assumes charge equilibrium, the Coulomb drag force can be approximated by

$$F_C \approx F_D \cdot I$$

where

$$I = \int_0^\infty dx \frac{1}{x} \exp{-x^2} \cdot \ln \frac{1 + (2\lambda_D/s)^2 x^4}{1 + 4x^4} \qquad (46)$$

with $x = (u - v)/c_i$. Setting $x_0 = \sqrt{s/2\lambda_D}$, the integral is divided into two parts and approximated by

$$I \approx \int_0^{x_0} dx \, \frac{1}{x} \, x_0^{-4} x^4 + \int_{x_0}^1 dx \, \frac{1}{x} \ln\left[1 + (x/x_0)^4\right]. \qquad (47)$$

We get

$$I \approx \frac{1}{4} - <\ln\left(1 + (x/x_0)^4\right) > \ln x_0 . \qquad (48)$$

Setting $<\ln[1+(x/x_0)^4]> \approx \frac{1}{2} \ln x_0^{-4}$ we obtain, for the typical values $s = 10^{-4}$ cm and $\lambda_D = 10^3$ cm, $I \approx 140$. Distant Coulomb collisions increase the drag force for subsonic grains in charge equilibrium with the plasma by a factor ~ 140. This factor is insensitive to parameter changes since it only depends on them logarithmically.

F. Mutual Collisions

Mutual collisions among dust particles can lead to grain destruction and fragment generation (high-velocity collisions) or grain growth (low-velocity collisions). By these effects dust particles are generated, destroyed, or modified inside planetary magnetospheres. Low-velocity collisions (relative speed $\lesssim 100$ m s^{-1}) occur among particles in planetary rings, whereas particles having largely different orbits will collide at much higher speed. Interplanetary meteoroids will collide with particles in orbit about a planet with typical speeds of tens of km s^{-1}. High-velocity collisions produce a large number of fragment particles, which couple differently to the magnetospheric environment than their parent particles do. In addition, high-velocity collisions provide a source for neutral gas and plasma inside magnetospheres. Both effects will be briefly described.

1. High-Velocity Collisions. From impact studies we obtain the following empirical relationships (see, e.g., Dohnanyi 1969; Gault and Wedekind 1969). When a small particle of mass m_p and velocity v collides with another body, we can get erosive collisions if the target is sufficiently large with respect to the projectile size, or we can get catastrophic disruption if the target is too small.

In *erosive collisions* the ejected mass is given by

$$m_e = \gamma m_p \qquad (49)$$

where $\gamma \sim 5 \times 10^{-10} v^2$ (v in cm s^{-1}). The largest fragment ejected has mass

$$m_L \sim 0.1 \, \gamma m_p \qquad (50)$$

and the size distribution of the ejected grains is

$$g(m) = Cm^{-\beta} \, dm \tag{51}$$

where $\beta = 1.8$ and C is determined from

$$\gamma m_p = \int_0^{m_L} m \, g(m) \, dm \ . \tag{52}$$

Catastrophic disruption occurs when the target mass is $\leqslant 100 \, \gamma m_p$. Again the size distribution of the fragments follows a power law distribution (Eq. 51) with $\beta \approx 1.8$ and C determined from

$$m_T = \int_0^{m_L} m \, g(m) \, dm \tag{53}$$

where m_T is the mass of the target particle. The mass of the largest fragment is, according to Fujiwara et al. (1977),

$$m_L = 1.66 \times 10^8 E_p^{-1.24} m_T^{2.24} \tag{54}$$

where E_p is the kinetic energy in c.g.s. units of the projectile with respect to the target particle.

Most fragment mass is ejected at a low speed (order of m s^{-1}); only a very small fraction of the fragments ($\sim 10^{-5}$ of the total mass, and only the smallest particles $m < m_p$) will be ejected at high speed (order of km s^{-1}) (cf. Gault and Heitowit 1963).

For recent applications of collision theory to ring systems see Morfill et al. (1980b), Burns et al. (1980) and Grün et al. (1980) for a discussion of the Jovian ring, and Morfill et al. (1983b) for collision effects in the Saturnian ring system. Morfill et al. (1983b) estimate that the erosion time of Saturn's rings is only $\sim 5 \times 10^6$ yr. However, for a large system like Saturn's rings, only a minute fraction of the ejecta is actually lost into the atmosphere or interplanetary space. The rest is redistributed over the rings and forms a regolith several cm deep on cores of larger ring particles (for more details see chapters by Durisen and by Weidenschilling et al.). This picture is consistent with the hypothesis put forward by Smith et al. (1982), which states that the inner Saturnian satellites (inside ~ 6 R$_S$) have been disrupted by impacts and reaccreted several times since the formation of the Saturnian system. High-velocity impacts also produce vapor and ions (see, e.g., Gault et al. 1972; Fechtig et al. 1978; Hornung and Drapatz 1981). For example, at impact speeds v = 30 km s^{-1}, the vapor mass produced roughly equals the projectile mass (actually a factor of ~ 3 higher according to Gault et al. [1972], and perhaps even higher than that for ice targets [Lange and Ahrens 1982]). Morfill et al. (1983a) suggest that the neutral gas emitted by impact vaporization is largely responsible for the observed neutral atmosphere above Saturn's rings (Broadfoot et al. 1981). The residual ionization of the impact-generated vapor cloud will be on the order of a few percent (Hornung and Drapatz 1981) for mm-sized projectiles and the speed considered above. These highly time-variable impact plasma clouds interact with the planetary magnetic field after their density is sufficiently reduced by expansion. Morfill and Goertz (1983) propose that a collection of such clouds produced, e.g., by the impact of a swarm of interplanetary meteoroids onto Saturn's ring, may trigger the formation of spokes. Ip (1983) suggest that inside a distance of 1.625 R_S from Saturn these impact-produced ions tend to move upward along the dipole field lines until they are lost into the ionosphere. Such a siphoning mechanism could lead to appreciable loss of ring material in this region. Northrop and Hill (1983b), on the other hand, claim that the inner edge of the B Ring has been caused by such a process (cf. Sec. III.A.2).

During the crossing of Saturn's ring plane the Voyager 2 plasma wave instrument detected impulsive noise (Scarf et al. 1982) which has been interpreted as micron-sized particles hitting the spacecraft and producing charge pulses recorded by the instrument (Gurnett et al. 1983).

2. *Low-Velocity Collisions.* We next discuss the effects of low-velocity collisions on the size distribution of E Ring particles and estimate the time scale for particle growth. There we consider small perturbations of grain orbits by fluctuations in the plasma drag. Other rings or ring systems may be quite different, e.g. Jupiter's ring where destructive collisions are dominant (see Morfill et al. 1980b; Burns et al. 1980; Grün et al. 1980; and chapter by Burns et al.).

The evolution of the size distribution f (s,t) in a spatially homogeneous system is given by

$$\frac{\partial f(s_1,t)}{\partial t} = - \int_0^\infty ds_2 f(s_1,t) f(s_2,t) \, \pi \, (s_1+s_2)^2 \, Q_{12} \langle \, (\Delta_{12}\delta v)^2 \rangle^{\frac{1}{2}}$$

$$(55)$$

$$+ \frac{1}{2} \int_0^{s_1} ds_3 \, f(s_2,t) f(s_3,t) \, \pi \, (s_2+s_3)^2 \, Q_{23} \left(\frac{S_1}{S_2} \right)^2 \langle (\Delta_{23}\delta v)^2 \rangle^{\frac{1}{2}} \, .$$

The first integral is a loss term due to collisions of particle 1 with any other (which removes it from the size range s_1 to $s_1 + ds_1$) and the second integral is the growth of particles into the size range s_1 to $s_1 + ds_1$ by two-body collisions between particles 2 and 3. Q is a sticking probability (≤ 1) and $\langle (\Delta_{ij}\delta v)^2 \rangle^{\frac{1}{2}}$

is the rms value of the stochastic relative velocities between particles i and j. The case of interest for us corresponds to a weak perturbing force (the plasma drag on micron-sized dust particles in Saturn's E Ring gives time scales for momentum exchange $\tau_f \approx 10$ yr [see Morfill et al. 1982b], in contrast to the correlation time of the perturbations τ_{K_0} which is short, i.e., the orbital period $\sim$ days). From Völk et al. (1980) we obtain for the case of stochastic gas drag forces $\langle (\Delta \delta v)^2 \rangle^{\frac{1}{2}} \approx 0.3$ km s^{-1} (for $\tau_f / \tau_{K_0} = 10^3$, nearly equal-sized grains, and a stochastic plasma velocity component 10 km s^{-1}). This is too small to cause fragmentation of most grains.

The collision rate between grains of similar size is

$$\nu = 4\pi s^2 n \langle (\Delta \delta v)^2 \rangle^{\frac{1}{2}} \tag{56}$$

with $n = \tau/4\pi s^2 h$ the spatial grain density, τ the normal optical depth, and scale height h. Hence Eq. (56) becomes

$$\nu = (4\tau/h) \langle (\Delta \delta v)^2 \rangle^{\frac{1}{2}} \tag{57}$$

i.e., the grain size cancels. For the E Ring, we use the numbers of Table I ($\tau = 10^{-6}$, $h = 10^9$ to 10^{10} cm). Then the collision time $\tau_{coll} = \nu^{-1} \sim 3000$ yr. This is also the grain growth time (doubling the mass). Of course, the value for $\langle (\Delta \delta v)^2 \rangle^{\frac{1}{2}}$ is uncertain, but it is not a rapidly varying function of particle size or τ_f (cf. Völk et al. 1980). The comparison of the grain growth time scale by coagulation with loss time scales (e.g. due to sputtering) will be discussed in Sec. III.B on radial transport.

III. KINETIC EFFECTS

A. Gravito-electrodynamics

While the motion of the plasma (both thermal and energetic) within a planetary magnetosphere is almost totally controlled by electromagnetic forces, with planetary gravitation playing only a subordinate role (e.g., causing slow azimuthal drifts), the motion of the larger bodies like satellites is overwhelmingly controlled by planetary gravitation, with gravitational perturbations by neighboring satellites playing a secondary role. Even for cm- and mm-sized grains that populate the rings, the electromagnetic effects are negligible compared to gravity for any acceptable values of the surface potential (even for those values close to their field emission or electrostatic disruption limits). It is only when we consider grains of micron size (0.1μm-5μm) that these two forces can become comparable for plausible values of the grain surface potential. The gravitational force is given by

$$\mathbf{F}_G = -\frac{\mu m}{r^2}\hat{r} \tag{58}$$

where $\hat{r}$ is the radial unit vector, m is the grain mass, and μ is G times the planet's mass; the Lorentz force is

$$\mathbf{F}_L = \frac{q}{c}(\mathbf{w} \times \mathbf{B}) \tag{59}$$

where q is given by Eq. (12) and c is the speed of light. With $|\mathbf{B}| = B_0 L^{-3}$ in the magnetic equatorial plane we obtain

$$\frac{F_L}{F_G} = k \frac{V_0 w_\perp}{s^2 L} \tag{60}$$

with the perpendicular (to the magnetic field) relative speed of the dust grain $w_\perp$ in cm s^{-1}, V_0 in V, s in cm, L in planetary radii, and the constant k. In the case of Jupiter k is 4×10^{-17}, and for Saturn it is 5×10^{-18}. For a dust grain of size $s = 10^{-5}$ cm (0.1 μm) at distance $L = 5$ from Jupiter which moves at a speed $w_\perp = 5 \times 10^6$ cm s^{-1} and is charged to a surface potential $V_0 = 10$ V, we get $F_L/F_G = 4$. For the values $s = 10^{-4}$ cm, $L = 2$, $V_0 = 10$ V and $w_\perp = 10^5$ cm s^{-1} we get, for Saturn, $F_L/F_G = 2.5 \times 10^{-4}$. The values of F_L/F_G for negatively charged grains in prograde circular orbits having different sizes and potentials at a distance $L = 5$ from Saturn are shown in Fig. 7. Clearly, electromagnetic forces are most important for submicron-sized grains (0.1 μm $\leqslant s_\mu \leqslant 0.5$ μm) in Saturn's E Ring (3.5 $\leqslant L \leqslant 9$). Closer to the planet, in the regions of the F Ring and the spokes, the dust particles must be much smaller to be dominated by electromagnetic forces, because $w_\perp$ decreases and the surface potential V_0 may be greatly reduced. At the corotation distance ($L = 1.86$), however, $F_L/F_G = 0$ even when the grain is charged.

Any attempt to bracket the range of F_L/F_G for the application of gravitoelectrodynamics is necessarily arbitrary, but the range 10^{-2} to 10^2 may be reasonable. Note that even when $F_L/F_G \sim 10^{-2}$, the electric force on the grains is still many orders of magnitude larger than, for instance, the typical gravitational perturbing force of a nearby satellite. Therefore, though the grain orbit is largely controlled by gravitation, in this case the perturbations produced by electromagnetic forces are sufficient to cause various subtle effects that may be observable. We next discuss the general solution of gravitoelectrodynamics and the stability of orbits.

1. General Solution. The equation of motion of a charged dust grain in the planet-centered inertial frame is given by

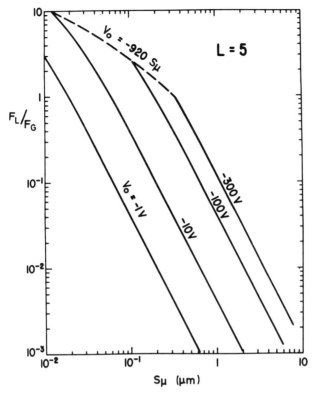

Fig. 7. The variation of F_L/F_G (ratio of electric to gravitational force) on grains of different sizes and potentials V_0 (in V) within the Saturnian magnetosphere at $L = 5.0$ (a position within the broad E Ring) (from Mendis et al. 1982b).

$$\ddot{\mathbf{r}} = \frac{q}{m}\ (\mathbf{E} + \frac{\dot{r}}{c}\ \times \mathbf{B}) - \frac{\mu}{r^3}\mathbf{r} + \mathbf{F} \tag{61}$$

where $\mathbf{F}$ represents forces associated with the collisions with photons, plasma and other grains.

Within the rigidly corotating regions of the planetary magnetospheres, where the ring systems are observed,

$$\mathbf{E} = -\ \frac{1}{c}\ (\Omega \times \mathbf{r}) \times \mathbf{B} \tag{62}$$

where Ω is the angular velocity of the planet. This is of course strictly applicable only when the Debye spheres of neighboring particles do not intersect, otherwise the electric fields of neighboring particles will also have to be included in $\mathbf{E}$. The single-particle approach of gravito-electrodynamics needs some essential modifications in this case. However, a comparison of the corotational electric force (due to the radial polarization of the corotating plasma) with the electric force due to neighboring grains indicates that the former is orders of magnitude larger in this case too. Consequently, it is possible that the predictions of the single-particle gravito-electrodynamic theory are not seriously invalidated, and may at least be correct to the first order. The main modification in this case is a decrease in the grain potential.

In the case of Saturn the magnetic moment and the spin vector are parallel (Connerney et al. 1982) within the observational uncertainties. It is easy to show (Mendis et al. 1982a; Northrop and Hill 1983a) that Eqs. (61) and (62) admit circular orbits in the equatorial plane moving with angular velocity ω_G, given by

$$\omega_G = \frac{\omega_{go}}{2} \left\{ -1 \pm \left[1 + 4 \left(\frac{\Omega}{\omega_{go}} + \frac{\omega_{Kep}^2}{\omega_{go}^2} \right) \right]^{\frac{1}{2}} \right\} \qquad (63)$$

where ω_{Kep} is the local Kepler angular velocity and ω_{go} is the gyrofrequency

$$\omega_{go} = -\frac{qB}{mc} . \qquad (64)$$

Equation (63) shows that two different motions are possible for a given grain. The plus sign in front of the square root corresponds to direct (prograde) motion, while the minus sign corresponds to indirect (retrograde) motion, for a negatively charged grain. For a positively charged grain, the minus sign gives a prograde motion while the plus sign gives a prograde or retrograde motion depending on the value of ω_{go}.

If a charged dust grain moving in a circular orbit in the equatorial plane of the planet (whose magnetic moment and spin are strictly parallel), is subject to a small perturbation in the plane (e.g., by the gravitational tug of a nearby satellite), it has been shown (Mendis et al. 1982a; Morfill et al. 1983b) that the grain will perform a motion that can be described as an elliptical gyration about a guiding center moving uniformly in a circle with angular velocity given by ω_G (Eq. 63). The gyration frequency ω about the guiding center is given by

$$\omega^2 = \omega_{go}^2 + 4\omega_{go} \omega_G + \omega_G^2 . \qquad (65)$$

Also, if a and b are the semimajor and semiminor axes of this ellipse,

$$\frac{b}{a} = -\frac{\omega}{2\omega_G + \omega_{go}} \qquad (66)$$

with the minor axis aligned in the radial direction.

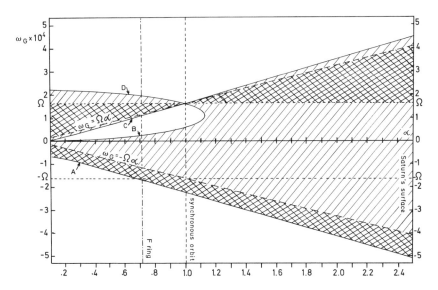

Fig. 8. The variation of ω_G with α ($= \omega_{\text{Kep}}/\Omega$). The curves marked A, B, C, D indicate the
values of ω_G when $\omega^2 = 0$, for various values of α. In the shaded regions, $\omega^2 > 0$, the
unshaded regions are where $\omega^2 < 0$. The dark shading corresponds to negative particles, and
the light shading to positive grains. The lines marked $\omega_G = \pm \Omega \alpha$ represent large particles
moving at Keplerian speed; the dashed line marked $\omega_G = \Omega$ represents the corotating
(small) particles. The values of $\alpha = 0.1, 0.7, 1,$ and 2.5 correspond to the limit of rigid
corotation in Saturn's magnetosphere, the F Ring, the synchronous orbit, and Saturn's
surface, respectively (from Mendis et al. 1982a).

2. Stability of Orbits. Not all the grain orbits are stable. Those which are
must satisfy the condition $\omega^2 > 0$. One can use Eqs. (63), (65) and (66)
together with this condition to obtain the stable orbits at any given distance
from the planet. The classes of stable and unstable orbits within the rigidly
corotating portion of the Saturnian magnetosphere are shown in a general
way in Fig. 8. Here $\alpha(=\omega_{\text{Kep}}/\Omega)$ is the independent variable and ω_G is the
dependent variable; α varies from 2.5, which corresponds to the Saturnian
surface, to 0.1, which corresponds to $r = 10$ R$_S$. Also, the positions of the
synchronous orbit ($\alpha = 1$) and the F Ring ($\alpha = 0.7$) are indicated.

We see that at the distance of the F Ring, for instance, negative particles
of all sizes from the smallest (corotating) particles to the largest (essentially
Keplerian) particles moving in the prograde sense are stably trapped, as expected. It is interesting that there are several other distributions of particles
that also can be stably trapped there. One of these is a set of retrograde
negative particles, from the largest (Keplerian) particles to moderately large
ones moving with angular velocity $\simeq -\Omega$. There are also three distinct
populations of positive prograde particles, and another population of positive
retrograde particles that can be stably trapped. One set consists of very small

(positive) particles moving slightly faster than the corotation speed, while a second set consists of very large (positive) particles moving prograde slightly slower than the Kepler speed. These are clearly to be expected. Somewhat less obvious is the existence of two stable populations of large to moderate-sized (positive) particles, one moving in the prograde sense and the other moving in the retrograde sense. The populations of large to moderate-sized (positive) prograde particles between curves B and C are unstable and are excluded.

Northrop and Hill (1982) have also studied the stability of charged grains in the equatorial plane of Saturn against perturbations normal to the ring plane. For a grain of given specific charge there is a critical distance R_c such that grains with $r < R_c$ are unstable. Northrop and Hill (1982b) have argued that the prominent change in the ring brightness, which seems to start at around 1.63 R_S, may be associated with the location of this stability radius at 1.625 R_S for grains of very large specific charge (q/m). $R_c = 1.625\ R_S$ corresponds strictly to "grains" with infinite specific charge, and is therefore more appropriate for plasma. This transition between stability and instability occurs because along a certain dipole field line the force acting on a charged particle may vary as a function of distance and latitude. This force can change from inward-pointing as gravity is the dominant force, to outward-pointing when centrifugal force takes over. Figure 9 shows these two regimes in the

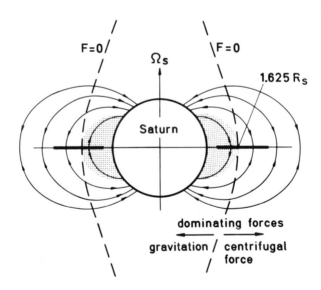

Fig. 9. Division of the rotating plasmasphere by the dashed curves $F = 0$ into two plasma regimes, according to the consideration of the balance of centrifugal and gravitational forces: (1) the upward siphon flow region denoted by the shaded area with equatorial distance $r \leqslant 1.6252\ R_S$; and (2) the equatorial confinement region with $r > 1.6252\ R_S$ (from Ip 1982).

Saturnian magnetosphere. Ip (1982) has argued that the decrease of optical depth by about a factor of 2 near $1.625 \, R_S$ is due to the field-aligned siphoning off of plasma formed inside this radius by the collision of interplanetary meteoroids with ring particles, whereas the collision-produced plasma outside this radius is confined to the equatorial plane and is not lost to the Saturnian ionosphere. Northrop and Hill (1983) pointed out that the stability limit at $1.625 \, R_S$ corresponds to highly charged particles that move in circular orbits with the local corotation speed. However, if highly charged ($|q/m| \gtrsim 7.5$ Coul g^{-1}) particles are launched in the ring plane at the local Kepler velocity then their stability limit is found almost exactly at the inner edge of the B Ring (Northrop and Hill 1983). Clusters of water molecules possibly created by micrometeoroid impacts onto the ring would satisfy this condition, according to these authors.

From Fig. 8 it is clear that negatively charged prograde grains outside the synchronous orbit move with an angular velocity ω_G larger than the Kepler angular velocity at that distance. Since ω_G depends on ω_{g0} and therefore on the grain size s, for a given potential there would be grains of a certain size that would move with the same angular speed Ω_{Sat} as a satellite interior to the grain orbit. This means that there is an exact 1:1 orbit-orbit resonance with such a satellite. Large positively charged grains outside the synchronous orbit move with ω_G, smaller than ω_{Kep}; these particles could have a 1:1 resonance with a satellite outside the grain orbit. A similar situation clearly does not arise in the purely gravitational case. This magneto-gravitational resonance would, for instance, arise between a certain size grain in the F Ring and its nearby satellites 1980S27 and 1980S26, depending on the charge state of the grains.

It must be stressed that, unlike a pure gravitational resonance which affects particles of all sizes equally, the magneto-gravitational resonance picks out a particular grain size s_c for a given potential V_0. Grains with sizes close to s_c (on either side) will also be strongly affected because their angular velocites will be close to that of the perturbing satellite, and so will remain in the vicinity of that satellite for long periods of time. If we consider such a particle with a gyration frequency ω about a guiding center moving with angular velocity ω_G, it can be shown (Mendis et al. 1982a) that the grain will move in an undulating orbit with a wavelength λ given by

$$\lambda = \frac{2\pi r}{\omega} (\omega_G - \Omega_{Sat}) \qquad (67)$$

where Ω_{Sat} is the angular velocity of the perturbing satellite. Furthermore, as each successive grain of the same size in the ring moves over the satellite it will be subject to the same perturbation and will therefore follow the same path as its predecessor in the reference frame of the satellite. Consequently, all the grains will move in phase to form a wavy pattern with wavelength λ in the frame of the perturbing satellite. It has been proposed by Mendis et al. (1982a) and Hill and Mendis (1982b) that the waves observed in the F Ring are formed this way. However, they require that the dust particles are charged to $V_0 = -38$ V. As mentioned earlier, collective self-shielding of the closely spaced dust population may rule out such large potentials.

Hill and Mendis (1982b) have also shown the existence of another type of resonance for charged dust grains, which they call gyro-orbital resonance. This is due to the modulation of the grain potential with orbital period as the grain moves in and out of the planet's shadow. Consequently, if the gyro-period of the grain is in resonance with the grain's orbital period, the orbital eccentricity grows until the grain is removed by collisions with its neighbors. Hill and Mendis (1982b) propose that this effect is responsible for clearing small isolated gaps in Saturn's B Ring from highly charged submicron-sized dust particles.

B. Radial Transport

Radial transport of dust particles can be described in two extreme ways: (1) large particles, basically in Keplerian orbits, which are gradually transported either systematically or stochastically; and (2) small particles, which are electromagnetically dominated, and are transported systematically or stochastically.

We use a diffusion-convection formalism to describe this transport. In terms of the integrals of motion I_i and their associated cyclic variables α_i, this yields for the evolution of the distribution function f in space and time

$$\frac{\partial f}{\partial t} + \frac{\partial}{\partial I_i}\left(\frac{dI_i}{dt} f - \frac{1}{2}\left\langle\frac{\Delta I_i \Delta I_j}{\Delta t}\right\rangle \frac{\partial f}{\partial I_j}\right) = 0 \qquad (68)$$

where dI_i/dt is a systematic change in I_i caused by a systematic interaction (e.g., momentum exchange with an ambient medium by particle friction or plasma drag). The transport coefficient in angled brackets requires knowledge of the power spectrum of perturbing forces. We are dealing with resonant phenomena; i.e., only those frequencies in the perturbations are picked out which are multiples of the particle's natural frequencies (e.g., orbital frequency, gyrofrequency).

From Morfill et al. (1983b) we obtain, e.g., a systematic radial drift velocity due to friction:

$$v_r \approx -\frac{2a_s}{\tau_f}\left(1 - \frac{\Omega}{\omega_{Kep}}\right) \qquad (69)$$

where a_s is the semimajor axis, Ω is the planetary rotation rate and ω_{Kep} is the local Keplerian angular velocity. The frictional coupling time scale is τ_f. In

TABLE III

Transport of Gravitationally Dominated Dust Grains[a]

Integral of Motion	Cyclic Variable
$I_1 = -\frac{1}{2}ma_s^2\omega_{Kep}$	$\alpha_1 = -\omega_{Kep}T$
$I_2 = m\omega_{Kep}a_s^2(1-e^2)^{\frac{1}{2}}$	$\alpha_2 = -\omega_p$
$I_3 = I_2\cos(i)$	$\alpha_3 = -\Omega_{asc}$
$I_4 = -\frac{1}{2}ma_s^2\omega^2_{Kep}$	$\alpha_4 = T$

[a] $\omega^2_{Kep} = GM/r_o^3$, where r_o is the initial particle position from the planet center, e = eccentricity, i = inclination, T = time of pericenter passage, ω_p = elongation of pericenter, and Ω_{asc} = longitude of ascending node. The integrals of motions are: I_1, action; I_2, total angular momentum; I_3, component of angular momentum; and I_4 total energy.

TABLE IV

Transport of Electromagnetically Dominated Grains[a]

Integral of Motion	Cyclic Variable
$I_1 = p_\perp^2/2mB$	
$I_2 = \oint p_\parallel\,ds$	$\alpha_i \equiv \dfrac{\partial W}{\partial I_i}$
$I_3 = \dfrac{g}{c}\phi$	

[a] $p_\perp$ is the component of particle momentum perpendicular to the magnetic field **B**, $p_\parallel$ is the parallel component, the integral $\oint ds$ extends along the field line to the mirror points, and ϕ is the magnetic flux enclosed by the drift shell. W is the total (kinetic + potential) energy of the particle. The integrals of motion are the three adiabatic invariants.

the case of large particles, we have the constants of motion I_i and the associated cyclic variables given in Table III (Barge et al. 1982; Hassan and Wallis 1982). One of the four integrals of motion is redundant.

1. Adiabatic Motion of Charged Dust Grains. In the case of small particles which are electromagnetically dominated, the integrals of motion and associated cyclic variables are given in Table IV (see e.g. Schulz 1975; Morfill 1978). The frequencies are the gyrofrequency (associated with I_1), bounce frequency (I_2) and longitudinal drift frequency (I_3).

Northrop and Hill (1983a) extended the usual adiabatic theory of charged particle motion (see Northrop 1963) to include the complication of a variable grain charge $q(t)$. The charge on a grain is modulated by variations of the relative speed between the grain and the plasma and by gradients in the plasma density and energy. Since the time scale for charging and discharging is finite, it introduces a phase lag in the charge with respect to the orbital position of the grain. The adiabatic theory developed is quite general, and is applicable to the case in which the magnetic moment and the spin vector of the planet are at any arbitrary angle. Converting the equation of motion to the frame rotating with the planet, Northrop and Hill (1983a) showed that the charged grain may be regarded as moving in an pseudo-magnetic field $\mathscr{B}$ given by

$$\mathscr{B} = \mathbf{B} + 2mc\,\mathbf{\Omega}\,/\langle q(t)\rangle \tag{70}$$

where $\langle q(t)\rangle$ is the value of $q(t)$ averaged over a gyrocycle and in an equivalent electric field $[m/q(t)]\,\nabla(\tfrac{1}{2}r^2\Omega^2 - \phi_g)$, where ϕ_g is gravitational potential and r is axial distance.

The charge $q(t)$ varies systematically with its gyrophase about the guiding center. Consequently the circle in which the grain gyrates rocks at the gyrofrequency. Since the variation of $q(t)$ is rather small, the angle through which the circle rocks is also small. The average plane of this circle is perpendicular to the pseudo-magnetic field $\mathscr{B}$. From this it is apparent, as expected, that electrons and ions with very small m/q_0 gyrate very nearly in a plane normal to $\mathbf{B}$, whereas bodies having large m/q_0 gyrate in a plane nearly normal to $\mathbf{\Omega}$.

In the intermediate particle size range, only direct orbit trajectory integrals can be used for a correct transport description. In conjunction with stochastic forces, this makes the calculation very tedious and time-consuming, but there seems to be no alternative (Hill and Mendis 1980). The two situations involving electromagnetic perturbations, which are amenable to Fokker-Planck type analysis are: (1) Io-produced "smoke" particles, which may escape into the magnetosphere of Jupiter (Johnson et al. 1980; Morfill et al. 1980a,b,c,d), and are probably magnetically-dominated; and (2) Saturn's E Ring (Morfill et al. 1982b) which is probably gravitationally dominated.

2. Systematic Drift. Two classes of processes have been considered so far: friction with the ambient plasma and radiation pressure drag on the one hand (Morfill et al. 1980a; Morfill et al. 1983b), and systematic charge variations caused by the dust grain motion and/or plasma density gradients on the other (Northrop and Hill 1983a).

In the case of *magnetically dominated particles*, one would expect systematic charge variations to occur at the particle's gyrofrequency. These charge variations can be caused by the dust grain's own velocity variation with respect to the ambient plasma, or by a radial plasma gradient or a radial gradient in the electron energy.

As shown by Northrop and Hill (1983a), a radial drift motion, always directed to the synchronous radius, is set up in the former case. The authors point out that the effect is relatively subtle, since it may only amount to a charge variation $\delta q/q \simeq 10^{-3}$ to 10^{-4} (i.e. ~ 1 electron charge or even less) per gyroperiod. In the Io plasma torus, where the plasma density is higher, the effect will be correspondingly more important.

A plasma gradient may lead to a more significant radial drift effect. The particle gyroradius is inversely proportional to the magnetic field strength (for constant particle velocity), or

$$r_g \approx R_g L^3 \tag{71}$$

where R_g is the gyroradius the particle would have at the planetary surface. The plasma density is generally described by

$$n \simeq N_0 L^m \tag{72}$$

where N_0 is the extrapolated value to the planet's surface. The plasma density fluctuation sampled by the dust grain during one gyroperiod is

$$\frac{\delta n}{n} = m \frac{2r_g}{R L} \tag{73}$$

where R is the planetary radius. Substituting Eq. (71) and writing $R_g = \epsilon R$ with $\epsilon \ll 1$, gives

$$\frac{\delta n}{n} = 2 m \epsilon L^2 . \tag{74}$$

For typical values (e.g., in Jupiter's inner magnetosphere) $m = -2$, $\epsilon = 10^{-2}$ (corresponding to ~ 0.1 μm dust grains) we get $\delta n/n = 0.04 L^2$, i.e., a few percent. The potential V of the dust particle is determined by the balance between plasma electron flux and photoelectron flux K (electrons cm^{-2} s^{-1}), at least in sunlight (cf. Eq. 6 and 9)

$$\pi s^2 n c_e \exp eV_0/E_e = \pi s^2 K . \tag{75}$$

Differentiating Eq. (75) yields

$$\frac{edV_0}{E_e} = \frac{dn}{n} \tag{76}$$

which shows that the surface potential fluctuates with the plasma density. The associated charge fluctuation over a gyroperiod is then $\delta q/q \approx \delta n/n$, i.e., it is significantly larger than the charge fluctuation induced by the dust motion itself.

As pointed out by Northrop and Hill (1983a), periodic charge fluctuations destroy the invariance of the grain's magnetic moment and induce a radial motion. The direction of the drift depends on the sign of the plasma density gradient. Friction with the ambient plasma and radiation pressure drag leads to a decay of the particle's gyroradius, as pointed out by Morfill et al. (1980a). This also leads to a decrease in the radial drift induced by both systematic and stochastic electromagnetic forces

In the case of *gravitationally dominated particles*, the drag forces have two effects. They circularize the orbits (plasma), and in the case of radiation pressure they induce large amplitude oscillations due to the daylight-shadow variations (Peale 1966; Burns et al. 1979; also chapter by Mignard). At the same time the orbital energy of the particles is changed (increased by the plasma drag outside the synchronous orbit, decreased by the radiation pressure drag). We use Saturn's E Ring to demonstrate this.

In the inner magnetospheres the plasma drag dominates considerably over radiation pressure. The associated systematic radial drift velocity is, in Saturn's plasma environment (Morfill et al. 1982):

$$v_r \, (4<L<7.5) \approx v_{ro} \, \frac{r}{\rho s} \tag{77a}$$

$$v_r \, (7.5<L<9) \approx v^*_{ro} \, \frac{r}{\rho s} \tag{77b}$$

where ρ is the particle's density (we take $\rho = 1 \text{ g cm}^{-3}$), v_{ro} is a constant $\approx 4 \times 10^{-13}$, and $v^*_{ro} \approx 4 \times 10^{-15}$. Distant Coulomb collisions have been taken into account. If we assume that the particle's inclinations are not strongly affected by the drag forces, we have to solve (in the absence of stochastic forces):

$$\text{div} \, (v_r n) + n/\tau_S = 0 \tag{78}$$

for the simplest scenario, where n is the dust particle spatial density and τ_S is the loss rate due to sputtering. Sputtering losses introduce a term $\partial[(dm/dt)m]/\partial m$, which we have simplified into a particle loss term because below a certain size the dust particles are essentially invisible.

The solution to Eq. (78) for a general $v_r \, (r)$, $\tau_S(r)$ is

$$n = n_0 \left(\frac{r_0}{r} \right)^2 \frac{v_r(r_0)}{v_r(r)} \exp\left(- \int_{r_0}^{r} \frac{dr}{\tau_S \, v_r} \right) \tag{79}$$

where the particle source is at $r = r_0$. From Eq. (77) this becomes

$$n \approx n_0 \left(\frac{r_0}{r} \right)^3 \exp\left(- \int_{r_0}^{r} \frac{dr}{\tau_S \, v_r} \right) . \tag{80}$$

The boundary condition at r_0 is

$$S = v_r n \mid_{r \to r_0} \tag{81}$$

where S (particles $\text{cm}^{-2} \, \text{s}^{-1}$) is the source strength and is presumed to be known. Then

$$n \approx \frac{S \rho s}{v_{ro} r_0} \left(\frac{r_0}{r} \right)^3 \exp \left(- \int_{r_0}^{r} \frac{dr \, s \rho}{\tau_S \, v_{ro} r} \right). \tag{82}$$

From Table II we find

$$\left. \begin{array}{c} \tau_S \, (4 \leqslant L \leqslant 5) = 1.8 \times 10^{11} \, \text{s} \\[2mm] \tau_S \, (5 \leqslant L \leqslant 7.5) = 4.5 \times 10^{9} \, \text{s} \\[2mm] \tau_S \, (L > 7.5) = 6 \times 10^{10} \, \text{s} \end{array} \right\} . \tag{83}$$

We then get

$$n = \frac{S \rho s}{v_{ro} r_0} \left(\frac{r_0}{r} \right)^{3 + s\rho/\tau_S \, v_{ro}} \tag{84a}$$

and changing to $L \equiv r/R_S$ as a convenient notation, we obtain for micron-sized grains:

$$\left. \begin{array}{ccccc} n \, (4 \leqslant L \leqslant 5) & \approx & 0.01S & \left(\dfrac{4}{L} \right)^{3 + 1/720} \\[4mm] n \, (5 \leqslant L \leqslant 7.5) & \approx & n \, (L=5) & \left(\dfrac{5}{L} \right)^{3 + 1/18} \\[4mm] n \, (L > 7.5) & \approx & n \, (L=7.5) & \left(\dfrac{7.5}{L} \right)^{3 + 5/12} \end{array} \right\} . \tag{84b}$$

This is a rather flat distribution, not compatible with the E Ring observations, although it is clear that the density drops off faster beyond $L = 7.5$. Diffusion due to fluctuating drag forces (e.g., caused by changes in the plasma density) will smear this distribution out even further. However, absorption by the satellites Tethys, Dione, and Rhea may play a role too, reducing the dust particle density as the grains are convected past, out of the Saturnian system. Absorption depends on the typical time scale t_e between distant encounters between a satellite (orbiting at radius $y = 1$) and the ring particles (orbiting at radius $1 \pm \Delta y$). For $\Delta y \ll 1$, we get

$$t_e \cong \frac{8\pi R_S^{3/2} L^{3/2}}{\sqrt{\mu_S}\,\Delta y} \qquad (85)$$

where $R_S\,L$ is the radial distance of the satellite from the planet center. This must be compared with the convection time across $2\Delta y$:

$$t_c \cong 2R_S L\,\Delta y/v_r \ . \qquad (86)$$

$R_S L\,\Delta y$ is a few satellite radii, denoting the sphere of influence of the particular satellite. The absorption by the satellite is then given roughly by

$$\eta \equiv \frac{n\,(1+\Delta y)}{n\,(1-\Delta y)} \approx e^{-t_c/t_e} \ . \qquad (87)$$

The ratio t_c/t_e in the region $4 < L < 7.5$ is

$$\frac{t_c}{t_e} = 2.27\times10^{-6}\ R_S^2\ s_\mu\ m^2\ L^{-7/2} \qquad (88)$$

where m is the number of satellite radii (R_{sat} in km) defining the spheres of influence ($m \sim 5$). In the region $L > 7.5$, t_c/t_e is a factor of 100 smaller. For Tethys this yields $\eta \approx 0.95$, for Dione we obtain $\eta \approx 0.97$, and for Rhea $\eta \approx 0.18$.

Combining the results of this analysis (see also Morfill et al. 1982b) we obtain the following qualitative picture, bearing in mind that there are large uncertainties in the sputter rates and in the somewhat arbitrary choice for satellite absorption:

1. Dust grains are created somehow on Enceladus at $L \cong 4$;
2. They are transported outwards mainly by plasma drag;
3. Sputtering and geometrical effects lead to a slowly decreasing grain density ($\sim L^{-3}$) as well as normal optical depth ($\sim L^{-2}$);
4. Absorption by Saturn's satellites leads to further grain reduction. This is particularly important for Rhea, because at $L = 8.7$ the radial drift due to plasma drag is small.

In principle the optical and trapped particle absorption signatures can be understood by invoking these processes.

   3. Stochastic Transport. Two processes leading to stochastic particle transport have been described in the literature. For magnetically-dominated particles, gyrocenter diffusion caused by stochastic charge variations is important (Morfill et al. 1980a,d). For gravitationally-dominated particles stochas-

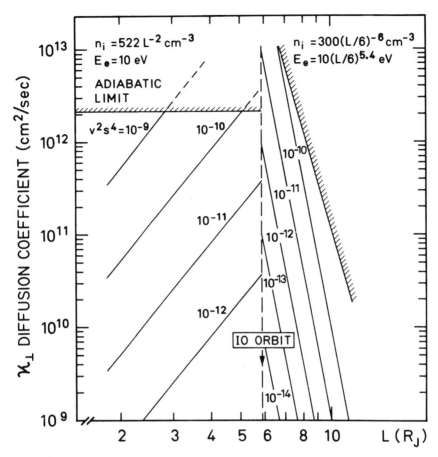

Fig. 10. Dust particle diffusion in the inner Jovian magnetosphere, showing the diffusion coefficient due to charge fluctuations as a function of position in the inner magnetosphere. The plasma characteristics (n_i, E_e) used to derive this result are indicated. The contours are given for different values of $v^2 s^4$ (v = particle velocity is cm s^{-1}; s = particle radius in cm), and the region where the guiding center diffusion theory is applicable shown by the hatched lines (from Morfill et al. 1980*a*).

tic forces lead to a diffusion of the particle orbits, in particular the orbital energy (Morfill and Grün 1979*a,b;* Consolmagno 1980; Barge et al. 1982; Hassan and Wallis 1982; Morfill et al. 1983*b*). Specific processes are magnetic fluctuations and plasma drag.

Figure 10 shows the calculated dust particle diffusion coefficient for *magnetically-dominated particles* in the Jovian magnetosphere (Morfill et al. 1980*a*) caused by random charge fluctuations. The plasma model (dependence on *L* of density n_i and electron energy E_e) used is indicated in the figure. The Io torus was not specially considered here. The quantity $v^2 s^4$, (v =

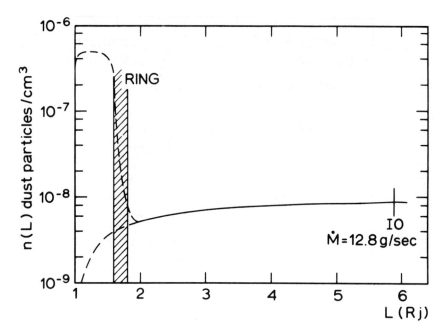

Fig. 11. Radial dependence of equatorial dust particle density inside the orbit of Io. A mass loss rate 12.8 g s^{-1} from Io in the form of submicron-sized particles ($\sim 10^{16}$ particles s^{-1}) was assumed. Inside the Jovian ring, at ~ 1.8 R$_J$ the dust population increases by a factor ~ 500 due to erosive collisions with parent bodies (from Morfill et al. 1980d).

particle injection velocity in a corotating frame; s = particle radius) was kept as a parameter.

Figure 11 shows the spatial dependence of Io injected ''smoke'' particles, with $v^2 s^4 \equiv 10^{-9}$, inside the Io radius (Morfill et al. 1980d). Losses by sputtering were shown to be relatively minor, and plasma drag was not considered (the plasma drag would be partially cancelled by the plasma gradient induced drift mentioned earlier, so that the net effect would be a slightly faster decrease of the particle density towards the planet than the value shown in the figure).

Derivation of the spatial diffusion coefficient is given by Morfill et al. (1980a). The physical reason for the diffusion is the following (cf. Fig. 12): During a gyration about a field line, a dust particle (carrying a surface charge of, e.g., 1000 electrons) undergoes normal statistical charge fluctuations on a time scale comparable to the gyroperiod (e.g., $\pm \sqrt{1000}$ in the time Δt that the particle needs to come into charged equilibrium). As a result, the gyroradius fluctuates statistically, the particle loses ''knowledge'' of its field line and the result is a gyrocenter diffusion across the magnetic field.

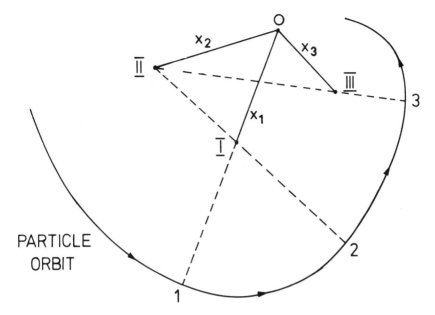

Fig. 12. Guiding center diffusion through charge fluctuations, showing the geometry of dust particle motion in the presence of charge fluctuations which leads to the guiding center (indicated by roman numerals) diffusion. x_i = distance from the original gyrocenter. Sudden changes in the charge occur at position 1, 2, and 3.

For *gravitationally-dominated particles*, short-period stochastic charge fluctuations are only of minor importance, since by definition electromagnetic forces are much smaller than gravity. This is even more so the case for perturbing electromagnetic forces, in particular when we consider that resonances at the particle orbital period, or a multiple thereof (but then with much reduced scatter efficiency), are required to evoke energy changes and real orbit diffusion. Long-period fluctuations in particle charge (and hence fluctuating electromagnetic forces) on the time scale of an orbital period may occur via long-period plasma variations. These plasma variations may be caused by solar wind interaction with the magnetosphere, growth and decay of ring currents, nightside plasma sheet changes, flux tube interchange instabilities, etc. Superimposed on the normal stochastic charge fluctuations is then another, also stochastic, long-period charge fluctuation. Relevant frequencies are those of the external (solar wind) fluctuation (periodicity ~ 100 hr, if we consider corotating interaction regions, solar flares, sector boundaries, etc.), the plasma azimuthal drift period, the corotation period (~ 10 hr) and the dynamical changes in the magnetosphere (ring currents, substorms, etc.) which could be ~ 100 hr (unfortunately, very little is known about this for Saturn's magnetosphere, and extrapolation from the Earth's magneto-sphere is certainly not straightforward). However, this discussion shows that we may expect plasma parameter variations over the relevant time scales (days); their strength, however, remains a subject for speculation.

Morfill et al. (1983b) have calculated the diffusion coefficient associated with such charge fluctuations. The result is

$$\kappa \approx \frac{<\Delta\omega_g^2>(\Omega-\omega_{\text{Kep}})^2<\Delta t>^4}{4(\tau_p+<\Delta t>)} \qquad (89)$$

where $<\Delta t>$ is the mean duration of a different plasma state (e.g., higher/lower density), τ_p is the mean period between such states, and $\Delta\omega_g$ is the induced fluctuation in the dust particle gyrofrequency,

$$\Delta\omega_g = B\Delta q/mc . \qquad (90)$$

The diffusion time scale

$$\tau_{\text{diff}} = r^2/\kappa \qquad (91)$$

is then evaluated:

$$\tau_{\text{diff}} \simeq 6\times10^3 \frac{s_\mu^4 L^6}{<\epsilon_i^2>V_0^2} \qquad (92)$$

(in yr) where V_0 is the grain potential in V, s_μ is the grain radius in μm, and $<\epsilon_i^2>^{\frac{1}{2}} \equiv \delta q/q$. As an example, in the E Ring at $L = 6$, with $<\epsilon_i^2>^{\frac{1}{2}} \equiv 0.1$ and $V_0 = 100$ V mean surface potential, we obtain $\tau_{\text{diff}} \simeq 3\times10^6$ yr. This is considerably larger than the convection time or the sputter time loss, so that a pure convection-loss description such as the one employed earlier seems justified, in spite of the large uncertainties in numerical values.

IV. OBSERVATIONS

The main emphasis of this chapter has been the detailed discussion of the individual effects comprised in dust-magnetosphere interactions. Occasional references to observations have already been made. In this section we sum-marize observations in the Earth's, Jupiter's and Saturn's magnetospheres which are related to dust-magnetosphere interactions. We should caution that electromagnetic phenomena have been inferred from most of these observa-tions because they could not easily be interpreted otherwise. Direct observa-tions, e.g. of charged dust grains, are scarce. In interplanetary space the dust experiment on board the Helios probe detected only 4 dust particles carrying a significant charge out of more than 200 observed particles (Grün et al. 1984). Also, as we will discuss below, direct observations of charged dust grains on the moon have been made (Berg et al. 1976). The next opportunity to observe directly charged dust particles will be via the dust experiments on board the Galileo and the International Solar-Polar Mission (ISPM) space probes, which will perform *in situ* measurements of dust particles and their charges in interplanetary space and in the Jovian magnetosphere.

Direct measurements of complete orbits of charged dust particles in planetary magnetospheres are not possible so their dynamics must be inferred from observations of an ensemble of dust grains. Remote sensing of dust particles by scattered or absorbed light or by high-energy particle absorption yields important but limited information on the integral cross-section area or the mass density of dust particles. Only the combination of remote observations of the large-scale distribution with *in situ* measurements of the dynamical state (mass, velocity, and charge) of individual particles can support theoretical predictions. Therefore some of the observations reported below have been only tentatively related to dust-magnetosphere interactions.

A. Dust in the Earth's Magnetosphere

In situ observations of dust particles near the Earth by the HEOS 2 micrometeoroid experiment (Hoffmann et al. 1975*a,b*) showed that the impact rate onto the sensor varied strongly within the Earth's magnetosphere inside 10 Earth radii (auroral zones). These short-term enhancements of the particle flux have been interpreted by Fechtig et al. (1979) as fragmentation products of fragile large meteoroids in the mass range 10 to 10^6 g. These bodies receive a large negative surface charge (corresponding to a surface potential of several 100 V) when they travel through the auroral zone at ~ 10 Earth radii. This leads to electrostatic fragmentation (see Sec.II.C.1) if the mechanical structure of the parent meteoroids is loose enough, and a swarm of small particles is produced. From the observed particle swarms Fechtig et al. were able to estimate the mass of the parent meteoroids and, from the viewing directions of the sensor, correlate them with type III fireballs (Ceplecha and McCrosky 1976). The masses and fluxes of the parent meteoroids are in agreement with the corresponding fireball values.

Electrostatic transport of lunar surface dust was first suggested by Gold (1955) on the basis of theoretical considerations. Observational evidence came from optical observations of horizon glow by the Surveyor spacecrafts which revealed that dust is elevated several tens of cm above the lunar surface (Rennilson and Criswell 1974). These authors conclude that surface particles are charged up and levitate due to intense electrostatic fields (~ 500 V cm^{-1}). They believe that electrostatic transport is the dominant local transport mechanism of lunar fines. *In situ* observation of levitated dust on the lunar surface has been reported by Berg et al. (1976). The Lunar Ejecta and Meteor-

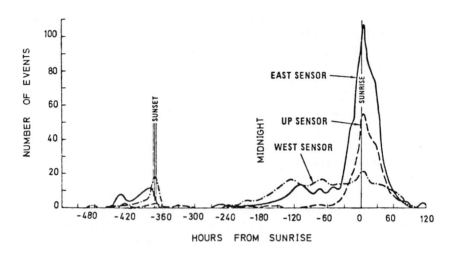

Fig. 13. Number of dust impacts onto the Lunar Ejecta and Meteorites (LEAM) experiment per 3-hr period, integrated over 22 lunar days (from Berg et al. 1976).

ites (LEAM) experiment placed on the Moon by the Apollo 17 astronauts recorded a strongly enhanced flux of charged dust particles near times when the terminator passes over the Apollo 17 site (see Fig. 13). The three sensors (facing up, to the east, and to the west) detected different impact rates during the terminator crossing, and these indicate a flux of charged particles directed across the terminator. A detailed model of the lunar dust transport which takes into account these observations has not yet been developed.

Dust has not only been observed in the immediate vicinity of the lunar surface; Severny et al. (1974) reported that the photometer on board the Lunokhod 2 vehicle measured scattered light extending up $\geqslant 260$ m above the surface. Visual observations of streamers and local horizon illuminations by the Apollo 8, 10, 15, and 17 astronauts before Apollo sunrise give evidence of high concentrations of dust up to the altitude of the spacecraft (120 km; cf. McCoy and Criswell 1974). Again, the physical implications of these observations are not fully understood.

B. The Jovian Ring

Voyager observations of the Jovian ring system have been described by Owen et al. (1979), Jewitt (1982), and Jewitt and Danielson (1981) and reviewed by Burns et al. (1980). See also the chapter by Burns et al. The Jovian ring system consists of three components: the bright ring, the faint disk and the even more tenuous halo. Many particles in the bright ring are micron-sized

(Jewitt and Danielson 1981; Grün et al. 1980). At surface potential 10 V (Morfill et al. 1980b) these particles are gravitationally dominated, and only plasma drag, sputter erosion, and mutual collisions are important for their dynamical evolution. The motion of particles in the faint disk, since they are highly concentrated towards the Jovian equatorial plane, must be dominated by gravitational forces. Therefore these particles cannot be much smaller than those in the outer bright ring; they are probably derived from this outer ring and drawn inward by plasma drag and radiation pressure drag (Morfill et al. 1980b; Burns et al. 1980). A tenuous ring envelops the other ring components, with a characteristic dimension of 10^4 km, normal to the plane of the bright ring. This out-of-plane material forms a broad lenslike halo, the outer limit being beyond 1.8 R_J and having a slight north-south asymmetry with respect to the bright ring (Jewitt and Danielson 1981). (For description of an alternative geometry see the chapter by Burns et al.) This vertical extent can easily be explained by assuming that its particles are quite small ($s \approx 0.1 \ \mu m$) (Morfill et al. 1980b; Consolmagno 1980; Grün et al. 1980). The motion of these particles is strongly influenced by electromagnetic effects, and the tilted Jovian magnetic field causes an out-of-plane force. We may regard these particles as being injected locally by collisional fragmentation and erosion of ring and disk particles and mirroring in the ''pseudo-magnetic field'' $\mathscr{B}$ given in Eq. (70). The average plane of their orbit is perpendicular to $\mathscr{B}$, the axis of which lies between the Jovian magnetic dipole axis $\mathbf{B}$ and the rotation axis $\mathbf{\Omega}$. Burns et al. (1980) pointed out that the halo seems not to be symmetric about the magnetic equator, but rather about the rotational equator. Also the lenslike shape of the halo, which is thinnest near the bright ring and thickest near the planet, is puzzling. To solve these questions, detailed studies of the evolution of ring, disk and halo particles must be carried out and improved photometric models developed which take spatially varying particle sizes into account (see chapter by Burns et al.)

Johnson et al. (1979) showed that the volcanic activity on the Jovian satellite Io not only leads to the ejection of copious amounts of gas into the Jovian magnetosphere, but can also be regarded as a source of small dust grains. It was suggested by Johnson et al. (1980) that these ''smoke'' particles are able to leave the satellite if their radii are $\lesssim 0.1 \ \mu m$. These particles become electrically charged by the ambient plasma and are removed from the gravitational field of Io by the influence of the Jovian magnetic field which sweeps past Io at a speed of ≈ 50 km s^{-1}. Morfill et al. (1980a) showed that, for such small particles and a typical particle surface potential of ~ 10 V, the subsequent motion in the magnetosphere is dominated by electromagnetic forces. The particle motion is practically adiabatic, i.e., the particles gyrate about their field lines, corotate, and bounce between their mirror points. Their main transport process in the magnetosphere is diffusive; the stochastic element in the particle motion is due to random charge fluctuations. Morfill et al. (1980b) and Grün et al. (1980) have proposed that these submicron-sized particles produce the visible ring particles by collisions with km-sized parent bodies (called "mooms" by Burns et al. 1980). However, these submicron Io particles have not yet been directly observed, and their existence has to be proven by the forthcoming Galileo mission.

C. Saturn's Ring System

1. Ring Structure and Small Particles. Micron- and submicron-sized grains have been identified in the dense A and B Rings, especially in the spokes region, and they dominate the populations in the outer F, G and E Rings. Dust-magnetosphere interactions may be important in determining the structure of Saturn's E, G, and F Rings. Plasma drag and sputtering, as well as diffusion of orbits due to plasma induced charge fluctuations, appear to be important for dust particles with radii $\leqslant 10$ μm (cf. Morfill al. 1983b). For the E Ring these processes may have caused the broad extent of the dust distribution within Saturn's magnetosphere (cf. Sec.II.D on sputtering and Sec.III.B. on radial transport for a more specific discussion). Any calculation of a detailed E Ring profile must take into account space variable loss of particles as well as transport and diffusion processes. Because of the short lifetimes of dust particles in that region due to sputtering and particle transport, an active source for E Ring particles is required, probably Enceladus.

During the crossing of the ring plane at a distance of 2.88 R_S, just at the edge of the G Ring, the plasma wave experiment of Voyager 2 detected intensive impulsive noise (Scarf et al. 1982). This noise has been attributed to micron-sized particles hitting the spacecraft, which produce charge pulses by impact ionization. Gurnett et al. (1983) were able to derive the masses of the dust grains from the observed signals. The results obtained show that the mass distribution varies as m^{-3}, and that most of the particles detected had radii in the range 0.3 to 3 μm. The observation of micron-sized particles in the G Ring which, like the E Ring particles, are subject to dispersive forces, and the narrowness of the G Ring imply either that this ring is a recent phenomenon ($\sim 10^3$ yr) or that it must be confined in some way, presumably by shepherding satellites; accordingly, it should perhaps have a structure resembling that of the F Ring.

The complicated, time variable, narrow structure (braids, kinks, multiple strands) of the F Ring is heavily influenced by the gravitational interaction of the shepherding satellites 1980S26 and 1980S27. It has been proposed by Morfill et al. (1983b) that coherent ring displacements, caused by plasma sheet induced charge variations on the grains, are observable in the F Ring in the form of waves which may subsequently evolve into kinks. Mendis et al. (1982a) and Hill and Mendis (1982b) suggest that the wave pattern observed in the F Ring originates from a magnetic-gravitational resonance between charged dust grains of a specific size and a shepherding satellite. Whether these effects will work depends on the plasma conditions at the F Ring and on the charge state that F Ring particles may acquire. Purely gravitational interaction has been considered by Showalter and Burns (1982) in order to explain the kinks and clumps observed by Voyager. However, the observed braiding is not understood at all.

2. Saturn Electrostatic Discharges. During both Voyager encounters with Saturn, the Planetary Radio Astronomy experiment detected strong discrete episodic bursts of radio emission, termed SED (Saturn electrostatic discharges) (Warwick et al. 1981, 1982; Kaiser et al. 1984). Although Evans et al. (1982) claim that the source for the SED is located in Saturn's B Ring, no physical process has been forwarded to explain the generation of SED in the ring. Recently Kaiser et al. (1983) have shown that SED are consistent with an extended lightning storm system in Saturn's atmosphere.

3. Spokes. Micron-sized particles cause the almost radial spokes in the B Ring. Smith et al. (1981) noted that spokes appear dark in back-scattered light but appear bright against the background B Ring in forward-scattered light, suggesting that micron-sized particles elevated above the ring plane are responsible for this phenomenon. A theory of spokes has to explain this and the following observed spoke characteristics (Smith et al. 1981, 1982; Proco and Danielson 1982; Grün et al. 1982): spokes have been observed in the B Ring between ∼ 100,000 km to 117,500 km from Saturn's center. They commonly appear wedge-shaped, with a vertex at a distance of 112,500 km (corotation distance). Their width at the base (towards Saturn) varies from about 2000 km to 20,000 km and their radial extension is about 3000 km to 12,000 km. However, narrow (typically 500 km in width) filementary spokes have been observed outside a radial distance of 110,000 km, mostly joined with a wider spoke farther in.

Several narrow spokes were observed during their formation along radial lines in the sunlit portion of the ring. The formation time is typically ≲ 5 min for a 6000 km long spoke. The rate of spoke formation is highest at the morning ansa outside Saturn's shadow. Spokes are nearly radial, or tilt away from radial in such a way that Keplerian motion will continue to tilt them further. From measurements of the angular velocity of spoke edges it was shown that in most cases both spoke edges revolve with Keplerian speed. However, a few spokes have been found where only one edge revolves with Keplerian speed whereas the other edge stays radial and corotates with Saturn, so their width increases with time. Spoke activity varies within the measurement errors of ± 22 min with a period of Saturn's rotation (10 hr 39.4 min). Several recurrent spoke patterns have been observed at that period. Spoke activity showed a pronounced peak in the quadrant centered around 115° SLS (Saturn longitude system, defined by Desch and Kaiser [1981]).

Spokes have been observed all around the illuminated side of the ring. However, they are visible with highest contrast relative to the underlying B Ring at the morning ansa. Bright spokes were also observed on the face of the B Ring which is not illuminated by the Sun but only by Saturn's shine. From edge-on views of the ring system an upper limit of 80 km for the height of spokes above the ring plane is derived.

4. *Spoke Formation*. In the following discussion, we briefly mention some theories on spoke formation and show their relation to the observations. Several theories (Gold 1980; Bastin 1981; Carbary et al. 1982) propose that spokes become visible because elongated grains are aligned due to polarization in weak radial electric fields. But Weinheimer and Few (1982) have argued against grain alignment theories. They showed that grain alignment theories require that grains are sufficiently conducting. However, unless ice (spoke particles, like the rest of the ring particles, are most likely to be composed of H_2O ice) becomes ferro-electric, the electric torque used to align the particles is many orders of magnitude too small, with the electric fields expected in the B Ring. Weinheimer and Few believe that it is very unlikely that the ice would become ferro-electric at the ring temperatures.

Most other theories proposed so far for spoke formation require charged dust particles which are elevated above the ring plane. Thomsen et al. (1982) interpret the deviation of the angular velocity of spoke features from the Keplerian value as being due to the orbital speed of negatively charged dust particles. The observed large deviations of the radial edge of active spokes (cf. Grün et al. 1982) from Keplerian speed would indicate, in their model, very high negative charges ($|q/m| > 100$ Coul kg^{-1}) on dust particles.

Hill and Mendis (1981*a*, 1982*a*) discuss dust particle dynamics in the electromagnetic environment. They calculate the orbits of grains of different sizes as the grains move around the planet, and find that high negative charges on dust particles should give rise to fine structure in wedge-shaped spokes. They predict that a number of almost straight, sharp "ribs" should radiate out from a point at the corotation distance, which revolve with time-varying angular speed. Finally, they show that the resettling of the grains on the larger bodies in the ring plane after their initial levitation results in a differential transport of grains across the ring plane. A consequence of this is the establishment of different multimodal size distributions of dust within the spokes at different planetocentric distances. None of these detailed predictions have yet been uniquely verified by the observations.

The most comprehensive theories of spoke formation are those of Goertz and Morfill (1983) and Morfill et al. (1983*c*). Goertz and Morfill (1983) show that relatively dense plasma ($n \sim 100$ cm^{-3}) is required near the rings in order to generate a strong enough surface electric field to lift dust particles off the ring. It is unlikely that the average plasma density near the rings is large enough to do this. Such a dense plasma cloud near the rings will contain slightly negatively charged dust, which due to the gravitational force, drifts relative to the plasma. This current causes a polarization of the plasma cloud and a radial $\mathbf{E} \times \mathbf{B}$ plasma drift at ~ 30 km s^{-1}. As long as the drifting plasma is dense enough, dust will be elevated marking the radial trail of the plasma.

The subsequent evolution of such a radially aligned, negatively charged dust cloud is discussed by Morfill et al. (1983c). The discharge of the fine dust by solar ultraviolet radiation produces a cloud of electrons which moves adiabatically in Saturn's dipolar magnetic field. The electron cloud is absorbed by the ring after one bounce, alters the local ring potential significantly, and reduces the local Debye length. As a result, more micron-sized dust particles may be elevated above the ring plane and the spoke grows in width. This process continues until the electron cloud has dissipated.

Both theories (Goertz and Morfill 1983; Morfill et al. 1983c) are able to account for the energetics of spoke formation, dust levitation off the ring plane, radial alignment of young spokes, spoke formation times, and the identification of structure (filamentary, narrow, and extended spokes).

Open questions in these theories include the origin of the dense plasma clouds proposed by Goertz and Morfill (1983), and the cause for the periodicity of spoke activity. While there is no satisfying explanation for the spoke periodicity, several proposals have been made for the origin of the dense plasma clouds. One possibility is meteoroid impacts (Morfill and Goertz 1983; Morfill et al. 1983a); another is sputtering of the rings caused by field aligned currents of accelerated keV particles (Morfill 1983b).

Acknowledgment: The authors are grateful to J. A. Burns for his helpful comments.

REFERENCES

Acuña, M. H., and Ness, N. F. 1976. Results from the GSFC fluxgate magnetometer on Pioneer 11. In *Jupiter,* ed. T. Gehrels (Tucson: Univ. Arizona Press), pp. 830–847.

Acuña, M. H., and Ness, N. F. 1980. The magnetic field of Saturn: Pioneer 11 observations. *Science* 207:444–446.

Ashworth, D. G. 1978. Lunar and planetary impact erosion. In *Cosmic Dust,* ed. J. A. M. McDonnell (New York: John Wiley and Sons), pp. 427–526.

Bagenal, F., and Sullivan, J. D. 1981. Direct plasma measurements in the Io torus and inner magnetosphere of Jupiter. *J. Geophys. Res.* 86:8447–8466.

Barge, P., Pellat, R., and Millet, J. 1982. Diffusion of Keplerian motions by a stochastic force. (I) A general formalism. *Astron. Astrophys.* 109:228–232.

Bastin, J. A. 1981. Note on the rings of Saturn. *Moon Planets* 24:467.

Belton, M. J. S. 1966. Dynamics of interplanetary dust. *Science* 151:35–44.

Berg, O. E., Wolf, H., and Rhee, J. 1976. Lunar soil movement registered by the Apollo 17 cosmic dust experiment. In *Interplanetary Dust and Zodiacal Light,* ed. H. Elsässer and H. Fechtig (Berlin: Springer-Verlag), pp. 233–237.

Bridge, H. S., Bagenal, F., Belcher, J. V., Lazarus, A. J., McNutt, R. L., Sullivan, J. D., Gazis, P. R., Hartle, R. E., Ogilvie, K. W., Scudder, J. D., Sittler, E. C., Eviatar, A., Siscoe, G. L., Goertz, C. K., and Vasyliunas, V. M. 1983. Plasma observations near Saturn: Initial results from Voyager 2. *Science* 215:563–570.

Bridge, H. S., Belcher, J. W., Lazarus, A. J., Olbert, S., Sullivan, J. D., Bagenal, F., Gazis, P. R., Hartle, R. E., Ogilvie, K. W., Scudder, J. D., Sittler, E. C., Eviatar, A., Siscoe, G. L., Goertz, C. K., and Vasyliunas, V. M. 1981. Plasma observations near Saturn: Initial results from Voyager 1. *Science* 212:217–224.

Bridge, H. S., Belcher, J. W., Lazarus, A. J., Sullivan, J. D., McNutt, R. L., Bagenal, F., Scudder, J. D., Sittler, E. C., Siscoe, G. L., Vasyliunas, V. M., Goertz, C. K., and Yeates, C. M. 1979. Plasma observation near Jupiter: Initial results from Voyager 1. *Science* 204:987–991.

Broadfoot, A. L., Sandel, B. R., Shemansky, D. E., Holberg, J. B., Smith, G. R., Strobel, D. F., McConnell, J. C., Kumar, S., Hunten, D. M., Atreya, S. K., Donahue, T. M., Moos, H. W., Bertaux, J. L., Blamont, J. E., Pomphrey, R. B., and Linick, S. 1981. Extreme ultraviolet observations from Voyager 1 encounter with Saturn. *Science* 212:206–211.

Brown, W. L., Lanzerotti, L. J., Poate, J. M., and Augustyaniak, W. M. 1978. Sputtering of ice by MeV light ions. *Phys. Rev. Letters* 40:1027.

Brown, W. L., Augustyniak, W. M., Brody, E., Cooper, B., Lanzerotti, L. J., Ramierez, A., Evatt, R., and Johnson, R. E. 1980. Energy dependence of the erosion of H_2O ice films by H and He ions. *Nucl. Instrum. Method* 170:321–325.

Burns, J. A., Lamy, P. L., and Soter, S. 1979. Radiation forces on small particles in the solar system. *Icarus* 40:1–48.

Burns, J. A., Showalter, M. R., Cuzzi, J. N., and Pollack, J. B. 1980. Physical processes in Jupiter's ring: Clues to its origin by Jove! *Icarus* 44:339–360.

Carbary, J. F., Bythrow, P. F., and Mitchell, D. G. 1982. The spokes in Saturn's rings: A new approach. *Geophys. Res. Letters* 19:420–422.

Carter, G., and Calligon, J. S. 1968. *Ion Bombardment of Solids* (New York: Elsevier).

Ceplecha, Z., and McCrosky, R. E. 1976. Fireball end heights: A diagnostic for the structure of meteoric material. *J. Geophys. Res.* 81:6257–6275.

Cheng, A. F., Lanzerotti, L. J., and Pirronello, V. 1982. Charged particle sputtering of ice surfaces in Saturn's magnetosphere. *J. Geophys. Res.* 87:4567–4570.

Connerney, J. E. P., Ness, N. F., and Acuna, M. H. 1982. Zonal harmonic model of Saturn's magnetic field from Voyager 1 and 2 observations. *Nature* 298:44–46.

Consolmagno, G. J. 1980. Electromagnetic scattering lifetimes for dust in Jupiter's ring. *Nature* 285:557–558.

Davis, D. R., Housen, K. R., and Greenberg, R. 1981. The unusual dynamical environment of Phobos and Deimos. *Icarus* 47:220–223.

Dermott, S. F., Gold, T., and Sinclair, A. T. 1979. The rings of Uranus: Nature and origin. *Astron. J.* 84: 1225–1234.

Dermott, S. F., Murray, C. D., and Sinclair, A. T. 1980. The narrow rings of Jupiter, Saturn and Uranus. *Nature* 284:309–313.

Desch, M. D., and Kaiser, M. L. 1981. Voyager measurement of the rotation period of Saturn's magnetic field. *Geophys. Res. Letters* 8:253–256.

Dobrovolskis, A. R., and Burns, J. A. 1980. Life near the Roche limit: Behavior of ejecta from satellites close to planets. *Icarus* 42:422–441.

Dohnanyi, J. S. 1969. Collisional model of asteroids and their debris. *J. Geophys. Res.* 74:2531–2554.

Evans, D. R., Romig, J. H., Hord, C. W., Simmons, K. E., Warwick, J. W., and Lane, A. L. 1982. The source of Saturn electric discharges. *Nature* 299:236–237.

Fechtig, H., Grün, E., and Kissel, J. 1978. Laboratory simulation. In *Cosmic Dust*, ed J. A. M. McDonnell (New York: Wiley, Chichester), pp. 607–669.

Fechtig, H., Grün, E., and Morfill, G. E. 1979. Micrometeoroids within ten Earth radii. *Planet. Space Sci.* 27:511–531.

Feuerbacher, B., and Fitton, B. 1972. Experimental investigation of photoemission from satellite surface materials. *J. Appl. Phys.* 43:1563–1572.

Fillius, W. 1976. The trapped radiation belts of Jupiter. In *Jupiter*, ed T. Gehrels (Tucson: Univ. Arizona Press), pp. 896–927.

Fillius, W., Ip, W. H., and McIlwain, C. E. 1980. Trapped radiation belts of Saturn: First look. *Science* 207:425–431.

Frank, L. A., Ackerson, K. L., Wolfe, J. H., and Mihalov, J. D. 1976. Observations of plasmas in the Jovian magnetosphere. *J. Geophys. Res.* 81:457–468.

Friichtenicht, J., F. 1964. Micrometeoroid simulation using nuclear accelerator techniques. *Nucl. Instrum. Meth.* 28:70–78.

Fujiwara, A., Kamimoto, G., and Tsukamoto, A. 1977. Destruction of basaltic bodies by high-velocity impact. *Icarus* 31:277–288.

Gault, D. E., and Heitowit, E. D. 1963. The partition of energy for hypervelocity impact craters formed in rock. In *Proceedings, Sixth Hypervelocity Impact Symposium*, Vol. 2, pp. 419–456.

Gault, D. E., and Wedekind, J. A. 1969. The destruction of tektites by micrometeoroid impact. *J. Geophys. Res.* 74:6780–6794.

Gault, D. E., Hörz, F., and Hartung, J. B. 1972. Effects of microcratering on the lunar surface. *Proc. Third Lunar Conf.* 3:2713–2734.

Goertz, C. K., and Thomsen, M. F. 1979. The dynamics of the Jovian magnetosphere. *Rev. Geophys. and Space Phys.* 17:731–743.

Goertz, C. K., and Morfill, G. E. 1983. A model for the formation of spokes in Saturn's ring. *Icarus* 53:219–229.

Gold, T. 1955. *Mon. Not. Roy. Astron. Soc.* 115:585–604.

Gold, T. 1980. Electric origin of the spokes seen in Saturn's rings. Contrib. *Roy. Soc. Conf. On Planetary Exploration*, London, Nov. 1980 (Transactions R. S.).

Grün, E., Morfill, G., Schwehm, G., and Johnson, T. V. 1980. A model of the origin of the Jovian ring. *Icarus* 44:326–338.

Grün, E., Morfill, G. E., Terrile, R. J., Johnson, T. V., and Schwehm, G. 1983. The evolution of spokes in Saturn's B ring. *Icarus* 54:227–252.

Grün, E., Fechtig, H., and Kissel, J. 1984. Physical and chemical characteristics of interplanetary dust: Measurements by Helios. In preparation.

Gurnett, D. A., Grün, E., Gallagher, D., Kurth, W. S., and Scarf, F. L. 1983. Micronsize particles detected near Saturn by the Voyager plasma wave instrument. *Icarus* 53:236–254.

Haff, P. K., Watson, C. C., and Tombrello, T. A. 1979. Ion erosion on the Galilean satellites of Jupiter. *Proc. Lunar Planet. Sci. Conf.* 10:1685–1699.

Hassan, M., and Wallis, M. K. 1983. Stochastic diffusion in inverse square field: General formulation for interplanetary gas. *Planet. Space Sci.* 31:1–10.

Hill, J. R., and Mendis, D. A. 1980. Charged dust in the outer planetary magnetospheres. II. Trajectories and spatial distribution. *Moon Planets* 23:53–71.

Hill, J. R., and Mendis, D. A. 1981a. On the braids and spokes in Saturn's ring system. *Moon Planets* 24:431–436.

Hill, J. R., and Mendis, D. A. 1981b. Electrostatic disruption of a charged conducting spheroid. *Can. J. Physics* 59:897–901.

Hill, J. R., and Mendis, D. A. 1982a. The dynamical evolution of the Saturnian ring spokes. *J. Geophys. Res.* 87:7413–7420.

Hill, J. R., and Mendis, D. A. 1982b. The isolated non-circular ringlets of Saturn. *Moon Planets* 26:217–226.

Hinteregger, H. E. 1965. Absolute intensity measurements in the extreme ultraviolet spectrum of solar radiation. *Space Sci. Rev.* 4:461–497.

Hoffman, H., Fechtig, H., Grün, E., and Kissel, J. 1975a. First results of the micrometeoroid experiment S215 on HEOS 2 Satellite. *Planet. Space Sci.* 23:215–224.

Hoffmann, H–J., Fechtig, H., Grün, E., and Kissel, J. 1975b. Temporal fluctuations and anisotropy of the micrometeoroid flux in the earth-moon system. *Planet. Space Sci.* 23:985–991.

Hornung, K., and Drapatz, S. 1981. Residual ionization after impact of large dust particles. In *The Comet Halley Probe Plasma Environment* (Washington, D. C.: ESA SP-155), pp. 23–37.

Hörz, F., Schneider, E., Gault, D. E., Hartung, J. B., and Brownlee, D. É. 1975. Catastrophic rupture of lunar rocks: A Monte Carlo simulation. *Moon* 13:235–258.

Ip, W–H. 1983. On plasma transport in the vicinity of the rings of Saturn: A siphon flow mechanism. *J. Geophys. Res.* 88:819–822.

Jewitt, D. C. 1982. The rings of Jupiter. In *Satellites of Jupiter*, ed. D. Morrison (Tucson: Univ. Arizona Press), pp. 44–64.

Jewitt, D. C., and Danielson, G. E. 1981. The Jovian ring. *J. Geophys. Res.* 86:8691–8697.

Johnson, T. V., Cook, A. F., Sagan, C., and Soderblom, L. A. 1979. Volcanic resurfacing rates and implications for volatiles on Io. *Nature* 280:746–750.

Johnson, T. V., Morfill, G. E., and Grün, E. 1980. Dust in Jupiter's magnetosphere: An Io source. *Geophys. Res. Letters* 7:305–308.

Kaiser, M. L., Connerney, J. E. P., and Desch, M. D. 1983. The source of Saturn electrostatic discharges: Atmospheric storms. *NASA Tech. Mem.* 84966, Goddard Space Flight Center.

Kaiser, M. L., Desch, M. D., Kurth, W. S., Lecacheux, A., Genova, F., and Pederson, B. M. 1984. Saturn as a radio source. In *Saturn*, eds. T. Gehrels and M. S. Matthews (Tucson: Univ. Arizona Press). In press.

Krimigis, S. M., Armstrong, T. P., Axford, W. I., Bostrom, C. O., Gloeckler, G., Keath, E. P., Lanzerotti, L. J., Carbary, J. F., Hamilton, D. C., and Roelof, E. C. 1981. Low energy charged particles in Saturn's magnetosphere: Results from Voyager 1. *Science* 212:225–231.

Krimigis, S. M., Armstrong, T. P., Axford, W. I., Bostrom, C. O., Gloeckler, G., Keath, E. P., Lanzerotti, L. J., Carbary, J. F., Hamilton, D. C. and Roelof, E. C. 1982. Low energy hot plasma and particles in Saturn's magnetosphere. *Science* 215:571–577.

Lange, M. A., and Ahrens, T. J. 1982. *Lunar Planet. Science* XIII:415–416.

Lanzerotti, J. L., Brown, W. L., Poate, J. M., and Augustyniak, W. M. 1978. On the contribution of water products from Galilean satellites to the Jovian magnetosphere. *Geophys. Res. Letters* 5:155–158.

Matson, D. L., Johnson, T. V., and Fanale, F. P. 1974. Sodium D-line emission from Io: Sputtering and resonant scattering hypothesis. *Astrophys. J.* 192:L43–L46.

McCoy, J. E., and Criswell, D. R. 1974. Evidence for a high altitude distribution of lunar dust. *Proc. Fifth Lunar Conf.* 3:2991–3005.

Mendis, D. A. 1981. The role of electrostatic charging of small and intermediate sized bodies in the solar system. In *Investigating the Universe*, ed. F. D. Kahn (Dordrecht: D. Reidel Publ. Co.), pp. 353–384.

Mendis, D. A., and Axford, W. I. 1974. Satellites and magnetospheres of the outer planets. *Ann. Rev. Earth Planet. Sci.* 2:419–474.

Mendis, D. A., Houpis, H. L. F., and Hill, J. R. 1982*a*. The gravito-electrodynamics of charged dust in planetary magnetospheres. *J. Geophys. Res.* 87:3449–3455.

Mendis, D. A., Houpis, H. L. F., and Hill, J. R. 1982*b*. Charged dust in Saturn's magnetosphere. *J. Geophys. Res.* 88:A929–A942 (*Suppl. Lunar Planet Sci. Conf.* XIII).

Meyer-Vernet, N. 1982. "Flip-flop" of electric potential of dust grains in space. *Astron. Astrophys.* 105:98–106.

Morfill, G. E. 1978. A review of selected topics in magnetosphere physics. *Rep. Progr. Phys.* 41:303.

Morfill, G. E. 1983*a*. Electromagnetic effects in planetary rings. *Advances Space Res.* 3:87–94.

Morfill, G. E. 1983*b*. The formation of spokes in Saturn's rings. Proceedings of *I.A.U. Colloqium 75 Planetary Rings,* ed. A. Brahic, Toulouse, France, Aug. 1982.

Morfill, G. E., and Grün, E. 1979*a*. The motion of charged dust particles in interplanetary space. I. The zodiacal cloud. *Planet. Space Sci.* 27:1269–1282.

Morfill, G. E., and Grün, E. 1979*b*. The motion of charged dust particles in interplanetary space. II. Interstellar grains. *Planet. Space Sci.* 27:1283–1292.

Morfill, G. E., and Goertz, C. K. 1983. Plasma clouds in Saturn's rings. *Icarus* 55-111–123.

Morfill, G. E., Grün, E., and Johnson, T. V. 1980*a*. Dust in Jupiter's magnetosphere: Physical processes. *Planet. Space Sci.* 28:1087–1100.

Morfill, G. E., Grün, E., and Johnson, T. V. 1980*b*. Dust in Jupiter's magnetosphere: Origin of the ring. *Planet. Space Sci.* 28:1101–1110.

Morfill, G. E., Grün, E., and Johnson, T. V. 1980*c*. Dust in Jupiter's magnetosphere: Time variations. *Planet. Space Sci.* 28: 1111–1114.

Morfill, G. E., Grün, and Johnson, T. V. 1980*d*. Dust in Jupiter's magnetosphere: Effect on magnetospheric electrons and ions. *Planet. Space Sci.* 28:1115–1123.

Morfill, G. E., Fechtig, H., Grün, E., and Goertz, C. K. 1983*a*. Some consequences of meteoroid impacts on Saturn's ring. *Icarus* 55:439–447.

Morfill, G. E., Grün, E., Johnson, T. V. 1983*b*. Saturn's E, G and F rings: Moduated by the plasma sheet. *J. Geophys. Res.* 88:5573–5579.

Morfill, G. E., Grün, E., Johnson, T. V., and Goertz, C. K. 1983c. On the evolution of Saturn's spokes: Theory. *Icarus* 53:230–235.

Müller, E. W. 1956. Field desorption. *Phys. Rev.* 102:618–624.

Mukai, T. 1980. Grain disruption by collisions with the solar energetic particles. In *Solid Particles in the Solar System,* eds. I. Halliday and B. MacIntosh (Dordrecht: D. Reidel Publ. Co.), pp. 385–389.

Northrop, T. G. 1963. *The Adiabatic Motion of Charged Particles* (New York: John Wiley and Sons).

Northrop, T. G., and Hill, J. R. 1983a. The adiabatic motion of charged dust grains in rotating magnetospheres. *J. Geophys. Res.* 88:1–11.

Northrop, T. G., and Hill, J. R. 1982. Stability of negatively charged dust grains in Saturn's ring plane. *J. Geophys. Res.* 87:6045–6051.

Northrop, T. G., and Hill, J. R. 1983. The inner edge of Saturn's B ring. *J. Geophys. Res.* 88:6102–6108.

Oechsner, H. 1975. Sputtering—A review of some recent experimental and theoretical aspects. *Appl. Phys.* 8:185–198.

Öpik, E. J. 1956. Interplanetary dust and terrestrial accretion of meteoritic matter. *Irish Astron. J.* 4:84–135.

Owen, T., Danielson, G. E., Cook, A. F., Hansen, C., Hall, V. L., and Duxbury, T. C. 1979. Jupiter's ring. *Nature* 281: 442–446.

Peale, S. J. 1966. Dust belt of the Earth. *J. Geophys. Res.* 71:911–933.

Pollack, J. B., Burns, J. A., and Tauber, M. E. 1979. Gas drag in primoridal circumplanetary envelopes: A mechanism for satellite capture. *Icarus* 37:587–611.

Porco, C. C., and Danielson, G. E. 1982. The periodic variation of spokes in Saturn's rings. *Astron. J.* 87:826–833.

Rennilson, J. J., and Criswell, D. R. 1974. Surveyor observations of lunar horizon-glow. *Moon* 10:121–142.

Rhee, J. W. 1967. Electrostatic potential of a cosmic dust particle. In *The Zodiacal Light and the Interplanetary Medium,* ed. J. C. Weinberg (Washington, D. C.: NASA SP-150), pp. 291–297.

Rhee, J. W. 1976. Electrostatic disruption of lunar dust particles. In *Interplanetary Dust and the Zodiacal Light,* eds. H. Elsässer and H. Fechtig (Berlin: Springer-Verlag), pp. 238–240.

Scarf, F. L., Gurnett, D. A., Kurth, W. S., and Poynter, R. L. 1982. Voyager plasma wave observations at Saturn. *Science* 215:287–294.

Schulz, M. 1975. Geomagnetically trapped radiation. *Space Sci. Rev.* 17:481–536.

Schulz, M., and Lanzerotti, L. J. 1974. *Particle Diffusion in the Radiation Belts* (New York: Springer-Verlag).

Scudder, J. D., Sittler, Jr., E. C., and Bridge, H. S. 1981. A survey of the plasma electron environment of Jupiter: A view from Voyager. *J. Geophys. Res.* 86:8157–8179.

Severny, A. B., Terez, E. I., and Zvereva, A. M. 1974. Preliminary results obtained with an astrophotometer installed on Lunokhod 2. *Space Research* XIV:603–605.

Shapiro, I. I., Lautman, D. A., and Colombo, G. 1966. The Earth's dust belt: Fact or fiction? 1. Forces perturbing dust particle motion. *J. Geophys. Res.* 71:5695–5704.

Showalter, M. R., and Burns, J. A. 1982. A numerical study of Saturn's F ring. *Icarus* 52:526–544.

Simpson, J. A., Bastian, T. S., Chenette, D. L., Lentz, G. A., McKibben, R. B., Pyle, K. R., and Tuzzolino, A. J. 1980. Saturnian trapped radiation and its absorption by satellites and rings: The first results from Pioneer 11. *Science* 207: 411–415.

Sittler, E. C., Scudder, J. D., and Bridge, H. S. 1981. Distribution of neutral gas and dust near Saturn. *Nature* 292:711–714.

Smith, B. A., Soderblom, L., Johnson, T. V., Ingersoll, A., Collins, S. A., Shoemaker, E. M., Hunt, G. E., Carr, M. H., Davies, M. E., Cook, II, A. F., Boyce, J., Danielson, G. E., Owen, T., Sagan, C., Beebe, R. F., Veverka, J., Strom, R. G., McCauley, J. F., Morrison, D., Briggs, G. A., and Suomi, V. E. 1979. The Jupiter system through the eyes of Voyager 1. *Science* 204:951–972.

Smith, B. A., Soderblom, L., Batson, R., Bridges, P., Inge, J., Masursky, H., Shoemaker, E., Beebe, R., Boyce, J., Briggs, G., Bunker, A., Collins, S. A., Hansen, C. J., Johnson,

T. V., Mitchell, J. L., Terrile, R. J., Cook, II, A. F., Cuzzi, J., Pollack, J. B., Danielson, G. E., Ingersoll, A., Davies, M. E., Hunt, G. E., Morrison, D., Owen, T., Sagan, C., Veverka, J., Strom, R., and Suomi, V. E. 1982. A new look at the Saturnian system: The Voyager 2 images. *Science* 215:504–537.

Smith, B. A., Soderblom, L., Beebe, R., Boyce, J., Briggs, G., Bunker, A., Collins, S. A., Hansen, C. J., Johnson, T. V., Mitchell, J. L., Terrile, R. J., Carr, M., Cook, II, A. F., Cuzzi, J., Pollack, J. B., Danielson, G. E., Ingersoll, A., Davies, M. E., Hunt, G. E., Masursky, H., Shoemaker, E., Morrison, D., Owen, T., Sagan, C., Veverka, J., Strom, R., and Suomi, V. E. 1981. Encounter with Saturn: Voyager 1 imaging science results. *Science* 212:163–191.

Smith, E. J., Davis, L., Jr., Jones, D. E., Coleman, P. J., Jr., Colburn, D. S., Dyal, P., Sonett, C. P., and Frandsen, A. M. A. 1974. The planetary magnetic field and magnetosphere of Jupiter: Pioneer 10. *J. Geophys. Res.* 79:3501–3513.

Smith, E. J., Davis, L., Jr., Jones, D. E., Coleman, P. J., Jr., Colburn, D. S., Dyal, P., and Sonett, C. P. 1980. Saturn's magnetic field and magnetosphere. *Science* 207:407–410.

Smoluchowski, R. 1980. Existence and role of amorphous grains in the solar system. In *Solid Particles in the Solar System*, eds I. Halliday and B. MacIntosh (Dordrecht: D. Reidel Publ. Co.), pp. 381–385.

Thomsen, M. F., Goertz, C. K., Northrop, T. G., and Hill, J. R. 1982. On the nature of particles in Saturn's spokes. *Geophys. Res. Letters* 9:423–426.

Van Allen, J. A. 1977. Distribution and dynamics of energetic particles in the Jovian magnetosphere. *Space Res.* 17:719–731.

Van Allen, J. A. 1982. Findings on rings and inner satellites of Saturn by Pioneer 11. *Icarus* 51:509–527.

Van Allen, J. A., Thomsen, M. F., Randall, B. A., Rairden, R. L., and Grosskreutz, C. L. 1980*a*. Saturn's magnetosphere, rings and inner satellites. *Science* 207:415–421.

Van Allen, J. A., Randall, B. A., and Thomsen, M. F. 1980*b*. Sources and sinks of energetic electrons and protons in Saturn's magnetosphere. *J. Geophys. Res.* 85:5679–5694.

Van Allen, J. A., Thomsen, M. F., and Randall, B. A. 1980*c*. The energetic charged particle absorption signature of Mimas. *J. Geophys. Res.* 85:5709–5718.

Vedder, J. F. 1963. Charging and acceleration of microparticles. *Rev. Sci. Instrum.* 34:1175–1183.

Vogt, R. E., Chenette, D. L., Cummings, A. C., Garrard, T. L., Stone, E. C., Schardt, A. W., Trainer, J. H., Lal, N., and McDonald, F. B. 1981. Energetic charged particles in Saturn's magnetosphere: Voyager 1 results. *Science* 212:231–234.

Völk, H. J., Jones, F. C., Morfill, G. E., and Röser, S. 1980. Collisions between grains in a turbulent gas. *Astron. Astrophys.* 85:316–325.

Warwick, J. W., Pearce, J. B., Evans, D. R., Carr, T. D., Schauble, J. J., Alexander, J. K., Kaiser, M. L., Desch, M. D., Pedersen, B. M., Lecacheux, A., Daigne, G., Boischot, A., and Barrow, C. H. 1981. Planetary radio astronomy observations from Voyager 1 near Saturn. *Science* 212:239–243.

Warwick, J. W., Evans, D. R., Romig, J. H., Alexander, J. K. Desch, M. D., Kaiser, M. L., Leblanc, Y., Lecacheux, A., and Pedersen, B. M. 1982. Planetary radio astronomy observations from Voyager 2 near Saturn. *Science* 215:582–587.

Wechsung, R. 1977. Industrial application of ion etching. Proc. 7th Intern. Vac. Congr. and 3rd Intern. Conf. *Solid Surfaces* pp. 1237–1244.

Wehner, G. K., Kenknight, C., and Rosenberg, D. L. 1963. Sputtering rates under solar wind bombarbment. *Planet. Space Sci.* 11:885–895.

Weinheimer, A. J., and Few, A. A. 1982. The spokes in Saturn's rings: A critical evaluation of possible electric processes. *Geophys. Res. Letters* 9:1139–1142.

Wyatt, S. P. 1969. The electrostatic charge of interplanetary grains. *Planet. Space Sci.* 17:155–171.

EFFECTS OF RADIATION FORCES ON DUST PARTICLES IN PLANETARY RINGS

F. MIGNARD

Centre d'Etudes et de Recherches
Géodynamiques et Astronomiques

The effects of radiation forces on a grain in planetary orbit are examined with particular emphasis on implications for planetary rings. The origin of radiation forces is briefly considered mainly to provide order of magnitude. The secular changes of orbital elements are derived and the effects of the planetary shadow assessed. We show that this shadow is of little importance in the dynamics of planetary rings, yielding but a small periodic oscillation in the semimajor axis. The periodical evolution of the other orbital elements is investigated, both for a spherical and for an oblate planet. We demonstrate the existence of a forced eccentricity that is strongly enhanced at a particular critical distance where resonance occurs. In the case of Saturn this critical distance is nearly coincident with the outer boundary of the E Ring, involving a depletion in very small particles.

Ever since Maxwell (1859) successfully demonstrated that the rings of Saturn must be made up of numerous independent bodies, dynamics of this cosmic jewel have attracted mathematicians and celestial mechanicians. How such a system might have evolved, and possibly may still be evolving is dependent upon the size of particles that make up the disk. During the 20th century, astronomers used propitious configurations (ring-plane crossings, occultations, well-opened rings) to learn more about particle properties, and bodies ranging from cm to km were reported. The two last decades have brought an impressive harvest of new observational materials thanks to both space exploration and the development of highly sensitive groundbased detectors. The classical three rings of Saturn have become a vast number of unnamed ringlets and Saturn has had to relinquish its privilege of being the only planet encircled with rings. Dynamics of planetary rings are governed by many factors and, to some extent each chapter of this book bears witness to the richness of the physics involved.

In the following pages we will discuss the possible role played by radiation forces on the tiny grains in certain rings. Are these forces more capable of shaping the rings than any other force? Can their dynamical consequences account for observational facts? To answer these questions we shall first examine the magnitude of radiation forces and the direction of their action. Next we will proceed with the secular change of the orbital elements which ultimately will inform us about the long-time stability of the swarms of small particles. Orbital evolution over intermediate timescales will be considered in connection with the planetary shadow and with the planet's oblateness.

More than an explanation of striking features observed in the ring system, this chapter aims at providing a framework for understanding some of the facts that can be ascribed to radiation forces. Indeed the subject pertains to more than just planetary rings even though nearly all numerical applications here specify ring parameters.

I. RADIATION FORCES IN GENERAL

A. Basic Physics

This section presents the basic ideas needed in order to work out the dynamical consequences of radiation forces. The goal of this section is not to give a comprehensive description of the interaction between grains and radiation, but only to outline the problem and to refer the reader to the original literature in this field for the details.

A particle of geometrical cross section A located at a point where the density flux (energy/area/time) of radiation is Φ would intercept an energy per unit time,

$$\Delta\Sigma = \Phi A \tag{1}$$

if it were a perfect absorber. At the same time a transfer of linear momentum occurs and the particle experiences a net force directed outward along the radiation source-particle line whose expression is given by

$$\mathbf{F} = \frac{\Phi A}{c}\,\mathbf{u}_r \tag{2}$$

where c is the speed of light and $\mathbf{u}_r$ is the unit vector radial to the source of the radiation. The absorbed energy heats the grain and if it is sufficiently small, it is subsequently isotropically reemitted bringing about no supplementary force. This simple picture must be refined to be more realistic. A particle is never a perfect absorber since diffraction implies the particle may interact with an electromagnetic wave in a region much larger than its geometrical cross section. Accordingly, the previous description must be amended and Eq. (2) becomes

$$\mathbf{F} = \frac{\Phi A}{c} Q_{pr} \mathbf{u}_r \tag{3}$$

where Q_{pr} is a dimensionless coefficient which can be evaluated for spherical and smooth particles with the help of Mie theory of light scattering (Fig. 1). Comprehensive discussion of the whole theory may be found in Born and Wolf (1975), McCartney (1976) and Van de Hulst (1981). Detailed computations and numerical results for materials of interest in the solar system have appeared in a series of papers by Lamy (1974, 1978) and Burns et al. (1979). The latter paper is recommended for its coverage of a broad range of dynamical effects involving radiation forces in planetary science. The knowledge of the scattering pattern has proven to be a powerful tool in determining the particle's size both in Jupiter's and Saturn's rings (see chapters by Cuzzi et al. and by Burns et al.). The physical origin of Q_{pr} is illustrated in Fig. 1.

B. Numerical Values of the Radiation Forces

When Q_{pr} is known from Mie scattering, the radiation force may be evaluated using Eq. (3). From the standpoint of dynamics it is preferable to use a new parameter instead of Q_{pr} that will immediately tell us how much the actual orbit can depart from the undisturbed orbit due to radiation forces.

Let F_G and F_{pr} respectively be the gravitational and the radiation force acting upon a grain in the solar environment. For a particle with radius r and density ρ and located at a distance R from the Sun we have

$$F_G = GM \frac{4}{3} \pi r^3 \rho / R^2 \tag{4}$$

$$F_{pr} = \frac{\Phi \pi r^2}{c} Q_{pr} = \frac{L}{4\pi R^2} \frac{\pi r^2}{c} Q_{pr} \tag{5}$$

where M and L are the solar mass and luminosity and G is the gravitational constant. Because both forces decrease as the inverse square of the distance to the Sun, their ratio depends only on the particle's properties. More precisely we have for this ratio

$$\beta = \frac{F_{pr}}{F_G} = \frac{3L}{16\pi GMc} \frac{Q_{pr}}{\rho r} \tag{6}$$

or numerically, $\beta = 0.6 \, Q_{pr}/\rho r$ with ρ in g cm^{-3} and r in μm. For the limiting case of geometrical optics $Q_{pr} = 1$, and by taking $\rho = 3$g cm^{-3} as an average value of density, $\beta = 0.2/r$. For micron-sized particles β is on the order of unity. If the geometrical optics were valid at this level, a particle

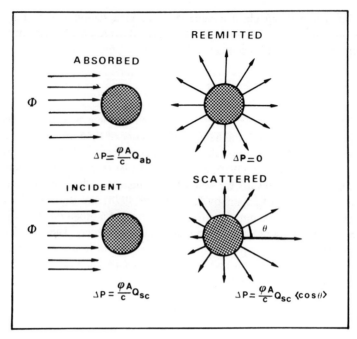

Fig. 1. Diagramatic illustration of the origin of the classical radiation force. The particle absorbs an amount of energy $\Phi A Q_{ab}$ while it scatters an energy $\Phi A Q_{sc}$. Each process involves an exchange of linear momentum ΔP. The absorbed energy is subsequently isotropically reemitted while scattered photons follow a well-defined pattern. The total budget of exchange of linear momentum leads to $Q_{pr} = Q_{ab} + Q_{sc} (1 - <\cos\theta>)$ and to Eq. (3) in the text.

smaller than 0.2 μm would be blown out of the solar system on an hyperbolic orbit. Such an escape constitutes the basis for the origin of the so-called β-meteoroids, first recognized by Berg and Grün (1973) from data collected by Pioneer probes and interpreted by Zook and Berg (1975).

Extensive numerical computations of β were carried out by Burns et al. (1979) and part of their results is reproduced in Fig. 2 for representative materials. It should be noted that all the curves have similar shapes and present a maximum in β for particle radius of $\sim$0.1 to 0.2 μm, with β close to unity. Geometrical optics is a fairly acceptable approximation for a particle whose radius is $>$0.5 μm, while it becomes meaningless for very small particles since β drops sharply when $r < 0.1$ μm. In Fig. 1 the ideal material is one that absorbs all radiation when $2\pi r/\lambda < 1$ (λ is wavelength) and is totally transparent elsewhere. For subsequent computations the reader should bear in mind that β is on the order of unity for micron-sized particles, i.e. the magnitude of the gravitational attraction due to the Sun is comparable to that of the radiation force.

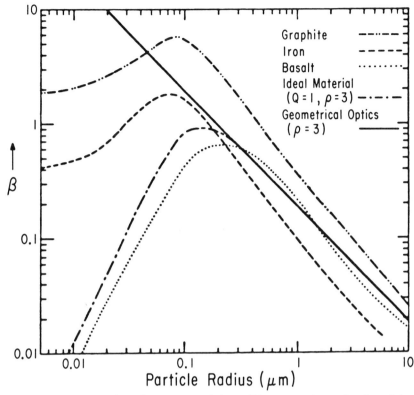

Fig. 2. A log-log plot of the β parameter (relative radition pressure) as a function of the particle size for three cosmically abundant materials and two comparison standards (from Burns et al. 1979).

In order to deal with tractable computations and to work with a preexisting physical model of interaction between matter and radiation Burns et al. were obliged to consider perfectly smooth spherical particles, which might raise doubts about the usefulness of their results in the real world. However, we have good reasons to rely on the numerical values they obtained. First, in a recent investigation, Mukai et al. (1981) have tackled the problem of light scattering by large (relative to the wavelength of the light) and rough spherical particles. Multiple reflections of the incident light yield a diffuse scattering but the whole pattern does not differ significantly from the one resulting from Mie scattering. Second, the continual jostling that causes a particle to be eroded is likely to maintain spherical shapes preferentially to other forms.

C. The Poynting-Robertson Component

The model of radiation forces developed above is not subtle enough to provide a complete picture of the interaction between light and matter. Some eighty years ago Poynting (1903) made public the results of his investigation about the consequences of the continuous absorption and reemission of solar radiation by particles, and claimed that a tangential drag should exist in addition to the well known radial radiation force (Eq. 2). Although his analysis was plagued by some shortcomings, in general he was right. The first correct analysis was performed by Robertson (1937) and generalized to extended sources by Guess (1962). The problem has been reexamined and updated by Burns et al. (1979) who pointed out the essentially non-relativistic character of the PR (Poynting-Robertson) drag, which had been overlooked for years.

The complete analysis is beyond the scope of this review; we restrict ourselves to a pictorial view displayed in Fig. 3. The reemission of the absorbed light is not isotropic when viewed in the solar frame. Accordingly, there exists a net tangential component in the radiation force and Eq. (1) must be rewritten as follows:

$$\mathbf{F} = \frac{\Phi A}{c} \, Q_{pr} \left[\left(1 - \frac{\dot{r}}{c} \right) \mathbf{u}_r - \frac{\mathbf{V}}{c} \right] \tag{7}$$

where $\dot{r} = dr/dt$ and $\mathbf{V}$ is the particle's velocity vector. Sometimes it is more convenient to express Eq. (7) in a slightly different form, namely

$$\mathbf{F} = \frac{\Phi A}{c} \, Q_{pr} \left[\left(1 - \frac{2\dot{r}}{c} \right) \mathbf{u}_r - \frac{r\dot{\theta}}{c} \mathbf{u}_\theta \right] \tag{8}$$

where $\mathbf{u}_\theta$ is the orthoradial unit vector. The term with $\mathbf{u}_r$ as a factor represents the corrected radial radiation force, while the second term may be named the drag component, So far, the terminology concerning radiation force has not been settled and in this chapter we shall call classical radiation force the part in Eqs. (7) and (8) that is of order zero in v/c and the remainder will be termed the PR force or PR drag even though it contains a radial component. It should now be apparent why the subscript "pr" was used in Sec. I. A.

D. Rings and Radiation Forces

The various points presented in Sec. A allow us to determine in which rings radiation forces are likely to play major or minor roles, or to have no role at all. First and foremost, some rings may be totally devoid of small particles; belonging to this group are the C Ring and the Cassini Division. On the other hand, micron-sized particles are present within the B Ring along with a population of larger bodies. Electrostatic forces are apparently responsible for the behavior of the small particles, as far as dynamics are concerned (Porco and Danielson 1983).

Photometric investigations show that a large fraction of the F Ring material is submicron-sized (Pollack 1981). But in the case of these ringlets, the dynamics is undoubtly governed by a shepherding mechanism generated by S27 and S26 (Showalter and Burns 1982; chapter by Cuzzi et al.; chapter by

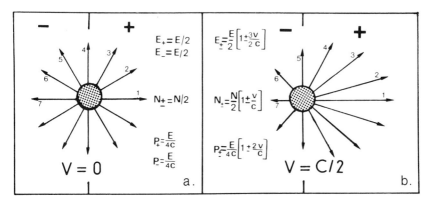

Fig. 3. The origin of the PR drag. The same radiation pattern is observed in the particle frame (a) and in a moving frame (b). *N, P, E* denote, respectively, the flow of photons, linear momentum and energy crossing a unit surface per unit time in the forward direction (+) or in the backward direction (−). In (a) the radiation being isotropic, the + and − flows are equal. A relativistic transform has been used to construct (b). The total budget shows that more linear momentum flows in the forward direction than in the backward. Accordingly, the particle will experience a drag force. Note that the energy and linear momentum flow are affected both because more numerous photons are emitted in the forward direction and because these photons are likewise the most energetic.

Dermott). Likewise, the Jovian rings, considered as a monodispersion, comprise particles of 3 to 5 μm in diameter (Jewitt 1982). Here the short lifetime due to sputtering and also the plasma drag prevent the radiation forces from shaping the ring. In addition, the ring and halo particles orbit the planet deep in its magnetosphere and become charged. Accordingly, electromagnetic forces overwhelm the radiation forces as demonstrated by Morfill and Grün (1980) and Burns et al. (1980). See also chapters by Grün et al. and by Burns et al. in this book.

The two remaining rings, the E Ring and the external extension of the G Ring, are worthy of paying more attention to in connection with radiation forces. The existence of an outermost ring of Saturn was first reported by Feibelman (1967) during the passage of the Earth in the ring plane in 1966. However, the reality of such a tenuous ring remained doubtful for more than a decade and was not confirmed until the recent ring plane crossing of 1979–1980. Photographic observations were made by Lamy and Mauron (1981) and Dollfus and Brunier (1982). More accurate data resulted from CCD observations made by Baum et al. (1981) and yielded edge-on profiles. Also images were obtained by Voyager 2 in forward scattering geometry at low elevation angles (Smith et al. 1982). Nearly all observations agree with a pronounced radial concentration of material at ~4 Saturn radii (R_S), in the neighborhood of the orbit of Enceladus. The inner edge of the ring is located at 3 R_S and the ring peaks abruptly at 4 R_S, then drops sharply to 5 R_S and spreads gradually to ~8 R_S. The spectral reflectance is accounted for if there are numerous scattering particles of some μm in diameter (Baum et al. 1981; Dollfus and Brunier 1982). The very small optical depth, $\tau \sim 10^{-6} - 10^{-7}$, shows that the ring surface is sparsely populated and that collisions are infrequent with typically $t_{col} = 10^4$ yr.

More recently, a concentration of small particles slightly outside the G Ring was detected during the Voyager 2 encounter (Gurnett et al. 1983), but the significance of this was not immediately recognized and actually came out as a by-product of the plasma wave experiment, during which an impulsive noise was recorded as the spacecraft crossed the ring plane. The detection took place at the unique distance of 2.88 R_S whereas the radius of the G Ring is 2.8 R_S (Smith et al. 1982). By modeling the noise production, a size and mass distribution is inferred and most of the particles that make up this concentration have a radius in the range of 0.3 to 3 μm. The vertical concentration decreases gradually with a characteristic thickness of 100 km. So far there is no groundbased observation of either this new ring or an outer G Ring appendage, and we do not know if there is a continuous, but tenuous, distribution of matter up to the E Ring. However, from low elevation-angle observations by Voyager, it can be stated that no ring of optical depth comparable to that of the E Ring lies farther out in the Saturn system (Smith et al. 1982).

Nothing is known about the size of the bodies that populate the ringlets of Uranus. However, some observational evidence suggests that, even if micron-sized particles are present, radiation forces are not a basic physical process in this system. The ring boundaries are sharply defined whereas interparticle collisions and radiation forces would produce diffuse boundaries. In addition, eccentric ringlets precess as rigid bodies (Nicholson et al. 1981; Elliot et al. 1981; chapter by Elliot and Nicholson), which involves either the action of unseen satellites or a compensation of the differential precession by the self-gravitation of the matter making up the rings (see chapter by Dermott). Whatever the actual processes may be, it is reasonable to rule out radiation forces as a possible explanation.

II. DUST IN PLANETARY ORBITS

A. Secular Changes of the Orbital Elements

Instantaneous Rate of Change. In addition to the Newtonian attraction of the planet, a grain experiences a supplementary force caused by solar radiation. According to its magnitude, this force will modify more or less the particle's orbital parameters. For a particle embedded within a ring, it is desirable to determine whether or not the radiation force may cause that particle to drift far from the place it was released. This drift can be either a secular change, which would alter to a large extent the distribution or only a periodical change, which would involve an excursion in the semimajor axis,

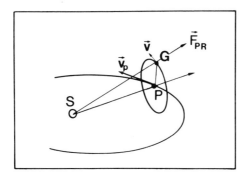

Fig. 4. A dust particle G orbits a planet P and is acted upon by the radiation coming from the Sun S. $\mathbf{V}_p$ denotes the planet's velocity with respect to the Sun while $\mathbf{V}$ is the particle's orbital velocity with respect to the planet. The figure is not to scale since $SP >> SG$. In the test vector $\mathbf{SP}$ is denoted by $\mathbf{r}_{SP}$.

the eccentricity or the inclination. In turn, such modifications would drastically affect the rate of collision and finally the ring structure.

The secular change of the orbit of a particle in motion around a planet stems from the energy removal by the Poynting-Robertson drag. Actually it is not obvious that an energy loss due to a drag force as viewed from the Sun likewise leads, on the average, to an orbital energy loss when measured from the planet's own frame. In order to derive analytical expressions for the different rate of change, we make the following assumptions:

1.   In any planetocentric motion we neglect the small variation of the distance between the Sun and the particle. Hence the solar flux is constant and equal to its value at the planet's level;
2.   The planet's orbit is assumed to be circular;
3.   No allowance is made for the planetary shadow, in a first approximation. This assumption will be reexamined later.

A vectorial expression of the radiation force was given in Eq. (7). By inserting the radius vector of the planet and that of the grain (Fig. 4) and by taking into account that the velocity $\mathbf{V}$ in Eq. (7) is relative to the Sun, we have

$$\mathbf{F} = \frac{\Phi A}{c} \; Q_{pr} \left\{ \left[1 - \frac{\mathbf{r}_{SP}}{r_{SP}} \left(\frac{\mathbf{V}_p}{c} + \frac{\mathbf{V}}{c} \right) \right] \frac{\mathbf{r}_{SP}}{r_{SP}} - \frac{\mathbf{V}_p + \mathbf{V}}{c} \right\} \qquad (9)$$

where $\mathbf{V}_p$ is the planet's orbital velocity and $\mathbf{V}$ is the velocity of the particle with respect to the planet, and $\mathbf{r}_{SP}$ is the sun-planet vector.

Let us consider a reference frame with its origin at the planet's center of mass and whose x-axis is directed toward the Sun. In this frame we can describe the particle's orbit with the six standard Keplerian elements, a, e, i,

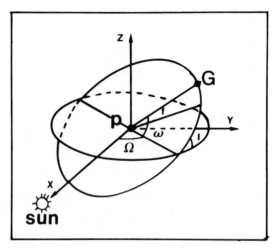

Fig. 5. The rotating coordinate system with its origin at the planet's center of mass. Ω is the
longitude of the node measured from a moving axis; the pericenter is located by $\tilde{\omega}$, and f
represents the particle's true anomaly.

Ω, ω, M. Since the Sun is moving, the longitude of the node is referred to a
moving axis (Fig. 5). The three components of the force in Eq. (9) are readily
obtained in our reference frame with the following notation

$$
\begin{aligned}
\mathbf{PG} = \mathbf{r} &= (x,y,z) \\
\mathbf{V} &= (\dot{x},\dot{y},\dot{z}) \\
\mathbf{V}_p &= (0,V_p,0) \\
\frac{\mathbf{r}_{SP}}{r_{SP}} &= (-1,0,0)
\end{aligned}
$$

$$
\mathbf{F} = -\frac{\Phi A}{c} \, Q_{pr} \left(1 + \frac{2\dot{x}}{c}, \frac{V_p}{c} + \frac{\dot{y}}{c}, \frac{\dot{z}}{c} \right) \, . \tag{10}
$$

In the right-hand side of Eq. (10) the term of zero order in v/c corresponds to
the classical radiation force and all the other terms to the PR components.
Finally the rate of change or orbital energy and angular momentum are given by

$$
\frac{dE}{dt} = \frac{\mathbf{F} \cdot \mathbf{V}}{m} \tag{11}
$$

$$
\frac{1}{2} \frac{dH^2}{dt} = \frac{(\mathbf{r} \times \mathbf{F}) \cdot \mathbf{H}}{m} \tag{12}
$$

where E denotes the orbital energy measured in the frame moving with the
planet and H the corresponding angular momentum, $\mathbf{H} \equiv \mathbf{r} \times \mathbf{V}$. Eqs. (11) and
(12) will eventually yield the rate of change of the semimajor axis a and of
the eccentricity e through the following equations of Keplerian motion

$$E = -\mu/2a \qquad (13)$$

$$H^2 = \mu a(1-e^2) \qquad (14)$$

with $\mu = GM_p$, as the product of the gravitational constant by the planet's mass.

Let us introduce a characteristic time t_{pr}, during which a particle intercepts from the radiation flux an energy equal to its own rest energy (Burns et al. 1979):

$$t_{pr} = \frac{mc^2}{3\Phi AQ_{pr}} \, . \qquad (15)$$

From Eqs. (10), (11) and (12) we have the instanteous rates of change

$$\frac{dE}{dt} = -\frac{c}{3t_{pr}}\left[\dot{x} + V_p\dot{y}/c + \dot{x}^2/c + V^2/c\right] \qquad (16)$$

$$\frac{dH}{dt} = -\frac{2}{3t_{pr}}\left[H^2 + H_zV_px - H_zy(c+\dot{x}) + H_yz(c+\dot{x}) - H_xV_pz\right]. \qquad (17)$$

Averaged Rate of Change. To obtain the secular changes from the two last equations we must average them first, over the orbital period, and second, over the precessional period of the node and of the pericenter. The order in which these averages are performed is irrelevant and will be determined from practical considerations. It is reasonable to assume that the two precessional motions are independent of each other and that they take place at a uniform rate. Over the orbital period we have, $<x> = <y> = <z> = <\dot{x}> = <\dot{y}> = <\dot{z}> = 0$. The notation $<>$ denotes the averaging over either the short period or long period. The nonvanishing contributions to the averages can be computed with the help of the following classical expressions (Brouwer and Clemence 1961) in the two-body problem

$$\begin{bmatrix} x \\ y \\ z \end{bmatrix} = \begin{bmatrix} P_x & Q_x \\ P_y & Q_y \\ P_z & Q_z \end{bmatrix} \begin{bmatrix} r\cos f \\ r\sin f \end{bmatrix} \qquad (18)$$

$$\begin{bmatrix} \dot{x} \\ \dot{y} \\ \dot{z} \end{bmatrix} = \begin{bmatrix} -P_x & Q_x \\ -P_y & Q_y \\ -P_z & Q_z \end{bmatrix} \begin{bmatrix} \dfrac{na^2\sin u}{r} \\ \dfrac{na^2(1-e^2)^{\frac{1}{2}}\cos u}{r} \end{bmatrix} \qquad (19)$$

where f and u are respectively the true and the eccentric anomaly, the mean motion n is determined by $n^2a^3 = \mu$, and the P's and Q's are given by

$$
\begin{aligned}
P_x &= & \cos\Omega\cos\omega & & -\cos i\,\sin\Omega\sin\omega \\
P_y &= & \sin\Omega\cos\omega & & +\cos i\,\cos\Omega\sin\omega \\
P_z &= & & & \sin i\,\sin\omega
\end{aligned} \tag{20}
$$

$$
\begin{aligned}
Q_x &= & -\cos\Omega\sin\omega & & -\cos i\,\sin\Omega\cos\omega \\
Q_y &= & -\sin\Omega\sin\omega & & +\cos i\,\cos\Omega\cos\omega \\
Q_z &= & & & \sin i\,\cos\omega
\end{aligned} \tag{21}
$$

Here ω and Ω are the argument of pericenter and longitude of the node. Then it follows that

$$
<\dot{x}^2> = \left\langle \frac{n^2a^4}{r^2} \left[P_x^2 \sin^2 u + Q_x^2(1-e^2) \right. \right.
$$

$$
\left. \left. \cos^2 u - P_x\,Q_x\,(1-e^2)^{\frac{1}{2}}\sin 2\,u \right] \right\rangle . \tag{22}
$$

This expression is easily averaged over the two precessional motions and leads to

$$
<\dot{x}^2> = < \frac{n^2a^4}{r^2}\,(\sin^2 u + (1-e^2)\cos^2 u)> \quad \frac{1+\cos^2 i}{4} \tag{23}
$$

and finally with $<V^2> = n^2a^2$ and with the average of Eq. (23) over the orbital period, the general result is

$$
\frac{dE}{dt} = -\frac{n^2a^2}{3t_{pr}}\,\frac{5+\cos^2 i}{4}. \tag{24}
$$

In Eq. (24) we have omitted the averaging brackets although a mean rate of change is understood.

For the averaging of Eq. (17) we need only to evaluate the nonzero terms, $<H_z\,y\dot{x}>$ and $<H_y\,z\dot{x}>$, and we have $<H_y\,z\dot{x}> = -Hna^2<\sin i\,\cos\Omega$ $[-P_z P_x\cos f\sin u + P_z Q_x(1-e^2)^{\frac{1}{2}}\cos f\cos u - Q_z P_x\sin f\sin u + Q_z Q_x$ $(1-e^2)^{\frac{1}{2}}\sin f\sin u]>$ and after the averages over the precessional motions are performed

$$
<H_y z\dot{x}> = Hna^2\,\frac{\sin^2 i}{4}\,<(1-e^2)^{\frac{1}{2}}\cos f\cos u + \sin f\sin u> \tag{25}
$$

and finally

$$
<H_y z\dot{x}> = Hna^2(1-e^2)^{\frac{1}{2}}\,\frac{\sin^2 i}{4}. \tag{26}
$$

Similarly we find

$$
<H_z y\dot{x}> = Hna^2(1-e^2)^{\frac{1}{2}}\,\frac{\cos^2 i}{2}. \tag{27}
$$

A straightforward combination of Eqs. (14), (17), (26) and (27) yields the time variation of the angular momentum

$$\frac{dH}{dt} = - \frac{n\,a^2(1-e^2)^{\frac{1}{2}}}{3t_{pr}} \frac{(1 + 5\cos^2 i)}{4} . \tag{28}$$

Time Evolution of the Semimajor Axis and of the Eccentricity. From Eqs. (13) and (14) the secular rate of change of the semimajor axis a and eccentricity e are

$$\frac{da}{dt} = - \frac{a}{t_{pr}} \frac{5 + \cos^2 i}{6} \tag{29}$$

and

$$\frac{de}{dt} = 0 . \tag{30}$$

The orbital eccentricity is secularly stable whereas the semimajor axis exhibits a secular decrease. In Eq. (29) t_{pr} appears as a characteristic decay time and can be expressed in terms of the coefficient β (Eqs. 6–15)

$$t_{pr} = \frac{1}{3\beta} \frac{a_p}{c} \frac{a_p}{GM/c^2} \tag{31}$$

or numerically, $t_{pr} = 530\,a_p^2/\beta$ in yr, where a_p is the Sun-planet distance expressed in astronomical units. This collapse time amounts to $5 \times 10^4/\beta$ yr for Saturn and to $10^4/\beta$ yr for Jupiter. Thus, even in planetary motion the PR drag causes the particle's orbit to shrink, but in a different way than for a heliocentric orbit. In the latter case the result is available in various forms in Robertson (1937), Wyatt and Whipple (1950), Burns et al. (1979), and Mignard (1983). For a circular orbit starting at $a = a_0$, it reduces to

$$\frac{da}{dt} = - \frac{2}{3t_{pr}} \frac{a_0^2}{a} \tag{32}$$

where t_{pr} is given by Eq. (31) by substituting a_0 for a_p. So, for spiraling motion about the Sun the shrinking speeds up as the particle comes close to the Sun, while the largest rate of variation of the semimajor axis is obtained far from the planet in the planetocentric case. As a consequence, t_{pr} in Eq. (32) is, to within an order of magnitude, the time of fall. In addition, the rate of change of a is independent of the eccentricity in Eq. (29) for planetocentric orbits, while an exact version of Eq. (32), valid for eccentric heliocentric orbits, shows that the time of fall becomes shorter as the initial eccentricity grows (Mignard 1983).

The impossibility of maintaining small particles in planetary orbit when coupled with the collisional spreading that occurs for larger particles (Brahic 1977), has prompted Goldreich and Tremaine (1979) to suggest that the narrow ringlets of Uranus must be either confined or very young structures.

Accordingly, several models have originated to constrain ringlets within a very narrow region (see chapter by Dermott).

Equation (29) allows us to draw some conclusions related to the equilibrium distribution of a stream of particles. Since the rate of change of a decreases with the distance to the planet, particles released at a constant rate from a source located at a distance $a_\circ$ will accumulate in the inner region. More quantitatively, for a disk-shaped stream, the radial number density $n(a)$ should be given by

$$n(a)a \, da/dt = \text{const.} \tag{33}$$

which expresses the constancy of the flux of particles in a sourceless region. Hence with Eq. (29), $n(a) \sim 1/a^2$.

The same reasoning holds for polydispersions, i.e., a range of particle sizes, provided that we consider the number density $n(r,a)$ in the vicinity of the semimajor axis a and in the range of the particle radius dr. In such a case, if we take $P(r)dr$ for the rate of creation of particles at a distance $a_\circ$ in the range dr, we find

$$n(r,a) \sim P(r)/a^2\beta(r) \tag{34}$$

and the optical depth at the same distance should be

$$\tau(a) \sim \int \frac{P(r) \, \pi r^2}{a^2 \, \beta(r)} \, dr \sim 1/a^2 . \tag{35}$$

Nothing that would resemble such a radial behaviour is observed in ring systems even in the inner part of the E Ring. However, the model is very crude because it assumes a conservation of the number of particles. If we keep within the approximation of the geometrical optics (Sec. I) $\beta \sim 1/r$ and if we assume that $P(r) \sim r^{-q}$, the smallest particles would be primarily responsible for the optical depth for $q > 3$; on the other hand, the largest particles would give the major contribution, if $q < 3$. However, as the small particles proceed inward they are progressively eroded and tend to disappear. Hence the right-hand side of Eq. (33) is no longer constant and the contribution of the small grains to the optical depth will not be as large as expected. Accordingly, any birth and death process within the swarm of particles requires that the radial dependence given by Eq. (35) be reconsidered.

Time Evolution of the Orbital Inclination. The mean rate of change of the orbital inclination can be obtained by inserting the component of the radiation force (Eq. 10) normal to the orbital plane in the perturbation equation (Brouwer and Clemence 1961). This procedure gives rise to rather lengthy calculations that we will not give in detail. Starting with the instantaneous rate of change

$$\frac{di}{dt} = -\frac{1}{3t_{pr}na(1-e^2)^{\frac{1}{2}}} \frac{r}{a} \cos(\omega+f)$$

$$\left[(c+2\dot{x})\sin\Omega\sin i - (V+\dot{y})\cos\Omega\sin i + \dot{z}\cos i\right] . \tag{36}$$

Here f is the true anomaly. Surprisingly, this equation when averaged over the orbital period and over the pericenter precession yields a very simple result that does not depend upon the eccentricity

$$\frac{di}{dt} = \frac{1}{6t_{pr}} \sin i \cos i \sin^2\Omega . \tag{37}$$

To derive this result, we have used the following nonvanishing mean values

$$
\begin{aligned}
&<\cos f> &&= -e \\
&<\cos u \cos f> &&= 1/2 \\
&<\sin u \sin f> &&= (1-e^2)^{\frac{1}{2}}/2 \\
&<\frac{r}{a}\cos f> &&= -(3/2)e .
\end{aligned}
\tag{38}
$$

In Eq. (37) i represents the inclination of the particle's orbital plane to the planet's orbital plane. Thus, for a ring particle lying close to the planet's equator, this inclination is on the order of the planet's obliquity, i.e. 3° for Jupiter and 27° for Saturn. The product $\sin i \cos i$ is 0.05 for Jupiter and 0.4 for Saturn. Consequently, the PR drag tends to generate an out-of-plane motion so that for a collisionless stream of particles orbiting a spherical planet, an equilibrium motion could occur in a plane perpendicular to the planet's orbital plane. In the actual situation, the orbital precession caused by the oblateness of the planet coupled with a collision regime forces the disk to spread in the planet's equator. The competition with the rate of change Eq. (37) must produce a certain thickness of the disk.

B. Dynamical Consequences of the Planetary Shadow

Overview of the Problem. In the early sixties scientists suddenly realized the importance of sunlight on the motion of an artificial satellite. More precisely this fact was particularly striking with the big balloon Echo 1, 30 m in diameter with a mass of 70 kg or $\beta = 0.01$. Echo 1 was a genuine β-particle although far from being micro-sized. At the same time it became clear that any prediction of the motion would be meaningless if the shadow were neglected. Even though the requirements for satellite prediction are very different from our present purpose, it is worth ascertaining whether or not the planetary shadow could change the conclusions we have drawn in the preceding section.

In their review Burns et al. (1979) rely on numerical integrations performed in connection with artificial bodies to discard the effect of the motion through the shadow on the orbit of small particles. Since ring particles lie primarily in the planet's equator and their orbits do not depart significantly from circular, it is possible to allow for this assumption on a more quantitative basis.

A particle spends a fraction of each revolution in the shadow cast by the planet; during this time the action of the radiation forces vanishes. For example, a particle on a circular orbit lying in the same plane as the planet's orbital plane passes the fraction

$$f = \frac{1}{\pi}\sin^{-1}(a/R) \tag{39}$$

of its orbital period in the shadow. In Eq. (39) R is the planet's equatorial radius. With $a/R = 2$ or 5 the Sun is occulted, respectively, during 17% and 6% of the time. We have established that as long as the PR forces are left out, the radiation force never causes secular change in the orbital elements. In other words, the energy loss and energy gain during the motion balance each other and average to zero. Obviously the result will remain true for a circular orbit even with the introduction of the shadow. In the case of an elliptic orbit, the symmetry will be broken and the balance of energy will not hold over one orbital period.

Periodical Change of the Semimajor Axis. Let us consider the net variation of orbital energy for a weakly elliptic orbit due to the classical radiation force. From Eq. (16) we have

$$\frac{dE}{dt} = -\frac{c}{3t_{pr}}\dot{x}\ . \tag{40}$$

The particle enters the shadow at point P_e and leaves it at point P_o (Fig. 6). Over one orbit the net exchange of energy amounts to

$$\Delta E = -\frac{c}{3t_{pr}}\int_{t_o}^{t_e}\dot{x}\ dt = -\frac{c}{3t_{pr}}(x_e - x_o) \tag{41}$$

where x_e and x_o are the x-coordinates of P_e and P_o. Then, there is a variation of orbital energy whenever the distance to the planet is different at inbound and the outbound. This corresponds to the change in potential energy according to the distance to the Sun. The net balance is zero for circular orbits and also for orbits when the pericenter longitude is 0 or π. Let f_e and f_o respectively be the true anomaly at P_e and P_o, and r_e and r_o be the corresponding distance to the planet's center. Hence, (Fig. 6)

$$x_e = r_e \cos(\omega + f_e) \tag{42}$$

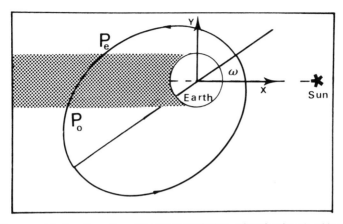

Fig. 6. The geometrical configuration of the planetary shadow, when the planet's orbit and the particle's orbit lie in the same plane. The particle enters the shadow at P_e and leaves it at P_o.

$$x_0 = r_0 \cos (\omega + f_0) \tag{43}$$

and for a small eccentricity, these expressions can be expanded to the first order in eccentricity, which yields

$$x_e - x_0 = a \cos (\omega + M_e) - a \cos (\omega + M_0) + a\, e/2 \left[\cos(\omega + 2M_e) - \cos (\omega + 2M_0) \right] \tag{44}$$

where M_e and M_0 are the mean anomalies. Inbound and outbound true and mean anomalies are determined by solving the two equations

$$r_e \sin(\omega + f_e) = R \tag{45}$$
$$r_0 \sin(\omega + f_0) = -R \tag{46}$$

which to the order e lead to

$$\sin (\omega + M_e) = \frac{R}{a} + e \left\{ \left[2 - \left(\frac{R}{a} \right)^2 \right] \right.$$
$$\left. \sin \omega - \frac{R}{a} \left[1 - \left(\frac{R}{a} \right)^2 \right]^{\frac{1}{2}} \cos \omega \right\} \quad , \tag{47}$$

$$\sin (\omega + M_0) = -\frac{R}{a} + e \left\{ \left[2 - \left(\frac{R}{a} \right)^2 \right] \right.$$
$$\left. \sin \omega + \frac{R}{a} \left[1 - \left(\frac{R}{a} \right)^2 \right]^{\frac{1}{2}} \cos \omega \right\} \quad , \tag{48}$$

and after some manipulations, we can compute to $O(e^2)$, $\cos(\omega + M_{e,0})$, to $O(e) \sin 2(\omega + M_{e,0})$ and $\cos 2(\omega + M_{e,0})$. By inserting the result in Eqs. (44) and (41) we obtain

$$\Delta E = -\frac{2ce}{3t_{pr}} \frac{R}{[1-(R/a)^2]^{\frac{1}{2}}} \sin \omega \tag{49}$$

and finally from the expression of E in terms of the semimajor axis, the mean rate of variation of a

$$\frac{da}{dt} = -\frac{2}{3\pi t_{pr}} \frac{e}{V_g} \frac{c}{V_g} \frac{R(a/R)^{\frac{1}{2}}}{[1-(R/a)^2]^{\frac{1}{2}}} \sin \omega \tag{50}$$

where V_g is the orbital speed of a grazing satellite

$$V_g = (GM/R)^{\frac{1}{2}} = \begin{cases} 25 \text{ km s}^{-1} \text{ for Saturn} \\ 43 \text{ km s}^{-1} \text{ for Jupiter.} \end{cases} \tag{51}$$

A more accurate calculation by Wyatt (1963) for $\omega = \pi/2$ is in agreement with the present calculation. It is surprising that, except for the slowly varying term $[1-(R/a)^2]^{\frac{1}{2}}$, the net budget of energy exchange does not depend on the size of the orbit, nor does there appear in Eq. (49) to be a way to recover the result $\Delta E = 0$ if the shadow is left out. From Eq. (49) we see that there is a loss of energy if ω is in the range $(0, \pi)$ and a gain otherwise. As expected, the maximum exchange occurs when the asymmetry is the most pronounced at $\omega = \pi/2$ and $3\pi/2$.

Implications for the Particles within the Rings. In the region of interest we take $[1-(R/a)^2]^{\frac{1}{2}} = 1$, which is an acceptable assumption even for $a/R \sim 2$. Within some planetary radii the pericenter precession is governed by the planet's oblateness and is uniform. The reader should remember that ω in Eq. (50) is actually measured from an axis directed towards the Sun. Thus, as long as the precessional rate due to the planet's flattening is larger than the mean motion of the Sun we can write

$$\frac{d\omega}{dt} = \frac{3}{2} J_2 \left(\frac{R}{a}\right)^2 n \tag{52}$$

where J_2 is the planet's quadrupole moment and n the particle's mean motion. Equation (52) will be valid in the inner region of the planet's environment; a more precise limit will be derived below.

Imagine a particle starting with $\omega = \pi$. The rate of variation of the semimajor axis would be positive from Eq. (50) as long as ω is between π and 2π. Therefore the semimajor axis would be increasing for half of the precessional period. At the same time, the rate of precession (Eq. 52) would decrease steadily and finally $\sin \omega$ would remain positive for more than half of the initial precessional period. Therefore, a would vary more than expected at first and, according to the value of e/t_{pr} in Eq. (50), we could obtain a large excursion in the semimajor axis caused by a sort of runaway amplification.

In order to show this runaway analytically, let us write Eq. (52) as follows:

$$\frac{d\omega}{dt} = \frac{2\pi}{t_{\omega g}} \left[\frac{R}{a} \right]^{\frac{7}{2}} \tag{53}$$

where $t_{\omega g}$ is the period of precession of the pericenter of a grazing satellite given by

$$\frac{2\pi}{t_{\omega g}} = \frac{3}{2} J_2 \frac{V_g}{R} \; . \tag{54}$$

For Saturn $t_{\omega g} = 7.2$ days; for Jupiter $t_{\omega g} = 5.4$ days. The differential system (Eqs. 50-53) possesses a first integral given by

$$\left[\frac{R}{a} \right]^3 = \left[\frac{R}{a_0} \right]^3 - \frac{e}{\pi^2} \frac{c}{V_g} \frac{t_{\omega g}}{t_{pr}} \cos\omega. \tag{55}$$

The initial condition is such that $a = a_0$ when $\omega = \pi/2$. Provided that the coefficient of $\cos\omega$ in Eq. (55) is large enough, this solution shows up the runaway behavior we have just described. Let K be this coefficient. If $K > (R/a_0)^3$, there exists ω such that a could approach infinity. A numerical evaluation of K proves that for Saturn and Jupiter the value of a_0 would have to be far beyond the external limits of the ring systems. Thus, the amplitude of oscillation in the semimajor axis remains very small in the actual system. Consequently, a linearized expression of Eq. (55) is good enough for our purpose:

$$\frac{a}{R} = \frac{a_0}{R} + \frac{e}{3\pi^2} \left(\frac{a_0}{R} \right)^4 \frac{t_{\omega g}}{t_{pr}} \frac{c}{V_g} \cos\omega \tag{56}$$

or numerically,

$$\frac{a}{R} = \frac{a_0}{R} + \begin{pmatrix} 2.4 \times 10^{-4} \text{ for Jupiter} \\ \text{or} \\ 1.5 \times 10^{-4} \text{ for Saturn} \end{pmatrix} e \, \beta \, (a_0/R)^4 \cos\omega \; . \tag{56a}$$

Thus, the radial excursion during a precessional period is hardly noticeable even in the outer regions of the E Ring, where for the very extreme conditions, $e = 0.1$, $\beta = 1$ and $a_0/R = 8$, the amplitude of variation of the semimajor axis amounts to 3700 km.

In Eq. (52) we have not taken into account the fact that the pericenter longitude is actually measured from a moving axis, although such an inclusion could be worked out easily. The more general solution risks not being very useful because in the resonance zone the orbital eccentricity will suffer large variations and cannot be kept constant. Thus, while the solution expressed in Eqs. (55) and (56) is well suited for planetary rings, its extension to other situations should be considered in connection with a more comprehensive study of the periodic perturbations.

To conclude this section, it can be stated that our results confirm that the planetary shadow is of little importance in shaping the orbit of a dust particle located at a few planetary radii. Moreover we have demonstrated that Burns et al. (1979) were right in extending the validity of numerical integrations designed for artificial bodies to the case of dust particles.

In the same way as we derived the rate of change of a, we could investigate the effect of the shadow on the other orbital elements, more particularly on the eccentricity. However, we shall show in the next section that, unlike the semimajor axis, there is a nonzero variation of the angular momentum when the particle remains in the sunlight during a complete orbital period. So the effects of the shadow would appear as a correcting term at least of the order of R/a.

C. Periodical Evolution

We have seen that the PR components of the radiation force are responsible for the secular evolution. Likewise the planetary shadow may possibly involve very important changes in the orbital energy, although this process is inefficient in planetary rings. We shall now consider orbital evolution over shorter timescales, typically on the order of the planet's orbital period.

For planetary orbits the classical radiation force is about four to five orders of magnitude larger than the PR component and, provided that it gives rise to a nonzero effect, we can discard the action of the PR drag. In the case of secular evolution, we were not allowed to make such an assumption since only the PR drag gives a nonvanishing contribution. Hence, in this section we restrict our study of orbital evolution to the mere action of the classical force (Eq. 3):

$$\mathbf{F} = \frac{\phi A}{c} \, Q_{pr} \mathbf{u}_r \,. \tag{57}$$

Spherical Planet: Equations of Motion. As before, we consider the particle's orbit to be small compared to the planet's orbit around the Sun. Under this assumption Eq. (57) is analogous to a force constant in magnitude and always directed along the Sun-planet line. Accordingly, it is simpler to use a rotating coordinate system which maintains the Sun at fixed coordinates and makes the equations of motion time-independent. A disturbing function may be derived from Eq. (57) and expressed in terms of the instantaneously fixed values of the Keplerian elements (Mignard 1982). Once this disturbing function is time-averaged over an orbital period, we obtain

$$R = \nu \left[\mu a (1-e^2) \right]^{\frac{1}{2}} \cos i + \frac{3F}{2m} e\, a\, (\cos\omega \cos\Omega - \cos i\ \sin\omega \sin\Omega) \quad (58)$$

where ν is the planet's orbital mean motion and $F = |\mathbf{F}|$. The first part in Eq. (58) results from the inertial forces in the rotating frame whereas the last term represents the additional potential energy of the particle in the radiation field.

By inserting Eq. (58) in the planetary equations (Brouwer and Clemence 1961, p. 289), we obtain

$$\frac{da}{dt} = 0 \quad (59)$$

$$\frac{d\phi}{dt} = \nu \tan\psi \, (\cos\Omega \ \sin\omega + \cos i \ \sin\Omega \ \cos\omega) \quad (60)$$

$$\frac{d\omega}{dt} = \nu \tan\psi \left[\frac{\cos\Omega \cos\omega}{\tan\phi} - \frac{\cos i \ \sin\Omega \ \sin\omega}{\sin\phi \ \cos\phi} \right] \quad (61)$$

$$\frac{d\Omega}{dt} = -\nu + \nu \tan\psi \ \tan\phi \sin\omega \sin\Omega \quad (62)$$

$$\frac{di}{dt} = \nu \tan\psi \ \tan\phi \ \sin i \ \sin\Omega \ \cos\omega \quad . \quad (63)$$

In Eqs. (59–63),

$$\sin\phi \equiv e \quad (64)$$

and

$$\tan\psi = \frac{3}{2} \frac{\beta V_p}{n\,a} = \frac{3}{2} \frac{F}{m\nu\,n\,a} = \frac{1}{2} \frac{1}{\nu t_{pr}} \frac{c}{n\,a} . \quad (65)$$

Other parameters have the same meaning as in the preceding section. We make the following comments concerning the above equations:

1. No allowance has been made for the shadow in deriving Eqs. (59)–(63). This fact accounts for the absence of periodical variation in the semimajor axis. Otherwise Eq. (50) would have been obtained again.
2. It was the variation in eccentricity (Eq. 60) that was invoked at the end of the Sec. B in the discussion of planetary shadow effects.
3. These equations represent the rate of orbital change in a frame rotating at a constant angular velocity ν, the planet's mean motion. Then, even for a negligible radiation force, $\psi = 0$, the orbital plane regresses at the rate $-\nu$. The general solution of Eqs. (59)–(63) is rather intricate (Mignard and Henon 1983); while it is valuable from the standpoint of theoretical

dynamics, it would be of little help to describe the main characteristics of the orbital evolution.

We restrict ourselves to the motion of a dust particle in the planet's orbital plane; observational evidence in planetary rings justify this assumption. In addition, it can be shown (Mignard 1982) that the inclination suffers only small oscillations provided that the eccentricity remains moderate. In this case, $i = 0$ and the pericenter longitude becomes $\tilde{\omega} = \omega + \Omega$. The two pertinent variables are now the eccentricity and the pericenter longitude. The initial set of differential equations proves to have a very simple form when expressed with the following variables,

$$X = \sin\phi \cos \tilde{\omega} \tag{66}$$
$$Y = \sin\phi \sin \tilde{\omega} \tag{67}$$
$$Z = \cos\phi \tag{68}$$

It should be noted that Z is a regularizing variable when the eccentricity approaches 1. In order to avoid the presence of nonphysical solutions, the phase space will be limited to the region $Z > 0$. This choice corresponds to the positive determination of $(1-e^2)^{\frac{1}{2}}$.

By inserting Eqs. (66)–(68) in (59)–(63) we have

$$\frac{dX}{d\bar{H}} = \cos\psi\, Y \tag{69}$$

$$\frac{dY}{d\bar{H}} = -\cos\psi\, X + \sin\psi\, Z \tag{70}$$

where

$$\frac{dZ}{d\bar{H}} = -\sin\psi\, Y \tag{71}$$

$$\bar{H} = \frac{\nu}{\cos\psi} t \;. \tag{72}$$

The solution of Eqs. (69), (70) and (71) is achieved by standard methods for linear differential systems, which yield

$$X = \cos\delta \sin\psi - \sin\delta \cos\psi \cos\bar{H} \tag{73}$$
$$Y = \sin\delta \sin\bar{H} \tag{74}$$
$$Z = \cos\delta \cos\psi + \sin\delta \sin\psi \cos\bar{H} \tag{75}$$

where δ is a constant of integration. The second constant of integration is determined by the choice of the origin of time. The physical meaning of δ is made clear with the first integral of Eqs. (69), (70) and (71),

$$\cos \delta = X \sin\psi + Z \cos\psi \tag{76}$$

which is proportional to the energy (Eq. 58).

The global behavior of the analytical solution given in Eqs. (73), (74) and (75) is better understood with a polar plot in the X–Y plane. The radius will represent $\sin\phi$ and $\tilde{\omega}$ will be the polar angle. From Eqs. (73) and (74), the equation of a trajectory labeled by δ is

$$\frac{(X - \cos\delta \, \sin\psi)^2}{\sin^2\delta \, \cos^2\psi} + \frac{Y^2}{\sin^2\delta} = 1 \tag{77}$$

i.e. the equation of a family of ellipses centered at

$$X_c = \cos\delta \, \sin\psi \tag{78}$$
$$Y_c = 0$$

and with semiaxes

$$a = \sin\delta$$
$$b = \sin\delta \, \cos\psi \, . \tag{79}$$

Two families are plotted in Fig. 7, one for a weakly disturbed particle, i.e. $\psi = 10°$, and the other for $\psi = 30°$. Larger values of ψ are unlikely in ring systems but they can occur around the terrestrial planets. Indeed $\tan\psi$ can be found from

$$\tan\psi = \beta(a/R)^{\frac{1}{2}}(\tan\psi)_g \tag{80}$$

where $(\tan\psi)_g$ denotes the value of $\tan\psi$ for a grazing particle with $\beta=1$. Values of $(\tan\psi)_g$ are listed for the different planets in Table I. The distinction between inner and outer planets is striking in the sensitivity of dust particles to radiation forces. The giant planets strongly attract particles, while at the same time radiation pressure is reduced in the outer solar system. The combination of these two facts accounts for the contrast disclosed in the table. From these figures we can conclude that orbiting terrestrial objects whose diameters range from 0.05 to 1 μm will undergo large variations in eccentricity so as eventually to strike the planet or be lost in its upper atmosphere. In contrast, around Jupiter or Saturn similar particles can be kept for a long time within 5 to 10 planetary radii, and affected only by the gradual orbital decay caused by the PR drag as shown in Eq. (29).

With these orders of magnitude in mind, we can proceed to the analysis of Fig. 7. We can see that according to the initial conditions, i.e. according to the value of δ, the pericenter may either circulate or librate. As long as $\delta < \psi$, the pericenter librates about $\tilde{\omega} = 0$ and the eccentricity oscillates between two extreme values given by

$$e_{\min} = \sin(\psi-\delta) \tag{81}$$

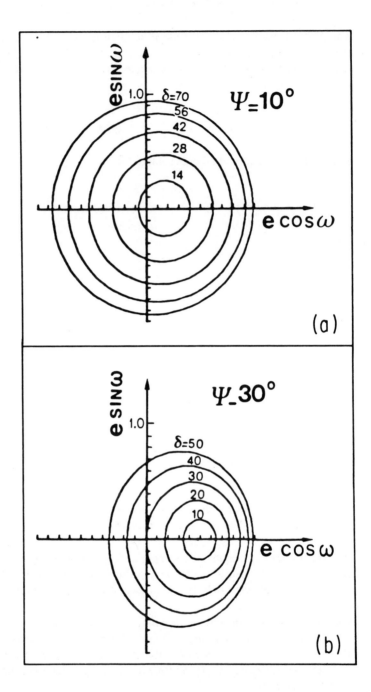

Fig. 7. Polar plots of the trajectories in the phase space $e - \tilde{\omega}$ for weakly (a) and strongly (b) disturbed particles. Each curve is labeled by δ in degrees. Libration corresponds to curves that do not encircle the origin. Otherwise, the pericenter has a circulatory motion.

TABLE I

Values of tan ψ for Grazing Particles[a]

	$(\tan\psi)_g$
Mercury	24
Venus	7.1
Earth	5.6
Mars	10
Jupiter	0.46
Saturn	0.58
Uranus	0.68
Neptune	0.49

[a]A grazing particle would orbit a planet in a circular orbit whose radius would be equal to the planet's equatorial radius.

$$e_{max} = \sin(\psi + \delta) \tag{82}$$

while the pericenter variations are restrained within $\pm\varpi_{max}$ with

$$\varpi_{max} = \sin^{-1}(\sin\delta/\sin\psi) . \tag{83}$$

In cases where the pericenter circulates, e_{max} is still given by Eq. (82) while e_{min} becomes

$$e_{min} = \sin(\delta - \psi) . \tag{84}$$

If the radiation force were to disappear slowly, the equilibrium solution $\delta=0$ would go to $\phi=0$ and $\varpi=0$. The array of ellipses would degenerate into circles of constant eccentricity $e=\sin\delta$, representing undisturbed Keplerian orbits. Thus δ may be considered the free eccentricity whose value is completely determined by the initial conditions. In contrast, for a disturbed orbit e cannot remain at zero even if the free eccentricity is zero. This permanent eccentricity can be called forced eccentricity and its value is given by $e_{forced} = \sin\psi$ which for small β is approximately equal to

$$e_{forced} \simeq \frac{3}{2} \beta \frac{V_p}{V_g} \left(\frac{a}{R}\right)^{\frac{1}{2}} = \begin{cases} 0.46\,\beta \left(\dfrac{a}{R}\right)^{\frac{1}{2}} \text{ for Juipter.} \\ 0.58\,\beta \left(\dfrac{a}{R}\right)^{\frac{1}{2}} \text{ for Saturn.} \end{cases} \tag{85}$$

This behavior is clearly analogous to behavior near orbital resonances (cf. Greenberg 1977).

Therefore if Saturn and Jupiter were spherical, the free eccentricity of a dust particle in the ring system would be significant. We will see in the next section to what extent this conclusion must be modified because of flattening. To appraise the real importance of the eccentricity variations, consider a particle in terrestrial orbit with $a/R = 3$ and $\beta = 0.05$. This particle is far enough from the Earth's center to allow the flattening to be disregarded. For initial conditions such that $e_{min} = 0.2$, i.e. $\phi = 11^\circ.5$, we have (Eqs. 80, 81, 82) $\psi = 26°$, $\delta = 14.5$ and $e_{max} = 0.65$. With these figures the pericenter distance amounts to $a(1 - e_{max}) = 1.05R$. Without any doubt the particle will be lost into the Earth's atmosphere, after it has described about half of the cycle or 164 days. More generally, a particle with a given semimajor axis will remain orbiting a planet as long as its orbital eccentricity is small enough to prevent it from entering the planet's atmosphere. Burns et al. (1979) pointed out that such a mechanism is more efficient to account for particle loss than ejections on hyperbolic orbit. Since H in Eq. (72) is a linear function of time, the solution described in Eqs. (73), (74) and (75) is periodical and the time required to complete one cycle is deduced from Eq. (72)

$$T = P \cos\psi \qquad (86)$$

where P is the planet's orbital period. For slightly disturbed particles T is essentially the same as P.

Oblate Planet. In order to extend the validity of the preceding solution and to make it useful for planetary rings it is desirable to take the planet's oblateness into account. The large variations that the eccentricity exhibits follow from a coupling between the eccentricity and the pericenter longitude expressed in Eqs. (69), (70) and (71). As we have shown, the typical period for the evolution is equal to the planet's orbital period and, for a spherical planet, the radiation force makes the pericenter precess or librate at a rate comparable to the planet's mean motion. On the other hand, oblateness will cause the apsidal line of a circular orbit in the equatorial plane to precess with an approximate rate given in Eq. (52). If the precession due to the flattening is faster than the planet's mean motion, the eccentricity will not have enough time to develop fully and the forced eccentricity will tend to disappear. In short, a rapid steady precession should break the coupling. Since the rate of precession due to J_2 decreases sharply with distance (Eq. 53), there exists a critical orbit dividing the planetary neighborhood into two regions: inside a critical distance $\dot{\varpi} > \nu$ and the reverse holds outside. The equality of the two rates of motion leads to the value of the critical distance a_c, given by the following equation

$$\frac{2\pi}{t_{wg}} \left(\frac{R}{a_c} \right)^{\frac{7}{2}} = \nu \qquad (87)$$

where t_{wg} is, as earlier, the period of precession of a grazing satellite. The pertinent data are collected in Table II. We have left out Uranus because it is

TABLE II

Pertinent Data for a Grazing Particle

	Quadrupole Moment $10^3 \times J_2$	Period of Precession $7_{\omega g}$ (days)	Critical Radius[a] a_c/R
Earth	1.08	36	1.94
Mars	0.2	230	1.37
Jupiter	14.7	5.4	6.69
Saturn	16.1	7.2	8.07

[a]Defined in Eq. (87), expressed in planetary radii.

meaningless to neglect the inclination of its equatorial plane to its orbital plane, while it is an acceptable approximation for the other planets.

From the figures in Table II we see that Saturn's and Jupiter's rings are easily inside the critical zone where flattening cannot be neglected. This critical distance was implicitly invoked to derive Eq. (55) by stating the rate of precession of the pericenter within the critical distance is nearly equal to the rate due to planetary oblateness alone. The transition between the two regions can be studied with

$$k = 1 - \tilde{\omega}/\nu \qquad (88)$$

and

$$\tilde{\nu} = k\nu \qquad (89)$$

so that $-\tilde{\nu}$ represents the mean rate of precession of the pericenter with respect to the planet-Sun line. Outside the critical distance k approaches 1 and $\tilde{\nu} = \nu$, while the classical rate of precession is restored within the critical distance. A situation of resonance will appear in the vicinity of the critical distance where the pericenter of a large particle would always be aligned with the Sun. This region is characterized by values of k close to zero. According to Eq. (87) we have

$$k = 1 - (a_c/a)^{\frac{7}{2}} \qquad (90)$$

which shows that the transition zone between the two regimes of precessional motion is very narrow, given the power $\frac{7}{2}$ that appears in Eq. (90). Hence we will almost always be in either of the two followings degenerate cases,

1. $a > a_c$ and the perturbations arising from the flattening could be disregarded since $k \sim 1$;
2. $a < a_c$ and from Eq. (88) $k \sim -(a_c/a)^{\frac{7}{2}}$. $\qquad (91)$

In order to characterize the magnitude of the perturbation consider an angle γ instead of ψ:

$$\sin \gamma \equiv \sin\psi /(\sin^2 \psi + k^2\cos^2\psi)^{\frac{1}{2}}$$
$$\cos \gamma \equiv k \cos \psi /(\sin^2\psi + k^2\cos^2\psi)^{\frac{1}{2}} \tag{92}$$
$$k \tan \gamma \equiv \tan\psi .$$

Outside the critical distance, ψ is recovered as a limiting value of γ. On the other hand, when $a < a_c$, γ is an angle which varies between $\pi/2$ and π, the smallest perturbations corresponding to γ close to π.

In the case of a spherical planet, the magnitude of the perturbation was mainly determined by β as it can be seen in Eq. (80). In the present situation we have two factors equally important that determine γ. First is β, obviously, and ultimately the particle's size, and second is the closeness of the resonance zone. In Eq. (92) we see that γ can be close to $\pi/2$ even for small β provided that $k \sim 0$. So the proximity of the resonance will amplify the effect of the radiation force, as usual in any resonance encountered in celestial mechanics (Greenberg 1977; see also chapter by Franklin et al.). Variations of γ with distance are plotted in Fig. 8 for Jupiter and Saturn for two values of β. We can see that $\tan\gamma$ drops sharply as soon as we enter the internal zone which implies that the perturbations will lessen as well.

We move back now to the differential system (Eqs. 60–63), with $i = 0$ and $\tilde{\omega} = \omega + \Omega$. Then, adding the contribution of the flattening to the rate of variation of the pericenter, we can derive two differential equations for the variation of e and $\tilde{\omega}$. With the same change of variables as in Eqs. (66), (67) and (68) we obtain the equivalent system

$$\frac{dX}{d\tilde{H}} = \cos\gamma \, Y \tag{93}$$

$$\frac{dY}{d\tilde{H}} = -\cos\gamma \, X + \sin\gamma \, Z \tag{94}$$

$$\frac{dZ}{d\tilde{H}} = -\sin\gamma \, Y \tag{95}$$

where

$$\tilde{H} = \tilde{\gamma} t/\cos\gamma . \tag{96}$$

This new system is similar to the system (Eqs. 69–71) derived for a spherical planet with the transpositions, $\psi \rightarrow \gamma$, $\nu \rightarrow \tilde{\nu}$. Outside the critical region γ is in the range $(0, \pi/2)$ and the complete solution (Eqs. 73–79) extends to the present situation. The inner region is the most interesting as far as rings are concerned. In this case, γ is in the range $(\pi/2, \pi)$ and the center of libration is at $\tilde{\omega} = \pi$ instead of $\tilde{\omega} = 0$. In addition, to fulfill the requirement

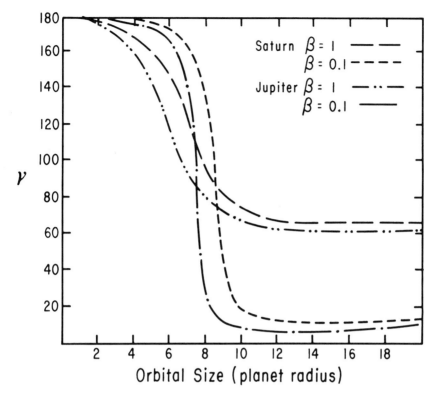

Fig. 8. Variation of γ with the distance to the center of the planet. The unity of distance is in Saturn radii in the curves for Saturn and otherwise in Jupiter radii.

$Z > 0$, the center of libration corresponds to a trajectory labeled with $\delta = \pi$. An increase of the free eccentricity corresponds to a decrease in δ and the semiaxes of the ellipsis remain defined by Eq. (79) and the center of libration by Eq. (78).

Let us try to derive some rules of thumb that can provide the reader with insight into the overall solution. By far, the most pertinent parameter is the forced eccentricity. We have

$$e_{\text{forced}} = \sin \gamma \qquad (97)$$

whose numerical value can be deduced from Fig. 8. Since the ring systems lie within critical distance (except for the outer part of the E Ring, which we will consider separately), we can use the approximate value of k given by Eq. (91) and Eq. (80) to obtain

$$e_{\text{forced}} = \beta (\tan \psi)_g \left(\frac{R}{a_c} \right)^{\frac{7}{2}} \left(\frac{a}{R} \right)^4 . \qquad (98)$$

This approximation is valid up to $(a/R) \sim 5$ for Saturn and up to $(a/R) \sim 4$ for Jupiter. With the figures contained in Tables I and II, the forced eccentricity becomes

$$e_{\text{forced}} = \begin{cases} 5.9 \times 10^{-4} \, \beta(a/R)^4 \text{ for Jupiter.} & (99) \\[2mm] 3.9 \times 10^{-4} \, \beta(a/R)^4 \text{ for Saturn.} & (100) \end{cases}$$

Such forced eccentricities are much smaller than those derived from Eq. (85) based on the assumption of a spherical planet. At the level of the outer bound of the Jupiter ring, $a/R = 1.8$ and the forced eccentricity is $e_f \sim 0.007 \, \beta$. In the case of Saturn it amounts to $0.024 \, \beta$ at the level of the G Ring where $a/R = 2.8$ and rises to $0.1 \, \beta$ at the level of Enceladus where the E Ring is most prominent. An exact evaluation of γ from Eq. (92) gives $e_f = 0.28 \, \beta$ at $a/R = 5$ and $e_f = 0.6 \, \beta$ at $a/R = 6$. The forced eccentricity keeps rising as we come closer to the resonance zone.

Eventually at the exact resonance, the array of ellipsis degenerates into straight lines

$$X = e \cos \varpi = e_0 \tag{101}$$

and

$$Y = e \sin \varpi = (1 - e_0^2)^{\frac{1}{2}} \cos (\nu \tan \psi t). \tag{102}$$

Equation (102) shows that the eccentricity may become equal to 1 and the characteristic time for such an evolution is on the order of $P/\tan\psi$ where P still denotes the planet's period of revolution. As a consequence of the orbital evolution in the resonance zone, we see that no particle can be kept in this region for a time longer than the characteristic time $(P/\tan\psi)$. This fact probably accounts for the outer boundary of the E Ring that Baum et al. (1981) set very close to 8 Saturn radii, i.e. exactly at the resonance distance.

This interpretation assumes that the remaining particles are moving about Saturn in weakly eccentric orbits, so that the observed distance can be taken as a semimajor axis. This implies the existence of an outward diffusion mechanism if Enceladus is the source to drive the particles from $a/R = 4$ to larger values. Mutual collisions may be responsible for such a diffusion. However, we should be careful in extending the results obtained by Brahic (1977) without modifications when colliding particles are micron-sized. Breaking or sticking are likely different from what they are for larger bodies.

As can be seen in Fig. 8 the width of the resonance zone becomes very narrow as β decreases. Then, the fact that we can observe matter orbiting Saturn in the vicinity of the critical distance implies that the forced eccentricity remains moderate. Hence from Eq. (100), or from a more accurate computation valid for $a/R \sim 8$, we can conclude that within the E Ring, or at least within its outer part, $\beta \ll 1$. This result implies that particles are some microns in diameter. In addition, the radiation forces tend to eliminate preferentially the smallest particles. This removal is very efficient because those particles are located close to the critical distance. Therefore, the outermost regions of the E Ring must be depleted in small grains compared to the innermost regions. At present the only available data have been obtained from edge-on profiles, and thus fail to provide us with reliable information concerning the size distribution within the E Ring, except for the fact that it comprises a large fraction of particles of 3 to 5 μm in radius. Likewise the photographic observations by Larson et al. (1981) carried out after the ring-plane crossing event were not sensitive enough to detect more than an unresolved line source smeared out by the motion of Saturn during the exposure.

Secular Effect Versus Periodical Evolution. So far the radiation pressure has been represented by the simple radial term directed along the Sun-particle line. From this assumption it turns out that the semimajor axis is secularly stable, while the other orbital parameters undergo various periodical changes. However, we have demonstrated in Sec. II.A, that the orbit is not stable over a period comparable to, or longer than, the characteristic time t_{pr} defined in Eq. (15). Since the semimajor axis decreases (Eq. 29), the relative magnitude of the disturbing radiation force lessens compared to the central gravitational attraction and both tan ψ and tan γ decrease.

On the other hand in the planetary environment a dust particle is continuously eroded either by impact with other particles or sputtered by protons and energetic ions (Burns et al. 1980; Morfill et al. 1980). These authors state that if ions of energy above some hundreds of eV hit a solid particle, molecules are ejected from the target and captured in the planet's magnetosphere. Therefore the particle's diameter would vary and eventually would lead to a variation of β as is seen in Fig. 2. According to the initial diameter β may either increase or decrease. Both processes, PR drag and erosion, yield a systematic change of β and of tan ψ over timescales much larger than the planet's orbital period which is the characteristic time for short periodical evolution. Then, the solution we have presented remains instantaneously valid but the diagrams in Fig. 7 will evolve adiabatically. It can be shown that the free eccentricity remains unchanged during the process, while the forced eccentricity is always adjusted to the instantaneous value of ψ or γ. Thus the forced eccentricity will be affected by systematic changes, which in turn will modify the rate of collision and the space distribution of particles.

When the PR drag alone is considered, the adiabatic change of the forced eccentricity is obtained from Eqs. (65)–(85)

$$\frac{d e_{\text{forced}}}{dt} = \frac{1}{4\nu t_{pr}} \frac{c}{V_g} \left(\frac{R}{a} \right)^{\frac{1}{2}} \frac{d\,(a/R)}{dt} \tag{103}$$

where da/dt is to be taken from Eq. (29). When the free eccentricity is small the forced eccentricity is representative of the mean eccentricity over intermediate timescales. Then Eq. (103) should replace Eq. (30) because of the PR drag on the eccentricity, and the orbital decay is accompanied by a gradual evolution of the orbit toward a circular one, as for heliocentric orbits. In the more general situation the disparition of the forced eccentricity leaves an elliptic orbit with eccentricity equal to the free eccentricity. When the PR drag and sputtering compete, each with its own direction of evolution, the situation is too complex to allow even general conclusions to be drawn.

The Out-of-Plane Motion. So far, we have limited our discussion to a situation that pertains to planetary rings, by only considering the motions of dust particles that take place in the planet's equator. Without going further in the general discussion, it is worth mentioning a result valid for small perturbations with the Sun lying in the planet's equatorial plane. In such a case, the orbital inclination of a grain accomplishes small oscillations around its mean value, whose quantitative expression is given by

$$i = i_0 + \frac{\tan \gamma}{2} \frac{e}{(1-e^2)^{\frac{1}{2}}} \sin i_0 \cos (\tilde{\nu} t / \cos \gamma). \tag{104}$$

As far as the amplitude of oscillation is concerned, this expression is in close agreement with the numerical integrations mentioned in Mignard (1982), even for values of $\tan \gamma$ on the order of unity. The occurrence of small oscillations does fail when the orbital eccentricity evolves towards very large values. The orbital plane becomes unstable whatever the initial conditions, provided that e grows rapidly.

D. Concluding Remarks

The most important conclusion of this chapter is actually somewhat negative: the implications of radiation forces in the dynamics of planetary rings are not far-reaching except for the E Ring and possibly for the diffuse matter which lies just outside the G Ring. But even in the case of the E Ring further investigations are needed to decide whether or not an orbital evolution controlled by radiation forces may produce the observed particle distribution.

A positive result belongs to the general development of planetary science. The discovery of numerous small particles in rings has revitalized the interest in their dynamics. Finally, the motion of a dust particle affected by radiation pressure is on the verge of being understood and this should benefit greatly other branches of science as well.

REFERENCES

Baum, W. A., Kreidl, T., Westphal, J. A., Danielson, G. E., Seidelman, P. K., Pascu, D., and Currie, D. G. 1981. Saturn's E-ring. I. CCD observations of March 1980. *Icarus* 47:84–96.

Berg, O. E., and Grün, E. 1973. Evidence of hyperbolic cosmic dust particles. *Planet. Space Sci.* 23:183–203.

Born, M., and Wolf, E. 1975. *Principles of Optics* (New York: Pergamon Press), Chapter 13.

Brahic, A. 1977. System of colliding bodies in a gravitional field. *Astron. Astrophys.* 54:895–907.

Brouwer, D., and Clemence, G. M. 1961. *Methods of Celestial Mechanics* (New York: Academic Press).

Burns, J. A., Lamy, P., and Soter, S. 1979. Radiation forces on small particles in the solar system. *Icarus* 40:1–48.

Burns, J. A., Showalter, M. R., Cuzzi, J. N., and Pollack, J. B. 1980. Physical processes in Jupiter's rings: Clues to its origin by jove! *Icarus* 44:339–360.

Dollfus, A., and Brunier, S. 1982. Observations and photometry of an outer ring of Saturn. *Icarus* 49:194–204.

Elliot, J. L., Frogel, J. A., Elias, J. H., Glass, I. S., French, R. G., Mink, D. J., and Liller, W. 1981. The 20 March 1980 occultation by the Uranian rings. *Astron. J.* 86:127–134.

Feibelman, W. A. 1967. Concerning the D-ring of Saturn. *Nature* 214:793–794.

Goldreich, P., and Tremaine, S. 1979. Towards a theory of the Uranian rings. *Nature* 277:97–99.

Greenberg, R. 1977. Orbit-orbit resonances in the solar system. *Vistas in Astronomy* 21:209–240.

Guess, A. W. 1962. Poynting Robertson effect for a spherical source of radiation. *Astrophys. J.* 135:855–866.

Gurnett, D. A. Grün, E., Gallagher, D., Kurth, W. S., and Scarf, F. L. 1983. Micron-size particles detected near Saturn by Voyager plasma wave instrument. *Icarus.* In press.

Jewitt, D. C. 1982. The rings of Jupiter. In *Satellites of Jupiter,* ed. D. Morrison (Tucson: Univ. of Arizona Press), pp. 44–64.

Lamy, P. 1974. Interaction of interplanetary dust grains with the solar radiation field. *Astron. Astrophys.* 35: 197–207.

Lamy, P. 1978. Optical properties of silicates in the far ultraviolet. *Icarus* 34:68–75.

Lamy, P., and Mauron, N. 1981. Observations of Saturn's outer ring and new satellites during the 1980 edge-on presentation. *Icarus* 48:181–186.

Larson, S. M., Fountain, J. W., Smith, B. A., and Reitsema, H. J. 1981. Observations of the E-ring and a new satellite. *Icarus* 47:288–290.

Maxwell, J. C. 1859. *On the Stability of Saturn's Rings* (London, 1859). Reprinted in *The Scientific Papers of James Clerke Maxwell* 2 vols. (Cambridge: Cambridge University Press, 1890), pp. 288–374.

McCartney, E. J. 1976. Optics of the atmosphere. In *Scattering by Small Particles* (New York: J. Wiley and Sons), Chapter 5.

Mignard, F. 1982. Radiation pressure and dust particle dynamics. *Icarus* 49:347–366.

Mignard, F. 1983. Dynamical consequences of radiation. Proceedings of *I.A.U. Colloquium 75 Planetary Rings,* ed. A. Brahic, Toulouse, France, Aug. 1982.

Mignard, F., and Henon, M. 1983. About an unsuspected integrable problem. Submitted to *Celestial Mechanics.*

Morfill, G. E., Grün, E., and Johnson, T. V. 1980. Dust in Jupiter's magnetosphere: Physical processes. *Planet. Space Sci.* 28:1087–1100.

Mukai, T., Giese, R. H., Weiss, K., and Zerull, R. H. 1982. Scattering of radiation by a large particle with a random rough surface. *Moon Planets* 26:197–208.

Nicholson, P. D., Matthews, K., and Goldreich, P. 1981. The Uranus occultation 10 June 1979. I. The rings. *Astron. J.* 86:596–606.

Pollack, J. B. 1981. Phase curve and particle properties of Saturn's F-ring. *Bull. Amer. Astron. Soc.* 13:727 (abstract).

Porco, C. C., and Danielson, G. E. 1983. The kinematics of spokes. Proceedings of *I.A.U. Colloquium 75 Planetary Rings,* ed. A. Brahic, Toulouse, France, Aug. 1982.

Poynting, J. H. 1903. *Phil. Trans. Roy. Soc.* 202:525; also collected in Scientific Papers, art. 20 p. 304. Cambridge 1920.

Robertson, H. P. 1937. Dynamical effects of radiation in the solar system. *Mon. Not. Roy. Astron. Soc.* 97:423–438.

Showalter, M. R., and Burns, J. A. 1982. A numerical study of Saturn's F-ring. *Icarus* 52:526–544.

Smith, B. A., Soderblom, L., Batson, R., Bridges, P., Inge, J., Masursky, H., Shoemaker, E., Beebe, R., Boyce, J., Briggs, G., Bunker, A., Collins, S. A., Hansen, C. J., Johnson, T. V., Mitchell, J. L., Terrile, R. J., Cook, A. F. II, Cuzzi, J., Pollack, J. B., Danielson, G. E., Ingersoll, A. P., Davies, M. E., Hunt, G. E., Morrison, D., Owen, T., Sagan, C., Veverka, J., Strom, R., and Suomi, V. E. 1982. A new look at the Saturn system. The Voyager 2 images. *Science* 215:504–537.

Van de Hulst, H. C. 1981. *Light Scattering by Small Particles* (New York: Dover Publ.), Chapter 29.

Wyatt, S. P. 1963. The effect of terrestrial radiation pressure on satellites orbits. In *Dynamics of Satellites,* ed. M. Roy (Berlin: Springer-Verlag), pp. 180–196.

Wyatt, S. P., and Whipple, F. L. 1950. The Poynting Robertson effect on meteor stream. *Astrophys. J.* 111:134–141.

Zook, H. A., and Berg, O. E. 1975. A source for hyperbolic cosmic dust particle. *Planet. Space Sci.* 23:183–203.

RING PARTICLES:
COLLISIONAL INTERACTIONS AND PHYSICAL NATURE

STUART J. WEIDENSCHILLING, CLARK R. CHAPMAN,
DONALD R. DAVIS, and RICHARD GREENBERG
Planetary Science Institute

Saturn's rings are composed of a myriad of individual particles. However, even in the post-Voyager era, no individual particle has yet been seen. Most treatments, therefore, consider the rings as a continuum, characterized by fluid-like properties such as viscosity. In this chapter, we concentrate on the properties of, and dynamical processes affecting, individual ring particles. Although the rings are inside the classical Roche limit, particles are generally gravitationally bound when they are located on the surfaces of larger ones. Furthermore, the net tidal stresses within ring particles are generally small. Thus, contrary to popular conception, particle collisions should produce accretion in Saturn's rings, especially in Ring A. The extremely rapid accretionary processes within the rings (time scales of days) are counterbalanced by tidal disruption of the larger accreted aggregates. We call these complex, evolving, house-sized objects, in quasi-equilibrium between accretion and disruption, dynamic ephemeral bodies (DEB's). This equilibrium may explain why the observed mass of the ring system is predominantly in particles with radii of at least 3 to 5 m. However, there are as yet no strong observational limitations against the existence of ring particles up to sizes exceeding a kilometer; indeed, it is not ruled out that most of the mass of the ring system could reside in such bodies. If large bodies exist, they might well be cohesive fragments from a satellite disruption that may account for the origin of the rings. On the other hand, accretionary processes acting on small primordial particles may be capable of growing some aggregates to dimensions of hundreds of meters or more, so the present size distribution does not constrain models for origin. The usual treatment of ring particles as ice spheres is implausible because such particles would accrete into aggregates (DEB's) or into regoliths surrounding cohesive cores. The coefficient of restitution is, therefore, probably very low, implying that the large particles containing most of the rings' mass are in a monolayer, although the small particles responsible for most of the rings' visible cross section form a layer many particles thick. Models of kinematic viscosity and interparticle erosive processes need to incorporate these properties.

[367]

I. INTRODUCTION

Since their discovery, Saturn's rings have appeared as a "system." While it has been recognized for more than a century that they must be composed of a myriad of particles, even in the post-Voyager era we have not yet seen a single ring particle. The Voyager pictures, of course, have revealed an almost overwhelmingly complex array of dynamical phenomena, the interpretation of which is well under way, as reflected in other chapters in this book. In the absence of direct information about what ring particles are like, idealized representations of them continue to be used, when necessary, for purposes of elucidating the dynamical behavior of the system as a whole and its resolvable elements (the ringlets, gaps, waves, spokes, etc.).

This chapter's purpose is to focus on the particles themselves. Collective dynamical phenomena depend on the interactions of individual particles. Were such interactions simply gravitational, further knowledge of the ring particle properties would be of little relevance. But to the degree that interparticle collisions are important, then the physical or even geological traits of these bodies must affect the observable behavior of the rings.

Occasionally it has been suggested that somehow the orbits of ring particles are phased so that there are practically no collisions among them. If such ideas were ever tenable, they seem less so now. Observations constrain the rings to be rather densely packed ($\sim$2% of the volume) with objects of a range of sizes from centimeters to many meters (see chapter by Cuzzi et al.). The forces of these bodies on one another, and of the planet and its satellites on the entire system, would seem to compel the particles to interact. While the behavior of the coorbital satellites demonstrates that particles on a collision course due to Keplerian shear need not necessarily collide, it is difficult to believe that within the crowded rings there exist avoidance trajectories that would prevent all collisions. The rings are not in static equilibrium. Even on a macroscopic scale, they have transient, rapidly evolving features that cannot all be explained by gravitational interactions. In the absence of convincing proof of a collisionless velocity pattern, we adopt the usual assumption that the rings are a collisionally interacting system.

Collisions must be considered in order to understand the origin and evolution of the rings. Studies of the rings as a dynamical system invoke parameters such as viscosity (see chapter by Stewart et al.), which actually reflect the integrated effects of all the individual collisions. Any scenario for the origin of the rings must consider a complex history of collisional phenomena, including impacts of comets and cometary meteoroids, impacts by satellite debris in Saturnicentric orbit, and collisions among the ring particles themselves. Also, ring particles are eroded on a microscale by processes discussed in more detail in the chapter by Durisen.

When two particles collide, the outcome depends on their physical nature. Their velocities are modified, depending on the coefficient of restitution.

If rebound velocities are sufficiently low, there is the possibility for accretion, even within the Roche limit, as we show in Sec. IV. Once particles accrete onto the surface of another or damage it, creating a regolith, the coefficient of restitution is much lower than on a solid block of the same material. Subsequent impacts into the regolith have outcomes different from impacts onto a bare surface (Hartmann 1978). Thus, impact experiments with smooth ice (see Stewart et al.'s chapter) can represent only a first step towards simulating ring processes.

Collisions also affect the size distribution of the particles. Accretion causes the development of larger bodies at the expense of smaller ones. Disruptive collisions produce small bodies at the expense of larger ones. Hypervelocity impacts by the external cometary flux of micrometeoroids and occasional larger objects are sufficient to disrupt and erode ring particles. Even very low-velocity collisions among ring particles themselves are potentially capable of erosion. Such modification of the size distribution through solar system history depends on the impact flux and on the material properties of the particles. Borderies et al. (see their chapter) argue that micro-erosion by low-velocity collisions among icy ring bodies can erode the rings very rapidly on the time scale of planetary history. If ring particles are of inherently weaker material than ice (e.g., snow, or an accretionary aggregate of ice particles), then interparticle collisions might be thought to be even more effective at disrupting or eroding such bodies. On the other hand, such materials have good damping properties, which inhibit the ejection of debris.

Despite the lack of direct observations of ring particles, it is important to understand what such objects are like. While it may be useful to adopt the physically simplest system (e.g., a single-sized swarm of ice spheres) for dynamical modeling, we must also attempt to evaluate the degree to which reality differs from the idealization. What are the "geological" properties of a ring particle? To what degree are chemical and mineralogical compositions the same everywhere in the ring system? Is there a component of silicate particles in addition to the predominant icy ones? What range of particle shapes exists? Are the particles smooth or angular? Are they homogeneous throughout their interiors? Are they spherically layered? Are they cracked or loosely bound aggregates? The answers to such questions are not generally known, but the physical outcome of each collision involving such bodies could depend strongly on the answers. In addition, the geological structure of a ring particle may not be invariant, but could change in response to the collisional process itself.

In considering the physical properties of ring particles, we must remain conscious that particles of different sizes manifest themselves in different ways. For example, optical observations of the rings tend to be dominated by particles of centimeter scales and smaller, while radio and radar observations are most sensitive to larger particles extending to many meters in size. Many of the properties of the rings inferred from the Voyager mission pertain to rather exceptional parts of the system. The radio occultation experiment (Marouf et al. 1983) yielded results on particle sizes, but in just a few locations in the rings (not including the B Ring), selected by the correct combination of geometry and of optical depth (neither too opaque nor too transparent). The photopolarimeter occultation experiment (Lane et al. 1982) yielded an upper limit to ring thickness, but just at the special locations at boundaries of the rare gaps. Estimates of ring thickness based on the opposition effect ($\sim$ 10 m) are widely regarded as the most powerful (see chapter by Cuzzi et al.), but those calculations are based on the assumption of a single particle size and clearly should be reconsidered in light of what we now know about size distributions. In fact, all observations to date, both Earth-based and from spacecraft, have been quite insensitive to any ring bodies with sizes of tens of meters to many kilometers; yet, such objects, if they exist, could contain a large fraction of the total mass of the system (see Sec. III), and their properties could have important effects on the evolution of the more visible system of smaller particles. Throughout this chapter, we emphasize the wide range of possible particle properties that are consistent with the observed optical, radio, and structural properties of rings.

A. Historical Perspective on Ring Particles

Influenced by both observation and theory, our concept of the physical nature of Saturn ring particles has evolved with time. J. D. Cassini and others suggested that the rings are composed of a swarm of very small satellites, moving separately (see van Helden's chapter). The idea was not widely accepted until Maxwell's proof in 1857. Only much later did specific concepts about the nature of the ring particles themselves evolve. The fact that the major rings are very bright, yet translucent, was recognized in the mid-twentieth century as implying that the rings must be of high albedo, with high backscattering efficiency. Kuiper (1952) interpreted his first infrared spectra suggesting an icy composition. During the 1950s and 1960s, Bobrov (1970) argued that the photometric properties of the rings require the particles to be centimeters to meters in scale with small-scale surface roughness. But there was little consensus on the composition, shapes, and sizes of ring particles until new types of data began to be acquired between 10 and 15 years ago.

Although Bobrov considered power-law size distributions, most of the pre-mid-1970s literature dealt with attempts to determine the mean size of ring particles and whether the dispersion of sizes is narrow or broad. The so-called typical size has tended to grow with time as longer wavelengths have been used to probe the rings. Early interpretations of optical data considered the possible roles of particles with scales of tens of microns to a few centimeters (Pollack et al. 1973). Later radiometric measurements of thermal behavior of particles suggested predominant sizes of at least several cm (Morrison 1974; Froidevaux et al. 1981). Passive microwave data and radar observations demonstrated that there was appreciable cross section in the rings of decimeter- to meter-scale particles. It became clear by the late 1970s (e.g., Greenberg et al. 1977; Cuzzi and Pollack 1978) that there must be comparable cross-section contribution by particles spanning at least two orders of magnitude in size (i.e., a power law-like size distribution with an incremental diameter index of roughly −3; see Sec. III. C). Finally, the Voyager radio occultation experiment (Marouf et al. 1983; also see chapter by Cuzzi et al. and Sec. III. C below) demonstrated that there is appreciable fractional area in particles spanning the range of cm to 10 m, with the predominant mass in house-sized bodies with radii of 3 to 5 m.

Two other related issues debated during the 1970s concerned, (a) the bulk composition and structure of the particles (ice/snow/silicate/metal), and (b) whether or not the ring particles are arrayed in a monolayer. For the sake of simplicity, models for icy compositions assumed quasi-spherical particles with the physical properties of solid ice (e.g., density $\rho \simeq 1$). In the 1970s, most analyses of collisional mechanics were based on the moderately high coefficients of restitution, material strengths, and other parameters appropriate for the assumed small spheres of solid ice. Such particles were not readily thought of as being capable of accreting each other. Most considerations of collisional evolution focused on the highly idealized erosion that may occur at microscopic points of contact when two ice particles slowly bump into each other (Goldreich and Tremaine 1978; chapter by Borderies et al.). The distribution of fine particles created by such erosion, and by hypervelocity impact with the external cometary flux, was treated as a transient state between creation and removal from the system by Poynting-Robertson and other effects (see Mignard's chapter).

A few articles in the pre-Voyager literature raised the possibility of irregular shapes, or of a snow-like structure. One suggestion from Cuzzi (1978) was that the rings were composed of very low-density particles ($\leqslant 0.1$ g cm^{-3}) having dimensions of many meters. Yet it awaited the Voyager radio occultation experiment to highlight the importance of many-meter-sized particles and to suggest the possibility that the ring particles are significantly less dense than ice. This last suggestion (cf. Marouf et al. 1983) is based on a comparison of the volumes of particles inferred from the experiment with the surface density of the rings implied by the observed density waves. Marouf et al. also note the inconsistency between the occultation results and the earlier upper limit to particle radius of about 1 m based on 3-mm scattered radiation (Epstein et al. 1980). That limit was based on the absorption coefficient of solid ice; presumably, the observations can be accounted for in terms of larger but fluffier particles.

The other issues debated in the 1970s concerned the question of a monolayer configuration for the ring particles. The concept of a monolayer was easy to understand when ring particles were thought to be all the same size. Early interpretation of photometric properties of the rings, especially the opposition effect, strongly implied a thickness of many particles, not a monolayer. However, the dynamical model of Goldreich and Tremaine (1978), involving particles of uniform size, indicates that such particles must collapse to a monolayer unless they have a high (>0.6) coefficient of restitution. The implication was that ring particles must have rather smooth, hard surfaces (cf. chapters by Harris and by Borderies et al.). We now realize that there is a broad particle size distribution and that the photometric data chiefly reflect properties of the smallest particles. The larger particles, containing most of the mass, might well be arranged in a monolayer configuration (straddling the ring plane) while being immersed in a layer of much smaller particles, perhaps straying only a large-particle-diameter, or so, from the ring plane but clearly *many* times the thickness of the small particles themselves. Cuzzi et al. (1979) first suggested combining elements of both a monolayer and a many-particle-thick layer. We will show that Voyager-derived ring thickness and particle size distributions (Sec. III) are compatible with such a combined model, involving inelastic particles, and with dynamical and photometric constraints on ring thickness.

B. Dynamic Ephemeral Bodies

It is too early in the process of studying Voyager's treasure of new data to be confident of the nature of Saturn's ring particles. Until an individual particle has actually been seen, we must rely on indirect and/or incomplete evidence, supplemented by theoretical considerations. Nevertheless, in this chapter we introduce a new concept of what a Saturn ring particle may be like. Of course, the traditional picture of ring particles as small, icy spheres is too idealized, but we will demonstrate, in addition, that it is qualitatively wrong. As we shall show in this chapter, physically realistic ring particles must go through a process of very rapid evolution involving accretion as well as tidal disruption. Particles tend to gather together, forming larger bodies in very short order. But the continuing presence of small particles demonstrates that the larger bodies are only temporary agglomerations. We call these constantly evolving aggregate objects "dynamic ephemeral bodies," or DEB's.

Traditionally, the rings have been regarded as a destructive environment where tidal stresses and collisions have worked together to tear particles apart. In that context, the size distribution has been thought of as the result of erosion and disruption of large particles info ever smaller ones. However, as we show in Sec. II, tidal stresses are surprisingly ineffective at disrupting particles; even very weak objects can remain intact. Moreover, for plausible physical properties and relative velocities (discussed in Sec. III), accretion, not fragmentation or erosion, may be a dominant outcome of collisions.

Such accretion is probably very rapid. Relative velocities among particles are ~ 0.1 cm s^{-1}, and coefficients of restitution are plausibly very low for low-density, regolith-covered particles. Even if coefficients of restitution are as high as some tens of percent, an impact involving a meter-sized or larger body can result in accretion by gravitational binding. The particles are quite

A 1982 conception of Saturn's rings showing rounded individual particles in a ring plane many particles thick. It differs from the model proposed by Weidenschilling et al. which would have a thick layer of very small particles (centimeter size), and a monolayer of irregular bodies of several meters in radius (painting by W. K. Hartmann; collection of A. Brahic). See also Color Plates 7 and 8.

densely packed in the rings. Particle-in-a-box calculations show, for a population like that proposed by Marouf et al. (1983; see also Cuzzi et al.'s chapter and Sec. III. C), that any given sub-3 m-size body hits and is accreted by a 3 to 5 m body in $\sim$ 10 days. A given 3 m body would double in mass during that time as it accretes smaller bodies. In Sec. IV, we confirm these crude estimates with numerical simulation of collisional evolution using a model incorporating a wide range of collisional outcomes, and based on the best estimates of relevant physical properties and parameters, as reviewed in Sec. III.

If such accretion represented the whole story, there would be no rings, or, at the very least, the small particles that provide most of the observable cross section would be eliminated by incorporation into larger bodies. Continued accretion would lead to substantial satellites rather than rings. However, we show in Sec. II that tidal disruption is possible under certain circumstances, and our numerical simulations described in Sec. IV incorporate such disruptive processes. After rapid growth to beyond several meters, ring particles become increasingly prone to disruption. Thus, at any time, the house-sized bodies that dominate the system may be only a few days or weeks old and may be on the verge of fragmenting into their much smaller components. The larger bodies in this model are indeed dynamic, and ephemeral. This picture contrasts sharply with the traditional view of a ring particle as a discrete object with a lifetime of perhaps billions of years. Still, the Voyager spacecraft have revealed unexpected dynamical structures on larger, observable scales, such as spokes, and kinks and braids in ringlets, which appear and disappear rapidly. Ephemeral ring particles are, in a sense, a conceptual extension of this behavior to smaller scales.

The processes and concepts that we introduce here are potentially relevant to all planetary ring systems. We apply them principally to the rings of Saturn, but do discuss, to a limited extent, implications for Jupiter and Uranus. Specific physical information about the latter two systems is too meager to make direct inferences about the properties of their larger constituent particles. The concepts we develop here for the case of Saturn should be taken into account in considerations of the evolution of the other ring systems.

II. DISRUPTIVE PROCESSES IN A TIDALLY DOMINATED ENVIRONMENT

Why does Saturn have rings where it does, rather than discrete satellites? The answer that the rings are within the Roche limit is inadequate. The mechanism by which tidal forces disrupt large bodies, or prevent their accretion, is not clear. There is at least one significant natural body inside a Roche limit, Mars' inner satellite Phobos. Phobos is subjected to a tidal pull equal to that in the inner part of Saturn's B Ring. Then why are Saturn's rings not composed exclusively of Phobos-sized or larger bodies? As we shall see, the much smaller size of Saturn ring particles limits their allowed density and implies that their material strength is extremely low.

A. The Classical Roche Limit

The only rigorously defined criterion for tidal disruption, the classical Roche limit, refers to the idealized problem of a synchronously rotating satellite without material strength (i.e., a liquid body). The idealized case offers useful insight into processes affecting real ring particles. But real ring particles differ from this model in important respects. Being solid, even if loose aggregates, they possess some finite strength; they may also have nonsynchronous rotation and various shapes. In this section, we review the meaning of the classical Roche limit, and then examine implications for tidal effects on the solid bodies in Saturn's rings.

The figure of hydrostatic equilibrium of a self-gravitating satellite was determined by Roche (1847; cf. Chandrasekhar 1969). The satellite was assumed to be rotating synchronously, and distorted by tidal and centrifugal forces. At sufficiently large distance, the equilibrium shape is a triaxial ellipsoid with the longest axis oriented toward the planet. Roche showed that no closed equipotential surface exists at orbital radii a less than the critical value given by

$$\frac{a}{R} = 2.456 \left(\frac{\rho_p}{\rho} \right)^{\frac{1}{3}} \qquad (1)$$

(see Table I for the definition of all symbols). Alternatively, for any given orbital radius, one can define a critical "Roche density" for a satellite, below which no equilibrium figure exists. If ring particles had the density of Mimas ($\rho = 1.44$ g cm^{-3}), the entire A Ring would be outside the Roche limit.

Within the Roche limit, a satellite cannot attain hydrostatic equilibrium (i.e., there is no shape for which the local surface gravity, including tidal and centrifugal forces, is everywhere perpendicular to the surface). Part of its surface is at lower potential than other parts. If the satellite cannot support a local slope, some of its mass must flow "downhill." A liquid body would become ever more elongated in the direction toward and away from the planet, eventually coming apart in some manner.

This result is often misinterpreted. The Roche limit is *not* the distance at which the planet's tidal force exceeds the satellite's gravitational attraction. It can be shown that the local surface gravity of the critical Roche ellipsoid (the limiting equilibrium figure) is nonzero and directed inward at all points of its surface. The same is true for a range of nonhydrostatic ellipsoids that are even more elongated and/or closer to the planet (Dobrovolskis and Burns 1980; Davis et al. 1981). Nor does the net binding energy, including tidal and centrifugal potentials, approach zero at the Roche limit; it is actually $\simeq 85\%$ that of an equivalent spherical body in free space.

TABLE I

Definition of Symbols Used in This Chapter

a	orbital radius	R	mean radius of Saturn
d	particle diameter	T	tidal and rotational stresses
g	surface gravity	v	mean random velocity
G	gravitational constant	v_e	escape velocity
H	ring thickness	v_i	impact velocity
k	coefficient of restitution	v_∞	approach speed of two particles
m	particle mass	μ	particle modulus of rigidity
P	particle spin period	ρ	particle density
P_o	orbital period	ρ_p	planet density
q	incremental diameter index	ω	particle spin rate
Q	inverse tidal dissipation	ω_c	spin rate for centrifugal
	parameter		force = surface gravity
r	particle radius	Ω	orbital frequency

How, then, is a satellite disrupted within the Roche limit? If the ellipsoid is elongated further along the direction of the orbital radius vector, the increase in the tidal potential exceeds the work done against its self-attraction. That is, the tidal field performs the additional work needed to pull the satellite apart. Equivalently, the satellite is unstable with respect to a fundamental mode of oscillation (Chandrasekhar 1969). This instability depends on the variation of gravitational and tidal potentials with gross shape. A fluid body would come apart, but a solid body would not unless its entire mass were mobilized (e.g., by a large impact). Tidal disruption at the classical Roche limit involves neither stripping of material from a satellite's surface nor its fracture by tensile stresses. (Such failure modes may occur, but much closer to the planet.)

B. Attraction Between Ring Particles

Because ring particles inside the classical Roche limit may have net inward surface gravity, they can accrete material in collisions and can retain regolith. The net attraction between two particles depends on several factors, including their shapes, relative sizes, and spin rates. It also depends on their orientation; it is minimized when their centers are aligned in the direction toward the planet so that the tidal gravity gradient opposes their mutual attraction. Considering only spherical particles in contact and so aligned, we can find the value at which their net attraction vanishes, for a few simple cases.

For two bodies of equal size and density with synchronous rotation (i.e., with their line of centers always pointing to the planet), the net attraction is zero at

$$\frac{a}{R} = 2.29 \; \left(\frac{\rho_p}{\rho}\right)^{\frac{1}{3}}, \tag{2}$$

well inside the Roche limit. If they are not rotating, the condition is

$$\frac{a}{R} = 2.0 \; \left(\frac{\rho_p}{\rho}\right)^{\frac{1}{3}}, \tag{3}$$

even further inside the Roche limit. A small particle resting on the surface of a large one in synchronous rotation feels no net attraction at

$$\frac{a}{R} = 1.44 \; \left(\frac{\rho_p}{\rho}\right)^{\frac{1}{3}}. \tag{4}$$

If the large particle is not rotating, the critical distance is

$$\frac{a}{R} = 1.26 \left(\frac{\rho_p}{\rho}\right)^{\frac{1}{3}}. \tag{5}$$

Faster spins result in zero net attraction at larger distances. Equations (1–5) define critical densities for the various cases at any distance (Fig. 1).

Inside the limits given by Eqs. (4) and (5), different parts of a ring particle's surface can have inwardly and outwardly directed surface gravity. Those regions with outward gravity cannot retain regolith. Such material might fall off the surface and escape, or simply migrate to other parts of the ring particle where gravity is inward. By definition, any solid body within the classical Roche limit has regions with local surface slopes relative to the potential field. These slopes can be quite steep, even where gravity has a net inward component (reaching 90° at the transition between inward and outward gravity). Noncohesive granular materials, having no tensile strength but able to bear shear stresses due to friction and self-compaction, are characterized by an angle of repose. Regolith of this nature on a ring particle could be highly mobile. For nonsynchronous rotation, the magnitude and direction of the local slope vary as the particle rotates, inducing material failure whenever and wherever the angle of repose is exceeded. Even on otherwise competent ring particles, regolith may migrate on the time scale of the rotation period.

C. Tidal Stresses and Disruption of Solid Bodies

What level of tidal stress can a body with finite strength withstand? The state of stress in a ring particle depends on its size, shape, density, and rotation rate. Aggarwal and Oberbeck (1974) analyzed the cases of nonrotating and synchronously rotating elastic spherical bodies. For our purposes, it is adequate to state that the maximum tensile and shear stresses due to tidal pull

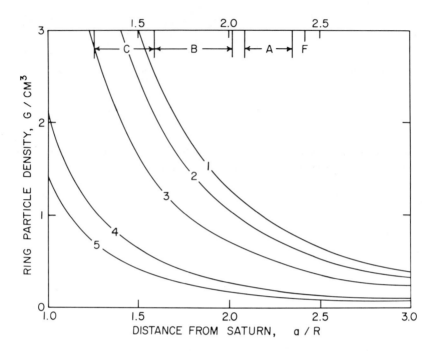

Fig. 1. Critical density of ring particles vs. distance from Saturn for phenomena discussed in
the text. The curves correspond to the equations with the same numbers. They are (1) the
classical Roche limit for a fluid body; and the condition for net attraction to vanish for; (2)
equal spheres in synchronous rotation; (3) equal spheres, nonrotating; (4) a small particle on
a large synchronously rotating sphere; and (5) a small particle on a large, nonrotating sphere.

are $\sim \rho \Omega^2 r^2$. Rotation induces additional stresses $\sim \rho \omega^2 r^2$. Rotational stresses
dominate if a particle's spin period is shorter than its orbital period; otherwise,
tidal stresses are more important. However, there is a qualitative difference
for nonsynchronous particles; rotational stress at any point is relatively constant, while the tidal stress varies as the particle rotates. Tidal stresses on
typical ring particles are small in any case; for $r = 10$ m, they are $\sim 10^{-2}$ dyne
cm^{-2}. If particles have their accretional growth limited by some form of tidal
disruption at such sizes, then they must be extraordinarily weak compared
with ordinary geological cohesive materials, which sustain stresses of order
$10^4 - 10^8$ dyne cm^{-2}.

Figures at lower right on the following pages show the effect of introducing a large (2.5 km
diameter) perturbing body within a ringlet at the distance of the F Ring. Arrows show
positions and velocities of the small particles relative to the large body (circle at center).
The view is downward from above the ring plane; Saturn is in the direction of the bottom of
the page. Dimensions are in km; the entire series covers one orbital period in 18 time steps.
For details, see p. 410. Figures appear in reverse order, so that flipping the pages produces
an animated display.

Note that stresses are proportional to ρ. This seems paradoxical, in that density greater than the Roche density would allow the existence of a stress-free configuration of specific shape (a Roche ellipsoid). Nevertheless, objects of nonequilibrium shape and in nonsynchronous rotation generally must resist stresses proportional to ρ in order to maintain their shapes. Shape can have an important effect on the peak stress level. For example, consider Phobos, which is an ellipsoidal object in synchronous rotation, with the longest axis pointing toward Mars. Dobrovolskis (1982) found stress levels in Phobos about one order of magnitude less than those given by Aggarwal and Oberbeck's formulae for a synchronously rotating sphere of equal size. However, if Phobos were not in synchronous rotation, stresses within it would vary with time, reaching much higher peak values. In the remainder of this discussion of tidal disruption mechanisms, we treat particles as roughly spherical, with the caveat that other shapes could be important.

Maximum tensile stress is developed across the plane central to the body and normal to the orbital radius vector. The particle's self-gravity tends to place the center under compression, so the peak tensile stress generally occurs at the surface. Hence, onset of tensile fracture at the surface does not necessarily result in disruption; the tidal pull must overcome self-compression for the fracture to propagate to the center. Consider a spherical ring particle split along a plane through its center. One can calculate directly the gravitational attraction between the two halves, and the tidal and centrifugal forces tending to separate them. The condition for zero net force between the two hemispheres is precisely the same as Eqs. (4) and (5) for synchronous and nonrotating bodies, respectively. These limits are very nearly the same as those for which fracture would propagate to the center of the body in the limit of zero strength for Aggarwal and Oberbeck's more elaborate analysis. This result suggests that a solid body under tidal stress may not break up into halves or into a few large pieces unless it is well within the limits of Eqs. (4) and (5). For example, this mode of failure could occur only for $\rho \lesssim 0.5$ g cm^{-3} in the inner B Ring or for $\rho \lesssim 0.1$ g cm^{-3} in the outer A Ring.

At higher densities, there may be a failure mode involving accumulation of tensional cracks over much of the surface, particularly if the body rotates to expose most of its area to the maximum stress. Eventually this would produce

18

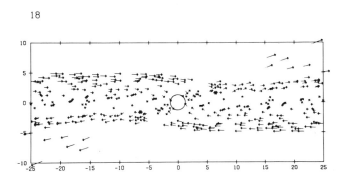

a weakened surface layer, exposed to significant local slope, which might give way in an "avalanche." If the mobilized material contains a significant fraction of the body's total mass, such failure would change its overall gravitational potential. In effect, the tidal field can do work on the body, as in the case of the classical instability of a liquid satellite. The end result might be disruption into many small particles, possibly leaving a residual core of unfractured material. A similar scenario may prevail for a competent body that gains mass by accretion. It may periodically build up a massive regolith, then lose it by such an avalanche process.

D. Spontaneous Escape From Ring Particles

Particle dynamics can be quite unusual in a tidally-modified environment. Particle motion near a satellite orbiting close to a planet has been studied by Greenberg et al. (1977), Dobrovolskis and Burns (1980), and Davis et al. (1981), who showed that impact speed, the effective surface gravity, and escape velocity could be greatly modified due to tidal and rotational effects. These effects could thus modify the rapid accretion that we envision among ring particles. Accretion may not be permanent in such an environment.

Consider an infinitesimal particle m_1 which is just touching the surface of a larger rotating ring particle m_2. The motion of m_1 relative to m_2 is governed by the gravitational attraction of m_2, and by the tidal and centrifugal acceleration. For the geometry shown in Fig. 2, where the obliquity of the spin axis is taken to be zero for simplicity, the net acceleration of m_1 is

$$g = 1 - 2 \left(\frac{\rho_p}{\rho} \right) \left(\frac{a}{R} \right)^{-3} \frac{\Delta a}{r} - \left(\frac{\rho_p}{\rho} \right) \left(\frac{a}{R} \right)^{-3} \frac{\omega}{\omega_c}^2 \tag{6}$$

where g is normalized to the nonrotating, free space gravity of m_2 and $g_2 = 2\pi G\rho_p d_2/3$. The characteristic frequency $\omega_c = \sqrt{4\pi G\rho/3}$ is the spin rate at which centrifugal force would equal the gravitational attraction at the large body's equator. For nonsynchronously rotating particles, the net acceleration on m_1 varies due to the varying tidal acceleration (i.e., Δa varies as the ring particle rotates). The minimum value of g, at the sub- and anti-Saturn points ($\Delta a = \pm r$), is

$$g_{\min} = 1 - (\rho_p/\rho) (a/R)^{-2} [2 + (\omega/\omega_c)^2] . \tag{7}$$

For a given ring particle density and distance from Saturn, $g_{\min}$ vanishes for a spin period

$$P_c = (G\rho/\pi) [1 - 2 (\rho_p/\rho) (a/R)^{-3}]^{-\frac{1}{2}} . \tag{8}$$

Ring particles spinning faster than P_c will have regions about the sub- and anti-Saturn points where the net force is *outward* from the ring particle.

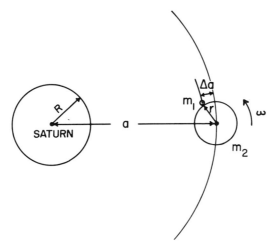

Fig. 2. Geometry for forces acting on a small ring particle m_1 located on the surface of a large particle m_2.

A particle such as m_1 would therefore lift off the surface and follow a ballistic trajectory.

We have numerically explored in three dimensions the motion of small particles that spontaneously leak off the surface of larger spherical Saturn ring particles rotating with arbitrary periods. Figure 3a depicts the trajectory of a particle which lifts off at the sub-Saturn point and reimpacts the surface of the larger body. Figure 3b shows a particle leaking off the surface and escaping from the larger particle. We refer to both types of behavior in Fig. 3 as *spontaneous leakage;* the latter type (Fig. 3b), which would tend to limit the accretionary growth of ring particles, we term *spontaneous escape.*

What then are the conditions required for a particle m_1 initially at rest on the surface of a larger particle m_2 to leak spontaneously away and escape? Consider the case where m_1 has finite size and mass. The effective surface

17

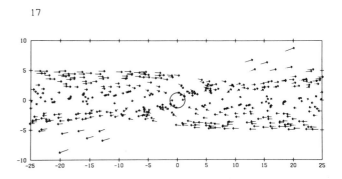

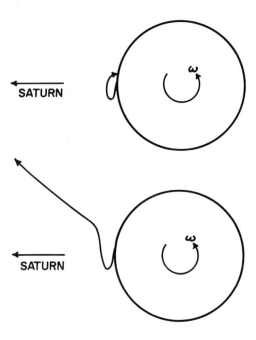

Fig. 3. Two types of trajectories followed by small particles that spontaneously leak off the
surface of large, rotating ring particles with $\rho = 0.5$ g cm^{-3} at $a/R = 1.75$. Motion is seen in
a rotating coordinate system in which the direction to Saturn is fixed. Top: spontaneous
leakage with immediate reimpact, $P = 6.0$ hr. Bottom: spontaneous leakage with escape,
$P = 5.95$ hr.

gravity must have an outward component for the particle to lift off the surface.
At the sub- and anti-Saturn points where the effective radial binding force is
at a minimum, the net acceleration between the two (assumed spherical)
particles is

$$g = \frac{1 + (d_1/d_2)^3}{1 + (d_1/d_2)^2} - \left(\frac{\rho_p}{\rho}\right)\left(\frac{a}{R}\right)^{-3}\left(2 + \frac{d_1}{d_2}\right)\left(1 + \left\lfloor\frac{\omega}{\omega_c}\right\rfloor^2\right) \tag{9}$$

where both particles have density ρ, and g is normalized as in Eqs. (6) and
(7). Note that for $d_1 = 0$, Eq. (9) reduces to Eq. (7). Figure 1 shows the
relations involving density, rotation period, and distance from Saturn that
yield zero effective gravity over some portion of a ring particle's surface.
However, as shown in Fig. 3, outward surface gravity does not guarantee
escape; some trajectories immediately return to other places on the surface.
We have numerically mapped the conditions for spontaneous escape from
spherical ring particles; Fig. 4 shows the relationships among particle density,

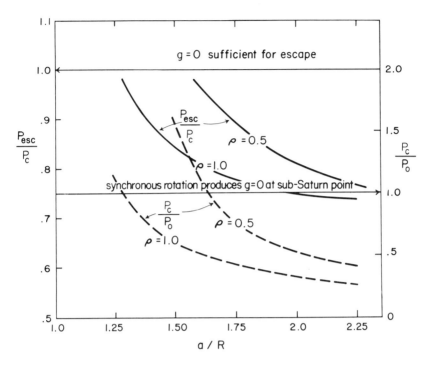

Fig. 4. Solid curves show rotation period P_{esc} required to first produce escape, normalized to P_c, the period at which $g = 0$ at the sub-Saturn point. Dashed curves show P_c for $\rho = 0.5$ g cm^{-3} ($P_c = 4.2$ hr) and $\rho = 1.0$ g cm^{-3} ($P_c = 3.3$ hr), normalized to orbit period P_0. P_0 ranges from 5.5 hr at $a/R = 1.25$ to 13.2 hr at $a/R = 2.25$.

distance from Saturn, and rotation period where spontaneous escape first occurs for zero obliquity. We expect qualitatively similar results for escape trajectories in three dimensions, with arbitrary orientation of the spin axis. We have assumed zero initial velocity. Collisions with a nonzero coefficient of restitution would give the small particle a finite radial velocity, tending to make escape easier. Actual trajectory integrations for such cases show that the

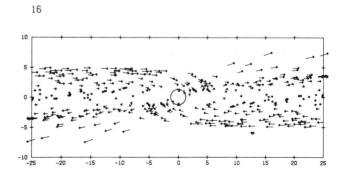

small particle may perform a series of bounces before either coming to rest or escaping. Larger ring particles may thus be surrounded by swarms of small ones in temporary associations.

The effectiveness of the leakage mechanism depends on location in the rings, particle densities, and rotation rates. Leakage is favored in the inner parts of the rings. The farther out in the rings, the more rapidly particles must rotate in order for leakage to be effective. Figure 4 shows that particles with $\rho = 0.5$ g cm^{-3} must rotate with periods less than 2 to 3 hr in the A Ring in order to have surface materials leak away. Apparently, evolution of the size distribution of ring particles depends on their rotation, which in turn is governed by collisions. In the next section, we discuss the rotational state as well as what is known about other traits of ring particles—their physical properties and relative velocities—relevant to their physical evolution.

III. PHYSICAL STATE OF RING PARTICLES

In this section, we discuss the constraints on the physical properties of the ring particles: their composition, physical state, and sizes. Other chapters in this book (especially that by Cuzzi et al.) treat these topics in greater detail. Here we are chiefly concerned with establishing basic parameters for numerical modeling of particle interactions. We emphasize the ranges of uncertainty associated with these parameters, and we elaborate on some possibilities that have tended to be ignored in the literature. We also summarize data pertinent to the question of relative velocities among particles and consider the responses of ring particles to potentially erosive collisions. Finally, we treat processes affecting the spin of a ring particle.

A. Chemical/Mineralogical Composition

The surfaces of Saturn's ring particles are predominantly composed of water ice (Pilcher et al. 1970), as revealed by diagnostic absorption features in the infrared reflection spectrum of the rings from 1 to 3 μm (Kuiper et al. 1970). The physical form of the ice, and the degree to which it is mixed with other materials (e.g., other ices or silicates) is only poorly known. Hydrated clathrates of methane or ammonia have reflection spectra indistinguishable from pure water frost throughout accessible wavelengths (Smythe 1975). Lebofsky et al. (1970) have pointed out that the coloration of the rings at visible wavelengths is inconsistent with pure water frost, but radiation damage to the crystal lattice or minor contaminants (e.g., silicates) could easily account for the coloration since ice is a very transparent material. The depth of the water absorption bands and the high inferred albedo for the ring particles demonstrate the predominance of the water ice; any surficial contamination of the ice must be slight.

In principle, there could be an intermixed population of low-albedo particles, provided they did not contaminate the ice particles. Microwave observations demonstrate, however, that the bulk material in the rings must be nearly lossless (like water ice) which rules out the presence of other types of materials contributing more than a few tens of percent of the cross section (Cuzzi et al. 1980). There is no constraint on composition for large particles, which might contain a substantial amount of mass but negligible cross section (see Sec. III.C below). Large particles with icy surfaces could have embedded cores of some different composition. While there is no observational requirement for particles to be anything other than slightly contaminated ices, the array of colorations observed in the rings—some with rather sharp boundaries (Smith et al. 1982)—advises us to keep an open mind about additional constitutents.

B. Physical Properties

There are observational constraints on the physical properties of the ring particles. The depths of the water-ice absorption bands are sensitive to particle size. Corrected for temperature effects, the band depths imply that the constituent ice particles have characteristic dimensions of tens of microns (Pollack et al. 1973), certainly indicative of the surface texture of larger particles, not a measure of particle size itself. Spacecraft polarimetry indicates rough, irregular, pulverized surfaces on icy ring particles (Esposito, et al. 1984; chapter by Cuzzi et al.). Telescopic observations of photometric phase effects, polarization, and infared eclipse radiometry also indicate that the surfaces of ring particles are more like snow or frost than solid ice in that they are rough and porous, at least on sub-centimeter scales.

The bulk density and internal structure of particles cannot be measured directly, but must be inferred from comparisons of ring cross section (optical depth) at various wavelengths, deduced size distributions, and independent estimates of ring surface density (or total mass). By combining independent estimates of ring surface density with particle sizes derived from Voyager radio occultation data, Marouf et al. (1983) estimate the bulk density of ring particles to be about ½ to ⅓ that of solid ice. This estimate applies chiefly to the larger particles of 3 to 5 m radius and is probably uncertain by at least a factor of 2.

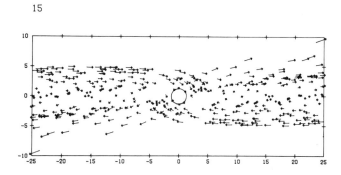

There is no definitive information concerning the shapes of individual ring particles or the configurations of clumps of particles. The Voyager radio occultation experiment (Tyler et al. 1983) does rule out very elongated particles. The constraints on physical properties of ring particles suggest to us that a typical ring particle is an irregularly shaped aggregation of snow or ice. Still, the observational data alone do not compel us to reject the traditional model of a spherical ball of ice, so long as it has a frosty surface (perhaps damaged by radiation or impact erosion).

C. Size Distribution

Most constraints on the size distribution of Saturn ring particles are satisfied fairly well by a power law for incremental number of radius r,

$$dN(r) = Ar^{-q} dr \qquad (10)$$

above some size r_{min}, where A and q are constants, with q somewhere in the range of 3.2 to 4.0. Greenberg et al. (1977) proposed a power law with $q = 3.3$ at all sizes larger than about a centimeter, based in part on expected q for a collisionally evolved system, but also consistent with available observations. Hénon (1981) proposed a much steeper power law ($q = 4.2$), arguing that a shallower slope would give a large number of particles bigger than 1 km inconsistent with the apparent thickness $\leqslant 1$ km. Some treatments of the problem over the past few years (Cuzzi et al. 1979; Goldreich and Tremaine 1982) have taken q to be exactly 3, in which case the observational evidence that most of the cross section in the rings is in particles of centimeter and meter scale would require that there be an upper bound r_{max} (~ 1 m) to the size range satisfied by the power law. Although such an upper bound of very roughly 30 cm was published by Epstein et al. (1980) based on microwave data, it is now known that the mass of the rings is dominated by particles at least an order of magnitude larger in radius (i.e., 10^3 times greater in mass; Marouf et al. 1983). Voyager data show that the size distribution has a more complex functional form than given by Eq. (10), and varies with location in the rings (Marouf et al. 1983). Still, the general trend is a power law, which can be approximated by a few power-law segments, modified by one or more humps.

Let us consider some critical properties of power laws with q in the range of 3 to 4 or more. First, if $q > 4$, then the bulk of the mass of the system resides in particles with sizes not much larger than r_{min}. On the other hand, for $q < 4$, the mass is dominated by the larger particles. However, most remote-sensing observations of the rings are insensitive to larger particles if $q > 3$ because the cross section is then dominated by smaller particles of sizes near r_{min}. This means that there is a range of possible values, $3 < q < 4$, in which the mass of the system could be dominated by large particles that contribute little to the cross section. So long as such particles are not so large as to be seen as distinct individual moonlets (which have not yet been detected in the rings), then there could be a consistent size distribution in which most of the mass of the ring system is, in a sense, invisible.

Even for Hénon's very steep power law, in which the bulk of the mass of the system is in small particles, the large ones in his distribution can be very important to a complete understanding of the rings. There are sufficient large particles in such a distribution that their edge-on projected area could be entirely responsible for the ~ 1 km thickness of the ring system observed (although with large uncertainties) from the ground. Indeed, such a distribution would include thousands of ring particles with radii of ~ 1 km, yet such bodies would contribute $<0.1\%$ of the mass of the total ring system and an infinitesimal fraction of the cross section. The largest particle in Hénon's size distribution has a radius of about 15 km, which itself may be within the bounds permitted by observation.

Let us consider further the observational constraints on the size distribution set by Voyager, including limits on the size of the very largest particles. Marouf et al. (1983) present size distributions derived from radio occultation data in four parts of the ring system, selected because they were relatively wide, homogeneous zones with moderate radio opacities. These regions were two parts of the C Ring, the outer Cassini Division, and a region near the center of the A Ring, spanning about one third of its width. The rest of the A Ring and the entire B Ring were too opaque to yield usable data.

Particle sizes were found by two different methods in distinct size ranges. In the range ~ 1 cm to 1 m, the differential extinction at wavelengths 3.6 and 13 cm was fitted to an assumed power law to derive the index q. The distribution in the range ~ 1 m to 15 m was derived by inversion of a model for forward scattering of the transmitted signal. A summary of the results is shown in Fig. 5 (for more details, see Marouf et al. [1983] and chapter by Cuzzi et al.). There are subtle differences in size distributions among the various regions, but they share some common characteristics. The most striking is the steepening of the distributions at sizes larger than a few meters. Since $q < 4$ for the smaller particles, most of the ring mass in the observable size range lies in bodies with radii of about 3 to 5 m. There may be a tendency for the effective size at the mass peak to increase with distance from Saturn.

14

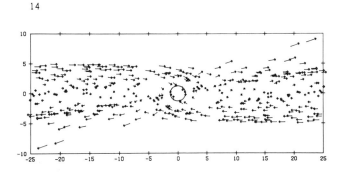

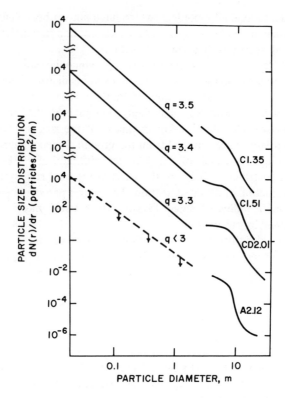

Fig. 5. Size distributions of Saturn ring particles measured by the Voyager 1 radio science experiment. The four locations include two parts of the C Ring, the outer Cassini Division, and inner A Ring (adapted from Marouf et al. 1983).

The inferred value of q for the sub-meter particles ranges from about 3.5 to $<$ 3. The differential extinction in the measured part of the A Ring was too small to do more than set an upper limit, but it is highly improbable that small particles are absent there. While the q variation appears to be monotonic in orbital radius for these selected regions, the amount of differential extinction increases again in the outer A Ring (Tyler et al. 1983).

There are several explanations for why the separately derived sub-and supra-meter size distributions do not join smoothly. (The details of the derived distributions are model-dependent.) The smaller particles may deviate from a true power law. Also, the q derived for them depends on the assumed refractive index; the q is smaller if the particles are porous, rather than solid ice. The scattering solution for the supra-meter size distribution depends on the assumed vertical structure of the rings. Marouf et al. assumed that such particles were in a many-particle-thick layer. Since the vertical distribution of the large particles may be more nearly a monolayer (ring thickness only a few times the largest particles' radii) (see Secs. V and VI), their absolute numbers could be several times smaller than the plotted curves. Recent calculations by Marouf and co-workers (personal communication) confirm that adoption of a monolayer model for the larger bodies does eliminate the discontinuity between the sub- and supra-meter size distributions. The refinement to their model reduces the prominence of the hump near 5 m, but the hump remains in the sense that most of the mass of the distribution resides in bodies of that size.

The radio occultation data, considered along with other data pertinent to ring particle size, present a picture intermediate between the simple power-law model introduced at the beginning of this section and a bimodal model with predominant particles of sizes near a few centimeters and again near a few meters. The size distribution apparently does not match the simple power-law form ($q = 2.8$ to 4.0) attributed by Hartmann (1969) to a variety of processes of rock fragmentation. However, the power law with superimposed bumps is qualitatively similar to the size distribution of the asteroids, which are certainly a collisionally evolving system (Zellner 1979).

Marouf et al. (1983) describe the steeper portion of their inferred size distributions (5 m $< r <$ 10 m) as an upper size cutoff. However, their analysis implies the existence of a "skirt" to the distribution at the largest sizes, which suggests the presence of a significant number of ring particles with dimensions at least as large as 15 m. The angular resolution of the sampling of the diffracted radiation makes their technique relatively insensitive to still larger particles. Thus, while their data appear to rule out a continuation of the curves in Fig. 5 to much greater sizes with a constant value of $q <$ 4 (Cuzzi et al.'s chapter), they cannot rule out the presence of a hump in the distribution at sizes ~ 1 km sufficient to dominate the mass of the system, provided the intermediate sizes (10 m to 1 km) are depleted relative to a pure power law. It is also important to bear in mind that size distributions have not been measured in the optically thick regions, including the entire B Ring. Hence, we must rely on indirect evidence or modeling to determine whether very large particles exist or not.

An upper limit to the sizes of moonlets embedded in the rings comes from

13

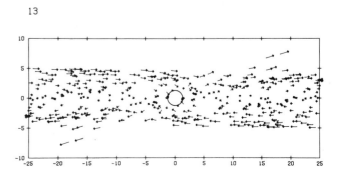

the failure to detect any such objects in the Voyager pictures. Several gaps have been carefully searched for embedded objects (Smith et al. 1982), but none have been found, giving an upper limit in those gaps ~ 10 km, depending on assumed albedo. Particles up to many tens of km in size could well have been missed in examination of the pictures if they were located within the brighter rings. Of course, such bodies might produce observable indirect effects, like gaps. Esposito et al. (1983) state that km-sized bodies "would clear gaps at least as large as their diameter." Therefore, constraints on the existence of large particles could be set by the photopolarimeter and ultraviolet spectrometer stellar occultation profiles, which did not reveal gaps that could be interpreted as the wakes of km-sized bodies. However, the large number of unexplained ring phenomena (chapter by Cuzzi et al.) suggest that such arguments against large bodies are not rigorous. The theory for gap formation is not yet adequate to specify the effects of numerous large bodies in the system. For example, the thousands of 1-km-scale bodies in Hénon's size distribution, whose integrated diameters span the ring system, could well stir up smaller particles sufficiently to close any gaps that might tend to form, without exceeding observational limits on the thickness of the rings.

Other possible effects of large particles on the ring system would include perturbing the observed density waves, but unless the mass of a particle approached the integrated mass in a density ripple, it would be unlikely to generate anything other than small, local effects. The fact that Holberg et al. (1982) report that some of the density waves are not "clean" could suggest the existence of massive particles.

Is it possible that the majority of the mass in the ring system could reside in particles > 100 m? It is commonly assumed that it does not, and values of several times 10^{22} g, based on surface densities derived from several density waves, are widely cited as the known value for the mass of the system. Actually, the most stringent direct upper limit on the mass of the rings is 10^{24} g, from the Pioneer 11 trajectory (Null et al. 1981). Depending on the degree to which the Voyager radio occultation experiment is sensitive to very large particles, it is easily possible that many times 10^{22} g exist in bodies with dimensions of tens of meters to kilometers. Such large particles could serve as reservoirs for replenishing ring material by periodic catastrophic disruption by hypervelocity impacts by objects from outside the Saturn system, much as Jupiter's inner satellites may be sources of its ring material (see chapter by Burns et al.).

For our modeling purposes, we adopt a size distribution similar to those of Marouf et al. (Fig. 5) as typical of the rings. The salient features are dominance (by mass) of particles in the size range of a few meters, and power law with $3 < q < 4$ for particles in the sub-meter range. It is important to bear in mind that this model is based on data that are representative of only a few special places in the rings and sensitive to only a limited size range, as discussed above.

D. Relative Velocities

There are no direct measurements of ring particle velocities, but several indirect methods yield fairly consistent estimates. The thickness of the rings yields an estimate of the out-of-plane (random) velocity $v \sim \Omega H$. Groundbased observations at edge-on appearances set an upper limit ~ 1 km for H, implying an upper limit to v of several cm s^{-1}. Burns et al. (1979) recognized that much of the apparent thickness could be due to warping of the ring plane. The Voyager 2 stellar occulation placed an upper limit on total thickness <200 m at the edges of gaps (Lane et al. 1982), implying $v \lesssim 1$ cm s^{-1}, if edges have typical thickness.

Another estimate is available from analysis of the bending wave excited at the 5:3 resonance with Mimas in the A Ring (see chapter by Shu). For kinematic viscosity 260 cm^2 s^{-1} inferred from the decay of this wave, Shu deduces $v \sim 0.4$ cm s^{-1}. This value may be higher than at most locations in the rings because of the energy input by numerous resonances in the A Ring.

As discussed above, the Voyager radio science data suggest that most of the ring mass may be in bodies of 3-5 m radius. Gravitational scattering theory (Safronov 1972) predicts that the maximum relative velocities should be of the order of the escape velocity of these dominant bodies. If their densities are that of ice, this approach implies $v \sim 0.1$ cm s^{-1}, which we take as the typical velocity. However, the classical scattering theory, based on 2-body encounters, may not be strictly applicable to the ring environment where many-body encounters are frequent (see chapter by Stewart et al.). Also, low relative velocities do not necessarily rule out the presence of much larger (km-sized) bodies in the rings. Their contribution to gravitational stirring is diminished, if they are not part of a continuous power-law size distribution (Levin 1978; Greenberg et al. 1978).

There is no direct evidence that v varies with particle size. Strict equipartition of kinetic energy, such that $v \propto m^{-\frac{1}{2}}$, can probably be ruled out. If bodies a few meters in size have $v \sim 0.1$ cm s^{-1}, equipartition would imply $v \sim 500$ cm s^{-1} for centimeter-sized particles, producing a ring thickness ~ 100 km. Even a modest amount of inelasticity in collisions is sufficient to prevent equipartition (Cuzzi et al. 1979).

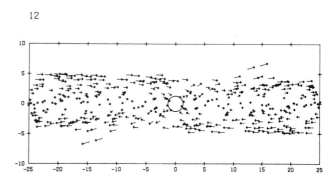

E. Microerosion by Interparticle Collisions

Goldreich and Tremaine (1978) proposed that ring particles are eroded by mutual low-velocity collisions. Although these ideas have been elaborated upon in the chapter by Borderies et al., the essential argument is simple. If two ice particles collide even at extremely low velocities ($\leq 10^{-2}$ cm s^{-1}), there will be localized stresses at the point of contact that exceed the strength of the ice. The eroded volumes are taken to be roughly equivalent to the volume of ice in which stresses exceed the critical value. Although relative velocities of ring particles are low by ordinary standards, they are much larger than the limiting velocities discussed by Goldreich and Tremaine, so in their model, erosion could proceed rapidly, given the frequent interactions of ring particles. The problem has been to understand why the ring particles have not eroded long ago into microscopic ice dust and disappeared. With the recent discovery (Marouf et al. 1983) that the dominant mass of the rings is in objects larger than previously thought, the net relative velocities must be even larger, so the problem is exacerbated.

However, this idealized model of erosion is not relevant, especially for very low relative velocities, for the following reasons. Even if fracture occurs, one must understand what happens to the fractured material. At the microscopic point of contact, the essentially planar surface of the impacting particle presses against the target particle and inhibits release of ejecta. It seems likely that intermolecular forces would combine with the external pressure of the impacting particle's surface to weld any fractured bits back together. Even if they were not welded, it is difficult to see why such ice-dust should vacate the parent particle. Sharp edges and projections might well be eroded, but after angularities are rounded off, it seems unlikely that significant erosion could continue.

Once a dusty regolith is formed on the particle by damage to the surface, subsequent impacts would not be into solid ice, but rather into the regolith. The material properties of ice would no longer be relevant to the interaction (Hartmann 1983). Any solid ice core that remains within a ring particle would be buffered from further erosion by the regolith. Being mobile, the regolith would tend to react to low-velocity impacts by redistributing itself, but the underlying core would be protected from further comminution. Thus, the erosion of the competent part of the ring particle is self-limiting. As we have discussed already, observational data indicate that the surfaces of ring particles are more like frost or snow than solid ice. While it is not clear how such materials might erode, calculations based on collisional behavior of solid ice seem inappropriate.

F. Impacts by External Flux

So far we have discussed the effects of low-velocity collisions among ring particles. Durisen (see his chapter) treats collisions between ring particles and an external flux, which involve high speed (tens of km s^{-1}) impacts that occur very infrequently and produce very different outcomes (e. g., cratering and catastrophic disruption) compared with mutual collisions among ring particles. The degree to which particle surfaces are regolith-covered is of major importance in determining the long-term evolution and collisional mixing in the rings due to bombardment by the external flux. Durisen argues that regolith should build up due to cratering and shattering by the external flux. However, given the greater frequency of collisions among ring particles, and the tidal dynamical effects we have discussed, the degree of regolith buildup on ring particles will be controlled by these rapid processes rather than by long-term bombardment by the external flux.

The external flux may be important, however, in fragmenting large ring particles and generating smaller particles. The much lower interparticle collision speed in the rings precludes cratering or disruption of cohesive ring particles, so only impacts with the external flux could break up larger cohesive ice (or silicate) bodies into multitudes of smaller ones (see Harris' chapter). We will return to these possibilities in Sec. V.

G. Rotation of Ring Particles

The spin state of ring particles is not known from observations. Their thermal radiation has been variously interpreted as implying rapid or slow rotation; such results are model-dependent (Esposito et al. 1984). The azimuthal brightness asymmetry of the A Ring could be explained either by synchronous rotation of particles or by collective gravitational effects resulting in asymmetric particle configurations (Colombo et al. 1976).

Pollack (1975) calculated the rate of tidal despinning and concluded that bodies > 1 km in size would not be despun over the age of the solar system. He assumed the ring particles had rigidity $\mu = 3 \times 10^{11}$ dyne cm^{-2}, characteristic of solid rock, and an inverse dissipation function Q of 7×10^4, which is appropriate to Saturn itself (Goldreich and Soter 1966). For solid ice, more plausible values are $\mu = 4 \times 10^{10}$ dyne cm^{-2} and $Q \sim 100$ (Cassen et al. 1982). The latter values imply more rapid despinning. In 4.5×10^9 yr, particles down to a few tens of meters would be despun in the A Ring and down to a few meters in the C Ring. (Larger bodies, if they exist, would be despun more rapidly, in $\sim 10^6$ to 10^7 yr for a kilometer-sized body.) Grain

11

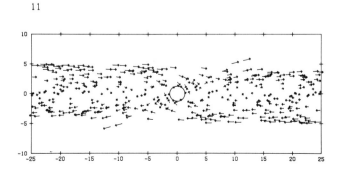

aggregates would have even lower μ and Q than solid ice; Q could be extremely low due to migration of regolith during rotation. Nevertheless, most ring particles must be spun up by collisions which occur at intervals much shorter than the tidal despinning times. Only large objects in gaps between ringlets (possibly the cause of such gaps by gravitational torques) are likely to rotate synchronously.

The spin state of a typical ring particle is determined by collisions with other particles. Both its rate of spin and the orientation of its axis vary randomly with time; this lack of preferred orientation supports treatment of ring particles as roughly spherical. The mean spin rate ω depends on a particle's size, the mass distribution of those it encounters, and the relative velocity. It also depends on the extent to which collisions are elastic. Highly elastic collisions would lead to equipartition of rotational and translational kinetic energies, with $\omega \propto r^{-\frac{3}{2}}$ (Colombo et al. 1976). However, as we discussed earlier, equipartition of translational energy and a high coefficient of restitution are ruled out, so rotational equipartition is unlikely.

For a system of equal-sized particles undergoing collisions due to Keplerian shear, where the collisions are inelastic but do not result in accretion, a simple dimensional argument implies $\omega \sim \Omega$. The situation is more complex for the case with accretion occurring with a power-law size distribution:

$$\omega = \frac{5(4-q)}{6} \left[\frac{3\,(1 + v^2/v_e^2)}{7-q} \right]^{\frac{1}{2}} \omega_c \tag{11}$$

based on Harris (1979), where $q < 4$. The derivation of Eq. (11) assumes the colliding particles follow 2-body trajectories, with gravitational focusing. This expression cannot be strictly correct inside the Roche limit, where all approach trajectories are strongly perturbed by the primary. However, 3-body trajectory integrations by Greenberg et al. (1977) and Wetherill and Cox (1983) show that impact rates and velocities are adequately represented by 2-body approximation over a wide range of conditions. We retain the form of Harris' equation with gravitational focusing, while recognizing that the numerical coefficient is uncertain.

As a numerical example, take reasonable values $\rho = 0.2$ g cm^{-3}, $q = 3.4$, and relative velocity v $= 0.1$ cm s^{-1}. For $r < 300$ cm ($v_e < v$), Eq. (11) gives $\omega \propto r^{-1}$ and spin periods P(hr) $\approx 0.05r$ (cm). The larger particles ($r \sim 3$ m) have periods ~ 15 hr. These values are sensitive to the assumed value of q; spins would be about twice as fast for $q = 2.8$. The strong q dependence is due to an assumption that the "target" is much larger than the accreted bodies that formed it. Near any lower size cutoff in the distribution, particles would have $\omega \sim v/r$, regardless of q; the smallest particles ($r \approx 1$ cm) have spin periods ~ 1 min.

For sizes larger than a few meters, the effective value of q is greater than 4 (cf. Fig. 5), so Eq. (11) is not valid. In that case, most of the mass and

angular momentum delivered to a large body is due to the meter-sized particles that dominate the mass distribution at sizes smaller than the large body.
Let the dominant bodies each have characteristic mass $m_1 \sim 10^7$ g, and let
m_2, r, and v_e be the mass, radius, and escape velocity of the large body.
As $v_e > v$, a typical impact velocity is $\sim v_e$, and the mean amount of
angular momentum delivered by a single impact is $m_1 v_e r/2$. It can be shown
that for a series of randomly oriented impacts, the expected rotation rate is

$$\omega \sim \left(\frac{m_1}{m_2}\right)^{\frac{1}{2}} \frac{v_e}{r} \simeq \left(\frac{2m_1}{m_2}\right)^{\frac{1}{2}} \omega_c \propto r^{-\frac{3}{2}}, \tag{12}$$

so larger bodies rotate slowly. For $r \gtrsim 30$ m, $P \gtrsim 100$ hr. This rotation is
much slower than synchronous. The direction of Saturn's tidal pull, relative
to a fixed point on the particle's surface would vary at about the orbital
frequency.

Note, however, that for any bodies much larger than the thickness of the
rings (~ 30 m), impacts on them may not be entirely random, but have a
systematic component of angular momentum. Suppose that the mean specific
angular momentum delivered is $\alpha v_e r$. The value of the parameter α must be
evaluated numerically by integrating over the range of possible approach
orbits resulting in impacts. The work of Guili (1968a,b) suggests that the
range of orbits contributes nearly equal positive and negative amounts of
angular momentum, so that $\alpha \ll 1$ (cf. Harris 1977). Integrating over the
accreted mass yields

$$\omega \simeq \frac{3}{2\sqrt{2}} \alpha \omega_c \tag{13}$$

thus, ω tends to a constant value. Since α is presumably small, such large ring
particles would probably be slow rotators. Most impacts would be near the
ring plane, so the obliquity of such particles would be small.

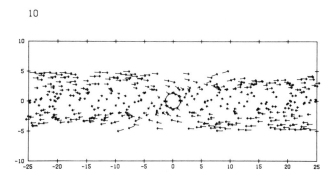

IV. PRESENT COLLISIONAL BEHAVIOR

We now discuss collisional behavior of ring particles in more detail by integrating the separate processes by means of numerical simulations. We begin with a simple numerical model of collisional outcomes involving small ice spheres to demonstrate that even this classical model of the rings results in very rapid accretion in the absence of tidal effects, which we have shown to be qualitatively negligible (Sec. II). We then augment our numerical simulation of ring particle interactions by incorporating parametrizations for ring particle rotation and for the tidal effects into subsequent numerical runs. Thus, we develop a reasonably complete and self-consistent picture of the present collisional behavior of Saturn ring particles.

We can learn several things using numerical simulations that cannot be inferred from the previous discussion of the basic processes alone. Our simulations not only provide a convenient method for calculating the cumulative effects of several simultaneous processes—collisions, accretion, tidal disruption, etc.—acting on particles of various size, but most importantly they enable us to trace the evolution of various hypothetical initial populations. The observed size distribution of the ring particles provides a real constraint which any model for ring particle evolution must satisfy. We use numerical simulations to test whether the physical processes that we believe are important in the ring system yield the observed size distribution as an equilibrium state. We also explore the range of other initial population distributions that can evolve into the present distribution.

A. Accretion

Earlier (Sec. I), we discussed reasons for believing that collisions must occur frequently in Saturn's rings, and we described a particle-in-a-box model showing that, in the dynamical environment of the ring system, accretion is a rapid ongoing process with bodies meters in size typically doubling in mass on time scales of several days. In this section, we give results from a numerical simulation, developed originally to study planetesimal growth (Greenberg et al. 1978) and asteroid collisional evolution (Davis et al. 1979).

Briefly, the numerical model simulates the collisional evolution of an arbitrary size distribution of bodies moving on orbits having small to moderate eccentricities and inclinations. A series of up to 30 diameter bins is used to represent the population, and each bin has a set of eccentricites and inclinations associated with it. The program computes the changes in the size and orbit distributions during a time step, considering collisions between all combinations of diameter bins as well as changes in orbits due to gravitational stirring. The outcome of a collision (catastrophic disruption, cratering, inelastic rebound) between two bodies depends on collisional energy and on the inherent strengths and sizes of the colliding bodies. For Saturn ring simulations, where the typical impact speed is <1 cm s^{-1}, inelastic rebound is the usual outcome of a collision between two particles. In this case, the coeffi-

TABLE II

Material Constants and Physical Parameters

	Solid Particles		Assemblages	
	Ice	Snow	Ice	Snow
Density, ρ (g cm^{-3})	0.9	0.2	0.45	0.1
Impact strength, (ergs cm^{-3})	3×10^5	10^3	0.01–10	0.01–10
Tensile strength, (dynes cm^{-2})	3×10^5	10^3	0.01–10	0.01–10
Coefficient of restitution, k	0.25–0.5	0.1	0.01	0.01

cient of restitution is the most important parameter for calculating whether the rebound leads to accretion or escape. The inputs to the simulation consist of the initial size and orbit distributions along with a set of physical and collisional parameters necessary to model collisional outcomes and orbital changes. During a run, the parameters are held constant, while the size and orbit distributions evolve with time.

In our models, we adopt bulk densities ranging from 0.1 to 0.9 g cm^{-3} and other physical properties characteristic of solid ice, snow, and their aggregates (assemblages with lumps of ice or snowballs as building blocks). The parameters listed in Table II represent our choice of values for numerical simulations of collisional evolution in the rings. The properties of very cold ices and snows responding to very low speed impacts have not been comprehensively measured in the laboratory. There are practical limitations; for example, a gravitationally-bound aggregate cannot even be formed on the surface of the Earth for experimental study. Impact strength is defined as the energy density in a collision that yields a largest fragment with half the mass of the target. While its dimensions are formally the same as for tensile strength, the two parameters are conceptually distinct (cf. Greenberg et al. 1978). In the absence of relevant experimental data, we assume them to be numerically equal. For the relative velocities and tidal forces in Saturn's rings, disruption is due to tidal stresses rather than impact energy.

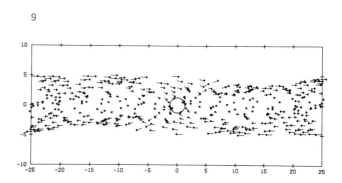

Our first simulation is for a power-law input distribution of ice particles and does not include tidal effects. (Subsequent numerical experiments will incorporate changes to simulate the tidally perturbed environment of Saturn.) At approach speeds $v_\infty \sim 0.1$ cm s^{-1}, collisions between ice particles result in inelastic rebound. The rebound outcome (i.e., whether the particles escape or eventually accrete) depends largely on the size of the largest body and on the coefficient of restitution k. A 3-m diameter ice body has a 2-body escape speed (v_e) of about 0.1 cm s^{-1} which, when combined with the approach speed, gives an impact speed (v_i) of ~ 0.14 cm s^{-1}, since $v_i = (v_e^2 + v_\infty^2)^{\frac{1}{2}}$. In this case, inelastic rebound with $k \gtrsim 0.7$ leads to escape, while $k \lesssim 0.7$ results in capture and accretion. Ice particles larger than ~ 3 m have an impact speed nearly the same as their escape speed, so that almost any coefficient of restitution less than unity will yield accretion. Particles smaller than 3 m have their impact speeds increasingly dominated by v_∞, and a smaller value of k is necessary to inhibit escape and yield accretion. For any given coefficient of restitution and ring particle density, there is a characteristic size above which, on average, collisions result in accretion and below which they result in inelastic rebound and escape. Gravitational stirring and resonances prevent collisional damping from further decreasing the mean approach speed; otherwise, v_∞ would drop, leading to accretion at all sizes.

We choose $k = 0.25$ as a reasonable value for impact of irregular ice spheres at low speeds. With this choice, bodies larger than ~ 0.7 m accrete smaller particles, while collisions involving smaller target particles result in rebound and escape. As ring particles accrete a regolith, a coefficient of restitution $\ll 0.25$ would become appropriate. Instead, we have kept the value fixed at 0.25 in this simulation, so our result conservatively underestimates the degree of accretion.

In this numerical simulation, accretion occurs rapidly, and after only 12 days the size distribution has changed dramatically (see Fig. 6). The largest bodies grow from the input value of 10 m up to 100 m and were still growing when the run was terminated. This simulation confirms our earlier conclusion (Sec. I) that rapid accretion would occur in the ring, given the range of collisional outcomes described by Greenberg et al. (1978).

B. Synthesis of Collisional Behavior

Next, we consider how to incorporate the tidal disaggregation mechanisms discussed in Sec. II (tidal disruption and spontaneous escape) and the effects of rotation discussed in Sec. III.G into our numerical simulation of ring particle size evolution. Tidal plus rotational stresses are calculated using

$$T = \rho r^2 (\Omega^2 + \omega^2) \tag{14}$$

(see Sec. II.B). Tidal stresses dominate over rotational stresses unless the rotation period is shorter than the orbit period. When the total stress exceeds

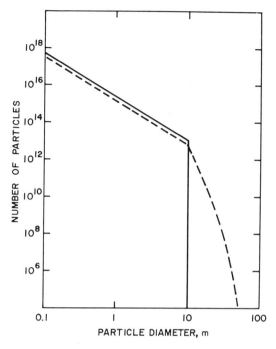

Fig. 6. Numerical simulation of accretion of ice particles in the A Ring without tidal disruption. Solid line is input distribution; dashed line is the distribution after 12 days.

the strength of the body, disruption occurs. Eq. (14) can be inverted to solve for the diameter at which the tidal stresses alone ($\omega = 0$) would be sufficient to disrupt bodies having the physical properties described in Sec. III. Table III(a) gives the largest body size that can withstand the tidal stresses occurring at two locations in the rings for the physical parameters adopted for the four types of ring particles considered. The large difference in limiting size between solid particles and assemblages arises from the much lower strength used for the assemblages, which are stripped apart at considerably smaller sizes than the relatively high-strength, solid bodies.

8

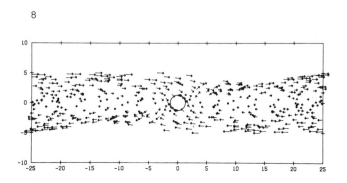

TABLE III
FOUR MATERIAL TYPES IN SATURN'S RINGS

	Solid Ice	Solid Snow	Ice Assemblages	Snow Assemblages
(a) Maximum Diameter Before Tidal Disruption				
Location (a/R)				
1.38 = mid-C Ring	45 km	5.5 km	12–120 m	25–250 m
2.10 = mid-A Ring	85 km	10 km	22–220 m	46–460 m
(b) Mean Rotation Period (hr)				
Particle diameter (m)				
1	1.1	2.4	1.6	3.4
10	11	24	16	34
100	350	740	500	1000
(c) P_c (hr)				
Location (a/R)				
1.38 = mid-C Ring	5.5	∞	∞	∞
2.10 = mid-A Ring	3.1	12	5	∞
(d) Size Range Within Which Accretion Occurs				
Mid-C Ring	0.5 m-45 km	None*	None*	None*
Mid-A Ring	0.3 m-85 km	5 m-10 km	2.5m-200 m	None*

*Due to spontaneous escape of surface particles, accretion does not occur at any size.

The numerical simulation employs the following relations for mean rotation rate as a function of particle size (based on the discussion in Sec. III.G):

$$\omega = 0.16 \frac{\sqrt{\rho}}{d} \quad \text{if } d < 10^3 \text{ cm} \tag{15a}$$

$$\omega = \frac{5.2 \sqrt{\rho}}{d^{3/2}} \quad \text{if } 10^3 < d < 10^4 \text{ cm} \tag{15b}$$

$$\omega = 5.2 \times 10^{-6} \sqrt{\rho} \quad \text{if } d > 10^4 \text{ cm} \tag{15c}$$

Table III(b) shows the rotation period for several size particles of different types. Since the mean rotation period exceeds the orbit period (which varies between 6 hr and 13 hr from the C Ring to the A Ring) for sizes larger than ~ 15 cm, we see that tidal stresses are the dominant mechanism for disruption; rotational stresses contribute little at the sizes at which disruption occurs.

In addition to the previously described algorithms for tidal and dynamical processes in the ring system, we need an algorithm for the size distribution of fragments of a tidally disrupted assemblage. In the absence of any information on the likely outcome of tidal disruption, we adopt several models spanning a range of possible outcomes:

   Model (1) places all the mass into a few large fragments.

   Model (2) assumes that all the disrupted mass goes into a shower of small particles having a power-law distribution with arbitrary index q.

   Model (3) is a combination of the first two; most of the disrupted mass goes into a number of intermediate-sized bodies with the rest going into a small-size power-law distribution.

Spontaneous escape is accounted for by parameterizing the numerical results of Sec. II (Fig. 4) as

$$\frac{P_c}{P_0} = \frac{3}{(a/R + 1.75)}\left[(\rho/\rho_p)\,(a/R)^3 - 2\right]^{-\frac{1}{2}} \qquad (16)$$

where P_c is the rotation period sufficient to produce spontaneous escape near the sub-Saturn point. ($P_c = \infty$ would indicate that the tidal acceleration alone is sufficient to strip small particles from the surfaces of nonrotating larger ones.) Table III(c) lists P_c for our adopted set of material parameters at two locations in the rings. Since our model for particle spins has bodies rotating more slowly the bigger they are, the limiting periods of Table III(c) provide a lower bound on the size at which accretion occurs. For example, an ice assemblage in the A Ring of size less than ~ 3 m could not accrete due to spontaneous escape; above this size accretion would occur until inhibited by tidal disruption.

   This scenario implicitly assumes that the particle density is independent of size for any given type of material. Alternatively, one might envision a hierarchical accretion process, in which basic constituent particles form aggregate bodies, which in turn form aggregates of aggregates, and so on, and that each generation would have a larger fraction of void space and lower density than the preceding one. In that case, spontaneous leakage would be more effective for larger bodies, despite their low rotation rates. However, we

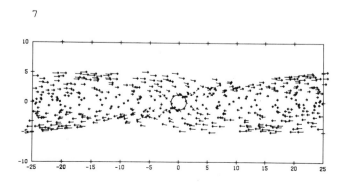

consider it unlikely that the constituent building blocks would retain their individuality much beyond the first generation; that is, ice aggregates would be less dense than solid ice, but probably would be fairly uniform in ρ among themselves. Hence, for our numerical simulations we assume constant density, independent of size, for each type of aggregate.

Bounds on the particle size at which accretion can occur may now be found by combining the limits due to different physical mechanisms. Table III(d) summarizes these bounds for the different types of ring particles. In the C Ring accretion occurs only for solid ice parameters; however, unless there were some annealing mechanism operating to maintain the effective properties of solid ice, such accretion would lead to formation of a lower-density assemblage, which would terminate the process due to spontaneous escape. Collisional damping could provide such an annealing mechanism, but it probably would not be effective on the short time scales of disruptive leakage. Thus, ring particle growth is not easy in the C Ring, which may explain its tenuous appearance if some process preferentially removes small particles.

In the A Ring, accretion occurs over a limited size range for all particle types except snow assemblages for which tidal leakage is still effective due to their very low density. The table shows that solid ice bodies would grow to sizes of 85 km, if (but only if) the strength of the accreted body were the same as that of its constituent pieces, which seems implausible. In this ring, aggregate bodies up to kilometers in size would accrete from smaller ice bodies, suggesting that low-density ice assemblages may well compose the large bodies in this part of the rings. Snow particles would accrete to form larger bodies unless the density of the aggregate bodies was <0.15 g cm^{-3}, in which case accretion would be stopped due to spontaneous escape, which would be effective even in the A Ring for the very low-density snow assemblages.

C. Numerical Simulations

Next we describe results of our numerical simulation program as modified to include tidal and rotational effects. We adopt as the nominal value for the tidal gravity gradient that for the middle of the A Ring. The objective is to find conditions that lead to the principal features of the measured size distributions, i.e., most of the mass in particles a few meters in size with a steep drop or cutoff toward larger sizes, and a power-law distribution with $3 \leqslant q \leqslant 4$ at sub-meter sizes (see Fig. 7). Understanding of more subtle features of the measured distributions, and their differences at various locations in the rings, will require a future generation of observations and modeling.

We first explored numerically the question of stability of the observed ring population. If we start with a size distribution like the observed one, does the evolution algorithm preserve with size distribution? The adopted nominal size distribution (Fig. 7) with ice-assemblage physical parameters was subjected to the three disruption-size models described in Sec. IV.B. We find that Model 1 (for a few large fragments) does not maintain the observed

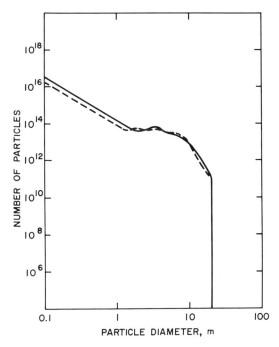

Fig. 7. Numerical investigation of the stability of the A Ring particle population. Physical parameters are for ice assemblage, with strength chosen so tidal disruption occurs at 20 m diameter. Solid line is nominal input size distribution; dashed line is the distribution after 0.5 yr.

population. Instead, the entire mass of the ring becomes concentrated in a narrow size range just below the size at which tidal disruption occurs; this spike results because the large particles are accreting smaller ones, while there is no source for resupplying the smaller sizes. For Model 2 (power law of small particles), the input population evolves to a new power law over the entire size range rather than a spike at the large diameter end. The index of the resulting power law is approximately that adopted for the fragments in each tidal disruption.

6

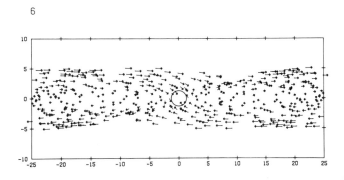

The size distribution does seem to be stable in the case of Model 3. Figure 7 compares the starting size distribution and population that evolves from it. This simulation spanned 0.5 yr in model time, much longer than the characteristic lifetime of any individual particle. The similarity between the initial population and the final one in this case suggests that the size distribution is in equilibrium. In this case, the steep falloff in population between 10 and 20 m is maintained by the high efficiency of tidal disruption at these larger sizes. Remember, for Model 3 the disruption process yields a swarm of intermediate-sized particles that produce a hump at sizes of a few meters, as well as many small particles that populate the power-law distribution in the centimeter to meter range. The intermediate and large particles accrete each other and also sweep up small particles, leading to the formation of more 10- to 20-meter-sized bodies, which are then tidally disrupted, thereby repopulating the smaller sizes. The overall population remains stable, but no individual body exists for more than a few days.

We next investigate the possibility that a radically different initial size distribution might evolve to the observed ring population using the same physical model that gave an equilibrium population. We find that an initial population with most of its mass in meter-sized bodies evolves to a distribution somewhat similar in overall shape to the observed population, but yields an excessive fraction of mass at the large-size end at the expense of the number of bodies in the small- and intermediate-size ranges. Further work is needed to explore the full range of initial populations and physical parameters that could lead to the observed ring particle size distribution.

In all the numerical experiments discussed above, encounter speeds were held fixed at 0.1 cm s^{-1} at all sizes. Other runs with relative velocities allowed to evolve along with sizes (see Greenberg et al. 1978) showed similar size evolution, but a strong tendency for random velocity to decrease with increasing particle size. This tendency is plausible, and has been suggested by Cuzzi et al. (1979; see also chapter by Cuzzi et al.). However, our simulations may not model random velocity evolution ideally for two reasons. First, our velocity stirring algorithm neglects resonances as a source of energy. Second, and perhaps more fundamental, the stirring algorithm is based on conversion of systematic Keplerian shear to random motions by gravitational encounters (which dominate in the planetesimal and asteroid cases for which the algorithm was originally developed), rather than due to collisions (which dominate in the rings). Nevertheless, the indication is that random velocities are less (certainly not greater), for bigger particles than small ones. Even a size-independent velocity of 0.1 cm s^{-1} implies out-of-plane excursions of only ~ 5 m, comparable to the size of the larger ring particles. Hence, the rings are essentially a monolayer, imbedded within a very low-mass component of small particles that make the rings look many particles thick.

Our numerical simulations are also not completely self-consistent in that they use an assumed distribution of spin rates to compute the particle size evolution. Actually, the spin rate and size distribution are coupled by a feedback mechanism. If the mean rotation rate is too low, small particles are accreted more efficiently, decreasing their numbers and lowering q. This raises the mean spin rate, releasing more small particles and tending to brake rotation. The system of particles will tend to reach an equilibrium value of q. Close to the planet, the greater tidal pull would allow regolith to escape more easily, allowing lower ω and larger q. This is consistent with the radio occultation results, which suggest a trend of shallower size distributions farther from Saturn. However, the situation is not simple; the relative number of small particles increases again in the outer A Ring, perhaps because of higher velocities due to the numerous resonances in that region.

In summary, our numerical simulations support the concept of Saturn's ring particles as short-lived dynamic ephemeral bodies (DEB's) in an ensemble steady state between accretion and tidal disruption. In order to construct models that match the observed size distribution, we must assume that tidal disruption of individual bodies yields a distribution of fragment sizes similar to the observed one. Unfortunately, we are unable to investigate such tidal breakup experimentally in terrestrial laboratories. The simulation shown in Fig. 7 assumes that tidal disruption occurs at a diameter 20 m, corresponding to a very low strength $\sim 10^{-2}$ dynes cm^{-2}. Another run assumed higher strength, which allowed kilometer-sized aggregates to form, and still assumed disruption primarily into bodies a few meters in size. That simulation yielded too many accreting intermediate-sized bodies (tens to hundreds of meters) to be compatible with the radio occultation data. While this result argues against the idea of a substantial number of large aggregate bodies in the rings, our models are still too primitive to draw a definite conclusion. Ultimately, *in situ* observations by spacecraft will be required.

V. IMPLICATIONS

The preceding discussion leads to the following picture of the present collisional behavior of the rings. Small particles of ice or snow are spun up by mutual collisions; due to their rapid rotation, accretion cannot occur, and any particles that find themselves touching spontaneously escape. There are also

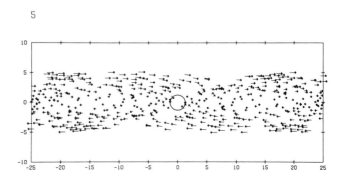

bodies would be tidally disrupted (Jeffreys 1947; Greenberg et al. 1977); they might then have been collisionally comminuted to the point that the largest cohesive fragments are somewhat or even considerably smaller (for example, 1 km). In any case, such cohesive fragments would serve as excellent "accretion nuclei" for the smaller particles. The processes we have described would then result in buildup of a deep regolith mantle on such a nucleus. When the mantle grows to sufficient depth (similar in scale to the sizes we have previously computed for tidal disruption of aggregates), the surface layers are tidally stripped, perhaps by avalanche processes operating as the large body's slow spin changes its orientation with respect to Saturn. Even in this case of fragmental origin, the sizes and shapes of the largest bodies are continually evolving.

Independent of the underlying size distribution of original cohesive material, the processes of accretion that we have described earlier would proceed, yielding aggregates with dimensions of several meters and larger. As a result, we doubt that the measured size distribution of Saturn's ring particles can be diagnostic of the mode of origin. Instead, the current size distribution should chiefly manifest the currently operating processes, retaining little signature of the origin of the particles. When we understand the processes better, more stringent limits might be placed on the form and size limits of the resulting distribution, in which case one could search for the signature of an underlying primordial distribution.

As we have noted, very large particles could be so widely distributed that they would not affect dynamical features (e.g., density waves) characterizing the continuous system of smaller particles. If a reservoir of such large bodies is necessary to explain the longevity of the rings (especially in view of the particles' higher relative velocities than previously expected, due to the prevalence of many-meter particles discovered by the radio occultation experiment), then the fragmental origin of the rings may be preferred. This is because the processes we have discussed would tend to yield a preferred characteristic size for aggregate particles at any radial distance in the ring. It is natural to identify this preferred size with the observed humps at a few meters radius. Indeed, there is a tendency in the data toward larger sizes at greater distances from Saturn, consistent with our model. Then it would be unlikely that aggregates much larger than the preferred size would contain enough large particles to dominate the mass of the system. A more likely origin for any large particles would be by fragmentation.

If the observed color variations within the rings are truly due to differences in composition (Smith et al. 1982; Cuzzi et al.'s chapter) rather than a particle-size effect, then mass reservoirs in the system are strongly implied. A primordial system of small particles, behaving in the ways we have described, could hardly establish narrow zones of compositional heterogeniety and maintain them throughout the age of the solar system. Even the dynamical confinement processes discussed in the chapter by Stewart et al. would be inadequate to prevent redistribution due to external impacts over the aeons. Much more likely, compositional differences represent relatively recent disruptions of large particles that had survived from the original fragmentation event.

If Saturn's rings are, indeed, the result of the fragmentation of a satellite, then ring particles today consist of a mixture of DEB's that are aggregates of smaller bodies and other DEB's with aggregate mantles surrounding cohesive cores. In addition, there is the ubiquitous population of centimeter-scale building blocks of the DEB's which constitute the chief component of observable cross-sectional area of the rings. There are no bare fragments. If the rings are primordial, the particles consist only of aggregate DEB's and their building blocks.

B. The F Ring and Shepherd Satellites

Because tidal effects decrease with distance from a planet, at sufficient distance they can no longer prevent accretion of discrete satellites. Saturn's system does not make a smooth transition; beyond the two innermost satellites is the F Ring, with another satellite just outside it. This coexistence is puzzling. If the satellites accreted *in situ,* why does the F Ring not form another satellite? Conversely, if spontaneous escape and tidal disruption prevent accretion of the F Ring, how did these shepherd satellites form? The satellites must then be competent bodies—chunks of ice, or at least with substantial solid cores beneath accreted regoliths. Such bodies could be a natural consequence of the scenario for origin of the rings by breakup of a parent body (cf. chapter by Harris). A few fragments, orbiting at the outer edge of the ring produced in this manner, would be driven outward by tidal/viscous torques, and protected from collisional grinding. (Similar bodies at the inner edge would be driven inward, facing ever-increasing tidal stresses and probable disruption.) These boundary satellites remain subject to external bombardment. The existing bodies might be fragments of one or more precursors that were disrupted, possibly rather recently in solar system history. The F Ring could be debris from that event.

Harris (personal communication) has suggested that the inner satellites could have accreted as viscous spreading of the ring system brought its outer

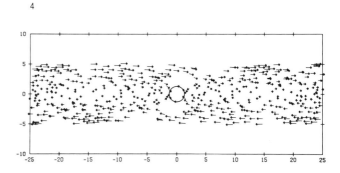

meter-scale bodies in the rings, and they rotate more slowly, permitting smaller particles to accrete onto them, forming thick regoliths. These large accretion nuclei may be remnant solid fragments from primordial times or they may be nothing but aggregates. The accretion of small particles onto larger ones is inhibited in the C Ring because ice aggregates are not sufficiently dense to resist the stronger tidal effects. But farther from the planet, continued accretion of small particles yields increasingly larger bodies, up to several meters in size and possibly approaching kilometer dimensions in some instances. Such aggregate bodies cannot accrete without limit, however. Tidal stresses increase with size; tidal disruption eventually limits their further growth. The precise mode of tidal disruption is not easy to characterize. Perhaps avalanches of loose material from the mantle of a body slough off in response to its slowly changing orientation with respect to Saturn's tides as it rotates. Our numerical simulations suggest that most of the larger house-sized objects fragment preferentially into relatively large (several meter) chunks plus a power-law-like tail of small particles, predominantly of centimeter scale. Relatively little mass is distributed in fragments with dimensions of tens of centimeters, nor are such sized particles readily produced by accretion because of spontaneous escape. These processes vary quantitatively from place to place in the rings, but the overall qualitative picture is the same. There is a steady-state ensemble of aggregate DEB's (dynamic ephemeral bodies) continually forming and disintegrating. Also, there is a significant population of centimeter-scale constituents shuffling from aggregate to aggregate. The largest bodies are arrayed in a monolayer, while the smaller constituents are perturbed into out-of-plane trajectories that correspond to a many-particle-thick layer. This multi-layer of small particles is only a few times thicker (tens of meters) than the monolayer of larger particles. For pictures that contrast the inferred close-up appearance of Saturn's rings as described here with the more traditional model of hard, smooth ice balls, see Color Plates 7 and 8.

A. Origin and Evolution of Saturn's Rings

How does this picture relate to scenarios for origin and evolution of the rings? Two classes of theories have been advanced (chapter by Harris). One class of theory for origin envisions the rings to be a primordial remnant of small particles that never grew into satellites. The size distribution implied by the primordial-origin model is not easy to specify. If the original constituent particles were all roughly centimeter-sized, the processes we have described in this chapter would tend to inhibit such particles from accreting, if the dynamical environment was similar to what exists in the present-day ring system. But relative velocities were likely lower in a primordial system of smaller particles, making collisional spinup less effective at preventing accretion, and allowing large bodies to grow. Also, instabilities could develop in such a primordial ring, resulting in gravitational clumping with direct formation of larger components (a process not represented in our models;

see chapter by Shu) in which case subsequent evolution could occur by the processes we have described.

An important conclusion, therefore, is that there is nothing inconsistent between the observed size distribution in the rings and a primordial origin model, contrary to statements like that by Marouf et al. (1983) that a primordial origin model would result in the absence of large particles in the rings. The details of the tidal disruption mechanisms we have discussed are too poorly understood to establish a limiting size to which bodies may accrete (see Table III). Such a limit could be as large as hundreds of meters to kilometers, especially in the outer parts of the ring system. While ring measurements have been interpreted as implying there is a cutoff in large-size particles, as we showed in Sec. III, there is no firm observational limit against bodies of hundreds of meters to possibly even 10 km sizes.

Another class of theory for origin has the rings produced by the breakup of a large, cohesive satellite, either by tidal breakup within the Roche zone or, more likely in our view, by collisional disruption due to impact with other bodies in either heliocentric or Saturnicentric orbit. Subsequent collisions with the cometary flux, or perhaps mutual collisions (although there is some question about the effectiveness of the latter process: cf. Harris 1976; Greenberg et al. 1977), serve to break up the bodies into a fragmental size distribution. This fragmental model for the origin of the rings is perhaps the more interesting of the two classes of theories for origin. First, one would expect a roughly power-law-like size distribution of small particles, representing the multigenerational fragmental debris from larger bodies. By small we mean down to centimeter scales, below which small particles may be removed from the ring system by radiation forces (see chapter by Mignard). In addition, unlike the primordial origin model, there is the possibility that numerous large, cohesive fragments remain in the rings. This depends on the degree of cumulative collisional fragmentation by the external flux. If it has completely ground the cohesive chunks down into small particles, further evolution would proceed as in the primordial-origin case. However, it would take an extremely large flux of cometary meteoroids to thoroughly smash up the ring particles. Thus, there might well have existed a distribution of large, cohesive fragments in the rings, possibly approaching the size (~ 100 km) above which even cohesive

3

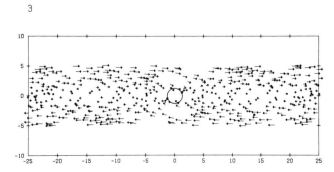

edge to the Roche limit. The rings do seem to extend very near to this limit. The irregular appearance of the F Ring is in marked contrast to the smooth outer boundary of the A Ring, although the magnitude of the tidal gravity gradient varies by $< 10\%$ between the two locations. In this scenario, unless we are viewing Saturn's system at the special epoch when accretion is beginning, the satellites should contain a nonnegligible fraction of the system's mass. (This argument assumes that the orbital evolution of the small satellites is not affected by resonances with the larger, more distant ones.) Since the viscous spreading of the B Ring is affected by the 2:1 Mimas resonance, it is more relevant to compare the masses of the shepherd satellites with that of the A Ring alone. Their total mass is $\sim 10^{-9}$ times Saturn's, assuming densities equal the local Roche density, ≈ 0.7 g cm^{-3}. This mass is $\sim 10\%$ that of the A Ring (Holberg et al. 1982). We regard their figure as a lower limit (see Sec. III). Hence, the shepherds may be only a few percent of the A Ring's mass, a value inconclusive regarding Harris' accretion suggestion. The shepherds may be made of very weak material, like accretional aggregate. They rotate synchronously with elongated shapes similar to Roche ellipsoids, and hence could be in nearly stress-free configurations.

The strongest argument against *in situ* formation is the F Ring, which is apparently not accreting (unless it is a transient feature due to a recent collision; cf. chapter by Cuzzi et al.). Weidenschilling and Davis (1983) note that the nominal orbits of the F Ring and its inner shepherd allow close approaches when their apsides are nearly 180° apart, which was the case at the time of the Voyager encounters. At such times, the tidal gravity gradient at the ring due to the satellite is comparable to that due to Saturn itself. This effect may disrupt condensations in the F Ring that would otherwise be stable. Unless the ring and satellites are locked in an unknown resonance, differential precession produces a series of close approaches at intervals ~ 20 yr. Weidenschilling and Davis suggest that the appearance of the F Ring may vary on this time scale, with alternating periods of accretion and disruption.

In the Voyager pictures, the F Ring appears to be on the verge of accretion, containing kinks and condensations that may be large embedded bodies (Smith et al. 1982; Terrile 1982). Our arguments in preceding sections indicate that such bodies, unless they have very low densities (≤ 0.1 g cm^{-3}), would readily accrete small particles (and each other). This behavior is confirmed by another sort of numerical simulation. We have performed simultaneous many-body integrations of systems consisting of one large body and many small ones, including their mutual gravity, tidal effects, and collisions. One such simulation, with tidal parameters appropriate to the F Ring, is shown in the series of small figures on the margin of this chapter (see pp. 379 to 413). At the assumed density of 0.5 g cm^{-3}, all particles hitting the large body are accreted, even though the coefficient of restitution is assumed to be high, 0.5. Some of the small particles also form gravitationally-bound pairs. The ring develops a kink similar to observed features. Spacecraft observations of the F Ring over a longer time base will show whether its structure varies periodically, and may offer a chance to observe formation and disruption of DEB's.

C. Rings of Jupiter and Uranus

The concepts developed here for Saturn's rings are also applicable to the rings of Jupiter and Uranus. In terms of planetary radii, both systems are at distances comparable to the B Ring (1.8 for Jupiter, 1.6 to 1.95 for Uranus). However, both planets are nearly twice as dense as Saturn. Their rings are deeper within the Roche limits, with tidal environments corresponding to that of Saturn's C Ring. The Roche density at Jupiter's ring is 3.5 g cm^{-3}, and in the Uranus system ranges from 2.5 to 4.5 g cm^{-3}.

The Jovian ring is extremely tenuous, with very low optical depth. The two small satellites near the ring's outer edge, and any other large particles within the ring itself, could retain regolith if their density is $\geqslant 1$ g cm^{-3}, as would plausibly be the case for silicate bodies. Such regolith might be a source for the small (μm-sized) particles dominating the ring's visible cross section (chapter by Burns et al.).

The Uranian rings have low albedo (chapter by Elliot and Nicholson), which has sometimes been taken to imply a silicate composition. One might argue that only silicates would be dense enough to avoid tidal breakup. On the other hand, low albedo could easily be explained by predominantly icy material, contaminated with a small amount of fine-grained, black carbonaceous material. Actually, as Elliot and Nicholson show, it is difficult to identify any candidate material dark enough. The rings are very narrow, presumably confined by small, unseen shepherd satellites orbiting between them (Dermott's chapter). Multiple stellar occultations have defined the orbital elements of the rings with high precision; the relative distances between them have uncertainties of only a few kilometers (Freedman et al. 1983). Such precision implies that any braiding or kinking is much less prominent than for Saturn's F Ring, which shows much larger excursions from an elliptical path. This property may place upper limits on the sizes of Uranian ring particles, either discrete bodies or DEB's (dynamic ephemeral bodies), and on the shepherd satellites.

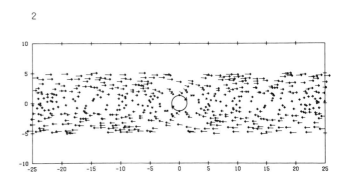

2

There is another way to determine an upper cutoff to the size distribution. Suppose the rings' low albedo is due to most of the cross section being in micron-sized particles (cf. Elliot and Nicholson). If the size distribution has a reasonable power-law index ~ 3.3, most of the cross section can indeed be in such small bodies. But an upper diameter cutoff at ~ 300 m is then necessary to limit the mass to the value ~ 20 g cm^{-2} based on the requirement for coherent ring precession (see discussion in chapter by Elliot and Nicholson). This cutoff is probably consistent with the kink-free nature of the rings.

VI. CONCLUSION

The dynamical processes we have described necessarily result in most of the mass of the observable system residing in bodies with an aggregate character. Collisions or impacts into regolith or balls of aggregate have very different effects from interactions involving solid ice. While the aggregate configurations may be greatly modified by collisions, removal of material by erosion into fine ice dust, which could then be removed from the system, is less effective than has been previously calculated. (This is especially important for impacts by the external flux.)

This is only one of several important implications of our thesis that solid ice is an inappropriate model for Saturn ring particles. As we have noted, the influential dynamical model of Goldreich and Tremaine (1978, 1982), which relates optical depth to coefficient of restitution in an equilibrium ring system, requires reconsideration. The usual presumption has been that the coefficient of restitution must be ≥ 0.63 for this model to apply, a value which may be too high even for solid ice and is certainly not applicable to a powdery surface or aggregate body. Our preferred interpretation is that the coefficients of restitution $\ll 0.63$, in which case the functional relationship to optical depth is not applicable, and the particles must collapse to a monolayer according to Goldreich and Tremaine. We regard it as quite plausible that the mass-dominant particles in the rings would be arranged roughly in such a monolayer, but with the smaller particles (which dominate optical observations) being scattered into a many-particle-thick layer. Such a model fits Voyager radio data quite well (see Sec. III.C).

Many observed large-scale features of Saturn's rings have been interpreted in the past by treating the rings as a self-gravitating fluid. Implicit in most of these models is a traditional view of uniform, cohesive, rebounding particles which control the fluid properties of the ensemble. As we have shown, however, ring particles may be very different from the traditional image. Perhaps for the macroscopic view, slight adjustments to such fluid parameters as viscosity could take this new ring-particle model into account. On the other hand, there may be fundamental qualitative changes in expected behavior.

A second iteration on such dynamical models is clearly needed. Most of the parameters (ring thickness, relative velocity, particle density, surface density) that have forced us to our new view of particle properties are based in part on inferences from the earlier dynamical models. Hence, a new generation of dynamical models may force us to modify our own physical model. Several iterations and perhaps injection of constraints from new data may be needed before we arrive at a consistent model for the dynamical behavior of the rings.

A number of implications of our model needs to be explored. How does the existence of DEB's affect the viscosity of the rings? Does the energy dissipation in frequent collisions imply unacceptably high spreading rates? If there are significant numbers of embedded large particles, can density waves propagate as modeled (see chapter by Shu)? Are the processes that we have described consistent with the theory of viscous instability invoked to explain ringlet formation (see chapter by Stewart et al.)? Do they affect shepherding and clearing of gaps by gravitational torques, and limits on the sizes of embedded particles inferred from the lack of such gaps? It is tempting to speculate that continual cycles of accretion followed by avalanche disruption might be in some way responsible for intermittent phenomena, such as spoke formation or electrostatic discharges (if the latter are indeed associated with the rings, rather than atmospheric in origin [Kaiser et al. 1984]). There is as yet no direct observational evidence for the new picture of ring particles presented in this chapter, but a consistent model of optical and dynamical phenomena would seem to require it.

Acknowledgments. This research has been supported by a contract from the National Aeronautics and Space Administration. We thank W. K. Hartmann for helpful discussions, and A. Harris and an anonymous reviewer for constructive criticisms of the original draft, as well as P. McBride for her valuable assistance with this chapter.

1

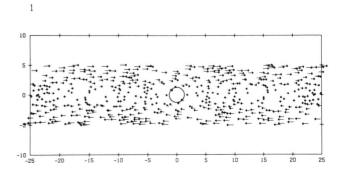

REFERENCES

Aggarwal, H. R., and Oberbeck, V. R. 1974. Roche limit of a solid body. *Astrophys. J.* 191:577–588.

Bobrov, M. S. 1970. Physical properties of Saturn's rings. In *Surfaces and Interiors of Planets and Satellites*, ed. A. H. Dollfus (New York: Academic Press), pp. 377–461.

Burns, J. A., Hamill, P., Cuzzi, J. N., and Durisen, R. H. 1979. On the "thickness" of Saturn's rings caused by satellite and solar perturbations and by planetary precession. *Astron. J.* 84:1783–1801.

Cassen, P. M., Peale, S. J., and Reynolds, R. T. 1982. Structure and thermal evolution of the Galilean satellites. In *Satellites of Jupiter*, ed. D. Morrison (Tucson: Univ. of Arizona Press), pp. 93–128.

Chandrasekhar, S. 1969. *Ellipsoidal Figures of Equilibrium*, New Haven, CT: Yale Univ. Press).

Colombo, G., Goldreich, P., and Harris, A. W. 1976. Spiral structure as an explanation for the asymmetric brightness of Saturn's A ring. *Nature* 264:344–345.

Cuzzi, J. N. 1978. The rings of Saturn: State of current knowledge and some suggestions for future studies. In *The Saturn System* (NASA Conf. Publ. 2068), pp. 73–104.

Cuzzi, J. N., and Pollack, J. B. 1978. Saturn's rings: Particle composition and size distribution as constrained by microwave observations. I. Radar observations. *Icarus* 33:233–262.

Cuzzi, J. N., Durisen, R. H., Burns, J. A., and Hamill, P. 1979. The vertical structure and thickness of Saturn's rings. *Icarus* 38:54–68.

Cuzzi, J. N., Pollack, J. B., and Summers, A. L. 1980. Saturn's rings: Particle composition and size distribution as constrained by observations at microwave wavelengths. II. Radio interferometric observations. *Icarus* 44:683–705.

Davis, D. R., Chapman, C. R., Greenburg, R., Weidenschilling, S. J., and Harris, A. W. 1979. Collisional evolution of asteroids: Populations, rotations, and velocities. In *Asteroids*, ed. T. Gehrels (Tucson: Univ. of Arizona Press), pp. 528–557.

Davis, D. R., Housen, K. R., and Greenberg, R. 1981. The unusual dynamical environment of Phobos and Deimos. *Icarus* 47:220–233.

Dobrovolskis, A. 1982. Internal stresses in Phobos and other triaxial bodies. *Icarus* 52: 136–148.

Dobrovolskis, A., and Burns, J. 1980. Life near the Roche limit: Behavior of ejecta from satellites close to planets. *Icarus* 42:422–441.

Epstein, E. E., Janssen, M., Cuzzi, J. N., Fogarty, W., and Mottman, J. 1980. Saturn's Rings: 3-mm observations and derived properties. *Icarus* 41:103–118.

Esposito, L. W., Cuzzi, J. N., Holberg, J. B., Marouf, E. A., Tyler, G. L., and Porco, C. C. 1984. Saturn's rings. In *Saturn*, eds. T. Gehrels and M. S. Matthews (Tucson: Univ. of Arizona Press), pp. 463–545.

Freedman, A. P., Tremaine, S., and Elliot, J. L. 1983. Weak dynamical effects in the Uranian ring system. *Astron. J.* 88:1053–1059.

Froidevaux, L., Matthews, K., and Neugebauer, G. 1981. Thermal response of Saturn's ring particles during and after eclipse. *Icarus* 46:18–26.

Goldreich, P. and Soter, S. 1966. Q in the solar system. *Icarus* 5:375–389.

Goldreich, P. and Tremaine, S. 1978. The velocity dispersion in Saturn's rings. *Icarus* 34:227–239.

Goldreich, P. and Tremaine, S. 1982. The dynamics of planetary rings. *Ann. Rev. Astron. Astrophys.* 20:249–284.

Greenberg, R., Davis, D. R., Hartmann, W. K., and Chapman, C. R. 1977. Size distribution of particles in planetary rings. *Icarus* 30:769–779.

Greenberg, R., Wacker, J. F., Hartmann, W. K., and Chapman, C. R. 1978. Planetesimals to planets: Numerical simulation of collisional evolution. *Icarus* 35:1–26.

Guili, R. T. 1968a. On the rotation of the earth produced by gravitational accretion of particles. *Icarus* 8:301–323.

Guili, R. T. 1968b. Gravitational accretion of small masses attracted from large distances as a mechanism for planetary rotation. *Icarus* 9:186–190.

Harris, A. W. 1976. Collisional breakup of particles in a planetary ring. *Icarus* 24:190–192.

Harris, A. W. 1977. An analytical theory of planetary rotation rates. *Icarus* 31:168–174.

Harris, A. W. 1979. Asteroid rotation rates. II. A theory for the collisional evolution of rotation rates. *Icarus* 40:145–153.

Hartmann, W. K. 1969. Terrestrial, lunar, and interplanetary rock fragmentation. *Icarus* 10:201–213.

Hartmann, W. K. 1978. Planet formation: Mechanism of early growth. *Icarus* 33:50–61.

Hartmann, W. K. 1983. Comments on collision mechanics in ring systems. In Proceedings of *IAU Colloquium 75, Planetary Rings*, ed. A. Brahic, Toulouse, France, Aug. 1982.

Hénon, M. 1981. A simple model of Saturn's rings. *Nature* 293:33–35.

Holberg, J. B., Forrester, W. T., and Lissauer, J. J. 1982. Identification of resonance features within the rings of Saturn. *Nature* 297:115–120.

Jeffreys, H. 1947. The relation of cohesion to Roche's limit. *Mon. Not. Roy. Astron. Soc.* 107:260–262.

Kaiser, M. L., Desch, M. D., Kurth, W. S., Lecacheux, A., Genova, F., Pedersen, B. M., and Evans, D. R. 1984. Saturn as a radio source. In *Saturn*, eds. T. Gehrels and M. S. Matthews (Tucson: Univ. of Arizona Press), pp. 378–415.

Kuiper, G. P. ed. 1952. Planetary atmospheres and their origins. In *Atmospheres of the Earth and Planets* (Chicago: Univ. of Chicago Press), pp. 306–405.

Kuiper, G. P., Cruikshank, D. P., and Fink, U. 1970. The composition of Saturn's rings. *Sky Telescope* 39:14.

Lane, A., Hord, C., West, R., Esposito, L., Coffeen, D., Sato, M., Simmons, K., Pomphrey, R., and Morris, R. 1982. Photopolarimetry from Voyager 2: Preliminary results on Saturn, Titan, and the rings. *Science* 215:537–543.

Lebofsky, L. A., Johnson, T. V., and McCord, T. B. 1970. Saturn's rings: Spectral reflectivity and compositional implications. *Icarus* 13:226–230.

Levin, B. J. 1978. Relative velocities of planetesimals and the early accumulation of planets. *Moon Planets* 19:289–296.

Marouf, E. A., Tyler, G. L., Zebker, H., Simpson, R., and Eshleman, V. R. 1983. Particle size distributions in Saturn's rings from Voyager 1 radio occultation. *Icarus* 54:189–211.

Morrison, D. 1974. Infrared radiometry of the rings of Saturn. *Icarus* 22:57–65.

Null, G. W., Lau, E. L., Biller, E. D., and Anderson, J. D. 1981. Saturn gravity results obtained from Pioneer II tracking data and earth-based Saturn satellite data. *Astron. J.* 86:456–468.

Pilcher, C., Chapman, C. R., Lebofsky, L. A., and Kieffer, H. H. 1970. Saturn's rings: Identification of water frost. *Science* 167:1372–1373.

Pollack, J. B. 1975. The rings of Saturn. *Space Sci. Rev.* 18:3–94.

Pollack, J. B., Summers, A., and Baldwin, B. 1973. Estimates of the size of the particles in the rings of Saturn and their cosmogonic implications. *Icarus* 20:263–278.

Roche, R. A. 1847. *Acad. des Sciences et Lettres de Montpelier, Mem. de la Section des Sciences* 1:243–262.

Safronov, V. 1972. *NASA TTF-677.*

Smith, B. A., Soderblom, L. A., Batson, R., Bridges, P., Inge, J., Masursky, H., Shoemaker, E., Beebe, R., Boyce, J., Briggs, G., Bunker, A., Collins, S. A., Hansen, C. J., Johnson, T. V., Mitchell, J. L., Terrile, R. J., Cook, A. F. II, Cuzzi, J. N., Pollack, J. B., Danielson, G. E., Ingersoll, A. P., Davies, M. E., Hunt, G. E., Morrison, D., Owen, T., Sagan, C., Veverka, J., Strom, R., and Suomi, V. E. 1982. A new look at the Saturn system: The Voyager 2 images. *Science* 215:504–537.

Smythe, W. D. 1975. Spectra of hydrate frosts: Their application to the outer solar system. *Icarus* 24:421–427.

Terrile, R. 1983. Structure and dynamical processes in Saturn's rings. In Proceedings of *IAU Colloquium 75*, ed. A. Brahic, Toulouse, France, Aug. 1982.

Tyler, G. L., Marouf, E., Simpson, R., Zebker, H., and Eshleman, V. R. 1983. The microwave opacity of Saturn's rings at wavelengths of 3.6 and 13 cm from Voyager I radio occultation. *Icarus* 54:160–188.

Weidenschilling, S. J., and Davis, D. R. 1983. Kinks, clumps, and accretion in Saturn's F Ring. *Lunar Planet Sci.* 14:840–841.

Zellner, B. 1979. I. The Tucson revised index of asteroid data. In *Asteroids*, ed. T. Gehrels (Tucson: Univ. of Arizona Press), pp. 1011–1013.

TRANSPORT EFFECTS DUE TO
PARTICLE EROSION MECHANISMS

RICHARD H. DURISEN
Indiana University

Various processes can erode the surfaces of planetary ring particles. Recent estimates for Saturn's rings suggest that a centimeter-thick surface layer could be eroded from an isolated ring particle in $<10^3$ yr by meteoroid impacts alone. The atoms, molecules, and chips ejected from ring particles by erosion will arc across the rings along elliptical orbits. For moderate ring optical depths, ejecta will be absorbed or inelastically scattered upon reintersecting the ring plane. Continuous exchange of ejecta between different ring regions can lead to net radial transport of mass and angular momentum, to changes in particle sizes, and to the buildup of chip regoliths several centimeters deep on the surfaces of ring particles. Because most of the erosional ejecta are not lost but merely exchanged over short distances, the net erosion rate of the surfaces of these ring particles will be much less than that estimated for an isolated particle. Numerical solutions for time-dependent ballistic transport under various assumptions suggest pile-up and spillover effects especially near regions of preexisting high optical depth contrast, such as the inner edges of A and B Rings. Global redistribution could be significant over billions of years. Other features in planetary ring systems may be influenced by ballistic transport.

I. INTRODUCTION

A. Evidence for Erosion

The action of erosion mechanisms in planetary rings can be inferred from the presence of the resultant gaseous or particulate ejecta. The first direct evidence came with the detection of Lyman $-\alpha$ radiation near the B Ring by sounding rocket and Earth-orbiting satellite experiments in the ultraviolet (Weiser et al. 1977; Weiser and Moos 1978; Barker et al. 1980). These observations and similar measurements by the Pioneer and Voyager space-craft imply a neutral hydrogen atmosphere extending one half to one Saturn radius above the ring plane with a number density of ~ 600 per cubic centimeter (Judge et al. 1980; Broadfoot et al. 1981). Because collisions with ring particles and formation of H_2 molecules would cause rapid loss of neutral hydrogen, the atmosphere must be replenished at a rate of $\sim 3 \times 10^{28}$ atoms per second (Carlson 1980; Ip 1983b), presumably by erosion of icy ring particles.

Pioneer and Voyager observations also show that micron-sized, forward-scattering particles dominate the visible part of Jupiter's ring and represent a significant or even dominant component in some regions of Saturn's rings. As discussed in chapters by Burns et al. and Grün et al., micron-sized particles are extremely short-lived in Jupiter's ring and in the E, F, and G Rings of Saturn, with lifetimes $\sim 10^2$ to 10^4 yr. Therefore on-going production by erosion of larger source particles seems required. In the main ring system of Saturn, the life span of micron-sized particles could be many orders of magnitude longer for two reasons: (1) these particles may reside most of the time in regoliths on the surfaces of larger ring particles; (2) the outer edge of the A Ring shields the rest of the rings from most trapped radiation and magnetospheric plasma. Nevertheless, the presence of micron-sized particles in Saturn's rings, even if relatively long-lived, suggests the action of a pulverizing, erosive mechanism.

B. Erosion Mechanisms and Rates

A variety of mechanisms can act to erode the surfaces of planetary ring particles: interparticle collisions; sputtering and thermal fatigue by cosmic rays or trapped radiation; sputtering due to neutralization of photoelectrons and ions from the ionosphere; sublimation of ices; photosputtering of ices by solar ultraviolet photons; and impacts by meteoroids. Some of these mechanisms, like sublimation or photosputtering, will produce only atomic or molecular ejecta. Others, like meteoroid impacts, will eject predominantly solid chips but also some gas, both neutral and ionized. Additional mechanisms can destroy ring particles that are already micron-sized. In this chapter we will concern ourselves primarily with mechanisms that produce copious ejections of neutral gas or chips from larger ring particles. Transport effects associated with micron- or submicron-sized charged particles or with plasmas are discussed in chapters by Burns et al. and Grün et al.

In 1967, Harrison and Schoen argued that absorption of solar ultraviolet photons in a surface monolayer of ice would cause the ejection of H and OH with high probability. This process is called photosputtering and is most effective for photons with a wavelength of ~ 1500 Å. According to their estimates, at 10 AU ice one centerimeter thick would be eroded in $\sim 3 \times 10^6$ yr. The Harrison and Schoen (1967) prediction resulted in several unsuccessful early searches for a ring atmosphere (e.g. Franklin and Cook 1969). Later

TABLE I

Gross Erosion Time Scales for Saturn's Rings

Mechanism	Time to Erode 1 cm Layer of Solid (1 g/cc) Ice (yr)	Reference
Sublimation at 100 K	3×10^9	Dennefeld (1974)
Ion Sputtering (E, F, G Rings)	10^8	Cheng et al. (1982)
Photosputtering	9×10^6	Carlson (1980)
Micrometeoroid impacts ($R_m < 100\mu$m)	2×10^5 1×10^5 1×10^5	Cook and Franklin (1970) Burns et al. (1980) Morfill et al. (1983)
Meteoroid impacts ($R_m > 100\mu$m)	3×10^3 3×10^2	Morfill et al. (1983) ($g = 4$) Morfill et al. (1983) ($g = 40$)

Blamont (1974) and Dennefeld (1974) considered a variety of erosional gas production mechanisms, including sublimation and meteoroid impacts, and predicted the density and distribution for a ring atmosphere. When this ring atmosphere was finally observed, Carlson (1980) again proposed photosputtering as the dominant gas production mechanism and revised the rate estimate (see Table I).

Bandermann and Wolstencroft (1969) and Cook and Franklin (1970) provided the first estimates of erosion rates for Saturn's rings due to micrometeoroid impacts by extrapolating the then existing inner solar system data to Saturn's orbit. Table I compares the Cook and Franklin estimate with later ones based on spacecraft data. This table gives gross erosion time scales, i.e., surface losses of ejecta are assumed to be uncompensated by gains of ejecta on the surface from elsewhere in the rings. The Burns et al. (1980) estimate is for Jupiter's ring and has been rescaled to Saturn by adjusting the gravitational focusing and yield to Saturn's gravitational field assuming that interplanetary micrometeoroid fluxes are approximately the same at 5 and 10 AU (Humes 1980). The most recent estimate in Table I given by Morfill et al. (1983) indicates that meteoroids with radii $R_m > 100\mu$m may be much more important than micrometeoroids. The factor g in the table is the enhancement factor applied to the interplanetary meteoroid flux in order to account for gravitational focusing by Saturn. According to Morfill et al., for the outer B Ring, $g \approx 4$ to 40 represents the range permitted by uncertainties in the distributions of the heliocentric orbital elements for meteoroids. The micrometeoroid ($R_m < 100\mu$m) erosion rates in the table are based on g-values of ~ 4.

Although erosion rates for the other mechanisms mentioned in Sec. I.B are often imperfectly known, meteoroid impacts and photosputtering are probably the dominant processes (cf. Table I of Carlson 1980).

C. Ejecta Properties for Photosputtering and for Meteoroid Impacts

Little information is available concerning the consequences of photo-absorption of ultraviolet radiation by water ice. A 1500 Å photon has an energy of about 8.3 eV. Carlson (1980) argues that most of this energy will be expended in dissociating the H_2O, in collisions with other surface molecules, and in excitation of the OH. Escape velocity at the outer edge of the A Ring is about 24 km s^{-1} so H and OH require kinetic energies of at least 3.0 eV and 51 eV, respectively, in order to escape. Carlson expects the total available kinetic energy to be <3 eV in most cases. Then, the ejected OH will have ejection velocities v_{ej} with respect to their parent ring particles of magnitude <6 km s^{-1}. Considering the number of possible energy loss mechanisms, even most H atoms are likely to photosputter with $v_{ej} \lesssim 10$ km s^{-1} and will probably avoid direct loss to escape. However, the uncertainties are considerable and determination of photosputtering erosion rates and v_{ej} distributions by laboratory experiments would be useful.

We know considerably more about ejecta due to hypervelocity meteoroid impacts. Hypervelocity in this context means an impact at a speed much greater than the sound speed in either the target or the projectile so that shock waves move into both from the point of impact. The shocks vaporize part (or all) of the projectile plus a small volume of the target, and they shatter and excavate a much larger volume of the target. Much of the detailed information comes from the now classic work by Gault and his collaborators (e.g. Gault et al. 1963; Gault and Wedekind 1969; Gault 1973; Stöffler et al. 1975). Let m and v_m be the mass and impact speed of the meteoroid, and let M_{ej} be the total mass of the fragments (chips) that are excavated and ejected. The experimental results may be crudely approximated, for our purposes, by

$$dM_{ej} \approx A\ v_{ej}^{-\alpha}\ dv_{ej} \quad \text{for } v_{ej} > v_0 \tag{1a}$$
$$\approx 0 \text{ for } v_{ej} < v_0$$

and

$$M_{ej} \approx 5 \times 10^3 m \left(\frac{v_m}{30 \text{ km s}^{-1}}\right)^2 \tag{1b}$$

where dM_{ej} is the mass of fragments ejected with speeds in the interval v_{ej} to $v_{ej} + dv_{ej}$.

Equations of the form (1a) have been used by Cook and Franklin (1970) with α varying between 2.6 and 4 and by Greenberg et al. (1978) with $\alpha = 3.25$. A comparison of an integral version of Eq. (1a) with data of Gault et al.

(1963) and Stöffler et al. (1975) can be found in Fig. 1 of Greenberg et al. (see also Fig. 3 of Housen et al. 1979). With $\alpha \approx 3$, Eq. (1a) is so steep that 75% of the ejecta have velocities between v_0 and $2\ v_0$. Below the characteristic speed v_0, the fraction of ejected mass is negligible. The value of v_0 depends on the strength and the surface composition of the target. For solid basalt, v_0 is ~ 100 m s^{-1} while v_0 for loose sand is only ~ 1 m s^{-1}. The values of α, however, are similar in both cases.

Equation (1b) is adopted from Grün et al. (1980) and is based on a summary of relevant experimental data for hard basaltic targets. For sand (Stöffler et al. 1975) and for hard ice targets (Lange and Ahrens 1982; Mizutani et al. 1982), the coefficient in Eq. (1b) is probably ~ 30 times larger. Croft (1982) reports that hypervelocity craters in snow are not much larger than those in solid ice, implying comparable ejecta yields for snow and ice. To be extremely conservative, Eq. (1b) was used to estimate the time scales of the gross meteoroid erosion in Table I without the extra factor of 30. Some caution is needed in applying Earth-based laboratory data for very weak or unconsolidated targets to the essentially zero-gravity conditions that hold for impact on centimeter- to meter-sized ring particles. Little data is available in the literature concerning the $\mathbf{v}_{ej}$ distributions for impacts on snow or ice.

In addition to fragments, hypervelocity impacts produce vapor and plasma. Summarizing the laboratory work of others, Morfill et al. (1983) cite vapor masses of $\lesssim 3m$ with v_{ej} less than but on the order of v_m. Only $\sim 1\%$ of the vapor is ionized. Meteoroid impacts are thus a significant source of gas and plasma as well as chips.

D. Consequences of Erosion for Ring Structure

Various lines of evidence suggest a surface mass density σ for Saturn's rings of $\lesssim 100$ g cm^{-2} (see chapter by Cuzzi et al.). If, for simplicity, we visualize the rings as a sheet of solid ice, they would have a thickness of 1 m. In this case, according to the Morfill et al. estimates for meteoroid impacts in Table I, the total gross erosion time for Saturn's rings is only 3×10^4 to 3×10^5 yr. If the rings are truly as old as the solar system, their total mass has been eroded 10^4 to 10^5 times over. Because, as described below, erosion results primarily in the trading of ejecta from the surfaces of ring particles over relatively short distances, net gains or losses in most ring regions will only occur on time scales several orders of magnitude longer than the gross erosion time. Nevertheless, if more than a few parts in 10^5 of the ejected mass has high enough velocity to be lost from the ring system by escape or by falling into the planet (e.g., impact vapor), then the gross erosion would be sufficiently large to raise questions about a primordial origin for the rings. Whether or not gross overall losses or gains prove to be this severe, the erosion time scales in Table I are sufficiently short that we can expect significant observable effects of erosion.

To understand these effects, let us first consider the kinematics of erosion ejecta. Most atomic and molecular ejecta are neutral and their time scales for ionization are considerably longer than an orbital period, at least in the Saturn environment (Carlson 1980). Most chip ejecta are larger than $1\,\mu$m in size and so do not experience notable non-Keplerian forces in one orbit period whether charged or not (see chapters by Burns et al. and Grün et al.). As a result, the vast bulk of erosion ejecta will follow Keplerian orbits determined by the vector sum $\mathbf{V} = \mathbf{v}_{ej} + \mathbf{v}_c$, where $\mathbf{v}_c$ is the circular orbit velocity of the parent ring particle. Typically only a total mass of chips comparable to m will have $V \geqslant$ escape velocity v_{esc} (Cook and Franklin 1970). Hence, the vast bulk of ejecta will arc across the rings (or, in some cases, into the planet) along elliptical orbits. Let r be the radial distance from the planet's center as measured in the ring plane. There will be a high probability that an ejectum will be absorbed or inelastically scattered if the ring optical depth is large at the radius r_{int} where the ejectum orbit reintersects the ring plane. Because r_{int} will in general be different from the orbit radius r_p of the parent ring particle, ejection of atoms, molecules, and chips by surface erosion can lead to net radial exchanges of material and angular momentum. We refer to this process as ballistic transport.

Local ($\mathrm{v}_{ej} \ll \mathrm{v}_c$) exchanges of ejecta between neighboring ringlets of similar properties will almost exactly cancel. But even when net radial transport is small for this reason, ring particles will build up steady-state regoliths of chips from meteoroid impacts, and the ring plane will be surrounded by a steady-state halo of ejecta which have not yet been reabsorbed. The regoliths and halo probably provide the reservoir of micron-sized particles which manifest themselves in spoke activity. In addition, the surfaces of ring particles will be contaminated by meteoroid material. Photosputtering and reabsorption of H and OH and impact vaporization will also modify surface composition and texture of ring particles.

Neighboring ring regions do not always have similar properties. Ring systems exhibit radial structure on many length scales, including sharp edges and abrupt optical depth gradients. Near such features, significant net effects can occur even locally. Although the fraction of ejecta causing global transport across a broad ring system will be rather small according to Eq. (1a), dramatic global redistribution is possible over billions of years, because the gross erosion rates are so high. Different ring regions could suffer net losses or gains of mass with consequent alteration of normal optical depth, particle surface properties, and particle sizes. Moreover, net changes in specific orbital angular momentum due to ejecta exchanges could lead to changes in the radius of a ringlet and could enhance (or blur) the sharpness of some ring features.

The remainder of this chapter deals with the quantitative description of ballistic transport and its application to planetary ring systems, with special emphasis on Saturn's rings.

II. BALLISTIC TRANSPORT

A. General Principles and an Illustrative Example

Most periodic or transient asymmetries, whether in a north-south or azimuthal sense, like shadowing, obliquity, density waves, or spoke activity, should average out over the erosion time scales in Table I. Therefore, for erosional studies, the system of parent ring particles can be viewed as a thin cylindrically symmetric sheet whose properties depend only on radius r and time t. Because the dispersion velocities of ring particles are small, typically $\ll v_0$, the orbits of parent ring particles can be treated as circular for computing ejecta orbits. The local vertical structure of the rings can influence gross erosion rates and ejecta velocity distributions through optical depth effects. A plane-parallel uniform slab approximation is usually adequate.

With these assumptions, the ballistic transport problem involves the description of how the properties of ring particles at all radii vary with time due to Keplerian exchange of ejecta. The gross ejection rate $R(r, t)$ is the mass ejected per unit ring area per unit time. It depends in part on the normal optical depth $\tau(r, t)$ of the rings, because the erosion is typically caused by an external flux of projectiles (photons or meteoroids). For given projectile influx, the properties of local ring particles and optical depth also determine the v_{ej} distribution and the yield, i.e., M_{ej}/m. Knowing these at each r, it is then possible, in principle, to calculate the orbits of all ejecta and, over an interval of time, to tally gains at r due to ejecta absorbed there from other regions against losses due to ejecta thrown from r and absorbed elsewhere. The calculation must account for the relative probability of absorption at r_{int} and reabsorption at r_p. Under some assumptions, ejecta may undergo inelastic scatterings prior to absorption (Ip 1983a). Overall then, ballistic transport has the general character of a diffusion or transfer problem, but one where the diffusing species follow elliptical or even parabolic and hyperbolic orbits.

To illustrate some features of ballistic transport, let us consider a drastically simplified hypothetical system of three narrow ringlets, which we call A, B, and C, with equal widths dr (see Fig. 1). (These ringlets must not be confused with Saturn's Rings A, B, and C.) Assume identical particle properties and projectile fluxes for the ringlets. Then, $\tau_A = \tau_B = \tau_C$ and $R_A = R_B = R_C = R$. Suppose there is a single value of $x \equiv v_{ej}/v_c \ll 1$ that is just sufficient to allow trading of ejecta by directly prograde and retrograde ejections, i.e., with v_{ej} nearly parallel or antiparallel to v_c. For such ejections, r_{int} corresponds to apocenter and pericenter respectively. Assume equal numbers of ejections fore and aft.

Keplerian orbit dynamics tells us that the orbital motion of an ejectum will satisfy

$$h^2 = G M a (1 - e^2) \tag{2}$$

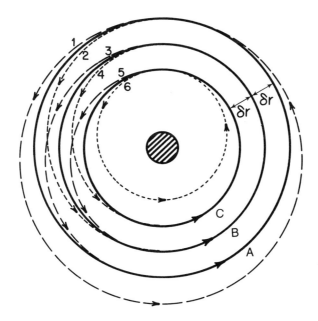

Fig. 1. Diagram illustrating a system of three narrow, coplanar, circular rings (heavy solid lines) exchanging ejecta by directly fore and aft ejections. The directions of orbital motions are indicated by arrows. Ejecta 1, 3, and 5 (long-dashed curves) are ejected in a prograde sense as viewed from the parent ring particle and ejecta 2, 4, and 6 (short-dashed curves) in a retrograde sense.

$$V^2(r) = G M \left(\frac{2}{r} - \frac{1}{a} \right) \tag{3}$$

where h is the specific orbital angular momentum, a the semimajor axis, e the eccentricity, G the gravitational constant, and M the mass of the central planet. By virtue of our assumptions, $V(r_p) = v_c(r_p) \pm v_{ej}$ and $h = Vr_p$. Recalling that apocenter and pericenter are just $a(1 + e)$ and $a(1 - e)$, then from Eq. (2) we find, to first order in x,

$$\delta r = r_{int} - r_p \approx \pm 4x r_p \tag{4}$$

$$h - h_p \approx \pm x h_p \tag{5}$$

$$h - h_{int} \approx \mp x h_{int} \tag{6}$$

where h_p and h_{int} refer to the specific orbital angular momenta for circular orbits at r_p and r_{int}, respectively. The upper signs correspond to prograde ejections, the lower signs to retrograde ejections. To lowest order in x, then, choosing $r_C = r_B (1 - 4x)$ and $r_A = r_B (1 + 4x)$ permits trading of ejecta.

Assume large τ's so that ejecta always collide with a ring particle at r_{int} and assume ejecta are not scattered but are promptly absorbed upon collision. Figure 1 shows sample trajectories for ejecta from each ringlet. Each ejectum is visualized as being emitted at the top of the diagram and initially follows an elliptical orbit tangent to its parent ringlet. Ejecta 2 and 5 are absorbed by B, 3 by A and 4 by C. All absorptions occur at the bottom of the diagram. Ejecta 1 and 6 are reabsorbed by their parent ringlets. The net rate of mass gain M_B by B is

$$\dot{M}_B = \pi R \; dr \, (r_A + r_C - 2r_B) \approx 0 \tag{7}$$

to first order in x. On the other hand,

$$\dot{M}_C = \pi R \; dr(r_B - r_C) \approx (4\pi R \; r_B \, dr)x \tag{8a}$$

$$\dot{M}_A = \pi R \; dr(r_B - r_A) \approx -(4\pi R \, r_B \, dr)x. \tag{8b}$$

Our system thus illustrates the following important conclusions. For ejection patterns which have fore-aft symmetry, the largest effects in erosional evolutions are first order in x and occur at the edges of a ring system. The effect of cylindrical geometry in an otherwise uniform ring system is to cause net erosion of the outer edge and net pile up at the inner edge. As the reader can verify, these conclusions are not affected to first order in x by a gradient in ring properties, e.g., $R = R_B + (r - r_B)d R /dr$, except that a sufficiently negative $d R /dr$ could reverse the signs in Eq. (8a,b).

Equations (2) and (3), solved exactly for prograde and retrograde ejections, give

$$\frac{r_{int}}{r_p} = \frac{1 \pm 2x + x^2}{1 \mp 2x - x^2} = 1 \pm 4x + 6x^2 + \mathcal{O}(x^3) \tag{9}$$

instead of Eq. (4). The symmetry causing $\dot{M}_B \approx 0$ to first order in x is broken to second order in x. Similarly, there are second order effects introduced by radial gradients and second derivatives. Over long enough times, these second order effects could accumulate even for small x.

To first order in x, Eq. (5) tells us that prograde ejecta decrease the specific angular momentum of the parent, while retrograde ejecta increase it. Furthermore, Eqs. (5) and (6) together say that the specific angular momenta of ejecta exchanged by neighboring ringlets are the same. This occurs because exchanged ejecta follow transfer orbits of the same a and e in Fig. 1. For instance, ejecta absorbed at A from B and ejecta absorbed at B from A both have $h = h_A (1 - x) = h_B (1 + x)$. Let J be the total angular momentum of ringlet $\mathcal{A}$. Then to lowest order $\dot{J}_A \approx h_A \dot{M}_A$. Hence the time derivative of $h_A = J_A /M_A$ is given by

$$\frac{\dot{h}_A}{h_A} = \frac{\dot{J}_A}{J_A} - \frac{\dot{M}_A}{M_A} \approx \left(\frac{h_A}{J_A} - \frac{1}{M_A} \right) \dot{M}_A \approx 0 \qquad (10)$$

to first order in x. Similarly $\dot{h} \approx 0$ to first order or higher for ringlets B and C. Thus, changes in specific angular momentum due to erosional transport are everywhere at most second order for fore-aft ejection symmetry. We need to focus attention on $\dot{h}$ not $\dot{J}$ because it is $\dot{h}$ which will cause the orbital radius of a ringlet to change.

We can easily generalize our three-ringlet example by adding more ringlets. $\dot{M}$ will remain second order for interior ringlets and first order for ringlets at or near the edges, while $\dot{h}$ remains at most second order everywhere. It should be noted that our assumption of directly prograde and retrograde ejections is unrealistic, because such ejecta see infinite slant optical depth and so do not readily escape their parent ringlet. However, only small deviations from fore and aft ejection can loft ejecta above a thin ring system. Therefore, we can multiply ringlets up to the limit of a continuous ring system, and generalize the distribution of ejecta directions (provided we preserve fore and aft symmetry) without affecting the conclusions stated above.

In realistic situations, the directional distribution of ejecta may not have fore-aft symmetry. As discussed in Sec. II.B, this could arise, for instance, due to anisotropic influx of meteoroids caused by the orbital motion of the ring particles. If the difference between the prograde and retrograde ejection rates is a significant fraction of R, then the reader can easily verify that the cancellations occurring in Eqs. (7), (8), and (10) are degraded by one order in the quantity x. A crucial implication is that, for ejection patterns with significant fore-aft asymmetry, the largest effects in mass redistribution are zero order in x and occur at the edges, while specific angular momentum changes become first order in x at the edges. Anisotropic ejections can also reverse the direction of pile up from inner to outer edges, if prograde ejections dominate. First order changes in specific angular momentum could play a role in ring dynamics by countering or enhancing viscous spreading.

Although the above illustrations do not cover all reasonable assumptions, they do provide a useful guide for understanding the results of more detailed calculations. The remainder of Sec. II describes two mathematical approaches which have been used so far to treat ballistic transport in planetary rings.

B. Mathematical Formulation

The elliptical orbits of ejecta have nodes at r_p and r_{int}. The optical depths $\tau(r_p)$ and $\tau(r_{int})$ and the slant path of an ejectum through the rings, which depends on $\dot{v}_{ej}$ together determine the relative probability that an ejectum will collide with a ring particle at one of these radii. A relatively simple and precise mathematical formulation of ballistic transport is possible in the limit of "prompt absorption" (Durisen et al. 1982, 1983). By this is meant that the collisions are so inelastic that the ejecta are immediately assimilated back into the ring particle population without further transport, presumably by incorporation into the regolith of a ring particle. The validity of this assumption will depend on the nature of the surfaces of ring particles and on the magnitude of v_{ej}. For the bulk of meteoroid ejecta, $v_{ej} \approx v_0 \approx 1$ to 100 m s^{-1} in Eq. (1a). Collisions with a ring particle that has a hard smooth surface are then mildly elastic, while collisions with a ring particle that has a regolith are highly inelastic (cf. Hartmann 1978). As argued in Sec. III.C, a principal effect of meteoroid erosion is likely to be regolith production. Hence, prompt reabsorption, as opposed to inelastic scattering, is probably a correct assumption for the bulk of chip ejecta, if regoliths do develop. For higher v_{ej}, the ejecta themselves become erosive projectiles. However, because the yields and v_{ej} of such secondary ejecta will be relatively small, they have so far been ignored in the treatment of ballistic transport.

For prompt absorption, net local changes in the surface density $\sigma(r,t)$ are readily expressed by a mass continuity equation

$$\frac{\partial \sigma}{\partial t} + \frac{1}{r}\frac{\partial}{\partial r}\left(\sigma r v_r\right) = \Gamma_m - \Lambda_m \tag{11}$$

where Γ_m and Λ_m are the rates of mass gains and losses, respectively, per unit area due to exchanges of ejecta with other ring regions. The left-hand side of the equation includes the possibility of slow radial drift v_r, because net changes in the specific angular momentum due to ballistic transport will cause a ringlet of material orbiting a central mass to change its orbital radius. Assuming that ring particles remain on nearly circular orbits, v_r can be determined via

$$v_r \frac{dh_c}{dr} = \frac{1}{\sigma}\left[\Gamma_h - \Lambda_h - h_c\left(\Gamma_m - \Lambda_m\right)\right] \tag{12}$$

where $h_c = \sqrt{GMr}$ is the specific orbital angular momentum for a circular orbit at r, and Γ_h and Λ_h are rates of angular momentum gains and losses per unit area due to ejecta exchanges. The left-hand side of Eq. (12) is the rate of change of specific angular momentum required by orbit mechanics in order for a circular orbit to drift radially at a speed v_r. The right-hand side of Eq. (12) is the instantaneous rate of change of the specific angular momentum of a ringlet due to ballistic transport. Equations (11) and (12) must be supplemented by equations describing the rate of change of local ring particle sizes and optical depth. For instance, in the simplest case of a single local particle radius $R_p(r, t)$ and internal density ρ_p, we get

$$\frac{\partial R_p}{\partial t} + v_r \frac{\partial R_p}{\partial r} = \frac{R_p}{3\sigma}\left(\Gamma_m - \Lambda_m\right) \tag{13}$$

and

$$\tau = \frac{3\sigma}{4R_p\rho_p} \tag{14}$$

if we assume that all particles grow or shrink simultaneously. Much of the difficulty in solving Eqs. (11) to (14) is hidden in the gain and loss terms, which are integrals over r and over ejection directions and speeds. Due to a high degree of cancellation between Λ's and Γ's, these integrals must be calculated accurately.

Figure 2 illustrates the geometry used to describe ejection direction distributions. The sphere in this figure is not a ring particle but represents the sphere of all possible directions for the vector $\mathbf{v}_{ej}$. The parent ring particle should be visualized as a very small sphere at the origin. In terms of θ and ϕ, Keplerian orbit mechanics gives

$$\frac{r_{int}}{r_p} = \frac{1 + 2x\cos\theta + x^2(\cos^2\phi + \cos^2\theta\sin^2\phi)}{1 - 2x\cos\theta - x^2(\cos^2\phi + \cos^2\theta\sin^2\phi)} \tag{15}$$

where $x = v_{ej}/v_c$ evaluated at r_p. The θ, ϕ which give the same r_{int}/r_p are indicated in Fig. 2 as curves on the sphere of all possible ejection directions for $x = 0.45$. Ejecta with θ, ϕ in the hatched cap escape from the system on hyperbolic orbits. As discussed in detail by Cook and Franklin (1970), ejecta can be lost by falling to the planet for ejection directions in the stippled region of Fig. 2, where the ejecta orbits have a pericenter less than a planetary radius. For many applications, neither hyperbolic escape nor fall into the planet is important, because most ejecta have $x \ll (\sqrt{2} - 1)$, i.e., ejection velocity much less than needed for escape. In regions of low optical depth near a highly oblate planet like Saturn, r_{int} may shift significantly due to apsidal motion prior to absorption. This effect has so far been ignored.

The mass loss integral in Eq. (11) is

$$\Lambda_m(r, t) = \mathcal{R}(r, t) \int_0^\infty dx \int_{-1}^1 d(\cos\theta) \int_0^{2\pi} d\phi\, Pf \tag{16}$$

where P is the probability that ejecta are truly lost from r through absorption at r_{int} or through loss to escape or the planet. In general, P depends on x, θ, ϕ, $\tau(r_{int})$, and $\tau(r)$. P depends on θ, ϕ because the probability of a collision at r or r_{int} is determined by the slant path of an ejectum through the rings as seen by a ring particle (see Durisen et al. [1983] for more details). The function f in Eq. (16) describes the fractional distribution of ejecta per unit x, $\cos\theta$, and ϕ

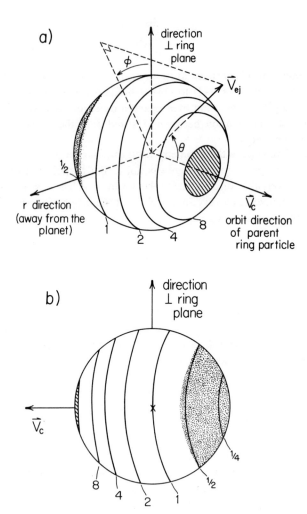

Fig. 2. The sphere of all possible ejection directions from a parent ring particle illustrated in two perspectives. The ring particle itself should be visualized as a small sphere at the origin. The case shown is for $x = 0.45$ and for r_p twice the planetary radius. The thin solid curves indicate the orientations of v_{ej} leading to particular values of r_{int}. The curves shown are for $r_{int}/r_p = 1/4, 1/2, 1, 2, 4, 8$ as labeled. In the cross-hatched region, $|V| = |v_c + v_{ej}| \geq v_{esc}$. In the stippled region, the periapse of the ejectum orbit is less than the planetary radius. A large value of x was chosen to illustrate all important behaviors clearly. The (a) view shows the spherical coordinate angles (θ, ϕ) used to describe the direction of v_{ej}. The (b) view looking radially outward from the planet (as indicated by the X) clearly shows the noncircular shape of the r_{int}/r_p curves and of the boundary for the planetary loss region (figure from Durisen et al. 1983).

interval. Frequently, f will be assumed to be separable in x and θ,ϕ and independent of r and t, so that

$$f(x,\ \theta,\ \phi) = f_x(x) f_{\theta\phi}(\theta,\phi). \tag{17}$$

The functions f_x and $f_{\theta\phi}$ are normalized so that their integrals over all x and all θ, ϕ are unity.

The gain term Γ_m is obtained by using orbit mechanics to determine which $d\theta,d\phi$ intervals at r' cause ejecta with given x to land in an interval dr around r. This is then integrated over x, ejection direction, and r'. (Casting Γ_m in a convenient form involves some subtleties beyond the scope of this chapter [see Durisen et al. 1983].) Γ_m also depends on R, P, and f. Λ_h and Γ_h have analogous forms with integrands $P f h_\perp (\theta,\phi)$, where $h_\perp$ is the component of an ejectum's specific angular momentum perpendicular to the ring plane. For both meteoroid impacts and photosputtering, net transport of other components of $\mathbf{h}$ should cancel over times longer than a few planetary orbits around the Sun.

It is convenient to divide the gross ejection rate $R(r,t)$ into two parts,

$$R(r,t) = R_e(r,t)\ R_\tau(r,t) \tag{18}$$

where R_e contains all dependences on the physical properties of the ring particles and the projectiles, while R_τ is the fraction of projectiles absorbed and depends only on ring optical depth. Using a uniform slab approximation for the local vertical ring structure, we find that

$$R_\tau = 1 - e^{-\tau/\cos\gamma} \tag{19}$$

if the projectiles are incident at an angle γ to the ring plane normal. For solar ultraviolet photons, the range of incidence angles γ will be limited by planetary obliquity. For meteoroid projectiles, R_τ will be complicated by aberration effects due to the orbital motion of ring particles (Cook and Franklin 1970). Calculations by Durisen et al. (1983), discussed in Sec. III.B of this chapter, simulate an extreme aberration effect by assuming that the intensity of meteoroid projectiles at incidence angle γ is proportional to sec γ. Then, for a uniform absorbing slab,

$$R_\tau = 1 - \int_1^\infty y^{-2} e^{-\tau y}\, dy = 1 - E_2(\tau) \tag{20}$$

where $E_2(\tau)$ is a function common to transfer problems and is tabulated in Appendix I of Kourganoff (1963).

To illustrate some general features of time-dependent solutions to Eq. (11), consider an idealized case with the following simplications: angular momentum transport is ignored so $v_r = 0$; σ and τ are initially uniform between two radii and zero elsewhere; $\tau \gg 1$ so $R_\tau = 1$ and $P = 1$ wherever $\sigma \neq 0$; all ejecta have the same x; R_e is a constant; losses to the planet are ignored; and ejections from the parent ring particle are isotropic, i.e., $f_{\theta\phi} = 1/(4\pi)$. Figure 3 presents the results in normalized units where the initial $\sigma = 1$. All evolutions shown are for one "total gross erosion time" $t_g \equiv \sigma/R_e$ (the time it would take a ring region with $\tau \gg 1$ to be eroded completely away if Γ_m were zero, i.e., if there were no compensating mass gains from other ring regions).

For $x = 0.01$ in Fig. 3a, we see exactly the behavior predicted from the 3-ringlet example in the previous section: net pile up at the inner edge and net erosion of the outer edge to first order in x and a second order net change in σ elsewhere. Figure 3a and b together illustrate the importance of geometry and distances thrown, through 4 special radii defined as follows:

1. $r_{\ell o}$ = the smallest r from which ejecta are lost to the outer edge;
2. r_{go} = the smallest r which gains ejecta from the outer edge;
3. $r_{\ell i}$ = the largest r from which ejecta are lost to the inner edge;
4. r_{gi} = the largest r which gains ejecta from the inner edge.

Only for small x, where $r_{gi} < r_{\ell o}$, does a central region exist for which net transport cancels to first order. For larger x, when r_{gi} and $r_{\ell o}$ cross over to $r_{\ell o} < r_{gi}$, as seen in Fig. 3b, the mass transport effects are no longer confined to edges but represent an overall inward transfer of mass. Figure 3c shows that the edge effects for small x preserve their structure even for narrow rings as long as $r_{gi} < r_{\ell o}$. The detailed structure in x = const. evolutions can of course be washed out with realistic x-distributions as shown below in Sec. III.B. Using $v_{ej}(r)$ = const. instead of $x(r)$ = const. changes only the details. Despite the relatively small effects in Fig. 3, $t_g = 1$ so that every ring region has been involved in mass exchange comparable to its initial mass. Ring particles at all r have thus developed regoliths of absorbed ejecta.

Figure 3d is identical to 3a except that $f_{\theta\phi} = (1 + 0.14 \cos \theta)/4\pi$, introducing a bias toward prograde ejections. Clearly, anisotropic ejection changes the sign and amplitude of the net effects. Unfortunately, $f_{\theta\phi}$ is one of the most uncertain parameters. Cuzzi and Durisen are currently elaborating earlier efforts by Cook and Franklin (1970) to determine $f_{\theta\phi}$ for meteoroid impacts. Meteoroids preferentially strike ring particles on their leading hemispheres due to aberration. For hypervelocity impacts on a compacted surface, there is a cone of ejecta roughly symmetric about the particle surface normal (Gault et al. 1963). Summing over such cones for an aberrated meteoroid flux gives a prograde sense to $f_{\theta\phi}$. Less is known about ejecta distributions resulting from oblique hypervelocity impacts onto a noncompacted regolith of ice

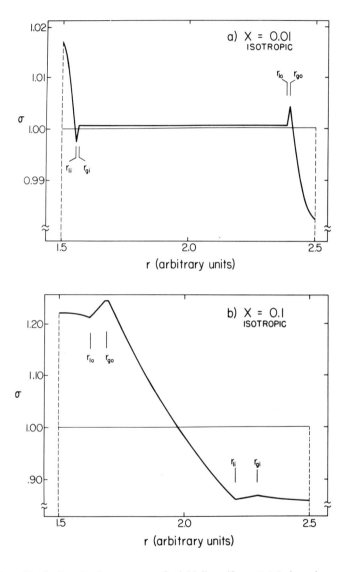

Fig. 3a and b. Surface density σ versus r for initially uniform $\tau \gg 1$ rings shown after one total gross erosion time by the heavy solid lines. The light solid lines indicate the initial σ normalized to unity. The simplifying assumptions made for these evolutions are listed in the text. The four special radii indicated in (a) and (b) are also defined in the text. (a) Ejections isotropic with $x = 0.01$; (b) Ejections isotropic with $x = 0.1$.

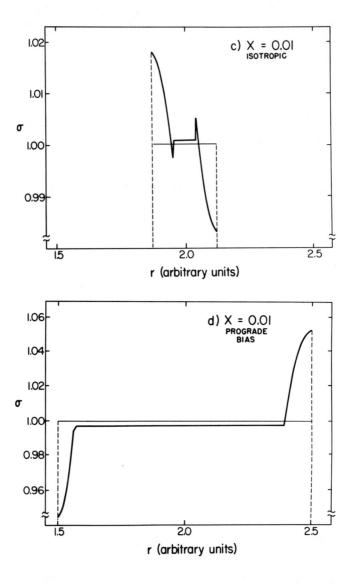

Fig. 3c and d. (c) Same as (a) except that the ring is initially narrower; (d) Same as (a) except the ejections are anisotropic with a prograde bias (figure from Durisen et al. 1983).

chips. The ejecta may preserve some sense of the projectile's direction of motion due to burrowing; in that case a net retrograde sense to $f_{\theta\phi}$ could result.

C. The Monte Carlo Method

The formalism presented above is susceptible to only limited generalization. For instance, high v_{ej} ejecta may produce a secondary shower of ejecta. This could be handled through extra terms in Eqs. (11) and (12), or by a two-step time integration. Unfortunately, the nature of collisions between ejecta and ring particles is fairly uncertain. If collisions are even moderately elastic, ejecta may undergo several orbit-changing scatterings prior to absorption. In the latter case, it is difficult to formulate tractable Γ's and Λ's for Eqs. (11) and (12). In the multiple scattering limit, the Monte Carlo approach taken by Ip (1983a) is much easier to implement.

A Monte Carlo simulation of ballistic transport involves following the history of a large number of ejecta sampled at random. With given values for elasticity and for scattering and absorption probabilities, a computer code can assign an outcome to each collision experienced by an ejectum until it is absorbed or effectively comes to rest. Clearly, such a technique is extremely flexible and may prove to be the only way to obtain results under complex assumptions. Disadvantages include some difficulty in understanding the results physically and the need to follow a large number of sample ejecta in each simulation. Monte Carlo techniques become especially cumbersome for time-dependent problems. However, the Λ's and Γ's in Eqs. (11) and (12) are also computationally cumbersome because of the multiple integrals (Eq. 16).

For his Saturn ring simulations, Ip (1983a) divides the main ring system into 20 radial bins with centers r_i. He follows 100 ejecta from each bin with a given v_{ej} and randomly oriented directions. For simplicity, he ignores the slant paths followed by ejecta through the ring plane and assumes instead that the probability of collision at r_{int} is $1 - \exp[-\tau(r)]$ (Eq. [19] with $\gamma = 0°$). When a particle collides, its relative speed is reduced by a factor β, chosen to be 0.5, and its direction is randomized. An ejectum is said to "come to rest" in Ip's simulations when its relative velocity is < 1 m s^{-1}. Losses to Saturn are included but usually are not very large.

The resulting sample ejecta trajectories can be used in a variety of ways. Ip assumes that ejecta come to rest without being absorbed and that they remain free particles for a finite "destruction" lifetime t_d after they come to rest. The free ejectum "destruction" mechanism is unspecified but could be absorption into a ring particle regolith or catastrophic fragmentation by a meteoroid. Under these assumptions, the steady-state surface number density n_j of free ejecta in bin j is

$$n_j r_j = \frac{t_d}{N} \sum_{i=1}^{\#\,bins} N_{ij} \; \mathcal{R}(r_i) r_i \qquad (21)$$

where N is the total number of sample ejecta and N_{ij} is the number of the sample ejecta from bin i that come to rest in bin j. The r_j and r_i's are present to account for the cylindrical geometry of the bins. Ip chooses $R \ (r_i) \sim \tau \ (r_i)$.

To generalize his calculations to include time-dependence, Ip could use $\Gamma_m = n_j/t_d$ and $\Lambda_m = R(r_j)$ in an equation like Eq. (11) for each bin. As σ and τ change, new sets of sample ejecta trajectories would have to be computed.

The reader will notice significant differences between Ip's (1983a) and Durisen et al.'s (1983) underlying assumptions about time-dependence, collision physics, ejection rates, optical depth effects, and the fate of ejecta. Unfortunately, these differences reflect the considerable uncertainties in our current understanding of these physical processes. However, these two mathematical formulations may prove fruitful in treating different problems.

III. APPLICATIONS

This section applies ballistic transport theory specifically to Saturn's rings. Although erosion is certainly important in Jupiter's ring and must also occur in the rings of Uranus, the nature and/or distribution of parent particles are poorly known in those cases, making it difficult to consider net transport. Nevertheless, some features of Saturn-oriented calculations may be generalized to the other ring systems, and these are noted below.

A. Steady-State Concentrations of Ejecta

So far, the only attempts to model global effects in Saturn's rings due to ballistic transport have been made by Ip (1983a) and by Durisen et al. (1983), using the methods described in the preceding sections. In their most detailed calculations, both groups use the step function form of initial $\tau(r)$ shown in Fig. 4. Unfortunately, because of differences in emphasis and techniques, the work by the two groups cannot be directly compared, although some features are common. This subsection concerns Ip's work with the Monte Carlo method.

Ip has used Eq. (21) with 20 radial bins to calculate the steady-state concentration n_j of multiply scattered free ejecta that have come to rest. As shown in Fig. 5, for all v_{ej}, Ip finds strong edge effects, primarily a pile up of materials at regions of preexisting large density contrast. The asymmetry somewhat favoring inner edge pile up is probably a consequence of cylindrical geometry, as in the three-ringlet example of Fig. 1. For Ip's multiple scattering calculation, however, outer edges are also regions of pile up even for isotropic initial ejections, because the density contrasts act as barriers to the diffusion of ejecta, i.e., a multiply scattering ejecta near a ring edge is more likely to come to rest on the higher τ side of the edge at the end of its damped random walk. Higher v_{ej} appears to favor trapping of ejecta in the B Ring, while lower v_{ej} causes more spillover of ejecta into neighboring ring regions.

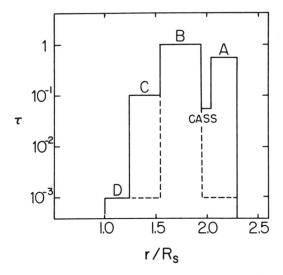

Fig. 4. The initial step-function $\tau(r)$ model for Saturn's rings as shown by the solid curve used by Ip (1983a) and by Durisen et al. (1983), except that Durisen et al. do not include the D Ring. R_S is the equatorial radius of Saturn. The dashed profile is used by Ip (1983a) for the calculations shown in Fig. 5d.

Ip concludes that the observed concentration of micron-sized fragments in the B Ring could be explained by a characteristic $v_{ej} \sim 0.7$ to 1.0 km s^{-1}. This seems considerably higher than the typical v_{ej} expected from meteoroid impacts.

It is important to remember that the profiles in Fig. 5 represent the steady-state distribution of ejecta only and not the surface mass density σ of the rings. If the free ejecta are eventually reabsorbed onto ring particles, then one can infer from Fig. 5 that σ will increase in regions where n_j is large and decrease where n_j is relatively small. If instead a significant fraction of the free ejecta is destroyed, then the net transport will be more complicated to characterize. Ip's techniques could be used to simulate such self-consistent time-dependent evolutions of σ but to date this has not been done.

Ip has also calculated steady-state n_j for other $\tau(r)$, for instance the dashed profile in Fig. 4 which simulates a relatively isolated high optical depth ringlet. Figure 5d shows a resulting profile that strongly resembles the structure of some narrow ringlets found in both Saturn's rings and the rings of Uranus (see chapters by Cuzzi et al. and Elliot et al.). Although several dynamical mechanisms have been proposed which could produce such double-peaked profiles with sharp edges (see chapter by Dermott), there is no single preferred explanation, and ballistic diffusion should be considered a possibility. The observed preference of small particles for the edges of the narrow C Ring "plateau" regions, as described in the chapter by Cuzzi et al.,

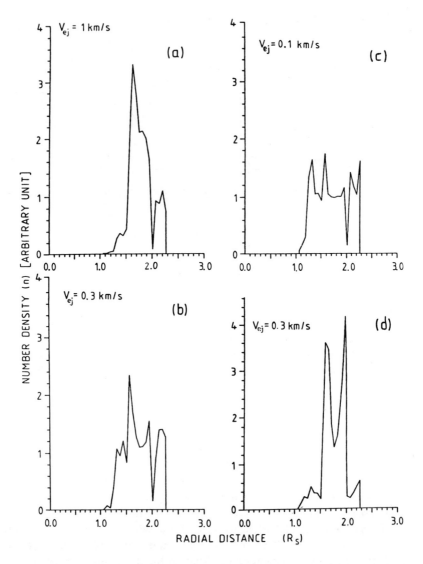

Fig. 5. The steady-state surface density of ejecta as a function of r in Saturn's rings as obtained by Ip shown for different $\tau(r)$ and v_{ej}. For (a), (b), and (c) the $\tau(r)$ is the solid curve of Fig. 4. For (d) the $\tau(r)$ is the dashed curve of Fig. 4 (from diagrams in Ip 1983a).

is suggestive of such a ballistic transport mechanism. Ip's calculations can be scaled to a narrow ringlet by keeping the ratio of v_{ej} to ring width the same. Ip's best case is shown in Fig. 5d, but his runs with other v_{ej} also give double-peaked structures.

B. Global Time-Dependent Evolutions

For the prompt absorption assumption of Durisen et al. (1983), similar general features of edge pile ups and spillovers are present but with notable differences in detail. These calculations are fully self-consistent evolutions of $\sigma(r)$ with time due to ballistic transport. Ejecta are only free between their initial ejection and their first collision with a ring particle.

Figure 6 shows evolved $\sigma(r)$ obtained by integrating Eq. (11) using 200 radial bins. The initial $\tau(r)$ for these evolutions is the same as for Ip's calculations except that the D Ring is omitted. Other assumptions include: only one initial particle size and internal density at each r; isotropic ejection; and $R_e(r, \tau) = $ const. Accumulated regoliths of ejecta are assumed to have the same internal mass density as the initial particles so that Eq. (14) gives $\tau \sim \sigma^{\frac{2}{3}}$ as R_p varies with time. For simplicity, σ is normalized so that $\tau = \sigma^{\frac{2}{3}}$ everywhere. Figure 6 gives the resulting $\sigma(r)$ after a time $t = 3t_g = 3\sigma/R_e$. Angular momentum transport has been ignored, and losses to Saturn are negligible. The choice $x = 0.1$ in Fig. 6c corresponds to $v_{ej} = 1.7$ km s^{-1} and 2.3 km s^{-1} at the outer A Ring and inner C Ring, respectively. Such velocities (or greater) are probably appropriate for gaseous or molecular ejecta. An $x \leq 0.01$ is more typical of chips ejected by meteoroid impacts. For Fig. 6a, c, and d, R_τ is $1 - E_2(\tau)$; for Fig. 6b, R_τ is $1 - \exp(-\tau)$. For Fig. 6d, $f_x \sim x^{-1}$ for $0.01 < x < 0.1$ and is zero otherwise. The x-integrals in Γ and Λ employ $20 x$-bins spaced evenly in $\ln x$.

Again, edge effects are prominent, as expected. Without multiple scattering, however, effects due to low x ejecta are extremely localized (see Eq. 4) and small in amplitude even after three gross erosion times. Notice especially the spillover features just inside the inner edge of the B Ring in Fig. 6a and b, indicated by arrows. As described in the chapter by Cuzzi et al., suggestively similar features exist in Saturn's rings near the inner A Ring and B Ring edges. In each case, a smooth slope of material is seen in $\tau(r)$ extending inward from a very sharp optical depth discontinuity. In a crude way, the $\tau(r)$ profiles of the observed sloping regions resemble a composite of the calculated spillover features in Fig. 6a and b. The C Ring and Cassini Division are bluer than the A and B Rings. In or near the observed sloping regions, there are variations in color between the extremes, suggestive of particle surface composition which is a mixture of the adjoining regions. The sloping region at the inner edge of the A Ring is ~ 1500 km wide, suggesting, through Eq. (4), a typical $x \approx 0.003$ or $v_{ej} \approx 70$ m s^{-1}, compatable with the v_{ej} expected from meteoroid impacts.

Comparison of Fig. 6a and b illustrates the unfortunate dependence of

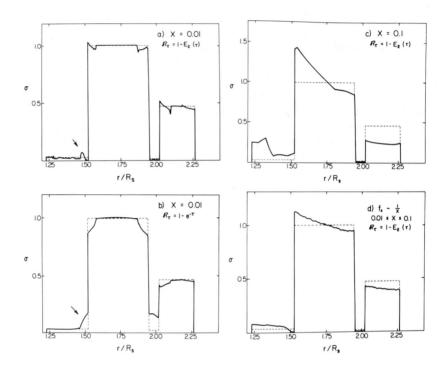

Fig. 6. Four surface density σ vs r plots for Saturn's rings shown after three total gross erosion times by the heavy solid lines. The dashed lines indicate the initial $\sigma(r)$. The different assumptions underlying each calculation are explained in the text. Note the spillover features indicated by the arrows (figure from Durisen et al. 1983).

details in the results on various poorly known quantities, such as R_τ. Because the dependence of P on τ is the same for both calculations, the low τ regions adjacent to the A and B Rings are just as good absorbers in both cases. However, $R_\tau = 1 - \exp(-\tau)$ makes low τ regions much weaker emitters of ejecta than does $R_\tau = 1 - E_2(\tau)$. Thus, in Fig. 6b, the low τ regions soak up material that would otherwise have piled up at the inner A and B Ring edges without ejecting much material back toward the high τ regions.

The power law relation in Eq. (1a) for meteoroid ejecta translates directly into $f_x \sim x^{-\alpha}$. The particular power law $f_x \sim x^{-1}$ used for Fig. 6d effectively gives equal weight to all x in the interval $(0.01, 0.1)$, because the important net transport effects are first order in x. Consequently, the evolved ring profile is relatively smooth, aside from ripples introduced by the finite number of x-bins. For $\alpha < 1$, global transport is dominated by the largest x-values, and evolutions resemble Fig. 6c. For $\alpha > 1$, the smallest x-values dominate, and evolutions resemble Fig. 6a and b. For meteoroid ejecta, α is almost certainly greater than one.

The calculations by both Ip (1983a) and Durisen et al. (1983) so far have effectively covered only a relatively short span of erosional history. As discussed earlier, t_g for Saturn's rings may be as short as 3×10^4 yr. Over millions or billions of years, even second-order effects have time to manifest themselves, and significant features could develop on all length scales even if $\alpha > 1$. Different mathematical approaches may be needed to treat such extensive evolutions.

From Figs. 3, 5, and 6, we conclude that erosional ballistic transport can produce prominent edge effects. These include enhancement of preexisitng density contrasts and spillovers from high to low τ regions with fairly distinctive shapes. Such features appear to exist in Saturn's rings and in the rings of Uranus. How these features evolve over long periods of time, say 10^2 to 10^5 t_g, remains to be addressed.

C. Angular Momentum Transport

All edges and narrow-ring structures are subject to viscous spreading unless confined by resonances or shepherding satellites (see chapters by Shu, Stewart et al., Franklin et al., and Dermott). To mention just a few features, the relatively narrow plateau regions in the C Ring with $\tau \approx \frac{1}{3}$ and the inner edges of the A and B Rings are not readily explained by known resonances or shepherds. Ballistic transport of angular momentum might play a role in confining these features.

As an illustrative example, consider the inner edge of a high τ region which is adjacent to a very low τ region and is undergoing ballistic transport with a strong bias toward retrograde ejections. Mass pile-up effects at an inner edge are zero order for a retrograde bias, and we expect $\dot{\sigma}/\sigma \approx 1/t_g$ there. This leads to $\dot{h}_c/h_c \approx x/t_g$ and a corresponding radial drift velocity $v_r \approx 2 r x/t_g$ for the edge. If the edge has a characteristic size scale $L \approx \sigma \mid d\sigma/dr \mid^{-1}$ and a kinematic viscosity ν, then it smears out with a speed $v_\nu \approx L/t_\nu$ where $t_\nu \approx L^2/\nu$ is the viscous spreading time. Comparing v_r and v_ν we see that ballistic angular momentum transport might confine or even sharpen ring features with characteristic sizes

$$L > L_{crit} \approx 15 \text{ km} \left(\frac{t_g}{10^5 \text{ yr}} \right) \left(\frac{10^{10} \text{cm}}{r} \right) \left(\frac{\nu}{10^2 \text{ cm}^2 \text{ s}^{-1}} \right) \left(\frac{10^{-2}}{x} \right) . \qquad (22)$$

Although the quantities t_g, ν, and x are subject to large uncertainties and the above estimate is admittedly very crude, L_{crit} is sufficiently small to be suggestive. Durisen et al. (1983) plan to include ballistic angular momentum transport in future detailed evolutions.

D. Regolith Accumulation and Particle Sizes

If $\alpha > 1$, as seems to be indicated, the vast bulk of ejecta have low x-values. Regardless of α, the evolutions in Fig. 6 demonstrate that the net rate of erosional transport is far less than the gross erosion rate. A meter-sized ring particle can build up several centimeters of regolith in a time $\ll t_g$. Thus, initial regolith accumulation is a relatively localized process and has been adequately treated in this limit by Morfill et al. (1983). Even in a ring region suffering slow net losses and hence decrease in R_p, the regolith will quickly achieve a sufficient steady-state depth so that craters produced by typical projectiles excavate mostly only previously accumulated chips. Applying results from laboratory cratering data, Morfill et al. argue for typical regolith depths of at least a few centimeters due to $R_m \gtrsim 100$ μm meteoroids. This depth should be attained in only a few thousand years according to Table I. In regions enjoying slow net gains and hence increasing R_p, the regoliths will grow in depth from this initial level without bond.

These arguments create some puzzles. If chip production is locally so effective everywhere in the ring system, why do we see abundant numbers of free micron-sized particles only in some ring regions? Moreover, the transport calculations in Figs. 5 and 6 often show even net gains in the C Ring, which in fact seems relatively devoid of small particles. Mechanisms other than ballistic transport are almost certainly involved, especially whatever mechanism causes the micron-sized particles to become released from regoliths and stay free of them in such large numbers. Their absence in the C Ring may reflect some process that preferentially removes the smallest particles from the rings, e.g., nongrativational forces (cf. Northrop and Hill 1982; chapter by Grün et al.). However, then the relatively short time scales for chip production in Table I would suggest that even the meter-sized particles now present in the C Ring should have been completely eroded down to submicron-sizes and lost long ago. Subcentimeter-sized particles can also be catastrophically fragmented by meteoroids and micrometeoroids. These small particles may survive somewhat longer in high τ regions because they are partially shielded from the destructive projectiles by the larger ring particles.

There are other processes that may influence the development of regoliths (cf. chapter by Weidenschilling et al.). Tidal forces and collision-induced spins could tend to strip loosely bound regoliths and make prompt absorption of chip ejecta more difficult. Oblique physical collisions between ring particles with deep regoliths would probably brush off some surface material. More nearly central collisions might actually compact and strengthen the regolith. The resulting effect of regoliths on the elasticity of ring particle collisions could have important consequences for ring dynamics and equilibrium (cf. chapters by Weidenschilling et al. and Stewart et al.). Moreover, loose surface grains of a regolith can be sputtered off if they become charged, a process that may be related to spoke formation (see chapter by Grün et al.). However, high and low τ regions may behave very differently in all these respects.

So far, no detailed evolutions of particle size distributions or regolith growth due to ballistic transport have been attempted. It appears that physical mechanisms in addition to transport must be included. Many important questions may prove best answered by local treatments where transport effects are ignored.

E. Chip Halos and Gaseous Atmospheres of Rings

During their out-of-plane motions between ejection and reabsorption or loss, ejecta will produce a steady-state concentration around a ring system. For instance, both photosputtering (Carlson 1980) and meteoroid impact vaporization (Morfill et al. 1983) can produce copious amounts of neutral hydrogen around the rings. The kinematics of ejecta are likely to be different in each case. Carlson argues that most photosputtered atoms or molecules will have velocities significantly below escape, move on elliptical orbits, and experience several scatterings prior to destruction. Morfill et al. expect v_{ej} to be more nearly comparable to v_m for impact vapor, so that this component would be moving closer to escape velocity and could be modeled near the rings by radial expansion away from the parent ring particle. Other yet undetected neutral gaseous components, such as OH, H_2O, and H_2 are also expected to be abundant in the ring atmosphere. Given reasonable assumptions about the v_{ej} distributions and source rates $R(r)$, maps for the steady-state gas density of various constituents could be made. So far, this has been done only crudely by Dennefeld (1974). A more detailed discussion of the ring atmosphere, including other source and sink mechanisms, has recently been provided by Ip (1983b).

The chip ejecta from meteoroid impacts will also produce a steady-state halo around the rings (Ip 1980). The heavy ejecta ($m_{ej} > m$), which are good backscatterers, tend to have very low x-values (Grün et al. 1980). Hence, as shown by Morfill et al. (1983), a steady-state chip halo of several kilometers vertical extent could exist with sufficient normal optical depth ($\sim 10^{-5}$) to add significantly to the apparent brightness of the rings at ring plane passage. The light ejecta ($m_{ej} < m$) will have considerably larger x-values and will produce a more diffuse halo of forward scatterers. With reasonably prompt reabsorption of ejecta, however, the steady-state surface density of chips on ballistic trajectories should be orders of magnitude smaller than the true surface densities implied by observations of the A and B Rings in forward-scattered light. As already noted in the previous section, a mechanism other than or in addition to erosional ballistic transport is probably needed to explain why there are so many free subcentimeter-sized particles in the A and B Rings.

Halos and atmospheres should exist around other ring systems as well. For Jupiter's ring, the faint disk is probably influenced primarily by non-Keplerian forces (see chapter by Burns et al.). However, the outer halo around the bright ring and the bright ring itself probably represent a steady-state distribution of ejecta around yet undiscovered parent ring particles (Burns et al. 1980; Grün et al. 1980). The rings of Uranus have so far exhibited no detectable diffuse component. However, it seems likely that one must exist at some level.

F. Overall Mass Balance and Contamination of the Rings and Planet

Cook and Franklin's (1970) pioneering work on the erosion of Saturn's rings by meteoroid bombardment was concerned primarily with whether the rings experience a net gain or loss of material overall. Gains could come about through retention of meteoroid mass. Losses could include chip ejecta lost to escape, to the planet, or possibly to the region exterior to the rings. In Fig. 2, we can see that even for large x-values loss to Saturn occurs over more θ,ϕ-space than does escape. Equation (1a) shows that only a relatively small fraction of ejecta have large x-values. For similar reasons, Cook and Franklin conclude that losses to Saturn dominate over losses to escape by a factor between 2 and 6 and that the rings lose at most about twice as much mass by chip loss as they gain in meteoritic mass.

The treatment by Cook and Franklin is still the most thorough in the literature and is worthy of careful reconsideration. The higher meteoroid influx adopted by Morfill et al. (1983) suggests a total meteoroid influx over a solar system age that is comparable to, if not greater than, the present ring mass. Therefore, in overall mass balance, the rings must have changed profoundly due to erosion over billions of years. Cook and Franklin conclude that impact vapor losses are negligible, but Morfill et al. attribute much higher velocity to impact vapor than do Cook and Franklin. Morfill et al. suggest a vapor mass $m_v \lesssim 3m$. If the vapor has a kinetic energy half that of the meteoroid, $v_{ej} \gtrsim 0.41 \, v_m$. In the rest frame of the ring particle, incoming meteoroids have v_m between $\sim(\sqrt{2} - 1) \, v_c$ and $(\sqrt{2} + 1) \, v_c$. So $v_{ej} \approx 0.41 \, v_m$ corresponds to a range of x-values between ~ 0.17 and 1.0. According to Fig. 2, about one third of the vapor with $x \approx 0.45$ would be lost immediately to escape or to Saturn. A subtantial fraction of the remainder might be lost by the rings later due to other processes. The fraction of the impact vapor which is ionized may also be lost (see Ip 1983b). Combining chip and vapor losses expected for Morfill et al.'s meteoroid parameters, it becomes difficult to understand how the present rings could have survived over 4.6×10^9 yr. Hypervelocity impact properties and meteoroid influx rates must be known with more confidence.

If most of the meteoroid vaporizes in the hypervelocity impact and the meteoroid vapor also has $v_{ej} \sim v_m$, there will probably be little net compensating gain of meteoroid mass. This could depend on composition. For instance, an iron meteoroid is probably much less susceptible to complete vaporization. Such meteoroids might be relatively uncommon in the outer solar system, but preferential accumulation could lead to detectable contamination over billions of years. For all compositions, it is again important to know confidently not only how much of the meteoroid is vaporized but how much of its original kinetic energy the vapor retains. Contamination by unvaporized traces of stony or iron meteoroids, combined with ballistic transport and regolith growth and destruction mechanisms, may help to explain the observed differences in color and albedo between ring regions.

In addition to the contamination of the rings by projectiles, erosion of ring particles causes contamination of Saturn's atmosphere by ejecta. The area of the rings is comparable to the surface area of Saturn. Using Cook and Franklin's estimates for losses to Saturn, we can see that chip ejecta from meteoroid impacts on the rings probably at least double the input of dusty aerosols into the upper atmosphere over the input expected by direct meteoroid influx alone. Shimizu (1980) has also suggested that OH photosputtered into Saturn could profoundly affect the ionization balance of the upper atmosphere. Free OH facilitates the recombination of H^+ and e^- through a charge exchange and tends to reduce the electron number density n_e substantially. In fact, spacecraft radio occultations do show n_e for Saturn to be much lower than predicted by conventional ionospheric models without OH (cf. Atreya and Waite 1981). For $v_{ej} \sim$ few km s^{-1}, OH can be photosputtered to midlatitudes. Impact vaporization could cause substantial contributions of OH at all latitudes, if $v_{ej} \sim v_m$. Other interactions between erosional ejecta from rings and planetary magnetospheres and ionospheres are discussed by Ip (1983b) and in the chapter by Grün et al.

IV. CONCLUSION

A. Summary

Erosion rates for the surfaces of icy ring particles due to photosputtering and meteoroid impacts are extremely high. Both these processes produce ejecta that will arc across the rings along Keplerian orbits and will eventually be reabsorbed into the ring system or lost from the ring system by escape, by falling into the planet, or by destruction. Between ejection and reabsorption or loss, atoms, molecules, and chips on ballistic trajectories will exist in steady-state concentrations around a ring system. The neutral hydrogen atmosphere around Saturn's rings, as evidenced by Lyman α, is probably produced in this manner. Chip halos of low optical depth ($\sim 10^{-5}$) should also exist around ring systems and, in some cases, like in Jupiter's ring, may actually dominate the appearance of the system (see chapter by Burns et al.). One important consequence of meteoroid bombardment should be the rapid buildup of several centimeter-thick chip regoliths on ring particles. In addition, losses of ejecta into the planet may have observable consequences in the upper atmosphere.

Exchanges of ejecta between ring regions can lead to significant net transport of mass and angular momentum. For prompt reabsorption of ejecta and isotropic ejection, mass transport effects are first order in v_{ej}/v_c near ring edges or near steep optical depth gradients, and they cancel to first order elsewhere. Net angular momentum transport is at most second order. When ejections are anisotropic or when ejecta scatter many times prior to reabsorption, transport effects can be larger in amplitude everywhere but remain greatest at the edges and near steep optical depth gradients. Saturn's rings appear to exhibit such edge effects. At the inner edges of the A and B Rings, material appears to have piled up to high optical depth and also spilled over into the adjacent low optical depth regions. Numerical simulations often show such optical depth pile ups and spillovers at the inner edges. Redistribution of mass in Saturn's rings has probably been extreme even on a global scale, considering the high erosion rates. Losses of ejecta to escape, to the planet, or to other processes and gains via meteoroid influx are comparable in amount but are unlikely to cancel. Consequently, large changes in overall ring mass and meteoritic contamination could result for Saturn's main ring system. If the meteoroid influx is as high as deduced by Morfill et al. (1983), it becomes somewhat difficult to understand the survival of Saturn's rings over the lifetime of the solar system.

B. Outstanding Problems

Details of erosional evolutions depend rather sensitively on the poorly understood physics of the erosion mechanisms themselves and of the interactions between ejecta and ring-particle surfaces. For both photosputtering and meteoroid impacts, we need more quantitative information on chip, gas, and plasma yields and on the distributions of ejecta sizes, speeds, and directions for realistic compositions and physical structures. The large influx rate for meteoroids estimated by Morfill et al. (1983) needs independent verification. Other important questions revolve around the structure of accumulated chip regoliths: How loose and fragile are they? Are they stripped by tidal, collisional, or electrostatic effects? Are they compacted by collisions? What effects do photosputtering and other atomic emission and absorption processes have on them? How does the regolith structure affect erosional yields, v_{ej} distributions, etc.?

The mathematical methods so far employed to treat ballistic transport are somewhat cumbersome and inelegant. In order to produce simulations of ring evolution over a great many gross erosion time scales, as seems desirable, new approaches incorporating techniques from more conventional transport problems may be required. Even within the existing mathematical framework, further refinements could be included, such as a particle size distributions, explicit modeling of regolith buildup, non-Keplerian motions, processes which destroy small particles, and radial drifts due to angular momentum transport. The possibility exists for understanding color and albedo differences and for explaining the maintainance of narrow ring features against viscous spreading in the absence of shepherding moonlets or resonances. Efficient techniques still must be developed to map out the spatial structure of steady-state ejecta atmospheres and halos. Mechanisms other than meteoroid impacts and the resultant ballistic transport seem to be needed to explain why there are large numbers of free subcentimer-sized particles in the A and B Rings but not in the C Ring nor in the Cassini Division. These mechanisms must be identified and included in erosional evolutions.

The importance of ring-particle erosion mechanisms, as evidenced by the short time scales in Table I, has only recently become widely realized. Further studies of ballistic transport and related processes could prove extremely fruitful for explaining many observed properties of planetary ring systems.

Acknowledgments. I am especially indebted to J. A. Burns, N. L. Cramer, J. N. Cuzzi, W.-H. Ip, A. Mecca, and T. Steiman-Cameron for their help with the preparation of this chapter. I am also grateful to D. R. Davis, L. W. Esposito, E. Grün, J. J. Lissauer, G. E. Morfill, M. R. Showalter, and G. R. Stewart for helpful comments, discussions, or correspondence. This work has been supported in part by the National Aeronautics and Space Administration.

REFERENCES

Atreya, S. K., and Waite, J. H. 1981. Saturn ionosphere: Theoretical interpretation. *Nature* 292:682–683.

Bandermann, L. W., and Wolstencroft, R. D. 1969. The erosion of particles in the rings of Saturn. *Bull. Amer. Astron. Soc.* 1:233 (abstract).

Barker, E. S., Cazes, S., Emerich, C., Vidal-Madjar, A., and Owen, T. 1980. Lyman-alpha observations in the vicinity of Saturn with *Copernicus. Astrophys. J.* 242:383–394.

Blamont, J. 1974. The "atmosphere" of the rings of Saturn. In *The Rings of Saturn*, eds. F. D. Palluconi and G. H. Pettengill (Washington, D.C.: NASA SP-343), pp. 125–129.

Broadfoot, A. L., Sandel, B. R., Shemansky, D. E., Holberg, J. B., Smith, G. R., Strobel, D. F., McConnell, J. C., Kumar, S., Hunten, D. M., Atreya, S. K., Donahue, T. M., Moos, H. W., Bertaux, J. L., Blamont, J. E., Pomphrey, R. B., and Linick, S. 1981. Extreme ultraviolet observations from Voyager 1 encounter with Saturn. *Science* 212:206–211.

Burns, J. A., Showalter, M. R., Cuzzi, J. N., and Pollack, J. B. 1980. Physical processes in Jupiter's ring: Clues to its origin by Jove! *Icarus* 44:339–360.

Carlson, R. W. 1980. Photo-sputtering of ice and hydrogen around Saturn's rings. *Nature* 283:461.

Cheng, A. F., Lanzerotti, L. J., and Pirronello, V. 1982. Charged particle sputtering of ice surfaces in Saturn's magnetosphere. *J. Geophys. Res.* 87:4567–4570.
Cook, A. F., and Franklin, F. A. 1970. The effect of meteoroid bombardment on Saturn's rings. *Astron. J.* 75:195–205.
Croft, S. K. 1982. Impacts on ice and snow: Implications for crater scaling on icy satellites. *Lunar Planet. Sci.* 13:135–136.
Dennefeld, M. 1974. Theoretical studies of an atmosphere around Saturn's rings. In *Exploration of the Planetary System,* eds. A. Woszczyk and C. Iwaniszewska (Dordrecht: D. Reidel), pp. 471–481.
Durisen, R. H., Cramer, N. L., Mullikin, T. L., and Cuzzi, J. N. 1982. The evolution of Saturn's rings due to particle erosion mechanisms. *"Saturn" Conf.,* held at Tucson, AZ.
Durisen, R. H., Cramer, N. L., Cuzzi, J. N., and Mullikin, T. L. 1983. Ballistic transport in planetary ring systems due to particle erosion mechanisms. *Icraus.* In preparation.
Franklin, F. A., and Cook, A. F. 1969. A search for an atmosphere enveloping Saturn's rings. *Icarus* 10:417–420.
Gault, D. E. 1973. Displaced mass, diameter, and effects of oblique trajectories for impact craters formed in dense crystalline rocks. *The Moon* 6:32–44.
Gault, D. E., Shoemaker, E. M., and Moore, H. J. 1963. *Spray Ejected from the Lunar Surface by Meteoroid Impact* (Washington, D.C.: NASA TN D-1767.)
Gault, D. E., and Wedekind, J. A. 1969. The destruction of tektites by micrometeoroid impact. *J. Geophys. Res.* 74:6780–6794.
Greenberg, R., Wacker, J. F., Hartmann, W. K., and Chapman, C. R. 1978. Planetesimals to planets: Numerical simulation of collisional evolution. *Icarus* 35:1–26.
Grün, E., Morfill, G. E., Schwehm, G., and Johnson, T. V. 1980. A model of the origin of the Jovian ring. *Icarus* 44:326–338.
Harrison, H., and Schoen, R. I. 1967. Evaporation of ice in space: Saturn's rings. *Science* 157:1175–1176.
Hartmann, W. K. 1978. Planet formation: Mechanism of early growth. *Icarus* 33:50–61.
Housen, K. R., Wilkening, L. L., Chapman, C. R., and Greenberg, R. 1979. Asteroidal regoliths. *Icarus* 39:317–351.
Humes, D. H. 1980. Results of Pioneer 10 and 11 meteoroid experiments: Interplanetary and near-Saturn. *J. Geophys. Res.* 85:5841–5852.
Ip, W.-H. 1980. Physical studies of the planetary rings. *Space Sci. Rev.* 26: 39–96.
Ip, W.-H. 1983a. Collisional interactions of ring particles: Ballistic transport process. *Icarus.* In press.
Ip, W.-H. 1983b. On planetary rings as sources and sinks of magnetospheric plasmas. In *Planetary Rings/Anneaux des Planetes: IAU Colloquium No. 75,* ed. A. Brahic (Toulouse: Centre National d'Etudes Spatiales). In press.
Judge, D. L., Wu, F.-M., and Carlson, R. W. 1980. Ultraviolet photometer observations of the Saturnian system. *Science* 207:431–434.
Kourganoff, V. 1963. *Basic Methods in Transfer Problems* (New York: Dover Publ.).
Lange, M. A., and Ahrens, T. J. 1982. Impact cratering in ice and in ice-silicate targets: An experimental assessment. *Lunar Planet. Sci.* 13:415–416.
Mizutani, H., Kawakami, S., Takagi, Y., Kato, M., and Kumazawa, M. 1982. Cratering experiments in sand. *Lunar Planet. Sci.* 13:530–531.
Morfill, G. E., Fechtig, H., Grün, E., and Goertz, C. K. 1983. Some consequences of meteoroid impacts on Saturn's rings. *Icarus.* In press.
Northrop, T. G., and Hill, J. R. 1982. Stability of negatively charged dust in Saturn's ring plane. *J. Geophys. Res.* 87:6045–6051.
Shimizu, M. 1980. Strong interaction between the ring system and the ionosphere of Saturn. *Moon Planets* 22:521–522.
Stöffler, D., Gault, D. E., Wedekind, J., and Polkowski, G. 1975. Experimental hypervelocity impact into quartz sand: Distribution and shock metamorphism of ejecta. *J. Geophys. Res.* 80:4062–4077.
Weiser, H., and Moos, H. W. 1978. A rocket observation of the far-ultra-violet spectrum of Saturn. *Astrophys. J.* 222:365–369.
Weiser, H., Vitz, R. C., and Moos, H. W. 1977. Detection of Lyman α emission from the Saturnian disk and from the ring system. *Science* 197: 755–757.

COLLISION-INDUCED TRANSPORT PROCESSES
IN
PLANETARY RINGS

GLEN R. STEWART
NASA Ames Research Center

D. N. C. LIN and PETER BODENHEIMER
Lick Observatory

The physics of collision-dominated particle disks around planets is analyzed from both the analytical and numerical point of view. The existence of Saturn's ringlet structure is analyzed in terms of a radial instability induced by viscous diffusion. The detailed model described is based on the assumption of uniform particle size. Modifications caused by finite particle size, gravitational effects, and a distribution of particle sizes are discussed.

I. GENERAL PHYSICAL PROPERTIES OF PLANETARY RINGS

Recently, data obtained from spacecraft and groundbased measurements have greatly refined our knowledge of the rings of Saturn (see chapter by Cuzzi et al.). In particular, the data indicate: (1) a vertical thickness of at most 150 m, and probably several times less (Lane et al. 1982), compared to a radial extent of about 10^{10} cm; (2) the rings are mostly composed of ice particles ranging from cm to m in size (Marouf et al. 1983); and (3) the rings are subdivided into a large number of ringlets with radial dimensions ranging from several hundred km down to the 10-km resolution of the Voyager spacecraft cameras (Smith et al. 1982). Many of these ringlets are observed in the B Ring where they show optical depth variations of 20 to 50% about a mean optical depth ranging from 0.6 to 3.0.

The investigation and explanation of the dynamical features of the rings have fascinated theoreticians over the past several centuries. Critical developments were provided by Maxwell (1859), who deduced that the rings were made up of a large number of orbiting particles, and by Jeffreys (1916, 1947) who was the first to realize that the particles in the rings frequently collide with each other. Jeffreys further showed that dissipation due to these collisions would rapidly flatten the system into a thin disk with a thickness of the order of a few particle diameters, and, on a longer time scale, would tend to spread the ring out in a radial direction. Contemporary researchers have concentrated on more detailed mathematical formulations of the collisional dynamics and the resultant transport processes. Such a description is essential if we are to deduce the basic mechanisms that determine the thickness, cause the radial structure, and regulate the overall evolution of the ring system. For example, we may ask whether the rings have a long enough evolution time scale to have existed for the lifetime of the solar system. A well-established model for the collisional dynamics of ring particles is also required when one considers the tidal interaction between Saturn's inner satellites and the rings. Some features in the rings such as density waves and sharp ring edges have been attributed to these interactions (see chapters by Borderies et al. and by Shu). A third example is the question of the stability of collisional evolution of the rings. Some features such as the density variations in the B Ring may be caused by a diffusion instability.

In this first section we discuss in an approximate way the important physical processes governing the evolution of a collisional particle ring. In later sections we discuss numerical simulations (Sec. II) and in a more rigorous fashion, the equations governing the evolution, first on the basis of kinetic theory (Sec. II) and then on the basis of the fluid approximation (Sec. III). In Sec. III we also analyze the energy budget and energy redistribution in an evolving particle disk. From these analyses, the evolution time scale of the rings as well as their thickness is deduced.

Having constructed a dynamical model for the rings, we examine in Sec. IV the stability of the system. In particular, we establish a general criterion that can be used to determine both the thermal and diffusive stability. This criterion is found to depend critically on the particular form of the coefficient of restitution as a function of impact velocity. We show that rings with optical depth of order unity are unstable and have a tendency to diffuse into ringlets. Qualitative comparisons between theoretical models and the observed radial structures in Saturn's rings are also presented. In Sec. V we discuss how the theory is modified if the finite size of the particles and interparticle gravitational effects are taken into consideration. Finally, in Sec. VI, we make some attempts towards incorporating the effects of a realistic distribution of particle sizes with a simple idealized model. We show that the density variations in large and small particles may be very different in the unstable regions. Section VII summarizes our conclusions.

A. The Role of Collisions

A planetary ring system, like Saturn's, consists mostly of small particles in nearly circular orbits about the central planet. Between successive collisions with other particles, the motion of an individual particle is described by Kepler's laws. For circular orbits the angular frequency of a particle is $\Omega = (GM/r^3)^{\frac{1}{2}}$ where r is the distance from the central mass M. We now estimate the importance of collisions in a disk system of *identical* particles. The collision frequency ω_c can be obtained in terms of the particle density (number per volume) N, the cross section πR_p^2 of a typical particle of radius R_p, and the velocity dispersion c, such that $\omega_c = R_p^2 Nc$. The scale height H of the disk is given by $NH = \sigma/m$, where σ is the surface density in g cm^{-2} and m is the mass of a particle. If the velocity dispersion, which characterizes the random motion of the particles with respect to the local circular orbit, is approximately isotropic, then $H = c/\Omega$ (derived in Sec. II.B.3). Then $\omega_c = R_p^2 \sigma \Omega/m = \tau\Omega/\pi$, where the optical depth $\tau = \pi R_p^2 \sigma/m$ by definition. Thus in regions where $\tau = 1$, as, for example, in portions of Saturn's rings, a typical particle has a collision with a neighboring particle at least twice per orbit. Since orbital periods in planetary-ring systems range from a few hours to a few days, even if τ is as low as 10^{-3}, a particle would still experience a billion collisions during the lifetime of the solar system. Thus collisions are clearly an important process (see chapter by Weidenschilling et al.).

Suppose we start with a set of particles in eccentric and mutually inclined prograde orbits about the central mass M. Collisions between particles tend to dissipate relative kinetic energy and thus to damp relative motion. In a few collision times this process should result in evolution towards coplanar, circular orbits (Jeffreys 1947). As a simple idealized example consider two coplanar particles with identical eccentricities e colliding at a distance r from M, when one particle is at pericenter and the other at apocenter. Because of angular momentum conservation the former is moving faster than the circular velocity at radius r while the latter is moving more slowly. In the case of a purely inelastic collision, the relative motion is completely damped so that the two particles move in the same circular orbit with the same total angular momentum as in the original orbits. Meanwhile, the total energy has decreased by the amount of their original relative kinetic energy, which is given by $\sim e^2 GM/(2r)$. Similarly, collisions between particles in mutually inclined orbits (inclinations $\pm i$) result in damping of relative kinetic energy ($\sim 2i^2 GM/r$) and establishment of coplanar orbits with the same total angular momentum.

Now consider the situation when the system has approached the disk state with nearly circular and coplanar orbits. The relative motion between particles is now dominated by the radial gradient in the circular orbital velocity. The system will continue to evolve toward a state of lower energy because energy is dissipated during collisions. To illustrate the form that this evolution takes, we consider two particles in nearly coplanar, circular orbits. The total energy is $\mathcal{E} = m(E_1 + E_2)$ and the total angular momentum per unit mass is $J = J_1 + J_2$ where

$$E = J_i^2/2r_i^2 - GM/r_i \tag{1}$$

and

$$J_i = (G M r_i)^{\frac{1}{2}} . \tag{2}$$

The change in the energy of a particle results from changes in both J_i and r_i. However, as already noted above, collisions on the average will tend to circularize orbits. Hence, for the purpose of calculating average energy changes, transitions due to collisions can be thought of as changes from one circular orbit to another. In this case a change in energy can be expressed solely in terms of a change in angular momentum:

$$\frac{dE}{dJ} = \frac{J}{r^2} = \Omega . \tag{3}$$

The change in E caused by a change in r can be ignored because for circular orbits

$$\left.\frac{\partial E}{\partial r}\right|_{J \,=\, \mathrm{const}} = -\frac{J^2}{r^3} + \frac{GM}{r^2} = 0 . \tag{4}$$

Now consider a small change in the individual angular momenta of the two particles, keeping total angular momentum constant. We have

$$d\mathcal{E} = m \left(\frac{dE_1}{dJ_1} \, dJ_1 + \frac{dE_2}{dJ_2} \, dJ_2 \right)$$

$$= m \, dJ_1 \, (\Omega_1 - \Omega_2) = m \sqrt{GM} \, dJ_1 \left(\frac{1}{r_1^{3/2}} - \frac{1}{r_2^{3/2}} \right) \tag{5}$$

since $dJ = dJ_1 + dJ_2 = 0$. Thus for $r_1 > r_2$, $d\mathcal{E} < 0$ if $dJ_1 > 0$. This argument was used by Lynden-Bell and Pringle (1974) to show that evolution toward a lower energy state is achieved by exchanging angular momentum in the direction such that particles in orbits with smaller semimajor axes give up angular momentum to those in orbits with larger semimajor axes. They use a similar argument to show that energy is also reduced by a net movement of mass to smaller radii.

The overall effect of collisions is thus first to circularize the orbits on a short time scale and then, on a much longer time scale to transport angular momentum from the inner to the outer region of the disk. The angular momentum transport results in transport of some mass toward the outer edge, but the bulk of the mass must drift inward as a result of the continual dissipation, which must be supplied from the gravitational energy of the disk.

This continual liberation and subsequent dissipation of energy will tend to maintain a small finite velocity dispersion. The minimum possible relative velocity of particles in nearby circular orbits, caused by the difference in angular velocity across one particle diameter $2R_p$, is ΩR_p. Since the orientation of contact of particles at collision will be nearly random, interchange between motions in different directions remains possible. As a result, a finite velocity dispersion in the vertical direction of order ΩR_p is unavoidable (Jeffreys 1947). If the velocity dispersion becomes much larger than ΩR_p, then the mean motion will no longer dominate the outcomes of the collisions and the rapid damping of random motions described at the beginning of this section will dominate. The disk will tend toward a steady state with a thickness on the order of a few particle radii. The efficiency of the angular momentum and mass transfer process in the disk is determined by the collision frequency and the mean free path of particles. Thus the structure and evolution of the system is determined by the detailed properties of the collisions.

B. Transport Processes

From the data mentioned above, we can deduce that each ringlet is composed of a large number of particles. Hence, any theory of the formation of these large-scale structures must ultimately abandon the detailed description of individual particle trajectories and adopt either a statistical mechanical or a fluid-dynamical formulation (see Sec. II.B.1) which describes the average behavior of a collection of particles as a whole. In this section (I.B) we estimate evolutionary time scales for significant radial spreading of the rings using these two approaches. We then (in Sec. I.B.3) consider the fundamental condition for energy balance and the role of the coefficient of restitution in establishing this balance. These concepts will be used in following sections to construct the steady-state vertical structure of the rings and to establish the stability of the steady state.

1. Statistical Properties of Particles in Planetary Rings. The random orientation of deviations from the local circular orbit effectively reduces the most important aspect of the particles' dynamical evolution to a 1-dimensional flow, i.e., a spreading in the radial direction. In a planetary ring with a moderate optical depth, a particle usually travels over a distance comparable to the circumference of an orbit between successive collisions. However, most of this journey is traveled in the tangential direction. Since we are primarily interested in efficiency of diffusion in the radial direction, the typical radial excursion between successive collisions may be regarded as the relevant scale for the mean free path.

If a typical particle experiences at least two collisions per orbit around the central planet (i.e., $\omega_c \geq \Omega/\pi$ or alternatively $\tau \geq 1$), this radial mean free path is approximately $\lambda = c/\omega_c$. If the number of particles in a ring is relatively sparse (i.e., $\tau < < 1$), the collision frequency may be small compared with the orbital frequency. In this case, the radial excursion of the particle is bounded by its orbital eccentricity e so that the mean free path approaches its upper limit of $ae \simeq c/\Omega$ where a is the semimajor axis of the orbit. For arbitrary τ both limiting values of λ may be included in a single prescription (Cook and Franklin 1964; Goldreich and Tremaine 1978) such that

$$\lambda^2 = c^2 \, \Omega^{-2} \, (1 + \tau^2)^{-1} \, . \tag{6}$$

The above form of the mean free path is entirely analogous to that for a plasma whose motion is confined by a magnetic field (Spitzer 1962). For a typical particle, the mean radial excursion after n such collisions may be approximated by the random-walk expression, $\Delta r = n^{\frac{1}{2}} \lambda$. Thus, the time scale t_{RW} required for a typical particle to randomly walk over a distance Δr is

$$t_{RW} = \frac{n}{\omega_c} = \frac{1 + \tau^2}{\tau} \Omega \left(\frac{\Delta r}{c} \right)^2 \, . \tag{7}$$

The above results indicate that a particle can diffuse over a distance > 10 ae on a time scale $100 \, (1 + \tau^2)/(\tau\Omega)$ which is longer than both the dynamical (orbital) and collisional time scales. So far, we have ignored the change in τ and c as a result of radial spreading. Thus the above "instantaneous" estimate may be applied to relatively small but not to arbitrarily large radial spreading.

2. Collective Nature of the Particles. In a typical planetary ring, the particles' mean free path λ is normally much smaller than the radial extent of the ring. Provided that the disk properties do not vary too rapidly with radius, we may choose a radial scale, Δa, which is both much smaller than L, the scale over which there is appreciable change in the physical properties of the ring, and much larger than λ. Over Δa, the particles have very similar kinematic properties so that their physical properties may be analysed in a collective manner. This aspect of the ring dynamics satisfies the requirements for a fluid description (Boyd and Sanderson 1969). In practice, since the mean free path in the radial direction is limited by the eccentricity of the orbit such that $\lambda \leq ae$, a collection of particles contained in a ring of width Δa may be regarded as a basic fluid element provided that $\Delta a > 10 \, ae$ and the collisionally induced dynamical evolution is stable on this scale.

The spreading rate may, therefore, also be obtained from fluid dynamics. This approach has been applied to the studies of gaseous accretion disks

(Lynden-Bell and Pringle 1974). In this case, the combined presence of viscosity and differential rotation induces a viscous stress that leads to an outward transfer of angular momentum and a general spread of a ring. The efficiency of angular momentum transfer is determined by the magnitude of the viscosity ν. If ν is a function of position only, the time scale for spreading in the radial direction is $t_v \simeq (\Delta r)^2/\nu$ (Lynden-Bell and Pringle 1974), as long as the radial spreading is small enough so there is not a substantial change in ν. The basic expression for the viscosity is $\nu = \omega_c \lambda^2$, so that

$$t_v \simeq \frac{1 + \tau^2}{\omega_c} \, \Omega^2 \left(\frac{\Delta r}{c}\right)^2 \qquad (8)$$

Since $\tau = \pi \omega_c/\Omega$, t_v is determined by the identical physical parameters as in the heuristic kinetic argument (see Eq. 7).

3. Energy Transport Processes. Both the kinetic and fluid treatments of diffusion processes indicate that ν is a function of τ and c. Thus, the determination of the magnitude of ν requires a prescription for the velocity dispersion which can be established through consideration of the detailed energy transport processes. Consider two neighboring particles with nearly circular orbits separated by a radial mean free path. The difference between their precollisional circular velocities is of order $\lambda \Omega$. After a collision they will move apart with a comparable velocity difference but their orbits will be reoriented. Consequently, they will each acquire an orbital eccentricity of the order $\lambda/a = c/\Omega a$. The energy change per unit mass associated with this gain in the eccentricity is of the order $(\lambda/a)^2 (\Omega a)^2 = \lambda^2 \Omega^2$. Since the kinetic energy per unit mass associated with eccentric motions is, in practice, the energy associated with dispersive random motion, i.e. c^2, its rate of increase may be determined in terms of the collision frequency ω_c such that $\dot{E}_t \simeq c^2 \omega_c \simeq \lambda^2 \Omega^2 \omega_c$. In the fluid approach, this tendency for increases in c is normally attributed to the effect of viscous stress. In a differentially rotating disk, the viscous stress continually converts energy, from that stored in the systematic shearing motion into that associated with random motion, at a rate

$$\dot{E}_t = \nu \, (r \, \partial \Omega/\partial r)^2 \qquad (9)$$

per unit mass, which is of the order $\nu \Omega^2$ for Keplerian orbits. With the substitution of the above formula for viscosity, the hydrodynamic treatment once again provides us with results indentical to those obtained from the kinetic approach.

In the absence of any dissipative process, the collection of particles would be "heated up" in the sense that the dispersion velocity would increase indefinitely. However, typical collisions between particles in the ring are partially inelastic. In the standard terminology for elasticity theory the amount of energy dissipation may be quantified with a coefficient of restitution ϵ

having values between zero (totally inelastic) and unity (perfectly elastic). Formally, ϵ may be expressed in terms of the relative velocities $\mathbf{g}_b \equiv \mathbf{v}_{1b} - \mathbf{v}_{2b}$ and $\mathbf{g}_a \equiv \mathbf{v}_{1a} - \mathbf{v}_{2a}$, before and after a collision between two particles. The central premise of the inelastic hard-sphere model states that the collisional change in relative velocity is proportional to the component of the relative velocity normal to the impact surface such that

$$\mathbf{g}_a - \mathbf{g}_b = - (1 + \epsilon) (\mathbf{g}_b \cdot \hat{\mathbf{k}}) \hat{\mathbf{k}} \tag{10}$$

where $\hat{\mathbf{k}}$ is the unit vector pointing from the center of particle 1 to the center of particle 2. In planetary rings, the typical value of the impact velocity (which we write as $v_c = |\mathbf{g}_b| = g$) is normally comparable to the dispersion velocity so that an amount $(1 - \epsilon^2) c^2$ of the kinetic energy per unit mass associated with random motion is dissipated into heat after each collision and is eventually radiated away from the system. The total number of collisions per unit time interval in the disk is ω_c so that the total rate of energy dissipation per unit mass is

$$\dot{E}_d = (1 - \epsilon^2) c^2 \omega_c \ . \tag{11}$$

At any radial position in the rings, the local kinetic energy associated with random motion may be changed by several other energy transport processes such as the advective transport caused by the bulk motion of the particles and the conductive transport induced by collisions between particles. However, these processes are normally much less efficient than the above local energy-generation and energy-dissipation mechanisms. Thus, a local energy equilibrium can only be established if these two processes are delicately balanced. The condition for energy balance is

$$\dot{E}_d = \dot{E}_t \ . \tag{12}$$

The above equation implies the following relationship between ϵ and τ for a disk with uniform particle size:

$$(1 - \epsilon^2) (1 + \tau^2) = b \tag{13}$$

where b is a constant of order unity (Cook and Franklin 1964; Goldreich and Tremaine 1978). This equilibrium $\epsilon - \tau$ relation does not explicitly depend on the velocity dispersion because both $\dot{E}_t$ and $\dot{E}_d$ have the same velocity dependence. In general $\dot{E}_t$ would have a different velocity dependence than $\dot{E}_d$ (if, e.g., we had included gravitational interactions between particles) and the energy balance would determine an explicit formula for the equilibrium velocity dispersion (Safronov 1969; see also chapter by Ward). In the present case, the equilibrium velocity may be determined from Eq. (13) once the dependence of ϵ on impact velocity is specified. Equation (13) also establishes a relation between ν and c, since $\nu = \omega_c \lambda^2$ and λ is given by Eq. (6).

If ϵ is independent of v_c, the rings may never establish an energy equilibrium and would either expand into a torus or collapse into a monolayer disk. This evolutionary tendency, which occurs because an increase in the velocity dispersion due to collisions is not compensated by an increase in dissipation, is somewhat analogous to thermal instabilities in accretion disks (Pringle et al. 1973; Shakura and Sunyaev 1976; Faulkner et al. 1983). However, ϵ for various rocky and metallic materials has been measured experimentally and found to be a monotonically decreasing function of v_c (Goldsmith 1960; see Sec. IV.B.1 for further discussion). If the ring particles have similar mechanical properties, the low-velocity collisions are much more elastic than high-velocity impacts. Consequently, for a ring with a particular τ, $\dot{E}_d$ increases faster than c^2 while $\dot{E}_t$ increases only as fast as c^2. Thus, for small c, the collisions would be mostly elastic so that $\dot{E}_t > \dot{E}_d$. In this case, the velocity dispersion of the particles would increase exponentially until ϵ were reduced to a small enough value such that $\dot{E}_t = \dot{E}_d$. Alternatively, for large c, $\dot{E}_d$ could be greater than $\dot{E}_t$ so that the kinetic energy stored in random motion would be continually drained until c were sufficiently small to establish an energy equilibrium. In the case that gravitational scattering is a significant energy source for random motion, a stable energy equilibrium can be established even if ϵ is independent of velocity because the effectiveness of gravitational scattering decreases with increasing relative velocity (chapter by Ward; also see Sec. V.A.1 below).

The above heuristic arguments indicate that, if ϵ is a decreasing function of v_c, not only is there an energy equilibrium, but this equilibrium is stable. In a particle ring, the energy equilibrium is established on the time scale for energy to diffuse through the thickness of the ring. For a semitransparent ring, this time scale is comparable to the orbital period (Goldreich and Tremaine 1978). Since the ringlet structure in Saturn's rings appears to be persistent over times much longer than the orbital period, it is reasonable to assume that these regions are in an energy equilibrium. Then, as shown above, the energy balance equation constrains $\epsilon(\tau)$. If $\epsilon(v_c)$ were known, this constraint would lead to $c(\tau)$ and $\nu(\tau)$ relationships which could be used to examine the evolution of the rings through solutions of the momentum equation alone.

II. MATHEMATICAL FORMULATION: KINETIC THEORY

A. Numerical Simulation

The collective physical processes, which were heuristically described above, occur on the length scale of a few mean free paths and on a time scale comparable to the orbital period. These length and time scales are much too small to be resolved by any current observations so that it is rather difficult to

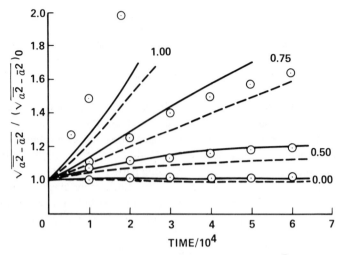

Fig. 1. The evolution of the distribution of particles' semimajor axes $(\bar{a}^2-\bar{a}^2)^{\frac{1}{2}}$. The subscript zero refers to the initial state. Curves are labeled with value of ϵ. *Open circles:* numerical calculation by Hämeen-Anttila and Lukkari (1980). *Solid curves* and *dashed curves:* results of two different analytical calculations by Hämeen-Anttila (1978). Times are given in units of the orbital period (figure from Hämeen-Anttila and Lukkari 1980).

directly compare theories with observations. However, the results of numerical simulation may be compared with theoretical arguments to verify the validity of various assumptions and to identify the key processes which dominate the evolution of the system. In the simulations of Trulsen (1972a), Brahic (1977), and Hämeen-Anttila and Lukkari (1980), the particles are modeled as identical rigid spheres which interact according to the hard-sphere collision model often used to simulate molecular dynamics. To simulate particle rings a degree of dissipation in the collisions must be included according to Eq. (10). The advantage of this idealized prescription is that comparisons between numerical and analytic solutions may be readily obtained.

The first step in a numerical simulation is to examine the simplest evolutionary tendencies, e. g., the radial spreading of a particle disk due to collisions. This effect is measured by the secular changes in the distribution of the particles' semimajor axes $(\bar{a}^2-\bar{a}^2)^{\frac{1}{2}}$ (Hämeen-Anttila and Lukkari 1980). According to the arguments in Sec. I, the rate of change in $(\bar{a}^2-\bar{a}^2)^{\frac{1}{2}}$ must be a function of ϵ. This qualitative expectation is indeed confirmed by the results of the numerical simulation. In Fig. 1, the circles denote simulations for several (constant) values of ϵ whereas the solid and dashed lines outline the results of two different analytic calculations by Hämeen-Anttila (1978). There is a general agreement between the numerical and analytic results.

As the evolution proceeds, the ring's thickness is automatically adjusted to establish an equilibrium velocity dispersion and scale height. The simula-

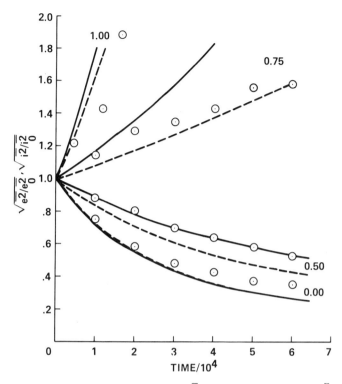

Fig. 2. The evolution of the mean eccentricity $(\overline{e^2})^{\frac{1}{2}}$ and the mean inclination $(\overline{i^2})^{\frac{1}{2}}$. Both sets of curves have been normalized to unity at $t = 0$ and are the same within the resolution of the plot. Notation is the same as in Fig. 1 (figure from Hämeen-Anttila and Lukkari 1980).

tions illustrate the rapid initial flattening and radial spreading of the system. The outcome is critically dependent on the prescription for ϵ that determines the rate of dissipation of orbital kinetic energy. Hämeen-Anttila and Lukkari's (1980) numerical simulations show how the ring settles into a flat-disk structure with root-mean-square orbital eccentricities and inclinations comparable to R_p/a if collisions are very inelastic (Fig. 2). Conversely, for highly elastic collisions, the ring evolves into a thick torus (Fig. 2). For moderate values of ϵ (e.g., $0.5 < \epsilon < 0.75$), the velocity dispersion may be maintained at an approximately constant value. Their analytic results also show similar evolutionary tendencies. Numerical simulations in which ϵ decreases with velocity of impact have also been carried out. Incorporating the prescription $\epsilon = \exp(-k v_c)$, where k is a constant, Hämeen-Anttila and Lukkari found that the velocity dispersion adjusted to allow energy equilibrium.

Regardless of the detailed prescription for ϵ, the orbital eccentricity and inclination quickly establish a quasi stationary equilibrium distribution of the form

$$F(x) = x^n \exp\left(-Cx^2/\overline{x^2}\right); x = e, i \qquad (14)$$

with the power index n slightly less than unity (Trulsen 1972a). One possible interpretation for this dynamical property is that there is a general tendency for F to relax quickly into a maximum entropy state allowed under various collisional constraints. In such a state the particles assume a Gaussian distribution of random kinetic energy such that

$$F(E) \sim \exp\left(-Cx^2/\overline{x^2}\right); x = e, i. \qquad (15)$$

The corresponding distribution for e and i is a Rayleigh distribution which has a functional form similar to Eq. (14) with $n = 1$. Since there are two degrees of freedom associated with the planar motion, the energy equipartition induces $(\overline{i^2}/\overline{e^2})^{\frac{1}{2}} = 1/\sqrt{2}$, if collisions are isotropic. Numerical simulations do in fact evolve to states near energy equipartition in e and i (see e.g., Trulsen 1972a). If ϵ is very small, particles settle down in a near-monolayer structure such that the orientation of colliding bodies (specified by $\hat{\mathbf{k}}$) becomes anisotropic with a preferred orientation in the plane of the rings. Hence, the random motion associated with orbital eccentricities is damped more efficiently than the motion due to orbital inclinations. Both numerical simulations and analytic solutions indicate that the ratio $(\overline{i^2}/\overline{e^2})^{\frac{1}{2}}$ decreases with increasing ϵ (Fig. 3).

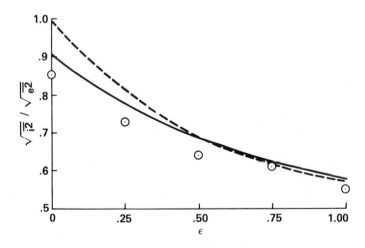

Fig. 3. The stationary value of $(\overline{i^2}/\overline{e^2})^{\frac{1}{2}}$ as a function of ϵ. Theoretical calculations (solid and dashed curves) and numerical simulations (open circles), as well as the figure itself, come from the work of Hämeen-Anttila and Lukkari (1980).

In contrast to the above studies of hard-sphere collisions, Trulsen (1972*b*) considered a more general collision model, the so-called snowflake model. Because of the assumed fluffy structure of the particles, this collision model led to an enhanced energy dissipation for grazing collisions, i.e., those which are most effective at producing random motions. When this model was used in a 2-dimensional simulation (all particles in the same plane) a small degree of radial focusing and clustering of pericenter arguments was observed. Brahic and Hénon (1977) made a preliminary test of a similar collision model in 3 spatial dimensions but did not report any results concerning radial clumping. Another mechanism, which can produce very definite local density enhancements, is the so-called diffusion instability, discussed in Sec. IV, which can occur even in the hard-sphere collision case as long as the coefficient of restitution is a decreasing function of impact velocity.

B. Analytical Approach.

The qualitative reasoning provided in Sec. I is too general to provide a deep appreciation of the detailed evolutionary characteristics. Due to the limitation of computers, numerical simulations are normally carried out with relatively coarse spatial resolution, limited number of particles, and for relatively short time scales. In order to make further progress towards a general understanding of the dynamics of planetary rings, rigorous mathematical analyses are necessary. In this section we discuss the dynamics from the point of view of the kinetic theory of gases.

1. Boltzmann Equation. The statistical properties of the particles may be expressed in terms of a probability density $f(m, r, v, t)$ for finding a particle of mass m at position r with velocity v at time t. Alternatively, f may be referred to as the distribution of particles in position-momentum space. In terms of f, the evolution of the system is described by the Boltzmann equation

$$\left(\frac{\partial}{\partial t} + v_i \frac{\partial}{\partial x_i} - \frac{\partial U}{\partial x_i} \frac{\partial}{\partial v_i}\right) f(m) = \sum_j \mathscr{C}[f(m), f(m_j)] \qquad (16)$$

where x and v are the position and velocity, respectively, expressed in some orthonormal coordinates. $U(x)$ is the gravitational potential due to the planet, its satellite system, any massive satellites that are imbedded in the rings and not included in the collection of particles described by f, and the self gravity of the particles included in the distribution function f. The right-hand side of the above equation represents the rate of change of f caused by collisions with particles of mass m_j.

The complexity of the collision integral generally prevents an exact solution of the Boltzmann equation. Considerable simplification can be achieved by replacing it with three sets of moment equations, obtained through multiplication, respectively, by 1, v_i, and $v_i v_j$, and integration over all velocities (e.g., Goldreich and Tremaine 1978):

$$\frac{\partial N}{\partial t} + \frac{\partial}{\partial x_i}(Nu_i) = \left(\frac{\partial N}{\partial t}\right)_c \tag{17}$$

$$\frac{\partial}{\partial t}(Nu_i) + \frac{\partial}{\partial x_j}(p_{ij} + Nu_iu_j) + \frac{N\partial U}{\partial x_i} = \left[\frac{\partial}{\partial t}(Nu_i)\right]_c \tag{18}$$

$$\frac{\partial}{\partial t}(p_{ij} + Nu_iu_j) + \frac{\partial}{\partial x_k}(p_{ijk} + u_jp_{ik} + u_kp_{ij} + u_jp_{jk})$$

$$+ Nu_iu_ju_k) + N\left(u_i\frac{\partial U}{\partial x_j} + u_j\frac{\partial U}{\partial x_i}\right) = \left[\frac{\partial}{\partial t}(p_{ij} + Nu_iu_j)\right]_c \tag{19}$$

where

$$N = \int f\,d^3v, \quad Nu_i = \int f v_i\,d^3v$$

$$p_{ij} = \int f(v_i-u_i)(v_j-u_j)\,d^3v . \tag{20}$$

In general, u_i consists of the local Keplerian circular velocity plus the radial flow generated by viscous evolution. These equations are entirely analogous to the moment equations of the kinetic theory of gases. This procedure of replacing the particle distribution function in the Boltzmann equation with its first three velocity moments is commonly referred to as the "hydrodynamic approximation." Note that this level of approximation is more general than the true "fluid" description given in Sec. III, which makes the further approximation that the diagonal components of pressure tensor p_{ij} are equal and thus may be replaced by a scalar pressure p.

In this treatment the physical processes which essentially determine the evolution of the rings are contained in the collision terms. For example, the term $(\partial N/\partial t)_c$ describes the evolution of the distribution function f due to coagulation and shattering whereas the term $[\partial(Nu_i)\partial t]_c$ includes, in addition, the effects of particle spin. The term $(\partial p_{ij}/\partial t)_c$ is due to contributions from gravitational scattering and inelastic collisions among the disk particles themselves. If all of these contributions are included at once, the mathematical complexity of the solution may easily obscure a general appreciation of the basic dynamical characteristics of the rings. Thus, it is advantageous to analyse the problem piecewise by considering only the dominant processes first. The simplest approach is to assume that the planetary rings are made of identical, hard, indestructible, nonspinning spheres, so that the collisional contribution to the zeroth- and first-moment equations (i.e., the right-hand sides of Eqs. (17) and (18) may be ignored). However, the effects of collisions in the second moment equation may not be ignored because these effects are essential for determining an energy equilibrium.

With the above assumption, the zeroth-moment equation is reduced to a simple continuity equation in which the evolution of the particle distribution function is determined by the gradient of the particle flux. The evolution of the flux may be readily obtained from the simplified first-order moment equation; it is determined by the gradient of the planet's gravitational potential and a collective stress due to the gradient in the pressure tensor **p**. Although the gravitational influence of the planet does not change significantly in time, the stress does.

The evolution of the stress tensor must be obtained from the second-order moment equation, which includes both the collisional effects and a contribution from a third-order pressure tensor p_{ijk}. The third order tensor must in turn be determined from a higher order moment equation. Since this mathematical procedure may be continued to arbitrarily high order, the hierarchy of velocity moments must be truncated at some stage. In the standard treatment of the problem (i.e., the hydrodynamic approximation used by Goldreich and Tremaine [1978]), the moment equations are closed by neglecting the third-order velocity moment of f occurring in the second-moment Eq. (19). Provided that the physical parameters such as the number density do not change significantly over a radial scale comparable to the vertical scale height, such an approximation is justified since the third-order velocity moment is smaller than the magnitude of all other contributions in the second-moment equation, by a factor approximately equal to the orbital eccentricity. Through such an approximation the velocity dependence of the probability density is totally constrained once a prescription for the collisional integral is specified.

2. Collisional Integrals. In general, the derivation for the collisional integral is rather complex since it contains contributions from various physical processes. Here, standard approximations used to evaluate collisional effects, such as the Fokker-Planck approximations, are not generally applicable because they are usually calculated for small-angle scattering whereas direct collisions in planetary rings result in large-angle scattering. One possible method adopted by Cook and Franklin (1964) is to use a Krook model for collision terms such that $(\partial f/\partial t)_c = \omega_c (f_o - f)$ where f_o is a Maxwellian which has the same number and energy density as f. The disadvantage of this method is that the energy loss due to inelastic collisions is not directly included in the above equations. In order to achieve an energy equilibrium, Cook and Franklin introduce an *ad hoc* prescription to account for the energy loss without altering the basic collision formula; thus the results of their calculations may be an artifact of their adopted assumptions. However, the evaluation of the collisional integral may be simplified considerably with the adoption of the following assumptions: (1) all particles are identical; (2) all collisions are hard-sphere collisions; (3) the particles' size is small compared with the interparticle spacing; (4) the epicyclic nature of the deviations from coplanar circular orbits is neglected so that the random velocities may be assumed to have a Gaussian distribution. The implications of this last approximation have not been thoroughly investigated, although Baxter and Thompson (1973) and Hämeen-Anttila (1978) have begun work in this direction. Under these assumptions, Trulsen (1971) derived the collisional integral associated with collision law (Eq. 10):

$$\mathscr{C}\,[f,f] \equiv \left(\frac{\partial f}{\partial t}\right)_c = 4R_p^2 \iint\limits_{(\mathbf{g}\cdot\hat{\mathbf{k}})>0} [f(\mathbf{v_1}^*)f(\mathbf{v_2}^*)/\epsilon^2$$

$$-f(\mathbf{v_1})\,f(\mathbf{v_2})](\mathbf{g}\cdot\hat{\mathbf{k}})\mathrm{d}\,\hat{\mathbf{k}}d^3\mathbf{v_2} \tag{21}$$

where

$$\mathbf{v}^*_{1,2} = \mathbf{v}_{1,2} \mp \frac{(1+\epsilon)}{2\,\epsilon}\,(\mathbf{g}\cdot\hat{\mathbf{k}})\,\hat{\mathbf{k}} \tag{22}$$

and R_p is the radius of a particle. Substituting this expression into the second-moment equation to replace the collisionally induced pressure tensor term, we find

$$\left(\frac{\partial p_{ij}}{\partial t}\right)_c = \frac{\pi}{12}R_p^2 \iint (1+\epsilon)\,[(1+\epsilon)g^2\,\delta_{ij} - 3\,(3-\epsilon)g_i g_j]$$

$$\times\, gf\,(\mathbf{v_1})f\,(\mathbf{v_2})d^3\mathbf{v_1}d^3\mathbf{v_2} \tag{23}$$

where δ_{ij} is the identity tensor. A similar expression for the above equation has been obtained by Goldreich and Tremaine (1978) for the special $i = j$ case. In the following discussion we shall use a cylindrical coordinate system (r, θ, z).

In order to evaluate the collision integral in Eq. (23) we must specify the velocity dependence of the distribution function $f(\mathbf{v})$ as well as of the coefficient of restitution ϵ. The results of the numerical simulations indicate that particle energies quickly establish an approximately Gaussian distribution (see Eq. 15). For most analytic treatments of the problem, it is usually assumed (Goldreich and Tremaine 1978; Hämeen-Anttila 1978) that f has a 3-dimensional Gaussian distribution and that it may be expressed in terms of a velocity dispersion tensor $T_{ij} = p_{ij}/N$ such that

$$f = N\,[(2\pi)^3\,\det\,\mathbf{T}]^{-\frac{1}{2}}\exp\,[-\tfrac{1}{2}v_i(T^{-1})_{ij}\,v_j]\,. \tag{24}$$

The trace of $\mathbf{T}$ (i.e., $\mathrm{Tr}\mathbf{T} = \mathbf{T}_{11} + \mathbf{T}_{22} + \mathbf{T}_{33}$) equals the mean square of the velocity dispersion, $\overline{c^2}$, due to orbital eccentricity and inclination. The basic motivations for adopting this particular expression are: (1) it satisfies the steady-state solution of the second-moment equation in the limit where collisions have a negligible effect on the evolution of the rings (i.e., $T_{rr} = 4T_{\theta\theta}$; $T_{r\theta} = 0$); (2) it reduces to a Maxwellian form in the isotropic collision dominated limit ($T_{ij} = \mathrm{Tr}\mathbf{T}\,\delta_{ij}/3$); and (3) it attains a nonvanishing $P_{r\theta}$ which induces angular momentum transport and matter redistribution. With minor alterations, a more general distribution function may be constructed to include the asymmetric distributions of pericenter arguments and ascending nodes (Hämeen-Anttila 1978).

Further progress can only be made with additional approximations. The standard procedure to follow at this point is to assume that ϵ has a weak dependence on v_c. Since the relative velocity has a Gaussian distribution, most of the collisions occur with $\overline{c^2} = \mathrm{Tr}\mathbf{T}$. Thus, it may be resonable to treat ϵ as a constant term in the integration and remove terms containing ϵ outside of the integral. The error introduced by this procedure may be simply treated as a deformation of ϵ since it is uncertain anyway (Hämeen-Anttila 1981).

Despite the above approximation, the anisotropy of the tensor $\mathbf{T}$ still prevents the exact evaluation of the integral in Eq. (23). However, it is possible to transform the integration variable into a diagonalized form so that the dimensionality of the integral may be reduced to one. The remaining integral can be evaluated easily over two of the principal axes. After some coordinate transformations, the integral over the third principal axis may also be evaluated if the velocity dependence in ϵ is ignored. The resultant integral can be expressed as

$$\left(\frac{\partial p_{ii}}{\partial t} \right)_c = 4\pi^{\frac{1}{2}} N^2 R_p{}^2 \; \frac{(1+\epsilon)}{(p_{jj} p_{kk})^{\frac{1}{2}}} \; P_{ii}^{\frac{3}{2}} \left[(1 + \epsilon) J_p^i + J_q^i \right] \quad (25)$$

where J_p^i and J_q^i are symmetric functions of p_{ii}/p_{jj} and p_{ii}/p_{kk}, respectively, and can be expressed in terms of elliptical integrals (Goldreich and Tremaine 1978). However, the resultant integrals are sufficiently complex so that it is usually more convenient to evaluate them numerically.

Alternatively, the integral may be evaluated analytically under the assumption that one factor of g in the integrand may be approximated by a constant value

$$g_0 = \frac{16}{3} \left(\frac{\mathrm{Tr}\mathbf{T}}{3\pi} \right)^{\frac{1}{2}} \quad (26)$$

(Hämeen-Anttila 1978, 1981). This approximation is analogous to the earlier approximation where terms containing ϵ are pulled out of the integral. The basis of the approximation is again the Gaussian form of the velocity distribution which implies that the dominant contribution of the integral may be attributed to a sufficiently narrow range of the integration variable that g may be treated as a constant. The numerical coefficient in g_0 was chosen to yield the exact result for the special case of an isotropic velocity distribution. The accuracy of this approximation is tested by a direct comparison with the numerical calculations below. With this approximation, the integral becomes

$$\left(\frac{\partial p_{ij}}{\partial t}\right)_c = \frac{8}{9} \, NR_p^2 (\pi \mathrm{Tr}\mathbf{T}/3)^{\frac{1}{2}} (1+\epsilon) \left[(1+\epsilon) \, \mathrm{Tr}\mathbf{p} \, \delta_{ij} - 3(3-\epsilon)p_{ij}\right] \qquad . (27)$$

3. Mass and Momentum Transfer. With the evaluation of the collisional integral, we now discuss the evolution of the mass distribution in the rings. In order to simplify this analysis, we examine the z-dependence of N with a thin-disk approximation. The basis of this approximation is to assume that the variation of gravitational potential in the vertical (z) direction is relatively small so that the dependence of physical parameters on the two independent spatial variables r and z, may be separated and the general physical properties of the disk may be averaged over the z direction. With this approximation, the dynamics of a disk or a ring may be characterized primarily by the evolution of the surface density $\sigma = m \int N(z)dz$. To determine $\sigma(r)$ we consider a thin, smoothly varying disk where the thickness $H \ll r$ and the radial gradient of H, dH/dr, $\ll 1$. Then the bulk motion in the vertical direction $u_z \approx (\partial H/\partial r)u_r$ is generally much less than u_r. With the neglect of contributions from u_z the z component of the first-moment equation is reduced to

$$\frac{1}{N} \frac{\partial}{\partial z}\left(NT_{zz}\right) = -\frac{GMz}{r^3} . \qquad (28)$$

If we assume T_{zz} is independent of z for a thin ring, we deduce from the above equation that the number density of the disk particles decreases exponentially with z such that

$$N(z) = N(0) \, \exp[-\Omega^2 z^2/(2T_{zz})] \qquad (29)$$

where $N(0)$ is the number density at the mid-plane. Consequently, the vertical scale height $H \cong (T_{zz})^{\frac{1}{2}}/\Omega$. Integrating $N(z)$ over z we find $\sigma = N(0) m(2\pi T_{zz})^{\frac{1}{2}}/\Omega$ which also implies $\tau = \pi R_p^2 \sigma/m \approx \omega_c/\Omega$.

If T_{ij} is independent of z, $\int p_{ij} \, dz = \sigma/m \, T_{ij}$ and the first-order moment equation reduces to

$$\frac{\partial u_r}{\partial t} + u_r \frac{\partial u_r}{\partial r} - \Omega^2 r = -\frac{GM}{r^2} + T_{\theta\theta} - \frac{1}{\sigma r} \frac{\partial}{\partial r} (r\sigma T_{\theta\theta}) \qquad (30)$$

and

$$\frac{\partial}{\partial t}(\Omega r) + \frac{u_r}{r}\frac{\partial}{\partial r}(r^2\Omega) = -\frac{1}{\sigma r^2}\frac{\partial}{\partial r}(r^2\sigma T_{r\theta}) . \tag{31}$$

The dominant terms in Eq. (30) are $\Omega^2 r$ and GM/r^2 unless either σ or $T_{\theta\theta}$ varies significantly over a characteristic length scale L comparable to H. For a typical, thin, smoothly varying disk, L is generally much larger than H so that Eq. (30) merely provides a requirement for the rotation law to be Keplerian, i.e. $\Omega = (GM/r^3)^{\frac{1}{2}}$ at all times. In this case, equation (31) reduces to

$$\sigma u_r r = \dot{m}/2\pi = \frac{2}{\Omega r}\frac{\partial}{\partial r}(r^2\sigma T_{r\theta}) . \tag{32}$$

The variable $\dot{m}$ may be interpreted as the mass transfer rate in the radial direction. Under the special condition of steady-state flow, i.e., when $\partial\sigma/\partial t = 0$, $\dot{m}$ is independent of radius and

$$u_r \simeq T_{r\theta}/(\Omega r) = T_{r\theta}^{\frac{1}{2}}\left(\frac{T_{r\theta}}{T_{zz}}\right)^{\frac{1}{2}}\frac{H}{r} << T_{r\theta}^{\frac{1}{2}} << \Omega r . \tag{33}$$

In general, the evolution of the particle distribution in the disk may be obtained from the zeroth-order moment equation:

$$\frac{\partial\sigma}{\partial t} = -\frac{1}{r}\frac{\partial}{\partial r}(r\sigma u_r) = \frac{2}{r}\frac{\partial}{\partial r}\left[\frac{1}{\Omega}\frac{\partial}{\partial r}(r^2\sigma T_{r\theta})\right] . \tag{34}$$

4. Evolution of the Pressure Tensor. For the solution of Eq. (34) a $(\sigma, T_{r\theta})$ relationship must be specified; it can be obtained from the second-order moment equation. If we ignore $(\partial N/\partial t)_c$ and $(\partial Nu_i/\partial t)_c$ we can write this equation in the simplified form

$$\frac{\partial p_{ij}}{\partial t} + p_{ik}\frac{\partial u_j}{\partial x_k} + p_{jk}\frac{\partial u_i}{\partial x_k} + \frac{\partial}{\partial x_k}\left(p_{ij}u_k\right)$$

$$+ \frac{\partial}{\partial x_k}\left(p_{ijk}\right) = \left(\frac{\partial p_{ij}}{\partial t}\right)_c \tag{35}$$

In a thin, axisymmetric disk, $p_{rz} = p_{z\theta} = 0$ so that there are only four nonvanishing components of the pressure tensor. However, even if we neglect contributions associated with u_z, vertically average the equations, and assume $L >> H$ (see Sec. II.B.1), the analytic solutions of partial differential Eq. (35) are still not readily transparent without further simplification. In principle the solution depends on the initial conditions in the ring. However, the results discussed above indicate that the ring tends to evolve into an energy equilibrium state on a relatively rapid time scale. In Sec. IV we shall rigorously demonstrate that such an energy equilibrium state is stable; for the moment we may assume that the pressure tensor is independent of time. Consequently, we may obtain an energy transport equation by summing up all the diagonal elements of T_{ij}:

$$-3\Omega\sigma T_{r\theta} = \left(\frac{\partial}{\partial t}\sigma\,\mathrm{Tr}\mathbf{T}\right)_c .\qquad(36)$$

The above equation implies that there is an overall balance between the rate of energy generation by viscous stress and energy dissipation by inelastic collisions. It is of interest to note that when collisions are totally elastic, so that the right-hand side of Eq. (36) is zero, the energy generation term nevertheless appears finite despite the absence of dissipation. However, in this case, steady-state solutions are not permitted so that Eq. (36) is not really applicable.

Equation (36) may be solved numerically with the aid of the numerically integrated values for the collisional integrals (Goldreich and Tremaine 1978). Using this approach, we can establish a relationship between ϵ and τ from the numerical solutions (see Fig. 4), which may be expressed as

$$\epsilon = [1 - b/(1 + \tau^2)]^{\frac{1}{2}}\qquad(37)$$

where b is a numerical constant equal to 0.61. This relationship is identical, within a constant factor of order unity, to that derived from the heuristic arguments in Sec. I.

Alternatively, we may utilize the simplified, analytical, approximate solutions for the collisional integrals (Eq. 27). In a thin disk, the second-order moment equation integrated over z becomes:

$$\left(\frac{\partial p_{rr}}{\partial t}\right)_c = \frac{\partial p_{rr}}{\partial t} - 4\Omega p_{r\theta} = \frac{\omega_0}{18}(1+\epsilon)\left[(1+\epsilon)\mathrm{Tr}\mathbf{p} - 3(3-\epsilon)p_{rr}\right]\qquad(38a)$$

$$\left(\frac{\partial p_{r\theta}}{\partial t}\right)_c = \frac{\partial p_{r\theta}}{\partial t} - \Omega(p_{rr}/2 + 2p_{\theta\theta}) = -\frac{\omega_0}{6}(1+\epsilon)(3-\epsilon)p_{r\theta}\qquad(38b)$$

$$\left(\frac{\partial p_{\theta\theta}}{\partial t}\right)_c = \frac{\partial p_{\theta\theta}}{\partial t} + \Omega p_{r\theta} = \frac{\omega_0}{18}(1+\epsilon)\left[(1+\epsilon)\mathrm{Tr}\mathbf{p} - 3(3-\epsilon)p_{\theta\theta}\right]\qquad(38c)$$

$$\left(\frac{\partial p_{zz}}{\partial t}\right)_c = \frac{\partial p_{zz}}{\partial t} = \frac{\omega_0}{18}(1+\epsilon)\left[(1+\epsilon)\mathrm{Tr}\mathbf{p} - 3(3-\epsilon)p_{zz}\right]\qquad(38d)$$

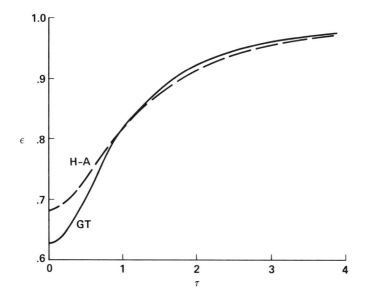

Fig. 4. Comparison between the $\epsilon - \tau$ relations calculated by Hämeen-Anttila (1978), marked H–A, and by Goldreich and Tremaine (1978), marked GT.

where the effective collisional frequency ω_0 can be defined in terms of the optical depth τ such that

$$\omega_0 = \frac{8\tau\Omega}{\pi} \, (\mathrm{Tr}\mathbf{T}/3T_{zz})^{\frac{1}{2}} \, . \tag{39}$$

In a steady state, the energy transport Eq. (36) becomes

$$9\Omega T_{r\theta} = \omega_0 (1 - \epsilon^2) \, \mathrm{Tr}\mathbf{T}. \tag{40}$$

Substituting this expression for ω_0 back into Eq. (38) we find

$$T_{rr} = \frac{9 - 7\epsilon}{3(3 - \epsilon)} \, \mathrm{Tr}\mathbf{T}$$

$$T_{zz} = \frac{1 + \epsilon}{3(3 - \epsilon)} \, \mathrm{Tr}\mathbf{T} \tag{41}$$

$$T_{\theta\theta} = \frac{3\epsilon - 1}{3(3 - \epsilon)} \, \mathrm{Tr}\mathbf{T}$$

so that

$$\frac{\omega_0}{\Omega} = \frac{8\tau}{\pi}\left(\frac{3-\epsilon}{1+\epsilon}\right)^{\frac{1}{2}} = \left[\frac{9(19\epsilon - 13)}{(1-\epsilon)}\right]^{\frac{1}{2}}/[(1+\epsilon)(3-\epsilon)] \qquad (42)$$

and

$$\tau = \frac{3\pi}{8}\left[\frac{19\epsilon - 13}{(1-\epsilon^2)(3-\epsilon)^3}\right]^{\frac{1}{2}}. \qquad (43)$$

Inserting this result back into Eq. (40) gives

$$T_{r\theta} = \frac{[(19\epsilon - 13)(1-\epsilon)]^{\frac{1}{2}}}{3(3-\epsilon)} \mathrm{Tr}\mathbf{T} \qquad (44)$$

and, indirectly, a relationship between $T_{r\theta}$ and τ.

The approximate analytic result for the $\epsilon - \tau$ relation in Eq. (43) closely agrees with the numerical result represented by Eq. (37) (see Fig. 4) and thereby provides an a posteriori justification for the approximation adopted by Hämeen-Anttila. The largest deviation between the two solutions occurs for small optical depth because the departure from isotropy in the velocity distribution is greatest at low collision frequencies. With some modifications and a more complex approximation for the collisional integral, a closer agreement between analytic and numerical results may be obtained (Hämeen-Anttila 1978).

5. *Collision-induced Viscous Evolution of the Disk.* In the previous section, we outlined two methods, both involving a rather restrictive set of assumptions, of obtaining relationships between τ and ϵ and between T_{ij} and $\mathrm{Tr}\mathbf{T}$. These relationships may be used to study the long-term viscous evolution of the disk. Provided that the disk is thermally stable, the thermal equilibrium solutions are fully applicable for this purpose since the time scale for establishing an energy equilibrium is much shorter than the diffusion time scale. The evolution is determined by the solution of the mass diffusion Eq. (34) which contains variables such as σ and $T_{r\theta}$. Thus, in addition to the above results a relationship between τ (or ϵ) and any component of the pressure tensor must be specified before the diffusion equation may be solved as an initial value problem.

If there is a one-to-one correspondence between ϵ and v_c (or $\mathrm{Tr}\mathbf{T}$) such that $\mathrm{Tr}\mathbf{T} = c_0^2 f(\epsilon/\epsilon_0)$, where f is some function to be determined experimentally and $f(1) = 1$, the diffusion equation may be written

$$\frac{\partial \tau}{\partial t} = \frac{2c_0^2}{r}\frac{\partial}{\partial r}\left\{\frac{1}{\Omega r}\frac{\partial}{\partial r}\left[r^2\tau\left(\frac{T_{r\theta}}{\mathrm{Tr}\mathbf{T}}\right)f(\epsilon/\epsilon_0)\right]\right\}. \qquad (45)$$

The analytic, approximate expression for $T_{r\theta}/\mathrm{Tr}\mathbf{T}$ is given by Eq. (44). Since there is a relationship between ϵ and τ (see Eqs. 37 or 43), the right-hand side of the diffusion equation contains a function, say h, of τ and the two independent variables, r and t. It seems likely that $h(\tau)$ is a nonlinear function of τ so that Eq. (45) will not in general yield a linear diffusion equation. In particular, since $T_{r\phi}$ is a function of τ, its magnitude may change with time such that the time scale for radial spreading of a ringlet is not necessarily proportional to the square of the width of a ringlet as was assumed in the heuristic arguments presented in Sec. I.B.

The $\epsilon - \tau$ and $T_{ij} - \mathrm{Tr}\mathbf{T}$ relationships obtained in Sec. II.B.4 and the $\epsilon - v_c$ relation to be determined experimentally may also be used to deduce a vertical scale height for the disk. In particular, from the observed τ we can determine a corresponding value of ϵ and then, from the $\epsilon - v_c$ relationship, a value for v_c or equivalently $\mathrm{Tr}\mathbf{T}$. Using the numerical value of $\mathrm{Tr}\mathbf{T}$ in the energy equilibrium solution (Eq. 41), we deduce a value for T_{zz}, and thus a value for H ($\approx T_{zz}^{\frac{1}{2}}/\Omega$).

Although a $\epsilon - v_c$ relationship is not yet available, we know that for most substances, very low-velocity collisions tend to be perfectly elastic. If the ring particles' ϵ remains unity up to an impact velocity v_1, there would be an absolute lower limit to the thickness of the rings such that $H \gtrsim H_1 = v_1/\Omega$. A ring cannot be thinner than H_1 because the ring particles do not dissipate their dispersion velocity below the critical value which will support such a thickness. At large v_c, ϵ decreases with v_c. However, according to the energy-equilibrium solution of the second-order moment Eq. (37), the ring particles' ϵ cannot be smaller than 0.62, corresponding to $(1-b)^{\frac{1}{2}}$ for $\tau = 0$. Below this value of ϵ, the collisions are too dissipative to be able to sustain an energy equilibrium. If the experimentally-determined ϵ for the ring particles is reduced to $(1-b)^{\frac{1}{2}}$ at some velocity, say v_2, there would be an upper limit to the thickness such that $H \lesssim H_2 = v_2/\Omega$. If, however, ϵ remains larger than $(1-b)^{\frac{1}{2}}$ for very larger values of v_c, there would be a limiting value in τ. A ring with a τ close to this limiting value can have a very large velocity dispersion such that the ring may be very thick and may evolve on a relatively short time scale. We show in Sec. IV that the disk is diffusively unstable for all τ in this case. The presence of regions of very low τ which are not influenced by external force fields such as resonant interactions with a satellite, will indicate that ϵ for the ring particles indeed reduces below $(1-b)^{\frac{1}{2}}$ at relatively small impact velocities.

The above discussions clearly indicate the importance of a theoretically determined $\epsilon - \tau$ law and the urgent need of an experimentally determined $\epsilon - v_c$ law. In particular, the limiting value of ϵ for arbitrarily small τ and the values of ϵ for very small and very large v_c basically determine the vertical structure and the evolutionary fate of the rings. Recall that the $\epsilon - \tau$ law was derived from the second-order moment equation so that its actual prescription may be sensitive to the approximations used in evaluating the collisional integral. In view of its importance and the numerous assumptions adopted for its evaluation, a more careful analysis of the $\epsilon - \tau$ relationship may be useful.

III. MATHEMATICAL FORMULATION: HYDRODYNAMICS

A. Basic Formulation

Although the kinetic treatment is designed to analyze rigorously the statistical properties of the particles in the ring, its complexity makes exact analytic solutions and their stability analyses unattainable. The above discussion is a clear indication that numerous approximations and simplifications are required to deduce and verify the most basic results. Overall, the mathematical complexity tends to obscure the central issue and the basic interpretation of the essential physical processes which dominate the structure, stability and evolution of the rings. However, to a considerable degree, these complexities may be reduced by the use of a fluid dynamical approximation. The essence of this approximation is that the dispersion velocity ellipsoid has a locally unique characteristic shape such that the pressure tensor may be divided into an isotropic part, represented by a scalar pressure, and a nonisotropic part represented by a stress tensor. The stress tensor is a product of locally determined transport coefficients, such as viscosity, and a globally determined shear tensor. This dependence on the global properties in the stress tensor is in contrast to the apparently total dependence on the local velocity dispersion ellipsoid in the pressure tensor in the kinetic treatment. As the analyses of the energy-equilibrium solutions of the second-order moment equations have already indicated, the pressure tensor is closely related to the global shear albeit indirectly. Through this approximation, all the complications associated with the evaluation of the dispersion-velocity ellipsoid are replaced by prescriptions for the transport coefficients. Consequently, the hydrodynamic equations have apparently somewhat simpler forms than the moment equations. In this case, the evolution of a ring may be determined from the hydrodynamical equations of continuity, momentum, and energy, which for a fluid of density ρ and velocity **u** can be written as

$$\frac{d\rho}{dt} + \rho \frac{\partial u_i}{\partial x_i} = 0 \tag{46}$$

$$\rho \frac{du_i}{dt} = -\frac{\partial P}{\partial x_i} - \rho \frac{\partial U}{\partial x_i} + 2 \frac{\partial}{\partial x_j} (\rho \nu e_{ij}) + \frac{\partial}{\partial x_i} \left[(\xi - \frac{2}{3}\nu) \rho \frac{\partial u_j}{\partial x_j} \right] \tag{47}$$

$$\frac{d}{dt} (\rho E) = -P \frac{\partial u_i}{\partial x_i} + 2 \rho \nu \left[e_{ij} e_{ij} - \frac{1}{3} \left(\frac{\partial u_i}{\partial x_i} \right)^2 \right]$$

$$+ \rho \xi \left(\frac{\partial u_i}{\partial x_i} \right)^2 + \frac{\partial}{\partial x_i} \left(\frac{\kappa \rho}{k_B} \frac{\partial E}{\partial x_i} \right) + \rho \dot{E}_c \qquad (48)$$

where v, ξ, and κ are the kinematic shear viscosity, bulk viscosity, and thermal conductivity, respectively, and $e_{ij} = 0.5(\partial u_i/\partial x_j + \partial u_j/\partial x_i)$ is the shear tensor. The pressure and internal energy (per unit mass) are denoted by P and E, respectively, and the rate of energy loss due to inelastic collisions is denoted by $\dot{E}_c$. The material time derivative, d/dt, contains both the local time derivative and the advective operator $(u_i \, \partial/\partial x_i)$ terms.

The leading terms of the somewhat simplified hydrodynamic equations have direct correspondence to leading contributions in the moment equations. From these corresponding terms, we may evaluate the magnitude and physical dependence of the transport coefficients. For example, after the vertical averaging, the continuity equation becomes

$$\frac{\partial \sigma}{\partial t} + \frac{1}{r} \frac{\partial}{\partial r} (r \sigma u_r) = 0 \qquad (49)$$

which is identical to the zeroth-order moment equation. Similarly, the leading terms in the momentum equation can also be identified with those in the first-order moment equation. In a thin, axisymmetric disk, the momentum equation has three components such that in cylindrical coordinates

$$\frac{\partial u_r}{\partial t} + u_r \frac{\partial u_r}{\partial r} - \Omega r^2 = -\frac{GM}{r^2} - \frac{1}{\sigma} \frac{\partial P'}{\partial r} + \frac{2}{\sigma} \frac{\partial}{\partial r} \left(\sigma v \frac{\partial u_r}{\partial r} \right)$$

$$+ 2 v \frac{\partial}{\partial r} \left(\frac{u_r}{r} \right) + \frac{1}{\sigma} \frac{\partial}{\partial r} \left[\sigma \left(\xi - \frac{2}{3} v \right) \frac{1}{r} \frac{\partial}{\partial r} (r u_r) \right] \qquad (50a)$$

$$\frac{\partial}{\partial t} (\Omega r) + \frac{u_r}{r} \frac{\partial}{\partial r} (r^2 \Omega) = \frac{1}{\sigma r^2} \frac{\partial}{\partial r} \left(v \sigma r^3 \frac{\partial \Omega}{\partial r} \right) \qquad (50b)$$

$$\frac{1}{\rho} \frac{\partial P}{\partial z} = -\frac{GMz}{r^3} \qquad (50c)$$

where $P' = \int P \, dz$ is the vertically integrated gas pressure. Equation (50c) is identical to the z component of the first-order moment equation if we identify P with mNT_{zz}. In our attempt to solve the moment equations, we have adopted an approximation in which T_{zz} is independent of z. This assumption corresponds to an isothermal equation of state in the hydrodynamic treatment. The solution for Eq. (50c) for an isothermal gas indicates that $N(z) = N(0) \exp [-z^2 GM/(r^3 C_s^2)]$ where C_s is the sound speed; this solution is, of course, identical to that obtained from the moment equations (see Eq. 29).

The analogy between Eqs. (50a) and (30) is more tenuous. Here, only the leading terms are identical whereas the interpretation of pressure, shear- and bulk-viscosity contributions is rather ambiguous. Fortunately, the contributions due to difficult-to-identify terms are negligibly small unless the disk flow is significantly noncircular for one of the following reasons: (1) the velocity dispersion on the ring is comparable to Ωr (or equivalently, the ring is very thick); (2) there is a very large pressure gradient in the radial direction due to sharp transitions; (3) there is a large bulk viscosity; and (4) the ring is strongly perturbed by external forces with very large radial gradients. In most regions of planetary rings these conditions are not realized so that Eq. (50a) is reduced to the standard prescription for Keplerian motion, $\Omega = (GM/r^3)^{\frac{1}{2}}$. However, near sharp edges of a typical ring, especially when the characteristic radial scale length is comparable to the vertical scale height, the ring flow may indeed become non-Keplerian and therefore it may be necessary to solve simultaneously the radial component as well as the θ-component of the equation of motion.

The θ-component of the equation of motion is useful for determining angular momentum transfer and mass flux in the radial direction. In those regions of the ring where particles follow Keplerian orbits Eq. (50b) becomes identical to Eq. (31) if we identify $T_{r\theta}$ with the quantity $-\nu r \partial\Omega/\partial r = 1.5\,\Omega\nu$. To dominant order, the radial mass flow becomes

$$\sigma u_r r = -\frac{3}{\Omega r} \frac{\partial}{\partial r} (r^2 \Omega \sigma \nu) .$$ (51)

The equivalent to Eq. (34), the mass diffusion equation, also can be obtained with this substitution. This equation may be solved, for various initial conditions once a prescription for ν is specified. Equation (6) provides an order-of-magnitude estimate from the standard formula $\nu \simeq \omega_c \lambda^2$. A more rigorous derivation for ν may be obtained if we identify $1.5\,\Omega\nu$ with $T_{r\theta}$ as above and use the numerical results obtained by Goldreich and Tremaine (1978):

$$\nu = K_1 c^2 \tau / [\Omega(1+\tau^2)]$$ (52)

where K_1 was calculated to be 0.15. An alternative expression for ν (Hämeen-Anttila 1978) is obtained from Eq. (44)

$$\nu = \frac{2[(19\epsilon - 13)(1-\epsilon)]^{\frac{1}{2}} \operatorname{Tr}\mathbf{T}}{9(3-\epsilon)\,\Omega}$$ (53)

and is in close agreement (see Sec. II.B.4). Furthermore, both relationships are closely analogous to an equivalent expression obtained by Cook and

Franklin (1964) who evaluated the collision integral with a Krook collision law. With the evaluation of these transport coefficients (or equivalently, different components of the pressure tensor), the mass diffusion equation may be solved numerically (or analytically under special circumstances) as an initial value problem once a relationship between τ and $\mathrm{Tr}\mathbf{T}$ is established.

In a fluid approximation, the energy equation is analogous to the second-order moment Eq. (35). Direct comparison between these two equations can be made by identifying E with the kinetic energy associated with random motion i.e. $\mathrm{Tr}\mathbf{T}$. The term $\partial(\rho E)/\partial t$ represents the rate of change in kinetic energy associated with the random motion which is analogous to $\partial(\sigma \mathrm{Tr}\mathbf{T})/\partial t$. The term $P\,\partial u_i/\partial x_i$ is the energy contribution due to $P\,dV$ work and is equivalent to the $P_{ij}\,\partial u_k/\partial x_k$ term. Viscous dissipation processes, described by the second and third terms on the right-hand side of Eq. (48), induce an energy transfer from the systematic shearing motion into the random motion and are related to the $P_{jk}\,\partial u_i/\partial x_k$ terms. The local heat flux due to conduction $\partial(\kappa\rho\,\partial E/\partial x_i)/\partial x_i$ is equivalent to the term $\partial p_{ijk}/x_k$. Finally, an energy loss rate, $\dot{E}_c$, is incorporated to account for the rate of energy loss due to inelastic collisions which is described by $(\partial p_{ij}/\partial t)_c$ in the kinetic theory approach. There is no simple way to estimate $\dot{E}_c$ without carrying out the evaluation of the collisional integral as we have done in the kinetic treatment presented in Sec. II.B.2. Therefore we shall utilize this result by setting $\dot{E}_c = K_2\,(1-\epsilon^2)\,\tau\,c^2\,\Omega$, where K_2 is a constant of order unity, in analogy with Eq. (11). If we compare this expression with the rigorous calculations referred to in Sec. II.B.3, we find that K_2 has a value ranging from 0.9 to 1.5, depending on the method of evaluation of the collisional integral. For convenience, we set $K_2 = 1$.

The energy equation for a thin disk may be written in cylindrical coordinates as follows:

$$\frac{\partial E}{\partial t} + u_r \frac{\partial E}{\partial r} + \left(\frac{\Gamma_3 - 1}{r}\right)\frac{\partial}{\partial r}(u_r r) = \frac{9}{4}\,\nu\,\Omega^2 - (1-\epsilon^2)\,\tau\,c^2\,\Omega$$

$$+ \frac{1}{r}\,\frac{\partial}{\partial r}\left(\frac{\kappa}{k_B}\,\frac{\partial E}{\partial r}\right) \tag{54}$$

where Γ_3 is the third adiabatic exponent. If the disk is smoothly varying, contributions due to the advective transport, PdV work, and thermal conduction are negligible compared with the energy generation process associated with viscous stress and energy dissipation due to inelastic collision.

Thus, in equilibrium, the energy equation becomes

$$\frac{9}{4}\,\nu\,\Omega^2 = (1-\epsilon^2)\,\tau\,c^2\,\Omega\ . \tag{55}$$

This equation is entirely analogous to Eq. (40). From Eq. (52) it follows that $(1 + \tau^2)(1 - \epsilon^2) = 0.6$. Substituting this $(\epsilon - \tau)$ relation, Eq. (51) and Eq. (52) into the continuity Eq. (49) and using $\tau = \pi R_p^2 \, \sigma/m$, we obtain a mass diffusion equation,

$$\frac{\partial \tau}{\partial t} = \frac{3K_1}{r} \frac{\partial}{\partial r}\left\{ \left(\frac{1}{\Omega r}\right) \frac{\partial}{\partial r}\left[\frac{\tau^2 r^2 c^2}{(1 + \tau^2)} \right] \right\} \tag{56}$$

where c is a function of ϵ and therefore of τ. Equation (56) is analogous to the corresponding Eq. (45) obtained from the kinetic approach.

B. The Energy Budget of Accretion-Disk Flow

Since ring particles undergo frequent inelastic collisions, the energy E associated with dispersive motion is continually drained from the system. This internal energy, in turn, is continually supplied by the kinetic energy of orbital motion via viscous stress. Ultimately, the kinetic energy must be replenished by the release of gravitational energy associated with a net inward drift of the ring particles. This general, qualitative description of the basic energy-flow pattern in a differentially rotating accretion disk, such as a planetary ring, may be examined rigorously with a fluid-dynamical approach which was first adopted by Lynden-Bell and Pringle (1974).

We begin by considering the rate of change of potential and kinetic energy carried by a fluid element with a fixed increment of mass dm and a variable physical volume, $dV = dm/\rho$. We first multiply u_i by the momentum Eq. (47), then add the result to the product of u_i and the continuity equation, and then integrate over dV such that

$$\frac{d}{dt} \int_V \left(\frac{u^2}{2} + U \right) dm = \int_V u_i \frac{\partial}{\partial x_j}\left[-P\delta_{ij} + 2\rho \nu e_{ij} \right.$$

$$\left. + \rho \left(\xi - \frac{2}{3}\nu \right) e_{kk} \delta_{ij} \right] dV. \tag{57}$$

In the derivation of the above equation, we have assumed $\partial U/\partial t = 0$. With the use of the divergence theorem, the above equation for a thin disk • between radii r_1 and r_2 becomes

$$\frac{d}{dt} \int_V \left(\frac{u^2}{2} + U \right) dm = [-2\pi r u_r P' + 2\pi \sigma \, \nu r^3 \Omega \, \partial\Omega/\partial r]_{r_1}^{r_2}$$

$$- \int_{r_1}^{r_2} 2\pi r^3 \, \nu\sigma \left(\frac{\partial\Omega}{\partial r} \right)^2 dr + \int_V P \frac{d}{dt}\left(\frac{1}{\rho} \right) dm \tag{58}$$

The left-hand side of the above equation represents the rate of change in kinetic and gravitational energy associated with ring material flowing into the potential well of the central planet. On the right-hand side, the first term represents the flux of internal energy advected into the region under consideration by the bulk motion of the ring particles. The second term is the rate of kinetic-energy transfer by viscous coupling. Note that even in the absence of any bulk motion, this viscous coupling can still provide an effective energy transfer across the disk. While potential energy may be converted into kinetic energy in one region of the ring, e.g., via spreading of the ring at the inner and outer boundaries, viscous coupling effects may cause the newly transformed kinetic energy to be transferred into another region. In fact, both of the first two terms are associated with global properties at the boundaries. They represent energy redistribution processes. The third term is associated with the rate of energy transfer, from kinetic into internal energy, due to viscous stress. Note that both the second and third terms depend only on the kinetic shear viscosity but not on the bulk viscosity. Thus, in a uniformly rotating disk, the local shear reduces to zero so that these terms vanish. Also of importance to this contribution is the presence of a nonvanishing frictional effect. The fourth term is associated with the rate of change in the internal energy associated with $P dV$ work. Both the third and fourth terms are referring to local effects. They represent the conversion between internal and kinetic energy. Actually, the last two terms may be replaced with a term corresponding to the rate of change of internal energy. If we integrate the internal energy Eq. (48) with respect to dV and then add it to Eq. (57) we obtain

$$\frac{d}{dt} \int_V \left(\frac{u^2}{2} + U + E \right) dm = \int \left[-P\delta_{ij} + 2\nu\rho e_{ij} + \rho \left(\xi - \frac{2}{3} \nu \right) \frac{\partial u_k}{\partial x_k} \delta_{ij} \right] u_i dS_j$$

$$+ \int \frac{\rho\kappa}{k_B} \frac{\partial E}{\partial x_i} dS_i + \int_V \dot{E}_c dm$$

$$= [-2\pi r u_r P' + 2\pi\sigma\nu r^3 \Omega \partial\Omega/\partial r]_{r_1}^{r_2} + \int_{r_1}^{r_2} \frac{\sigma k}{k_B} \frac{\partial E}{\partial r} dr + \int E_c dm \qquad (59)$$

where the integration variable S refers to the surface of a cylinder concentric with the axis. Although the rate of change of internal energy, due to motion in the potential well of the planet, on the left-hand side of Eq. (59) has replaced terms three and four on the right-hand side of Eq. (58), the internal and kinetic energy can still be transferred by advective and viscous-coupling effects, respectively. In addition, the internal energy may be redistributed in a ring by collisionally induced conduction. The internal energy is drained by inelastic collisions.

The above equation clearly indicates that the loss of internal energy, by inelastic collisions, at any region of the ring is roughly balanced by a local gain in which the viscous stress transforms kinetic energy into internal energy. The loss of kinetic energy may be countered by global transport processes which redistribute the kinetic energy from elsewhere. The overall energy flow is maintained by the ring's motion into deeper parts of potential well of the planet. This characteristic can be best appreciated in the limit of small E and κ where only the viscous coupling and the inelastic collision terms are of importance. It is their balance which determines the evolutionary fate of the ring.

Under certain specialized conditions it is possible for a disk to evolve to a quasi-steady state in which the local time derivatives are small (Lynden-Bell and Pringle 1974; Lin and Bodenheimer 1982). From the continuity Eq. (49) we find that $\dot{m} = 2\pi\sigma u_r r$ is independent of r. Thus we may integrate the θ-component of the momentum Eq. (50b) to obtain

$$\sigma u_r r^3 \Omega = \sigma \nu r^3 \, \partial\Omega/\partial r + \dot{J}_c/2\pi \; . \tag{60}$$

The integration constant $\dot{J}_c$ may be interpreted as the Lagrangian angular-momentum flux across any radius. Although $\dot{J}_c$ is independent of r, it is a linear combination of two radially-varying functions: an outwardly directed angular-momentum flux induced by viscous transfer and an inwardly directed Eulerian angular momentum flux carried by the inwardly drifting particles. Note that we have defined $\dot{m}$ and $\dot{J}_c$ in such a manner that negative values imply mass and angular momentum transfer onto the planet. At the inner boundary of a freely expanding ring there is no viscous coupling so that $\dot{m}$ and $\dot{J}_c$ would have the same sign. Alternatively it may be possible to maintain a planetary ring in a steady state, in which $\dot{m} = 0$, by a strong viscous coupling at the rings' two boundaries. Such coupling may be caused by a variety of physical processes such as tidal effects from shepherding satellites or a boundary-layer interaction between the ring and the accreting planet. In this case, the energy loss due to inelastic collision is continually supplied by the viscous coupling process. For a ring with a finite mass flux, the particles steadily drift inward with a radial velocity

$$u_r = \frac{\nu}{r} \frac{\partial \ln\Omega / \partial \ln r}{\left(1 - \dfrac{\dot{J}_c}{\dot{m}\,\Omega r^2}\right)} \; . \tag{61}$$

Away from boundaries of the ring, the quantity $\dot{J}_c/(\dot{m}\Omega r^2)$ is usually positive and much smaller than unity so that in a Keplerian disk $u_r = -3\nu/2r$.

In the quasi-steady state the physical significance of the dominant terms in the energy equation is vividly demonstrated. In this case

$$\frac{d}{dt} \int_V \left(\frac{u^2}{2} + U + E \right) dm = \int 2\pi r \, \sigma u_r \frac{\partial}{\partial r} \left(\frac{u^2}{2} + U + E \right) dr$$

$$= \dot{m} \left[\frac{u^2}{2} + U + E \right]_{r_1}^{r_2} \tag{62}$$

so that Eq. (59) becomes

$$\dot{m} \left[\frac{u^2}{2} + U + E \right]_{r_1}^{r_2} = [2\pi r^3 \sigma \nu \Omega \partial \Omega / \partial r]_{r_1}^{r_2} + \int_V \dot{E}_c \, dm \tag{63}$$

where we have neglected the $P dV$ work and the thermal conduction terms. The viscous coupling contribution (first term on the right-hand side of Eq. (63) may be expressed in terms of the mass flow,

$$[2\pi r^3 \sigma \nu \Omega \, \partial \Omega / \partial r]_{r_1}^{r_2} = [\dot{m} \Omega^2 r^2]_{r_1}^{r_2} = [\dot{m} \, u^2]_{r_1}^{r_2} \tag{64}$$

to show that it is twice the contribution arising from the orbital kinetic energy of the inwardly drifting particles. Thus, Eq. (63) becomes

$$[\dot{m} E]_{r_1}^{r_2} = \left[\frac{3}{2} \dot{m} \, \Omega^2 \, r^2 \right]_{r_1}^{r_2} + \int_V \dot{E}_c \, dm \ . \tag{65}$$

Since the difference in internal energy ($E = c^2/2$) between r_1 and r_2 is small, Eq. (65) indicates a balance between internal energy growth caused by viscous evolution and internal energy damping by collisions. The total orbital energy difference between particles at r_1 and r_2 is

$$\left[\frac{1}{2} u^2 + U \right]_{r_1}^{r_2} = - \left[\frac{1}{2} \Omega^2 r^2 \right]_{r_1}^{r_2} \tag{66}$$

which is three times smaller than the power liberated between radii r_1 and r_2. The balance is supplied by the energy transport due to the viscous couple (Lynden-Bell and Pringle 1974). This simplified steady-state analysis clearly illustrates the energy redistribution processes that occur in a viscously evolving particle disk. Qualitatively, a similar energy redistribution process must occur in general nonsteady-state disks.

IV. THERMAL AND VISCOUS STABILITY OF A PLANETARY RING

There are three important reasons for examining the hydrodynamic stability of a planetary ring:

1. Both the moment equations in the kinetic theory and the transport equations in fluid dynamics are established on the basis of the hydrodynamic approximation that radial gradients are small. As we have indicated in Sec. I.B.2, in order to justify this approximation, there must exist a basic fluid element within which the ring particles' dynamics may be studied collectively. This justification may be rigorously satisfied by a demonstration that the ring is indeed stable against perturbations with wavelength comparable to H.

2. The derivation of transport coefficients, such as viscosity, is based on an energy-equilibrium assumption. In order for this assumption to be satisfied, a planetary ring must be thermally stable, i.e., it can return to a state of energy-equilibrium, through inelastic collisions and viscous stress, after a small perturbation is imposed. Although we have already qualitatively outlined a condition for thermal stability in Sec. I, a rigorous stability analysis is still worthwhile.

3. The steady-state assumption produces an optical depth which varies smoothly with r, i.e., $d \ln \tau / d \ln r$, should be of order unity. However, Voyager data on Saturn's rings indicate that the ring system is divided into thousands of ringlets in which the optical depth varies rapidly with radius. Although some of the gaps and ringlets may be caused by resonant interactions with Saturn's numerous inner satellites (see chapters by Shu and by Franklin et al.), many features seem to be intrinsic to the ring's internal dynamics. One possible interpretation of the phenomenon is that it is caused by a viscous instability somewhat analogous to the traffic jam problem. This possibility may be verified with a rigorous stability analysis.

The above three issues may be addressed with different degrees of sophistication. The third issue, i.e., the origin of ringlets, may be most simply addressed with a viscous-stability analysis based on a perturbation calculation of the mass diffusion equation. The second issue requires a simultaneous perturbation analysis of the mass as well as energy transport equations whereas the first requires, in addition, a perturbation analysis of the momentum equations. We shall discuss these issues in order of increasing sophistication, i.e., in reverse order. In Sec. IV.A.3. we also give a qualitative discussion of the physical mechanisms which give rise to diffusion instabilities and contrast them with the earlier ideas of jet-stream formation.

A. Viscous Stability of Planetary Rings and the Origin of the Ringlets

The first stability analysis is to be carried out with the simplest possible approach, which is somewhat analogous to that performed by Lightman and Eardley (1974) in a different astrophysical context. In this analysis, the ring is assumed to always be in a state of energy-equilibrium so that (1) there is a $\epsilon - \tau$ relationship (see Eq. 37) and (2) ν may be expressed in the form of Eq. (52). Both assumptions may be justified later with a thermal stability analysis provided ϵ decreases with v_c. Since an $\epsilon - v_c$ relationship is not yet available, we adopt the *ad hoc* assumption $c \propto \sigma^\delta$ and investigate both analytically and numerically, under what conditions the system is stable.

1. An Analytic Stability Treatment. For computational convenience, the physical dimensions in the diffusion equation may be eliminated by use of the variable $\xi = r/r_0$ where r_0 is taken to be the point of unit optical depth, i.e., $\tau(r_0) = \tau_0 = 1$. Thus $\sigma/\sigma_0 = \tau$ where $\sigma_0 = \sigma(r_0)$. Now let $X = \xi^{\frac{1}{2}}$ and $T = t\Omega_0/R$ where $\Omega_0 = \Omega(r_0)$ and $R = \Omega_0 r_0^2/\nu_0$ is the Reynolds number; then Eq. (56) reduces to

$$\frac{\partial \tau}{\partial T} = \frac{3K_1}{X^3} \frac{\partial^2}{\partial X^2} \left[X^4 \tau^{2\delta} (1 + \tau^{-2})^{-1} \right] . \tag{67}$$

Following standard procedures in linear stability analyses, we introduce a small-amplitude, short-wavelength, sinusoidal perturbation τ_2 upon a general solution τ_1, in the neighborhood of some radius r_1, so that

$$\tau(X,T) = \tau_1(X,T) \left[1 + \tau_2(X,T) \right] \tag{68}$$

where

$$\tau_2 = Ae^{-sT} \sin k(X - X_1) . \tag{69}$$

By assumption, we may choose $A \ll 1$ and $kX_1 \gg 2\pi$, where the dimensionless perturbation wavelength is $\lambda/r = 2\pi/k$. The criteria for choosing the appropriate limits for wavelength are (1) it should be sufficiently short so that the mass-diffusion equation may be linearized, and (2) it should be sufficiently long so that the pressure-gradient effects, on momentum and energy transport, remain negligibly small.

Introducing Eqs. (68) and (69) into (67) and neglecting terms second-order in A, we obtain

$$s = 6k^2 \left[\delta + (1 + \tau_1^2)^{-1} \right] X \tau_1^{2\delta} / (1 + \tau_1^2) . \tag{70}$$

Equation (70) indicates that $s > 0$ for $\delta > 0$ and therefore these disks are stable and perturbations on them are always damped. For $\delta < -1$, s is always negative and perturbations would grow if the $c \propto \sigma^\delta$ law remains valid. For $-1 < \delta < 0$, s is positive if $\tau < \tau_c = [-(1+\delta)/\delta]^{\frac{1}{2}}$ and negative if $\tau > \tau_c$. Provided $-1 < \delta < 0$ we would expect the instability to develop into regions of very high optical depth, separated by gaps with optical depth $\leq \tau_c$, since these regions are stable. Note, however, if $\delta < -1$, all regions of the disk are unstable, whether optically thin or optically thick. One would expect the instability to develop into regions of alternately very high or very low τ.

The above linear stability analysis was based on work by Lin and Bodenheimer (1981). A similar analysis was performed by Ward (1981), who found the essentially equivalent criterion that instability occurs if

$$\frac{\partial}{\partial \sigma} (\nu \sigma) < 0 . \tag{71}$$

Assuming that the viscosity coefficient is given by Eq. (52) with $K_1 = 0.15$ and the coefficient of restitution is related to c by $(1 - \epsilon^2) \propto (c^2)^{1/\alpha}$, where α is expected to be fairly large, he finds that criterion (71) is satisfied if $\tau > \tau_c = \alpha^{-\frac{1}{2}}$. Thus, if τ is larger than some critical value of perhaps 0.3 to 0.5, instability occurs. Note that this $\epsilon - c$ relation replaces the assumption $c \propto \sigma^\delta$ used by Lin and Bodenheimer.

2. *Numerical Confirmation*. The occurrence of the instability can be verified and studied in more detail by the use of numerical techniques which can follow the evolution into the nonlinear regime. Such models have been presented by Lin and Bodenheimer (1981) with fluid-dynamical techniques and by Lukkari (1981) with N-body particle techniques. Lin and Bodenheimer (1981) consider the full time-dependent nonlinear solution of Eq. (67). This diffusion equation for τ has the same general form as the simple heat conduction equation which can be solved, as described by Richtmyer (1957), by an implicit numerical technique. The initial conditions were taken to be steady-state solution calculated under the assumption that $c \propto \sigma^\delta$, modified by short-wavelength sinusoidal perturbations in τ of the form for Eq. (69). Parameters investigated included δ, wave number k, amplitude A, and location X_1. Typical examples of stable and unstable solutions are given in Figs. 5 and 6. In all cases investigated, the stability criterion deduced from Eq. (70) was verified. Since no cutoff on the $c \propto \sigma^\delta$ law was imposed, perturbations grew indefinitely in the unstable regime. In the region where $-1 < \delta < 0$, the surface density decreased, in the regions where it was initially perturbed downwards, until $\tau = \tau_c$, as expected from the stability condition. Although the calculations could not be continued beyond one e-folding time, because numerical perturbations of very short wavelengths began to grow rapidly, some calculations were carried into the nonlinear regime by a choice of a relatively large initial value for A. In these cases also, there was good agreement between the numerical results and the linear stability analysis.

The growth time for the instability can be deduced from the quantity s (Eq. 70) and the scaling relation for T. The result is

$$t_g \simeq \frac{R\tau}{3Xk^2 \Omega} \simeq \frac{R\lambda^2 \tau}{r^2 \Omega} . \tag{72}$$

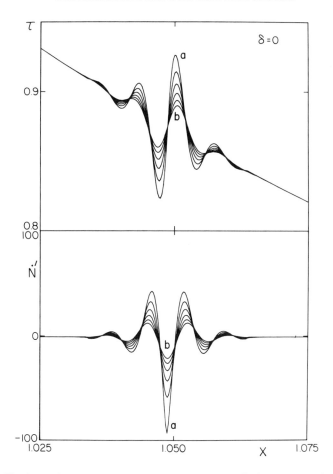

Fig. 5. The decay of a perturbation on a stable steady-state disk for $\delta = 0$, $X_1 = 1.05$, $A = 0.07$, and $k = 8.4 \times 10^2$ (see Eq. 69). *Upper portion,* optical depth; *lower portion,* normalized mass transfer rate $\dot{N}' = \dot{N}/(3\pi\sigma\nu)$ where $\dot{N}$ is calculated from Eq. (75). The six curves are separated by equal time intervals starting with initial condition (a) and ending after one *e*-folding time (b). Numerical simulations are from Lin and Bodenheimer (1981).

The diffusion time for the overall spreading of the disk (Lynden-Bell and Pringle 1974) is $t_d \approx R\,\Omega^{-1}$. The simple argument that t_d must be $> 5 \times 10^9$ yr yields the condition $R \gtrsim 3 \times 10^{13}$. Thus for ringlets with wavelength 100 km, $t_g \gtrsim 10^4\tau$ yr. Alternatively, by setting $R \cong \Omega_0 r_0^2/\nu_0$ we obtain $t_g \simeq \lambda^2/\nu_0$, thus for $\lambda = 100$ km and measured estimates for $\nu \approx 260$ (Lissauer et al. 1983; chapters by Shu and by Cuzzi et al.) the growth time is $\sim 10^4$ yr at $\tau = 1$. The numerical results give very good agreement with these estimates from the linear stability analysis. Note that there is no preferred length scale for the instability; however, the growth time decreases sharply with decreasing wavelength.

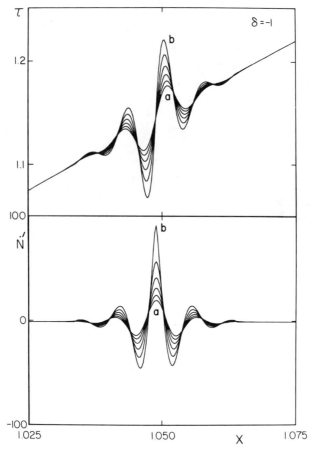

Fig. 6. The growth of a perturbation on an unstable steady-state disk for $\delta = 1$, $X_1 = 1.05$, $A = 0.06$, and $k = 8.4 \times 10^2$. The symbols and the source of the figure are the same as for Fig. 5.

An alternative time-scale estimate can be obtained from the linear stability analysis of Ward (1981):

$$t_g \simeq 5.4 \times 10^3 \left(\frac{\lambda_2}{H'} \right)^2 \frac{(1 + \tau^2)^{2+\alpha}}{(\alpha\tau^2 - 1)} \; \text{yr} . \tag{73}$$

Here $\lambda_2 = \lambda/500$ km where λ is the wavelength and $H' = H_0/10$ m where H_0 is the scale height at zero optical depth. For typical values ($\lambda_2 = H' = 1$) at optical depth 1, $t_g = 10^5$ yr for $\alpha = 3$ and 10^6 yr for $\alpha = 10$. Thus for $\lambda_2 = 100$, $t_g = 4 \times 10^3$ yr and 4×10^4 yr, respectively, in reasonable agreement with the results from Lin and Bodenheimer and much shorter than the age of the solar system.

A second approach to the numerical problem was carried out by Lukkari (1981) who explicitly calculated the evolution arising from collisions of a system of 250 orbiting particles. The coefficient of restitution was assumed to obey the condition.

$$\epsilon = \left[1 + K \left(\tau^3 + \tau_a^3 \right) c \right]^{-1} \tag{74}$$

where K and τ_a are constants. With this relation, instability sets in for $\tau \geqslant 0.75$. An initially assumed broad density maximum in a small interval of distance from the planet increased in amplitude by a factor of 4 and evolved into a narrow spike after 33,000 collisions. This amplification of the density confirmed earlier analytical work on collisional Keplerian disks by Hämeen-Anttila (1978) who apparently was the first to predict that such amplification could occur and that one might expect the Saturn ring system to be broken up into ringlets. An analogous instability had been investigated earlier in the context of gaseous accretion disks around neutron stars and black holes where turbulent viscosity is responsible for angular momentum transport and radiation pressure plays an important role (Lightman and Eardley 1974; Shakura and Sunyaev 1976).

   3. Physical Explanation for the Instability. On a Keplerian particle disk the collisions normally induce a viscous stress such that mass transfer occurs with a flux

$$m\dot{N} = \frac{-2}{\Omega r} \frac{\partial g}{\partial r} \tag{75}$$

where $g = 3\pi \Omega r^2 \nu \sigma$ is the viscous couple (Lynden-Bell and Pringle 1974). In the case of quasi-steady state, $\sigma \nu$ is constant so that $\dot{N}$ is a negative constant and the direction of mass flow is inward. However, in an unstable disk this assumption is not valid and the direction of mass transfer depends on the sign of $\partial g / \partial r$. It turns out that the local mass flow is towards a local density maximum from both directions. To see this, write

$$m\dot{N} = \frac{-2}{\Omega r} \frac{\partial g}{\partial \sigma} \frac{\partial \sigma}{\partial r} . \tag{76}$$

If, in a density perturbation, σ increases outwards, the mass flow is outwards (toward the σ-maximum) if $\partial g / \partial \sigma < 0$. Similarly, if σ increases inwards in the perturbation, the mass flow is inwards (again towards the density maximum) if $\partial g / \partial \sigma < 0$.

   To examine the sign of $\partial g / \partial \sigma$ more carefully, we assume, as above that $c \propto \sigma^\delta$ and we consider a short-wavelength perturbation $\lambda \ll r$. Then we can write

$$\frac{\partial g}{\partial \sigma} \propto \frac{\partial}{\partial \sigma}\left[\sigma^{2\delta}(1 + \tau^{-2})^{-1}\right]. \qquad (77)$$

Clearly, if $\delta < 0$ and $\tau > 1$, the disk is unstable since $\partial g/\partial \sigma < 0$, and if $\delta < -1$ it is unstable for $\tau \ll 1$, since $\partial g/\partial \sigma \propto 2\ (\delta+1)$. Note that the stability criterion is modified for small τ in this model, and in the models of Ward and Lukkari it does not occur at all for τ less than some critical value.

The numerical diffusion calculations (Figs. 5 and 6) illustrate the direction of mass flow. For the stable solution, in the region where σ increases outwards $\dot{N}$ is sharply negative, that is, viscosity tends to decrease the amplitude and spread out the perturbation. Similarly, where σ decreases outward, $\dot{N}$ is positive, away from the density maximum. On the other hand, in the unstable case where $\partial\sigma/\partial r > 0$, $N > 0$ so the density enhancement is reinforced. Similarly when $\partial\sigma/\partial r < 0$, $N < 0$ and mass transfer again is toward the density maximum.

In summary, the criterion that determines whether a particle disk spreads or clumps depends on the relationship between velocity dispersion and surface density, since this relationship controls the form of the viscous stress given in Eq. (77). Our assumption of energy equilibrium also establishes a relationship between τ and ϵ (Eq. 37). Thus, the criterion for diffusion instability ultimately depends on how ϵ varies with the velocity dispersion. The fact that ϵ is a decreasing function of velocity does not guarantee a diffusion instability; the stability criterion depends in a complicated way on the particular form of the $\epsilon-v_c$ relation. For the particular power-law relation assumed by Ward (1981), $(1-\epsilon^2) = K(c^2)^{1/\alpha}$, the exponent $1/\alpha$ determines the critical optical depth above which there is instability. If α were too small, the instability would require an impossibly large τ.

It should be emphasized that our explanation of the diffusion instability is fundamentally a *fluid* description caused by a stress imbalance between neighboring fluid elements of particles. It has yet to be rigorously described by a kinetic theory (see Sec. IV.B.2). The diffusion instability is distinctly different from Alfvén and Arrhenius' (1976) "jet stream" hypothesis. They argue that the inelasticity of collisions would dissipate relative motion to such an extent that particle orbits would cluster into a narrow stream with correlated orbital elements (i.e., arguments of pericenter and ascending nodes as well as semimajor axes would cluster). This idea does not require a particular $\epsilon-v_c$ relation; only that collisions be very inelastic. Indeed, their scenario seems to require an initial state out of energy equilibrium such that the excess velocity dispersion can be efficiently damped during the jet-stream formation process. In contrast, the diffusion instability can start from a state of energy equilibrium. Baxter and Thompson (1973) attempted to verify the jet-stream hypothesis with a kinetic theory, but their favorable conclusions are based on overly restrictive assumptions for the distribution of orbital elements and their relative rates of change. Trulsen (1972b) succeeded in demonstrating jet-stream formation in a 2-dimensional N-body simulation, but only for an extremely dissipative collision model (see Sec. II.A). Hämeen-Anttila and Lukkari (1980) found a suggestion of a jet-stream effect in a 3-dimensional hard-sphere simulation only when they assumed a restitution coefficient ϵ that was too small (~ 0.25) to allow an energy equilibrium to be established (see Fig. 2). Under realistic conditions in planetary rings, it seems unlikely that the jet-stream mechanism plays an important role.

B. Intermediate-Wavelength Thermal and Viscous Stability Analysis.

The above viscous stability analysis is based on the assumption that the ring is always in a state of energy equilibrium. We now derive the conditions under which this assumption may be satisfied.

1. The Fluid Approximation. The thermal stability of an accretion-disk flow is closely related to the viscous stability (Shakura and Sunyaev 1976). In the stability analysis, a perturbation must be introduced to offset the delicate energy balance between inelastic collisions and viscous stress. As in Sec. IV.A, we concentrate our attention on intermediate wavelength perturbations for simplicity. In principle, the stability analysis should be carried out with the kinetic approach since such a perturbation may distort the general shape of the dispersion-velocity ellipsoid. The evolution of individual elements of the pressure tensor can only be obtained through the 3 components of the second-order moment Eq. (19) and the mass diffusion equation. In practice, the most important aspects of the stability analysis, namely the evolution of perturbations on σ and TrT, may be carried out with a fluid-dynamical approach. In this case, perturbations are imposed on τ and $E = c^2/2$ such that

$$
\tau = \tau_1 (1 + \tau_2 e^{\lambda \tau} \sin kr)
$$
$$
E = E_1 (1 + E_2 e^{\lambda \tau} \sin kr)
$$

(78)

where τ_1 and E_1 are the initial value of τ and E at r_1. The evolution of these perturbations may be analyzed with the mass diffusion Eq. (56) and the energy Eq. (54). In this analysis, a prescription for ν of the form

$$
\nu = \frac{2K_1 E \tau}{(1+\tau^2)\,\Omega}
$$

(79)

may be adopted (see Eq. 52). Such a viscosity law may be justified, even in a perturbed state, by the heuristic arguments in Sec. I. In essence, only the effects of TrT and τ on ν are included whereas the effect of the distortion of the dispersion-velocity ellipsoid has been neglected. The equations then become

$$\frac{\partial \tau}{\partial t} = \frac{6K_1}{\sqrt{GM}} \frac{1}{r} \frac{\partial}{\partial r}\left[r^{\frac{1}{2}} \frac{\partial}{\partial r}\left(\frac{E\tau^2 r^2}{1+\tau^2} \right) \right] \tag{80}$$

$$\frac{\partial E}{\partial t} = E\tau\Omega\left[\frac{9}{4} \frac{2K_1}{(1+\tau^2)} - (1 - \epsilon^2) \right]. \tag{81}$$

For intermediate wavelength perturbations, these equations retain all the important terms for mass, angular momentum and energy transport.

We start the stability analysis with an initial condition that the ring is in a state of energy equilibrium such that τ_1 and E_1 at some radius r_1 obey the normal $\epsilon-\tau$ relationship for energy equilibrium (Eq. 37). The linearized resulting equations become

$$\lambda\tau_2 + \frac{6K_1k^2}{\Omega}\left[\frac{2\tau E}{(1 + \tau^2)^2} \tau_2 + \frac{\tau E}{(1 + \tau^2)}E_2 \right] = 0 \tag{82}$$

$$E_2\left(\lambda - \tau\Omega E \frac{\partial \epsilon^2}{\partial E} \right) + \frac{9K_1 \tau^3 \Omega \tau_2}{(1 + \tau^2)^2} = 0 .$$

These two simultaneous equations for E_2 and τ_2 can be solved to give

$$\lambda^2 + A\lambda + B = 0 \tag{83}$$

$$\text{where } A = \frac{12K_1k^2\tau E}{\Omega(1 + \tau^2)^2} - \tau\Omega E \frac{\partial \epsilon^2}{\partial E} \tag{84a}$$

$$\text{and } B = \frac{-12K_1 k^2 \tau^2 E^2}{(1 + \tau^2)^2}\left(\frac{\partial \epsilon^2}{\partial E} \right) - 54\frac{K_1^2 k^2 \tau^4 E}{(1 + \tau^2)^3} . \tag{84b}$$

The two roots are

$$\lambda_{\pm} = \frac{-A \pm \sqrt{A^2-4B}}{2} . \tag{85}$$

The negative root corresponds to thermal stability if $A^2 > 4B$ (which a little algebra shows is always true) and $A > 0$. Thus, thermal stability is guaranteed as long as $\partial \epsilon^2/\partial E < 0$ while thermal instability occurs ($A < 0$) if the condition

$$k^2 < \frac{\Omega^2 (1+\tau^2)^2}{12 K_1} \frac{\partial \epsilon^2}{\partial E} \tag{86}$$

is satisfied, which is impossible unless $\partial \epsilon^2 / \partial E > 0$.

The positive root corresponds to viscous instability if $\lambda_+ > 0$, which occurs if $A > 0$ and $B < 0$. The first condition is satisfied if $\partial \epsilon^2 / \partial E < 0$ so that the disk is thermally stable. In that case we can employ the condition for collisional equilibrium:

$$\frac{9}{2} \frac{K_1}{(1+\tau^2)} = (1 - \epsilon^2) . \tag{87}$$

Substituting this expression into the equation for B, we obtain the instability criterion

$$E \frac{\partial \epsilon^2}{\partial E} + (1 - \epsilon^2) \tau^2 > 0. \tag{88}$$

Thus the analysis gives the conditions for both viscous and thermal instability, and Eq. (88) defines a general relation between ϵ, c, and τ that is required for the occurrence of a viscous diffusion instability. Given a relation between ϵ and τ, this expression provides an estimate of the minimum value of τ that is required for instability. Setting the left-hand side of Eq. (88) equal to zero, substituting Eq. (37) and plotting the resulting curve in Fig. 7, we find that the ($\partial \ln \epsilon / \partial \ln c$) plane can be divided into regions which indicate where both thermal and viscous instability occur.

When an experimentally determined $\epsilon - v_c$ relationship becomes available, Fig. 7 may be used to map out the region of instability. For the purpose of illustration, we construct four hypothetical experimental results in Fig. 7. Curve (a) represents thermal instability over the entire range of ϵ; curve (b) represents viscous instability over a certain range of ϵ and marginal thermal and viscous instability otherwise; and curves (c) and (d) represent viscous instability over a certain range of ϵ and stability for $\epsilon < \epsilon_c$ or ϵ_d, respectively. Curve (d) corresponds to the power-law $\epsilon - v_c$ relation assumed by Ward (1981), with $\alpha = 10$. When these results are analysed together with the $\epsilon - \tau$ relationship (Eq. 37), each value of ϵ corresponds to an optical depth. Physically, curve (a) would cause a ring to evolve into a torus or a thin sheet on dynamical time scales. Curve (b) would cause a ring system to break into many ringlets separated by gaps with arbitrarily small τ. Curves (c) or (d) would cause a ring system to break into ringlets separated by gaps with finite optical depth. Voyager's data on Saturn's B Ring indicate that the ringlets are separated by gaps with moderate optical depth (chapter by Cuzzi et al.). If the ringlet structure is caused by a viscous instability, these data suggest

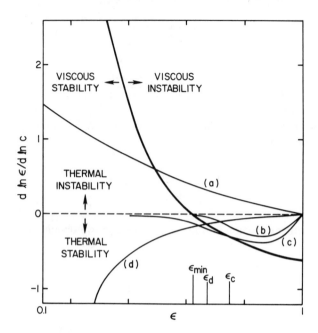

Fig. 7. The $\epsilon-c$ relation ($\epsilon-v_c$ relation) plotted as a function of ϵ. The critical line for viscous instability, defined by Eq. (88) is given by the heavy solid curve. Hypothetical $\epsilon-c$ relations, as explained in the text, are plotted as curves a, b, c, and d. Equilibrium solutions are allowed only to the right of ϵ_{min} and below the dashed line.

that the intrinsic properties of the ring particles may indeed be similar to those portrayed in curve (c).

Both preliminary experimental data (Bridges et al. 1984) and theoretical impact models (chapter by Borderies, et al.) give results similar to curve (c). However, both of these results are based on ring particles that are modeled as solid ice spheres. It is entirely possible that some degree of fragmentation or aggregation of ring particles actually occurs during collisions. In that case substantial regoliths may be built up on the larger ring particles which could greatly alter their restitution coefficient (see chapter by Weidenschilling et al.). For particles which are not perfect spheres, the restitution coefficient may depend on the orientation of the collision and particle spin as well as on the impact velocity. Although these effects may be important, they are much too complex to be considered here.

2. Kinetic Approach. In the above fluid-dynamical analyses, only the trace of the velocity-dispersion ellipsoid is analyzed. In reality, we have assumed $T_{r\theta}/\mathrm{Tr}\,\mathbf{T}$ only depends on τ as expressed in Eq. (52). This assumption may be relaxed when the stability analysis is carried out with the kinetic approach. In this case, medium-wavelength perturbations must be introduced to T_{rr}, $T_{r\theta}$, $T_{\theta\theta}$, and T_{zz} so that we must solve for the linearized mass diffusion equation and all four components of the second-order moment equation. Such a computation may be carried out numerically along the same lines that Goldreich and Tremaine (1978) used to deduce the transport coefficient (see Sec. II.B.3). Alternatively, the approximate analytic results obtained by Hämeen-Anttila (1978) (see also Sec. II.B.3) may be used to make somewhat further analytical advances. Applying the perturbation into Eqs. (38) yields five linearlized equations. Preliminary eigenvalue results indicate that the condition for viscous instability is not changed. However, the analysis for thermal instability becomes much more complex, and it may be necessary to solve it numerically. The above results clearly indicate the advantages and the limitation of the fluid-dynamical approach. While it does simplify stability analyses considerably, it must be based on certain assumptions.

C. Thermal and Viscous Stability at Very Short Wavelength

The viscous instability criterion deduced in Sec. IV.B (i.e., Eq. 88) does not contain any wavelength dependence. Literal interpretation of this result implies that when the ring becomes unstable, it may be divided into ringlets with arbitrarily small width. If this is the case, the validity of the hydrodynamic approximation, on which the entire rigorous analyses are based, becomes questionable. However, at very small wavelength several physical effects become important. For example, the pressure gradient effect, due to large gradients in τ and c, may cause the flow to depart significantly from Keplerian rotation and to induce a large radial velocity which would enhance the previously negligible advective transport processes. In addition, the conductive process also becomes important. Thus, some basic assumptions in standard analyses of ring structure, such as the stationary and Keplerian nature of Ω and the magnitude of u_r, must be relaxed. These effects can only be taken into account if short-wavelength perturbations are also introduced to Ωr and u_r and their evolution is analyzed with the momentum equation, in the fluid approximation, or the first-moment equation, in the kinetic approach. In the fluid case, four equations, i.e., Eqs. (49), (50a), (50b), and (53) must be solved simultaneously. Preliminary eigenvalue analysis of the linearized equations indicate no simple analytical criteria for thermal or viscous instability.

Although the fluid approximation for instability analysis is very attractive, due to its simplicity, there are several limitations to its applicability to planetary rings. First, as we have stated in Sec. IV.B, it is based on the assumption that the transport coefficients, such as the viscosity, are not changed by the perturbations. Second, a functional form for the conductive coefficient has to be introduced. Third, it is difficult to find unambiguous correspondence between certain terms in the fluid dynamical transport equations and others in the kinetic moment equations. These problems may be obviated with the kinetic treatment. In this case, seven linearized perturbation equations must be solved simultaneously. Besides the difficulties associated with obtaining and interpreting the eigenvalue solutions of a 7×7 matrix, the truncation of the third-order pressure tensor becomes problematic. Obviously, we cannot proceed on this topic without considerable additional complications. Therefore, we leave discussion of this aspect of stability analyses to further progress in the field.

V. VISCOUS STABILITY AT LARGE AND SMALL OPTICAL DEPTH

One of the most important applications of the stability analyses presented in Sec. IV is to provide a scenario for the origin of the ringlets. If the ringlet structure is indeed caused by the diffusion instability, it is relevant to discuss the evolution of the ring after the onset of the instability. In the preliminary discussion in Sec. IV.A we argued that, when an infinitesimal perturbation is imposed on a ring system with moderate τ, ring particles continually drift from regions with a deficit surface density into regions with an enhanced surface density so that the contrast between the ringlets and the gaps increases with time. However, Voyager data on Saturn's B Ring indicate that variations in optical depth between the ringlets and the gaps are relatively moderate and that transitions between those regions can be resolved. These data suggest that if the ringlet structure is indeed due to a viscous instability, there must be stabilizing effects at both high and low optical depth limits. There are several stabilizing mechanisms:

1. At sufficiently large τ, c becomes so small that gravitational attraction between particles may become important;
2. Also at large τ, the mean free path of the particles cannot be reduced below the size (R_p) of a typical particle, so that there is a lower limit to the magnitude of ν.
3. At small τ, the form of the $\epsilon - v_c$ relation may lead to a viscous stress that increases with c for τ less than some critical value (see Sec. IV.B.1), in which case the disk is stabilized.

In this section we first discuss stabilizing effects at large τ and then at small τ.

A. Self-gravitating Effects at Large τ

Depending on the values of c and σ, gravitational attraction between ring particles can produce at least three significant effects:

1. It induces gravitational scattering through which a minimum c may be maintained (Cuzzi et al. 1979; Ward and Harris 1983);

2. It increases the frequency of mid-plane crossings as a result of additional gravitational pull towards the mid-plane of the ring (Salo and Lukkari 1982);
3. It may lead to collective gravitational instability.

1. Gravitational Scattering, Thermal Stability and Minimum c. Gravitational scattering between ring particles becomes important when their relative velocity g = max ($\sqrt{2}c$, ΩR_p) is less than the surface escape velocity from a particle, $V_{esc} = (8\pi G\rho/3)^{\frac{1}{2}}R_p$. For ring particles made of ice (density 1 g cm^{-3}), gravitational scattering between ring particles becomes important if g is sufficiently small that the ring's thickness is less than

$$H_{esc} \simeq V_{esc}/\Omega = 1.8 \, (r/R_s)^{\frac{3}{2}}R_p \qquad (89)$$

where R_s is the radius of Saturn. At the inner edge of Saturn's A Ring where $r = 2.0\,R_s$, and for a particle of 5 m radius, $H_{esc}/\sqrt{2} = 18$ m. Thus gravitational scattering between particles only plays an important role if the scale height of the ring is less than a few particle radii (Goldreich and Tremaine 1982). If there is a broad distribution of particle sizes, however, gravitational scattering more strongly influences the smaller particles because they can pass closer to a larger particle without suffering a collision (see Sec. VI.C.).

Gravitational scattering is a purely elastic process ($\epsilon = 1$) which merely stirs random motion without causing dissipation. The increase in velocity dispersion caused by a scattering event between two particles with separation $r_{12} = |\mathbf{r}_1 - \mathbf{r}_2|$ is on the order of $\Delta c \sim Gm/r_{12}c$. This mechanism will tend to increase c until $(\Delta c)^2$ is reduced to the point where the total energy input rate due to both gravitational scattering and direct collisions balances the energy dissipation rate due to the inelasticity of collisions. The inverse relation between Δc and c for gravitational scattering will always allow the system to adjust to a thermal equilibrium state provided there is a finite level of energy dissipation which does not rapidly decrease with increasing c.

Thus, gravitational scattering may effectively maintain a velocity dispersion of the same order as the surface escape velocity from the particle size which represents most of the mass of the rings. Further discussion of this process may be found in Ward's chapter. This lower limit on c implies that the $\epsilon - \tau$ relationships derived in Sec. I (Eq. 13) and Sec. II (Eq. 43) must be somewhat modified to take into account the fact that only a fraction of the collisions are dissipative. Similarly, the criteria for thermal and diffusion instability (Eqs. 86 and 88) must also be modified.

Gravitational scattering plays a predominant role in the context of the protoplanetary accretion disk (see Ward's chapter). However, its contribution is of marginal importance in typical planetary rings because the scattering cross section, which is limited to H_{esc}^2 by the thin disk geometry, never greatly exceeds the physical cross section of a typical ring particle. In addition, typical planetary rings usually lie within two planetary radii. At this distance, the central planet can exert a gravitational influence on the surface of a typical ring particle that has a strength comparable to that due to the ring particle itself unless the internal density of the ring particle considerably exceeds that of the planet. Thus, the tidal radius of a typical particle does not extend much beyond its surface. In this case, the gravitational interaction between neighboring particles tends to occur inside the Roche radius of the central planet itself. This planetary tidal influence tends to decrease the efficiency of gravitational scattering between particles. Mathematically, the planetary tidal influence may be analyzed numerically by integration of the 3-body orbits (see chapter by Weidenschilling et al.).

Although gravitational scattering may only play a limited role in the dynamics of a planetary ring, it can stabilize the viscous instability at large τ. In particular, a lower limit on c implies a lower limit on ν. If a ring is viscously unstable, c would decrease as τ increased in a ringlet. Eventually, τ would become sufficiently large and c sufficiently small so that gravitational scattering would become important. Then ν would no longer decrease as τ increased and the ring would become stabilized. Finally, it should be pointed out that at large τ, the interparticle spacing may become so small that the gravitational potential energy is comparable to the kinetic energy of relative motion. In this case, the kinetic theory described in Sec. II.B breaks down and a more complicated approach analogous to the theory of liquids must be applied.

2. Enhancement in Plane-crossing Frequency. The increase in the frequency of passages through the mid-plane of the ring results in an increased collisional frequency and thus an increased shear stress at large τ (Salo and Lukkari 1982). The stability analyses in Sec. IV.A indicate that a ring is stabilized against viscous diffusion if viscous stress increases with τ.

In the limit of small orbital inclination, the out-of-plane motion of a ring particle can be described by

$$\ddot{z} = -\Omega^2 z + f_z \tag{90}$$

where f_z is the force due to self gravity of the ring which may be obtained from Gauss's law:

$$f_z = -2\pi G \sigma(z) . \tag{91}$$

The quantity $\sigma(z) = m \int_{-z}^{z} N(z')dz'$ is that fraction of the ring's surface density contained between z and $-z$. Let us assume that $N(z)$ has the Gaussian form of Eq. (29), except that $\overline{z^2}$ is now depressed below its unperturbed value, $\overline{z_0^2} = T_{zz}/\Omega^2$, by the force f_z. Equation (91) may be expanded for small z such that

$$f_z = -2\pi \, GmN \, (0) \int_{-z}^{z} \left[1 - z'^{\,2}/(2\overline{z^2}) + \dots \right] dz'$$

$$= -2\sqrt{2\pi} \, G\sigma \left[z/(\overline{z^2})^{-\frac{1}{2}} - \left(z^3/6 \right) \left(\overline{z^2} \right)^{-\frac{3}{2}} + \dots \right].$$

(92)

Keeping only terms linear in z, the equation of motion becomes

$$\ddot{z} = -\Omega_*^2 z \quad \text{where} \quad \Omega_*^2 = \Omega^2 \left[1 + 2 \left(\frac{2\pi}{\overline{z^2}} \right)^{\frac{1}{2}} \frac{G\sigma}{\Omega^2} \right].$$

(93)

The above result indicates that the self-gravity of the ring effectively increases the frequency of vertical oscillations across the mid-plane of the ring by a factor Ω_* / Ω. The reduced scale height $\overline{z^2}$ is given by the equation

$$\overline{z^2} + 2\sqrt{2\pi} \, G\sigma \left(\overline{z^2} \right)^{\frac{1}{2}} / \Omega^2 - T_{zz}/\Omega = 0$$

(94)

which results from setting T_{zz}/Ω_*^2 equal to $\overline{z^2}$. Solving for $\overline{z^2}$ one finds

$$\Omega_* / \Omega = \left[(1 + \Delta^2)^{\frac{1}{2}} + \Delta \right] \quad \text{where} \quad \Delta = \left(\frac{2\pi}{T_{zz}} \right)^{\frac{1}{2}} \frac{G\sigma}{\Omega}.$$

(95)

Both the increase in the frequency of oscillation across the mid-plane and the decrease in the scale height induce increases in the collisional frequency and shear viscosity. One may expect that this effect increases the collision frequency by a factor of 2 at most. The net result is to enhance the stabilizing effect of finite particle size (see Sec. V.B below) at large optical depth (Salo and Lukkari 1982).

3. *Gravitational Instability.* Toomre (1964) has shown that, in the context of galactic disks, a collisionless particle ring may become gravitationally unstable against axisymmetric perturbations if the radial component of the dispersion velocity ellipsoid drops below the value

$$(T_{rr})_{\text{min}}^{\frac{1}{2}} = 3.36 \, G\sigma/\Omega .$$

(96)

Since internal energy associated with dispersive motion cannot be dissipated in a collisionless stellar system, such as a spiral galaxy, this gravitational instability acts to virialize the particles in the sense that it provides a stirring mechanism which effectively prevents the velocity dispersion from falling below $(T_{rr})_{\text{min}}^{\frac{1}{2}}$. In a gravitationally unstable planetary ring, internal energy may still be lost through dissipative collisions at a rate which may reduce somewhat the critical minimum dispersion velocity (Goldreich and Tremaine 1982). We have already stated that typical ring particles have physical sizes comparable to their tidal radii. Nevertheless, gravitational instability can occur since it is caused by a collective interaction of a large number of particles. The axisymmetric instability leads to a ring-like density enhancement which at close range acts on nearby particles with an attractive gravitational force that decreases as the inverse of the distance from the ring rather than the inverse square.

Harris and Ward (1983) have suggested that the transient collective motion induced by the gravitational instabilities may result in an effective kinematic viscosity ν_e which may be defined in an expression analogous to Eq. (12):

$$\nu_e \, (3\Omega/2)^2 = (1-\epsilon^2) \, \omega_c \, c_{min}^2 . \tag{97}$$

If the minimum velocity dispersion c_{min}, has the approximate form of Eq. (96), the effective kinematic viscosity would be

$$\nu_e = K(1-\epsilon^2) \, G^2 \, \sigma_0^2 \, \tau^3/\Omega^3 \tag{98}$$

where $\sigma_0 = \sigma/\tau$ is the surface density at unit optical depth and K is a constant ~ 0.1 to 10. With such an effective viscosity, the viscous stress increases with τ^4 such that the ring is stable against viscous diffusion. Since gravitational instability can only occur for large τ, the above result may be interpreted as a stabilizing effect which acts to prevent τ from increasing beyond a certain level.

B. Effect of Finite Size of Particles at Large τ

The finite size of the particles induces and maintains a residual relative velocity, ΩR_p, that can not be damped out regardless of the efficiency of the collisional dissipation (Brahic 1977). This residual relative velocity, which is comparable to the difference in Keplerian velocity between particles with orbits separated by $2R_p$, provides a lower limit to the magnitude of the kinematic viscosity (Goldreich and Tremaine 1982):

$$\nu_{min} \simeq \omega_c R_p^2 \simeq \Omega R_p^2 \tau . \tag{99}$$

This minimum viscosity exceeds ν_e (Eq. 98) for particles with size larger than $R_{p,min} \simeq KG\sigma/\Omega^2$. In Saturn's rings, for example, $\sigma \approx 50$ g cm^{-2} and $\Omega \approx 2 \times 10^{-4}$ s^{-1} so that $R_{p,min}$ is about K m. Unless K is much larger than unity, direct physical collisions among particles larger than meter size can be more efficient than gravitational instability in preventing c from decreasing below a certain critical value at large τ.

The above qualitative deductions may be substantiated with a more rigorous mathematical analysis. For simplicity, such a calculation may be based on the assumption that the ring is composed of uniform hard spheres with a collision law as prescribed by Eq. (10). The second velocity moment of the collisional integral (Eq. 21) may be written in the form (Trulsen 1971)

$$
\left(\frac{\partial p_{ij}}{\partial t} \right)_c = 2R_p^2 \iiint \left[\mathbf{v}_{1a}\,\mathbf{v}_{1a} + \mathbf{v}_{2a}\,\mathbf{v}_{2a} - \mathbf{v}_{1b}\,\mathbf{v}_{1b} - \mathbf{v}_{2b}\,\mathbf{v}_{2b} \right]
$$

(100)

$$
\times \, (\mathbf{g} \cdot \hat{\mathbf{k}}) f\,(\mathbf{v}_{1b}) f\,(\mathbf{v}_{2b})\, \mathrm{d}\hat{k}\, d^3\,v_{1b}\, d^3\,v_{2b}
$$

where the post-collisional velocity is related to the pre-collisional velocity by

$$
\mathbf{v}_{1,2,a} = \mathbf{v}_{1,2,b} \mp 0.5\,(1+\epsilon)\,(\mathbf{g} \cdot \hat{\mathbf{k}})\hat{\mathbf{k}} \ .
$$

(101)

In the limit of vanishing velocity dispersion Eq. (100) becomes

$$
\left(\frac{\partial p_{ij}}{\partial t} \right)_c = N^2 R_p^2\,(1+\epsilon)^2 \int (\mathbf{g} \cdot \hat{\mathbf{k}})^3 \hat{\mathbf{k}}\,\hat{\mathbf{k}}\, \mathrm{d}\hat{\mathbf{k}} \ .
$$

(102)

The evaluation of the above integral depends on the direction of the relative velocity $\mathbf{g}$.

Now consider a collision between two neighboring particles, which are originally on circular Keplerian orbits. It is mathematically convenient to analyze the collision in a rotating Cartesian coordinate system $(r_1,\ \theta_1,\ z_1)$ which is centered on particle 1 (see Fig. 8) and which corotates with the position vector of particle 1. Suppose that it is on a collision course with particle 2. Due to the difference in their positions in the ring, the direction as well as the magnitude of the orbital velocities, v_1 and v_2, of the two particles may differ (see Fig. 8). To first order in R_p/r, $\mathbf{g}$ may be expressed in the newly defined coordinates such that

$$
\mathbf{g} = (-v_2 \sin\delta,\ v_2 \cos\delta - v_1,\ 0) \simeq [-\Omega r_2 \delta,\ -\tfrac{1}{2}\Omega\,(r_2 - r_1),\ 0] \quad (103)
$$

where δ is the angle between the position vectors of the colliding particles. The directional vector $\hat{\mathbf{k}}$, pointing to the center of particle 2, is

$$
\hat{\mathbf{k}} = (\sin\psi\,\cos\phi,\ \cos\psi,\ \sin\psi\,\sin\phi) \tag{104}
$$

where ψ is the angle between the θ_1-axis and $\hat{\mathbf{k}}$ and ϕ is the angle between the r_1 axis and the projection of $\hat{\mathbf{k}}$ onto the $r_1 - z_1$ plane. For finite values of ϕ, Eqs. (103) and (104) actually describes collisions between particles in circular orbits distributed in parallel planes a distance $z \neq 0$ above the mid-plane. Although such a situation is not possible in reality, the inclusion of such orbits provides

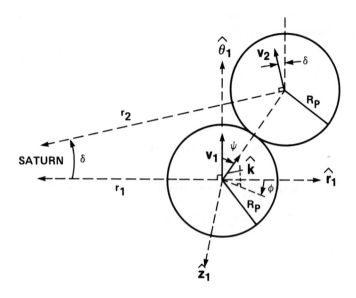

Fig. 8. Geometry of collisions between hard spheres in circular orbits of radius r_1 and r_2 with velocities $\mathbf{v}_1$ and $\mathbf{v}_2$.

a useful channel for interpolating between the two extreme physical situations of large velocity dispersion ($c \gg \Omega R_p$) and strictly coplanar, concentric, circular orbits ($c = 0$) (Hämeen-Anttila 1982). In terms of this geometry,

$$\mathbf{g} = -\Omega R_p (2\cos\psi, \sin\psi\cos\phi, 0). \qquad (105)$$

By substitution of Eqs. (104) and (105) into the integral in Eq. (102), ($\partial p_{ij}/\partial t)_c$ may be evaluated in the restricted range ($0 < \psi < \pi/2$ and $-\pi/2 < \phi < \pi/2$ for $r_1 < r_2$; $\pi/2 < \psi < \pi$ and $\pi/2 < \phi < 3\pi/2$ for $r_1 > r_2$). The result of the integration is

$$\left(\frac{\partial p_{ij}}{\partial t}\right)_c = \frac{8}{35} \frac{N\tau(1+\epsilon)^2\Omega^3 R_p^3}{\pi(\pi\overline{z^2})^{\frac{1}{2}}} \begin{pmatrix} 4 & 9\pi/8 & 0 \\ 9\pi/8 & 4 & 0 \\ 0 & 0 & 1 \end{pmatrix}. \qquad (106)$$

This expression is entirely analogous to a similar expression obtained by Hämeen-Anttila (1982).

In reality, collisions due to overlapping circular orbits always induce a true velocity dispersion that is of random nature. This effect can be incorporated into the evaluation of ($\partial p_{ij}/\partial t)_c$. Instead of simply adding the two contributions (Eq. 27) and (Eq. 106) for ($\partial p_{ij}/\partial t)_c$, Hämeen-Anttila (1982) suggests an interpolation formula of the form

$$\left(\frac{\partial p_{ij}}{\partial t}\right)_c = \omega_0^* \left\{ \frac{(1+\epsilon)}{18}\left[(1+\epsilon)\,\mathbf{Tr}\,\mathbf{p}\,\delta_{ij} - 3(3-\epsilon)p_{ij}\right]\right.$$

$$\left. + \frac{(1+\epsilon)^2}{35} N(\Omega R_p)^2 \begin{pmatrix} 4 & 9\pi/8 & 0 \\ 9\pi/8 & 4 & 0 \\ 0 & 0 & 1 \end{pmatrix}\right\} \tag{107}$$

where

$$\omega_0^* = \frac{8\tau\Omega}{\pi}\left[\frac{\mathrm{Tr}\mathbf{T} + (3/\pi)\,(\Omega R_p)^2}{3\Omega^2\,z^2}\right]^{\frac{1}{2}} \tag{108}$$

is the collision frequency. The advantage of the above expression is that it is approximately valid for both large and small velocity dispersion. If a planetary ring is in an energy equilibrium, the energy transport Eq. (36) reduces to

$$T_{r\theta}(3\Omega/2) = \omega_0^*\left[\frac{1-\epsilon^2}{6}\mathrm{Tr}\mathbf{T} - \frac{9}{70}(1+\epsilon)^2\,(\Omega R_p)^2\right]. \tag{109}$$

This result differs from Eq. (40) since it contains an extra energy source term due to the finite size of the particles.

Following the same procedure used in Sec. II.B.4, we can deduce a generalized $\epsilon - \tau$ relationship such that

$$\tau = \frac{\pi\omega_0^*}{8\Omega}\left(\frac{1+\epsilon}{3-\epsilon}\right)^{\frac{1}{2}}\left[\frac{1 + (18/35)\,(R_p\,\Omega)^2/\mathrm{Tr}\mathbf{T}}{1 + (3/\pi)\,(R_p\,\Omega)^2/\mathrm{Tr}\mathbf{T}}\right]^{\frac{1}{2}}\left[(1+\Delta^2)^{\frac{1}{2}} - \Delta\right] \tag{110}$$

where

$$\Delta = \frac{G\sigma}{\Omega}\left[\frac{6\pi\,(3-\epsilon)}{(1+\epsilon)\,[\mathrm{Tr}\mathbf{T} + (18/35)\,(\Omega R_p)^2]}\right]^{\frac{1}{2}} \tag{111}$$

and

$$\frac{\omega_0^*}{\Omega} = \frac{-B + \sqrt{B^2 - 4\,AC}}{2A} \tag{112}$$

where

$$A = \left[\frac{(1+\epsilon^2)}{9}\mathrm{Tr}\mathbf{T} - \frac{3}{35}\,(1+\epsilon)^2\,(\Omega R_p)^2\right]\frac{(1+\epsilon)^2(3-\epsilon)^2}{36} \tag{113a}$$

$$B = \frac{-3\pi}{560}\,(1+\epsilon)^3\,(3-\epsilon)\,(\Omega R_p)^2 \tag{113b}$$

$$C = - \frac{(19\,\epsilon - 13)\,(1+\epsilon)}{36}\,\mathrm{Tr}\mathbf{T} - \frac{18}{35}\,(1+\epsilon)^2\,(\Omega R_p)^2 . \qquad (113c)$$

Although the above expression reduces to the original $\epsilon - \tau$ relationship (see Eq. 43) in the limit of very small particle size, there are some fundamental differences between the above formula and our previous results. For example, the new expression depends explicitly on $\mathrm{Tr}\mathbf{T}$. Therefore, an $\epsilon - v_c$ relationship is needed to deduce a unique $\epsilon - \tau$ relationship. Moreover, there is an explicit dependence on ΩR_p. The factor depending on Δ accounts for the increased collision frequency due to the vertical component of self-gravity discussed in Sec. V.A.2. These additional terms all contribute to stabilization against viscous diffusion at large τ since they provide a lower limit for v. Mathematically, once an $\epsilon - \tau$ relationship is established, the shear stress may be calculated as a function of τ from Eq. (109), using Eq. (112) for ω_0^*. This procedure leads to stress curves qualitatively similar to those in Fig. 9. Unfortunately, detailed quantitative analyses cannot be carried out until an $\epsilon - v_c$ relationship is available.

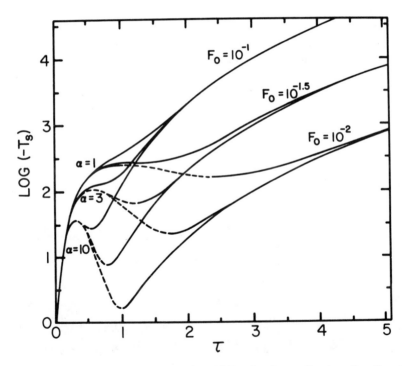

Fig. 9. The viscous torque $T_s = 2\pi r^3\,\sigma v \partial\Omega/\partial r$ plotted as a function of τ. Curves are parameterized by the value of α and F_o as explained in the text (figure from Harris and Ward 1983).

C. Stabilizing Effect at Small τ

If a ring is thermally stable, c must decrease with τ. The condition for an energy equilibrium (Eq. 13) indicates that ϵ increases with τ. Thus, in a thermally stable ring low optical depth corresponds to large velocity dispersion. For relatively large dispersion velocities, gravitational scattering occurs less frequently than direct collision so that the self-gravitating effects discussed in Sec. V.A are negligibly small. Similarly, the finite size of the particles does not influence the dynamics of the ring except that it provides a cross section for direct physical impact. Consequently, stabilization against the viscous diffusion instability at small τ must occur because of the intrinsic collisional properties of the particles themselves. As we have discussed in Sec. V.B.1, an $\epsilon - v_c$ relationship similar to curve (c) in Fig. 7 can stabilize a ring at $\tau < \tau_c$.

D. Comparisons with Observations

In order to verify that the viscous diffusion instability scenario is a reasonable explanation for the origin of the ringlets, several observable parameters need to be deduced. Voyager's data contains a vast amount of information on the optical depth variation in the ringlet systems (chapter by Cuzzi et al.). If we combine the above results on the condition for viscous instability and the stabilizing effects at large and small τ, we can make a direct comparison between the theory and the Voyager data. The above discussion indicates that the stability criterion may depend sensitively on the $\epsilon - v_c$ relationship. However, due to the lack of an experimentally determined $\epsilon - v_c$ law, most of the previous work is based on *ad hoc* assumptions. For example, in order 'o obtain a $v - \tau$ relation valid for both large and small τ, Harris and Ward (1983) simply added Eq. (98) to the viscosity formula (52). Figure 9 shows the logarithm of the resultant viscous stress as a function of τ for various *ad hoc* $\epsilon - v_c$ laws parameterized by $\alpha = d \ln c^2 / d \ln (1 - \epsilon^2)$ and for various filling factors $F_0 = \Omega R_p / c_0$, where c_0 is the equilibrium velocity dispersion obtained from Eq. (12) in the limit of very small τ. The viscous stress typically exhibits a local minimum value at τ near unity. Once a viscous diffusion instability occurs, the local optical depth may grow until the ringlet is stabilized at the optical depth corresponding to this minimum stress value S_{min}. The adjacent gaps would be stabilized at the smaller (but nonzero) τ where the viscous stress again attains the value S_{min}. Thus, ringlet formation may induce a stress value considerably lower than that deduced for a uniform τ. The precise value of τ at which the ring is stabilized may eventually be determined once all of the stabilizing mechanisms discussed in this section are included in the theory and an $\epsilon - v_c$ law is available. However, an important further modification discussed in the next section may arise from the observed fact that there is a broad distribution of particle sizes.

VI. THE EFFECT OF A PARTICLE SIZE DISTRIBUTION

A. Implications of Observational Data

So far, the analyses have been limited to planetary rings with uniform particle size. We adopted this case to study first since it simplifies the mathematical procedures considerably. However, Voyager's radio occultation measurement of Saturn's A Ring (Marouf et al. 1983; chapter by Cuzzi et al.) can be approximated by a power-law distribution in the number density $N(R_p)$ of the form

$$dN(R_p) = \begin{cases} A R_p^{-q} dR_p, & \text{for } 1 \text{ cm} < R_p < 5 \text{ m} \\ 0, & \text{for } R_p < 1 \text{ cm or } R_p > 5 \text{ m} \end{cases} \tag{114}$$

where $2.8 < q < 3.4$. A similar power law is observed in the asteroid belt (Anders 1965) and may be generally representative of collisional particle disks (chapter by Weidenschilling et al.). For simplicity, we adopt $q = 3$ for the discussion below.

Particles with radii between $R_{p,1}$ and $R_{p,2}$ contribute an optical depth

$$\tau_{12} = \int_{R_{p,1}}^{R_{p,2}} \pi R_p^2 dN(R_p) = \tau \frac{\log(R_{p,2}/R_{p,1})}{\log(500)} \tag{115}$$

where τ is the total optical depth due to all the particles. This result implies that the observed optical depth represents comparable contributions from particles in each size range if $q \sim 3$. However, particles with $R_{p,1} < R_p < R_{p,2}$ contain a fractional mass

$$\Delta M = \int_{R_{p,1}}^{R_{p,2}} \frac{4}{3} \rho \pi R_p^3 dN(R_p) = \frac{4\rho\tau (R_{p,2} - R_{p,1})}{3 \log(500)} \tag{116}$$

which implies that 80% of the total ring mass is contained in particles with $R_p > 1$ m.

It is also of interest to estimate the mean spacing d between particles within a given size range, a quantity which may be deduced once the scale height of the ring is determined from the condition for an energy equilibrium. Such a calculation requires the generalization of the second-order moment Eqs. (38) consistent with this size distribution along with an experimentally determined $\epsilon - v_c$ law. These analyses remain to be carried out. Nonetheless, an order of magnitude value for d may be estimated in a qualitative manner. For example, suppose that the scale height for all particles is comparable to the size of the largest particles and τ is on the order of unity. In this case, the largest meter-sized particles in the distribution (Eq. 114) would be separated by a distance comparable to their own diameters, while the smallest cm-sized particles would be separated by ten times their own diameters. The implications are two fold. First, successive gravitational encounters with the largest m-sized particles, by either small or large particles, may not be well-separated events. Thus, mathematical analyses based on simple 3-body gravitational scattering (see Sec. V.A.1) may be inadequate. Second, the large m-sized particles have more frequent encounters with smaller cm-sized particles than with themselves. The collisional frequency $\omega_{i \to j}$ between particles with different sizes, say $R_{p,i}$ and $R_{p,j}$ can be estimated from

$$\omega_{i \to j} = \Omega \tau_j \left(1 + R_{p,i}/R_{p,j}\right)^2 . \tag{117}$$

The inferences from the above equation are: (1) a cm-sized particle collides on the average with four other cm-sized particles before colliding with a m-sized particle, and (2) a m-sized particle collides with several thousand cm-sized particles between collisions with another m-sized particle.

Although $\omega_{2 \to 1}/\Omega \gg 1$ for $R_{p,1} \ll R_{p,2}$, the presence of small cm-sized particles hardly influences the viscous evolution of the large m-sized particles since most of the ring's mass is contained in the large particles. Therefore, the stability analyses based on uniform particle size are applicable to the large m-sized particles. In contrast, small cm-sized particles collide as frequently with much larger particles as with themselves. Thus, the viscous evolution of the small particles is strongly coupled to the number density and the velocity dispersion of particles of all sizes. Consequently, viscous instability for the small cm-sized particles may be distinctly different from that in the case of uniform particle size.

B. Collisional Evolution of the Small Particles in a Bimodal Size Distribution.

1. Qualitative Analysis. For simplicity, the evolution of a ring with nonuniform particle sizes may be first illustrated with a simple qualitative analysis of a bimodal size distribution where $m_1 \ll m_2$ and τ_1/τ_2 is a constant of order unity. According to the argument presented in Sec. VI.A, the dynamics of the large particles are not influenced by the motion of the small particles. Since small particles collide often with the large particles, their evolution is strongly influenced by the kinetic state of the large particles. For example, if the large particles have zero dispersion velocity, then their main effect will simply be to limit the mean free path of the small particles. An effective mean free path may be obtained from Eq. (6) if we replace τ with $(\omega_{1 \to 1} + \omega_{1 \to 2})/\Omega$. The corresponding effective viscosity acting on the small particles has the form

$$\nu_1 = \frac{c_1^2}{\Omega} \frac{(\tau_1 + q\,\tau_2)}{1 + (\tau_1 + q\,\tau_2)^2} \tag{118}$$

where c_1 is the dispersion velocity of the small particles and $q \sim 2$ for $R_1 << R_2$ (Stewart 1984).

In the case that the large particles have a finite velocity dispersion, the mean free path of the small particles will be somewhat larger than the above estimate. The velocity change of particle 1 resulting from a collision with particle 2 is

$$\mathbf{v}_{1a} - \mathbf{v}_{1b} = - \frac{m_2\,(1+\epsilon)}{m_1 + m_2}\,(\mathbf{v}_{1b} - \mathbf{v}_{2b}) \cdot \hat{\mathbf{k}}\,\hat{\mathbf{k}} \;. \tag{119}$$

Since the velocity components of different particle sizes are uncorrelated, Eq. (119) requires the mean square velocity change of particle 1 due to non-zero dispersion velocity of the large particles c_2 to be proportional to c_2^2. In the limit of large collision frequency the additional contribution to the mean free path of the small particles is just $c_2/(\omega_{1\to1} + \omega_{1\to2})$. The total ν_1 in this limit therefore becomes

$$\nu_1 = \frac{c_1^2 + c_2^2}{\Omega\,(\tau_1 + q\,\tau_2)} \quad \text{for } \tau_1 \text{ and } \tau_2 >> 1 \;. \tag{120}$$

In the low-τ limit the calculation of the mean free path is complicated by the epicyclic motion of the particles. Consider the approximate equation of motion of particle 1 in the reference frame of its circular orbital motion:

$$\frac{\partial \mathbf{v}_1}{\partial t} + 2\mathbf{v}_1 \times \hat{\mathbf{z}}\,\Omega = -\omega_{1\to1}\,\mathbf{v}_1 - \omega_{1\to2}\,(\mathbf{v}_1 - \mathbf{v}_2) \;. \tag{121}$$

The right-hand side of the above equation estimates the time-averaged change in particle 1's velocity caused by infrequent collisions with particles 1 and 2. For $\omega_{1\to1}$, $\omega_{1\to2} << \Omega$ this equation may be solved with a series expansion $\mathbf{v}_1 = \mathbf{v}_1^0 + \mathbf{v}_1^1 + \ldots$ where $\mathbf{v}_1^0$ satisfies the collisionless approximation to Eq. (121).

$$\frac{\partial \mathbf{v}_1^0}{\partial t} + 2\mathbf{v}_1^0 \times \hat{\mathbf{z}}\,\Omega = 0 \;. \tag{122}$$

The small collision-induced average perturbation $\mathbf{v}_1^1$ is given by

$$2\mathbf{v}_1^1 \times \hat{\mathbf{z}}\,\Omega = - (\omega_{1\to1} + \omega_{1\to2})\,\mathbf{v}_1^0 + \omega_{1\to2}\,\mathbf{v}_2^0 \;. \tag{123}$$

$\partial v_1'/\partial t$ does not appear in the above equation because v_1' is the average change in v_1 after many orbital periods. This relation indicates how small perturbations from the collisionless motion are of order ω_c/Ω times the relative velocity of the colliding bodies. Since v_1^0 and v_2^0 are uncorrelated, the resultant change in the mean square velocity of the small particles has just two contributions, $(\omega_{1\rightarrow1} + \omega_{1\rightarrow2})^2 c_1^2/\Omega^2$ and $\omega_{1\rightarrow2}^2 c_2^2/\Omega^2$. The first of these contributions tends to decrease the velocity dispersion of the small particles and thereby decrease the radial mean free path. Thus, if $c_2 = 0$, the resultant shear viscosity for infrequent collisions is

$$\nu_1 = (\omega_{1\rightarrow1} + \omega_{1\rightarrow2}) \frac{[c_1^2 - c_1^2 (\omega_{1\rightarrow1} + \omega_{1\rightarrow2})^2/\Omega^2]}{\Omega^2}$$

$$\approx \frac{c_1^2}{\Omega} \left[\tau_1 + q\tau_2 - (\tau_1 + q\tau_2)^3 \right]. \tag{124}$$

Note that Eq. (124) is just the low-τ expansion of Eq. (118). The second contribution to the small particles' velocity dispersion, $\omega_{1\rightarrow2}^2 c_2^2/\Omega^2$, adds to c_1^2 because a finite c_2 can increase the post-collisional velocity and hence the mean free path of the small particles. The generalization to Eq. (124) is thus

$$\nu_1 = (\omega_{1\rightarrow1} + \omega_{1\rightarrow2}) [c_1^2 - c_1^2 (\omega_{1\rightarrow1} + \omega_{1\rightarrow2})^2/\Omega^2 + c_2^2 \omega_{1\rightarrow2}^2/\Omega^2]/\Omega^2$$

$$\approx \frac{1}{\Omega} \left\{ (\tau_1 + q\tau_2) \left[1 - (\tau_1 + q\tau_2)^2 \right] c_1^2 + (\tau_1 + q\tau_2) (q\tau_2)^2 c_2^2 \right\} \tag{125}$$

for τ_1 and $\tau_2 \ll 1$.

The more rigorous calculation presented in the following section (VI.B.2) suggests a rough interpolation between Eqs. (120) and (125) of the form

$$\nu_1 \approx \frac{(\tau_1 + q\,\tau_2) [c_1^2 + q^2\,\tau_2^2 c_2^2/(1 + q^2\,\tau_2^2)]}{\Omega[1 + (\tau_1 + q\,\tau_2)^2]} \tag{126}$$

The energy of random motion of the small particles, $\tfrac{1}{2}c_1^2$, increases at the rate $(9/4)\Omega^2\nu_1$ due to the above shear viscosity. In addition, the large particles transfer some of their energy of random motion to the small particles at the rate

$$\dot{E}_{tr} \approx \omega_{1\rightarrow2} \left(c_2^2 - \frac{m_1}{m_2} c_1^2 \right). \tag{127}$$

This tendency toward energy equipartition would occur even in the absence of the systematic shear of the mean orbital motion. Because of the inelasticity of the collisions, however, the energy transfer rate is modified from that given in Eq. (127). The kinetic theory described below yields an actual energy transfer rate of the form

$$\dot{E}_{tr} = \frac{\omega_{1\rightarrow2} m_2}{3(m_1 + m_2)} \; (1+\epsilon) \left[\frac{m_2(1+\epsilon)(c_1^2 + c_2^2)}{m_1 + m_2} - 2c_1^2 \right]. \qquad (128)$$

This equation would reduce to the form of Eq. (127) in the perfectly elastic limit, $\epsilon = 1$. The energy balance equation for the small particles is now

$$\frac{9}{4} \, \Omega^2 \, \nu_1 + \dot{E}_{tr} = (1 - \epsilon^2) \, \omega_{1\rightarrow1} \, c_1^2 \qquad (129)$$

where ν_1 and E_{tr} are given by Eqs. (126) and (128), respectively. A solution for the steady-state c_1 can be obtained if Eq. (129) is supplemented by the energy balance equation for the large particles. If we neglect the effect of the small particles on c_2, the equation for a single particle size applies:

$$\frac{9}{4} \, \Omega^2 \nu_2 = (1 - \epsilon^2) \, \omega_{2\rightarrow2} \, c_2^2$$

$$\qquad (130)$$

$$\nu_2 = \frac{c_2^2 \tau_2}{\Omega \, (1 + \tau_2^2)} \; .$$

Equations (129) and (130) may be used to find c_1^2 as a function of τ_1 and τ_2. It should be noted that the quantity ϵ, which appears in Eq. (128), should be evaluated at $v_c = (c_1^2 + c_2^2)^{\frac{1}{2}}$; however, in Eqs. (129) and (130) ϵ should be evaluated at $v_c = \sqrt{2}\,c_1$ and $\sqrt{2}\,c_2$, respectively. In general, one will derive a steady-state $m_1 c_1^2 / m_2 c_2^2 \ll 1$ because ν_1 results in nonequipartition. Substituting the steady-state value of c_1^2 back into Eq. (126), one obtains a shear stress acting on the small particles proportional to $\tau_1 \nu_1$.

We find that the effect of large particles on small particles depends sensitively on the kinetic state of the large particles and the ratio τ_1/τ_2. In the case $c_2^2 < c_{cr}^2$, where c_{cr}^2 is some critical velocity between $(m_1/m_2)\,c_1^2$ and c_1^2, large particles merely reduce the mean free path which causes the maximum viscous stress, acting on the small particles, to attain a smaller value and to occur at a smaller $\tau = \tau_1 + \tau_2$. According to arguments in Sec. IV.C, a ring may be stable against viscous diffusion at sufficiently small τ if the $\epsilon - v_c$ law for the ring particles resembles that of curve (c) in Fig. 7. The above results suggest it is possible for the small particles to become viscously unstable even when the optical depth for each population may be smaller than the critical value for instability in the case of a single particle size. This selective instability would tend to generate ringlets and gaps of small particles against a uniform background of large particles. As the density-enhanced region of the small particles grows, the effect of the large particles would diminish. However, the density enhancement in the small particles may continue owing to viscous instability among the small particles themselves. In the small-particle gaps, the large particles dominate and eventually stabilize the evolution of the small particles.

In the case $c_2^2 > c_{cr}^2$, collisions with large particles increase the velocity dispersion of the small particles more effectively than they limit the small particles' mean free path. The resultant viscous stress, therefore, increases with τ_2/τ_1. Thus, the large particles induce a stabilizing effect against viscous diffusion among the small particles. In this case, provided $\tau_1/\tau_2 \approx 1$, the small particles can only become viscously unstable when the large particles are also unstable.

Since the kinetic properties of the large particles are not affected by the small particles, c_2 may be deduced in terms of τ_2 from the energy equilibrium condition (Eq. 130) and $\epsilon - v_c$ law. Hence, there is a unique solution for a given value of τ_2. Furthermore, when an $\epsilon - v_c$ law is available, the evolution of viscous instability may be deduced for the two populations.

2. Procedures for Rigorous Analysis. The heuristic discussion in the preceding section may be verified with a rigorous analysis. We generalize the kinetic theory approach for hard-sphere collisions as presented in Sec. II.B.2. Let p_{ij} and P_{ij} be the pressure tensors associated with the probability densities $f(m_1, r_1, v_1, t)$ and $F(m_2, r_2, v_2, t)$. For the same reasons as discussed above, the condition for energy equilibrium for the large particles does not change from Eq. (40). For the small particles, the second-order moment equation contains two collisional contributions such that

$$\left(\frac{\partial P_{ij}}{\partial t}\right)_c = \int (v_i - u_i)(v_j - u_j)[C(f,f) + C(f,F)]d^3v \tag{131}$$

where

$$\int (v_i - u_i)(v_j - u_j)C(f,F)d^3v \equiv I \tag{132}$$

$$I = (R_{p,1} + R_{p,2})^2 \iiint (v_{1i}^* v_{1j}^* - v_{1i} v_{1j})f(v_1)F(v_2)(g \cdot \hat{k})d\hat{k}d^3v_1 d^3v_2$$

and

$$v_1^* = v_1 - \frac{(1+\epsilon)m_2}{m_1 + m_2}(g \cdot \hat{k})\hat{k} . \tag{133}$$

If the dispersion-velocity distributions are analogous to that expressed in Eq. (24), the integral I may be transformed into a 6-dimensional integral as follows:

$$I = \frac{\pi(R_{p,1} + R_{p,2})^2 N_1 N_2}{4(2\pi)^3 \sqrt{\det \mathbf{Q}}} \int \frac{(1+\epsilon)m_2}{m_1 + m_2} \left[\frac{(1+\epsilon)m_2}{m_1 + m_2} \right. \tag{134}$$

$$\times \left(\frac{g^2}{3}\delta_{ij} + g_i g_j \right) - (g_i h_j + h_i g_j) - 2g_i g_j \bigg]$$

$$\times g \, \exp\left(-\tfrac{1}{2}\mathbf{q}^T \cdot \mathbf{Q}^{-1} \cdot \mathbf{q}\right) d^6 q$$

where

$$\mathbf{Q} = \begin{pmatrix} \mathbf{t+T} & \mathbf{t-T} \\ \mathbf{t-T} & \mathbf{t+T} \end{pmatrix}, \quad \mathbf{q} = (g_x, g_y, g_z, h_x, h_y, h_z), \tag{135}$$

$$\mathbf{t} = \mathbf{p}/N_1 \quad \text{and} \quad \mathbf{T} = \mathbf{P}/N_2$$

where $\mathbf{g} = \mathbf{v}_1 - \mathbf{v}_2$ and $\mathbf{h} = \mathbf{v}_1 + \mathbf{v}_2$. The anisotropy of the velocity distribution prevents an exact evaluation of the above integral. However, if we adopt the same approximation techniques as in Sec. II.B.2, by replacing a factor of g under the integral with a constant value $g_0 = (16/3) [\mathrm{Tr}\,(\mathbf{t+T})/6\pi]^{\frac{1}{2}}$, the integral becomes

$$I = \frac{4}{3} \; (R_{p,1} + R_{p,2})^2 N_1 N_2 \left[\frac{\pi}{6} \, \mathrm{Tr}\,(\mathbf{t+T}) \right]^{\frac{1}{2}} \frac{[1 + \epsilon(g_0)]m_2}{m_1 + m_2}$$

$$\times \left\{ \frac{[1 + \epsilon(g_0)]m_2}{m_1 + m_2} \left[\frac{1}{3} \, \mathrm{Tr}(\mathbf{t+T})\delta_{ij} + t_{ij} + T_{ij} \right] - 4t_{ij} \right\}. \tag{136}$$

In the above integration, we also approximated $\epsilon(v_c)$ by $\epsilon(g_0)$. The value of g_0 is so chosen as to provide the exact integral for the isotropic limit. Applying the above result into the second-order moment equation for the small particles, we find that t_{ij} is closely coupled to T_{ij}. Note that the impact velocity between small and large particles may be generally different from that among the two populations themselves. Therefore ϵ may assume a different value for different types of collisions even if ϵ is independent of particle size.

The solutions of the above equations may be obtained numerically for a given $\epsilon - v_c$ law (Stewart 1984). Preliminary results for the ad hoc prescription $\alpha = d \log (v_c^2) / d \log (1-\epsilon^2)$ are illustrated in Fig. 10 for different values of c_1/c_2 and τ_2. These results are in general agreement with those obtained from the heuristic discussion in Sec. VI.B.I. This agreement provides an a posteriori justification for the adopted approximation.

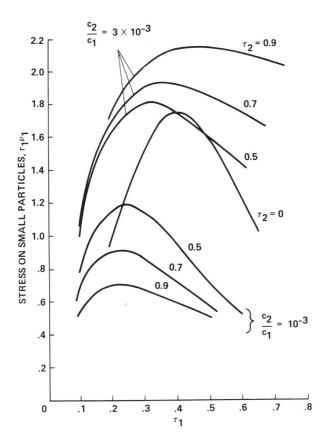

Fig. 10. Shear stress acting on small particles as a function of small particle optical depth τ, for various values of large particle optical depth τ_2 and two values of the velocity ratio c_2/c_1. For the smaller value of c_2/c_1, the finite value of c_2 has negligible effect. The velocity dependence of ϵ was assumed to be of the form $\alpha = d \log (v_c^2)/d \log (1 - \epsilon^2)$ with $\alpha = 10$.

C. Gravitational Effects on Small Particles

Rigorous analyses for gravitational scattering among ring particles would be considerably more complex than the above analyses for direct collisions. We mentioned some of the complications associated with gravitational scattering in Sec. V.A.1. A full-scale analysis would require a description modeled after Stewart and Kaula (1980) modified for the physical conditions appropriate to planetary rings. Nevertheless, a heuristic discussion, in the spirit of the preceding section, may be of interest (Cuzzi et al. 1979).

In a bimodal mass distribution where $m_1 \ll m_2$, the velocity deflection of a small particle passing a distance d from a much larger particle m_2 is of order $\Delta c_1 = 2 Gm_2/(c_1 d)$. The corresponding change in the kinetic energy is $c_1 \Delta c_1$

$= 2\,Gm_2/d$. The frequency of such encounters is $\omega_{\mathrm{enc}} = \pi\,d^2\,N_2\,c_1$ so that the rate of energy transfer between the two populations is $E_{\mathrm{grav}} = 2\pi\,Gm_2\,N_2\,c_1\,d$. For a thin ring, most gravitational encounters occur over a distance $R_{p,2} < d < c_1/\Omega$. The cumulative effects of nearby and distant encounters may be approximated by the effect of one single encounter at a distance $d_{\mathrm{eff}} = 2\,Gm_2/c_1^2$ so that $E_{\mathrm{grav}} = 4\pi\,G^2\,m_2^2\,N_2/c_1$, (Cuzzi et al. 1979).

In addition to the energy change, gravitational scattering can also change the mean free path of the particles. To lowest order, the resultant change in the shear viscosity may be estimated in the same way as perfectly elastic collisions with collision frequency

$$\omega_{\mathrm{enc}} \simeq 4\pi\,G^2\,m_2^2\,N_2\,c_1^{-3} \approx \left[v_{\mathrm{esc}}^4\,c_1^{-3}/(c_2/\Omega + R_{p,2}) \right]\tau_2 \ . \tag{137}$$

The straightforward extension of Eq. (118) is thus

$$\nu_1 = \frac{(\tau_1 + q\,\tau_2 + \omega_{\mathrm{enc}}/\Omega)c_1^2}{\Omega\,[1 + (\tau_1 + q\,\tau_2 + \omega_{\mathrm{enc}}/\Omega)^2]} \ . \tag{138}$$

If we insert this expression back into the energy equilibrium Eq. (55) and substitute ω_{enc} from Eq. (137) we obtain the following form of the energy equation:

$$\frac{(9/4)\,\Omega\,[\tau_1 + q\,\tau_2 + 4\pi\,G^2\,m_2^2\,N_2/(c_1^3\Omega)]\,c_1^2}{1 + [\tau_1 + q\,\tau_2 + 4\pi\,G^2\,m_2^2\,N_2/(c_1^3\,\Omega)]^2} = \omega_{1\to 1}\,(1-\epsilon^2)\,c_1^2 \ . \tag{139}$$

In the low-τ limit this expression agrees with the corresponding equation given by Cuzzi et al. (1979), who simply added $\dot E_{\mathrm{grav}}$ to Eq. (55). In the limit of $\tau_1,\ \tau_2,\ \omega_{\mathrm{enc}}/\Omega \gtrsim 1$, Eq. (139) must be used. This equation can be solved for the steady-state c_1; then the stress acting on the small particles is obtained from $\tau_1\nu_1$. This stress differs from that given by Ward and Harris (1983) who neglected to include the effect of ω_{enc} in the shear viscosity.

Gravitational scattering can also cause energy transfer from the large particles' random motion into the small particles' random internal motion. This dynamical friction effect was recently described in the context of the solar system accretion disk by Stewart and Kaula (1980). The resultant energy transfer rate will be negligible compared to the collisional energy transfer rate, Eq. (128), unless the velocity dispersion of the small particles falls below the escape velocity from the larger particles, $v_{\mathrm{esc}} = (2Gm_2/R_{p,2})^{\frac{1}{2}}$. As in the case of direct collisions, the underlying energy source is derived from the systematic shearing motion. Therefore this source of energy must again be equivalent to an additional viscous stress. The above formula does not include this effect; the calculation could be done with the methods of Stewart and Kaula (1980).

In an energy equilibrium, a small optical depth corresponds to a large velocity dispersion so that gravitational scattering is relatively ineffective. For a fixed ratio of τ_1/τ_2, the viscous stress due to gravitational scattering increases with τ since ω_{enc} increases with τ and c_1 decreases with τ. Therefore, gravitational scattering between the two populations again provides a stabilizing effect for the small particles. More detailed calculations are necessary to clairfy this effect, as it may depend sensitively on the validity of the 2-body scattering formula for ω_{enc}.

The above analysis for a bimodal distribution is a highly simplified attempt to account for a mass distribution of the form given by Eq. (114). Meaningful comparisons with observations may require a more general analysis where a realistic particle-size distribution is included.

VII. CONCLUSION

In this chapter, we have outlined the general formalism for calculating the structure, stability and viscous evolution of planetary rings. We have attempted to demonstrate the basic underlying physics with an idealized collision model for ring particles. For simplicity, much of the discussion is based on a uniform particle-size distribution and direct physical collisions. We also provided a limited discussion of additional effects due to nonuniform size distribution and gravitational interaction among the ring particles. The main conclusions that can be reached at the present time are as follows:

1. The steady-state structure of a particle disk is highly dependent on the degree of dissipation that occurs in collisions, i.e., the $\epsilon - v_c$ relation.

2. For the simple case of nongravitating hard spheres, thermal stability of a disk is assured provided that $d\epsilon^2/dE < 0$, where E is the energy of random motion. Thermal stability implies that the disk can reach an energy equilibrium at a finite scale height.

3. Equation (88) gives the conditions under which a viscous diffusion instability occurs, which may cause a ring system to break into many small ringlets separated by gaps with moderate optical depth. The occurrence of instability depends on the relationships between ϵ and v_c and between ϵ and τ. Preliminary experimental results for the $\epsilon - v_c$ relation for ice particles indicate that this mechanism could operate in Saturn's B Ring.

4. In regions of planetary rings with high optical depth, both the finite size of the particles as well as their mutual gravitational interaction produce important deviations from the hard-sphere model. These effects tend to enhance thermal stability and therefore to modify the requirement $d\epsilon^2/dE < 0$. The criterion for diffusion instability is also modified to enhance stability at high τ.

5. A distribution of particle sizes can lead to more complicated size-selective diffusive instabilities. Preliminary analysis of a simple model with two particle sizes suggests that the diffusion instability for the smaller particles may occur at a smaller optical depth than for the larger

particles. This effect arises from the strong influence of the large particles on the dynamics of the small particles as opposed to the nearly negligible response of large particles to impacts of small particles. The resulting size segregation could lead to observable radial structure in the rings.

We can identify two directions for future investigations. First, it is important to analyze carefully and verify the several assumptions required for the kinetic theory of hard-sphere particle disks described in Sec. II. The traditional Boltzmann-equation formalism employed by Goldreich and Tremaine (1978) is based upon the approximate representation of Keplerian orbits as a random velocity dispersion about coplanar, circular orbits; the detailed dynamics of individual orbit changes as well as the systematic nature of the epicyclic motion is neglected in favor of collective conservation laws for total angular momentum and energy.

A more rigorous approach should formulate a Boltzmann equation directly in terms of the orbital elements of the particles (Hämeen-Anttila 1978). Since a particle's semimajor axis as well as the eccentricity and inclination is changed by a typical collision, the zeroth-order moment of Hämeen-Anttila's kinetic equation already describes the radial spreading of a particle disk. This treatment contrasts with the more traditional approach (described in Sec. II) in which the continuity equation must be supplemented by the first and second velocity moment equations so that a closed equation for the surface density evolution can be obtained. Nevertheless, by use of a Taylor expansion about the collisional change in semimajor axis and several additional approximations, Hämeen-Anttila obtains a coefficient of shear viscosity identical to our Eq. (53). More work is needed to relax some of the assumptions of previous work and to rederive our basic results in a more fundamental way. Such work should also lead to generalizations of the present results to more complicated situations, such as steep density gradients, where the hydrodynamic approximation breaks down.

Once the physics of the hard-sphere collision model in particle disks has been satisfactorily established, further investigation of more realistic collision models should be pursued in parallel with laboratory experiments on the collisional properties of candidate ring particle materials. Some of the complications which can arise in the event of particle aggregation and fragmentation, for example, are indicated by recent work on the protoplanetary disks of planetesimals (see chapters by Ward and by Weidenschilling et al.).

Future observations and further data analysis of the Voyager observations are needed to better constrain the size distribution of particles in Saturn's rings and the degree of size segregation associated with observed structure in the rings. Many component kinetic theories should help to elucidate these phenomena. The most complex structure appears in the denser portions of Saturn's B Ring where the effects of finite particle size and self-gravitation are expected to be important. A complete incorporation of these effects into the kinetic theory will undoubtedly lead to complicated many-body effects. The wealth of phenomena known to occur in the theory of liquids compared to that of the ideal gas suggests many possible transport processes yet to be discovered in Saturn's rings.

Acknowledgments. This work was supported in part by National Science Foundation grants to the University of California at Santa Cruz. GRS was a NAS/NRC Resident Research Associate at NASA-Ames Research Center. The authors wish to thank J. Lissauer and J. Cuzzi for useful comments.

REFERENCES

Alfvén, H., and Arrhenius G. 1976. *Origin and Evolution of the Solar System.* (Washington, D.C.: NASA SP-345).

Anders, E. 1965. Fragmentation history of asteroids. *Icarus* 4:399–408.

Baxter, D. C., and Thompson, W. B. 1973. Elastic and inelastic scattering in orbital clustering. *Astrophys. J.* 183:323–336.

Boyd, T. J. M., and Sanderson, J. J. 1969. *Plasma Dynamics* (New York: Barnes and Noble), p. 51.

Brahic, A. 1977. Systems of colliding bodies in a gravitational field: I. Numerical simulation of the standard model. *Astron. Astrophys.* 54:895–907.

Brahic, A., and Hénon, M. 1977. Systems of colliding bodies in a gravitational field: II. Effect of transversal viscosity. *Astron. Astrophys.* 59:1–7.

Bridges, F. G., Hatzes, A., and Lin, D. N. C. 1984. On the structure, stability and evolution of Saturn's rings: Preliminary measurements of the coefficient of restitution of ice-ball collisions (preprint).

Cook, A. F., and Franklin, F. A. 1964. Rediscussion of Maxwell's Adams prize essay on the stability of Saturn's rings. *Astron. J.* 69:173–200.

Cuzzi, J. N., Durisen, R. H., Burns, J. A., and Hamill, P. 1979. The vertical structure and thickness of Saturn's rings. *Icarus* 38:54–68.

Faulkner, J., Lin, D. N. C., and Papaloizou, J. 1983. On the evolution of accretion disc flow in cataclysmic variables. I. The prospect of a limit cycle in dwarf nova systems. *Mon. Not. Roy. Astron. Soc.* 205:359–375.

Goldreich, P., and Tremaine, S. 1978. The velocity dispersion in Saturn's rings. *Icarus* 34:227–239.

Goldreich, P., and Tremaine, S. 1982. The dynamics of planetary rings. *Ann. Rev. Astron. Astrophys.* 20:249–284.

Goldsmith, W. 1960. *Impact* (London: Arnold).

Hämeen-Anttila, K. A. 1978. An improved and generalized theory for the collisional evolution of Keplerian systems. *Astrophys. Space Sci.* 58:477–519.

Hämeen-Anttila, K. A. 1981. Quasi-equilibrium in collisional systems. *Moon Planets* 25:477–506.

Hämeen-Anttila, K. A. 1982. Saturn's rings and bimodality of Keplerian systems. *Moon Planets* 26:171–196.

Hämeen-Anttila, K. A., and Lukkari, J. 1980. Numerical simulations of collisions in Keplerian systems. *Astrophys. Space Sci.* 71:475–497.

Harris, W. R., and Ward, A. W. 1983. On the radial structure of planetary rings. Proceedings of *I.A.U. Colloquium 75 Planetary Rings,* ed. A. Brahic, Toulouse, France, Aug. 1982.

Jeffreys, H. 1916. On certain possible distributions of meteoric bodies in the solar system. *Mon. Not. Roy. Astron. Soc.* 77:84–112.

Jeffreys, H. 1947. The effects of collisions on Saturn's rings. *Mon. Not. Roy. Astron. Soc.* 107:263–267.

Lane, A. L., Hord, C. W., West, R. A., Esposito, L. W., Coffeen, D. L., Sato, M., Simmons, K. E., Pomphrey, R. B., and Morris, R. B. 1982. Photopolarimetry from Voyager 2: Preliminary results on Saturn, Titan, and the rings. *Science* 215:537–543.

Lightman, A. P., and Eardley, D. M. 1974. Black holes in binary systems: Instability of disk accretion. *Astrophys. J.* 187:L1–L3.

Lin, D. N. C., and Bodenheimer, P. 1981. On the stability of Saturn's rings. *Astrophys. J.* 248:L83–L86.

Lin, D. N. C., and Bodenheimer, P. 1982. On the evolution of convective accretion disk models of the primitive solar nebula. *Astrophys. J.* 262:768–779.

Lissauer, J. J., Shu, F. H., and Cuzzi, J. N. 1983. Viscosity in Saturn's rings. Proceedings of *I.A.U. Colloquium 75 Planetary Rings,* ed. A. Brahic, Toulouse, France, Aug. 1982.

Lukkari, J. 1981. Collisional amplification of density fluctuations in Saturn's rings. *Nature* 292:433–435.

Lynden-Bell, D., and Pringle, J. E. 1974. The evolution of viscous discs and the origin of the nebular variables. *Mon. Not. Roy. Astron. Soc.* 168:603–637.

Marouf, E. A., Tyler, G. L., Zebker, H. A., Simpson, R. A., and Eshleman, V. R. 1983. Particle size distributions in Saturn's rings from Voyager 1 radio occultation. *Icarus* 54:189–211.

Maxwell, J. C. 1859. On the stability of the motions of Saturn's rings. (Macmillan and Company, Cambridge and London). Also printed in *Scientific Papers of James Clerk Maxwell.* (Cambridge: Cambridge University Press, 1890), vol. 1.

Pringle, J. E., Rees, M. J., and Pacholczyk, A. G. 1973. Accretion onto massive black holes. *Astron. Astrophys.* 29:179–184.

Richtmyer, R. D. 1957. *Difference Methods for Initial Value Problems.* (New York: Interscience), ch. 6.

Safronov, V. S. 1969. Evolution of the protoplanetary cloud and formation of the Earth and planets. Nauka, Moscow, USSR, Transl. Israel Program for Sci. Transl., 1972, NASA TTF-677.

Salo, H., and Lukkari, J. 1982. Self-gravitation in Saturn's rings. *Moon Planets* 27:5–12.

Shakura, N. I., and Sunyaev, R. A. 1976. A theory of the instability of disk accretion onto black holes and the variability of binary X-ray sources, galactic nuclei and quasars. *Mon. Not. Roy. Astron. Soc.* 175:613–632.

Smith, B. A., Soderblom, L., Batson, R., Bridges, P., Inge, J., Masursky, H., Shoemaker, E., Beebe, R., Boyce, J., Briggs, G., Bunker, A., Collins, S. A., Hansen, C. J., Johnson, T. V., Mitchell, J. L., Terrile, R. J., Cook, A. F. II, Cuzzi, J., Pollack, J. B., Danielson, G. E., Morrison, D., Owen, T., Sagan, C., Veverka, J., Strom, R., and Suomi, V. 1982. A new look at the Saturn system: The Voyager 2 images. *Science* 215: 504–537.

Spitzer, L., Jr. 1962. *Physics of Fully Ionized Gases* (New York: Interscience).

Stewart, G. R. 1984. Dynamical effects of nonuniform particle size in Saturn's rings. Submitted to *Icarus.*

Stewart, G. R., and Kaula, W. M. 1980. A gravitational kinetic theory for planetesimals. *Icarus* 44:154–171.

Toomre, A. 1964. On the gravitational stability of a disk of stars. *Astrophys. J.,* 139:1217–1238.

Trulsen, J. 1971. Towards a theory of jet streams. *Astrophys. Space Sci.,* 12: 329–348.

Trulsen, J. 1972a. Numerical simulation of jet streams. I: The three-dimensional case. *Astrophys. Space Sci.* 17:241–262.

Trulsen, J. 1972b. Numerical simulation of jet streams. II: The two-dimensional case. *Astrophys. Space Sci.* 18:3–20.

Ward, W. R. 1981. On the radial structure of Saturn's rings. *Geophys. Res. Letters* 8:641–643.

Ward, W. R., and Harris, A. W. 1983. Diffusion instability in a bi-modal disc. In *Proceedings of IAU Colloquium 75, Planetary Rings,* ed. A. Brahic, Toulouse, France, Aug. 1982.

WAVES IN PLANETARY RINGS

FRANK H. SHU

Institute for Advanced Study

The Voyager spacecraft revealed the rings of Saturn to have an unexpected richness of structure. Many of the observed features have now been identified as collective effects arising from the self-gravity of the ring material. These effects include spiral density waves and spiral bending waves, the main topics of this review chapter. Both kinds of waves were first discussed in the astronomical literature in connection with the dynamics and structure of spiral galaxies, and our discussion contrasts the similarities and differences between the disks of galaxies and planetary rings. After developing the theory of free and forced waves of both types, we discuss how the observed waves can be used as diagnostics to obtain crucial parameters that characterize the physical state of the rings.

I. INTRODUCTION

A. Similarities and Differences Between Galactic Disks and Planetary Rings

Superficially, apart from a difference in scale of a factor of a trillion, the disks of spiral galaxies and of Saturn's rings would seem to have many similarities; they are both spatially thin structures supported primarily by centrifugal equilibrium; they are both made of innumerable discrete objects whose random motions are small compared to their circular speeds; they both have considerable internal structure. Indeed, it is the premise of this review that collective gravitational effects explain much of the internal structure of both objects.

There is, however, a crucial difference in the relative scale of the collective processes which operate in disk galaxies and in Saturn's rings. The natural scale of self-gravitational disturbances in flattened distributions of matter with surface mass density σ and angular rotation speed Ω is roughly given by (cf. Toomre 1964)

$$L = \frac{2\pi G \sigma}{\kappa^2} \tag{1}$$

where the epicyclic frequency κ is related to Ω through Eq. (4a) below. In spiral galaxies, the scale L is comparable to the radius r of the disk, while in planetary rings, $L \ll r$. Indeed, in Saturn's rings, $\sigma \sim 50$ g cm^{-2}, $\kappa \sim 2 \times 10^{-4}$ s^{-1}, and $L \sim 500$ cm, which is much less than $r \sim 10^{10}$ cm. The reason for the difference in the ratio of L to r, of course, is that the ratio of mass in the disk to that in the rest of the system is much smaller for Saturn than for spiral galaxies.

A heuristic derivation of Eq. (1) exhibits the physical ideas. Consider an axisymmetric perturbation that results in a ring of mass per unit length $\sigma \pi l$ over half a radial wavelength πl. For $l \ll r$, the ring may be modeled locally as a straight wire which produces, at a distance l, an excess gravitational acceleration toward the center of the ring determined by Gauss's law:

$$2G (\sigma \pi l)/l = 2\pi G \sigma .$$

Notice that the collective attraction of a narrow ring of given surface density σ is independent of its radial width l, a result that makes our consideration different from the usual Roche limit arguments. Resisting the collapse of the ring is a Coriolis acceleration directed radially away from the center of the ring, 2Ω v, which results because the tendency to conserve angular momentum produces perturbational tangential velocities v. To estimate v, note that the excess (or deficit) of specific angular momentum ΔJ of material moved a small radial distance l inward (or outward) is

$$\Delta J = l \frac{d}{dr} (r^2 \Omega) .$$

The excess velocity v associated with ΔJ is v $= \Delta J/r$. If we make use of Eq. (4a), we now find the Coriolis acceleration to be

$$2\Omega \text{ v} = \kappa^2 l$$

which increases linearly with l. The destabilizing effect of self-gravity, $2\pi G \sigma$, and the stabilizing effect of rotation (which acts similarly to tidal forces) $\kappa^2 l$, are exactly balanced if l has the critical value L given by Eq. (1). Axisymmetric disturbances of (inverse wavenumber) scale smaller than Eq.

(1) require more than rotation alone to stabilize them against the self-gravity of the disk.

The agent that stabilizes galactic disks against short-scale disturbances is random motions; a different agent probably works in planetary rings because of the difference in the physical nature of the constituent bodies. The particles constituting Saturn's rings are solid particles whose effective radii R are of order 5 m (Marouf et al. 1983; see also chapter by Cuzzi et al.), i.e., comparable to the value of L computed above. In our opinion, the rough equality $R \sim L$ is no coincidence, but may reflect the operation of self-gravitational instabilities in Saturn's past. The idea is analogous to Goldreich and Ward's (1973) suggestion for the formation of asteroid-sized bodies in the early solar system.

Imagine that the distribution of ring-particle sizes was such that initially all the particles' radii were less than L. Then, inelastic collisions reduce the level of random velocities beneath the minimum required for gravitational stability of axisymmetric disturbances (cf. Safronov 1960; Toomre 1964). Collective instability would set in, and many particles would grow to size L, or perhaps even larger. At that point, however, the system would be stabilized even if the random velocities were reduced to zero, for all disturbances of scale larger than L would be stabilized by rotation, and all disturbances smaller than L would be stabilized by the effects of finite particle size. (It is meaningless to contemplate disturbances of inverse wavenumbers less than the size of a solid particle.) Without collective instabilities, the particles may be unable to agglomerate further, on a particle-particle basis, because the rings of Saturn are, of course, within the Roche limit of Saturn. However, the Roche-limit obstacle would not have applied to the growth of planets in the solar nebula, or indeed to the growth of the satellites of Saturn exterior to its rings.

Left to themselves, then, the rings of Saturn are incapable of exhibiting resolvable collective gravitational behavior. [Of course, the influence of relatively large embedded bodies within the ring system may be enhanced by the effects of the collective self-gravity. For example, it has been proposed (Colombo et al. 1976; Franklin and Colombo 1978) that trailing "density wakes" of the type studied by Julian and Toomre (1966) may explain the azimuthal asymmetries which have been observed in Saturn's A Ring (Carmichel 1958; Esposito et al. 1980). Such wakes may be regarded as superpositions of many density waves produced by a localized source (Hunter 1973).] Indeed, unlike models of disk galaxies, the rings of Saturn have no known gravitational instabilities; all the observed collective gravitational phenomena are driven waves. These waves turned out to be visible by Voyager spacecraft because they are excited by resonant interaction with various external satellites, and the wavelength of density waves or bending waves near their resonant source can exceed the corotating scale $2\pi L$ by a few orders of magnitude. Before the Voyager encounters with Saturn, driven density waves had already been invoked by Goldreich and Tremaine (1978b) as a mechanism by which the influence of resonances could be extended to clear wide gaps. The basic idea is simple. Spiral density waves launched from inner Lindblad resonance (see Sec. II.C) carry negative angular momentum. As these waves propagate outward and dissipate, their angular momentum deficit is transferred to the basic rotational motion of the ring particles, causing them to drift slowly inwards toward the position of the inner Lindblad resonance. If this tendency is sufficiently strong, a gap in front of the inner Lindblad resonance may be opened. Except for the outer edges of the A and B Rings of Saturn, the gap-producing ability of resonances remains unverified, but the waves that they do excite have yielded a powerful diagnostic tool for investigating the physical state of planetary rings.

B. The Role of Physical Collisions

Physical collisions play an important role in galaxies only for the dynamics of the interstellar gas, a minor fraction of the total mass of a galaxy. On a large scale, such collisions make processes such as galactic shocks possible (Fujimoto 1968; Roberts 1969; Shu et al. 1973; Woodward 1975), and thereby they accentuate the observability of spiral structure. But their absence in the stellar component necessitates that the bulk of the matter in the disk acquire nonnegligible random motions (Toomre 1964,1977).

In planetary rings, inelastic collisions between ring particles affect the large-scale structure more directly (Goldreich and Tremaine 1982). First, in the absence of perturbing effects, they make the collection of ring particles ultimately settle into the equatorial plane of the planet. Second, they quickly dissipate the random motions of all but the smallest ring particles (which may be subject to strong electromagnetic forces). Third, they lead to friction, which generally works to destroy structure in the ring system (Cook and Franklin 1964; see, however, chapter by Stewart et al. in this book). Because of inelastic collisions, the random velocities would tend toward zero, were they not offset by the input of kinetic energy from viscous spreading or from gravitational stirring by external satellites.

The level of shear friction in the rings is characterized by the kinematic viscosity ν. For ring material of normal optical depth τ not very different from the order of unity, and made of particles with characteristic size R that possess rms random speed c in a typical direction, kinetic theory yields the following approximate formulae for the kinematic viscosity (Brahic 1977; Goldreich and Tremaine 1978a):

$$\nu = \Omega R^2 \tau \text{ for } c < 2\Omega R \ . \tag{2a}$$

$$\nu = \left(\frac{c^2}{2\Omega} \right) \left(\frac{\tau}{1 + \tau^2} \right) \text{ for } c > 2\Omega R \ . \tag{2b}$$

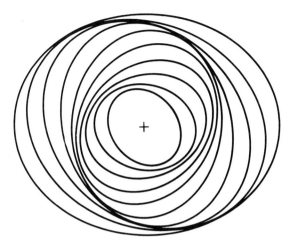

Fig. 1. Schematics of spiral density waves. The actual wrapping is much tighter for real density waves in ring systems.

Since the typical vertical displacement of a particle from the central ring plane is c/Ω, Eq. (2a) gives the value of the "monolayer viscosity," because it applies when the vertical excursions of a ring particle (the characteristic ring thickness) become comparable to or smaller than the size of the (largest plentiful) particles. In planetary rings, particles probably come in a broad distribution of sizes (Marouf et al. 1983), and Eqs. (2a, b) apply only in an average sense (Henon 1981; Goldreich and Tremaine 1982).

In Saturn's rings, it has been found (Cuzzi et al. 1981; Lane et al. 1982; Lissauer et al. 1982) that the random speed c is so small that the noncircular excursions of ring particles are at most tens of meters, much less than the scale of resolvable collective phenomena. Thus, to a high degree of approximation, we may ignore the effects of the noncircular motions in their contribution to a force of "pressure" in our discussion of the collective dynamics, thereby greatly simplifying the conventional treatments (cf. Lin and Shu 1968; Toomre 1969; Goldreich and Tremaine 1978b; Bertin 1980). In the context of density waves, this means that we restrict ourselves to the study of the "long waves," and ignore the "short waves" that are so prominent in theories of the spiral structure of disk galaxies (e.g., Shu 1970; Lin and Lau 1979). We take account of the nonzero random speeds only insofar as they contribute to the gradual damping of density waves and bending waves through the viscosity (Eq. 2b).

C. Schematics of Density Waves and Bending Waves

Before we begin our formal analysis, we will give a brief descriptive account of how density waves and bending waves are initiated in planetary rings by resonant interaction with external satellites. Consider first the excita-

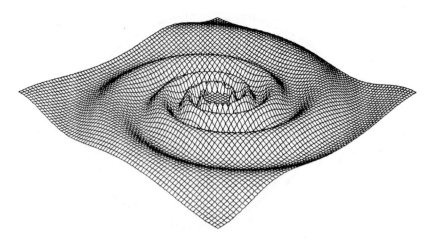

Fig. 2. Schematics of spiral bending wave.

tion of density waves. In the absence of external forcing, the unperturbed orbit of a ring particle is a perfect circle if we ignore the small random motions. When disturbed by the Fourier component of a satellite's gravity which is m-times periodic in ϑ, the steady-state response of the ring particle will contain m outward excursions and m inward excursions, which try to follow the rotation rate of the disturbing force field. These inward and outward excursions will be most severe for ring particles near a resonant condition. Beyond an inner Lindblad resonance, the outward excursions will be less than the outward excursions of the ring particles closer to exact resonance; thus the intervening material will be compressed. The corresponding material between the inward excursions will be rarefied. The excess and deficit of material will exert gravitational accelerations on neighboring pieces of matter, whose nonaxisymmetric density response communicates the disturbance further. This is the beginning of a density wave. Because of the differential rotation, the crests and troughs will trace a spiral pattern (Fig. 1).

A similar scenario holds for the excitation of bending waves. A satellite circling in an inclined orbit with respect to the ring plane will exert periodic vertical forces on the ring particles. The upward and downward motions will be most severe near resonance. As these particles move up and down, their gravitational attraction on neighboring regions will cause the latter also to oscillate. The disturbance therefore propagates away from the resonance region as a bending wave, much as shaking a bedsheet at one end will put flaps into the sheet. In the case of Saturn's rings, a steady state is soon reached where the vertical input of energy (and angular momentum) at the resonance is carried off by a spiral bending wave (Fig. 2). This drainage of the secular input at resonance prevents the vertical excursions of the ring particles from becoming so large as to invalidate the linear theory.

II. ORBITS AND RESONANCES

A. The Epicyclic Theory and Its Relation to the Keplerian Problem

Adopt cylindrical coordinates (r, ϑ, z), and consider a free particle orbiting in a circle of radius r with angular speed Ω (r) in the equatorial plane $z = 0$ of an axisymmetric body (the planet) with an associated gravitational potential $\varphi_p(r,z)$. Centrifugal equilibrium requires

$$r\Omega^2(r) = \left[\frac{\partial \varphi_p}{\partial r} \right]_{z=0} . \tag{3}$$

If this test particle is displaced by an arbitrary small amount, it will oscillate freely in the horizontal and vertical directions about the reference circular orbit with epicyclic frequency κ (r) and vertical frequency μ (r) given by Lindblad's theory of epicyclic motion (Chandrasekhar 1942):

$$\kappa^2(r) = r^{-3} \frac{d}{dr}\left[(r^2\Omega)^2 \right] . \tag{4a}$$

$$\mu^2(r) = \left[\frac{\partial^2 \varphi_p}{\partial z^2} \right]_{z=0} . \tag{4b}$$

When the field point is outside the body of Saturn, φp satisfies Laplace's equation

$$\frac{1}{r} \frac{\partial}{\partial r}\left[r \frac{\partial \varphi_p}{\partial r} \right] + \frac{\partial^2 \varphi_p}{\partial z^2} = 0 . \tag{5}$$

Equation (5) applied to $z = 0$ yields the following useful relation between the vertical frequency μ, the radial frequency κ, and the tangential frequency Ω:

$$\mu^2 + \kappa^2 = 2\Omega^2 . \tag{6}$$

If the planet were spherical, μ, κ, and Ω would all be equal; however, because Saturn is oblate, μ is slightly greater than Ω, and κ is slightly less.
   The relation of the above discussion to classical notions in celestial mechanics is simple. Because $\mu \neq \kappa \neq \Omega$, a noncircular orbit in inertial space is generally not closed. With $\mu > \Omega$ and $\kappa < \Omega$, we have nodal regression and apsidal advance. In terms of an expansion in multipole moments, it is conventional to write the planetary potential in the equatorial plane as

$$\varphi_p (r, 0) = - \frac{GM_p}{r}\left[1 - \sum_{n=1}^{\infty} J_{2n} (R_p/r)^{2n} P_{2n}(0) \right] , \tag{7}$$

where M_P is the planet's mass, R_P is its radius, J_{2n} is its $2n$-th multipole moment, and

$$P_{2n}(0)\left(-\frac{1}{4}\right)^n \frac{(2n)!}{(n!)^2} \tag{7a}$$

is the value of the Legendre polynomial of order $2n$ at the origin. Given the parameters for the planet, Eqs. (3), (4a), and (6) allow us to calculate $\Omega(r)$, $\kappa(r)$, and $\mu(r)$.

B. Forcing by an Orbiting Satellite

Consider now a satellite of mass M which orbits the planet with a small eccentricity e_M in a plane inclined by a small angle i_M with respect to the planet's equatorial plane $z = 0$. Let (r_M, ϑ_M, z_M) be the time-dependent cylindrical coordinates of this satellite. The (direct) contribution of the satellite to the total gravitational potential is

$$\varphi_M(r,\vartheta,z,t) = -GM\left[r_M^2(t) + r^2 - 2r_M r\cos\left[\vartheta_M(t) - \vartheta\right] + \left[z_M(t) - z\right]^2\right]^{-\frac{1}{2}}. \tag{8}$$

Strictly speaking, one should also include the effects of the usually less important "indirect" term, which arises because the presence of the satellite accelerates the center of the planet, where we have placed the origin of our coordinate system (see Appendix A).

What is important for us here is that $r_M(t) - a_M$, $\Theta_M(t) \equiv \vartheta_M(t) - \Omega_M t$, and $z_M(t)$ represent small quantities and, moreover, are periodic functions. To a sufficient degree of approximation, the quantities $r_M(t)$ and $\Theta_M(t)$ have period $2\pi/\kappa_M$, while $z_M(t)$ has period $2\pi/\mu_M$. Thus, not only can we expand for small e_M and i_M, but since $\varphi_M(r,\vartheta,z,t)$ is periodic in time t and angle ϑ, we may also perform Fourier analysis in these variables. The result consists of a series of expressions ordered in e_M and $\sin i_M$, each one of which can be decomposed, by means of trignometric identities, into elementary terms of form

$$\text{Re}\left[\Phi_M(r,z)e^{i(\omega t - m\vartheta)}\right], \tag{9}$$

where the disturbance frequency ω is given as a sum of integer combinations of Ω_M, μ_M, and κ_M,

$$\omega = m\Omega_M \pm n\,\mu_M \pm p\kappa_M, \tag{10}$$

where m, n, p are nonnegative integers. In Appendix A, we give explicit techniques and formulae for calculating the coefficient $\Phi_M(r, z)$ and its derivatives in the plane $z = 0$. For the present discussion, we merely state that n is even for horizontal forcing and odd for vertical forcing. Moreover, the forcing amplitude associated with the disturbance frequency (Eq. 10) is proportional to $e_M^{|p|}\sin^{|n|}i_M$ (Goldreich and Tremaine 1978b; Shu et al. 1983).

C. Classification of Horizontal and Vertical Resonances

Consider the response of a test particle placed in a circular orbit in the ring plane. For small perturbations from a satellite, the particle responds as a multidimensional, forced, linear-harmonic oscillator. In particular, it may suffer resonances if the relative frequency at which it experiences the disturbance of the satellite, referred to its local rotation rate, is equal to any of the natural frequencies of its free oscillations (cf. Franklin and Colombo 1970; chapter by Franklin et al. in this book). In the present notation, it will undergo horizontal (Lindblad) resonance if it is placed at $r = r_L$ where r_L satisfies (Lin and Shu 1966):

$$\omega - m\,\Omega(r_L) = \pm\kappa(r_L), \tag{11}$$

with ω given by Eq. (10). On the other hand, it will suffer vertical (inclination) resonance if its radial position r_V satisfies (Shu et al. 1983):

$$\omega - m\Omega(r_V) = \pm\,\mu\,(r_V). \tag{12}$$

When condition (11) holds for the lower (upper) choice of sign, we refer to r_L as an inner (outer) Lindblad resonance or horizontal resonance. When condition (12) holds for the lower (upper) choice of sign, we refer to r_V as an inner (outer) inclination or vertical resonance. Because large satellites exist exterior to the main ring systems, $\Omega_M < \Omega$, and inner resonances are generally more important than outer ones. Of all inner Lindblad resonances, apsidal resonances, where

$$\Omega_M = \omega = \Omega\,(r_L) - \kappa\,(r_L),$$

are especially interesting in the context of collective effects, since planetary rings can be in near resonance, $\Omega\,(r) - \kappa\,(r) \approx \Omega_M$, with a distant satellite for a large range of radii r, making the density waves driven by such resonances more easily resolved (Cuzzi et al. 1981).

To fix ideas for a more typical case, let us give an example. The disturbance frequencies associated with the 4:2 inner horizontal and vertical resonances of Mimas are given by

$$\omega_{\text{horiz}} = 3\Omega_M + \kappa_M,$$

$$\omega_{\text{vert}} = 3\Omega_M + \mu_M,$$

while the positions of the corresponding resonances are given through

$$3\Omega\,(r_L) - \kappa\,(r_L) = \omega_{\text{horiz}},$$

$$3\Omega\,(r_V) - \mu\,(r_V) = \omega_{\text{vert}}.$$

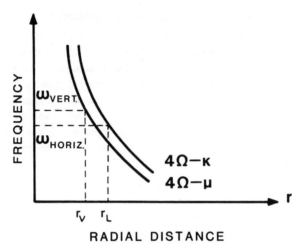

RADIAL DISTANCE

Fig. 3. Schematic drawing which shows how the oblateness of Saturn results in different radial locations r_L and r_V for the 5:3 inner Lindblad resonance and inner vertical resonance of Mimas.

If Ω, μ, κ were all equal, then r_L and r_V would coincide and equal a position r where the ring particle circles Saturn four times for each two times Mimas circles Saturn: $\Omega (r)/\Omega_M = 4/2$. This rational number gives the resonances their names. However, because $\mu > \Omega > \kappa$ due to Saturn's oblateness, the degeneracy $r_L = r_V$ is split, and the 4:2 vertical resonance lies somewhat inside the 4:2 horizontal resonance, which is itself somewhat inside the 2:1 horizontal resonance of Mimas (the outer edge of the B Ring), where

$$\omega_{\text{horiz}} = 2\Omega_M,$$

$$2\Omega (r_L) - \kappa (r_L) = \omega_{\text{horiz}}.$$

Figure 3 demonstrates the analogous situation for the 5:3 resonances.

Clearly, if one were willing to contemplate all possible rational combinations, one could have more than enough resonances to do anything one wanted. But our comments after Eq. (10) imply that the satellite forcing associated with resonances characterized by a ratio of integers which contain a difference N between numerator and denominator must be proportional to a product of $(N - 1)$ factors of the small quantities, e_M and $\sin i_M$. Thus, the strongest inner horizontal resonances correspond to the ratio $m:m-1$, involving only the circular part of the satellite's motion, while the strongest inner vertical resonances correspond to the ratio $m+1:m-1$, involving one factor of the inclination of the satellite's orbit.

III. WAVES AS PROCESSES

We now begin our formal dynamical analysis. We wish to calculate the collective response of a planetary ring to horizontal and vertical forcing by an external satellite. In what follows, we develop the inviscid theories and incorporate the effects of friction as small corrections to the final results. We limit our discussion to waves of small amplitude. In the linear approximation, horizontal and vertical motions decouple, so density waves and bending waves may be treated separately. Nonlinear effects do play a role in many of the density waves observed in Saturn's rings, but those effects were still incompletely evaluated at the time this review was written.

A. Spiral Density Waves

It is known, from studies of the collective behavior of galactic disks, that if random velocities of the constituent bodies can be ignored, the (nondisspative) behavior calculated from a kinetic treatment (using an encounterless Boltzmann equation) is identical to a fluid treatment (using the gas-dynamical equations with zero pressure and viscosity). Since the fluid treatment is much simpler, we adopt it here.

1. Basic Equations. Consider the horizontal dynamics of a disk of matter, which we shall idealize as being infinitesimally thin. Let the surface mass density of this disk be $\sigma\,(r,\vartheta,t)$. If $\varphi_D(r,\vartheta,z,t)$ is the self-gravitational potential associated with this disk, it satisfies Poisson's equation

$$\frac{1}{r}\frac{\partial}{\partial r}\left(r\,\frac{\partial\varphi_D}{\partial r}\right) + \frac{1}{r^2}\frac{\partial^2\varphi_D}{\partial\vartheta^2} + \frac{\partial^2\varphi_D}{\partial z^2} = 4\pi G\sigma\delta(z)\ , \qquad (13)$$

where $\delta\,(z)$ is the Dirac delta function.

Let $u(r,\vartheta,t)$ and $v\,(r,\vartheta,t)$ be the r and ϑ components of the fluid velocity. The equations of compressible fluid dynamics for purely horizontal motions in a pressureless inviscid disk read

$$\frac{\partial\sigma}{\partial t} + \frac{1}{r}\frac{\partial}{\partial r}\,(r\sigma u) + \frac{1}{r}\frac{\partial}{\partial\vartheta}(\sigma v) = 0, \qquad (14)$$

$$\frac{\partial u}{\partial t} + u\,\frac{\partial u}{\partial r} + \frac{v}{r}\frac{\partial u}{\partial\vartheta} - \frac{v^2}{r} = -\frac{\partial}{\partial r}\left[\varphi_P + \varphi_D + \varphi_M\right]. \qquad (15)$$

$$\frac{\partial v}{\partial t} + \frac{u}{r}\frac{\partial}{\partial r}\,(rv) + \frac{v}{r}\frac{\partial v}{\partial\vartheta} = -\frac{1}{r}\frac{\partial}{\partial\vartheta}\left[\varphi_D + \varphi_M\right]. \qquad (16)$$

2. Equilibrium State and Linear Perturbations. We take the total state of the system to be one of centrifugal balance plus small perturbations. Neglect-

ing terms of second order in the perturbations, and looking ahead to calculating the linear response for satellite forcing of the form given by Eq. (9), we
write

$$\sigma(r,\vartheta,t) = \sigma_0(r) + \mathrm{Re}\left[S(r)e^{i(\omega t - m\vartheta)}\right], \tag{17a}$$

$$u(r,\vartheta,t) = 0 + \mathrm{Re}\left[U(r)e^{i(\omega t - m\vartheta)}\right], \tag{17b}$$

$$v(r,\vartheta,t) = r\Omega(r) + \mathrm{Re}\left[V(r)e^{i(\omega t - m\vartheta)}\right], \tag{17c}$$

where, strictly speaking, we should take $\Omega(r)$ to satisfy

$$r\Omega^2(r) = \left[\frac{\partial}{\partial r}(\varphi_P + \varphi_{Do} + \sum\varphi_{Mo})\right]_{z=0}, \tag{18}$$

with φ_{Do} being the potential that corresponds to the equilibrium disk of surface
density σ_0, and $\sum\varphi_{Mo}$ the time-independent axisymmetric part of the potential
due to all of the planet's satellites (see Eq. A13 of Appendix A). In practice,
since the masses of the satellites and known planetary rings are very small
compared to the Jovian planets, Eq. (3) provides a sufficiently accurate approximation to Eq. (18), except possibly if we want very precise locations for
apsidal resonances. Our treatment of waves in this chapter is sufficiently
general to accommodate the replacement of Eq. (3) by the more accurate Eq.
(18), desirable for some applications (e.g., close binary stars, the solar
nebula).

The substitution of Eqs. (17) and (18) into Eqs. (14) to (16) yields the
linearized set

$$i(\omega - m\Omega)S + \frac{1}{r}\frac{d}{dr}(r\sigma_0 U) - \frac{im}{r}\sigma_0 V = 0, \tag{19}$$

$$i(\omega - m\Omega)U - 2\Omega V = -\frac{\partial}{\partial r}(\Phi_D + \Phi_M), \tag{20}$$

$$i(\omega - m\Omega)V + \frac{\kappa^2}{2\Omega}U = \frac{im}{r}(\Phi_D + \Phi_M). \tag{21}$$

In Eqs. (20) and (21), the functions $\Phi_D(r,z)$ and $\Phi_M(r,z)$ are evaluated in the
plane $z = 0$. The function Φ_M is the appropriate Fourier coefficient in expression (9), while Φ_D satisfies the linearized form of the Poisson Eq. (13) for the
self-consistent field of the disk.

$$\frac{1}{r}\frac{\partial}{\partial r}\left(r\frac{\partial \Phi_D}{\partial r}\right) + \frac{\partial^2 \Phi_D}{\partial |z|^2} - \frac{m^2}{r^2}\Phi_D = 0 \text{ for } |z| > 0, \qquad (22a)$$

$$S(r) = \frac{1}{2\pi G}\left(\frac{\partial \Phi_D}{\partial |z|}\right)_{|z|=0+}. \qquad (22b)$$

To derive Eq. (22b), integrate Eq. (13) from $z = 0-$ to $z = 0+$ under the assumption that Φ_D is continuous across $z = 0$, but that its z-derivative reverses signs.

If we solve Eqs. (19) to (21) algebraically for S, U, V, we obtain

$$U = i\,\Lambda_U\,(\Phi_D + \Phi_M), \qquad (23a)$$

$$V = \Lambda_V\,(\Phi_D + \Phi_M), \qquad (23b)$$

$$S = \Lambda_S(\Phi_D + \Phi_M), \qquad (23c)$$

where Λ_U, Λ_V, Λ_S are the real differential operators,

$$\Lambda_U = \frac{1}{D}\left[-(\omega - m\Omega)\frac{d}{dr} + \frac{m}{r}2\Omega\right], \qquad (24a)$$

$$\Lambda_V = \frac{1}{D}\left[\frac{\kappa^2}{2\Omega}\frac{d}{dr} - \frac{m}{r}(\omega - m\Omega)\right], \qquad (24b)$$

$$\Lambda_S = \left(\frac{1}{\omega - m\Omega}\right)\left(-\frac{1}{r}\frac{d}{dr}r\sigma_0\Lambda_U + \frac{m}{r}\sigma_0\Lambda_V\right), \qquad (24c)$$

and where D is the determinant of the coefficient matrix associated with Eqs. (20) and (21):

$$D = \kappa^2 - (\omega - m\Omega)^2 . \qquad (25)$$

Clearly, D is a discriminant for the distance (in frequency) from Lindblad resonances (cf. Eq. 11).

Given Φ_M, Eqs. (22) and (23c) form our basic set to solve for $S(r)$ and $\Phi_D (r,0)$. Having obtained $\Phi_D (r,0)$, we recover from Eqs. (23a) and (23b) the two components of fluid velocity $U(r)$ and $V(r)$. We now formulate an asymptotic approximation which greatly simplifies our task.

3. *Asymptotic Theory.* We are interested in solutions where the self-consistent part of the disturbance gravitational potential in the disk plane,

$$\Phi_D(r,0) \equiv F(r), \qquad (26)$$

varies rapidly with r (because the scale length of collective effects is small). With this assumption, Eq. (22a) has the approximate form (for more accurate treatments, see Shu [1970] and Bertin and Mark [1979]):

$$\frac{\partial^2 \Phi_D}{\partial r^2} + \frac{\partial^2 \Phi_D}{\partial |z|^2} = 0 \text{ for } |z| > 0, \qquad (27)$$

Laplace's equation in rectangular coordinates. The solution to Eq. (27), subject to the condition (26), is

$$\Phi_D(r,z) = F(r + is\,|z|) , \qquad (28)$$

where $s = \pm 1$, and the correct choice of sign is to be made on the basis of physical arguments (vanishing of Φ_D at large $|z|$). With the solution (28), Eq. (22b) becomes

$$S(r) = \frac{is}{2\pi G} \frac{dF(r)}{dr} = \frac{is}{2\pi G} \frac{d\Phi_D}{dr} \qquad (29)$$

where Φ_D in the last expression is evaluated in the plane $z = 0$. Apart from the assumption that $d\Phi_D/dr$ is large in comparison with Φ_D/r, we allow for the possibility that the derivative of σ_0/D might be large. Then Eqs. (24a,b,c) allow us to approximate

$$\Lambda_S \Phi_D = \frac{d}{dr} \left(\frac{\sigma_0}{D} \frac{d\Phi_D}{dr} \right) , \qquad (30a)$$

$$\Lambda_S \Phi_M = \frac{d}{dr} \left\{ \frac{\sigma_0}{D} \left[\frac{d\Phi_M}{dr} - \frac{m 2\Omega \Phi_M}{r(\omega - m\Omega)} \right] \right\} . \qquad (30b)$$

Note that the dominant contribution to Λ_S comes from the effect of the radial motion in the equation of continuity (Eq. 19). Substituting Eqs. (29) and (30) into Eq. (23c) allows us to integrate once in r. Rearranging terms, we obtain

$$r \frac{d\Phi_D}{dr} - \frac{isrD}{2\pi G\sigma_0} \Phi_D = -r \frac{d\Phi_M}{dr} + \frac{m 2\Omega}{(\omega - m\Omega)} \Phi_M . \qquad (31)$$

With Φ_M given, Eq. (31) becomes the fundamental differential equation of our system to solve for the radial dependence $\Phi_D(r,0)$ of the disk potential. Although only one spatial derivative of Φ_D appears in Eq. (31), the presence of i in front of the undifferentiated term makes Eq. (31), as we shall soon see, an inhomogeneous wave equation.

3a. Free Spiral Density Waves. Ignore, for the time being, the in-homogeneous terms proportional to Φ_M in Eq. (31), and consider free solutions with the WKBJ form

$$\Phi_D(r,0) = A(r) \exp\left[i \int k \, dr\right],\qquad(32)$$

where k is real width $|k| \, r >> 1$, so the phase of $\Phi_D \, (r, 0)$ varies much more rapidly than its amplitude $A(r)$. The requirement that the expression on the right-hand side of Eq. (28) decays with increasing $|z|$ (as $\exp\left[-|kz|\right]$) requires us to identify $s = \pm 1$ to be

$$s = \text{sign}(k)\qquad(33)$$

With this identification, Eq. (29) becomes

$$S(r) = -\frac{|k|\Phi_D(r,\,0)}{2\pi G},\qquad(34)$$

showing that density maxima correspond to potential minima. The loci of surface density maxima (cf. Eq. 17a) correspond to the curves

$$\omega t - m\vartheta + \int k \, dr = n2\pi, n = 0,1,\ldots,\, m-1,$$

which, at any fixed time t, describe an m-armed spiral. These spirals trail with the direction of rotation if $k < 0$, and lead if $k > 0$. For $m \neq 0$, the pattern of spiral arms rotates as a rigid body at the angular pattern speed

$$\Omega_p \equiv \omega / m.$$

The substitution of Eq. (32) into Eq. (31), with the right-hand side set equal to zero, yields, to lowest asymptotic order, the dispersion relation for long spiral density waves (Lin and Shu 1964):

$$(\omega - m\Omega)^2 = \kappa^2 - 2\pi G \sigma_0 |k|,\qquad(35)$$

which has an associated turning point (where k is zero) at $D = 0$. For regions where σ_0/D is not rapidly varying, the associated solution for the dyamical response (Eq. 30a) of the surface density is

$$S(r) = -\sigma_0 \frac{k^2}{D} \Phi_D .\qquad(36)$$

Comparison of Eqs. (34) and (36) lends insight into why self-sustained density waves can only exist in regions where $D > 0$. In the regions where

$D < 0$, the density response (Eq. 36) is 180° out of phase compared to the value (Eq. 34) required to sustain the gravitational potential that drives the response. Thus, density waves are evanescent in the regions where $D < 0$.

For axisymmetric disturbances, $m = 0$, Eq. (35) formally predicts instability, $\omega^2 < 0$, for large wavenumbers. However, such potential instabilities take place only for wavelength scales where it is inappropriate to ignore the effects of the finite random motions and sizes of the particles. For example, if we include acoustic effects and the finite thickness of the disk, we obtain a dispersion relationship of the form (Lin and Shu 1968; Vandervoort 1970):

$$(\omega - m\Omega)^2 = \kappa^2 - 2\pi G \sigma_0 T |k| + c^2 k^2, \qquad (37)$$

where c is the rms velocity dispersion in the radial direction and $T = T(|k| z_0)$ is a thickness reduction factor associated with a finite equivalent vertical thickness (see Appendix B):

$$2 z_0(r) \equiv \frac{\sigma_0(r)}{\rho_0(r, 0)} . \qquad (38)$$

where $\rho_0(r, 0)$ is the average volume mass density in the plane $z = 0$. The function $T(\zeta)$ has a value of unity for $\zeta = 0$, and it monotonically approaches zero as $1/\zeta$, for ζ very large. If we set $T = 1$ in Eq. (37), we easily show that all local instabilities are suppressed if c exceeds the critical value (Safronov 1960; Toomre 1964):

$$c_{crit} = \frac{\pi G \sigma_0}{\kappa} .$$

In the other limit, where ζ is very large, Eq. (37) becomes the dispersion relationship appropriate for density waves propagating in a cylinder of infinite (relative) height (cf. Chandrasekhar 1961):

$$(\omega - m\Omega)^2 = \kappa^2 - 4\pi G \rho_0(r, 0) + c^2 k^2.$$

This limit is, however, of little practical astronomical interest.

In the present context, it is more interesting to imagine inelastic collisions reducing the actual c to zero. The quantity z_0, as defined by Eq. (38), would not be zero, for the particles would still have finite size. Indeed, in this situation we expect the disk to form a monolayer of particles with $z_0 \sim R$, where R is a characteristic particle size. For $m = 0$, Eq. (37) can now be written in the form

$$\omega^2 = \kappa^2 - 2\pi G \frac{\sigma_0}{z_0} \zeta T(\zeta) . \qquad (39)$$

Since $\zeta T(\zeta)$ is always less than unity, the above shows that instabilities of all wavelengths are still suppressed if

$$z_0 > \frac{2\pi G \sigma_0}{\kappa^2}, \tag{40}$$

where the right-hand side represents nearly the same length scale as Eq. (1).

Equation (39) can also be used to express an idea complementary to the Roche limit. Begin by noting that $\rho_0 (r, 0)$ cannot exceed the internal volume density, ρ_I of the individual particles. Note also that $|k|$ cannot meaningfully be taken much larger than the reciprocal of R; thus, the product $\zeta T(\zeta)$ is also likely to be a small fraction. Setting

$$\frac{\sigma_0}{2z_0} \zeta T(\zeta) = \rho_I \delta,$$

where δ is a small fraction, and writing

$$\kappa^2 \approx \frac{GM_p}{r^3} = \frac{4\pi}{3} G \rho_P \left(\frac{R_P}{r} \right)^3$$

where ρ_P is the planet's mean density, we may now express the stability criterion, $\omega^2 > 0$, as

$$\frac{r}{R_P} < \left(\frac{\rho_P}{3\rho_I \delta} \right)^{\frac{1}{3}}.$$

The physical interpretation of the above is that a monolayer of solid particles, which attract each other only by mutual gravitation, cannot agglomerate to larger sizes if they orbit in a ring too close to the planetary surface. Thus, if all particles are initially very small and the surface density is too large for them to form a monolayer, then inelastic collisions would bring the ring to a condition of gravitational instability. The particles would grow to sizes such that the surface density could be more or less accommodated by a monolayer. The ring system would then be stable to collective gravitational effects even if the velocity dispersion were reduced to zero. Tremaine (personal communication) has independently reached similar conclusions. This does not imply that planetary rings *are* monolayers, for there may be external inputs of random kinetic energy. Such inputs exist in Saturn's A Ring, for example, in the form of many driven density waves (Lissauer et al. 1983). Moreover, the largest particles may form a monolayer, yet the smaller particles may be kept many particles thick by gravitational scattering from the large ones (for observations bearing on this point, see e.g., Cuzzi et al. 1980).

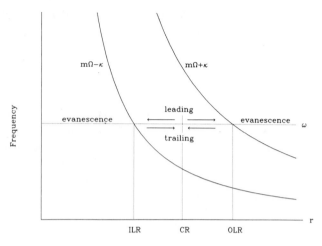

Fig. 4. Schematic drawing showing that long trailing spiral density waves of frequency ω and angular mode number m propagate away from inner and outer Lindblad resonances, while long leading spiral density waves propagate toward them. If long waves reach the corotation circle where $\omega = m\,\Omega$, they reflect and transmit as short waves (see Lin and Lau 1979), a process irrelevant for planetary rings. The regions beyond the inner and outer Lindblad resonances are regions of evanescence.

For waves observable by the Voyager spacecraft, however, we always deal with structures whose radial scales are many orders of magnitude larger than scales at which we have to contend with corrections to Eq. (35), or its more general predecessor, Eq. (31). The physical interpretation of Eq. (35) is simple. In the absence of self-gravitation, i.e., in the absence of the last term, a disk with angular periodicity m can perform free horizontal oscillations at a frequency referred to the local rotation rate, $\omega - m\,\Omega$, only equal to the natural frequency $\pm\kappa$. Since the self-gravity of compressed regions slows down the reexpansion of such regions, the actual square of the Doppler-shifted frequency must be smaller than κ^2 by the amount $2\pi G\sigma_0|k|$. Long density waves of frequency ω therefore, can exist only in the regions between the inner and outer Lindblad resonance (Fig. 4). Of course, a planetary ring may be truncated before the outer Lindblad resonance is reached. The group velocity of long density waves is given (Toomre 1969) by:

$$c_g = - \left[\frac{\partial \omega}{\partial k}\right]_r = \frac{\pi G\sigma_0}{\omega - m\,\Omega}\, s \ . \tag{41}$$

Thus, long training waves ($s = -1$) propagate outward from inner Lindblad resonance (where $\omega - m\,\Omega = -\kappa$) and inward from outer Lindblad resonance (where $\omega - m\,\Omega = +\kappa$). Long leading waves ($s = +1$) propagate toward Lindblad resonances, not from them; therefore they cannot be excited at

Lindblad resonances without entering a condition of evanescence. Such considerations probably prompted Goldreich and Tremaine (1978b) to expect that orbiting exterior satellites would excite trailing spiral density waves that travel outward from the inner resonant location in a planetary disk. This expectation is reinforced by noting that Eq. (35) formally predicts free density waves to have zero radial wavenumber k at Lindblad resonance (where $D = 0$), so such waves have the right spatial structure near resonance to couple to the smoothly varying forcing of the satellite's gravitational field (represented by the right-hand side of Eq. (31). We now proceed to rederive Goldreich and Tremaine's theory of resonantly forced spiral density waves.

3b. Forced Spiral Density Waves. To calculate how density waves are excited near Lindblad resonances, we assume that the fractional distance from resonance

$$x \equiv (r - r_L)/r_L \tag{42}$$

is small, and expand the coefficients in Eq. (31) in a Taylor series about $r = r_L$. In particular,

$$D = \mathcal{D} \, x \text{ where } \mathcal{D} \equiv \left(r \, \frac{dD}{dr} \right)_{r_L}. \tag{43}$$

The quantity $\mathcal{D}$ is positive for inner Lindblad resonances; negative for outer ones. If we choose $s = -1$, Eq. (31) now becomes

$$\frac{d\Phi_D}{dx} + \frac{ix}{\epsilon} \, \Phi_D = \Psi_L, \tag{44}$$

where

$$\Psi \equiv \left(- r \, \frac{d\Phi_M}{dr} \pm m \, \frac{2\Omega}{\kappa} \, \Phi_M \right)_{r_L}. \tag{45a}$$

$$\epsilon \equiv \left[\frac{2\pi G \sigma_0}{r \, \mathcal{D}} \right]_{r_L}. \tag{45b}$$

In Eq. (45a), the lower (upper) sign applies to inner (outer) Lindblad resonances. The quantity ϵ is positive (negative) for inner (outer) Lindblad resonances, and it generally has a magnitude of about 10^{-8} in Saturn's rings. This means that the nondimensional length scale of the waves near resonance, $|\epsilon|^{\frac{1}{2}}$, is fairly small. The square of the quantity Ψ_L is proportional to the resonant torque exerted by the satellite (Goldreich and Tremaine 1978b).

Expressions for the angular momentum density and energy density carried by spiral density waves may be found in Goldreich and Tremaine (1978b) or in Shu (1970). The important point is that density waves found inside of corotation (where $\omega - m\Omega = 0$) carry negative angular momentum (and energy), while density waves found outside of corotation carry positive angular momentum (and energy). Thus, when these waves damp, they cause matter on either side of corotation to move away from corotation (toward the Lindblad resonances). Thus, the net effect of density waves excited in planetary rings by satellite forcing is usually to cause the ring material to flow away from the satellite.

Since $\exp(ix^2/2\epsilon)$ is an integrating factor for Eq. (44), and since $\Phi_D \to 0$ as x heads toward the region where density waves are evanescent, the solution may be written as

$$\Phi_D(r,0) = A_L H_q(\xi), \tag{46a}$$

$$A_L = q(2\pi|\epsilon|)^{\frac{1}{2}}\Psi_L, \tag{46b}$$

where we have defined

$$q \equiv \mathrm{sgn}(\epsilon), \tag{47a}$$

$$\xi \equiv \frac{qx}{\sqrt{(2|\epsilon|)}}, \tag{47b}$$

$$H_q(\xi) \equiv \frac{1}{\sqrt{\pi}}\exp(-iq\xi^2)\int_{-\infty}^{\xi}\exp(iq\eta^2)d\eta . \tag{47c}$$

Notice that H_- is the complex conjugate of H_+:

$$H_-(\xi) = H_+^*(\xi) . \tag{48}$$

Plotted in Fig. 5 is the function $H_-(\xi)$.

It is illuminating to consider the asymptotic forms taken by the solution (Eqs. 46a,b). For ξ large and negative (far into the region of evanescence for waves excited either at inner or outer resonance), we use

$$\exp(iq\eta^2)\, d\eta = -\frac{iq}{2\eta}\, d[\exp(iq\eta^2)] , \tag{48a}$$

and integrate Eq. (47c) by parts to obtain

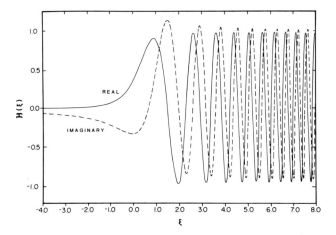

Fig. 5. The real and imaginary parts of the wavefunction $H_-(\xi)$ defined by Eq. (47c).

$$\Phi_D \rightarrow -i\epsilon \, \frac{\Psi_L}{x} \text{ as } \xi \rightarrow -\infty \,, \tag{49}$$

which gives the nonwavy part of the disk response, the noncircular distortion of streamlines purely by the gravity of the satellite. This part of the solution corresponds to the neglect of the effects of self-gravitation, i.e., the neglect of the first term on the left-hand side of Eq. (44) in comparison with the second. For ξ large and positive (far into the region of wave propagation), we can use the identity

$$\int_{-\infty}^{\xi} \exp(iq\eta^2) d\eta \equiv \int_{-\infty}^{+\infty} \exp(iq\eta^2) d\eta - \int_{\xi}^{+\infty} \exp(iq\eta^2) d\eta,$$

to show

$$H_q(\xi) \rightarrow \exp[iq(-\xi^2 + \pi/4)] - \frac{iq}{2\xi\sqrt{\pi}} \text{ as } \xi \rightarrow +\infty \,.$$

This now yields for Eq. (46)

$$\Phi_D \rightarrow A_L \exp\left[i(-x^2/2\epsilon + q\pi/4)\right] - i\epsilon\Psi_L/x \text{ as } \xi \rightarrow +\infty \,. \tag{50}$$

The first term represents a trailing spiral density wave with amplitude

$$|A_L| = (2\pi|\epsilon|)^{\frac{1}{2}} |\Psi_L| \,, \tag{51}$$

and phase (cf. Eq. 32)

$$\int_{\tau_L}^{\tau} k\,dr + \frac{\pi}{4} + \arg(q\,\Psi_L) \ . \tag{52}$$

The second term in Eq. (50) is again the nonwavy part of the disk response. Note that the extrapolation to $x = 0$ (the position of the resonance) of the asymptotic form of the nonwavy disk response misleadingly suggests divergent behavior. In fact, the actual solution (Eq. 46a) remains finite at $\xi = 0$:

$$\Phi(r_L,0) = \frac{1}{2} A_L e^{iq\pi/4} \ . \tag{53}$$

The rest of the solution, S, U, V, may be obtained from Eqs. (29), (23b), and (23c) by differentiation of Φ_D. However, when we are in the wave zone, it is easier to obtain the perturbational surface density (which is usually assumed to be proportional to the observed optical depth variations) by using Eqs. (34) and (35) (cf. Cuzzi et al. 1981; Lane et al. 1982). Because Eq. (35) implies that $|k| \approx x/\epsilon\, r_L$ near resonance, the surface-density variation has fractional amplitude in the near wave-zone (Goldreich and Tremaine 1978b):

$$\frac{|S|}{\sigma_0} = \left[\frac{2\pi}{|\epsilon|^3}\right]^{\frac{1}{2}} \left[\frac{|\Psi_L|}{r_L{}^2|\,\mathcal{D}\,|}\right]|x| \ . \tag{54}$$

The linear increase of $|S|$ with distance x from resonance reflects the conservation of wave energy and angular momentum (Toomre 1969; Shu 1970). Since $\Psi_L/r_L{}^2|\mathcal{D}| \propto M/M_p$ for the strongest horizontal resonances, and since ϵ is so small in planetary rings, Eq. (54) implies that density waves excited in this manner can become nonlinear within a small fractional distance x of resonance, often within one wavelength of the resonance source. Lissauer and Cuzzi (1982) have tabulated the values of the characteristic distance for waves to become nonlinear,

$$x_{NL} \equiv \left[\frac{|\epsilon|^3}{2\pi}\right]^{\frac{1}{2}} \frac{r_L{}^2|\,\mathcal{D}\,|}{|\Psi_L|} \ , \tag{55}$$

for all known horizontal resonances of importance in Saturn's rings.

3c. Viscous Damping of Density Waves. According to the linear inviscid theory, once spiral density waves are launched, the crests and troughs should gather greater and greater contrast as the waves propagate from their resonant source. None of the density waves observed in Saturn's rings behave in this manner, and we need to ask why not.

A plausible answer in many cases is that the observed waves have such large amplitudes that nonlinear effects enter. The theory of self-consistent nonlinear density waves that include dissipative agents requires development. The work done in the galactic context on this problem (galactic shocks) may be adaptable to the present situation.

In the case in which the wave amplitudes remain small enough for linear theory to apply, the theory of viscous dissipation is relatively simple and well developed. When shear and bulk viscosities are both present and their spatial derivatives are small in comparison with those of the fluid velocities, then we need to add to the right-hand sides of the equations of motion (Eqs. 15 and 16) the terms (Landau and Lifshitz 1959, p. 49):

$$\nu\nabla^2\mathbf{u} + (\zeta+ \frac{1}{3}\nu)\nabla(\nabla\cdot\mathbf{u}) \tag{56}$$

where ν and ζ are the shear and bulk viscosities divided by the fluid volume density. In particulate disks of roughly unit optical depth, the concept of bulk viscosity is somewhat dubious (see chapter by Borderies et al.), but we shall continue with the fluid treatment for historical continuity. Substituting in the WKBJ form (Eq. 32), so that each radial derivative brings down a (large) factor ik, where k may now be complex, and using Eq. (29) to eliminate Φ_D, where s is now defined to be sign[Re(k)], we can reduce the homogeneous verison of Eqs. (19)–(21) to the approximate set:

$$i(\omega-m\Omega)\, S + i\, \sigma_0 k\, U = 0, \tag{57a}$$

$$i(\omega-m\Omega)\, U - 2\Omega V - is\, 2\pi\, GS + (\zeta+ \frac{4}{3}\nu)k^2 U = 0, \tag{57b}$$

$$i(\omega-m\Omega)\, V + \frac{\kappa^2}{2\Omega} + \nu k^2 V = 0. \tag{57c}$$

Since the above is a set of linear homogeneous equations, in order to have nontrivial solutions, the determinant of the coefficient matrix must be zero. Discarding quantities quadratic in the small coefficients of viscosity, this requirement leads to the dispersion relation,

$$(\omega-m\Omega)\, [D-2\pi G\, \sigma_0 sk]+i[(\omega-m\Omega)^2 k^2(7\nu/3 + \zeta)$$
$$+ 2\pi G\, \sigma_0 sk^3\nu] = 0. \tag{58}$$

Write $k = k_R + ik_I$; then, for small ν and ζ, the real part of k satisifes the usual dispersion relation (Eq. 35), while the imaginary part is given by

$$k_I = \left[\left(\frac{\omega - m\Omega}{2\pi G \sigma_0} \right) s k_R^2 (7\nu/3 + \zeta) + \frac{k_R^3}{\omega - m\Omega} \nu \right] . \qquad (59)$$

Near a Lindblad resonance, $(\omega - m\Omega) \approx -q\kappa$, and

$$|k_R| \approx \frac{\mathcal{D} x}{2\pi G \sigma_0} . \qquad (60)$$

For small $|x|$, we may therefore further approximate Eq. (59) by (Goldreich and Tremaine 1978b)

$$k_I = -qs \frac{\kappa \mathcal{D}^2 x^2}{(2\pi G \sigma_0)^3} (7\nu/3 + \zeta) . \qquad (61)$$

This small imaginary part of the wavenumber will lead to a viscous damping factor

$$\exp \left(-\int_{r_L}^{r} k_I \, dr \right) \approx \exp \left[-\left(\frac{|x|}{x_{\text{vis}}} \right)^3 \right], \qquad (62)$$

where the nondimensional viscous damping length is given by

$$x_{\text{vis}} \approx |\epsilon| \left[\frac{3r^2 |\mathcal{D}|}{\kappa(7\nu/3 + \zeta)} \right]^{\frac{1}{3}} . \qquad (63)$$

For our treatment of the viscous effects as a small correction to be valid, the combination $7\nu/3 + \zeta$ must be small enough so that x_{vis} is much greater than $|\epsilon|^{\frac{1}{2}}$, the natural nondimensional scale of the waves near resonance. At first sight, this appears to be a requirement that might be satisfied or violated by large margins in different planetary rings. However, Goldreich and Tremaine (1982) have summarized why there are good reasons for supposing that the viscosity in planetary rings is not very different from the monolayer value (Eq. 2a). If, moreover, the normal optical depth τ is of order unity (and in a monolayer, it cannot greatly exceed unity), if the characteristic size R is given by Eq. (1), and if $|\mathcal{D}| \sim \Omega\kappa$, then we have

$$R \sim \epsilon r \text{ and } 7\nu/3 + \zeta \sim \epsilon^2 r^2 \Omega ,$$

which yields for Eq. (63)

$$x_{\text{vis}} \sim |\epsilon|^{\frac{1}{3}} .$$

For $|\epsilon|$ very small x_{vis} would then be appreciably (but not overwhelmingly) larger than $|\epsilon|^{\frac{1}{2}}$. In Saturn, where $|\epsilon|$ is typically 10^{-8}, we expect to see on the order of tens of wave crests before the viscous damping kills the waves. This expectation is roughly borne out by the actual observations, suggesting that the viscosity is, indeed, not many orders of magnitude larger than its monolayer value. More precise estimates will be discussed in Sec. IV.

Here, we merely summarize that Eq. (60) can be used to determine the surface density σ_0 of planetary rings from measurements of the observed wavelength behavior of spiral density waves and Eq. (62), to determine the viscosity combination $\nu + 3\zeta/7$ from measurements of the amplitude behavior (see Cuzzi et al. 1981; Lane et al. 1982; Holberg et al. 1982; Lissauer et al. 1982). We move now to consider the theory of bending waves.

B. Spiral Bending Waves

Since bending waves and long density waves have many analogous properties, we shall develop the theory of bending waves using much of the same notation that we have introduced for density waves. Apart from a substitution of μ for κ, r_V for r_L, and G for $-G$, the dispersion relations for bending waves and long density waves turn out to be identical (cf. Eqs. 35 and 84). The frequency μ replaces κ because μ is the natural frequency of free vertical oscillations; G replaces $-G$ because the resulting self-gravity in bending motions is in the opposite direction as the displacement, whereas it is in the same sense as the displacement for horizontal compressions. Other symbols, such as D, have different meanings in this section (III.B) from their meanings in the section on density waves (III.A). The usage is closely analogous, however. With these preliminary comments, there should be no confusion.

1. Basic Equations. Consider the small vertical displacement $Z(r,\vartheta,t)$ from the equilibrium plane $z = 0$ of an otherwise infinitesimally thin axisymmetric disk of surface density $\sigma(r)$. If the noncircular motions in the plane of the disk are ignored, the vertical displacement satisfies the equation of motion (Hunter and Toomre 1969):

$$\left(\frac{\partial}{\partial t} + \Omega \, \frac{\partial}{\partial v} \right)^2 Z = g_z , \tag{64}$$

where g_z is local gravitational acceleration in the vertical direction. In the present context, g_z has three contributions

$$g_z = g_P + g_M + g_D , \tag{65}$$

where g_P is from the planet, g_M is from some satellite, and g_D is from the self-gravity of a bent disk.

For $|Z| \ll r$, the term g_P has the form of the restoring force due to a linear harmonic oscillator:

$$g_P = -\mu^2 Z . \tag{66}$$

With regard to g_M, we concentrate on one Fourier component (Eq. 9) of the satellite's gravitational potential:

$$g_M = \mathrm{Re}\left[f_M(r)e^{i(\omega t - m\vartheta')}\right] , \tag{67a}$$

$$f_M(r) = -\frac{\partial \Phi_M}{\partial z}(r,0) , \tag{67b}$$

where the evaluation of the right-hand side at $z = 0$ instead of at $z = Z(r,\vartheta,t)$ implies that we restrict ourselves to a linear theory. The gravitational attraction of a warped disk on itself at a field point (r,ϑ,Z,t) is given by

$$g_D = -G\int r'dr'\int d\vartheta' \frac{\sigma(r')\,(Z-Z')}{[r^2+r'^2-2rr'\cos(\vartheta-\vartheta')+(Z-Z')^2]^{\frac{3}{2}}} \tag{68}$$

where Z refers to $Z(r,\vartheta,t)$ and Z' means $Z(r',\vartheta',t)$. In the linear approximation, we expect the vertical response to the forcing (Eq. 67) to take the form

$$Z(r,\vartheta,t) = \mathrm{Re}[h(r)e^{i(\omega t - m\vartheta)}] . \tag{69}$$

The linearized form of Eq. (68) then becomes

$$g_D = \mathrm{Re}[f_D(r)e^{i(\omega t - m\vartheta)}] , \tag{70a}$$

$$f_D(r) = G\int \sigma(r')r'dr' \int_{-\pi}^{+\pi} d\psi \frac{[h(r')\cos(m\psi)-h(r)]}{(r'^2+r^2-2rr'\cos\psi)^{\frac{3}{2}}} . \tag{70b}$$

With the above development, Eq. (64) becomes the following inhomogeneous integral equation to solve for $h(r)$ (Hunter and Toomre 1969):

$$(\Lambda_D + D)h(r) = f_M(r) , \tag{71}$$

where D is the quantity

$$D = \mu^2 - (\omega - m\Omega)^2 , \tag{72}$$

and where the linear integral operator Λ_D is defined by $\Lambda_D h(r) \equiv -f_D(r)$, with $f_D(r)$ given by Eq. (70b). With the help of the half-angle formula,

$\cos(m\psi) = 1-2\sin^2(m\psi/2)$, we may write $\Lambda_D h(r)$ in terms of its axisymmetric and nonaxisymmetric contributions:

$$\Lambda_D h\,(r) = G\int\Big\{K_0\,(r,r')\,[h\,(r)-h\,(r')\,]+N_m\,(r,r')\,h\,(r')\Big\}\sigma\,(r')\,r'\mathrm{d}r'\;, \qquad (73)$$

where

$$K_0(r,r') = \int\limits_{-\pi}^{+\pi}\frac{\mathrm{d}\psi}{(r^2+r'^2-2rr'\cos\psi)^{\frac{3}{2}}} \qquad (74a)$$

$$N_m(r,r') = \int\limits_{-\pi}^{+\pi}\frac{2\sin^2(m\psi/2)\mathrm{d}\psi}{(r^2+r'^2-2rr'\cos\psi)^{\frac{3}{2}}}\;. \qquad (74b)$$

It is possible to express K_0 in terms of the Laplace coefficient $b_{3/2}^{(o)}$, and N_m in terms of $b_{3/2}^{(o)}-b_{3/2}^{(m)}$, familiar to celestial mechanicians (Brouwer and Clemence 1961; Appendix A), but we shall not do so here. What is more important for us is to note that, taken alone, $N_m(r,r')$ is integrable near $r'=r$ while $K_0(r,r')$ is not:

$$K_0(r,r') \approx \frac{2}{r'|r-r'|^2} \quad \text{for } r' \approx r, \qquad (75a)$$

$$N_m(r,r') \approx \frac{m^2}{r^3}\ln\left[\frac{r}{|r-r'|}\right] \quad \text{for } r' \approx r\;. \qquad (75b)$$

Equation (71) is the basic equation that governs the linear dynamics of forced bending motions. In what follows, we take advantage of the properties expressed in Eqs. (75a,b) to develop an asymptotic technique to turn the integral Eq. (71) into a simple differential equation.

2. Asymptotic Theory. We are interested in solutions where the radial part $h(r)$ of the vertical displacement is a rapidly varying function, with approximately equal numbers of positive and negative excursions for both its real and imaginary parts. Thus, we expect that the contribution of the term $N_m(r,r')h(r')$ to the integral (Eq. 73) will largely cancel out; moreover, we expect that the largest contribution from the term $K_0(r,r')[h(r)-h(r')]$ will be due to local effects. Using this reasoning to approximate $K_0(r,r')$ by Eq. (75a), we now obtain

$$\Lambda_D h(r) = 2GP\int\frac{[h(r)-h(r')]}{(r'-r)^2}\,\sigma(r')\mathrm{d}r' \qquad (76)$$

where the P in front of the integral means that we are to take its principal value. If one does not like this *ad hoc* mathematical procedure, one can introduce the physical argument that completely flattened disks have apparent singularities not relevant to real disks of finite thickness. For a disk of equivalent thickness $2z_0$ (see Eq. 38), one would then heuristically replace Eq. (76) by

$$\Lambda_D h(r) = 2G \int \frac{[h(r) - h(r')]}{[(r'-r)^2+z_0^2]} \sigma(r')dr' \ .$$

What we mean by Eq. (76) is, then, the mathematical idealization in which we evaluate the above in the limit $z_0 \to 0$.

There is a simple physical interpretation to Eq. (76). For vertical displacements of radial scale much smaller than the azimuthal scale, we can approximate each local piece of the disk as a straight wire in the tangential direction (see Borderies et al. 1982). For a straight wire of density per unit length $\sigma(r')dr'$, at a height $h(r')$, the gravitational field directed perpendicular toward the wire at a distance $d = [(r - r')^2+(h - h')^2]^{\frac{1}{2}}$ is given by Gauss's law as $2G\sigma(r')dr'/d$. The z component is smaller by a factor $(h-h')/d$, which, for $|h-h'|$ much less than d (the linear approximation), yields Eq. (76).

For disturbances where $h(r')$ varies much more rapidly than $\sigma(r')$, we may pull $\sigma(r')$ out of the integral (Eq. 76) as $\sigma(r)$. We may then integrate once by parts to obtain

$$\Lambda_D h(r) = - 2G\sigma(r) P \int \frac{dh/dr'}{r'-r} \ dr' \ . \tag{77}$$

The quickest and most elegant route to the desired result (Eq. 80) makes use of residue calculus, an idea suggested to me by S. Balbus (personal communication). We analytically continue the function dh/dr' into the entire complex plane $z = x + iy$, defining it to be zero on the negative real axis. For r real and positive, the integral

$$\int_0^\infty \frac{dh/dr'}{r'-r} \ dr' = \int_{-\infty}^\infty \frac{dh/dz}{z-r} \ dz \tag{78}$$

may be evaluated by considering a contour which detours in a small semicircle about the simple pole at $z = r$ on the real line, and which closes in the upper or lower half plane depending on where dh/dz is exponentially decaying. Since we are interested in bending motions which asymptotically acquire the WKBJ form (cf. Eq. 32)

$$h(r) = A(r) \exp\left[i \int kdr\right], \tag{79}$$

we should close the contour in the upper half plane if k is positive, and in the lower half plane if k is negative. Residue calculus then provides the result,

$$\Lambda_D h(r) = i2\pi\, G\,\sigma(r)s\,\frac{dh}{dr}\,(r)\,,\tag{80}$$

where again we use the notation

$$s \equiv \mathrm{sign}\,(k)\,.\tag{81}$$

For $h(r)$ given by Eq. (79), the self-gravity of a sinusoidally bent disk produces a restoring force of radial variation (Hunter and Toomre 1969; Bertin and Mark 1980):

$$f_D(r) \equiv -\Lambda_D h(r) \approx -2\pi\, G\,\sigma(r)|k|h(r).\tag{82}$$

With the substitution of Eq. (80) into Eq. (71), we obtain the central differential equation for forced bending waves (Shu et al. 1983):

$$-i2\pi\, G\,\sigma s\,\frac{dh}{dr}+Dh = f_M\,.\tag{83}$$

The above bears a strong resemblance to the first-order ordinary differential Eq. (31), especially when we remember that the right-hand sides of both equations are known inhomogeneous terms. The formal mathematical relationship between long density waves and bending waves is thereby established.

2a. Free Spiral Bending Waves. Ignore, for the time being, the inhomogeneous term f_M in Eq. (83), and consider free solutions of the WKBJ form (Eq. 79). The substitution of Eq. (78) into the homogeneous version of Eq. (83) then yields, to lowest asymptotic order,

$$(\omega-m\Omega)^2 = \mu^2+2\pi\, G\,\sigma|k|\,,\tag{84}$$

which is the desired dispersion relationship (Hunter and Toomre 1969, pp. 760–761; see also Eq. (D7) of Bertin and Mark 1980).

Equation (84) may be directly compared with the dispersion relationship for long density waves, Eq. (35). We see, as advertised, that the two relationships are identical except for the substitution of μ for κ and G for $-G$. Unlike density waves, bending waves like to oscillate at a Doppler-shifted frequency which is faster than the natural frequency, because the force of self-gravity in bending motion opposes the direction of the bending distortion (see Eq. 82),

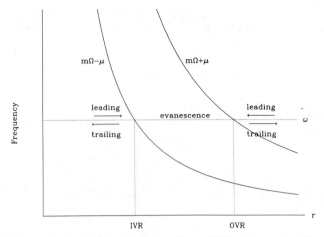

Fig. 6. Schematic drawing showing that trailing spiral bending waves of frequency ω and angular mode number m propagate away from inner and outer vertical resonances, while leading spiral bending waves propagate toward them. The region between the inner and outer vertical resonance is a region of evanescence.

making the disk return toward the equilibrium position faster than if we had ignored the self-gravity.

Because of the difference in sign of the restoring force, bending waves have the opposite sense of propagation as long density waves. Thus, bending waves can exist only in the range outside the two vertical resonances (where $D < 0$); in the region between the inner and outer vertical resonances (where $D > 0$), bending waves are evanescent (see Fig. 6). The group velocity of bending waves is given by

$$c_g = - \left[\frac{\partial \omega}{\partial k} \right]_\tau = - \frac{\pi G \sigma}{(\omega - m\Omega)} s \, . \tag{85}$$

Trailing bending waves ($s = -1$) propagate inward from inner vertical resonance (where $\omega - m\,\Omega = -\mu$) and outward from outer vertical resonance (where $\omega - m\,\Omega = +\mu$). Leading bending waves ($s = +1$) propagate toward vertical resonances, not from them; therefore they cannot be excited at vertical resonances without entering a condition of evanescence. In particular, we might expect that exterior satellites whose orbits are inclined with respect to the mean ring plane (the equatorial plane of the planet) would excite trailing spiral bending waves that travel inward from the inner resonant location in the planetary disk. In particular, since the satellite Mimas in the Saturn system has both an inclined and an eccentric orbit, we might expect to see that spiral bending waves and spiral density waves it excites come in pairs, with the degeneracy in resonance location being split by the oblateness of Saturn (see

Fig. 3), and with the rest of wavetrains not interfering with each other because trailing bending waves and trailing density waves have opposite senses of propagation. This is indeed the observed situation (Shu et al. 1983).

2b. *Forced Spiral Bending Waves.* To calculate how bending waves are excited near vertical resonances, let us assume again that the fractional distance from resonance

$$x \equiv (r - r_V)/r_V \tag{86}$$

is small, and expand the coefficients in Eq. (83) in a Taylor series about $r = r_V$. We define

$$\mathcal{D} \equiv \left[r \frac{dD}{dr} \right]_{r_V}, \tag{87a}$$

$$\epsilon \equiv \left[\frac{2\pi G\sigma}{r \, \mathcal{D}} \right]_{r_V}, \tag{87b}$$

$$L_V \equiv \frac{f_M(r_V)}{\mathcal{D}} . \tag{87c}$$

The quantities $\mathcal{D}$ and ϵ are positive when r_V refers to an inner vertical resonance and negative when r_V refers to an outer vertical resonance.

With these definitions (Eqs. 87a,b,c), a Taylor series expansion of the coefficients of Eq. (83) yields

$$i\epsilon \frac{dh}{dx} + xh = L_V . \tag{88}$$

The solution to Eq. (88) may be written as

$$h = A_V H_{-q}(\xi) , \tag{89}$$

where we have defined

$$q \equiv \text{sign} (\epsilon) , \tag{90a}$$

$$\xi \equiv - qx/(2|\epsilon|)^{\frac{1}{2}} , \tag{90b}$$

$$A_V \equiv \left(\frac{2\pi}{|\epsilon|} \right)^{\frac{1}{2}} iL_V , \tag{90c}$$

and where the functions $H_{\pm}(\xi)$ are defined by Eqs. (47c) and (48).

For ξ large and negative (far into the region of evanescence),

$$h \to \frac{L_V}{x} \text{ as } \xi \to -\infty , \tag{91}$$

which corresponds to the nonwavy part of the disk response. For ξ large and positive (far into the region of wave propagation),

$$h \to A_V \exp[i(x^2/2\epsilon - q\pi/4)] + \frac{L_V}{x} \text{ as } \xi \to +\infty . \tag{92}$$

The first term represents a trailing spiral bending wave with amplitude $|A_V|$ and phase (cf. Eq. 78):

$$\int_{\tau_L}^{\tau} k\,dr + q\,\frac{\pi}{4} + \arg[f_M(r_V)] . \tag{93}$$

The second term in Eq. (92) is again the nonwavy part of the disk response.

The maximum slope $|dh/dr|$ of the disk reached locally in the near wave-zone is given by $|kA_V|$. Because Eq. (84) implies that $|k| \approx -x/\epsilon r_V$ near resonance, the maximum local slope has a value in the near wave-zone

$$|kA_V| = \left[\frac{2\pi}{|\epsilon|^3} \right]^{\frac{1}{2}} \left[\frac{|L_V|}{r_V} \right] |x| . \tag{94}$$

Since $L_V/r_V \propto (M/M_p)\sin i_M$ for the strongest vertical resonances, the small inclination of a typical satellite's orbit helps to keep bending waves from becoming as nonlinear as the density waves excited by the strongest horizontal resonances (cf. Eqs. 54 and 94). Indeed, viscous dissipation of the waves as they propagate implies that even the very strongest bending wave observed in Saturn's rings (associated with the 5:3 vertical resonance of Mimas) never becomes highly nonlinear (Lissauer et al. 1982).

2c. Viscous Damping of Bending Waves. In the presence of viscous shear stresses, we need to add to the right-hand side of Eq. (64) the term $\nu\partial^2 w/\partial r^2$ where $w \approx \partial Z/\partial t + \Omega\partial Z/\partial\vartheta$ is the vertical component of the fluid velocity (cf. Eq. 56). With the WKBJ form (Eq. 79), and $s = \text{sign}[\text{Re}(k)]$, this results for the viscous but homogeneous version of Eq. (83):

$$2\pi G\sigma sk + D = -iv(\omega - m\Omega)k^2 . \tag{95}$$

Writing $k = k_R + ik_I$, we obtain for small ν (Shu et al. 1983):

$$|k_R| = -D/2\pi G\sigma \tag{96a}$$

$$k_I = -\nu\,\frac{(\omega-m\Omega)D^2}{(2\pi G\sigma)^3}\ . \tag{96b}$$

As the bending wave propagates (over distances $|x| \ll 1$) the amplitude suffers viscous damping by a factor

$$\exp\left(-\int_{\tau_L}^{\tau} k_I\,dr\right) \approx \exp\left[\left(-\frac{|x|}{x_{\text{vis}}}\right)^3\right] \tag{97}$$

where the dimensionless viscous damping length is defined by

$$x_{\text{vis}} \equiv |\epsilon|\left(\frac{3|\mathcal{D}|r^2}{\mu\nu}\right)^{\frac{1}{3}}\ . \tag{98}$$

This expression should be compared with Eq. (63), which is appropriate for the viscous damping length of density waves. Notice in particular that because linear bending waves involve no compression, the bulk viscosity does not enter into Eq. (98). For this reason, among others, the damping of bending waves allows a cleaner analysis of the kinematic viscosity ν in Saturn's rings than the damping of density waves.

IV. WAVES AS DIAGNOSTICS

So far, our interest has centered on developing the theory of collective waves in planetary rings as a physical phenomenon. Since the processes are relatively well understood now, another viewpoint is possible, to apply the theory to observed waves so as to obtain information about the physical properties of the rings and their environment. At the time this chapter was written, the only set of rings observed to have either density or bending waves is the Saturn system. Thus, our use of waves as diagnostics is necessarily confined to Saturn's rings. The discussion which follows is merely a brief summary of the results; for details, the reader is referred to the chapter by Cuzzi et al. in this book and to the references cited therein.

A. Surface Densities of Ring Systems

The first density wavetrain discovered in Saturn's rings was associated with the apsidal ($m = 1$ inner Lindblad) resonance of Iapetus. Because the quantity $\mathcal{D}$ of Eq. (43) is especially small for an apsidal resonance, the separations of wavecrests and troughs near resonance (cf. Eq. 60) were long enough to be resolved on Voyager 1 images of the unlit face of Cassini's Division (see Fig. 1 of Cuzzi et al. 1981). The crest-to-crest and trough-to-trough spacings of the brightness undulations, seen in diffuse transmission as a function of radial distance from exact resonance, were observed to satisfy a $1/x$ relationship, allowing, from the proportionality factor in Eq. (60), a determination of the surface mass density σ_0 in the region of wave propagation. The derived surface density in the outer part of the Cassini Division turned out to be 16 g cm^{-2}.

The same procedure was followed to reduce the stellar occultation observations of a set of (nonlinear) density waves excited in the inner B Ring by a 2:1 resonance with Saturn's coorbital satellite 1980S1 (Lane et al. 1982). The derived surface density, $\sigma_0 = 60$ g cm^{-2}, was in good agreement with the Cassini Division measurement in that the ratio of normal optical depth τ in this part of the B Ring is observed to be a factor of 4 higher than the optical depth in the outer Cassini Division.

Measurements of many more examples have confirmed the rough constancy of the ratio τ/σ_0 in the regions of density-wave propagation (Holberg et al. 1982). From such studies, the view has developed that the transport of angular momentum and material by density waves in Saturn's rings—which Goldreich and Tremaine (1978b) had anticipated would open up a gap (Cassini Division) in the strongest case—has, in somewhat weaker situations, actually enhanced the optical depth of the propagation regions just outside the inner Lindblad resonances. On the order of fifty density wavetrains have now been detected and identified, and measurements of their wavelength behaviors has allowed the derivation of a surface density model of Saturn's rings (Esposito et al. 1982). The total mass of Saturn's rings is found in this manner to be $\sim 5 \times 10^{-8}$ that of the planet.

In a similar way, the surface mass density in the neighborhood of the bending waves excited by 5:3 and 8:5 vertical resonances of Mimas has been analyzed via Eq. (96) (Shu et al. 1982). In the middle A Ring, $\sigma_0 \approx 5$ g cm^{-2}. In other words, if all the material there were condensed out as solid ice, it would form a layer ~ 45 cm thick. In fact, we shall see from Sec. IV.B below that there is reason to suspect that the actual local scale height is ~ 3000 cm, implying that the ring particles are not densely packed.

The analysis of the Mimas 5:3 bending-wave profiles was especially interesting because absolute phases (see Eq. 93) could be found. The observed wave profiles varied as theoretically predicted with ring longitude (relative to Mimas's location) at a given radius. The good fit with theory demonstrated conclusively that the observed brightness variations did indeed have a tightly-wrapped spiral waveform, and were not merely a series of alternating circular ringlets with an appropriate radial distribution to mimic the behavior of collective waves.

B. Viscosities in Ring Systems

There have been two attempts to measure the viscosity in Saturn's rings by using the formulae (Eqs. 62 and 63) for frictional damping of spiral density waves. Analysis of the amplitude behavior of the density waves driven by apsidal resonance with Iapetus led to an estimate of $\nu + 3\zeta/7 = 170$ cm^2 s^{-1} for the outer Cassini Division (Cuzzi et al. 1981). A similar analysis of the density waves driven by the 2:1 resonance with the coorbital satellite 1980S1 gave $\nu + 3\zeta/7 = 20$ cm^2 s^{-1} (Lane et al. 1982). The latter value is close to the monolayer viscosity predicted by Eq. (2a). Unfortunately, the first measurement is subject to uncertainty because some or all of the observed amplitude decay might have been caused by wave propagation up an increasing density profile, and the second is doubtful because the linear theory of waves was applied to a distinctly nonlinear context (see Lissauer et al. [1982] for a more detailed discussion). If the "monolayer" value holds up for the B Ring under a more rigorous analysis of the B Ring density waves, it would cast severe doubt on any simple form of the "viscous instability" idea for ringlet structure, because "monolayer viscosity" has the wrong dependence on optical depth for the proposed instability to proceed (cf. chapter by Stewart et al.).

A more reliable estimate for the viscosity has been extracted from an analysis of the 5:3 bending waves excited by Mimas (Fig. 7). The sag of the overall brightness of the bending wave region results from shadowing, because the local slopes of the wave profiles in these regions exceed the critical value 0.20 needed to produce shadows for the given Sun-ring geometry (Shu et al. 1983). The important observation is that the general sag is seen to disappear at a distance of ~ 200 km from exact resonance. Thus, although individual wave crests and troughs cannot be resolved at such locations, we may infer from the lack of shadowing that the local slope cannot exceed 0.20 anywhere in these regions. But the inviscid theory says that the maximum value of the local slope produced by bending waves should increase indefinitely as the waves propagate inward from inner vertical resonance (see Eq. 94). To avoid a contradiction, it seems reasonable to attribute the required reduction of wave amplitude to viscous damping.

If we assume various values for the kinematic viscosity ν, Eqs. (97) and (98) predict the theoretical behavior indicated in Fig. 8. For shadowing to end somewhat < 200 km inside exact resonance, the kinematic viscosity must be roughly $\nu = 260$ cm^2 s^{-1}. For such a value of ν, we expect that shadowing begins at ~ 50 km inside resonance, and that maximum slopes of ~ 0.34 (slope angles of $\sim 19°$) are reached over a relatively broad range of radii. The first expectation is realized by a detailed analysis of the observed wave profiles; the second, by an examination (Shu et al. 1982) of the ring longitude of brightness contrast-reversal seen in wide-angle Voyager 1 shots of the bending wave region (Smith et al. 1981). The basic idea behind contrast-reversal is that the bending wave regions look dark relative to their surroundings when

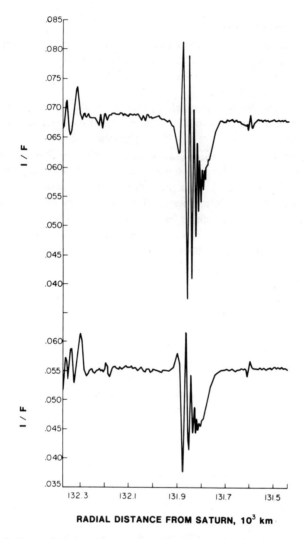

Fig. 7. Two radial scans of reflectivity I/F (geometric albedo) for the 5:3 Mimas bending wave. Exact resonance occurs at 131,900 km from the center of Saturn (from Shu et al. 1982).

the Sun shines across the furrows (producing shadows), while the same regions look bright when the Sun shines along the furrows.

The value $\nu = 260$ cm^2 s^{-1} is an order of magnitude larger than what we would expect for the "monolayer viscosity" of Eq. (2a). If we suppose ν is given by Eq. (2b), the implied random velocity c is ~ 0.4 cm s^{-1}. The corresponding scale height c/Ω is 30 m, appreciably greater than the radii R of the largest numerous ring particles. Why does c take on an unexpectedly large

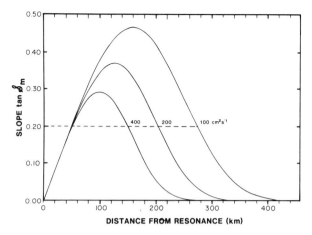

Fig. 8. The theoretical value of the slope where the ring plane is most bent as a function of radial distance from the 5:3 vertical resonance of Mimas. The curves are drawn for kinematic viscosites ν taken to be 400, 200, and 100 cm^2 s^{-1}. Values of the slope > 0.20 correspond to shadowed regions in Fig. 7 (from Lissauer et al. 1982).

value in the A Ring? The answer seems to be the input of kinetic energy from numerous density waves driven by resonances with known satellites. If this input is balanced in a steady state by dissipation in inelastic collisions, then we expect a mean random velocity of ~ 0.5 cm s^{-1} in the middle of the A Ring (Lissauer et al. 1982). Given the crudeness of these estimates, the latter value is in excellent agreement with the value of c derived from our viscosity measurement.

C. Masses of Satellites Around Rings

The absolute amplitudes of driven density waves or bending waves depends on the mass of the satellite which is doing the forcing (cf. Appendix A). Excellent agreement exists between the observed and predicted amplitudes of the bending waves excited by Mimas, whose mass is known with fair accuracy from celestial mechanics (Shu et al. 1983). In particular, the full up-and-down displacement of 1.4 km near the 5:3 vertical resonance of Mimas probably explains the groundbased measurements of 1 to 2 km for the apparent edge-on thickness of Saturn's rings (Brahic and Sicardy 1981). This indicates that the theory of forced waves developed so far has no glaring deficiencies. Thus, in principle one could turn the procedure around and use the measured amplitudes of density waves or bending waves to derive gravitational values for satellites whose masses cannot be directly measured. Unfortunately, density waves are by far the more numerous of the waves excited in Saturn's rings, and the analysis of their amplitude information is incomplete both theoretically and observationally. Theoretically, we face the issue that the most easily observed waves have large amplitudes, so there is an urgent need to develop the theory of forced, self-consistent, nonlinear density waves in the presence of viscous dissipation. Observationally, we face the challenge of reliably transforming measured brightness variations to inferred surface density variations. Both of these obstacles can be surmounted, and we look forward to the use of wave diagnostics for satellite mass determinations in the near future.

Acknowledgments. It is a pleasure to thank J. Bahcall and the Institute for Advanced Study at Princeton for support and hospitality during the period of the writing of this review while I was on sabbatical leave from the University of California at Berkeley. Discussions with S. Balbus, J. Cuzzi, P. Goldreich, C. C. Lin, J. Lissauer, J. Mark, A. Toomre, and S. Tremaine have been most enlightening. The research associated with this work was funded in part by consortium grants from the NASA Ames Research Center and a grant from NSF.

APPENDIX A — CALCULATION OF FORCING STRENGTHS

In this appendix we discuss the technique used to calculate the horizontal and vertical forcing strengths at resonances. We begin by noting that we need to calculate the various Fourier coefficients of Eq. (8) and its z derivative in the plane $z = 0$; i.e., we need to consider how to Fourier decompose

$$- GM \left\{ r_M^2 (t) + r^2 - 2r_M (t)r \cos[\vartheta_M (t) - \vartheta] + z_M^2 (t) \right\}^{-\frac{1}{2}} , \qquad (A1a)$$

$$GM z_M (t) \left\{ r_M^2 (t) + r^2 - 2r_M (t)r \cos[\vartheta_M (t) - \vartheta] + z_M^2 (t) \right\}^{-\frac{3}{2}} . \qquad (A1b)$$

For small departures from circular motion with radius a_M and circular speed Ω_M, we may expand Eqs. (A1a and b) in a power series in the small quantities

$$R_M (t) \equiv r_M (t) - a_M = -e_M a_M \cos[\kappa_M (t - t_0)] , \qquad (A2a)$$

$$\Theta_M (t) \equiv \vartheta_M (t) - \Omega_M t = e_M \frac{2\Omega_M}{\kappa_M} \sin [\kappa_M (t - t_0)] , \qquad (A2b)$$

$$Z_M (t) \equiv z_M (t) = a_M \sin (i_M) \sin (\mu_M t), \qquad (A2c)$$

where the last expressions in Eqs. (A2a,b,c) correspond to Lindblad's theory of epicyclic motion (see e.g., Chandrasekhar 1942). The quantities e_M and i_M are the eccentricity and inclination of the satellite's orbit, and are assumed to be small. In Eqs. (A2a,b,c), we have defined $t = 0$ to be the instant of ring-plane crossing (ascending node) and $t = t_0$ to be the instant of periapse passage of the satellite M. To first order in the departure from circular motion, the expansion of expressions (A1a,b) leads to the expressions

$$-\frac{GM}{\Delta^{\frac{1}{2}}} + \frac{GM}{\Delta^{\frac{3}{2}}}\left\{R_M[a_M - r\cos(\Omega_M t - \vartheta)] + \Theta_M a_M r\sin(\Omega_M t - \vartheta)]\right\} , \quad \text{(A3a)}$$

$$\frac{GMZ_M}{\Delta^{\frac{3}{2}}} - \frac{3GMZ_M}{\Delta^{\frac{5}{2}}}\left\{R_M[a_M - r\cos(\Omega_M t - \vartheta)]\right.$$
$$\left. + \Theta_M a_M r\sin(\Omega_M t - \vartheta)]\right\} , \quad \text{(A3b)}$$

where we have defined

$$\Delta \equiv a_M^2 + r^2 - 2a_M r\cos(\Omega_M t - \vartheta) . \quad \text{(A4)}$$

Since $\Delta^{-\frac{1}{2}}$, $\Delta^{-\frac{3}{2}}$, $\Delta^{-\frac{5}{2}}$, are periodic even functions of $(\Omega_M t - \vartheta)$, they can be Fourier analyzed as cosine series in this variable

$$\frac{a_M}{\Delta^{\frac{1}{2}}} = \frac{1}{2}b_{\frac{1}{2}}^{(0)} + \sum_{m=1}^{\infty} b_{\frac{1}{2}}^{(m)}\cos[m(\Omega_M t - \vartheta)] , \quad \text{(A5a)}$$

$$\frac{a_M^3}{\Delta^{\frac{3}{2}}} = \frac{1}{2}b_{\frac{3}{2}}^{(0)} + \sum_{m=1}^{\infty} b_{\frac{3}{2}}^{(m)}\cos[m(\Omega_M t - \vartheta)] , \quad \text{(A5b)}$$

$$\frac{a_M^5}{\Delta^{\frac{5}{2}}} = \frac{1}{2}b_{\frac{5}{2}}^{(0)} + \sum_{m=1}^{\infty} b_{\frac{5}{2}}^{(m)}\cos[m(\Omega_M t - \vartheta)] , \quad \text{(A5c)}$$

where the b's are defined by the integrals

$$b_\gamma^{(m)}(\beta) \equiv \frac{2}{\pi}\int_0^{\pi} \Gamma^{-\gamma}\cos(m\psi)\,d\psi, m = 0,1,2,\dots \quad \text{(A6a)}$$

$$\beta \equiv r/a_M, \quad \text{(A6b)}$$

$$\Gamma \equiv 1 + \beta^2 - 2\beta\cos\psi . \quad \text{(A6c)}$$

These b's are called Laplace coefficients in celestial mechanics (Brouwer and Clemence 1961). Their derivatives have the properties

$$\frac{db_\gamma^{(m)}}{d\beta} = \gamma[-2\beta b_{\gamma+_1}^{(m)} + b_{\gamma+_1}^{(m+1)} + b_{\gamma+_1}^{(m-1)}], m = 0,1,2,\ldots \tag{A7}$$

with $b_\gamma^{(-1)} = b_\gamma^{(1)}$. Given r_L/a_M or r_V/a_M, it is straightforward to compute numerically all the necessary b's that eventually enter in Eqs. (45a) and (87c).

For reference, we note some useful relationships that may be derived as follows. Integration by parts of Eq. (A6a), plus the use of the trignometric identity

$$2\sin(m\psi)\sin\psi = \cos[(m-1)\psi] - \cos[(m+1)\psi],$$

yields

$$b_\gamma^{(m)} = \frac{\gamma}{m}\beta[b_{\gamma+_1}^{(m-1)} - b_{\gamma+_1}^{(m+1)}], m = 1,2,3,\ldots \tag{A8}$$

On the other hand, substituting the identity

$$\Gamma^{-\gamma} = (1 + \beta^2 - 2\beta\cos\psi)\Gamma^{-(\gamma+1)}$$

into Eq. (A6a) gives

$$b_\gamma^{(m)} = (1+\beta^2)b_{\gamma+_1}^{(m)} - \beta[b_{\gamma+_1}^{(m+1)} + b_{\gamma+_1}^{(m-1)}], m = 0,1,2,\ldots \tag{A9}$$

The three Eqs. (A7), (A8), and (A9) allow us to express $b_{\gamma+_1}^{(m-1)}$, $b_{\gamma+_1}^{(m)}$, and $b_{\gamma+_1}^{(m+1)}$ in terms of $b_\gamma^{(m)}$ and its derivative.

Putting Eqs. (A5) and (A2) into expressions (A3), we see that $\varphi_M(r,\vartheta,0,t)$ can be written as a series of terms of the form

$$\text{Re}[\Phi_M(r,0)\,e^{i(\omega t - m\vartheta)}]. \tag{A10}$$

while $-\partial\varphi_M/\partial z$ evaluated at $(r,\vartheta,0,t)$ can be expressed as a series of terms of the form

$$\text{Re}[f_M(r)\,e^{i(\omega t - m\vartheta)}], \tag{A11}$$

where ω is given by Eq. (10).

The strongest horizontal resonances are associated with $\omega = m\Omega_M$,

$$\Phi_M(r,0) = -\frac{GM}{a_M}b_{\frac{1}{2}}^{(m)}, \tag{A12}$$

which holds for all m except $m = 0$ and $m = 1$. For $m = 0$, the expression (A12) needs a factor of ½, but this is a pedantic point since the resonance condition (Eq. 11) cannot be satisfied anywhere in the ring system for $\omega = 0$ and $m = 0$. The axisymmetric part of the satellite's potential can contribute only to changing slightly the circular frequency Ω and the epicyclic frequency κ. These changes are effected through Eqs. (18) and (4a), where we take

$$\varphi_{M0}(r,0) = -\sum \frac{GM}{2a_M} b_{\frac{1}{2}}^{(o)} , \qquad (A13)$$

where the sum is taken over all satellites. The satellite system aggravates the precession of the periapses of ring particles, and this influences accurate calculations of the locations of apsidal resonances. Equation (A13) implies that the contribution of each satellite to this effect is as if it were a circular wire of radius a_M and mass per unit length $M/2\pi a_M$.

For $m = 1$, we need to include the indirect contribution of the planet's motion to the forcing at apsidal resonances. The instantaneous acceleration felt by the planet due to the attraction of the satellite is $-M/M_P$, times the acceleration $-\nabla\varphi_P(\mathbf{r}_M)$ felt by the satellite at its vector position r_M. By the principle of equivalence, this instantaneous acceleration of our coordinate frame is equivalent to the action of a gravitational field

$$\mathbf{g} = - \frac{M}{M_P} \nabla\varphi_p(\mathbf{r}_M) .$$

But the above gravitational field is independent of location $\mathbf{r}$ and is derivable as the negative gradient of the (indirect) potential

$$\frac{M}{M_P} \mathbf{r} \cdot \nabla\varphi_P (\mathbf{r}_M) . \qquad (A14)$$

If we use Eq. (3), then to linear order in M/M_P, the above expression in the plane $z = 0$ has the form

$$\frac{M}{M_P} rr_M \Omega^2 (r_M) \cos[\vartheta_M - \vartheta] , \qquad (A15)$$

which contributes only to $m = 1$ (**apsidal resonance**). Note that Goldreich and Tremaine (1978b, 1980) have misprints in their starting expressions for the indirect potential.

The physical interpretation of expression (A15) is simple in the case of a circular orbit. For a circular orbit of radius a_M and angular speed $\Omega_M \equiv \Omega(a_M)$, we may go into the frame which rotates with speed Ω_M. If we define ψ

$\equiv \vartheta - \Omega_M t$, apart from the stationary potentials of the planet and the satellite, there will also be a centrifugal potential of form (cf. the discussion of Lubow and Shu [1975] on accretion disks in close binary stars, which also uses a coordinate system centered on one of the massive bodies):

$$-\frac{1}{2} \Omega_M^{2}(r^2 + r_{CM}^{2} - 2rr_{CM}\cos\psi) ,$$

where r_{CM} is the radial position of the center of mass. Only the nonaxisymmetric part of the above contributes to the indirect potential in an inertial frame. But to linear order in M/M_P and for a circular orbit, $r_{CM} = (M/M_P)a_M$, $\vartheta_M = \Omega_M t$, and the nonaxisymmetric part of the above is identical to expression (A15). It is now easy to generalize Eq. (A12) to take into account the above comments. Thus, for $\omega = m\Omega_M$

$$\Phi_M(r,0) = -\frac{GM}{(1+\delta_{m0})a_M} [b_{\frac{1}{2}}^{(m)} - \delta_{m1}f\beta], \quad m = 0,1,2\ldots . \tag{A16}$$

where δ_{mn} is the Kronecker delta and where $f \equiv \Omega_M^{2}\alpha_M^{3}/GM_p$ is unity in the Keplerian approximation.

The next strongest horizontal resonances are associated with $\omega = m\Omega_M \pm \kappa_M$. For $\omega = m\Omega_M + \kappa_M$ and $m \neq 1$, we have

$$\Phi_M(r,0) = -\frac{GMe_M\, e^{-i\kappa_M t_o}}{(1+\delta_{m0})2a_M} B_{\frac{1}{2}}^{(+)} , \tag{A17a}$$

$$B_{\frac{1}{2}}^{(+)} \equiv \left\{ b_{\frac{1}{2}}^{(m)} - \frac{\beta}{2}\left[\left(1 - \frac{2\Omega_M}{\kappa_M}\right)b_{\frac{1}{2}}^{(m-1)} + \left(1 + \frac{2\Omega_M}{\kappa_M}\right)b_{\frac{1}{2}}^{(m+1)} \right] \right\}. \tag{A17b}$$

The use of Eqs. (A7), (A8), and (A9) allows the conversion of Eq. (7) of Goldreich and Tremaine (1980) into the form (A17). For $\omega = \Omega_M + \kappa_M$, with r_M and ϑ_M given by Eqs. (A2a) and (A2b), the indirect potential (A15) contributes an additional term

$$\frac{GMe_M}{2a_M} e^{-i\kappa_M t_o}\left(3 - \frac{\kappa_M^{2}}{\Omega_M^{2}} + \frac{2\Omega_M}{\kappa_M}\right)f\beta , \tag{A18}$$

to Eq. (A17a). For $\omega = m\Omega_M - \kappa_M$ and $m \neq 1$, we have

$$\Phi_M(r,0) = -\frac{GMe_M\, e^{i\kappa_M t_o}}{(1 + \delta_{m0})2a_M} B_{\frac{3}{2}}^{(-)} , \tag{A19a}$$

$$B_{\frac{3}{2}}^{(-)} \equiv \left\{ b_{\frac{3}{2}}^{(m)} - \frac{\beta}{2} \left[\left(1 + \frac{2\Omega_M}{\kappa_M} \right) b_{\frac{3}{2}}^{(m-1)} + \left(1 - \frac{2\Omega_M}{\kappa_M} \right) b_{\frac{3}{2}}^{(m+1)} \right] \right\}. \quad \text{(A19b)}$$

For $\omega = \Omega_M - \kappa_M$, the indirect potential (A15) contributes an additional term,

$$\frac{GMe_M}{2a_M} e^{i\kappa_M t_0} \left(3 - \frac{\kappa_M^2}{\Omega_M^2} - \frac{2\Omega_M}{\kappa_M} \right) f\beta, \quad \text{(A20)}$$

to Eq. (A19a). Although they are written differently, the indirect terms in Eqs. (A18) and (A20) are identical to those contained in Eqs. (8) and (9) of Goldreich and Tremaine (1980).

Let us now give explicit expressions for the coefficient $f_M(r)$ in Eq. (A11) as evaluated from the expression (A3b). The strongest vertical resonances are associated with $\omega = m\Omega_M \pm \mu_M$,

$$f_M = \pm \frac{GM \sin i_M \, e^{-i\pi/2}}{(1+\delta_{m0}) 2 a_M^2} b_{\frac{3}{2}}^{(m)}, \quad \text{(A21)}$$

which holds for all m. The contribution of the indirect potential (see A14) is of order ZM/M_p and is therefore ignored in a linearized treatment.

The next strongest resonances are generally associated with $\omega = m\Omega_M \pm \mu_M \pm \kappa_M$. For $\omega = m\Omega_M \pm \mu_M + \kappa_M$, we have

$$f_M = \pm \frac{3GMe_M \sin i_M}{(1+\delta_{m0}) 4 a_M^2} e^{-i(\kappa_M t_0 + \pi/2)} B_{\frac{5}{2}}^{(+)}, \quad \text{(A22a)}$$

$$B_{\frac{5}{2}}^{(+)} \equiv \left\{ b_{\frac{5}{2}}^{(m)} - \frac{\beta}{2} \left[\left(1 - \frac{2\Omega_M}{\kappa_M} \right) b_{\frac{5}{2}}^{(m-1)} + \left(1 + \frac{2\Omega_M}{\kappa_M} \right) b_{\frac{5}{2}}^{(m+1)} \right] \right\}. \quad \text{(A22b)}$$

For $\omega = m\,\Omega_M \pm \mu_M - \kappa_M$, we have

$$f_M = \pm \frac{3GMe_M \sin i_M}{(1+\delta_{m0}) 4 a_M^2} e^{-i(\kappa_M t_0 - \pi/2)} B_{\frac{5}{2}}^{(-)}, \quad \text{(A23a)}$$

$$B_{\frac{5}{2}}^{(-)} \equiv \left\{ b_{\frac{5}{2}}^{(m)} - \frac{\beta}{2} \left[\left(1 + \frac{2\Omega_M}{\kappa_M} \right) b_{\frac{5}{2}}^{(m-1)} + \left(1 + \frac{2\Omega_M}{\kappa_M} \right) b_{\frac{5}{2}}^{(m+1)} \right] \right\}. \quad \text{(A23b)}$$

The derivation of Eqs. (A16)–(A23) completes our discussion of the calculation of forcing strengths.

APPENDIX B—EFFECTS OF FINITE THICKNESS

In this appendix we consider the modifications introduced for the properties of density waves by removing the assumption that the disk of matter is infinitesimally thin. If the three-dimensional fluid equations are written in their conservation form (mass and momentum) with volume density ρ replacing surface density σ, and if the perturbational equations are appropriately integrated over all z under obvious symmetry requirements, then we recover Eqs. (19)–(21), except that the quantities U, V, Φ_D, Φ_M must be replaced by their averages over the vertical mass distribution of the equilibrium state. For example, Φ_D should be replaced by

$$<\Phi_D> \equiv \frac{1}{\sigma_0(r)} \int_{-\infty}^{+\infty} \rho_0 (r,z)\Phi_D (r,z) \ dz \ , \tag{B1}$$

where

$$\sigma_0(r) \equiv \int_{-\infty}^{+\infty} \rho_0 (r,z)dz \ . \tag{B2}$$

The perturbational surface density $S(r)$ now has the interpretation

$$S(r) \equiv \int_{-\infty}^{+\infty} P \ (r,z) \ dz \ , \tag{B3}$$

where $P(r,z)$ contains the radial and vertical variation of the perturbational volume density

$$\rho_1 \ (r,z) = \text{Re} \ [P(r,z)e^{i(\omega - m\vartheta)}] \ . \tag{B4}$$

In the wave zone, we assume the WKBJ form

$$\Phi_D \ (r,z) = A(r,z)\exp\left(i \int k dr\right) \ , \tag{B5}$$

yielding as the generalization of Eq. (36),

$$S(r) = - \sigma_0 \frac{k^2}{D} <\Phi_D> \ . \tag{B6}$$

On the other hand, the substitution of Eq. (B5) into the generalization of Eq. (27) (i.e., into Poisson's equation) gives

$$\frac{\partial^2 \Phi_D}{\partial z^2} - k^2\Phi_D = 4\pi \ GP \ (r,z). \tag{B7}$$

At this point, the methods of Shu (1968) and Vandervoort (1970) diverge. Shu (as presented in Lin and Shu 1968) proceeded by noting that the Green's function method yields the following formal solution for Eq. (B7):

$$\Phi_D (r,z) = - \frac{2\pi G}{|k|} \int_{-\infty}^{+\infty} P(r,z') \, e^{-|k| |z-z'|} \, dz' . \tag{B8}$$

To go further, we need to specify how to obtain the volume density response $P(r,z)$ when all we have so far is a formal solution (Eq. B6) to the surface density response $S(r)$. Rather than start all over with the full dynamical equations, Shu (1968) argued that the vertical variation of the perturbational volume density is approximately proportional to the product of the unperturbed volume density and the perturbation in the potential. Since the interesting parameter regime is $|k|z_0 \ll 1$, we could ignore the z variation of Φ_D in comparison with that of ρ_0, that is, in the evaluation of the integral (Eq. B8), we could use the approximation

$$P(r,z) = S(r) \, \frac{\rho_0(r,z)}{\sigma_0(r)} . \tag{B9}$$

The equilibrium structure of a thin disk can be taken to be of the form

$$\rho_0(r,z) = \frac{\sigma_0 (r)}{z_0(r)} f(z/z_0) , \tag{B10}$$

where z_0 can be chosen so that $f(x)$ is a form function with normalization properties (cf. Eqs. [38] and [B2]):

$$f(0) = \frac{1}{2} \quad \text{and} \quad \int_{-\infty}^{+\infty} f(x)dx = 1 . \tag{B11}$$

In any case, the substitution of Eqs. (B8–B10) into Eq. (B6) now yields the modified dispersion relation

$$D = 2\pi G \, \sigma_0 |k| T(\zeta) , \tag{B12}$$

where $\zeta \equiv |k|z_0$ and T is defined as

$$T(\zeta) \equiv \int_{-\infty}^{+\infty} \int_{-\infty}^{+\infty} f(x) f(y) e^{-\zeta|x-y|} dxdy . \tag{B13}$$

Equation (39) is the axisymmetric version of Eq. (B12).

Clearly, $T(\zeta)$ is a monotonically decreasing function of its argument with $T(0) = 1$. Since $f(y) \leq \frac{1}{2}$, it is easy to show that $T(\zeta) \leq 1/\zeta$, i.e., that the product $\zeta T(\zeta)$ is always less than unity, a property we used to establish Eq. (40). For monolayers, it is interesting to consider the homogeneous slab:

$$f(x) = \frac{1}{2} \quad \text{for } -1 \leq x \leq +1; f(x) = 0 \text{ for other cases.} \tag{B14}$$

With this choice, Eq. (B13) has the analytic expression,

$$T(\zeta) = \frac{1}{\zeta}(1 - \zeta^{-1}e^{-\zeta}\sinh\zeta) \tag{B15}$$

which possesses explicitly the properties $T(0) = 1$, and $\zeta T(\zeta) \to 1$ as $\zeta \to \infty$.

A somewhat more accurate method was introduced by Vandervoort (1970). In the present context, his method amounts to replacing the ansatz (trial form) Eq. (B9) with the ansatz

$$P(r,z) = -\frac{k^2}{D}\rho_0(r,z)\Phi_D(r,z) \tag{B16}$$

which restores the z-variation of Φ_D and satisfies the integral relations (B6), (B3), and (B1). The substitution of Eq. (B16) into Eq. (B7) results in the second-order differential equation:

$$\frac{\partial^2\Phi_D}{\partial z^2} + k^2\left(4\pi G\frac{\rho_0}{D} - 1\right)\Phi_D = 0. \tag{B17}$$

For the vertical structure (Eq. B10) and (Eq. B14), the solution of Eq. (B17), which is even in z and vanishes at infinity, reads

$$\Phi_D(r,z) = \Phi_D(r,0)\cos(\alpha|kz|) \quad \text{for } |z| < z_0$$

$$\Phi_D(r,z) = Ae^{-|kz|} \quad \text{for } |z| > z_0$$

where A is to be determined and α is defined by

$$\alpha \equiv \left[\frac{2\pi G\sigma_0}{z_0 D} - 1\right]^{\frac{1}{2}}. \tag{B18}$$

The requirement that Φ_D and its z derivative are continuous at $z = \pm z_0$ yields A, and the requirement

$$\alpha \tan(\alpha |k| z_0) = 1 \qquad (B19)$$

which is a generalized dispersion relationship.

For axisymmetric disturbances, $m = 0$, the curve of marginal stability, $\omega^2 = 0$, satisfies the relationship (Eq. B19) where α is now given by (see Eq. 25):

$$\alpha = \left[\frac{2\pi G \sigma_0}{z_0 \kappa^2} - 1 \right]^{\frac{1}{2}} . \qquad (B19a)$$

With σ_0 and κ held fixed in the above equation, the relationship (Eq. B19) then yields the minimum value of z_0 required to stabilize axisymmetric density waves of radial wavenumber k. Clearly, axisymmetric waves of even the highest wavenumbers are marginally stabilized if z_0 had the critical value

$$z_{0 \text{ crit}} = \frac{2\pi G \sigma_0}{\kappa^2} \qquad (B20)$$

giving once again the stability criterion of expression (40).

For z_0 not much greater than the critical value (Eq. B20), the quantity α in Eq. (B18) will be very large near Lindblad resonances (where $D \approx 0$). When $\alpha \gg 1$, Eq. (B19) has the approximate solution $|k| z_0 = \alpha^{-2}$, which can be put into the form (Eq. B12) with

$$T(\zeta) = \frac{1}{(1+\zeta)} . \qquad (B21)$$

It is easy to verify numerically that Eqs. (B15) and (B21) are very similar functions.

REFERENCES

Bertin, G. 1980. On the density wave theory for normal spiral galaxies. *Phys. Rept.* 61:1–69.

Bertin, G., and Mark, J. W.-K. 1979. Local density-potential relations for spiral density waves in galaxies. *S.I.A.M. J. Appl. Math.* 36:407–420.

Bertin, G., and Mark, J. W.-K. 1980. On the excitation of warps in galaxy disks. *Astron. Astrophys.* 88:289–297.

Borderies, N., Goldreich, P., and Tremaine, S. 1983. Precession of inclined rings. *Astron. J.* 88:226–228.

Brahic, A. 1977. Systems of colliding bodies in a gravitational field: I. Numerical simulations. *Astron. Astrophys.* 54:895–907.

Brahic, A., and Sicardy, B. 1981. Apparent thickness of Saturn's rings. *Nature* 289:447–450.

Brouwer, D., and Clemence, G. M. 1961. *Methods of Celestial Mechanics* (New York: Academic Press).

Camichel, H. 1958. Mesures photométriques de Saturne et de son anneau. *Ann. Astrophys.* 21:231–242.

Chandrasekhar, S. 1942. *Principles of Stellar Dynamics* (Chicago: Univ. Chicago Press).

Chandrasekhar, S. 1961. *Hydrodynamic and Hydromagnetic Stability* (Oxford, U.K.: Oxford Univ. Press).

Colombo, G., Goldreich, P., and Harris, A. W. 1976. Spiral structure as an explanation for the asymmetric brightness of Saturn's A ring. *Nature* 264:344–345.

Cook, A. F., and Franklin, F. A. 1964. Rediscussion of Maxwell's Adams prize essay on the stability of Saturn's rings. *Astron. J.* 69:173–200.

Cuzzi, J. N., Pollack, J. B., and Summers, A. L. 1980. Saturn's rings: Particle composition and size. *Icarus* 44:683–705.

Cuzzi, J. N., Lissauer, J. J., and Shu, F. H. 1981. Density waves in Saturn's rings. *Nature* 292:703–707.

Esposito, L. W., Dilley, J. P., and Fountain, J. W., 1980. Photometry and polarimetry of Saturn's rings from Pioneer Saturn. *J. Geophys. Res.* 85:5948–5956.

Esposito, L. W., O'Callaghan, M., and West, R. A. 1983. The structure of Saturn's rings: Implications from the Voyager stellar occultation. *Icarus*. In press.

Franklin, F. A., and Colombo, G. 1970. A dynamical model for the radial structure of Saturn's rings. *Icarus* 12:338–347.

Franklin, F. A., and Colombo, G. 1978. On the azimuthal brightness variations of Saturn's rings. *Icarus* 33:279–287.

Fujimoto, M. 1968. In *Non-stable Phenomena in Galaxies, Proc. I.A.U. Symp. No. 29*, pp. 453–463. In Russian.

Goldreich, P., and Tremaine, S. 1978*a*. The velocity dispersion in Saturn's rings. *Icarus* 34:227–239.

Goldreich, P., and Tremaine, S. 1978*b*. The formation of the Cassini division in Saturn's rings. *Icarus* 34:240–253.

Goldreich, P., and Tremaine, S. 1980. Disk-satellite interactions. *Astrophys. J.* 241:425–441.

Goldreich, P., and Tremaine, S. 1982. The dynamics of planetary rings. *Ann. Rev. Astron. Astrophys.* 20:249–283.

Goldreich, P., and Ward, W. R. 1973. The formation of planetesimals. *Astrophys. J.* 183:1051–1061.

Hénon, M. 1981. A simple model of Saturn's rings. *Nature* 293:33–35.

Holberg, J. B., Forrester, W. T., and Lissauer, J. J. 1982. Identification of resonance features within the rings of Saturn. *Nature* 297:115–120.

Hunter, C. 1973. Patterns of waves in galactic disks. *Astrophys. J.* 181:685–705.

Hunter, C., and Toomre, A. 1969. Dynamics of the bending of the galaxy. *Astrophys. J.* 155:747–776.

Julian, W. H., and Toomre, A. 1966. Non-axisymmetric response of differentially rotating disks of stars. *Astrophys. J.* 146:810–830.

Landau, L. D., and Lifshitz, E. M. 1959. *Fluid Mechanics* (London: Pergamon Press).

Lane, A. L., Hord, C. W., West, R. A., Esposito, L. W., Coffeen, D. L., Sato, M., Simmons, K., Pomphrey, R. B., and Morris, R. B. 1982. Photopolarimetry from Voyager 2: Preliminary results on Saturn, Titan, and the rings. *Science* 215:537–543.

Lin, C. C., and Lau, Y. Y. 1979. Density wave theory of spiral structure of galaxies. *Studies Appl. Math.* 60:97–163.

Lin, C. C., and Shu, F. H. 1964. On the spiral structure of disk galaxies. *Astrophys. J.* 140:646–655.

Lin, C. C., and Shu, F. H. 1966. On the spiral structure of disk galaxies. II. Outline of a theory of density waves. *Proc. Nat. Acad. Sci.* 55:229–234.

Lin, C. C., and Shu, F. H. 1968. Density wave theory of spiral structure. In *Astrophysics and General Relativity,* eds. M. Chretian, S. Deser, and J. Goldstein (New York: Gordon and Breach), 2:236–329.

Lissauer, J. J., and Cuzzi, J. N. 1982. Resonances in Saturn's rings. *Astron. J.* 87:1051–1058.

Lissauer, J. J., Shu, F. H., and Cuzzi, J. N. 1982. Viscosity in Saturn's rings. *Icarus*. In press.

Lubow, S. H., and Shu, F. H. 1975. Gas dynamics of semi-detached binaries. *Astrophys. J.* 198:383–405.

Marouf, E. A., Tyler, G L., Zekbar, H. A., Simpson, R. A., and Eshleman, V. R. 1983. Particle size distributions in Saturn's rings from Voyager 1 radio occultation. *Icarus* 54:189–211.

Roberts, W. W. 1969. Large-scale shock formation in spiral galaxies and its implications on star formation. *Astrophys. J.* 158:123–143.

Safronov, V. S. 1960. On the gravitational instability in flattened systems with axial symmetry and non-uniform rotation. *Ann. Astrophys.* 23:979–982.

Shu, F. H. 1968. *The Dynamics and Large-Scale Structure of Spiral Galaxies.* Ph.D. Dissertation, Harvard University, Cambridge.

Shu, F. H. 1970. On the density-wave theory of galactic spirals. II. The propagation of the density of wave action. *Astrophys. J.* 160:99–112.

Shu, F. H., Milione, V., and Roberts, W. W. 1973. Nonlinear gaseous density waves and galactic shocks. *Astrophys. J.* 183:819–841.

Shu, F. H., Cuzzi, J. N., and Lissauer, J. J. 1983. Bending waves in Saturn's rings. *Icarus* 53:185–206.

Smith, B. A., Soderblom, L., Beebe, R., Boyce, J., Briggs, G., Bunker, A., Collins, S. A., Hansen, C. J., Johnson, T. V., Mitchell, J. L., Terrile, R. J., Carr, M., Cook, A. F., Cuzzi, J. N., Pollack, J. B., Danielson, G. E., Ingersoll, A., Davies, M. E., Hunt, G. E., Masursky, H., Shoemaker, E., Morrison, D., Owen, T., Sagan, C., Veverka, J., Strom, R., and Suomi, V. E. 1981. Encounters with Saturn: Voyager 1 imaging science results. *Science* 212:163–191.

Toomre, A. 1964. On the gravitational stability of a disk of stars. *Astrophys. J.* 139:1217–1238.

Toomre, A. 1969. Group velocity of spiral waves in galactic disks. *Astrophys. J.* 158:899–913.

Toomre, A. 1977. Theory of spiral structure. *Ann. Rev. Astron. Astrophys.* 15:437–478.

Vandervoort, P. 1970. Density waves in a highly flattened, rapidly rotating galaxy. *Astrophys. J.* 161:87–102.

Woodward, P. R. 1975. On the nonlinear time development of gas flow in spiral density waves. *Astrophys. J.* 195:61–73.

RING PARTICLE DYNAMICS IN RESONANCES

FRED FRANKLIN, MYRON LECAR
Harvard-Smithsonian Center for Astrophysics

and

WILLIAM WIESEL
Air Force Institute of Technology

We present a general calculation of parameters describing resonant orbits. Development of a time-averaged Hamiltonian provides an evaluation of the range in semimajor axis where resonant orbits occur, their forced eccentricities and characteristic times for development. Then we apply some of these results and develop a qualitative argument to show that a sufficiently strong resonance can produce a broad region of low particle density, i.e. a 'gap', just outside it. We estimate roughly what 'sufficiently strong' means by obtaining a lower limit on the mass of the satellite responsible for the resonance. The mechanism that we discuss involves the ability of resonances to establish a net diffusion of ring particles toward smaller semimajor axes; it operates even when self-gravitation of (concentrations of) ring particles is ignored. Thus, it is a property of resonance that is distinct from the excitation of spiral density waves.

I. INTRODUCTION

Our intent in this chapter is to analyse in a general way the characteristics of gravitational perturbations that arise when a resonance exists between the orbital frequencies of a satellite and a ring particle of negligible mass. The concept of resonance implies exact, or very near, equality between an external forcing frequency and the natural frequency of the responding system. Examples of such a case exist in the solar system. The mean orbital frequency of the Trojan asteroids is precisely that of Jupiter and the periodic variation of the orbital elements of those bodies—even their stability—can only be explained by the existence of the 1:1 resonance. Thanks to careful searches in the Saturn system during 1980, we now know that this example is no longer unique. Two Trojan-type satellites resonate with Tethys (S3) and at least one with Dione (S4) (Reitsema 1981a, b). The coorbital satellites, 1980S1 and 1980S3, provide a novel example of 1:1 resonance in which the masses of the two bodies are comparable (Yoder et al. 1983). In an analysis of the effects of a perturbing secondary on a ring of material, the 1:1 resonance is not of particular concern, except for the case of a satellite imbedded in the ring. More interesting is the case in which the forcing frequency is a fraction (denoted by $(i-j)/i$, where $i > j$) of the local orbital frequency. On the basis of these remarks, we claim that a resonance exists between a ring particle and an external satellite (subscript s) when $n_s \simeq [(i-j)/i]n$, where as usual the mean motion n is the time derivative of the mean longitude λ. A principal concern of this chapter (Sec. II) centers on deriving the maximum induced eccentricity $e_{\max}$ of resonant orbits together with the range of semimajor axis Δa in which increased eccentricities occur and the time scales T_ℓ for their development as a function of the integer indices i and j. Values of j denote the order of resonance so that the first order case $j=1$ includes resonances where the forcing frequency is $\frac{1}{2}$, $\frac{2}{3}$, etc., of the local value (i.e. where the local mean motion is 2, $\frac{3}{2}$, etc. times that of the secondary). We shall see that, although resonant strengths measured in terms of $e_{\max}$ and Δa depend weakly on i (there is little difference between $\frac{1}{2}$ and $\frac{2}{3}$), the dependence on j is critical; the $\frac{1}{2}$ resonance ($j=1$, $i=2$) is much stronger than $\frac{1}{3}$ ($j=2$, $i=3$). The quantity that principally distinguishes between the various orders is the exponent to which the secondary-to-primary mass ratio is raised. Its monotonic increase with j for resonant orbits coupled with the small masses of the Saturn satellites means that resonances of high order, $j \geq 3$ are inconsequential in that environment.

Although the mean motions are the most important terms, two other slower motions are required for a complete description of the conditions for resonance. In all cases of resonance in the solar system, either the oblateness of the primary and/or the presence of perturbing bodies produces a slow motion of the apsidal and nodal lines at rates $\tilde{\omega}$ and $\dot{\Omega}$ that are normally but not always (cf. Ward 1981) a progression of the former and a regression of the latter. Thus a complete and general condition for exact orbital resonance between two bodies, which replaces the approximate form involving only the mean motions, can be written as

$$i_1 n_s + i_2 n + i_3 \tilde{\omega}_s + i_4 \tilde{\omega} + i_5 \dot{\Omega}_s + i_6 \Omega = 0 \ , \tag{1}$$

where the i's are integers. (We have chosen to maintain the link with celestial mechanics in this chapter by using its conventions and language to describe behavior at resonance. Those more at home with the nomenclature of galactic dynamics will find the equivalence of the two schemes described by Green-berg [1984].) Although we shall quote results pertaining to the forced inclination of a particle, we confine the analytical development in this chapter to the planar case. With this simplification we shall find in Sec. II that this relation takes the form (cf. Peale 1976):

$$(i+j')n_s + (-i+j)n - j'\dot{\tilde{\omega}}_s - j\dot{\tilde{\omega}} = 0 \ , \tag{2}$$

which reduces to the approximate relation when the $\tilde{\omega}$'s and j' are set equal to zero.

Inclusion of the terms in $\dot{\tilde{\omega}}$ (and $\dot{\Omega}$) is a matter both of refinement and necessity. The former because, given the remarkable resolution achieved by Voyager, positions of features in Saturn's ring are now known to nearly km accuracy. The identification of observed structure in the ring with predicted locations of satellite resonances (Lissauer and Cuzzi 1982) requires a knowledge of $\dot{\tilde{\omega}}$ and $\dot{\Omega}$ whose values in the C Ring exceed one percent of n. Their consideration is also necessary because the resonance condition shows that allowed multiples of the $\tilde{\omega}$'s (and $\dot{\Omega}$'s) lead to a group of adjacent resonances rather than a single one. Consider the example of the 8:5 commensurability, $8n_{\mathrm{Mimas}} \simeq 5\,n_{\mathrm{local}}$, noted as a density wave by Voyager imaging. Associated with this commensurability are a number of individual resonances, e.g., $8n_s - 5n - \tilde{\omega}_s - 2\tilde{\omega}$, $8n_s - 5n - 2\tilde{\omega}_s - \tilde{\omega}$, etc., limited by the requirement of rotational invariance of the coordinate system, namely that the coefficients must sum to zero. As mentioned above, it is expected (to be verified in Sec. II) that the first order or $j=1$ resonance (with $j'=2$) dominates over the higher order cases, $j \geq 2$. It will not, however, induce perturbations that rival the $j=1$, $j'=0$ case (an example of which is $2n_s - n - \tilde{\omega}$) because, as we shall show in Sec. II, $e_s^{|j'|}$ appears as a multiplier in the expression for the forced eccentricity. These weaker, but still first-order resonances that contain powers of e_s or I_s (satellite's eccentricity and inclination) are often called eccentricity or inclination resonances.

Studies of the 3-body problem show that stable periodic orbits exist at resonance; i.e., after one synodic period T_s, which is defined by the orbital periods of the particle and satellite, all positions and velocities repeat exactly. Transformation to a coordinate system that rotates with the mean angular velocity of the massive bodies provides a useful technique to obtain these periodic solutions numerically (Colombo et al. 1968). Such a treatment shows that (1) although the orbital elements of the massless body can vary appreciably during the short period T_s, since we are dealing with periodic orbits, no effects related to longer periods are present; and (2) the eccentricities $\bar{e}$ of periodic resonant orbits, when averaged over T_s are many times greater than values for nonresonant orbits. Even Mimas, whose mass relative to Saturn is only 6.7×10^{-8}, induces eccentricities at first-order resonances that are a factor of $> 10^4$ above nonresonant values. If the number density of bodies were comparable throughout a ring, then the collision frequency between resonant and nonresonant particles would rise by the same amount over the value corresponding to collisions between nonresonant particles. The behavior of periodic orbits near a first-order resonance helps to guarantee a high-collision frequency even among resonant particles because they possess the following property. At a semimajor axis a somewhat less than the resonant value a_r, periodic orbits are such that the massless body lies at pericenter when at (inferior) conjunction with the two massive objects. On the other hand, bodies beyond resonance, so that $a > a_r$, are at apocenter when in conjunction. Pictured in the (rotating) frame so that Mimas and Saturn are stationary, resonant periodic orbits for $a < a_r$ are represented by ovals whose long axes are perpendicular to the line joining Mimas and Saturn but by ovals aligned with the two for $a > a_r$. Collisions must therefore occur even for a "nested set" of periodic orbits.

For the general case in which a particle does not move in a periodic orbit, its orbital elements oscillate, or librate, with a period T_ℓ that is much longer than T_s about those of the periodic solution. In particular, when particles diffuse into resonance as the result, for example, of collisions, their eccentricities, inclinations and, to a lesser degree, their semimajor axes will, during the period T_ℓ, vary considerably. In Sec. II we evaluate these quantities at various resonances and in Sec. III use them to describe how gaps in Saturn's rings can be formed. Note that at all resonances there are orbits having semimajor axes at and/or near the resonant value that will not execute stable librations. Comets that librate temporarily about various resonances with Jupiter (Marsden 1970) are a good example. For the case in which two massive bodies move in circular orbits (i.e., for the restricted 3-body problem), Hénon (1969) and Jefferys (1971) have, for various secondary-to-primary mass ratios, provided diagrams indicating where stable solutions exist. In general, for first-order resonances, stable librations occur at resonance unless particle eccentricities become too large (actual values depend on the mass ratio) while, in higher order cases, orbits of low-eccentricity are also unstable (cf. Wisdom 1982).

The theory for the production of density waves at resonance, as developed and summarized by Goldreich and Tremaine (1982) and by F. Shu (see his chapter and the chapter by Borderies et al.), uses the epicyclic approximation to obtain the radial perturbation near resonance. Their formulae show that the eccentricity at a distance Δx from the resonance that forces it is proportional to $m/\Delta x$, where m is the perturbing satellite's mass. This relation has a singularity at resonance and is of unknown accuracy close to it. The treatment in Sec. II is not limited to this approximation. We follow methods laid down by Poincaré and Brouwer to obtain a Hamiltonian that, after averaging, leaves the action variables as constants of the motion. These procedures enable us to obtain analytical expressions, not widely available, for the behavior of orbits arbitrarily close to resonance. (Characteristics of resonant asteroids provide convenient comparisons for certain of these results.) We can also establish limits within which the linear approximation just mentioned is appropriate. Considered in this sense, Sec. II provides material bearing directly on the applicability and/or extension of the density wave theory (see also chapter by Borderies et al.).

Much of Sec. III is given over to a largely qualitative argument, drawing upon numerical studies, leading to a description of how gaps (i.e., extended regions of very low optical thickness) might be produced by the heightened collision frequency at strong resonances. We are expressly interested in the possibility of accounting for the Cassini Division. Although the presence of predicted density waves (Goldreich and Tremaine 1978) has been clearly established at many resonances (Esposito et al. 1983), it is far from clear that they can be held responsible for producing the entire Cassini Division whose width is ~ 4500 km. Because the discussion in Sec. III considers effects of collisions induced by resonant perturbations on a particle-by-particle basis and does not include the self-gravitation of developing concentrations of ring material, it differs from the density wave analysis. Future studies should benefit from the incorporation of both effects.

II. PERTURBATIONS AT RESONANCE

Motivated by the Kirkwood Gap problem, Brouwer (1963) obtained an integral (Hamiltonian) that is valid both at resonance and when the perturbing satellite moves in an eccentric orbit. With the semimajor axis and mean motion of the perturber set equal to one, it has the remarkably concise form:

$$\frac{1}{2} L^{*-2} + \frac{p+q}{q} L^* + R^* = \Gamma_B \ . \tag{3}$$

L is a Delaunay variable, equal to the square root of the semimajor axis, $\overset{\circ}{R}$ is the disturbing function which we shall consider presently and Γ_B is a constant. The asterisks mean that the quantity in question has been averaged over a synodic period in order to eliminate short-period terms. This technique assumes that no elements vary appreciably during one period. The integers p and q provide a way of cataloguing resonances by describing the mean motion ratio, $n/n_s = (p+q)/p$ where the subscript refers to the perturbing secondary and its absence implies the massless body. We have rederived Eq. (3) and use another notation. Readers wishing to compare this work with Brouwer's may find the compilation of equivalents in Table I useful. In terms of Keplerian elements, the Poincaré variables of Table I are

$$\Lambda \equiv a^{\frac{1}{2}}; \quad \Gamma \equiv a^{\frac{1}{2}} [1-(1-e^2)^{\frac{1}{2}}] \ . \tag{4}$$

TABLE I

Relations Between Parameters for Case of Zero Inclination

	Brouwer (1963)	This Chapter
Delaunay variables:	L	Λ
	$L-G$	Γ
Integers:	p	i-j
	q	j
Combinations of above:	$X_1^* \equiv [(p+q)L-pG]^*/q$	U
	$X_2 \equiv (L-G)/q$	S
Giving this identity:	L^*	$U+(-i+j)S$

Recall that we restrict our analysis to objects moving in a plane, but will quote analogous results for inclination resonances.

The remaining term required for Eq. (3) is the disturbing function R. Newcomb (1895) obtained, for the case of a massless body in the field of a satellite, the potential in the form:

$$V_s = -\frac{Gm}{a_s} \sum_{i,j,j',k,k'} e^k\, e_s^{k'}\, P_{jj'}^{kk'}\left(\frac{a}{a_s}\right) \times$$

$$\cos\,[(i+j')\lambda_s + (-i+j)\lambda - j'\varpi_s - j\varpi] \tag{5}$$

where the λ's are mean longitudes and the ϖ's longitudes of pericenter. The summation indices k and k' are positive; j assumes the values $k, k-2 \ldots -k$ and j' depends similarly upon k'. $P_{jj'}^{kk'}$ (a/a_s) is expressible as a Newcomb operator (cf. Izsak et al. 1964) and depends only on the ratio of the semimajor axes. Equation (5) shows V_s to be an infinite series, whose leading term, i.e. the one with the lowest power of both e's, is

$$V_s = -\frac{-Gm}{a_s}\, e^{|j|} e_s^{|j'|}\, P_{jj'}^{|j|\,|j'|}\left(\frac{a}{a_s}\right)\cos\,\sigma, \tag{6}$$

where σ is the argument of the cosine term in Eq. (5). Near a given resonance, specified by the various indices, σ is a slowly varying function that oscillates (librates) between finite limits. At some critical distance from resonance (which we determine later) it begins to circulate continuously. This critical distance, Δa, provides a meaningful definition of resonant width, although others (Lissauer and Cuzzi 1982) have been used. When all the variables contained in σ are measured from the same origin, rotational invariance requires that their coefficients sum to zero. (Had we retained the inclination, the added restriction that the sum of the coefficients of the nodal longitudes must be even also applies.)

Brouwer (1963) using a technique due to Poincaré, chose a set of canonical variables that neatly isolates one critical term of σ at each resonance. Because, at any given resonance, all other angular arguments in the expansion of R have much shorter periods, Brouwer's technique is the equivalent of averaging over these short-period terms. As long as the orbital elements do not vary during the averaging time (which is measured in a few orbital periods) these terms will vanish. We form R^* from V_s of Eq. (6), once we have expressed the a's and e's in terms of the appropriate canonical variables, by ignoring all terms except the critical one of lowest order in e at any resonance.

Replacement of a and e in the disturbing function by U and S is readily accomplished from the relations given in Table I. For small e, we have $e^2 \simeq 2jS/a^{\frac{1}{2}}$ and, to $\mathcal{O}(e^2)$, $a^{\frac{1}{2}} = U$ so that

$$e = (2jS/U)^{\frac{1}{2}} . \tag{7}$$

Equation (3), with the semimajor axis of the satellite, $a_s = 1$, now takes the form

$$\frac{1}{2}[U + (-i+j)S]^{-2} - iS + m\, e_s{}^{|j'|} P(U)\, (2js/U)^{|j|/2} \cos \sigma \tag{8}$$

$$= \Gamma_B - \left(\frac{i}{i-j}\right) U$$

where m is the secondary-to-primary mass ratio and $P(U)$ are exactly the polynomials involving Laplace coefficients and their derivatives tabulated by Brouwer and Clemence (1961). As errors exist for certain $P(U)$ given by Newcomb (1895), we recommend the tabulation by Izsak et al. (1964) for any $P(U)$ not provided by Brouwer and Clemence. The right-hand side of Eq. (8) is a constant that we shall denote by K.

For the case $j = 0$, Eqs. (7) and (8) no longer apply because S is undefined. However, a treatment is possible for $j = 0$ and it has the novel aspect that neither e nor i appears in the forcing team. Thus only the semimajor axis varies in this type of resonance which we label as being of zeroth order.

In planetary rings (as we shall explain in Sec. III) we are particularly interested in the fate of particles that diffuse into resonance from orbits that are initially almost circular and of low inclination. For all resonances other than $j = 0$, a particle's orbit becomes eccentric as the resonance dynamics force S, which is not a constant of the motion (Brouwer 1963), away from zero.

In a study of the effects of resonance, we require first a measure of the range Δa of semimajor axis in which librating orbits occur. Second, particles within $\sim \pm \Delta a$ of exact resonance ($a = a_r$) can have large forced eccentricities whose characteristic maximum value e_{max} are obtainable once S is determined. The technique of using an averaged Hamiltonian, Eq. (3), to investigate the dependence of S (hence the square of the forced eccentricity) on σ has been used since the time of Poincaré (1902) and more recently by Schubart (1964). Schubart's studies are numerical (which we have found useful) but no analysis appears to have exhausted the information that can be obtained analytically. We follow Poincaré by introducing the rectangular canonical coordinates,

$$x \equiv \sqrt{2S} \cos \sigma$$

$$y \equiv \sqrt{2S} \sin \sigma \, , \tag{9}$$

so that $S = \frac{1}{2}(x^2 + y^2)$, into Eq. (3). For resonances described by the argument $\sigma = (i + j')\lambda_s + (-i + j)\lambda - j'\varpi_s - j\varpi$, the mean motion λ in terms of that of the satellite is $\dot{\lambda} = (i + j')/(i - j)\dot{\lambda}_s$, for $i > j$ and $j,j' = 0,1,2\ldots$. Therefore, in the units adopted here, at a resonance r,

$$a_r^{\frac{3}{2}} \simeq U_r^3 = \frac{i-j}{i+j'} \, . \tag{10}$$

Upon expansion of its leading term to $\mathcal{O}(S^2)$ and substitution of Eq. (9), the Hamiltonian Eq. (8) becomes

$$K + \frac{1}{2U^2} = -\frac{3}{8}(-i+j)^2 \frac{(x^2+y^2)^2}{U^4} + \frac{[(i+j')U^3 + (-i+j)]}{2U^5}(x^2+y^2)$$

$$-mP(U)e_s^{|j'|}(j/U)^{|j|/2} x(x^2+y^2)^{(|j|-1)/2} \, . \tag{11}$$

Exactly at resonance, Eq. (10) shows that the term in Eq. (11) in square brackets vanishes. When expanded to first order in ΔU about resonance, this term becomes $(3/2)(i-j)\,\Delta a/a$. For the 3 orders of resonance $j = 1, 2$, and 3 we can write Eq. (11) first as

$$K + \frac{1}{2U^2} = -\frac{3}{8} \frac{(i-j)^2}{U^4}(x^2+y^2)^2 + \frac{3}{4} \frac{(i-j)}{U^3} \frac{\Delta a}{a}(x^2+y^2)$$

$$-me_s^{|j'|}P(U)\frac{x}{U^{\frac{1}{2}}} \, , \text{ for } j = 1, \tag{12}$$

where the final term is replaced by $\pm 2me_s^{|j'|}P(U)[x(x^2+y^2)^{\frac{1}{2}}]/U$ for $j = 2$ and $3\sqrt{3}\,me_s^{|j'|}P(U)[x(x^2+y^2)]/U^{\frac{3}{2}}$ for $j = 3$.

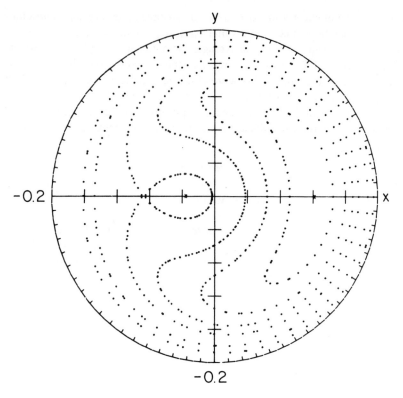

Fig. 1. A typical surface of section, after Schubart (1964), of the restricted 3-body problem, drawn for the first-order resonance, $\sigma = 2\lambda_{\text{Jupiter}} - \lambda - \tilde{\omega}$, with Jupiter, $m_{\text{Jupiter}}^{-1} = 1047.355$, as the secondary.

We call a resonance for which $|j| = 1$ (whence, for example, $\sigma = 2\lambda_s - \lambda - \tilde{\omega}$) "first order." Since x and y both vary as $\sqrt{2S}$ which in turn is proportional to e, the initial condition $e_0 = 0$ implies $x_0 = y_0 = 0$. In this case the two constants of the motion K and U are related by $K = -1/2U^2$. Solving Eq. (12) for this and other values of K provides the "surface of section" of a resonance, obtained numerically by Schubart (1964). Figure 1 is taken from his work. Here we examine the behavior of Eq. (12) at $y = 0$ (a quartic equation in x). Equation (12) has a root at $x = 0$ (cf. Fig. 1). Exactly at resonance, defined by $\Delta a = 0$, Eq. (12) reduces to

$$x = \left[-\frac{8mP(U)}{3(i-1)^2} U^{\frac{7}{2}} \right]^{\frac{1}{3}} \qquad (13)$$

for the case $j' = 0$. We have already seen that $U = a^{\frac{1}{2}}$ to $\mathcal{O}(e^2)$, so from Eqs. (7) and (9), $e \cong xa^{-\frac{1}{4}}$. Equation (13) therefore becomes

$$e \simeq \left[-\frac{8maP(a)}{3(i-1)^2} \right]^{\frac{1}{4}} , \tag{14}$$

where $P(a)$ is negative for all $j=1$ resonances. (The mass m and semimajor axis a are measured in terms of that of the primary and secondary, respectively.) A more general treatment shows that the forced inclination differs from the form of Eq. (14), and its analogues at other resonances, only by factors of order unity. Equation (14) provides the total range of eccentricity forced upon a particle exactly at a $j=1$, $j'=0$ resonance. (Written in terms of x, it gives the positive intercept of the x axis in Fig. 1.) For the 2:1 resonance $\sigma = 2\lambda_s - \lambda - \tilde{\omega}$, Eq. (14) reduces to $e = (2m)^{\frac{1}{4}}$, which is equal to 0.1241 for the Jupiter-Sun case. For Mimas, with $m = 6.7 \times 10^{-8}$, $e = 0.0051$ and the characteristic radial excursion, $2ae$ is 1200 km.

A more general solution of Eq. (12) shows that even larger eccentricities are forced on a body at a semimajor axis slightly displaced from exact resonance. The requirement that Eq. (12) considered as a cubic has 3 real roots, 2 of which are equal—which is the condition for the extreme libration shown by the contour passing through $x = y = 0$ in Fig. 1—is

$$\frac{1}{4} \left[\frac{8}{3} \frac{mP(U)}{(i-1)^2} U^{\frac{7}{2}} \right]^2 + \frac{1}{27} \left[\frac{-2U}{i-1} \frac{\Delta a}{a} \right]^3 = 0 . \tag{15}$$

The single positive root x_1 and the repeated negative roots $x_2 = x_3$ imposed by Eq. (15) are factors of $2^{\frac{2}{3}}$ and $2^{-\frac{1}{3}}$ times the value provided by Eq. (13), viz., $x_1 = 0.1754$ and $x_2 = x_3 = -\frac{1}{2}x_1 = -0.0877$ for an asteroid at the 2:1 resonance with Jupiter. These values agree very well with those obtained numerically by Schubart and shown in Fig. 1. Whereas positive x corresponds to the usual libration of σ about 0° (conjunction at pericenter of the asteroid), negative x implies libration about 180° (conjunction at apocenter). Apocentric libration has been detected for 7 asteroids (Franklin et al. 1975) and the maximum observed eccentricity for several of them agrees well with the value derived here, $e = |x_2|a^{-\frac{1}{4}} = 0.0985$. For Mimas, the limiting excursion $2ae$ now reaches 1900 km.

Solution of Eq. (15) yields a characteristic value for the resonance half-width $\Delta a/a$. More precisely, it provides the amplitude of the variation in semimajor axis of a single body whose libration approaches $\pm 180°$ about $\sigma = 0°$. Since amplitudes of this amount are unlikely to be stable, the solution of Eq. (15)

$$\frac{\Delta a}{a} = \left\{ \frac{6}{(i-1)} [amP(a)]^2 \right\}^{\frac{1}{3}} \tag{16}$$

provides an upper limit. At the 2:1 resonance, $\Delta a/a = 3/2\, m^{\frac{2}{3}}$, so with Jupiter as the perturber $\Delta a = 0.048$ AU for the somewhat poorly defined Kirkwood

Gap whose observed half-width is perhaps slightly greater (Brown et al. 1967). A better comparison (because any correspondence between Δa and the width of a Kirkwood Gap is complex) is provided by the Hilda asteroids that form an isolated group librating at the 3:2 resonance. In an analysis of 21 members, Schubart (1968) has shown that the largest Δa encountered, corresponding to a libration of $\sim 100°$, is 0.055 AU. Equation (16) yields $\Delta a = 0.075$, so the overestimate is not severe. Because $\Delta a \sim m^{\frac{2}{3}}$, scaling to Mimas' mass enormously reduces Δa. At the 2:1 resonance Δa is only 2.9 km, so the source region for particles whose eccentricity variations are considerable is itself remarkably narrow.

We return to Eq. (12) with $j=1$, $j'=0$ one more time to obtain the linear approximation by dropping the first term on its right-hand side, so that

$$x \frac{\Delta a_\ell}{a} = \frac{4}{3} \frac{mP(U)}{(i-1)} U^{\frac{5}{2}},$$

or

$$e = \frac{4}{3} \frac{amP(a)}{(i-1)} \frac{a}{\Delta a_\ell}, \tag{17}$$

where, for the 2:1 resonance, $P(a) = -\frac{1}{2}[4b_{\frac{1}{2}}^{(2)} + a(db_{\frac{1}{2}}^{(2)}/da)]$ and, as noted earlier, the a's are measured in terms of the semimajor axis of the perturber. The $b_{\frac{1}{2}}^{(2)}$'s are Laplace coefficients (Brouwer and Clemence 1961). The linear approximation is accurate and useful only outside the librating orbit region of width Δa. We indicate this fact here by using Δa_ℓ when dealing with the linear regime. Because we have used Kepler's Third Law in the form $\Omega_0^2 a^3 = 1$, the radial perturbation r_1 at a distance Δa_ℓ from this resonance ($i=2$) is

$$r_1 = ae = -\frac{2m}{3\Delta a_\ell \Omega_0^2}\left[4b_{\frac{1}{2}}^{(2)} + a \frac{db_{\frac{1}{2}}^{(2)}}{da} \right]. \tag{18}$$

Equation (18) has the same form as, but is a factor of 2 larger than, the expression given by Goldreich and Tremaine (1982; their Eq. 45). The factor of 2 results because the two studies consider slightly different orbits; we shall clarify this remark toward the end of this section. At a distance $\Delta a_\ell = \Delta a$ from the exact 2:1 resonance Eq. (17) underestimates e by 35%, but at $2\Delta a$ the discrepancy has fallen to only 2%; thus, the linear approximation is quite accurate for Δa_ℓ larger than the Δa given by Eq. 16.

Equation (12) also applies to the cases $j=2$ and $j=3$, and any discussion is sufficiently similar to that for $j=1$ resonances that we will only quote results. However, resonances for which $j \geq 2$ do differ in one regard from the first-order case. The equations of motion of these types have as factors powers of the eccentricity and/or inclination, so that orbits initially circular or uninclined would seem to remain so. Yet, evolution from these states does occur because in a region near resonance such orbits are unstable (Wiesel 1974). For $j=2$ and $j'=0$, the analogue of Eq. (14) is

$$e = \frac{4}{3(i-2)} \, [6amP(a)]^{\frac{1}{2}}, i = 3, 4, \ldots \tag{19}$$

where $P(a) > 0$ for all $j = 2$ resonances. At a 3:1 resonance ($i = 3$ or $\sigma = 3\lambda_s - \lambda - 2\varpi$), Eq. (19) gives $e = 0.056$ for m_{Jupiter} and $e = 0.00047$ for m_{Mimas}. The former matches the observed forced eccentricity of 887 Alinda (Marsden 1970) while the latter gives $2ae = 83$ km. A realistic lower limit on the resonance width follows from Eq. (15):

$$\frac{\Delta a}{a} \gtrsim \frac{8}{3(i-2)} \, amP(a) \, , \tag{20}$$

which for the above cases respectively yields $a \gtrsim 0.004$ AU and 10 m.

The $j = 2$ resonance just considered is not the only one lying in the region where $\dot{\lambda} \simeq 3\dot{\lambda}_s$. As the cosine term in Eq. (5) indicates, near any mean motion commensurability there exists a number of resonances described by the critical angle $\sigma = (i+j')\lambda_s + (-i+j)\lambda - j'\varpi_s - j\varpi$. When $j=j'=1$, the forced eccentricity becomes (cf. Eqs. 12 and 14):

$$e = \left[-\frac{8}{3} \, \frac{amP(a)}{(i-1)^2} \, e_s \right]^{\frac{1}{3}} \tag{21}$$

where $P(a) < 0$ for all i. For Mimas, Kozai (1957) finds $e_s = 0.0202$ so that $e = 0.0016$ when $i=2$. Thus, this so-called eccentricity resonance, which is displaced inward by 610 km because of Saturn's oblateness, dominates over the previously considered $j=2$, $j'=0$ case because both e and in particular Δa are larger. The latter rises to 0.006 AU for Jupiter ($e_s = e_{\text{Jupiter}} = 0.048$) and 0.21 km for Mimas when e_s is included in Eq. (16). Clearly other values of j, j', and i produce additional weaker resonances near, for example, $\dot{\lambda} = 3\dot{\lambda}_s$, which might be detected in the outer regions of Ring C where the optical thickness is low. The question arises whether weak features could be of third order $j=3$ which we now examine.

For third-order resonances, the power of the mass ratio present in expressions for the forced eccentricity (inclination) and resonant width continues to increase. Equation (12) for $j=3$ yields

$$e = -\frac{24}{(i-3)^2} \, amP(a), i = 4, 5, \ldots \tag{22}$$

where $P(a) < 0$. At the 4:1 and 8:5 resonances with Mimas, Eq. (22) gives

$e = 1.64 \times 10^{-7}$ and $e = 5.18 \times 10^{-7}$, or linear excursions, $2ae$, of 24 and 140 m. The resonant width,

$$\Delta a = \frac{24}{(i-3)^3} \, [amP(a)]^2 \, a \tag{23}$$

shrinks to $\sim 10^{-3}$ cm at the 8:5 resonance. Since two density waves have been recorded in the vicinity of $5\dot{\lambda} = 8\dot{\lambda}_s$, whose characteristic patterns make it clear that one is associated with the eccentricity and the other the inclination (cf. chapters by Shu, and by Cuzzi et al.), resonances described by $j=1, j'=2$ or $j=2, j'=1$ must be much more important than $j=3, j'=0$. For the eccentricity resonances the two candidates are therefore: (1) $8\lambda_s - 5\lambda - 2\tilde{\omega}_s - \tilde{\omega}$, i.e., $j=1, j'=2, i=6$; or (2) $8\lambda_s - 5\lambda - \tilde{\omega}_s - 2\tilde{\omega}$, i.e., $j=2, j'=1, i=7$. Equations (14) and (16), (19) and (20) with $e_s^{|j'|}$ included give: for Case (1) $e = 0.00049$, $2ae = 133$ km, and $\Delta a = 0.15$ km; while for Case (2) $e = 0.00013$, but $\Delta a \geqslant 3$ m. Evidently, of all cases thus far considered, the $j=1$ resonances dominate over others when the driving satellites have small masses characteristic of the Saturn system.

We defer treatment of inclination resonances for a more detailed account (Wiesel et al. in preparation) but it is clear that the analogue of case (1) above, viz. $8\lambda_s - 5\lambda - \tilde{\omega}_s - \Omega_s - \Omega$ is the strongest. (Because $\sin i_s$ and e_s for Mimas are nearly equal, the forced inclination, $I_f \sim [me_s \, I_s]^{\frac{1}{3}}$ and resonant width, $\Delta a/a \sim [me_s \, I_s]^{\frac{2}{3}}$ for this resonance hardly differ from case (1).) The distance of this inclination resonance from Saturn, for $J_2 = 0.016299$ and $J_4 = -0.000917$ (Null et al. 1981) is 135618 km, while the dominant eccentricity resonance lies at 135798 km. The observed separation between the density waves driven by each resonance is 190 ± 10 km (J. Cuzzi, personal communication). A similar pair of stronger $j=1$ resonances with Mimas near $3\dot{\lambda} \simeq 5\dot{\lambda}_s$ appear as well-developed density waves (cf. chapter by Shu) in the A Ring. The outer, defined by $\sigma = 5\lambda_s - 3\lambda - \tilde{\omega}_s - \tilde{\omega}$ lies at 132275 km from the planet, the inner, $\sigma = 5\lambda_s - 3\lambda - \Omega_s - \Omega$ at 131877 km.

For resonances of zeroth order $j = 0$, the slowly varying angle σ equals $(i + j')\lambda_s - i\lambda - j'\tilde{\omega}_s$ in the planar case. Since $\tilde{\omega}_s$ has replaced $\tilde{\omega}$, the position of a $j = 0$ resonance is slightly displaced (characteristically by a few hundred km toward Saturn in the Cassini Division region) from that of a nearby $j = 1$ resonance when perturbing terms like the oblateness are present. Because for $j = 0$ neither the longitudes of pericenter $\tilde{\omega}$ nor node Ω appear in the Hamiltonian, the conjugate momenta are constants of the motion. Therefore, the resonance only produces oscillations Δa_0 in the semimajor axis. Equation (12) is not immediately applicable for the $j = 0$ case, but Brouwer's treatment can be generalized to include it. The companion to Eq. (11) can be written as

$$K_0 = - \frac{1}{2(-iU_0)^2} + (i+j')U_0 - me_s^{|j'|} P(a) \cos \sigma \tag{24}$$

where, apart from a constant, U_0 has replaced U for this case. As before, we expand the first two terms of Eq. (24) about the value of U_0 at resonance, viz. $U_0(a_r) = a_r^{\frac{1}{2}}/(-i)$, and absorb a constant into K_0 to yield

$$K'_0 = \frac{1}{2} \frac{\Delta a_0}{a_r} \frac{1}{U_0^2(a_r)} \left[\frac{1}{i^2} + (i+j')U_0^3(a_r) \right]$$

$$-\frac{1}{8}\left(\frac{\Delta a_0}{a_r}\right)^2 (i+j')U_0(a_r) + m e_s^{|j'|} P(a) \cos \sigma . \quad (25)$$

The term in square brackets vanishes at resonance $a_r = [i/(i+j')]^{\frac{2}{3}}$. For zero-order resonances, librations of the longitude of conjunction of the satellite and a ring particle occur about the apocenter of the satellite's orbit $\sigma = 180°$ (cf. Fig. 2). We can evaluate K'_0 by setting $\Delta a_0 = 0$ when $\sigma = 0°$ so that at $\sigma = 180°$ Eq. (25) yields

$$\frac{\Delta a_0}{a} = 4 \left[a m e_s^{|j'|} P(a) \right]^{\frac{1}{2}} \quad (26)$$

One caveat remains on the general application of Eq. (26) because our development of the disturbing function R has ignored the indirect term. The latter arises because we have referred all coordinates to the planetocentric rather than center of mass frame. We can readily allow for this effect by including an appropriate numerical correction to the $P(a)$'s (Brouwer and Clemence 1961, p. 493). The 2:1 resonance ($i=1$, $j'=1$) is the only $j'=1$ resonance possibly present in Saturn's rings requiring this modification. For this case with Mimas as the perturber, we obtain $\Delta a_0 = 11$ km (cf. Fig. 2). Ignoring the indirect correction increases Δa_0 to 18 km.

Thus far we have obtained expressions for the forced eccentricity and range of semimajor axis where librating orbits exist. We turn now to the question of the time scale for the variation of these quantities, which is the libration period T_ℓ. Because energy and time are canonical conjugates, one expects that Δa and T_ℓ^{-1} will be similarly dependent on critical quantities such as the mass ratio m. Our analysis uses standard techniques (cf. Goldstein 1953) to obtain the frequency ω that describes oscillations arising from slight displacements about the equilibrium solutions of the Hamiltonian (Eq. 12). The first partial derivative K_x yields an equation for the equilibrium solutions $\sigma = 0$ of Eq. (12):

$$(K_x)_{y=0} = 0 . \quad (27)$$

The second partials K_{xx} and K_{yy} provide the secular equation which, in the cases studied here, has the form

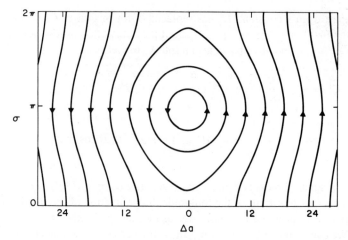

Fig. 2. A surface of section for the zero-order resonance, $\sigma = 2\lambda_s - \lambda - \bar{\omega}_s$, with Mimas as the secondary. Horizontal coordinate, given in km, measures the distance from exact resonance, $\Delta a = 0$.

$$\omega^2 = \left[K_{xx} K_{yy} \right]_{y=0} . \tag{28}$$

If we let the coefficients of the terms containing the x's and y's as they appear on the right-hand side of Eq. (12) be A, B, and C, Eq. (28) for first-order $j = 1$, $j' = 0$ resonances becomes

$$\omega^2 = 4\,(6\,Ax^2 + B)\,(2\,Ax^2 + B) . \tag{29}$$

Well away from resonance, Eq. (29) reduces to

$$\omega \cong 2B = \frac{3}{2}\,\frac{(i-1)}{a^{\frac{3}{2}}}\,\frac{\Delta a}{a} \tag{30}$$

which has the expected form of the drift frequency of particles moving in Keplerian orbits.

How best to employ Eq. (29) at resonance presents a slight quandary because we do not know the precise value of Δa that corresponds to the equilibrium solution given by Eq. (27). We commit no serious error (as we shall see presently in a numerical example) by setting $\Delta a = 0$ so that $B = 0$. Equation (29) then yields

$$\omega \cong 4\sqrt{3}\,Ax^2 = \left(\frac{3}{a} \right)^{\frac{3}{8}} \left[\frac{-(i-1)\,mP(a)}{2^{\frac{1}{2}}} \right]^{\frac{3}{4}} , \tag{31}$$

thanks to Eq. (27) written as

$$x = \left(- \frac{C}{4A} \right)^{\frac{1}{4}} = \left[\frac{-2\,mP\,(U)}{3(i-1)^2} U^{\frac{1}{2}} \right]^{\frac{1}{4}}. \tag{32}$$

Equations (31) and (32) apply only for the first-order resonances, $j=1, j'=0$. As provided by Eq. (32) for the equilibrium solution, x (which equals $ea^{\frac{1}{4}}$) is a factor of 2 smaller than the value obtained for that librating orbit which, during the range of its resonant oscillations, passes through $x=e=0$ and which was given earlier in this section by Eqs. (13) and (14). The different orbits considered would appear to reconcile the factor of 2 discrepancy between the linear approximation developed here (Eq. 18) and that of Goldreich and Tremaine (1982). A final point should be made on a related matter: x, as given by Eq. (32) ($x = 0.070$ for $m=m_{\text{Jupiter}}$ at the 2:1 resonance) is not equal to the equilibrium solution shown by the cross near $x = 0.12$ in Fig. 1 because Schubart's solution does not correspond, as Eq. (32) requires, to $\Delta a=0$.

For the 2:1 resonance and $m=m_{\text{Jupiter}}$, Eq. (31) yields $\omega^{-1} \equiv T_\ell = 32$ Jovian orbital periods. T_ℓ equals 40 when, instead of setting $\Delta a=0$, we calculate T_ℓ one resonant width (Eq. 16) from exact resonance. Both Schubart (1964) and Lecar and Franklin (1973) obtained $T_\ell \cong 38$ in different types of numerical studies for small e. In the case of 1362 Griqua ($\bar{e} = 0.3$, $\Delta\sigma = \pm 100°$) Marsden (1970) found $T_\ell \cong 32$. For the same resonance, but with Mimas' mass, $T_\ell = 1.86 \times 10^4$ periods of Mimas or ~ 48 yr.

The discussion for other resonances is sufficiently similar that we need only provide expressions analogous to Eq. (31). For $j=0, j'=1$

$$\omega = \frac{1}{2} \left[\frac{me_s^{|j'|}P(a)}{a} \right]^{\frac{1}{2}}, \tag{33}$$

which, for the 2:1 resonance with Mimas yields $T_\ell = 69$ yr. For $j=2, j'=0$,

$$\omega = 4\sqrt{6} \left[\frac{mP(a)}{a^{\frac{1}{2}}} \right]. \tag{34}$$

Equation (34), and all expressions developed here for ω, apply for the condition of both small free eccentricity and libration amplitude that are likely to be appropriate in planetary rings. The only well-observed body at a resonance of this type ($\sigma = 3\lambda_{\text{Jupiter}} - \lambda - 2\bar{\omega}$) is 887 Alinda, for which the free eccentricity is 0.5 and $\Delta\sigma$ is 114°. Therefore, it is not surprising that its observed period, $T_\ell \cong 360$ yr (Marsden 1970) is a factor of nearly four shorter than the value $T_\ell = 1390$ yr given by Eq. (34). (The decrease of T_ℓ with increasing e was established numerically by Schubart (1964) for the 2:1 resonance.) In the case of Mimas, $T_\ell = 4340$ yr at the $j= 2$ resonance ($3\lambda_s - \lambda - 2\bar{\omega}$), but Eq. (31) with $e_s = 0.0202$ included yields $T_\ell = 210$ yr for the $j=1$ resonance ($3\lambda_s - \lambda - \bar{\omega}_s - \bar{\omega}$). Finally, for $j=3, j'=0$ we have

TABLE II

Dependence of the Libration Amplitude (Described by Variation in Semimajor Axis Δa, Libration Period T_ℓ and Forced Eccentricity) on Mass m_s and Eccentricity e_s of the Secondary

Resonance Order $\|j\| + \|\ell\|$	j'	Form of σ (Example)[a]	Δa or $(T_\ell)^{-1}$	Forced Eccentricity[b]
0	1	$(i + j')\lambda_s - i\lambda - j'\varpi_s$	$m_s^{\frac{1}{2}}$	—
1	0	$i\lambda_s + (-i+1)\lambda - \varpi$	$m_s^{\frac{2}{3}}$	$m_s^{\frac{1}{3}}$
2	0	$i\lambda_s + (-i + 2)\lambda - 2\varpi$	m_s	$m_s^{\frac{1}{2}}$
1	1	$(i + 1)\lambda_s + (-i + 1)\lambda - \varpi_s - \varpi$	$(m_s e_s)^{\frac{2}{3}}$	$(m_s e_s)^{\frac{1}{3}}$
3	0	$i\lambda_s + (-i + 3)\lambda - 3\varpi$	m_s^2	m_s
2	1	$(i + 1)\lambda_s + (-i + 2)\lambda - \varpi_s - 2\varpi$	$m_s e_s$	$(m_s e_s)^{\frac{1}{2}}$
1	2	$(i + 2)\lambda_s + (-i + 1)\lambda - 2\varpi_s - \varpi$	$(m_s e_s)^{\frac{2}{3}}$	$(m_s e_s)^{\frac{1}{3}}$

[a]In its most complete form, a resonance is described by $\sigma \equiv (i + j'+k)\lambda_s + (-i + j + k)\lambda - j'\varpi_s - j\varpi + (-2k + \ell)\Omega_s - \ell\,\Omega$.
[b]Dependence of the forced inclination is analogous to the forced eccentricity with I_s replacing e_s.

$$\omega = 54\,(3a)^{\frac{1}{2}} \left[\frac{mP\,(a)}{(i-3)} \right]^2 , \qquad (35)$$

which gives, for asteroids possibly librating near $2\dot\lambda = 5\dot\lambda_{\text{Jupiter}}$, $T_\ell = 5 \times 10^5$ yr. For the nearby first-order $j=1, j'=2$ resonance, $T_\ell = 1800$ yr which makes a search for such bodies feasible.

Table II compiles some of the results of this section for resonances of order equal to and less than 3. The dominant resonances are the $j=1$ and $j=0$ cases, both of which are observed in Saturn's rings. Are $j=2$ resonances likely to be important? If a first-order, $j=1$, $j'=1$ resonance is significant, then the adjacent second-order one ($j=2$, $j'=0$) will be equally so if the mass and eccentricity of the driving secondary are such that $m_s \cong (m_s e_s)^{\frac{2}{3}}$, or $m_s \cong e_s^2$. Thus the asteroid belt is the best hunting ground for resonances with $j=2$ although their existence as minor features in regions of low optical thickness is not excluded in the Uranus and Saturn ring systems.

III. DEVELOPMENT OF GAPS AT RESONANCE

Our discussion of resonance phenomena must address the question of how many of the conspicuous ring features can be attributed to various satellite resonances, and how are such features produced. This broad question is nearly as difficult to answer as it is obvious to propose. In this section we argue that not only do resonances excite spiral density waves (cf. chapter by Shu) but, in addition, they generate, by a distinctly different process also related to resonant perturbations, regions of very low optical thickness that appear as dark gaps in the ring. Our approach, which relies on the background material of Sec. II, sets an approximate lower limit on the satellite mass necessary for gap production. There are also many gaps in our knowledge, and we shall mention several topics for further study.

A qualitative picture describing a process by which resonances can deplete the amount of material in their vicinity has been given by Franklin et al. (1980). That paper concentrated on the truncation of the asteroid belt and the development of a deep gap at the 2:1 resonance. As a prelude to the Saturn ring problem, we review its arguments, refining some of them to apply to the questions at hand. Consider particles drifting into resonance from either direction as the result of viscous diffusion. The eccentricities of such objects are in general very much less than the eccentricity, $e_{\max}$, forced by the resonance. Now, assume that the perturbing satellite moves in a circular orbit so that, for these particles whose semimajor axes equal a_0, we can write the Jacobi constant

$$J \simeq \frac{1}{2a_0} + a_0^{\frac{1}{2}} + \mathcal{O}(m) \tag{36a}$$

where m is the perturber's mass. At a first-order resonance $\delta a/a \sim \delta e^2 \sim m^{\frac{2}{3}}$ (see Sec. II), so that we are justified in dropping terms of order m. As resonant perturbations develop a larger eccentricity e_1, the Jacobi constant becomes

$$J \simeq \frac{1}{2a_1} + \left[\, a_1(1-e_1^2) \, \right]^{\frac{1}{2}} \tag{36b}$$

where e_1 is comparable to $e_{\max}$. Collisions are more likely at an eccentricity e_1 than at values near zero. Since J is conserved, the two versions of Eq. (36) show that $a_1 < a_0$; i.e., when the probability of a collision is highest, a is smallest. As collisions are partially inelastic, a_1 is even further reduced so that after leaving resonance via a collision, a_1 is reduced below what it was before entering resonance. There is, therefore, a net drift to smaller values of a. Franklin et al. (1980) likened resonances to semi-permeable membranes in the sense that collisions assist particles trying to diffuse inwards but impede bodies on a net outward passage, the latter being preferentially returned to their region of origin.

Taking the variation of an equation with the form of Eq. (36b) provides a useful restatement and extension of this discussion:

$$\delta J = 0 = \frac{[a(1-e^2)]^{\frac{1}{2}}}{2} \left\{ \frac{\delta a}{a} \left[\, 1 - \frac{1}{a^{\frac{3}{2}}(1-e^2)^{\frac{1}{2}}} \right] - \frac{\delta e^2}{(1-e^2)} \right\} . \tag{37}$$

For a resonance located interior to a satellite so that $a < 1$, Eq. (37) shows that δa is negative when δe^2 is positive. It also makes clear that particles are losing angular momentum because $\delta[a^{\frac{1}{2}}(1-e^2)^{\frac{1}{2}}] = \delta a/2a^2$ is less than zero. The angular momentum lost by particles is transferred to the perturber which spirals outward.

When applied to the asteroid belt, this argument which Franklin et al. (1980) checked by a series of numerical simulations can explain the rapid decline in the asteroid population just outside the 2:1 resonance, and give an accurate width for the Kirkwood Gap at the 2:1 resonance itself. In the context of the asteroidal distribution, these results lose some interest because they apply only to bodies with eccentricities $\lesssim m^{\frac{1}{3}} \simeq 0.02$. For larger eccentricities, viscous diffusion of particles is more rapid than angular momentum transport so that neither a sharp truncation nor clear gaps have the opportunity to develop. Thus, to be a viable explanation, an additional mechanism (cf. chapter by Ward) is required to raise the average eccentricity and inclination subsequent to this sculpturing of the asteroidal belt material. The eccentricity limit poses no problem in Saturn's rings where, as we shall show, characteristic values are $\sim 10^{-7}$, but a potential problem for gap formation exists because of the small masses of the inner satellites and, specifically, of Mimas. How can so small a satellite be responsible for a broad gap is a question that has haunted the search for an explanation of the Cassini Division for some time. The discussion in Sec. II shows, for example, that resonant perturbations are large over a range of semimajor axis that is only ~ 3 km wide. In the hope of establishing a causal connection between the resonance and the Division, the latter part of this section introduces some heuristic arguments that are suggestive though not yet complete. We claim that the narrowness of the region itself in no way prevents a sufficiently strong resonance from functioning as a barrier against the outward migration of ring particles. Consequently, what determines the presence of a gap (and its eventual width) are the details of viscous diffusion at and near resonance owing to the enhanced frequency of collision there. One immediate prediction, verified in the case of the Cassini Division, follows. The position of the (narrow) resonance, because it acts as a one-way barrier, defines the sharp inner edge of the gap it produces. The location and appearance (e.g. slope or sharpness) of the outer boundary is governed by the behavior of particles induced by partially inelastic collisions to diffuse inwards. In this sense, the gap develops or eats its way outward in time from resonance.

An alternative approach to gap development that makes the same prediction with regard to the location of its inner boundary was first proposed quantitatively in the context of planetary rings by Goldreich and Tremaine (1978). Their mechanism, based on effects of spiral density waves produced at resonance, differs from the considerations offered here. Both operate toward the same end, and are two aspects of the response of a particle disk to resonant perturbations, by transferring angular momentum from the region of resonance to the perturber. There is, however, a significant difference: ours produces gaps even when the self-gravitation of nonaxisymmetric material concentrations is neglected, while the density wave theory clearly requires self-gravitation. A pictorial representation of the development of density waves near the 2:1 resonance has been provided by Lin and Papaloizou (1979). In Sec. I, we pointed out that, in the frame rotating with the massive bodies, periodic orbits are represented by ovals about which other orbits librate and that the family of smaller ovals interior ($a < a_r$) to resonance is orthogonal to those of the larger exterior one. In the presence of dissipation accompanying partially inelastic collisions, the entire field of orbits would be altered so that successively larger ovals would be aligned at intermediate angles. Put in other terms, in the region beginning inside resonance and moving outside it, one can represent periodic orbits by a set of ovals (the librating orbits serve to make individual ovals fuzzier) each of whose long axis is rotated by an increasing angle. The totality of such orbits appears as a spiral, shown schematically by Lin and Papaloizou (1979, their Fig. 3) and more clearly, though in a strictly illustrative context, by A. J. Kalnajs (see Toomre 1977, his Fig. 3). The nonaxisymmetric spiral pattern produces a torque which Goldreich and Tremaine (1978) determined analytically and applied as a mechanism to generate the Cassini Division. Clearly, the self-gravitation of the developing wave pattern is the source of the torque. Although the existence of very well-developed spiral patterns (but not deep gaps) at many resonances provides a strikingly beautiful confirmation of the Goldreich-Tremaine theory for the production of density waves, its role as the principal explanation of the Cassini Division is placed in doubt by the numerical simulations of Franklin et al. (1980) and Schwarz (1981). The models developed in these papers, which ignore self-gravitation and presumably for that reason do not exhibit density waves at resonance, require only partially inelastic collisions to develop gaps. Franklin et al. show that the effects of diffusion as we have sketched them here closely conform to the behavior of their numerical models. Schwarz argues that the gap produced at the 2:1 resonance with Mimas is no more than ~ 300 km wide. The accuracy of this numerical value is somewhat compromised both by the limited duration of the integrations ($\sim$ one libration period) and by the drastic scaling to Mimas' mass used to reduce computing time. The two simulations employ mass ratios m of 10^{-2} and 2×10^{-3}, each of which is a factor close to 10^5 greater than the Mimas-Saturn value. This in turn makes the ratio of nonresonant to resonant eccentricity larger by a factor of $\sim 10^3$ in the models than in Saturn's rings. We suspect that in such a "noisy" ring, the full width of any gap produced by a resonance will not be apparent.

We can safely conclude that gaps are formed at sufficiently strong resonances and that they can occur independently of the self-gravitation produced by nonaxisymmetric disturbances. Rather they owe their development to the enhanced collision frequency associated with the relatively high eccentricity orbits at resonance and the capacity of the latter to encourage only the inward diffusion of particles. Just how wide a gap can grow in the real environment of Saturn's rings remains unknown. Therefore, it cannot yet be claimed that the *entire* Cassini Division width is the responsibility of the 2:1 resonance with Mimas. A new generation of models that reduce the scaling factor and include self-gravitational effects seem valuable in addressing this problem. It would be useful to consider the behavior at other first-order resonances in such models, even if their relevance to Saturn's rings is hypothetical, because it is not clear whether the orthogonal nature of periodic orbits interior and exterior to a_r is a property common to all of them.

Having reviewed what is reasonably well-known, we offer some remarks and speculations on how collisions operate to define the inner boundary and possibly also enlarge the gap at the 2:1 Mimas resonance beyond Schwarz' (1981) estimate which, taken at face value, is more than one order too small to account for the Cassini Division. A companion topic is why gaps are not seen at other resonances, e.g., those corresponding to satellites less massive than Mimas. In the next paragraphs we consider how resonant perturbations affect particle diffusion.

Random-walk arguments (cf. Franklin et al. 1980) lead to an estimate of the amount by which a particle will diffuse in the radial direction as the result of collisions. Each collision produces an incremental step ℓ_p so that, after a time t, a body undergoing many collisions will have traveled a distance Δr_d given by

$$\Delta r_d = \ell_p \left(\frac{t}{t_c} \right)^{\frac{1}{2}} , \tag{38}$$

where t_c is the average time between collisions, and is equal to the mean free path λ_p divided by the random velocity v_p of the particles. The random velocity is related to what we shall call the "free eccentricity" e_{fr} of ring particles by $v_p = e_{\mathrm{fr}} r \Omega_0$, where Ω_0 is the local angular velocity of rotation. Estimates reviewed in the chapter by Cuzzi et al. place the 1-dimensional random velocity at ~ 0.4 cm s^{-1}. Because this number is uncertain by a total factor of ~ 2, we shall use this numerical value here whenever a random velocity is required, i.e., no correction is applied to the 1-dimensional value. Near the Cassini Division, $r = 1.17 \times 10^{10}$ cm so that we obtain $e_{\mathrm{fr}} \simeq 2.3 \times 10^{-7}$ and $e_{\mathrm{fr}} r \simeq 2.7 \times 10^3$ cm. The step length ℓ_p taken at each collision is the smaller of the two quantities $e_{\mathrm{fr}} r$ or $\lambda_p = (n_p \sigma_p)^{-1}$, where n_p is the number density of particles and σ_p their average cross section. To evaluate λ_p we use the value for surface mass density $\sigma_p \sim 100$ g cm^{-2} as characteristic of the outer part of the B Ring (see chapter by Cuzzi et al.). For σ_p we have the relation

$$\sigma_p = 2h \, m_p \, n_p \tag{39}$$

where h is the half-thickness of the ring and m_p a typical particle mass. Both Saturn's field and the self-gravitation of ring material contribute to yield for h an equation that is well represented by

$$h = h_o \left[1 + \left(\frac{2\pi G\sigma}{\Omega_o^2} \right) \frac{1}{h} \right]^{-\frac{1}{2}} \tag{40}$$

where h_o, the value of h when only Saturn's gravity operates, is $(\pi/2)^{\frac{1}{2}} v_p/\Omega_o$. For $\sigma = 100$ g cm^{-2} the solution to Eq. (40) gives $h = 16.5$ m (consistent with observations described in Cuzzi et al.'s chapter) and shows that $h = 0.695\,h_o$, where the coefficient is the correction due to self-gravitation. Combining these numerical estimates with an assumed average particle radius of 10 cm (and a mean density of 1), we find from Eq. (39) that $n_p = 7.2 \times 10^{-6}$ cm^{-3} and $\lambda_p = 440$ cm. (Smaller particle radii decrease λ_p linearly.) Since $\lambda_p < e_{fr} r$, it is the appropriate quantity, $\ell_p = \lambda_p$, for use in Eq. (38) which, with $v_p = e_{fr}\, r\Omega_o$ and $\Omega_o = 2\pi/T$, can be written

$$\frac{\Delta r_d}{r} = \left(2\pi e_{fr} \frac{\lambda_p}{r} \frac{t}{T} \right)^{\frac{1}{2}} \tag{41}$$

where T is the local orbital period. In terms of the numerical estimates discussed above, Eq. (41) becomes

$$\frac{\Delta r_d}{r} = 2.4 \times 10^{-7} \left(\frac{t}{T} \right)^{\frac{1}{2}}. \tag{42}$$

This relation provides that the radial distance particles will diffuse in a time t if their mean random velocities are 0.4 cm s^{-1}.

In Sec. II we obtained a value for the eccentricity, e_{max}, forced at the 2:1 resonance, together with the time T_ℓ for its development from a much smaller value. These quantities combine to give an expression for the induced radial perturbation Δr_r of

$$\frac{\Delta r_r}{r} \simeq e_{max} \sin \left(\frac{t}{T_\ell} \right). \tag{43}$$

For the 2:1 Mimas resonance, $e_{max} = (2 \text{ m})^{\frac{1}{2}} = 0.005$ and, since $T_\ell = 48$ yr, $T_\ell/T = 3.6 \times 10^4$ so that, as long as $t \ll T_\ell$, Eq. (43) becomes

$$\frac{\Delta r_r}{r} \simeq 1.4 \times 10^{-7} \left(\frac{t}{T} \right). \tag{44}$$

A comparison of Eqs. (42) and (44) shows that the magnitude of the radial excursions of resonant bodies dominates those resulting from diffusion after only some ten orbital periods. We interpret this result and summarize the process of density depletion as follows: consider Saturn's rings initially to be a nearly uniform medium in which a strong resonance has just been established. Just interior to the resonance, particle motions are characterized by a small free eccentricity e_{fr} determined by their random motions. Mutual collisions cause some particles to diffuse outward into resonance. Once in resonance, nearby particles experience perturbations that are relatively large but coherent, so that the mean random velocity is not greatly altered. That is why we can still use e_{fr} in Eq. (42) to estimate the amount of diffusion in the resonant region, as long as the number density of particles in resonance is still comparable to the value outside it. (The analogy comes to mind of a dense crowd of commuters exiting through a narrow stairwell; their elbowing is largely independent of how fast or in what direction the stream of people moves.) As the resonant perturbations develop, the jostling particles that have diffused into resonance are moved as a body over a wide range which is, deep in resonance, measured by e_{max}. We have described early in this section how particles subjected to resonant perturbations preferentially collide with non-resonant bodies interior to resonance because that is where the collision frequency is largest. The dominance of the resonant return of particles over their outward diffusion implied by Δr_r being greater than Δr_d on a short time scale acts to reduce the amount of material in resonance.

Formally, Eq. (42) indicates that, after $t/T \simeq 10$, Δr_d has penetrated only a few percent of the total resonant width $\Delta a = 2.9$ km of the 2:1 Mimas resonance. In fact the time t/T is longer and the penetration deeper because, as the particle density in resonance falls, the mean free path λ_p increases so that the coefficient in Eq. (42) rises. We would require a solution of the diffusion equation to understand the details precisely. But the important point of this heuristic example is that, if the resonance width were narrower, which goes hand-in-hand with the slower development of resonant perturbations, then particles would diffuse through resonance so rapidly that no pronounced gap would arise. We have, therefore, in principle the means of establishing a criterion that sets the minimum satellite mass necessary to inhibit outward particle diffusion. This criterion imposes the condition that, for narrow gaps to occur at resonance, the diffusion time t_d must be greater than the time T_e which measures the growth of the forced eccentricity. Slightly recast, Eq. (41) gives the time for a particle to diffuse through the resonant width Δa:

$$t_d = \left(\frac{\Delta a}{a}\right)^2 \frac{T}{2\pi e_{fr}} \frac{a}{\lambda_p} \tag{45}$$

where, from Eq. (16) in Sec. II, $\Delta a/a = 3/2 \, m^{\frac{2}{3}}$ for the 2:1 resonance. Equation (31) yields

$$T_\ell = \left(\frac{a}{3}\right)^{\frac{3}{8}} \frac{2^{\frac{1}{3}}T}{[mP(a)]^{\frac{2}{3}}} .$$ (46)

The condition $t_d \gtrsim T$ and the numbers quoted above yield the limiting mass ratio, $m \gtrsim 8.7 \times 10^{-8}$. Increasing the mean free path in resonance raises the limit and reducing e_{fr} (cf. chapter by Borderies et al.) would lower it. Formally, Mimas with $m = 6.7 \times 10^{-8}$ fails, but improvements in the model as well as revised numbers could place it safely on the gap-forming side of the limit. Results from Voyager 2 (Tyler et al. 1982) suggest that Mimas' mass may be 20% larger than the value we have used here.

Other satellites of Saturn are much less likely to form gaps in its ring. A ranking by mass of the newly discovered inner satellites (see Goldreich and Tremaine 1982), which is somewhat approximate because only their sizes have been measured, places the larger of the two coorbital satellites, 1980S1 (Janus), next in mass to Mimas but smaller by a factor of at least ten. Within the boundaries of the ring all other first and higher order resonances, whether associated with Mimas or other satellites, have smaller values of Δa and longer times T_ℓ than the 2:1 Mimas resonance. We conclude from this pilot study that in all likelihood only this resonance has the possibility of generating a gap by establishing a barrier to outward particle diffusion. However, we should not dismiss the reasonable chance that the outer boundary of the A Ring is determined by an application of the arguments developed here to the 7:6 resonance with Janus. Within the confines of the ring it is the next strongest resonance. The nagging question remains: even if we have a possible mechanism to establish an inner boundary to the Cassini Division, what determines the total division width, i.e. what fixes its outer boundary? To this question we cannot now provide a satisfactory answer, but the generation of a gap as large as ~ 4500 km by the inward diffussion of particles to the 2:1 resonance does not appear hopeless. The effect of a resonance on ring dynamics, as given by Eq. (17), will be negligible if the induced eccentricity $e_r < e_{fr}$. The converse is likely to be true: particle motions and diffusion will be affected if $e_r > e_{fr}$. These two values of the eccentricity are equal at a distance from the 2:1 resonance of $\sim 3 \times 10^4$ km, some 7 times the observed division width. Perhaps the division is still slowly expanding into the A Ring. We urge that solving the diffusion equation for this problem be given priority.

The foregoing paragraphs have argued that the strongest resonance present within the ring system may be responsible for the Cassini Division. Many resonances that are weaker, either because the driving satellite has a lower mass or because of the nature of the resonance as explored in Sec. II, have a clear association with density waves. Esposito et al. (1983) have identified 13 definite density waves and numerous possible ones. On images of good resolution, there is a definite matching between first-order $j=1, j'=0$ resonances and small variations in optical thickness throughout much of the A Ring (cf.

Holberg et al. 1982). Resonances involving 1980S26 (outer F Ring shepherd) with i as great as 18 are clearly visible. Slightly more conspicuous is the converging series associated with 1980S27 (inner F Ring shepherd) with i as large as 35, i.e., $\sigma = 35\lambda_s - 34\lambda - \varpi$. The coorbital satellites, 1980S1 (Janus) and 1980S3 (Epimetheus), assumed to have a single mean motion of $518°24/day$, contribute resonances in the A and C Rings, where the $i=7$ case closely corresponds to the outer boundary of the former. Other $j=1$ resonances with Mimas (but with $j'=1$ and $j'=2$ so that $e_s^{|j'|}$ is present in various parameters; cf. Table II) have been detected as density waves at the three commensurabilities, $3\lambda_s \simeq \lambda$, $5\lambda_s \simeq 3\lambda$ and $8\lambda_s \simeq 5\lambda$. Wiesel (1982) has searched in the Cassini Division, where the optical thickness $\tau \lesssim 0.1$, for evidence of higher order resonances and concludes that the $j=2$ and $j=3$ (i.e. $\sigma = 2\lambda_s - \lambda + 2\varpi_s - 3\varpi$) cases are possibly present.

A question raised by these remarks is: Do all resonances greater than a given strength produce density waves—or gaps? Any answer requires some qualification because the response to a resonance depends on the local optical thickness which may be, but need not be, determined by the resonance. With the caveat in mind, our opinion is that there is no evidence for strong missing resonances. Esposito et al. (1983) provide some useful quantitative remarks in reaching the same conclusion.

If the answer to the question just posed is cautious, the response to its converse is definite: there are several large gaps, and many narrow features, that are not linked to any known resonance. The attempt (Franklin et al. 1982) to associate several conspicuous gaps in the outer parts of the A and C Rings with resonances excited by longitude dependent harmonics of the planet seems to have been unsuccessful given the more accurate positions now available, unless good reason can be found for supposing that the density inhomogeneities responsible for the effect rotate with a ~ 5 minute longer period than Saturn's magnetic field. Even were this explanation a possible one, the second most conspicuous ring feature (after the Cassini Division), the Encke Gap, remains unaccounted for by any resonant effect linked to known satellites external to the A Ring.

In ring systems other than Saturn's (and the Sun's asteroid belt) resonances appear to be of no importance. The population of the tenuous Jovian ring is dominated by small particles (cf. chapter by Burns et al.) whose motions are also affected by the planet's magnetosphere. No strong resonances with any of the Galilean satellites lie within this ring system. At present, there is no evidence that any of the known satellites of Uranus directly affect its rings. But in the case of Saturn as this and other chapters attest, resonances make a significant, if still incompletely understood, contribution to the structure and evolution of its rings.

Acknowledgments. We appreciate the comments of several readers on an earlier draft and thank the editors for their considerate prodding.

REFERENCES

Brouwer, D. 1963. The problem of the Kirkwood gaps in the asteroid belt. *Astron. J.* 68:152–159.

Brouwer, D., and Clemence, G. 1961. *Methods of Celestrial Mechanics* (New York: Academic Press).

Brown, E., and Shook, C. 1933. *Planetary Theory* (London: Cambridge Univ. Press).

Brown, H., Goddard, I., and Kane, J. 1967. Qualitative aspects of asteroid statistics. *Astrophys. J. Suppl.* 14:57–123.

Colombo, G., Franklin, F., and Munford, C. 1968. On a family of periodic orbits of the restricted three-body problem. *Astron. J.* 73:111–123.

Cuzzi, J., Lissauer, J., and Shu, F. 1981. Density waves in Saturn's rings. *Nature* 292:703–707.

Esposito, L., O'Callaghan, M., and West, R. 1983. The structure of Saturn's rings: Implications from the Voyager stellar occultation. *Icarus.* In press.

Franklin, F., Colombo, G., and Cook, A. 1982. A possible link between the rotation of Saturn and its ring structure. *Nature* 295:128–130.

Franklin, F., Lecar, M., Lin, D., and Papaloizou, J. 1980. Tidal torques on infrequently colliding particle disks. *Icarus* 42:271–280.

Franklin, F., Marsden, B., Williams, J., and Bardwell, C. 1975. Minor planets and comets in libration about the 2:1 resonance with Jupiter. *Astron. J.* 80:729–746.

Goldreich, P., and Tremaine, S. 1978. The formation of the Cassini Division in Saturn's rings. *Icarus* 34:240–253.

Goldreich, P., and Tremaine, S. 1982. The dynamics of planetary rings. *Ann. Rev. Astron. Astrophys.* 20:249–283.

Goldstein, H. 1953. *Classical Mechanics* (Cambridge, MA: Addison-Wesley Press), Chap. 10.

Greenberg, R. 1984. Orbital resonances in the Saturn system. In *Saturn,* eds. T. Gehrels and M. S. Matthews (Tucson: Univ. of Arizona Press). In press.

Hénon, M. 1969. Numerical exploration of the restricted problem. V. *Astron. Astrophys.* 1:223–238.

Holberg, J., Forrester, W., and Lissauer, J. 1982. Identification of resonant features within the rings of Saturn. *Nature* 297:115–120.

Izsak, I., Gerard, J., Efimba, R., and Barnett, M. 1964. Construction of Newcomb operators on a digital computer. *Smith. Astrophys. Obs. Spec. Rep.* 140.

Jefferys, W. 1971. *An Atlas of Surface of Sections of the Restricted Problem of Three Bodies* (Austin, TX: Publ. of Dept. of Astron.), Series II, Vol 3, No. 6.

Kozai, Y. 1957. On the astronomical constants of the Saturnian satellites. *Ann. Tokyo Astron. Obs.* 5:73–127.

Lecar, M., and Franklin, F. 1973. On the original distribution of the asteroids. *Icarus* 20:422–436.

Lin, D., and Papaloizou, J. 1979. Tidal torques on accretion disks in binary systems with extreme mass ratios. *Mon. Not. Roy. Astron. Soc.* 186:799–812.

Lissauer, J., and Cuzzi, J. 1982. Resonances in Saturn's rings. *Astron. J.* 87:1051–1058.

Marsden, B. 1970. On the relation between comets and minor planets. *Astron. J.* 75:206–217.

Newcomb, S. 1895. Development of the perturbing function. *Astron. Pap. of Amer. Eph.* V:1–48.

Null, G., Lau, E., Biller, E., and Anderson, J. 1981. Saturn gravity results obtained from Pioneer 11 and earth-based Saturn satellite data. *Astron. J.* 86:456–468.

Peale, S. 1976. Orbital resonances in the solar system. *Ann. Rev. Astron. Astrophys.* 14:215–246.

Poincaré, H. 1902. Sur les planètes du type d'Hécube. *Bull. Astron.* 19:289–310.

Reitsema, H. 1981a. The libration of Saturn's satellite Dione B. *Icarus* 48:23–28.

Reitsema, H. 1981b. Orbits of the Tethys' Lagrangian bodies. *Icarus* 48:140–142.

Schubart, J. 1964. Long-period effects in nearly commensurable cases of the restricted three-body problem. *Smith. Astrophys. Obs. Spec. Rep.* 149.

Schubart, J. 1968. Long-period effects in the motion of Hilda-type planets. *Astron. J.* 73:99–103.

Schwarz, M. 1981. Clearing the Cassini Division, *Icarus* 48:339–342.

Sinclair, A. 1969. The motions of minor planets close to commensurabilities with Jupiter. *Mon. Not. Roy. Astron. Soc.* 142:289–294.

Toomre, A. 1977. Theories of spiral structure. *Ann. Rev. Astron. Astrophys.* 15:437–478.

Tyler, G. L., Eshleman, V. R., Anderson, J. D., Levy, G. S., Lindal, G. F., Wood, G. E., and Croft, T. A. 1982. Radio science with Voyager 2 at Saturn: Atmosphere and ionosphere and the masses of Mimas, Tethys, and Iapetus. *Science* 215:553–558.

Ward, W. 1981. Solar nebula dispersal and the stability of the planetary system. *Icarus* 47:234–264.

Wiesel, W. 1974. *A Statistical Study of the Kirkwood Gaps.* Ph.D. Dissertation, Harvard Univ., Cambridge, MA.

Wiesel, W. 1982. Saturn's rings: Resonances about an oblate planet. *Icarus* 51:149–154.

Wisdom, J. 1982. The origin of the Kirkwood gaps: A mapping for asteroidal motion near the 3:1 commensurability. *Astron. J.* 87:577–593.

Yoder, C., Colombo, G., Synnott, S., and Yoder, K. 1983. Theory of motion of Saturn's coorbital satellites. *Icarus* 53:431–443.

DYNAMICS OF NARROW RINGS

STANLEY F. DERMOTT
Cornell University

The ring models described here were developed to account for the dynamical problems posed by the narrow rings of Uranus. Some of these rings are now known to be eccentric, inclined, nonuniform in width, optically thick, and narrow, with very sharp edges. The eccentric rings have common pericenters and large, positive eccentricity gradients. The theory of shepherding satellites successfully accounts for most of these features and can also account for some features of the narrow Saturnian rings, in particular, waves, kinks, and periodic variations in brightness. Outstanding problems include the putative relation between eccentricity and inclination displayed by eight of the nine Uranian rings, and the magnitudes of the tidal torques acting on the shepherding satellites. The horseshoe-orbit model, although viable, probably has more application to the narrow rings from which the Saturnian coorbital satellites formed. The angular momentum flow rate due to particle collisions is a minimum at the Lagrangian equilibrium points L_4 and L_5, and we can expect accretion to be rapid at these points.

The Voyager 2 image (Fig. 1) of the Saturnian F Ring showing the two "guardian" satellites, 1980S26 and 1980S27, leaves little doubt that the theory of shepherding satellites proposed by Goldreich and Tremaine (1979a) for the Uranian rings is essentially correct. There is no direct evidence that satellites confine the nine Uranian rings, but this must be regarded as very likely. The horseshoe-orbit model of Dermott et al. (1979), postulates that each narrow ring *contains* a small satellite. While viable, this model cannot account for some of the observed features of the Uranian rings. However, I will argue that the model can be applied to the narrow rings from which the coorbital Saturnian satellites presumably were formed. The aim of this chapter

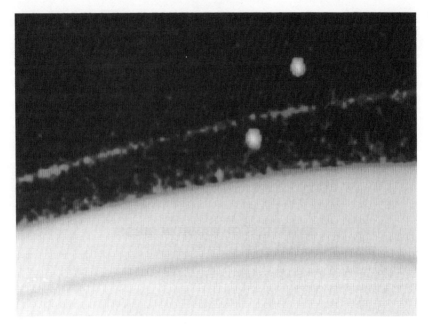

Fig. 1. Image taken by Voyager 2 of Saturn's A Ring, showing the narrow F Ring bracketed by its two shepherding satellites. Because the inner one (1980S27) orbits the planet slightly faster than the outer (1980S26), the satellites lap each other every 25 days. When this picture was taken, the shepherds were < 1800 km apart; they passed each other ~ 2 hr later. Marked azimuthal variations in the brightness of the F Ring are evident. (Image courtesy of JPL/ NASA.)

therefore will be to give an account of the dynamical processes involved in both types of ring-satellite gravitational interaction.

In Sec. I, I describe the problems posed by the narrow Uranian and Saturnian rings. The initial outstanding problem was that of particle confinement, and a good portion of this chapter is devoted to that problem alone. In the model of Goldreich and Tremaine (Sec. II), particle confinement is achieved by tidal torques exerted on the ring by small nearby satellites. The torque arises from the second-order change in semimajor axis experienced by each ring particle at each close encounter with the shepherding satellite. However, for the cumulative change in semimajor axis to be finite, the ring particle must lose some orbital energy between consecutive encounters. This loss is achieved either by collisions, which result in eccentricity damping, or by energy transfer to other ring particles through the excitation of spiral density waves. I give a simple derivation of the torque equation and explain why this torque is correctly described as a tidal torque. In Sec. II, I also discuss wave formation in very narrow rings and show how these waves could account for the kinks, braids, and periodic variations in brightness displayed by the Saturnian F Ring.

The linear relation between ring width and orbital radius displayed by the
ϵ Ring of Uranus and some other narrow rings shows that the pericenters of
the particle orbits have a common precession rate, and that the pericenters are
closely aligned (Sec. III). Such alignment may be maintained by the ring's
self-gravitation (Goldreich and Tremaine 1979a,b), in which case the ring
mass and surface density can be deduced from observable ring parameters.
However, it is notable that all the measured eccentricity gradients are close to
unity; this prompted Dermott and Murray (1980) to suggest that the rings
are close-packed at pericenter and that apse alignment is maintained by
particle collisions. This problem has since been complicated by the observation that some Uranian rings have small inclinations, and by a possible
relationship between a ring's eccentricity and its inclination (French et al.
1982). Node alignment is easily accommodated by the self-gravitational
model (Borderies et al. 1982a; Yoder 1983), but at present there is
no theory to account for the putative relation between eccentricity and inclination. In Sec. III, I also discuss how recent developments of the
shepherding-satellite model of Goldreich and Tremaine account for the
existence of rings with very sharp edges.

In the horseshoe-orbit model (Sec. IV), confinement is achieved by the
gravitational action of a satellite embedded in the ring; this satellite also acts
as a source of ring particles. The dynamics of this problem are of particular
interest when the satellite/planet mass ratio, m/M, is very small $(m/M)^{\frac{1}{3}} \ll 1$,
for only then are the horseshoe-orbit solutions of the equations of motion
dominant. I discuss the stability of this type of orbit and the formation of
coorbital satellites. A question of interest here is the complete absence of
coorbital satellites in the Jovian system, in contrast to their relative abundance around Saturn.

The most important recent review to emphasize the dynamics of narrow
rings is by Goldreich and Tremaine (1982). Other reviews include those of Ip
(1980a, b), Dermott (1981a) and Brahic (1982).

I. PROBLEMS

A. Observations.

The models described in this chapter were inspired by the discovery of
the rings of Uranus (Elliot 1977, 1979), so it is appropriate that I describe
these rings first. Here I emphasize only those features which constrain the
models; for a more complete description of the Uranian rings see the chapter
by Elliot and Nicholson in this book.

All the radii of the nine Uranian rings lie in a comparatively narrow range
of radial width $< 10,000$ km. The innermost ring is 16,000 km above the
planet and the outermost ring is 78,000 km lower than the orbit of Miranda.
The rings are very narrow and optically thick (optical depths $\tau \lesssim 1$). In fact,

only three of the rings have been fully resolved. These have been found to have nonuniform widths. The width of the ϵ Ring (the outermost one) increases from 21 km at pericenter to 96 km at apocenter and has a mean normal optical depth $\tau \approx 1$ at its widest point; the optical depth at the narrowest point is too high to be measured reliably. The width of the α Ring increases from 5 km at pericenter to 10 km at apocenter and the corresponding range of τ is 1.4 to 0.7. The figures for the β Ring are similar; the range in width is 5 to 11 km and the corresponding range in τ is ~ 1.5 to 0.35. All the other rings except the η Ring have widths < 4 km (see chapter by Elliot and Nicholson; also Nicholson et al. 1982).

The edges of some of the rings are remarkably sharp; witness the Fresnel diffraction spikes exhibited by, for example, the γ Ring. The Voyager 2 photopolarimeter occulation profiles (Lane et al. 1982) have revealed that the outer edge of Saturn's A Ring, located at the 7:6 resonance with the larger coorbital satellite (1980S1), is sharp on a distance scale < 1 km. The outer edge of Saturn's B Ring, at the 2:1 resonance with Mimas, is comparably sharp. In Sec. III, I discuss the role of resonance in the formation of these sharp edges (Borderies et al. 1982b). The Uranian rings are very dark and particulate (gaseous rings [Van Flandern 1979] can be discounted for the reasons given by Fanale et al. [1980], Gradie [1980] and Hunten [1980]). There appears to be very little material between the rings (Matthews et al. 1982).

Most of the resolved rings show some structure. That of the ϵ Ring has been described as undulating and appears to be stable, i.e., time independent. Near apoapse, the optical depth of the α Ring is a minimum near its center: the so-called "double-dip" structure (see chapter by Elliot and Nicholson). A similar structure is displayed by the narrow ring in Saturn's C Ring at 1.29 Saturn radii (R_s) (Sandel et al. 1982) and by the narrow, optically thick spike discovered by the Voyager 2 PPS in Saturn's F Ring (Lane et al. 1982). The Uranian η Ring is broad (width ≈ 50 km) and diffuse with a sharp, unresolved spike at its inner edge, just inside a region of low optical depth (see Fig. 6 in chapter by Elliot and Nicholson). The δ Ring also shows a broad section of diffuse material.

That the rings may be eccentric was suggested by Elliot et al. (1977), developed by Lucke (1978), and proved by Nicholson et al. (1978) who discovered the linear relation between the radii and the radial widths of sections of the ϵ Ring (see Sec. III). All the Uranian rings, with the possible exceptions of the η and ϵ Rings, are now known to be both eccentric and inclined to the equatorial plane of the planet (French et al. 1982). It also appears that all the rings except the ϵ Ring have $e \approx \sin i$ (to within a factor < 5), where e is eccentricity and i inclination, and that both e and i increase with decreasing semimajor axis a (see Fig. 14 of Elliot and Nicholson, in this book). If this relation has some physical significance, then the rate of increase of e with decreasing a is so marked that the inner boundary of the ring system may be that region where $2ae$ would be comparable with the mean ring separation; this would occur at $a \approx 38,000$ km.

Most of the structure observed in Saturn's rings is probably due to either diffusion instabilities (Lin and Bodenheimer 1981; Lukkari 1981; Ward 1981) or spiral density waves (Goldreich and Tremaine 1978b; Cuzzi et al. 1981). However, in almost every case where clear gaps appear in the rings, eccentric ringlets are found (Smith et al. 1981, 1982; Stone and Miner 1982; Lane et al. 1982). Narrow eccentric rings exist at 1.29 R_s (associated with the Titan apsidal resonance [Porco et al. 1982]), at 1.45 R_s, and at 1.95 R_s. The latter is just outside the 2:1 Mimas resonance, but appears to be a Keplerian ellipse precessing under the influence of Saturn's oblateness (Smith et al. 1982). In each case, like the eccentric Uranian rings, these rings are widest at apocenter and narrowest at pericenter.

At least two narrow discontinuous rings, or arcs, exist within the Encke Division near 2.21 R_s. Two separate rings were seen in the Voyager 1 images, whereas the Voyager 2 images of the gap each show only one ring, but in different images these arcs appear at more than one radial location suggesting that the ring or rings may be discontinuous. Both arcs show large azimuthal variations in brightness on a length scale of 3000 km, and one arc shows kinks or waves tens of km in amplitude (peak to peak) and 1000 km apart (Smith et al. 1982; chapter by Cuzzi et al.).

The strangest narrow ring in the solar system is undoubtedly the F Ring of Saturn discovered by Pioneer 11 (Gehrels et al. 1980). Voyager 1 images of this ring revealed "braids," regions where the ring is split into two separate components (see Fig. 2), clumps, kinks, and large azimuthal variations in brightness (Smith et al. 1981). Braiding was also seen by Voyager 2, but only in one image (Fig. 3) (Smith et al. 1982). The braids or loops have lengths between 7000 and 10,000 km; the initial report of a length of 700 km (Smith et al. 1981) was a mistake (Smith et al. 1982). The spacing of the clumps is similar, but ranges from 5000 to 13,000 km (Smith et al. 1982). These clumps appear fuzzy on most images, but at least one clump is sharply defined and may be a small satellite embedded in the ring (Smith et al. 1982). The widths of the loops shown in the Voyager 1 images are between 30 and 40 km (Smith et al. 1981, 1982).

In the Voyager 2 images, the F Ring appeared 500 km wide and consisted of one bright component and at least four faint components. The Voyager 2 PPS detected a ring of width about 60 km with a sharp, 1 km wide, optically thick ($\tau \approx 1$) component (Lane et al. 1982).

B. Confinement.

Interparticle collisions, Poynting-Robertson light drag and plasma drag cause an unconstrained narrow ring to gradually spread (Goldreich and Tremaine 1979a, 1982). Brahic (1977) has shown that a narrow ring of uniform surface density, mass m_r, mean radius r and width W ($<<r$) has an

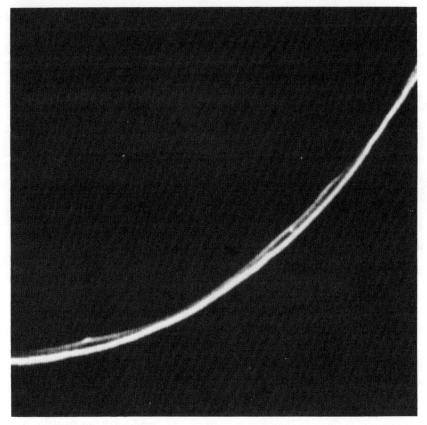

Fig. 2. Three separate components of Saturn's F Ring, seen in a Voyager 1 image. Two prominent bright strands appear twisted and kinked, giving a braided appearance to the ring; the fainter innermost strand largely lacks such nonuniformities. (Image courtesy of JPL/ NASA.)

energy E (at fixed angular momentum) that varies with W as

$$E \approx - \frac{GMm_r W^2}{32r^3} + \text{constant}, \tag{1}$$

where M is the mass of the planet and G the gravitational constant. Thus E is a maximum when W is a minimum, and any loss of energy will result in spreading on a timescale t_d given by

$$t_d = W/\dot{W} = - m_r \, \Omega^2 W^2/(16 \, \dot{E}) \tag{2}$$

where $\Omega \equiv (GM/r^3)^{\frac{1}{2}}$ is the mean angular velocity of the ring. On the microscopic scale, we can relate $\dot{E}$ to the particle collision frequency ω_c by

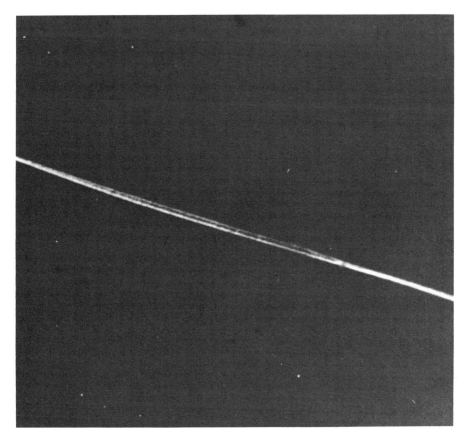

Fig. 3. The only Voyager 2 image of the F Ring that shows braiding. (Image courtesy of JPL/NASA.)

$$\dot{E} \approx - 3v^2 m_r \, \omega_c \, (1 - \epsilon^2) \qquad (3)$$

where v is the one-dimensional random velocity and ϵ is the coefficient of restitution of the particles (Goldreich and Tremaine 1982). An alternative approach is to treat the ring as a differentially rotating fluid of density ρ in which a shear stress

$$S \equiv \rho \nu r \, \frac{d\Omega}{dr} \qquad (4)$$

generates a torque which transfers angular momentum with direction and rate determined by the angular velocity gradient $d\Omega/dr$ (Safronov 1969; Lynden-Bell and Pringle 1974). $\dot{E}$ is then the work done by the torque. For effective kinematic viscosity ν we have

$$\nu = \frac{v^2}{\Omega} \frac{0.46\tau}{1+\tau^2} \tag{5}$$

(Cook and Franklin 1964; Goldreich and Tremaine 1978a). Both approaches yield the result that t_d is comparable to the time a particle needs to random walk across the ring (Brahic 1977; Goldreich and Tremaine 1978a),

$$t_d \approx \frac{1}{\tau\Omega} \left(\frac{W}{d} \right)^2 \tag{6}$$

where d is the characteristic radius of the particles. For discussion of the meaning of "characteristic" see Hénon (1981, 1983), Goldreich and Tremaine (1982), and the chapter by Weidenschilling et al. For rings located at about 2 planetary radii,

$$t_d \approx \frac{2\times 10^{-4}}{\tau} \left(\frac{W}{d} \right)^2 \text{ yr.} \tag{7}$$

Poynting-Robertson light drag causes the orbit of a particle of density ρ_r and radius d, moving in a circular orbit about a planet of mean orbital radius a_p, to decay on a time scale

$$t_{\text{pr}} \approx \frac{8\rho_r dc^2}{3(L_\odot/4\pi a_p^2)\, Q_{\text{pr}} (5+\cos^2 i)} \tag{8}$$

where $L_\odot$ is the solar luminosity, c is the velocity of light, i is the inclination of the particle orbit relative to the ecliptic and

$$Q_{\text{pr}} = Q_{\text{abs}} + Q_{\text{sca}} (1-<\cos\alpha>) \tag{9}$$

where Q_{abs} is the absorption coefficient (1 for a perfect absorber), Q_{sca} is the scattering coefficient, and α is the scattering angle (Burns et al. 1979); see also the chapter by Mignard. The orbits of particles of different sizes will decay at different rates and, despite the effects of collisions which will average out the decay rates, if the surface density of the ring is not radially uniform, then we must expect a narrow ring to broaden on a time scale

$$t_{\text{spread}} \approx t_{\text{pr}} (W/r) . \tag{10}$$

For the Uranian rings

$$t_{\text{spread}} \approx 2 \times 10^5\, Wd \text{ yr}, \tag{11}$$

where W is in km and d is in cm. Since t_d increases and t_{spread} decreases if d

decreases, and vice versa if d increases, it follows that regardless of particle size narrow rings could not remain narrow for more than about 10^7 yr. Therefore, if the observed narrow rings are not young, then they must be confined (Goldreich and Tremaine 1979a).

If a planet has a magnetic field which maintains a corotating magnetosphere, and recent observations of auroral hydrogen Lyman-α emission (Clarke 1982) indicate that this is the case for Uranus as well as Saturn, then absorption of this plasma by the ring particles leads to orbital decay on a timescale

$$t_p \approx \frac{2\,dp_r}{3\,r\rho_p}\,\frac{\Omega}{(\Omega-\Omega_p)^2} \tag{12}$$

where Ω_p is the angular velocity of the planet and ρ_p is the plasma density (Burns et al. 1980). Since the planet is the source of the angular momentum, then plasma drag, like tidal drag, acts to push the particles away from the synchronous orbit (see chapter by Grün et al.). For the F Ring of Saturn

$$t_p \approx 3 \times 10^7 d \text{ yr,} \tag{13}$$

where d is in cm. However, since the F Ring partially clears out its flux tubes, the above estimate is a lower limit (Goldreich and Tremaine 1982).

C. Corotational Resonance.

Shortly after the discovery of the narrow Uranian rings, it was suggested that the observed occultations may have been produced by arcs of particles librating in stable, corotational resonances (Dermott and Gold 1977). The dynamics are illustrated in Fig. 4 (see also Greenberg 1984 and the chapter by Franklin et al.). In a resonance of the type shown, we have

$$(p+q)\lambda' - p\lambda - q\varpi' = \phi \tag{14}$$

where λ and λ' are, respectively, the mean longitudes of a particle and of the perturbing satellite, ϖ' is the pericenter of the perturbing satellite's eccentric orbit, and p and q are integers. In the stable configuration ϕ librates (i.e. oscillates) about π; if $q=1$ (first-order resonance), then all conjunctions of the satellite and the particle take place near the apocenter of the satellite's orbit. Thus, the existence of the resonance guarantees that the separation of particle and satellite at conjunction is close to a maximum. (If $q=2$ (second-order resonance), then only every other conjunction takes place near apocenter, and so forth.) Resonances can involve the motions of nodes or even combinations of the motions of nodes and of pericenters, but these more complex cases are not discussed here (see chapters by Franklin et al., and by Shu; also Greenberg 1984).

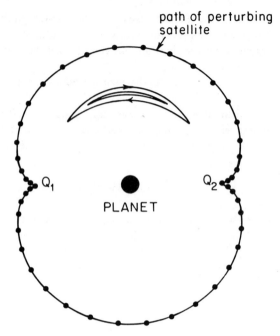

Fig. 4. Libration of particles in a corotation resonance about a longitude which is stationary in a frame corotating with the pattern speed of the perturbing potential. In this frame, the path of the perturbing satellite is closed and stationary. The example shown here is the 3:2 ($\approx$ n/n') resonance. Points marked on the path of the perturbing satellite denote positions at equal time intervals. The motion of the perturbing satellite in this frame is slow near the points marked Q_1 and Q_2.

Differentiating Eq. (14) with respect to time and rearranging, we obtain

$$\frac{n' - \dot{\bar{\omega}}'}{n - \dot{\bar{\omega}}'} = \frac{p}{p+q} . \tag{15}$$

Thus, the mean motions relative to the motion of the pericenter are exactly commensurate. It follows that, in a frame rotating with the mean motion n of the particle, the path of the perturbing satellite is closed (see Fig. 4). The gravitational influence of the perturbing satellite on the orbit of the particle can now be modeled by spreading the mass of the satellite along its closed path in such a way that the line density at any one point is proportional to the time spent in that part of the path. In Fig. 4, the positions of the satellite are marked at equal time intervals; thus the spacing of these marks is a measure of the line density. The line density distribution represents the disturbing potential, and we can say that in a corotational resonance the resonant particle corotates with the pattern speed of the disturbing potential. The example shown in the 3:2 resonance ($[p+q]/p=3/2$), for which the line density is a maximum at the points marked Q_1 and Q_2. Different values of p and q give rise to different distributions of mass, but in all cases, if the orbital eccentricity of the satellite is not zero, then the line density distribution is not uniform. It is this azimuthal nonuniformity which gives rise to the forces that stabilize the resonance. Consider a particle displaced from the equilibrium point. If the particle is displaced towards the planet, then its mean motion will be greater than the resonant mean motion n and, as shown in Fig. 4, it will drift in the prograde sense. The force on the particle due to the mass distribution at Q_1 will then have a greater effect than that at Q_2, and will act to increase the angular momentum of the particle. But, since mean motion decreases with increasing angular momentum, the net effect of the force is to reverse the sense of drift. Thus, a displaced particle can librate about a longitude that is fixed in the rotating reference frame. Particles moving in nested librating paths (see Fig. 4) will form a compact arc of particles, that is, a narrow discontinuous ring (Dermott and Gold 1977).

Particle arcs associated with corotational resonances have some interesting properties. Weak external drag forces, such as Poynting-Robertson light drag, do not necessarily destroy their stability; the perturbing satellite can supply the angular momentum removed by the drag force in such a way that the exact resonance is maintained, and the equilibrium longitude in the rotating reference frame is merely displaced (Goldreich 1965). Internal dissipation due to particle collisions still leads to ring spreading, but, since for small displacements the libration period is independent of the libration amplitude, the shear forces and the net angular momentum flow rates due to particle collisions are very much less than in an unconstrained ring of similar proportions (Dermott et al. 1983).

The weakness of this ring model is that for most resonances the arcs are very narrow (Aksnes 1977; Goldreich and Nicholson 1977). The maximum width W of an arc of librating particles is determined by the strength of the average perturbing force and is given by

$$W = 8 \left(\frac{a|R|}{3GM} \right)^{\frac{1}{2}} a \qquad (16)$$

where a is the semimajor axis of the ring particles, and R is the term in the expansion of the disturbing function associated with the resonant argument (Goldreich and Nicholson 1977; Dermott and Murray 1983). For two-body resonances of the type described by Eq. (14),

$$\frac{a|R|}{GM} = \frac{\alpha f(\alpha) m' e'^q}{M} \qquad (17)$$

where the primed quantities refer to the perturbing satellite and $\alpha = a/a'$.

S. F. DERMOTT

TABLE I

Arc Widths of Corotational Resonances

Satellite	m'/M	e	$p+q{:}p$	a (Planetary Radii)	$\alpha f(\alpha)$	W (km)
Mimas	$7\ 10^{-8}$	0.0201	2:1	1.95	0.750	18
			5:3	2.20	2.329	5
1980S1	$8\ 10^{-9}$	0.0070	5:4	2.17	3.145	8
			6:5	2.23	3.946	9
			7:6	2.27	4.747	10
Miranda	$10^{-6?}$	0.012	4:1	1.97	0.097	0.1

Values of $\alpha f(\alpha)$ and W for a few of the corotational resonances in Saturn's rings near Encke's Gap are given in Table I. The arcs associated with these particular resonances are wide and should give rise to observable phenomena. However, the two-body resonances in the region of the Uranian rings are high-order resonances ($q > 3$), so the perturbing forces are weak and the arcs are narrow ($W < 1$ km). Three-body corotational resonances involving two satellites of masses m_1 and m_2 and a ring particle are also possible, but, since $|R| \propto m_1 m_2 / M^2$, these resonances tend to be even weaker (Aksnes 1977; Goldreich and Nicholson 1977).

The corotational resonance model is a viable ring model in that it can account for narrow discontinuous arcs of confined particles which are to some extent stable against the disruptive effects of Poynting-Robertson light drag and interparticle collisions. However, it was soon realized that the model could not account for the widths of the Uranian rings, and other models were sought. We now know that the Uranian rings are not discontinuous arcs, but arcs associated with known satellites should exist in the Saturnian ring system.

II. SHEPHERDING SATELLITES

A. Shepherd Dynamics.

The dynamics of the shepherding satellite model of Goldreich and Tremaine (1979a) are not easily described in a few lines. The impulse approximation of Lin and Papaloizou (1979), the usual simple approach, has been rejected by Hénon (1983). There is also the problem of explaining the role of dissipation, since an eccentricity damping term is curiously absent from the "standard formula" (Eq. 25) for the torque (Hénon 1983; Greenberg 1983).

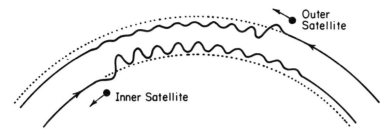

Fig. 5. Schematic diagram showing the action of the shepherding satellites. Arrows on the ring particle paths show the direction of motion of the particles with respect to the perturbing satellite. For clarity I assume here that the outer strand of particles is perturbed only by the outer satellite; in fact both satellites sometimes act on the same particles at the same time, and at all times each satellite acts on all the strands. At conjunction, a satellite changes the eccentricity and the semimajor axis of each ring particle. Eccentricity damping by particle collisions results in a further change in the semimajor axis, but this change is negligible in comparison with that produced by the satellite interaction.

A qualitative description of the mechanism is given in Fig. 5. On encounter with a nearby satellite, a ring particle briefly experiences an attractive force in the direction of the satellite. For a particle initially moving in a circular orbit, this causes the excitation of a small eccentricity e and a change δa in its semimajor axis in such a direction that the particle appears to have been repelled by the satellite. The change δh in the angular momentum h of the particle is given by

$$2\,\frac{\delta h}{h} = \frac{\delta a}{a} - e^2 \;. \tag{18}$$

However, even though we usually have $ea \gg \delta a$, $\delta a/a \gg e^2$ (see Eq. 21) and δh is effectively determined by δa alone. But, paradoxically, $\delta a/a$, which is of second order in m'/M, can be calculated from e, which is of first order in m'/M, using the Jacobi integral for the circular restricted three-body problem, or equivalently Tisserand's relation:

$$\frac{a'}{a} + 2\left(\frac{a}{a'}\right)^{\frac{1}{2}} (1-e^2)^{\frac{1}{2}} = C + \mathcal{O}\,(m'/M) \tag{19}$$

where C is a constant (Goldreich and Tremaine 1982). Substituting $\Delta a = a - a'$ into Eq. (19) and expanding binomially, we obtain (Dermott and Murray 1981a)

$$\frac{3}{4}\left(\frac{\Delta a_n}{a'}\right)^2 - e_n^2 \simeq C - 3 \tag{20}$$

where the subscript n refers to values after the nth encounter with the perturbing satellite. Thus, an increase in e always produces an increase in the separation $|\Delta a|$ of the semimajor axes of the ring particle and the satellite.

For the configuration shown in Fig. 5, $\delta a = \Delta a_1 - \Delta a_0$, and substitution into Eq. (20) yields

$$\frac{\delta a}{a} = \frac{2}{3} \frac{a}{\Delta a_0} e^2 . \tag{21}$$

e can be estimated from Gauss's form of the perturbation equation:

$$\frac{de}{dt} \simeq \frac{1}{na} (D \sin f + 2T \cos f) \tag{22}$$

where D and T are the radial and tangential forces (per unit mass) on the particle due to the satellite, and f is the particle's true anomaly. If we neglect T and assume that a radial force $Gm'/\Delta a_0^2$ acts for a time $2\Delta a_0/Ua$ (as in the impulse approximation), where $U = 3n\Delta a_0/2a$ is the relative angular velocity of the particle and the satellite, and that during this brief interval ($0.2P$ where P is the orbital period of the particle), the particle is always close to quadrature and $\sin f \approx 1$, then we obtain

$$e = \frac{4}{3} \frac{m'}{M} \left(\frac{a}{\Delta a_0}\right)^2 \tag{23}$$

(cf. Lin and Papaloizou 1979). A more accurate calculation by Julian and Toomre (1966) shows that the coefficient in this equation is 2.24 rather than 4/3. That our approximation underestimates the coefficient is partly due to neglecting the tangential force. Although the sign of T changes during the encounter, the sign of $\cos f$ also changes and $< T \cos f >$ is actually far from negligible.

To calculate the mean torque Γ exerted on the ring, we must now make an assumption about the effects of repeated encounters between a ring particle and a satellite. In Fig. 5 I assume that collisions between the ring particles act to damp the excited eccentricity, and that at the next encounter with the same satellite the particle is again moving in a circular orbit and experiences the same exchange of angular momentum δh. Since the time between encounters is $2\pi/U$, Eqs. (18), (21) and (23) (with coefficient 2.24 rather than 4/3) yield

$$\Gamma = \dot{h} = \frac{\delta h}{2\pi/U} , \tag{24}$$

$$= 0.399 \left(\frac{Gm'}{n\Delta a_0^2}\right)^2 m_r . \tag{25}$$

In this case, a damping factor does not appear in the equation for the torque because it is assumed that between encounters the excited eccentricity is totally damped.

At the other extreme, if the excited eccentricity is completely undamped, then the evolution of Δa due to repeated encounters is described by the Jacobi integral, or more approximately by Eq. (20). Using first-order perturbation theory it can be shown that e does not increase indefinitely but merely oscillates about some mean, and it follows from this and Eq. (20) that to second-order in m'/M, in accord with Poisson's theorem on the invariability of semimajor axes, Δa must also oscillate about some mean and that the torque on average is zero.

Between these two extremes, we must expect the torque to depend on the rate of eccentricity damping. The following argument makes it clear that this is the case. The existence of a torque Γ implies that work is performed and that the total mechanical energy E of the system decreases, i.e., is dissipated as heat, at a rate

$$\dot{E} = -U\Gamma . \tag{26}$$

If ΔE is the energy dissipated in the time $2\pi/U$ between encounters, then

$$\Delta E = 2\pi\Gamma \tag{27}$$

and, from Eqs. (18), (21), and (23), we have

$$\Delta E = \frac{1}{2} m_r (ean)^2 . \tag{28}$$

That is, ΔE may be thought of as the kinetic energy associated with the radial or eccentric motion of the ring particles. The magnitude of the torque is related to the rate at which this energy is dissipated.

B. Wave Formation.

In a frame corotating with the perturbing satellite, all particles initially moving in circular orbits must follow identical paths after encounter. It follows that each satellite generates a standing wave of amplitude $A = ea$ and wavelength

$$\ell = \frac{aU}{n/2\pi} = 3\pi\Delta a_0 \tag{29}$$

(see Fig. 5). In the inertial frame, each particle moves in an independent Keplerian ellipse, but the pericenters of these elliptical orbits and the phases of the particles on the orbits are such that the locus of the particles is a sinusoidal wave that moves through the ring with the angular velocity of the perturbing satellite (Dermott 1981b). For the F Ring of Saturn

$$A = 6 \times 10^{15} \, (m'/M) \, (\Delta a_0)^{-2} \, \text{km} \tag{30}$$

where Δa_0 is in km, and

$$l = 9.4 \, \Delta a_0 \, . \tag{31}$$

Just as the Moon is attracted by the tidal wave that it raises on the Earth, the satellite is attracted by the tidal wave that it raises on the ring, and the calculation of this force gives us another way of estimating the torque Γ. For heuristic purposes, the ring particles are shown in Fig. 6 as unperturbed ($x=0$) before encounter ($y<0$), and with a displacement

$$x = ea \, \sin(2\pi y/ \, l \,) \tag{32}$$

after encounter ($y>0$). The torque can easily be shown to be

$$\Gamma = 2.24 \, \frac{3I}{2\pi} \left(\frac{Gm'}{n\Delta a_0^2} \right)^2 m_r \tag{33}$$

(cf. Eq. 25), where

$$I = \int\limits_0^\infty \frac{(y/\Delta a_0) \, \sin \, (2y/3\Delta a_0) \, \mathrm{d} \, (y/\Delta a_0)}{[1 + (y/\Delta a_0)^2]^{\frac{3}{2}}} \approx \frac{1}{6} \, . \tag{34}$$

Thus, in this approximation the coefficient in Eq. (33) is 0.18 rather than 0.399. This discrepancy is due to my oversimplified representation of the particle path. The wave is only truly sinusoidal when y is large ($>> \, l \,$). Also, the phase of the wave in Fig. 6 is not an accurate representation of the true phase, but the basic physics is correct. Note that the mean torque is finite only if the wave maintains a constant phase with respect to the satellite. For this reason, a satellite only interacts with its own wave; the waves raised by other satellites have no direct effect, although effects may arise if the eccentricity damping mechanism is nonlinear.

C. Lindblad Resonance.

In any narrow section of a ring, the perturbations of the ring particle orbits are highly coherent, and, although particle collisions will act to damp the waves, the process may be slow. For this reason, the magnitude of the torque given by Eq. (25) may in some circumstances be an overestimate. If

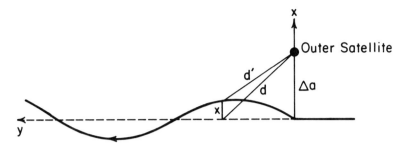

Fig. 6. The tidal torque estimated from the force on the satellite due to the wave. The integral (Eq. 34) is dominated by contributions from those particles with $|y| < \ell/2$. Since $d' < d$, those particles with $y > 0$ exert a greater force on the satellite than corresponding unperturbed particles with $y < 0$. Thus, the resultant force accelerates the motion of the satellite and retards the motion of the ring particles. The arrow on the wave shows the direction of motion of the ring particles with respect to the outer perturbing satellite.

the waves survive from one encounter to the next, then we must consider the possibility of resonance (see chapter by Franklin et al.). This will occur wherever the local ring circumference is an integral number of wavelengths, that is, wherever

$$2\pi a/\ell = p+1 \tag{35}$$

where p is an integer ≥ 0. Consecutive perturbations will then be in phase and a wave with amplitude significantly $> A$ may result.

The approximate resonance condition can also be written

$$\frac{n'}{n} \approx \frac{p}{p+q} \text{ or } \frac{p+q}{p}, \tag{36}$$

depending on whether the perturbing satellite is outside or inside the ring, respectively. For a satellite outside the ring $(n' < n)$ the exact resonance condition is

$$pn - (p+q)n' + q\tilde{\omega} = 0 \tag{37}$$

where $\tilde{\omega}$ is the pericenter of the ring particle orbit. For Lindblad resonance, the order of the resonance q must be unity. (There are also Lindblad resonances with terms in $\tilde{\omega}'$, but those are not discussed here; see Greenberg 1984 and chapters by Franklin et al., and by Shu.) If the particle is locked in resonance and the forced eccentricity is small, then the pericenters of the ring particle orbits are not aligned; rather, $\tilde{\omega}$, $\tilde{\omega}$, and the mean longitude or phase of each particle are such that conjunctions of the particle and the satellite always occur at an apse of the particle's orbit. The magnitude of the forced eccentricity, in the absence of damping, is given by

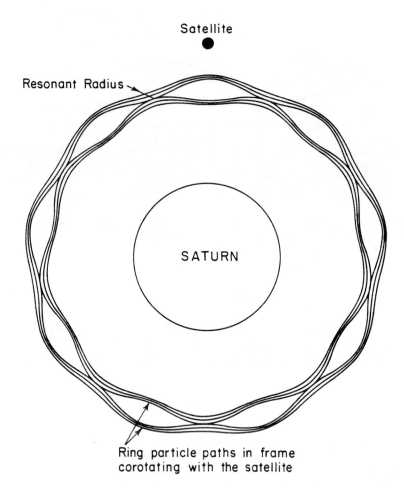

Fig. 7. Ring particle paths in a frame corotating with the perturbing satellite, illustrating the dynamics of Lindblad resonance. Resonant gravitational interactions at radii in the ring where the ratio of the satellite and ring particle mean motions are close to $p/(p+1)$, where p is an integer, generate a system of waves which are stationary in the rotating frame. The waves on opposite sides of the exact resonance have a phase difference of 180°; the resultant wave pattern consists of $p+1$ equally spaced loops. The variations of surface density associated with this pattern can act on the other ring particles to excite a spiral density wave (see Fig. 1i). (Figure copyright of *Nature*, MacMillan Journals Ltd.)

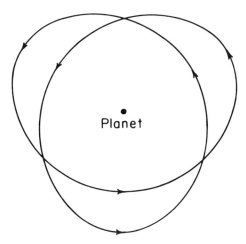

Fig. 8. Path of a particle in a reference frame rotating with the perturbing satellite. The particle is trapped in a 3:1 resonance for which $(p+q)n'-pn-q\dot{\omega}=0$, $p=1$, and $q=2$. In this frame, all paths for which $q \neq 1$ are self-intersecting. Thus, in a densely populated ring of particles, Lindblad resonances for which $q \neq 1$ cannot be established.

$$e = \left| \frac{1}{2} \frac{(m'/M)f(p)n}{(p+1)n'-pn} \right| \qquad (38)$$

(Greenberg 1973). Thus e increases markedly as the exact resonance is approached. At the exact resonance, the phase of the response changes by 180°. Similar behavior is observed in any driven harmonic oscillator. For particles outside the exact resonance, conjunction always occurs at apocenter, whereas for particles inside the exact resonance, conjunction always occurs at pericenter. Thus, the satellite excites a wave pattern of $p+1$ equally spaced loops which corotate with the perturbing satellite (see Fig. 7).

The particle path pattern shown in Fig. 7 is one of streamline flow for which interparticle collisions are a minimum. Such nonintersecting nested paths are possible only if $q=1$ and every conjunction takes place at the same point in the orbit. If $q>1$, then, no matter how small the forced eccentricity, each particle path always intersects itself (see Fig. 8). Obviously, in a densely populated ring of particles, resonant orbits that intersect can neither be established nor maintained.

Even if $q=1$, then, because of the phase change at exact resonance, orbits close enough to exact resonance can still intersect. In Fig. 7, these orbits have been eliminated. The edges of the empty loops are then defined by those orbits

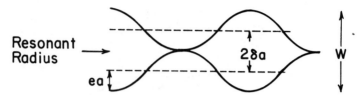

Fig. 9. Particle paths displaced $\delta a (=W/4)$ from exact resonance. Since $ea = \delta a$ and $e = 0.546(m/M)/(\delta a/a)$, we have $W = 2.96\,(m/M)^{\frac{1}{4}} a$.

that are displaced from exact resonance by a distance δa, such that the resultant forced eccentricity is $\delta a/a$ (see Fig. 9). If p is large, then

$$e = 0.546\,\frac{(m'/M)}{\delta a/a} \tag{39}$$

and e is independent of p. Hence, since $ea = \delta a$,

$$0.546\,\frac{(m'/M)}{\delta a/a^2} = \delta a \tag{40}$$

and

$$W = 4\delta a = 2.96\,(m'/M)^{\frac{1}{4}} a. \tag{41}$$

Thus, W is approximately the same for all first-order resonances. Approximate values of these loop widths for Lindblad resonances in Saturn's rings are given in Table II. If $ea = \delta a$, then

$$\frac{d(ea)}{d|\delta a|} = -1 \tag{42}$$

and where the loop width is a maximum the streamlines converge on common points. This occurs at apocenter or pericenter, depending on whether the particle orbits are outside or inside exact resonance. The finite size of the ring particles would prevent such close packing of the particles from occurring, and thus the above value for W should be regarded as an underestimate.

The particle path pattern shown in Fig. 7 represents the undamped configuration. The lag angle between the equilibrium, or steady-state, tide and the tide-raising satellite is zero and there is no torque. Any dissipation of the energy stored in the tidal wave due to interparticle collisions results in a lag in the tidal response of the ring. The resultant torque is proportional to the magnitude of that lag. Figure 10 is a schematic representation of the configuration at the outer edge of the Saturnian B Ring, which is just inside the 2:1 resonance with Mimas.

TABLE II

TABLE II

Loop Widths of Lindblad Resonances

Satellite	m'/M	W^a (km)
Mimas	7×10^{-8}	94
1980S1	8×10^{-9}	32
1980S3	10^{-9}	11
1980S26	6×10^{-10}	9
1980S27	10^{-9}	11
1980S28	2×10^{-11}	2

[a]We assume that p is large and that $a \approx 2$ planetary radii.

D. Spiral Density Waves.

If a ring has sufficiently high surface density, then the azimuthal variation of the gravitational potential associated with the $p + 1$ loops at a Lindblad resonance can act on the orbits of the other ring particles, to generate a spiral density wave with $p + 1$ arms. The theory of this important ring phenomenon has been described in great detail by Goldreich and Tremaine (1978b,c, 1979c, 1980, 1981, 1982) and is reviewed in the chapter by Shu. Here, I content myself with a few qualitative remarks.

Outside the region of exact resonance, the particle paths in a frame corotating with the perturbing satellite are closed and contain the same number of waves. However, each closed path is displaced azimuthally with respect to its neighbor, and it follows from geometrical considerations that a spiral density wave must result (see Figs. 11 and 12). The whole pattern is stationary in the corotating reference frame. Thus, the gravitational potential associated with the spiral arms acts on the ring particles with the same frequency as the disturbing potential, and the pattern is self-enhancing.

The magnitude of the torque that now arises from the force between the satellite and the spiral arms is still determined by the rate of energy dissipation, but this energy is now dissipated at locations well-removed from the exact resonance. This process has been well described by Harris and Ward (1982) who compare the spiral density waves with "water waves [that] propagate away and leave the site of the original disturbance calm . . . ready for the next impulse." Consistent with this view, detailed calculations by Goldreich and Tremaine (1978b,c, 1979c) show that the magnitude of the torque is given by Eq. (25), that is, by the formula for waves in very narrow rings in which the wave energy is completely dissipated between impluses.

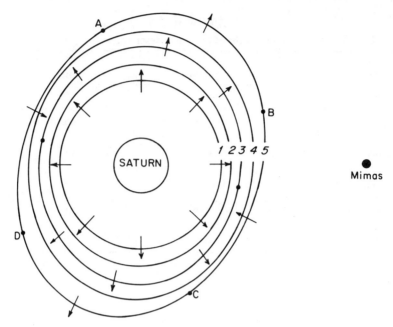

Fig. 10. Particle paths (streamlines) near the 2:1 Lindblad resonance associated with Mimas (located at the outer edge of the Saturnian B Ring). The radial arrows show the direction of angular momentum flow, determined by the local angular velocity gradient. Streamline #3 is critical, having two azimuthal positions (marked with dots) where the angular velocity gradient is zero. Streamline #4 has a limited azimuthal domain in which angular momentum flows inward. For streamline #5, the outward angular momentum flow between A and B and between C and D is exactly balanced by the inward flow between B and C and between D and A, and the net flow across the streamline is zero. According to Borderies et al. (1982b), this streamline marks the ring boundary; they have shown that such boundaries can be remarkably sharp. (Figure by P. Goldreich, personal communication.)

For an inner Lindblad resonance, for which (according to Eq. 37) the perturbing satellite is outside the ring and $n/n' \approx (p+1)/p$, the spiral density wave is only present outside the resonance and the torque on the ring is negative. The density wave carries negative energy and angular momentum and propagates towards the satellites. This wave is damped by particle collisions, and those particles involved in the damping lose energy and angular momentum and move inwards towards the planet and away from the perturbing satellite. Consequently, just outside the exact resonance a broad gap may open up (Goldreich and Tremaine 1978b, 1982). Gaps and density waves associated with inner Lindblad resonances have been observed in Saturn's rings (Cuzzi et al. 1981). For an outer Lindblad resonance, for which the perturbing satellite is inside the ring and $n/n' \approx p/(p+1)$, the torque on the ring is positive and damping of the density waves causes the particles to move away from the planet and away from the perturbing satellite. In both cases, the perturbing satellite acts to repel the ring particles.

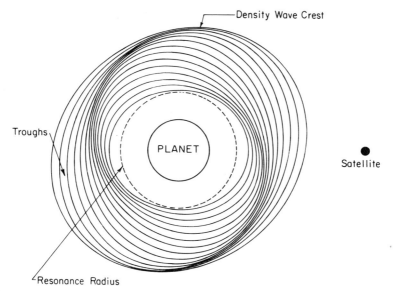

Density Wave Crest

Troughs

PLANET

Satellite

Resonance Radius

Fig. 11. Schematic diagram of the particle path pattern and the associated trailing spiral density wave generated by the $p=1$ Lindblad resonance. The pattern is stationary in a frame co-rotating with the perturbing satellite.

Ring particles trapped in a corotational resonance form a compact arc of particles (see Fig. 4), and the azimuthal variation of the gravitational potential associated with this arc can also act on the other ring particles to generate a spiral density wave (Goldreich and Tremaine 1979c, 1982). Unlike Lindblad resonances, corotational resonances are not confined to first-order ($q=1$) resonances, so the number of possible resonant locations in a ring can be large. However, the width of a corotational arc is less than the width of a Lindblad loop by a factor $\sim e'^{q/2}$ (see Eqs. 16, 17, and 41) and, since the eccentricities of satellites in the solar system tend to be small, the gravitational influence of a corotational resonance tends to be less than that of a Lindblad resonance.

E. Goldreich-Tremaine Model.

If a wide, diffuse ring is bounded by two satellites, then the repulsive action of the satellites will reduce the ring width until the confining torques just counteract the tendency for the ring to spread (Goldreich and Tremaine 1979a). (This shepherding action will not occur if the motion of the ring particles is retrograde with respect to that of the satellites. In that case, the two torques would act on the ring in the same sense and the space between the satellites would be swept clear of particles.)

In equilibrium, the net external torque on the ring is zero. Hence, from Eq. (25)

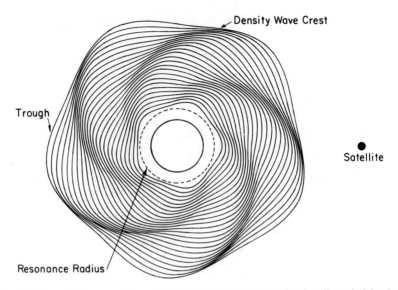

Fig. 12. Schematic diagram of the particle path pattern and the associated trailing spiral density wave generated by the $p=4$ Lindblad resonance. If the ratio of the resonant mean motions is $p/(p+1)$, then $p+1$ spiral arms are generated.

$$\frac{m_{\text{out}}}{\Delta a_{\text{out}}^2} = \frac{m_{\text{in}}}{\Delta a_{\text{in}}^2} \qquad (43)$$

where the subscripts 'out' and 'in' refer to the outer and inner satellites respectively. It follows from Eq. (23) that the amplitudes of the two waves raised on the ring are equal, although their wavelengths may be quite different.

Collisions between the ring particles generate a torque

$$2\pi a^3 \sigma v \left| \frac{d\Omega}{da} \right| = 3\pi a^2 \sigma v n \qquad (44)$$

where σ is the surface density of the ring. This is the magnitude of the torque that must be applied to the ring to maintain an equilibrium width W. The excess torque on each half of the ring due to variations of satellite torques across the ring is of magnitude

$$W <\Delta a^{-1}> |\Gamma| \qquad (45)$$

where $|\Gamma|$ is the magnitude of the total torque exerted on the ring by each satellite. Equating Eqs. (44) and (45), and assuming for convenience that $\Delta a_{\text{out}} = \Delta a_{\text{in}} = \Delta a$ and that $v \approx n d^2 \tau$, we obtain

$$W = \left(\frac{15\tau}{4}\right)^{\frac{1}{4}} \frac{M}{m} \left(\frac{\Delta a}{a}\right)^{\frac{3}{2}} d \tag{46}$$

where d is the "characteristic" size of the ring particles (Goldreich and Tremaine 1979a; see also discussion in chapter by Weidenschilling et al.). Plausible shepherd satellites, for example with $m/M \sim 10^{-10}$ and $\Delta a \sim 500$ km, would give the observed width for the Uranian rings, assuming cm-sized particles. However, the observed width of Saturn's F Ring is ~ 500 times that given by Eq. (46). Perhaps the problem is that the torque formulae do not strictly apply when the orbits are eccentric. Showalter and Burns (1982) argue that the F Ring particles are appreciably "stirred" at each close encounter with the shepherding satellites and that this could account for the excessive width of the ring.

The torque Γ exerted by the ring on each satellite pushes the satellites away from the ring at rates given by

$$\Gamma = h = \frac{1}{2} mna\dot{a} . \tag{47}$$

The timescale for the rate of change of Δa

$$T_{\text{sep}} = \frac{\Delta a}{\dot{a}} = \frac{5}{4} \frac{M^2}{mm_r} \left(\frac{\Delta a}{a}\right)^5 n^{-1} \tag{48}$$

can be used to place bounds on the surface densities of the rings. Using $\sigma = \tau \rho_r d$ and Eq. (46), we can eliminate mm_r/M^2 from Eq. (48) to obtain

$$T_{\text{sep}} = 3 \times 10^{14} \tau^{\frac{1}{2}} \left(\frac{\rho_r}{\sigma}\right) \left(\frac{\Delta a}{a}\right)^{\frac{5}{2}} \text{yr} \tag{49}$$

where ρ_r and σ are given in cgs units. If we demand that $T_{\text{sep}} > 5 \times 10^9$ yr, then, with cgs units,

$$\sigma < 200 \tau^{\frac{1}{4}} \rho_r^{\frac{1}{2}} (\Delta a/a)^{\frac{5}{4}} . \tag{50}$$

Thus, for the Uranian rings we must have $\sigma < 1$ g cm^{-2}, a value which in itself is not objectionable. However, Goldreich and Tremaine (1979a,b) consider that the apse alignment of the ϵ Ring is maintained by self-gravitation (see Sec. III), in which case $\sigma \approx 25$g cm^{-2}. If $\Delta a/a \approx 10^{-2}$ (and in this argument Δa should be regarded as the smaller of Δa_{out} and Δa_{in}), then their value for σ yields $T_{\text{sep}} \approx 10^7$ yr. If this were the case, we would have to conclude that the ϵ Ring is very young.

The easiest way out of this dilemma is to allow that σ is < 1 g cm^{-2} and that apse alignment is not maintained by self-gravitation (Dermott and Murray

1980); this is discussed further in Sec. III. Goldreich and Tremaine (1979a,b) suggest that the guardian satellites may be trapped in corotational resonances with the known Uranian satellites. Since the torque exerted by the ring would have to increase the angular momentum of all the satellites involved in the resonance, T_{sep} would be larger by a factor $\geqslant m*/m$, where $m*$ is the sum of the masses of the resonant satellites. However, most of the possible resonances in the vicinity of the rings are three-body resonances for which

$$\dot{\phi} = qn - (p+q)n_B + pn_A \tag{51}$$

where the subscripts B and A denote the inner and outer satellites respectively (Dermott and Gold 1977; Goldreich and Nicholson 1977; Freedman et al. 1983). I find that these resonances may be too weak to trap the satellites.

The equation of motion of the resonant argument ϕ is

$$\ddot{\phi} = -\omega^2 \sin\phi + q\dot{n}_{drag} \tag{52}$$

where $\dot{n}_{drag}$ is the rate of change of the mean motion of the guardian satellite due to the ring torque:

$$\frac{n}{\dot{n}_{drag}} = \frac{a}{\Delta a} T_{sep} . \tag{53}$$

The libration frequency ω is related to the width W of the corotation arc (see Eq. (16) by

$$\frac{\omega}{n} = \frac{3W}{8a} = \left(\frac{3a|R|}{GM}\right)^{\frac{1}{2}} . \tag{54}$$

For stability, the sign of $\ddot{\phi}$ must reverse (Goldreich 1965), hence

$$|q\dot{n}_{drag}| < \omega^2 . \tag{55}$$

For the guardian satellites of the Uranian rings, we need

$$T_{sep} > 5 \times 10^{-9}q \left(\frac{GM}{\Delta a|R|}\right) \text{ yr.} \tag{56}$$

If we allow that $\sigma \approx 1$ g cm^{-2}, then for resonant trapping we need $a|R|/GM > 10^{-16}$; there are indeed many 3-body resonances that satisfy this requirement (Goldreich and Nicholson 1977). However, if $\sigma \approx 25$ g cm^{-2}, as is perhaps the case for the ϵ Ring, then $T_{sep} \approx 10^7$ yr and for resonant trapping we need $a|R|/GM > 10^{13}$. It so happens that there are two, and only two, resonances close to the ϵ Ring that satisfy this requirement. The $p = 7$ Miranda-Ariel resonance has a strength $\approx 2 \times 10^{-13}$ and the Miranda 4:1 resonances have strengths $\approx 10^{-11}$ ($e \sin^2 \frac{1}{2} i$ term) and $\approx 6 \times 10^{-13}$ (e^3 term). However, these resonances have semimajor axes of 50887 km, 51470 km, and 51520 km respectively (data and formulae from Freedman et al. 1983). Thus, the Miranda-Ariel resonance lies ≈ 100 km outside the pericenter of the ϵ Ring at 50782 km, and the Miranda 4:1 resonances lie ~ 100 km inside the apocenter at 51595 km. Unless the pericenters of the satellites and the ring orbits are permanently aligned (and this possibility is easily dismissed), these resonances cannot be the locations of the guardian satellites.

I conclude that, unless resonances involving unseen satellites that orbit between the rings and Miranda and have masses $> 10^{-8} M$ anchor the guardian satellites, we must have $\sigma \sim 1$ g cm^{-2}, and that the apse alignment of the ϵ Ring is not maintained by self-gravitation. But there is another possibility: Eq. (25) may greatly overestimate the magnitude of the ring torque. That the actual torques are probably much less than those given by Eq. (25) is suggested by the observation (Smith et al. 1981, 1982) that comparatively massive satellites exist close to the Saturnian A Ring. Goldreich and Tremaine (1982) estimate that

$$T_{\text{sep}} = 1.5 \times 10^{-6} \sigma^{-1} R_s^{-3} \Delta a^4 \text{ yr} \tag{57}$$

where R_s is the satellite radius in cm; Δa is now the separation in cm of the satellite from the outer edge of the A Ring, and σ is in g cm^{-2}. T_{sep} for 1980S1 and 1980S27 may be as small as 7×10^7 and 6×10^5 yr, respectively. The orbits of these satellites are well determined, and they are not stabilized by any known resonance (Goldreich and Tremaine 1982). In the case of 1980S27, if we require $T_{\text{sep}} > 5 \times 10^9$ yr, then $< 10^{-4}$ of the energy associated with the excited eccentricities must be dissipated between encounters. If resonances in the A Ring associated with this satellite are strong enough to open gaps, then T_{sep} may be larger, but only by a factor ≈ 4 (Goldreich and Tremaine 1982). Perhaps we should consider the possibility that the rings are young.

F. Kinks, Braids and Eccentric Rings.

The theory that I have described so far is appropriate for near-circular satellite and ring orbits. It may be applicable to some of the Uranian rings, but for eccentric rings, particularly the Saturnian F Ring, modifications are necessary.

All aspects of the F Ring appear somewhat extreme. The amplitudes A of the waves on the ring are $\approx 6 (1000 \text{ km}/\Delta a)^2$ km (see Eq. 30) and thus are comparable with the 30 km width of the main ring. For the Uranian rings, with plausible values $m/M \sim 10^{-10}$ and $\Delta a \sim 500$ km, the wave amplitudes are probably very much less than the ring widths.

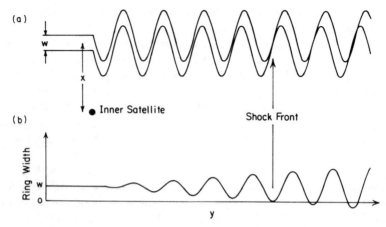

Fig. 13. (a) Wave pattern for a satellite-generated wave of length $3\pi\Delta a$, where Δa is the mean distance of the satellite from the ring particles. There is a small variation in wavelength across a ring of finite width. The resultant particle path pattern could lead to the formation of shock fronts and, if the ring contains gaps, loops. This could account for some of the features seen in the Saturnian F Ring. (b) Radial variation of ring width associated with the wave pattern shown in (a).

For a ring of finite width, there is a variation of wavelength across the ring and this can lead to the formation of shocks (see Fig. 13). The shock front will form at a distance

$$y = \frac{3\Delta a^2}{2A} \tag{58}$$

from the perturbing satellite. For $y < 2\pi a$, we require

$$\Delta a < 300 \left(\frac{m/M}{10^{-10}}\right)^{\frac{1}{4}} \text{ km} \tag{59}$$

for the Uranian rings, and

$$\Delta a < 1400 \left(\frac{m/M}{10^{-9}}\right)^{\frac{1}{4}} \text{ km} \tag{60}$$

for the F Ring. Thus, in both cases shocks may form between encounters with the perturbing satellites. The shocks will have radial widths $< A$. For the Uranian rings this may be very small (< 0.1 km), whereas for the F Ring the radial width of the shocked region could be far from negligible.

Shock fronts are likely sites for the formation of temporary particle clumps, thus I would expect these to have a mean azimuthal separation of one wavelength l. These shocks and the associated variations of ring width (see Fig. 13) may also contribute to the marked periodic variations in brightness observed in the F Ring and in narrow rings in Encke's gap (Gehrels et al. 1980; Smith et al. 1981, 1982).

The lines drawn in Fig. 13 represent the boundaries of a narrow ring, but they could equally well represent the boundaries of a narrow gap within a ring. In that case, near the shock the gap would degenerate into a series of loops or braids of width $2W$ and length l, where W is the unperturbed width of the gap. The width of the loops could not exceed $4A$ ($\approx$ 24 km for the F Ring). The gap would be expunged at the shock and thus would have to be regenerated by some agency. A natural suggestion is that small satellites (or large particles) exist in the ring.

The width of the Lindblad loops is

$$W = 1.5 \left(\frac{m/M}{10^{-9}} \right)^{\frac{1}{2}} \tag{61}$$

for the Uranian rings, and

$$W = 13 \left(\frac{m/M}{10^{-9}} \right)^{\frac{1}{2}} \tag{62}$$

for the F Ring. Thus, we could only expect to observe these loops in the F Ring. It was in fact suggested that these loops could account for the braided appearance of the ring (Dermott 1981b). However, first-order resonances are separated by a distance

$$s = \frac{3\Delta a^2}{2a} . \tag{63}$$

For the F Ring $s = 7$ km, which is $\ll W$. Thus, even if the F Ring were circular the resonant configuration could not be established. For the Uranian rings, assuming $\Delta a \sim 500$ km, s is large enough that the resonances are probably well separated. Borderies et al. (1982b) consider that Lindblad resonances define the edge of each narrow ring. This is certainly possible, but for some of the narrow rings the interior of the ring may be free of resonances. Perhaps s defines the minimum width of a sharp-edged ring.

If a ring is appreciably eccentric, then there is no direct application of our discussion of corotational and Lindblad resonance; the only resonance that can exist is an "eccentric" resonance (Goldreich and Tremaine 1981). However, the perturbations can be divided into corotational terms that perturb the mean longitude and Lindblad terms that perturb the eccentricity. Both terms act to change the ring eccentricity e_r. If $e_r \ll \Delta a/a$, then the resultant time scale $(e_r/\dot{e}_r)$ is

$$T_e = \frac{2}{cn} \left(\frac{M}{m_s} \right)^2 \left(\frac{\Delta a}{a} \right)^5 \tag{64}$$

where the subscript s refers to the guardian satellite (Goldreich and Tremaine 1982). Using Eq. (46) for the ring width W and $\sigma = \tau \rho_r d$ to eliminate m_s/M and $\Delta a/a$, we obtain

$$T_e = 8 \times 10^{-5} \, (\tau/c) \, (W\rho_r/\sigma)^2 \, \text{yr} \tag{65}$$

for the Uranian rings, where quantities on the right are given in cgs units. If no gaps are present in the ring, then the corotation terms dominate, $c = -0.148$, and the eccentricity damps. If gaps open at the first-order resonances in the ring, then the Lindblad terms dominate, $c = +1.52$, and the eccentricity grows (Goldreich and Tremaine 1981, 1982). Both timescales are uncomfortably short. We now know that only the η Ring appears to be truly circular (French et al. 1982).

The perturbation of a narrow, eccentric ring by a nearby satellite has been studied numerically by Showalter and Burns (1982). They do not take account of interparticle collisions which, particularly in the case of the F Ring, can have a major influence on the particle flow patterns even on timescales as short as the period between encounters. Nevertheless, their methods are well suited for studying the dynamics of encounter, and they have revealed a number of interesting phenomena.

Their satellite-ring configuration is shown in Fig. 14. Only the range of variation of the ring-satellite separation is significant. If the eccentricities are small, then for a given value of this range the separate eccentricities and orientations of the satellite and ring orbits do not matter. The encounter dynamics can be accurately modeled by the simplest configuration, that of a circular ring and a nearby satellite in an eccentric orbit (Showalter and Burns 1982).

There are large, qualitative differences between this case and that of a satellite in a circular orbit. At encounter, there is now a large (first-order in m/M) change δa in the semimajor axis of the ring particle orbit. For all encounter phases, we estimate that

$$|\delta a| \approx \frac{4}{3} \, a^2 e_s \left| \frac{\delta e}{\Delta a} \right| + 0(m/M)^2 \tag{66}$$

where δe is the forced eccentricity still given by Eq. 23 and e_s is the eccentricity of the satellite orbit. However, the actual magnitude and the sign of δa vary systematically with the phase of the encounter. The change in sign of

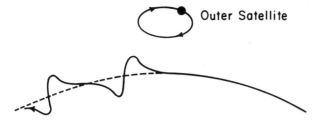

Fig. 14. Wave pattern for a wave, also of length $3\pi\Delta a$, where Δa is mean distance of the satellite from the ring, generated by a satellite in an eccentric orbit. This wave is not sinusoidal. In a frame rotating with the mean motion of the perturbing satellite, the path of the satellite is an ellipse with semimajor and semiminor axes $2ae$ and ae, respectively.

δa is particularly important, since it follows that there is a tendency for gap formation, or at least appreciable azimuthal variations in the particle number density on a scale of one wavelength ℓ. These variations could be largely responsible for the marked variations in brightness observed in the F Ring, although as I have discussed there could be other contributing factors.

The long-term variations of the eccentricities of narrow rings have been studied by Borderies et al. (1983a). In their model, both the ring and the nearby satellite are replaced by one dimensional elliptical wires of line densities $m_r/2\pi a$ and $m_s/2\pi a$, respectively. The radial force between these wires determines the variation of the eccentricity and apse precession rate $\tilde{\omega}_r$ of the ring (see Sec. III). The relative influence of the satellite and the quadrupole moment of the planet J_2 on $\tilde{\omega}_r$ is determined by the ratio

$$\Gamma_e = \frac{21\pi}{2} \frac{M}{m_s} J_2 \left(\frac{B}{a}\right)^2 \left(\frac{\Delta a}{a}\right)^3 \tag{67}$$

where B is the radius of the planet. For the Uranian rings, we can again use Eq. (46) to obtain

$$\Gamma_e = \frac{1}{4} \tau^{\frac{1}{2}} (W\rho_r/\sigma) (\Delta a/a)^{\frac{1}{2}} . \tag{68}$$

Thus, for these rings, $\Gamma_e \gg 1$ and the influence of J_2 is dominant. However, for the inner guardian satellite of the F Ring, $\Gamma_e = 16.8$ and there are small but appreciable variations in both $\tilde{\omega}_r$ and e_r on a timescale of 18 yr (Borderies et al. 1983a).

The configuration of the F Ring and the inner guardian, 1980S27, at the time of the Voyager 2 encounter is shown in Fig. 15. The distance of closest approach was then ≈ 483 km. The corresponding radial separation Δr of the apocenter and pericenter distances was $+ 134 \pm 142$ km, and the separation Δa of the semimajor axes was 832 ± 30 km (Synott et al. 1983). 1980S27 has

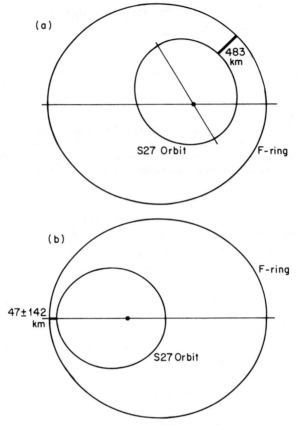

Fig. 15. (a) Configuration of the Saturnian F Ring and inner shepherding satellite (1980S27) orbits, at the time of Voyager 2 encounter. (b) The configuration expected in 1993.

a long semiaxis of 70 km (Smith et al. 1982). Thus, even if the orbital elements were to remain constant, the satellite and the ring must periodically experience very close encounters. In fact, perturbations by 1980S27 act to increase the eccentricity of the F Ring and Δr is reduced to $+ 47 \pm 142$ km (Borderies et al. 1983a). Here, the uncertainty in Δr has been estimated from the uncertainties in the orbital elements and no account has been taken of the effect of these uncertainties on the variation of the eccentricity.

The period between close encounters is largely determined by J_2 and is ≈ 18 yr. Taking the figures for Δr at face value, it would appear that every 18 yr the satellite may enter the ring. I consider this to be unlikely. Neglecting the gravitational field of the satellite, I calculate that the relative velocity of the satellite and the ring particles at closest approach is 39 ± 14 m s^{-1}. This is comparable with the escape velocity (38 m s^{-1}) of the isolated satellite (the satellite is close to the Roche limit and so the actual escape velocity will vary with position on the surface of the satellite, and in places will be considerably less than this value; see chapter by Weidenschilling et al.). If the satellite and the ring orbits intersect, then particles would be swept onto the satellite surface intermittently for a period as long as 2 yr, and this catastrophe would be repeated every 18 yr. If the satellite has an energy absorbing regolith, then I would not expect those particles that impinge on the satellite to survive. However, it is probably significant that Δr is comparable with the satellite radius. I would guess that either the mechanism that acts to increase the eccentricity of the ring switches off when Δr is small, or the ring is eventually destroyed. We might ask what is special about the F Ring and its inner guardian. The values of m_s/M and $\Delta a/a$ are not markedly different from those values suggested for the Uranian rings (Goldreich and Tremaine 1979a). Perhaps the eccentricities of the Uranian rings are limited by close encounters with their guardians.

III. APSE AND NODE ALIGNMENT

Several types of eccentric rings are now known to exist in the solar system. Rings with forced eccentricities associated with Lindblad resonances have been found in Saturn's rings (Smith et al. 1981, 1982). If $p = 0$ in Eq. (37), then $\tilde{\omega} = n'$ and the line of apses rotates with the perturbing satellite. The eccentric Saturnian ring at $1.29R_s$ is locked in a $p = 0$ Lindblad resonance with Titan (Porco et al. 1983). In terms of the wave description (see Fig. 7), the ring contains a single wave and the resultant shape is an ellipse with Saturn at one focus. If $p = 1$, then the ring contains two waves and the resultant shape is a non-Keplerian ellipse centered on the planet (see Figs. 10 and 11 in Sec. II.C). As predicted by Goldreich and Tremaine (1978b), particles in the outer edge of the Saturnian B Ring are observed to be locked in a $p = 1$ Lindblad resonance with Mimas (Smith et al. 1981, 1982).

The eccentric Uranian rings, the eccentric Saturnian rings at 1.45 R_s and in the Maxwell gap, and the Saturnian F Ring have free eccentricities, and their precession rates are largely determined by the dynamical oblateness of the planet J_2. The rings which are wide enough to be resolved, and whose geometry is well determined (the α, β and ϵ Uranian rings and the Saturnian rings at 1.29 R_s and 1.45 R_s), all have markedly nonuniform widths. In all cases, the widths are a minimum at pericenter and a maximum at apocenter (see Fig. 16).

The variation in width of these rings implies either that there is a variation in eccentricity across the ring (Nicholson et al. 1978) or that the pericenters of the particle orbits are systematically misaligned, or both (Dermott and Murray 1980). We define the mean eccentricity gradient of the ring g_r by

$$g_r = a \frac{\delta e}{\delta a} \tag{69}$$

Fig. 16. Composite image of Saturn's C Ring. The horizontal line through the center marks the border between the two images; at the top is shown the trailing ansa of the rings, and at the bottom the leading ansa. The dark gap in the center of both images contains a narrow ring which is clearly both eccentric and of nonuniform width. (Image courtesy of JPL/NASA.)

where δe and δa are the differences in the eccentricities and the semimajor axes of the Keplerian orbits that define the inner and outer edges of the ring. If the pericenters are aligned, then the variation of the radial ring width W with the true anomaly f is given by

$$W = \delta a \left[1 - (g_r + e) \cos f \right] . \tag{70}$$

Since, if $e \ll 1$, the orbital radius r varies as

$$r = a(1 - e \cos f) , \tag{71}$$

it follows that the harmonic variations of W and r are in phase and that W varies linearly with r (Nicholson et al. 1978). Any departure from linearity would imply a misalignment of pericenters. For the Uranian ϵ Ring, at least, this misalignment must be $< 0°2$ (Dermott and Murray 1980). Since W cannot

TABLE III

Eccentric Rings of Nonuniform Width

Planet	Ring	$<W>$ (km)	e $\times 10^4$	$J_2(B/a)^5$ $\times 10^4$	g_r
Uranus[a]	α	7.5	7.2	2.3	0.35
Uranus[a]	β	7.8	4.5	2.1	0.35
Uranus[a]	ϵ	58	79.2	1.2	0.65
Saturn[b]	$1.29\,R_s$	25	2.7	46.9	0.35
Saturn[c]	$1.45\,R_s$	69	3.9	26.1	0.55

[a]Data from chapter by Elliot and Nicholson in this book.
[b]Data from Porco et al. (1983).
[c]Data from Esposito et al. (1983).

be negative (the particle orbits cannot intersect), we must have $|g_r + e| < 1$. However, it is interesting and probably significant that all the observed values of g_r are positive and all are close to the critical value of unity (see Table III).

If the pericenters precessed under the influence of J_2 alone, then the differential precession rate across a ring would be

$$\left(\frac{d\tilde{\omega}}{da}\right)_J = \frac{-21}{4}\frac{n}{a}J_2\left(\frac{B}{a}\right)^2 . \tag{72}$$

Thus, if no other forces acted, the pericenters would rapidly disperse.

A. Self-gravitation.

Goldreich and Tremaine (1979a,b) consider that apse alignment is maintained by the self-gravitation of the ring particles. The contribution of the self-gravitation of the ring to $d\tilde{\omega}/da$, which arises from the variation of the radial forces acting on the particles with true anomaly, is given by

$$\left(\frac{d\tilde{\omega}}{da}\right)_r = \beta\frac{na^3\sigma_r}{M<W>^2} \tag{73}$$

where, for most mass distributions,

$$\beta \approx \frac{3\pi<W>g_r}{ae} . \tag{74}$$

If

$$\left(\frac{d\tilde{\omega}}{da}\right)_J + \left(\frac{d\tilde{\omega}}{da}\right)_r = 0 , \tag{75}$$

then

$$g_r \approx + 2.3\,e\!<\!W\!>\!J_2 \left(\frac{B}{a}\right)^5 \frac{\rho_b}{\sigma_r} \tag{76}$$

where ρ_b is the density of the planet. Thus, knowledge of the ring geometry yields an estimate of the mean surface mass density σ_r of the ring. Note that self-gravity only acts to align the pericenters if the eccentricity gradient g_r is positive and the radial forces are a maximum at pericenter. This is in accord with the observations. The surface mass densities of the Saturnian rings estimated from Eq. (76) are also in accord with independent estimates based on the relation between optical depth and surface mass density established from both the radial wavelengths of density waves and the scattering of the Voyager 1 radio signal (Borderies et al. 1982a; see also the chapter by Elliot and Nicholson in this book).

The self-gravitation model of apse alignment has now been extended to node alignment (Borderies et al. 1982a; Yoder 1982). The analysis predicts that if the self-gravitation of the ring alone acts to align the apses and the nodes, then we must have

$$\frac{\delta i}{i} = \frac{\delta e}{e} \tag{77}$$

where δi is the variation of the inclination of the particle orbits across the ring. Yoder (1982) has also shown that the configuration of aligned pericenters and nodes is stable against small displacements. An eccentric ring is stable even if its inclination is zero, but a ring with an appreciable inclination is only stable if it is also eccentric (Yoder 1982).

B. Embedded Rings.

The eccentric Saturnian rings at 1.29 R_s and 1.45 R_s have special locations. In both cases, the rings are very close to the outer edges of the clear gaps in which they lie (Porco et al. 1983; Esposito et al. 1983). In the case of the 1.45 R_s ring, the particles in the outer edge of the eccentric ring may even brush the surrounding disk at their apocenters (Esposito et al. 1983). These configurations may not be fortuitious. The closeness of the rings to the surrounding disk leaves little room for shepherding satellites. However, the ring particle orbits will be perturbed at each close encounter with the disk. Perhaps this interaction results in shepherding action.

The location of the rings may also have some other significance. The contribution of a circular disk to the differential precession of a nearby narrow eccentric ring is given by

$$\left(\frac{d\dot{\tilde{\omega}}}{da}\right)_d = \frac{(1+\gamma)\,na^3\sigma_d}{M\Delta a^2} \tag{78}$$

(cf. Eq. 73) where

$$\gamma = \frac{3}{4} g_d^2 - \frac{1}{2} g_d g_r + \frac{5}{8} g_d^4 - \frac{1}{2} g_d^3 g_r + \dots \tag{79}$$

where g_d ($= ae/\Delta a$) is the eccentricity gradient between the ring and the inner edge of the circular disk, $\Delta a (<0)$ is the corresponding difference in the semimajor axes and σ_d is the surface mass density of the disk. Regardless of the sign of g_r, $(d\tilde{\omega}/da)_d$ is only positive if the eccentric ring is close to the inside edge of the disk. That this is the case for the Saturnian rings suggests that the disk may have a role in their apse alignment. Using Eqs. (72), (78) and (79), I calculate that even if the rings were massless ($\sigma_r = 0$), $d\tilde{\omega}/da$ would be zero where

$$\Delta a = -56 \left(\frac{\sigma_d}{100 \text{ g cm}^{-2}} \right)^{\frac{1}{2}} \text{km} \tag{80}$$

in the case of the 1.45 R_s ring, and where

$$\Delta a = -42 \left(\frac{\sigma_d}{100 \text{ g cm}^{-2}} \right)^{\frac{1}{2}} \text{km} \tag{81}$$

in the case of the 1.29 R_s ring (neglecting the contribution of Titan to $d\tilde{\omega}/da$). Thus, the disk may partly determine the local eccentricity gradients of the outer edges of the rings; these local gradients may even be negative. This might imply that the eccentricity gradients at the edges may differ appreciably from the observed mean eccentricity gradients, and may be equal to those critical values for which the angular momentum flux rates are zero and the edges are sharp (Borderies et al. 1982b, 1983b).

C. Precessional Pinch.

Dermott and Murray (1980) pointed out that by itself self-gravitation does not explain why all the observed values of g_r are close to unity. If g_r is determined by three presumably independent parameters of the ring (e, $<W>$ and σ_r), then it is unreasonable to expect those quantities to be always such that $g_r \approx 0.5$ (see Table III). They argued that the ring particles may be close-packed at pericenter and that close-packing may prevent differential precession. Figure 17 gives an heuristic description of how differential precession, particle collisions, and self-gravitation acting together always transform a narrow eccentric ring of uniform width into a ring with a large positive eccentricity gradient and aligned pericenters. Equilibrium (Fig. 17d) is only reached when the particles are so close together at pericenter that the ring width there cannot be reduced any further. Since differential precession always acts to reduce the ring width, it must then cease.

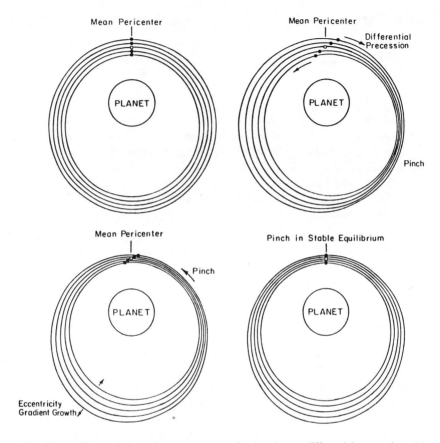

Fig. 17. Possible evolution of a narrow eccentric ring due to differential precession. (a) Initially the pericenters are aligned and the ring width is uniform. (b) Differential precession produces a harmonic variation of width with the pinch (W_{min}) located before pericenter ($f \simeq - \pi/2$). The pericenters of the eccentric orbits are denoted by filled circles; the mean pericenter of the ring by an open circle. As the pericenters separate, W_{min} decreases. Dermott and Murray (1980) contend that close packing of particles at the pinch prevents W_{min} from being reduced to zero. (c) The further evolution of the ring is now determined by self-gravitation, resulting in the growth of a positive eccentricity gradient. As the eccentricity gradient increases, the pinch moves in a prograde sense towards pericenter. (d) Only at pericenter is the pinch in stable equilibrium.

The only alternative to this argument is to allow that e, $<W>$, σ_r and g_r are not independent parameters (Dermott and Murray 1980). The coupled evolution of these parameters has now been solved by Borderies et al. (1983c). In their model, self-gravitation is always responsible for the alignment of the pericenters, and evolution of the parameters only ceases when close-packing of the particles at pericenter limits the growth of the mean eccentricity e.

D. Sharp Edges.

The particle dynamics of an eccentric ring with a large eccentricity gradient are quite different from those of a circular ring, because both the relative velocities of the particles and their collision frequencies can vary markedly with true anomaly (Dermott and Murray 1980). The radial variation of angular frequency $\dot{f}$ for a ring with aligned pericenters is given by

$$\frac{d\dot{f}}{dr} = -\frac{3}{2}\frac{n}{a}\frac{\left(1 - \frac{4}{3}g_r\cos f\right)}{(1 - g_r\cos f)} \tag{82}$$

and, if $g_r > 3/4$, then the angular velocity gradient at pericenter is positive. Borderies et al. (1982b) have shown that this change of sign has a profound effect on the angular momentum transfer rate due to particle collisions, and probably accounts for the existence of rings with very sharp edges. Their model is shown in Fig. 10 (see Sec. II.C).

Using the fluid approximation (Eq. 4), Borderies et al. have shown that the net angular momentum flow rate is zero and the ring edge is sharp when the magnitude of the eccentricity gradient at the ring edge is $(3/4)^{\frac{1}{2}}$. [Note that in their model the fundamental quantity is the azimuthal variation of the radial width of adjacent streamlines, and that this is determined by the eccentricity gradient g_r alone only when the pericenters are aligned (see Dermott and Murray 1980, and Eq. 70). Their more general treatment also allows for the azimuthal variation of radial width associated with misalignment of pericenters (this effect is shown in Fig. 17). For very narrow rings, the latter effect can be dominant.] A more sophisticated analysis, using the Boltzmann equation, shows that the critical eccentricity gradient varies with the mean optical depth of the ring τ (Borderies et al. 1983b). For $\tau > 3$, the critical gradient is close to the fluid limit 0.866, but the critical gradient decreases with τ and for $\tau \approx 0.25$ it is as low as 0.5. The local eccentricity gradient, ade/da, is strongly radially dependent near any Lindblad resonance (see Eq. 38) (and, perhaps, near the inner edge of a disk). For this reason, Borderies et al. (1982b) consider that the sharp edges of all rings in the solar system are probably associated with Lindblad resonances.

IV. HORSESHOE ORBITS

Dermott et al. (1979) proposed that each narrow ring contains a small satellite that maintains solid particles in stable, horseshoe orbits about its Lagrangian equilibrium points (Fig. 18). The case for which the ring-satellite has zero eccentricity can be studied by considering the Jacobi integral. In a rotating reference frame in which the satellite is fixed, this integral is (Brown 1911)

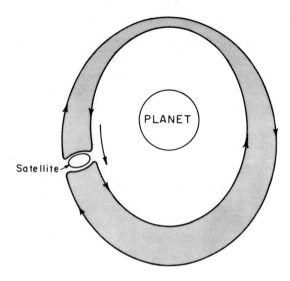

Fig. 18. Ring model of Dermott et al. (1979); each ring contains a small satellite which maintains particles in horseshoe orbits. Loose solid particles leave the satellite surface and enter orbits closely similar to that of the satellite (which can be both eccentric and inclined to the equatorial plane of the planet). The gravitational force of the satellite in a 1:1 resonance with the ring particles provides the critical phenomenon needed to define a narrow ring. (Figure copyright of *Nature*, MacMillan Journals Ltd.)

$$\frac{2}{r} + r^2 + \frac{m}{M}\left(\frac{2}{\Delta} + \Delta^2\right) = V^2 + C \tag{83}$$

where r is the distance of the particle from the center of the planet, Δ is the distance from the satellite, and V is the speed in the rotating reference frame (Fig. 19). The unit of distance is the separation of the satellite from the center of the planet, and the unit of time is chosen such that the mean motion n of the satellite is given by

$$n^2 = 1 + m/M \ . \tag{84}$$

This requires $GM = 1$. The curves $V^2 = 0$ for a range of values of the Jacobi constant C are called zero-velocity curves, and define regions of the plane within which the particle is confined to move.

The shapes and widths of these curves are very good guides to the geometry of the actual particle paths. Dermott and Murray (1981a) have shown that, if $(m/M)^{\frac{1}{3}} \ll 1$, there is a close correspondence between a particle's path and its associated zero-velocity curve. If a particle is moving in a

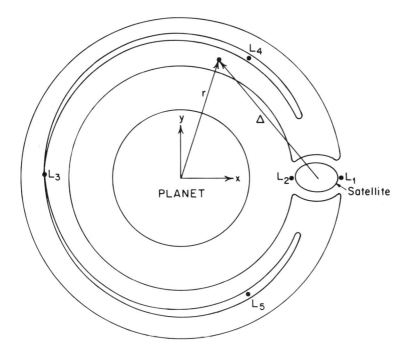

Fig. 19. Schematic diagram showing the Lagrangian equilibrium points and the critical zero-velocity curves. The critical horseshoe curve actually passes through L_1 and L_2 and the critical tadpole curve passes through L_3. Horseshoe orbits will exist between these two extremes. The rectangular coordinate frame is centered on the planet and corotates with the satellite.

near-circular orbit, and the radial displacement of its associated zero-velocity curve from the unit circle is W_z, then the radial displacement of the particle path at the same longitude is $2W_z$; if the particle orbit has a small eccentricity, then this statement applies to the motion of the guiding center.

Tadpole orbits encompass either L_4 or L_5 alone, whereas horseshoe orbits encompass L_3, L_4 and L_5 (Fig. 19). In most of the past work on the three-body problem, emphasis has been placed on the tadpole solutions, since these describe the motion of the Trojan asteroids with respect to Jupiter. Until the Voyager encounters with Saturn, examples of horseshoe orbits in the solar system were unknown. Dermott et al. (1979, 1980) pointed out that for two reasons we must expect horseshoe orbits in the solar system to be associated only with very small satellites. First, the ratio of the widths of those regions where, respectively, tadpole alone and tadpole and horseshoe orbits are possible is $\approx (m/M)^{\frac{1}{6}}$, and it follows that the horseshoe orbit region is only dominant if $(m/M)^{\frac{1}{6}} \ll 1$. The second reason obtains from a study of orbital stability.

For motion in a horseshoe orbit it is convenient to write the Jacobi constant C as

$$C = 3 + \alpha \left(\frac{m}{M} \right)^{\frac{2}{3}}$$ (85)

where α is a constant $\leq 3^{\frac{1}{3}}$. If we write

$$a = a_s + \Delta a$$ (86)

where a and a_s are the semimajor axes of a ring particle and a ring satellite respectively, then

$$\Delta a = 2 \left(\frac{\alpha}{3} \right)^{\frac{1}{2}} \left(\frac{m}{M} \right)^{\frac{1}{3}} a_s$$ (87)

and the distance of closest approach of a particle (or its guiding center) to the satellite is

$$y = \frac{2}{\alpha} \left(\frac{m}{M} \right)^{\frac{1}{3}} a_s$$ (88)

(Dermott and Murray 1981a).

Dermott et al. (1980, 1981a) have found by numerical integration of particular cases that the type of path followed by a particle depends on the value of α. For large values of $\alpha (>1)$ the particles either strike the satellite or are scattered, but for small values of α (<1) the particles are apparently repelled by the satellite and motion in horseshoe orbits is possible (for a simple discussion of the dynamics involved see Dermott et al. 1979). The nature of the path changes dramatically as α is reduced, and for very small values of α the horseshoe paths are almost perfectly symmetric with respect to the unit circle.

If we write

$$\left| \frac{\Delta a_0}{a_s} \right| - \left| \frac{\Delta a_j}{a_s} \right| = \pm \left(\frac{m}{M} \right)^{n}$$ (89)

where the subscript j refers to the number of consecutive encounters with the satellite that the particle experiences, then we find that n increases to values >0.7 ($j=1$) and ≈ 1.2 ($j=2$) as α decreases to <0.2 (Fig. 20). Thus, for small values of α the orbits are near-periodic and Δa_0 and Δa_2 are equal to order m/M. This symmetry is not a property of the circular orbit case alone. Fig. 20 shows that there is no substantial difference between the cases of circular and eccentric orbits.

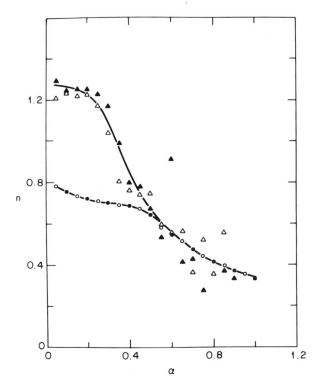

Fig. 20. Summary of horseshoe paths. n is a measure of the symmetry of the horseshoe path (see Eq. 89). α is the impact parameter, proportional to Δa^2. The circular points refer to changes in the semimajor axis a of the particle orbit after a single encounter with the satellite: filled circles refer to the circular orbit case ($e_s = 0$, where e_s is the eccentricity of the satellite orbit); open circles refer to the elliptical orbit case with $e_s = 0.01$. The triangular points refer to the total change in a after two consecutive encounters: filled triangles, $e_s = 0$; open triangles, $e_s = 0.01$.

The effect on the horseshoe orbit of a ring particle by an external force due, for example, to Poynting-Robertson light drag can now be understood. If α is small, then Δa_0 and Δa_2 are always equal; drag forces have little influence on the encounter dynamics. Therefore, if the ring were very narrow and the magnitudes of the drag forces acting on the particle were the same in both halves of the horseshoe path, then the orbital decay of the particle achieved in one half of the path would in effect be cancelled by that achieved in the other half (Fig. 21). Since the drag force extracts angular momentum from the system, some orbital decay would of course occur, but the satellite would supply angular momentum to the particle to maintain the 1:1 resonance and the orbit of the ring particle and the satellite would decay together at some rate r times less than that of an unconstrained particle, where

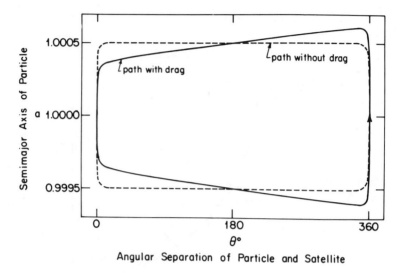

Fig. 21. Path of a particle moving in a horseshoe orbit around a ring-satellite of mass ratio $m/M = 10^{-9}$ (circular orbit case). The dashed line refers to the particle path in the absence of drag, and the solid line shows the effect of an external drag force. Since $\alpha < 0.2$, both paths are highly symmetric about the line $a = 1$ and both paths are closed (to order m/M). Thus, even in the presence of drag, the particle orbit is stable; the ring-satellite provides the energy and angular momentum needed to maintain the 1:1 resonance.

$$r \approx \frac{m_r}{m_s + m_r} \tag{90}$$

(Dermott et al. 1979, 1980; Dermott 1981a). Thus, even if a ring consisted of very small particles ($d \ll 0.1$ cm), if $m_s \gg m_r$ then Poynting-Robertson light drag acting over times comparable with the age of the solar system would not result in significant orbital decay or ring spreading. However, second-order effects associated with the variation of the magnitude of the mean drag force with distance from the planet may not be negligible.

From Fig. 21, we see that orbital decay on the inside of the horseshoe acts to increase the width W of the path, whereas that on the outside acts to decrease it. If $\dot{a} < 0$ and $d|\dot{a}|/da < 0$, then one might expect W to increase with time. However, since the particle's orbital period decreases with increasing a (Kepler's third law), the particle spends a greater time on the outside than on the inside of the horseshoe path, and the sign of $\dot{W}$ is found to depend on the magnitude of $d|\dot{a}|/da$. If $\dot{a} = -k^2 a^n$, where k is a constant, then $\dot{W}/W = (3+n)(\dot{a}/a)$ and, if $n > -3$, then $\dot{W} < 0$ and the particles are driven towards L_4 and L_5. From Eq. (8), we see that this is the case for the Poynting-Robertson light drag.

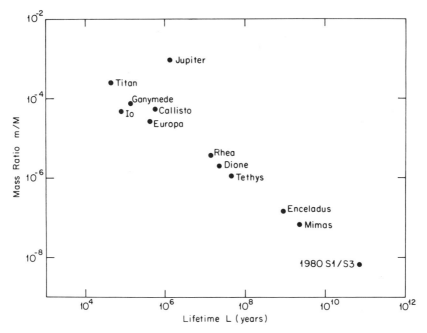

Fig. 22. Values of L for solar system bodies. If the evolution of the semimajor axis of a small satellite moving in a horseshoe orbit due to encounters with a primary or more massive coorbital satellite can be described by a random-walk process, then the lifetime L of the small satellite can be estimated from the mass ratio m/M, where m is the mass of the primary satellite and M is the mass of the planet. Horseshoe orbits may be associated only with very small or young primary satellites.

If Eq. (89) were sufficient to describe the effects of particle and satellite encounters, then I would expect particles to be lost from horseshoe orbits due to a random walk of the quantity $|\Delta a_0| - |\Delta a_2|$. If $n = 1$ in Eq. (89), then this would occur on a time scale

$$L \approx \frac{T}{(m/M)^{\frac{5}{3}}} \tag{91}$$

where T is the orbital period of the satellite (Dermott et al. 1980). Values of L for various bodies in the solar system are shown in Fig. 22. On this basis, Dermott et al. (1980) suggested that very small satellites which lie outside the Roche zone may be associated with narrow rings of primordial material that they have yet to accrete. Their numerical investigations and those of Dermott and Murray (1981a) were not very extensive. Thus Eq. (91) cannot be regarded as well supported. Nevertheless, it is encouraging that those satellites since discovered to be associated with companions in horseshoe orbits, have lifetime $L > 5 \times 10^9$ yr; these are the coorbital satellites 1980S1 and 1980S3

(Smith et al. 1981, 1982) and Mimas (Simpson et al. 1980; Van Allen et al. 1980; Stone and Miner 1982; Van Allen 1982).

It is natural to suggest that the coorbital satellites 1980S1 and 1980S3 are direct collision products, but I consider this unlikely. Dermott and Murray (1981a) have shown that high-α orbits are very unstable. A satellite could not remain in such an orbit for long without either being scattered away from the more massive satellite or colliding with it. Since any objects which leave the surface of the main satellite must, if they move in horseshoe orbits at all, move in high-α orbits, it is improbable that the present orbital configuration, characterized by a very low α value (0.025), was formed in that way. Similar arguments apply to the formation of the coorbital satellites of Mimas, Dione, and Tethys. If collisions have had a role in the formation of coorbital satellites, then I consider that the formation of a narrow ring of coorbital debris out of which the coorbital satellites then accrete is a necessary intermediate step. I can see no other way of placing large quantities of material in the low-α orbits necessary for orbital stability and satellite survival (Dermott and Murray 1981b; Yoder et al. 1983).

The existence of satellites in horseshoe orbits lends support to the horseshoe orbit ring model. However, other observations argue against it. In particular, occultation data show that the structure of some of the Uranian rings is highly complex. Such asymmetric profiles could not possibly be modeled by a ring which contains a single satellite. Whether small satellites exist in Saturn's rings that maintain rings of particles on horseshoe paths is not known, but I would argue that they probably do.

Acknowledgment. This research was supported by the National Aeronautics and Space Administration.

REFERENCES

Aksnes, K. 1977. Quantitative analysis of the Dermott-Gold theory for Uranus's rings. *Nature* 269:783.

Borderies, N., Goldreich, P., and Tremaine, S. 1982a. Precession of inclined rings. *Icarus* (in press).

Borderies, N., Goldreich, P., and Tremaine, S. 1982b. Sharp edges of planetary rings. *Icarus* (in press).

Borderies, N., Goldreich, P. and Tremaine, S. 1983a. The variations in eccentricity and apse precession rate of a narrow ring perturbed by a close satellite. *Icarus* 53:84–89.

Borderies, N., Goldreich, P. and Tremaine, S. 1983b. Perturbed particle disks. *Icarus* 55:124–132.

Borderies, N., Goldreich, P. and Tremaine, S. 1983c. The dynamics of elliptical rings. *Icarus* (in press).

Brahic, A. 1977. Systems of colliding bodies in a gravitational field. I. Numerical simulation of the standard model. *Astron. Astrophys.* 54:895–907.

Brahic, A. 1982. The rings of Uranus. In *Uranus and the Outer Planets,* ed. G. Hunt (Cambridge: Cambridge Univ. Press), pp. 211–236.

Brown, E. W. 1911. On a new family of periodic orbits in the problem of three bodies. *Mon. Not. Roy. Astron. Soc.* 71:438–454.

Burns, J. A., Lamy, P. L., and Soter, S. 1979. Radiation forces on small particles in the solar system. *Icarus* 40:1–48.

Burns, J. A., Showalter, M. R., Cuzzi, J. N., and Pollack, J. B. 1980. Physical processes in Jupiter's ring: Clues to its origin by Jove! *Icarus* 44:339–360.

Clarke, J. T. 1982. Detection of auroral hydrogen Lyman-alpha emission from Uranus. *Astrophys. J.* 263:L105–L109.

Cook, A. F., and Franklin, F. A. 1964. Rediscussion of Maxwell's Adams prize essay on the stability of Saturn's rings. *Astron. J.* 69:173–200.

Cuzzi, J. N. 1982. Mysteries of the ringed planets. *Nature* 300:485–486.

Cuzzi, J. N., Lissauer, J. J., and Shu, F. H. 1981. Density waves in Saturn's rings. *Nature* 292:703–707.

Dermott, S. F. 1981a. The origin of planetary rings. *Phil. Trans. Roy. Soc. London.* A303: 261–279.

Dermott, S. F. 1981b. The braided F ring of Saturn. *Nature* 290:54–57.

Dermott, S. F., and Gold. T. 1977. The rings of Uranus: theory. *Nature* 267:590–593.

Dermott, S. F., Gold, T., and Sinclair, A. T. 1979. The rings of Uranus: Nature and origin. *Astron. J.* 84:1225–1234.

Dermott, S. F., and Murray, C. D. 1980. Origin of the eccentricity gradient and the apse alignment of the ε ring of Uranus. *Icarus* 43:338–349.

Dermott, S. F., and Murray, C. D. 1981a. The dynamics of tadpole and horseshoe orbits. I. Theory. *Icarus* 48:1–11.

Dermott, S. F., Murray, C. D., and Sinclair, A. T. 1980. The narrow rings of Jupiter, Saturn and Uranus. *Nature* 284:309–313.

Dermott, S. F., and Murray, C. D. 1981b. The dynamics of tadpole and horseshoe orbits. II. The coorbital satellites of Saturn. *Icarus* 48:12–22.

Dermott, S. F., and Murray, C. D. 1983. Kirkwood gaps in the distribution of the asteroids. I. The 3:1 resonance with Jupiter. In preparation.

Dermott, S. F., Murray, C. D., and Williams, I. P. 1984. Drag forces in the three-body problem and the formation of coorbital satellites. In preparation.

Elliot, J. L. 1979. Stellar occultation studies of the solar system. *Ann. Rev. Astron. Astrophys.* 17:445–475.

Elliot, J. L., Dunham, E. W., and Mink, D. J. 1977. The rings of Uranus. *Nature* 267:328–330.

Esposito, L. W., Borderies, N., Goldreich, P., Cuzzi, J. N., Holberg, J. B., Lane, A. L., Pomphrey, R. B., Terrile, R. J., Lissauer, J. J., Marouf, E. A., and Tyler, G. L. 1983. The eccentric ringlet in the Huygens gap at 1.45 Saturn radii: Multi-instrument Voyager observations. *Science* 222:57–60.

Fanale, F. P., Veeder, G., Matson, D. L., and Johnson, T. V. 1980. Rings of Uranus: Proposed model is unworkable. *Science* 208:626.

Freedman, A. P., Tremaine, S., and Eliot, J. L. 1983. Weak dynamical effects in the Uranian ring system. *Astrophys. J.* In press.

French, R. G., Elliot, J. L., and Allen, D. A. 1982. Inclinations of the Uranian rings. *Nature* 298:827–829.

Gehrels, T., Baker, L. R., Beshore, E., Bleman, C., Burke, J. J., Castillo, N. D., Dacosta, B., Degewij, J., Doose, L. R., Fountain, J. W., Gotobed, G., Kenknight, C. E., Kingston, R., McLaughlin, G., McMillan, R., Murphy, R., Smith, P. H., Stoll, C. P., Strickland, R. N., Tomasko, M. G., Wijesinghe, M. P., Coffeen, D. L., and Esposito, L. 1980. Imaging photopolarimeter on Pioneer Saturn. *Science* 207:434–439.

Goldreich, P. 1965. An explanation of the frequent occurrence of commensurable mean motions in the solar system. *Mon. Not. Roy. Astron. Soc.* 130:159–181.

Goldreich, P., and Nicholson, P. 1977. The revenge of tiny Miranda. *Nature* 269:783–785.

Goldreich, P., and Tremaine, S. 1978a. The velocity dispersion in Saturn's rings. *Icarus* 34:227–239.

Goldreich, P., and Tremaine, S. 1978b. The formation of the Cassini Division in Saturn's rings. *Icarus* 34:240–253.

Goldreich, P., and Tremaine, S. 1978c. The excitation and evolution of density waves. *Astrophys. J.* 222:850–858.

Goldreich, P., and Tremaine, S. 1979*a*. Towards a theory for the uranian rings. *Nature* 277:97–99.

Goldreich, P., and Tremaine, S. 1979*b*. Precession of the ϵ ring of Uranus. *Astron. J.* 84:1638–1641.

Goldreich, P., and Tremaine, S. 1979*c*. The excitation of density waves at the Lindblad and corotation resonances by an external potential. *Astrophys. J.* 233:857–871.

Goldreich, P., and Tremaine, S. 1980. Disk-satellite interactions. *Astrophys. J.* 241:425–441.

Goldreich, P., and Tremaine, S. 1981. The origin of the eccentricities of the rings of Uranus. *Astrophys. J.* 243:1062–1075.

Goldreich, P., and Tremaine, S. 1982. The dynamics of planetary rings. *Ann. Rev. Astron. Astrophys.* 20:249–283.

Gradie, J. 1980. Rings of Uranus: Proposed model is unworkable. *Science* 208:625–626.

Greenberg, R. 1973. Evolution of satellite resonances by tidal dissipation. *Astron. J.* 78:338–346.

Greenberg, R. 1983. The role of dissipation in shepherding of ring particles. *Icarus* 53: 207–218.

Greenberg, R. 1984. Resonances in the Saturn system. In *Saturn,* eds. T. Gehrels and M. S. Matthews (Tucson: Univ. Arizona Press). In press.

Harris, A. W., and Ward, W. R. 1982. Dynamical constraints on the formation and evolution of planetary bodies. *Ann. Rev. Earth Planet. Sci.* 10:61–108.

Harris, A. W., and Ward, W. R. 1983. On the radial structure of planetary rings. Proceedings of *I.A.U. Colloquium 75 Planetary Rings,* ed. A. Brahic, Toulouse, France, Aug. 1982.

Hénon, M. 1981. A simple model of Saturn's rings. *Nature* 293:33–35.

Hénon, M. 1983. A simple model of Saturn's rings-revisited. Proceedings of *I.A.U. Colloquium 75 Planetary Rings,* ed. A. Brahic, Toulouse, France, Aug. 1982.

Holberg, J. B., Forrester, W. T., and Lissauer, J. J. 1982. Identification of resonance features within the rings of Saturn. *Nature* 297:115–120.

Hunten, D. M. 1980. Rings of Uranus: Proposed model is unworkable. *Science* 208:625–626.

Ip, W. H. 1980*a*. Physical studies of planetary rings. *Space Sci. Rev.* 26:39–96.

Ip, W. H. 1980*b*. New progress in the physical studies of planetary rings. *Space Sci. Rev.* 26:97–109.

Julian, W. H., and Toomre, A. 1966. Non-axisymmetric responses of differentially rotating disks of stars. *Astrophys. J.* 146:810–832.

Lane, A. L., Hord, C. W., West, R. A., Esposito, L. W., Coffeen, D. L., Sato, M., Simmons, K., Pomphrey, R. B. and Morris, R. B. 1982. Photopolarimetry from Voyager 2: Preliminary results on Saturn, Titan and the rings. *Science* 215:537–543.

Lin, D. N. C., and Bodenheimer, P. 1981. On the stability of Saturn's rings. *Astrophys. J.* 248:L83–L86.

Lin, D. N. C., and Papaloizou, J. 1979. Tidal torques on accretion discs in binary systems with extreme mass ratios. *Mon. Not. Roy. Astr. Soc.* 186:799–812.

Lucke, R. L. 1978. Uranus and the shape of elliptical rings. *Nature* 272:148.

Lukkari, J. 1981. Collisional amplification of density fluctuations in Saturn's rings. *Nature* 292:433–435.

Lynden-Bell, D., and Pringle, J. E. 1974. The evolution of viscous discs and the origin of the nebular variables. *Mon. Not. Roy. Astron. Soc.* 168:603–637.

Matthews, K., Neugebauer, G., and Nicholson, P. D. 1982. Maps of the rings of Uranus at a wavelength of 2.2 microns. *Icarus* 52:126–135.

Nicholson, P. D., Matthews, K., and Goldreich, P. (1982). Radial widths, optical depths and eccentricities of the Uranian rings. *Astron. J.* 87:433–447.

Nicholson, P. D., Persson, S. E., Matthews, K., Goldreich, P., and Neugebauer, G. 1978. The rings of Uranus: Results of the 1978 10 April occultation. *Astron. J.* 83:1240–1248.

Porco, C., Borderies, N., Danielson, G. E., Goldreich, P., Holberg, J. B., Lane, A. L., and Nicholson, P. D. 1983. The eccentric ringlet at 1.29 R_s. Proceedings of *I.A.U. Colloquium 75 Planetary Rings,* ed. A. Brahic, Toulouse, France, Aug. 1982.

Safronov, V. S. 1969. *Evolution of the Protoplanetary Cloud and Formation of the Earth and Planets.* Moscow: Nanka. Transl. Israel Program for Scientific Translations, 1972. NASA TTF-677.

Sandel, B. R., Shemansky, D. E., Broadfoot, A. L., Holberg, J. B., Smith, G. R., McConnell, J. C., Strobel, D. F., Atreya, S. K., Donahue, T. M., Moos, H. W., Hunten,

D. M., Pomphrey, R. B., and Linick, S. 1982. Extreme ultraviolet observations from the Voyager 2 encounter with Saturn. *Science* 215:548–553.

Showalter, M. R., and Burns, J. A. 1982. A numerical study of Saturn's F Ring. *Icarus* 52:526–544.

Sicardy, B., Combes, M., Brahic, A., Bouchet, P., Perrier, C., and Courtin, R. 1983. The 15 August 1980 occultation by the Uranian system: structure of the rings and temperature of the upper atmosphere. *Icarus* 52:454–472.

Simpson, J. A., Bastian, T. S., Chenette, D. L., McKibben, R. B., and Pyle, K. R. 1980. The trapped radiations of Saturn and their absorption by satellites and rings. *J. Geophys. Res.* 85:5731–5762.

Smith, B. A., Soderblom, L. A., Johnson, T. V., Ingersoll, A. P., Collins, S. A., Shoemaker, E. M., Hunt, G. E., Carr, M. H., Davies, M. E., Cook, A. F., Boyce, J., Danielson, G. E., Owen, T., Sagan, C., Beebe, R. F., Veverka, J., Strom, R. G., McCauley, J. F., Morrison, D., Briggs, G. A., and Suomi, V. E. 1981. Encounter with Saturn: Voyager 1 imaging science results. *Science* 212:163–191.

Smith, B. A., Soderblom, L., Beebe, R., Boyce, J., Briggs, G., Bunker, A., Collins, S. A., Hansen, C. J., Johnson, T. V., Mitchell, J. L., Terrile, R. J., Carr, M., Cook, A. F., Cuzzi, J., Pollack, J. B., Danielson, G. E., Ingersoll, A., Davies, M. E., Hunt, G. E., Masursky, H., Shoemaker, E., Morrison, D., Owen, T., Sagan, C., Veverka, J., Strom, R., and Suomi, V. E. 1982. A new look at the Saturn system : The Voyager 2 images. *Science* 215:504–537.

Stone, E. C., and Miner, E. D. 1982. Voyager 2 encounter with the saturnian system. *Science* 215:499–504.

Synott, S. P., Terrile, R. J., Jacobson, R. A., and Smith, B. A. 1983. Orbit's of Saturn's F Ring and its shepherding satellites. *Icarus* 53:156–158.

Van Allen, J. A. 1982. Findings on rings and inner satellites of Saturn by Pioneer 11. *Icarus* 51:509–527.

Van Allen, J. A., Thomsen, M. F., and Randall, B. A. 1980. The energetic charged particle absorption signature of Mimas. *J. Geophys. Res.* 85:5709–5718.

Van Flandern, T. C. 1979. Rings of Uranus: invisible and impossible? *Science* 204:1076–1077.

Vogt, R. E., Chenette, D. L., Cummings, A. C., Garrard, T. L., Stone, E. C., Schardt, A. W., Trainor, J. H., Lal, N., and McDonald, F. B. 1982. Energetic charged particles in Saturn's magnetosphere: Voyager 2 results. *Science:* 215:577–582.

Ward, W. R. 1981. On the radial structure of Saturn's ring. *Geophys. Res. Lett.* 8:641–643.

Yoder, C. F. 1983. The gravitational interaction between inclined, elliptical rings. Proceedings of *I.A.U. Colloquium 75 Planetary Rings*, ed. A. Brahic, Toulouse, France, Aug. 1982.

Yoder, C. F., Colombo, G., Synnott, S. P., and Yoder, K. A. 1983. Theory of motion of Saturn's coorbiting satellites. *Icarus* 53:431–443.

PART IV

Origins

THE ORIGIN AND EVOLUTION OF PLANETARY RINGS

A. W. HARRIS
Jet Propulsion Laboratory

Saturn's rings have often been regarded as the uncoagulated remnant of a circumplanetary disk from which the regular satellites formed. In 1847, Roche suggested that they may instead be the fragments of a disrupted satellite. More than a century later, this central issue of whether ring particles are primordial or evolved remains unresolved. In this chapter, I review a number of dynamical processes which act on a gas/solid disk in Keplerian motion: viscous spreading, gas drag, coagulation of particulates, and the effect of further infall of matter from heliocentric orbit onto the planet/disk. From these considerations, I conclude that the survival of a remnant ring is unlikely, and turn to the possibility that rings were created by the disruption of sizable satellites, which were less sensitive to the destructive processes present during planet formation. This hypothesis, while appearing at first more complex than the other one, seems upon closer examination to have fewer unresolved difficulties. It offers a natural explanation for the presence of shepherd satellites (i.e., large collision fragments) coexisting in the same orbital range as ring particles, where coagulation into satellites is apparently not possible. This feature of the hypothesis is particularly attractive with regard to the origin of the Uranian rings.

There is little disagreement that the planetary rings which we see today are basically products of the same formation process that gave rise to the regular satellite systems surrounding each of the ringed planets. The primary issue debated in theories of ring origin is whether they represent a failure of the innermost portion of a circumplanetary disk to accumulate into satellites, or whether they are the result of the disruption of preexisting satellites. The principal difficulty of the former scenario is the matter of survival of such small particles under the conditions of planetary formation. In order to avoid this same difficulty, the latter scenario must provide for a delay from the time of satellite (and planet) formation to the time of disruption of the satellite(s) under more quiescent conditions. Furthermore, a breakup scenario must provide for the radial migration of the satellite(s) from the site of origin to the site of breakup (since satellite formation appears to be impossible at the present ring locations), and for the subsequent grinding and spreading of fragments to the present size and orbital distributions. In spite of the apparently greater difficulties, I favor the breakup scenario, and will attempt to describe and defend one such model in the following sections.

In this chapter I take the Roche limit or Roche zone to mean that orbital range near the planet within which accumulation of a particulate disk into discrete bodies fails to occur on account of the planet's tide. (See the chapter by Weidenschilling et al. for a discussion of tides which can affect a particle's disruption and accretion.) I do not review in detail the broader topic of planet and satellite formation. Such a review is in preparation by Stevenson et al. (1984). Section I of this chapter is a brief overview of the aspects of planet formation that affect ring formation. In Sec. II I consider the various processes operating during planet formation that lead to rapid evolution of small ring particles. The severe constraints imposed by these processes lead me to favor a scenario in which the ring material is preserved in the form of satellites until after the process of planet formation is essentially complete. In Section III I outline such a formation model. In the final section, the interrelation of the various known ring systems, and the implications of my ring formation model for the larger question of planetary formation, are discussed.

I. THE FORMATION OF PLANETS AND SATELLITE SYSTEMS

Models of giant planet formation can be grouped in two categories. If the planets formed by large scale gravitational instabilities in the solar nebula (essentially Jeans instability with rotational energy included), then the entire mass of the planet would collapse into a giant gaseous protoplanet on a time scale on the order of the orbit period (Cameron 1978). Following a time of 10^5 to 10^7 yr, during which the mass would be in hydrostatic equilibrium but in a very distended state, hydrodynamic collapse due to the dissociation of hydrogen would occur in $\leqslant 1$ yr, resulting in a planet not much hotter and larger than at present (Bodenheimer et al. 1980). Pollack et al. (1976, 1977) have extrapolated the cooling histories of Jupiter and Saturn backward in time, assuming hydrostatic equilibrium. Figure 1 shows the time required for Saturn to cool enough to allow ices to condense at various orbital radii. From the later work of Bodenheimer et al. (1980), it appears that the hydrodynamic collapse model merges with the hydrostatic one at $\sim 10^5$ yr, hence that point on the timeline in Fig. 1 should be taken as zero and everything to the left disregarded. The important conclusion with regard to satellite and ring formation is that ices could not condense at the present ring radii for $\sim 10^6$ to 10^7 yr. Therefore the ring mass must have been stored as a gaseous ring for that long,

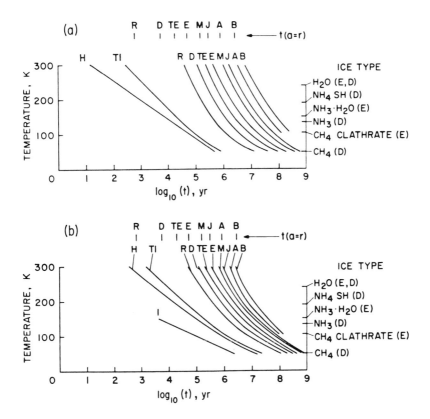

Fig. 1. The temperature within a circumplanetary disk about Saturn as the planet cooled and contracted. The two profiles represent assumed optically thin (a) and optically thick (b) disks. Each curve refers to a constant orbital distance and is identified by the initial of the satellite or ring currently occupying that orbital distance. (J for Janus may be regarded as the distance of the coorbital satellites.) The scales across the top identify the times at which the radius of Saturn equaled the orbital distances indicated. See Pollack et al. (1976) for further details. (Figure from Pollack et al. 1976.)

or else condensed at a larger radius and later transported into the present ring radius.

The second general class of models of giant planet formation require solid cores of 1 to 10 Earth masses which trigger gas accumulation about them (Safronov 1972; Harris 1978a; Mizuno et al. 1978; Mizuno 1980). The advantage of these models is that a less massive solar nebula is required, since gravitational instability of the gas phase is not required. On the other hand, nucleated gas accretion only occurs about fairly massive cores, which are likely to require $\sim 10^5$ to 10^6 yr to grow from the condensates of the cooling nebula. Even if a core of several Earth masses grew more rapidly, the gas envelope surrounding it would require $\sim 10^5$ yr to cool sufficiently for hydrodynamic collapse to occur (Harris 1978a). Following the initial hydrodynamic collapse, further gas accumulation would occur as rapidly as gas could be delivered to the accumulation zone. This time scale could be very short for a vigorously turbulent solar nebula; however, it is unlikely that it was $\lesssim 10^5$ yr, or the solar nebula would have been dissipated before the cores were formed. Unfortunately, thermal history calculations analogous to those discussed above have not been done for a nucleated growth model of planet formation, so it is not clear whether ice could condense about such a protoplanet while it was still growing. Based upon the hydrostatic models (Fig. 1), one might guess that if the time scale of gas accumulation were $\lesssim 10^6$ yr, the protoplanet would be too hot to allow condensed ice to exist as close to Saturn as the present rings, and the ring material would have had to be preserved as a gaseous disk for $\sim 10^6$ yr after formation, or else be transported in from a greater radius after that period of time, as in the previous model. On the other hand, if gas accumulation took $\gtrsim 10^6$ yr, ice may have been stable at the ring radius during the time of planet growth, and the storage problem would not exist.

In summary, it appears that the time scale of planet formation and/or ice condensation is probably $\gtrsim 10^5$ yr, so that one should be concerned about dynamical processes within a gas/solid protoplanetary disk which tend to disrupt it more rapidly than that. Furthermore, it is not clear whether or not the planets were still accumulating matter during the course of satellite/ring formation, so the effect of continuing mass infall onto the disk should be considered.

In order to illustrate (in Sec. II) the possible importance of various dynamical processes present in a gas/solid protosatellite disk, I shall make use of the numerical values given in Table I. For each planet with a known ring, a minimum mass disk is estimated at several orbital radii. The surface density of solids is estimated by smearing out the present ring/satellite masses over the disk area available. The temperature is taken to be the condensation temperature for the dominant solid species present. (Ice is assumed for Uranus since the lower temperature results in more conservative estimates of the importance of the various processes.) The equivalent thickness of the gas disk is (see Safronov 1972, p. 25) $h \approx 1.5\, v_T/\Omega$, where $v_T \approx 10^{-4}\, T^{\frac{1}{2}}$ cm s^{-1} is the thermal velocity of the gas molecules and Ω is the orbit frequency. The gas density is obtained by first estimating the mass of gas of the planet's composition necessary to yield the observed mass of solids in the disk. Others (e.g., Weidenschilling 1982) have assumed reconstitution to solar composition, which implies a more massive protosatellite disk. The assumed reconstitution factors, σ_g/σ_s, are tabulated. The solid surface density is mutiplied by this factor and divided by h to obtain the midplane gas density.

In evaluating the results based on this table, it must be remembered that if satellite formation is an inefficient process, then the disk masses may have

TABLE I

Models of Primitive Circumplanetary Disks

Planet Zone	Type of Solids	Temperature (K)	Orbit Radius (10^{10}cm)	Mass of Solids (g)	Surface Density of Solids, σ_s ($g\ cm^{-2}$)	Equivalent Thickness of Gas Disk (10^{10}cm)	Gas/Solid Ratio, σ_g/σ_s	Mid-planet Gas Density ($g\ cm^{-3}$)
Jupiter								
rings-V	rock	1000	1.8	10^{22}	10	0.3	100	3×10^{-5}
I-II	rock	1000	7	10^{26}	10^4	1.5	100	7×10^{-5}
III-IV	ice	250	20	3×10^{26}	3×10^3	3	25	3×10^{-6}
Saturn								
rings	ice	250	1.5	10^{24}	10^3	0.1	20	2×10^{-5}
I-V	ice	250	5	5×10^{24}	10^3	0.6	20	3×10^{-6}
Titan	(methane)	100	12	1.4×10^{26}	3×10^3	3	5	5×10^{-7}
Uranus								
rings	(ice)	250	0.5	10^{19}	10^{-1}	0.1	3	3×10^{-10}
I-IV	(ice)	250	6	3×10^{24}	3×10^2	1.5	3	7×10^{-8}

been much greater. On the other hand, if satellite formation was concurrent with slow growth of the planets, then the instantaneous mass of the disk may have been much less. Thus Table I represents a middle-of-the-road model of circumplanetary disks. Also, the low densities in the Jupiter and Uranus ring zones may well be due to loss of solid matter from those zones. If this is the case, then the surface densities of the inner satellite regions may be more representative of the original ring zone densities than those inferred from the amount of solid material presently there.

II. DYNAMICAL EVOLUTION OF A PROTOSATELLITE DISK

There are several dynamical processes which act to dissipate a solid/gas disk about a planet. Most of the same processes would occur in the solar nebula as well, but because of the smaller length scale, generally higher surface density, and shorter orbit period of a circumplanetary disk, the corresponding time scales are very much shorter than for the solar nebula.

A protosatellite disk is subject to viscous shear, just as are the present planetary rings (e.g., see chapter by Stewart et al.). If the gas phase of the ring were not turbulent, then the time scale of viscous spreading would be very long ($\sim 10^9$ yr). However, since even a minimum mass circumplanetary disk, with a surface density $\gtrsim 10^4$ g cm^{-2} (cf. Table 1) would be optically thick, it is likely that the disk would be convectively unstable and hence turbulent (e.g., Lin and Papaloizou 1980). The time scale for an element of gas to migrate a distance of the order of its orbit radius by viscous shear is

$$t \sim \frac{a^2}{\nu_e} . \tag{1}$$

The effective viscosity is $\nu_e \sim \Delta v \delta / \text{Re}$, where Re is the effective Reynolds number. The turbulence velocity Δv is on the order of the product of the eddy scale δ times the radial gradient of the orbital velocity. For Keplerian motion, this is just half of the orbit frequency Ω, so $\nu_e \sim \frac{1}{2} \Omega \delta^2 / \text{Re}$. If the turbulence is driven by thermal convection in the vertical direction, then δ will be on the order of the vertical scale height: $\delta \sim h \approx v_T / \Omega$. The effective Reynolds number is expected to be $\sim 10^3$ (e.g., Goldreich and Ward 1973; Lin and Papaloizou 1980). Thus, combining all these dimensional estimates, the time scale in years of viscous spreading of a turbulent disk becomes:

$$t \sim \text{Re} \frac{\Omega a^2}{v_T^2} \sim 10^5 \left(\frac{m_p}{m_\odot} \frac{a}{a_\oplus} \right)^{\frac{1}{2}} \left(\frac{100 \text{ K}}{T} \right) \tag{2}$$

where $m_p / m_\odot$ is the mass of the central body in solar units, $a/a_\oplus$ is the orbit radius in AU, and T is the temperature of the gas disk. The dissipation time for the solar nebula itself is thus on the order of 10^5 yr; for a circumplanetary disk about Jupiter or Saturn, $t \sim 100$ yr. One consequence of this difference in time scales is that, if the giant planets grew by nucleated instability (see e.g., Mizuno 1980), then mass infall onto a circumplanetary accretion disk would occur on a time scale of $\sim 10^5$ yr, whereas viscous dissipation of the disk would occur on a scale of $\sim 10^2$ yr. Thus it is questionable whether a protoplanetary accretion disk could exist as a steady state feature under these circumstances. Regardless of the mode of formation of the giant planets, the above time scale of dissipation imposes a serious constraint on models of condensation and growth of satellites from a circumplanetary disk. It has been argued (see e.g., Coradini et al. 1981) that a circumplanetary disk would not be turbulent and hence the diffusion time scale would be very long. I consider the arguments either for or against turbulence to be poorly developed at present; and the time scale of dissipation of the gaseous disk remains an important problem.

Another dissipation process that occurs as solids condense from the gaseous disk is aerodynamic drag (see e.g., Whipple 1964; Goldreich and Ward 1973; Weidenschilling 1977). If a radial pressure gradient exists in the gaseous disk, then the balance between gravitational and centrifugal forces will be altered by an additional gas pressure term, resulting in a slightly different circular orbit velocity for the gas than for solids at the same orbital radius. If the gas pressure gradient is of the order of the pressure divided by the orbit radius, then the difference between the Keplerian orbit velocity and the gas orbit velocity is (see e.g. Goldreich and Ward 1973):

$$v_w \sim \frac{1}{6} \frac{v_T^2}{\Omega\, a} . \tag{3}$$

For a typical circumplanetary disk at the temperature of ice condensation, $v_w \sim 30$ m s^{-1}. Small particles ($\leqslant$ meter sized) would be carried along with the gas and would experience relatively little radial drift before settling to the midplane (Weidenschilling 1982). If the particulate disk is optically thick (i.e., like the present rings of Saturn), the gas drag is probably best estimated as a shear stress on the surfaces of the disk (see e.g. Goldreich and Ward 1973). The stress per unit area of the disk (both sides) is:

$$S \sim 2\rho_g v_w^2/\text{Re} \tag{4}$$

where ρ_g is the gas density and Re is the Reynolds number. For laminar flow, the Reynolds number would be $\geqslant 10^5$, hence it is likely that the shear zone will be turbulent, with an effective Reynolds number Re $\sim 10^3$. The torque per unit area on the disk is $S\, a$. This can be equated to the rate of change of orbital angular momentum as follows:

$$\frac{2\rho_g v_w^2 a}{Re} = \frac{\sigma_s}{2} \Omega a \dot{a} \tag{5}$$

where σ_s is the surface density of the particulate disk. The gas density can be expressed in terms of the gas surface density σ_g, divided by the equivalent thickness of the gas disk $h \approx 1.5 \, v_T / \Omega$ (Safronov 1972, p. 25):

$$\rho_g \approx \frac{2}{3} \frac{\sigma_g \Omega}{v_T} . \tag{6}$$

By substituting this expression, along with the expression for wind speed Eq. (3), into Eq. (5), the time scale $t \sim a/\dot{a}$ for drag decay of the solid ring can be written in a simple form:

$$t \sim \frac{27}{2} \, Re \, \frac{\sigma_s}{\sigma_g} \frac{Gm_p}{v_T^3}$$

$$t \approx 6 \times 10^4 \, yr \, (Re) \left(\frac{\sigma_s}{\sigma_g} \right) \left(\frac{m_p}{m_\odot} \right) \left(\frac{100°K}{T} \right)^{\frac{3}{2}} . \tag{7}$$

Note that t is not dependent on the actual mass of the disk, nor on its radial extent, but only on the mass ratio of solids to gas. For an effective Reynolds number $\sim 10^3$, the decay time for silicate condensates ($T \sim 10^3$ K, $\sigma_s/\sigma_g \sim 10^{-2}$) is $\sim 10^4$ yr for the solar nebula or ~ 10 yr for a circumplanetary nebula about Jupiter or Saturn. For ices ($T \sim 250$ K, $\sigma_s/\sigma_g \sim 10^{-1}$), the lifetimes are $\sim 10^6$ yr or $\sim 10^3$ yr for solar or planetary nebulae, respectively.

For larger solid aggregates, the disk approximation above may not be appropriate. Rather, one should consider drag about individual bodies. For a circumplanetary nebula, the Reynolds number for a sphere ≥ 1 m moving at a velocity v_u, given by Eq. (3) through even a thin gas nebula ($\rho_g \sim 10^{-5}$ g cm^{-3}) would be ≥ 1000, hence the appropriate drag law is (cf. Weidenschilling 1977):

$$F_D \approx 0.7 r^2 \rho_g v_w^2 . \tag{8}$$

The time scale of orbital decay from aerodynamical drag on an isolated sphere of density ρ_s and radius r is thus:

$$t \sim 100 \frac{\rho_s}{\rho_g} \frac{\Omega^3 a^3}{v_T^4} r . \tag{9}$$

For the minimum mass circumplanetary disk about Saturn, the time scale of aerodynamic decay is ~ 200 (r km^{-1}) yr in all zones. For Jupiter, the time scale is also ~ 200 (r km^{-1}) yr in the ring/Amalthea and Ganymede/Callisto zones, but only ~ 5 (r km^{-1}) yr in the Io/Europa zone. Weidenschilling (1982) derives considerably shorter time scales because he assumes a higher gas density and a steeper pressure gradient. For Uranus, $t \sim 600$ (r km^{-1}) yr in the satellite zone, or $\sim 2 \times 10^6$ (r km^{-1}) yr in the ring zone. The large value in the ring zone is due to the low mass of the present ring system. If the same density is assumed as is present in the satellite zone, then the time scale of decay in the ring zone would be $\sim 10^4$ (r km^{-1}) yr.

If the planet is experiencing infall of matter from the solar nebula, then as a result both the gas and solid components of a circumplanetary disk will experience drag (Harris 1978b). The infalling matter is expected to arrive with very little angular momentum with respect to the planet, otherwise the accumulation of such matter would greatly overspin the planet (see e.g., Harris 1977). This being the case, the matter that falls onto the disk will cause it to decay into the planet as it gains mass. The same will be true of satellites in orbit about the planets, even if they do not retain the gas component of the influx of heliocentric matter. Harris (1978b) derived the rate of inward spiraling of a satellite, for the case of matter arriving with no preferred angular momentum with respect to the planet, and infalling from the solar nebula with originally negligible velocity relative to the planet. Since the velocity of the infalling matter exceeds that of the satellite (escape velocity versus orbit velocity), Epstein drag was assumed. This assumption is perhaps not perfect for gas infall, but should yield a dimensionally valid estimate. The rate of inward spiraling is

$$\frac{da}{a} \approx -3 \frac{\rho_p}{\rho_s} \frac{r_p^2}{r_s a} \frac{dm_p}{m_p} \tag{10}$$

where ρ, r, and m are the density, radius, and mass of the planet (subscript p) or satellite (subscript s). The flux of mass on the satellite is

$$\frac{dm_s}{m_s} \approx -\frac{2}{5} \frac{da}{a} \tag{11}$$

hence even if the mass sticks, the infall of the satellite is rapid compared to its growth. Figure 2 is a plot of the evolution of the Saturnian satellites as a function of mass gain of the planet. Since satellites could not survive evolution across neighboring orbits, it can be inferred that the present satellites, particularly the small ones, were formed very near the end of the growth of the planet, if not after the growth was completed.

A similar pair of equations can be derived for the rate of mass gain and orbital evolution of an optically thick ring (subscript r) subject to mass infall (Harris 1978b):

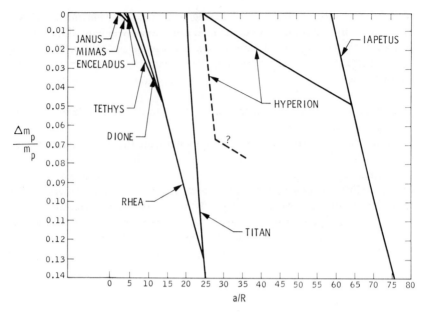

Fig. 2. Evolutionary tracks of the satellites of Saturn due to the growth of the planet. $\Delta m_p/m_p$ is the mass remaining to be accumulated and a/R is the orbit radius in current Saturn radii. The end points of the tracks ($\Delta m_p/m_p = 0$) are at the present orbital radii. (The track labeled Janus may be regarded as that of the coorbital satellites.) Hyperion may have been carried along with Titan for some time after the establishment of the present resonance (dashed line). (Figure from Harris 1978*b*.)

$$\frac{dm_r}{dm_p} \approx 1 \tag{12}$$

$$\frac{da}{a} \approx 2 \frac{dm_r}{m_r} \approx 2 \frac{dm_p}{m_r}. \tag{13}$$

If such a ring existed throughout the growth of the planet, then approximately half of the total planet's mass would have arrived via ring infall. This mass would have arrived with orbital angular momentum, which would greatly overspin the planet. Thus the scenario of a growing planet continuously surrounded by an optically thick ring is contradictory to the observed spins of the planets. From Eq. (13), it is clear that disks of the mass of those outlined in Table I would be extremely sensitive to decay by infall of heliocentric matter. Even the "reconstituted" masses of the disks are $\sim 10^{-3}$ of the planet masses, hence such disks would not survive even the last 1% of growth of their respective planets.

TABLE II

Coagulation Time Scales in Model Circumplanetary Disks

Planet/Zone	Time scale (yr) $(r_s \, \mathrm{km}^{-1})$
Jupiter	
Rings/V	3
I-II	0.05
III-IV	0.3
Saturn	
Rings	0.02
I-V	0.2
Titan	0.3
Uranus	
Rings	100
I-IV	2

One last time scale which should be compared with the above ones is that of coagulation of the solid particles through mutual collisions. The time scale of growth, assuming all particles stick on contact, is (Safronov 1972, ch. 9):

$$t \sim \frac{m_s}{\dot{m}_s} \sim \frac{2\rho_s \, r_s}{\sigma_s \, \Omega} \; . \tag{14}$$

The coagulation time scales in the various zones of the disk models of Table I are given in Table II. In general, these time scales are very much shorter than any of the preceding ones, hence one should expect the particulates in a circumplanetary disk to coagulate into $\gtrsim 100$ km bodies before any of the other processes are effective.

The various time scales discussed above are summarized in Table III. If the disk is turbulent, coagulation may be stalled, but viscous spreading will dissipate both the gas and the solids rapidly. Even if the gas is quiescent, aerodynamic drag would quickly remove a particulate disk from inside the Roche limit, while small particles outside the Roche limit would probably coagulate into satellites on an even shorter time scale. Thus it seems unlikely that solid particles as small as those of the present Saturnian or Uranian rings could have survived either within or outside the Roche zone for as long as $\sim 10^5$ yr, as required by the arguments in Section I. For this reason I favor a model of ring origin in which the ring mass is preserved in the form of satellites until after planet growth is complete and any gaseous disk has been dissipated. In Sec. III, this scenario is outlined.

TABLE III

Summary of Dynamical Time Scales in a Gas/Solid Circumplanetary Disk

Process	Time Scale (yr)
Viscous spreading (gas)	$\geqslant 100$
Gas drag on particulate disk	10-1000
Gas drag on satellites	$\sim 200\ (r_s\,\mathrm{km}^{-1})$
Infall of circumsolar matter on disk	$\geqslant 10^{-3}$ (planet growth time)
Infall of circumsolar matter on satellites	$\geqslant r_s/r_p$ (planet growth time)
Coagulation/fragmentation	$\sim 0.1\ (r_s\,\mathrm{km}^{-1})$

III. A SATELLITE BREAKUP MODEL OF RING FORMATION

The primary motivation for this scenario of ring formation is to avoid the difficulty presented by the very short orbital decay times of small particles during planet/satellite formation. Since the time scale of coagulation of particles outside the Roche limit is shorter than their orbital decay times, it is unlikely that rings were formed farther out and then evolved inward without first coagulating into satellites. Therefore, I suggest that satellites formed rapidly, or were captured, into orbits about the giant planets, and then spiraled inward due to the processes discussed in the previous section. When planet growth was completed, and/or any circumplanetary gas was dissipated, the inward migration ceased, leaving the presently observed satellite systems, plus additional small satellites in the regions presently occupied by rings. Such satellites would not necessarily be tidally disrupted (see e.g., Jeffreys 1947; Aggarwal and Oberbeck 1974), but may have persisted intact for some time, much as is the case for Phobos today (Dobrovolskis 1982). However, if subsequently disrupted by large meteoritic collisions, these satellites would be unable to reaccumulate into satellites, due to the planetary tides, and could instead evolve collisionally into the presently observed ring systems.

The first critical test of this hypothesis is whether the rate of infall of satellites into the planet poses an inconsistency. If the planets formed by hydrodynamic collapse, then circumplanetary gas disks must have persisted for $\sim 10^6$ yr in order to deliver satellites into the Roche zone after the planet had cooled enough to allow them to survive there. The persistence of a gas disk for so long itself poses difficulties due to the level of turbulence likely to have been present (cf., Table I). Ignoring this difficulty for the moment, the relevant time scale of decay is that of grown satellites in response to gas drag. Note that this time scale is $\sim 10^4$ yr for Saturn's ring parent body at its present radius. Thus, in the course of $\sim 10^6$ yr ~ 100 such bodies (comparable in size to a typical inner satellite) should have been lost in order to assure that a ring parent body would have a high likelihood of being left behind in the Roche zone. This is $\sim 10^{26}$ to 10^{27} g, not an impossible amount of solid mass to have been lost into Saturn.

The alternate hypothesis, that satellites were both created and driven inward by the infall of matter from heliocentric orbit during the course of slow planet growth, also leads to plausible limits (cf. Fig. 2 and Eq. 10), of several times the present masses of the satellite systems. Thus either mode of planet/satellite formation might lead to one or more satellites being left within the Roche limit of a planet after the conclusion of planet/satellite formation.

The next step in the scenario, collisional disruption, was probably accomplished by the same meteoritic bombardment which left a record of large craters on some of the older surfaces of Callisto and most of the Saturnian satellites. By extrapolating the crater density on Iapetus to the innermost satellites and rings of Saturn, Smith et al. (1982) estimate that these bodies have experienced enough collisions to have disrupted them several times over. Indeed, if a problem exists at all, it is one of an embarrassment of riches: it might be difficult for a satellite, once inside the Roche limit, to avoid disruption for the requisite time until formation processes have ceased.

Once a single ring parent body has been disrupted, further disruptions will occur as a result of mutual collisions between large fragments. The time scale of this process is given by Eq. (14), where r_s should be taken to be on the order of that of the parent body (~ 200 km for Saturn or ~ 10 km for Jupiter and Uranus), and σ_s the value for the respective ring zones from Table I. The time scales are ~ 30 yr, 4 yr, and 1000 yr, respectively, for Jupiter, Saturn, and Uranus. This process is limited in that as the fragments become smaller, the collisions become more gentle and less effective in causing further fragmentation. Since the impact velocity between two particles is at least on the order of the surface escape velocity of the larger fragment, the kinetic energy of such an impact can be equated with the limiting energy for fragmentation (Harris 1975; Greenberg et al. 1977) to define the size to which the largest fragments will be reduced:

$$ r \sim \left(\frac{3E_c}{4\pi G} \right)^{\frac{1}{2}} \approx 20 \left(\frac{E_c}{\rho_s} \right)^{\frac{1}{2}} \qquad \text{meters} \qquad (15) $$

where E_c is the critical specific energy for fragmentation in ergs g^{-1}. Greenberg et al. (1977, 1978) suggest values for E_c of $\sim 10^7$ and 2×10^5 for solid rock and ice, respectively, but also suggest that values as low as $\sim 10^5$ to 10^6 may be appropriate for hydrated mineral assemblages that may have been prevalent in the outer solar system during formation (i.e., Uranian ring material). For these values of E_c, rapid collisional fragmentation should cease in an ice ring when the largest fragments are ~ 5 km in radius, and for a rocky ring, fragments should be no larger than ~ 40 km, but probably much smaller, perhaps $\leqslant 10$ km.

Following the rapid reduction of the rings to particles no larger than ~ 10 km, the gentle bumping of particles should continue to result in slow erosion, much as rocks in a stream bed are slowly eroded by frequent collisions which are individually much too gentle to cause large scale fracturing. Borderies et al. (see their chapter) estimate that the time scale of this erosion is $\sim 10^5/\tau$ yr, where τ is the normal optical thickness of the ring. Weidenschilling et al. (see their chapter) argue that as particles develop a regolith erosion will not occur in the usual sense; however, they agree that mass exchange between ring bodies will occur on a time scale very short compared to the age of the solar system, so that in any case the particle size distribution should be highly evolved to an equilibrium state.

What is the equilibrium particle size, and why haven't they ground themselves to dust? Weidenschilling et al. (see their chapter) suggest that ring particles may in fact be gravitationally bound aggregates, which grow as large as they can before tidal stress disrupts them. I would suggest instead the following hypothesis. Aggregation is only favored as long as the collision velocity is less than the surface escape velocity of the particles, which in turn is the case if $\tau \gtrsim 1$ (Goldreich and Tremaine 1978). As particles aggregate, τ becomes less and the dispersion velocity increases, which hinders further aggregation. Thus the equilibrium that may be established is not a specific particle size, but rather whatever size results in $\tau \sim 1$. A tantalizing observational fact favoring this hypothesis is that both the Saturnian and Uranian rings have $\tau \sim 1$, in spite of the apparent difference of 2 to 3 orders of magnitude in ring particle size (cf. chapters by Elliot and Nicholson and by Cuzzi et al.). Furthermore, one would expect a tendency at increasing ring radii toward lower τ and larger particles, until finally aggregation wins even at zero optical depth and accumulation into satellites becomes possible.

In order for the above hypothesis to be tenable, ring particles must somehow evolve in size, by erosion and/or reannealing of dust onto larger particles, to achieve the above defined equilibrium, and eventually achieve the high degree of elasticity required by the dynamical constraints (see e.g., Goldreich and Tremaine 1978). The latter requirement is a difficulty for both this model and that of Weidenschilling et al.

Returning to the problem of the early evolution of collision fragments, a disrupted satellite should be reduced, on a time scale of a few years or less, to a swarm of debris in which the largest few fragments are ~ 10 km in diameter, but much of the mass is contained in much smaller fragments typical of comminution outcomes (see e.g., Dohnanyi 1969; Fujiwara et al. 1977). Subsequent erosion should occur on a much longer time scale, $\sim 10^5$ yr. Before this could happen, the narrow ring of fragments would be altered by viscous spreading, and possibly also by tidal truncation of the fluid ring by the largest fragments. Consider first viscous spreading. After a time t a narrow ring should diffuse to a width Δa (e.g., see chapter by Borderies et al.):

$$\Delta a \sim \left(\nu t \right)^{\frac{1}{2}} \gtrsim \left[\tfrac{1}{2}\, \Omega t \left(\frac{\tau}{1+\tau^2} \right) \right]^{\frac{1}{2}} r_s \qquad (16)$$

where the viscosity $\nu \approx 0.5\, v^2\, \tau/\Omega(1+\tau^2)$ and the random velocity is at least that due to Keplerian shear for a body of radius r_s, $v \gtrsim \Omega r_s$. In the few years required for rapid fragmentation, the debris ring should spread to only ~ 100 times the size of the characteristic particle size, r_s. However, on the time scale of further erosion, $\sim 10^5$ yr, spreading should occur to $\sim 10^4$ times the width of the characteristic particle size. Thus if $\sim 10^5$ yr is really the appropriate time scale for reduction of the largest particles in Saturn's ring from ~ 10 km to the present size, then spreading to the present radial extent may have occurred in the same amount of time. In any case, radial spreading to the current width from an initially narrow debris ring would appear to be an expected outcome, even for the very low interparticle collision velocities in the present ring.

If the dispersion velocity were to become low enough while some large fragments remained, then those fragments might clear gaps in the rings. If this were to occur, then the large fragments would not suffer continued erosion from mutual collisions and might persist for a much longer time. The condition for tidal truncation is that the torque transferred to the disk by the large fragment (satellite) must exceed that of viscous spreading, out to a distance from the satellite at least as great as the distance to the Lagrange points L_1 or L_2 (see e.g. Lin and Papaloizou 1979):

$$\frac{8}{27}\ \frac{G^2 m_s^2\, a\ \sigma_s}{x^3} \gtrsim \pi a^2\, \sigma_s\, v^2\ \frac{\tau}{1+\tau^2} \tag{17}$$

where $x = (m_s/3m_p)^{\frac{1}{3}}\, a$ is the minimum half-width of the gap (distance to the Lagrange points) for truncation to occur. Equation (17) can be rearranged to define the minimum particle size for which truncation occurs:

$$\frac{m_s}{m_p} \gtrsim \frac{9}{8}\ \pi \left(\frac{v}{\Omega a}\right)^2\ \frac{\tau}{1+\tau^2}\ . \tag{18}$$

For the present rings of Saturn, $v \sim 1$ cm s^{-1}, hence $m_s/m_p \sim 10^{-12}$, corresponding to a minimum diameter of an icy satellite of ~ 10 km. Thus if an initially narrow debris ring spreads radially, so that the dispersion velocity can fall below ~ 1 to 10 cm s^{-1} more rapidly than erosion reduces the largest fragments below ~ 10 km in diameter, then those fragments may open gaps in the ring. In an extreme case of many large fragments, clear gaps may be the dominant feature of the ring system, as in Uranian-type rings.

One final process which continues to reduce shepherd satellites even after gaps have been cleared is meteoritic bombardment. If the meteoritic bombardment of the ring zone was sufficient to cause the disruption of a ring parent body, then it probably was great enough to cause further disruptions of the largest fragments. Consider a power-law number density of bombarding particles $N(>m_1) \propto m_1^{1-q}$. The power-law index q is expected to lie in the range 1.5 to 2.0 (Dohnanyi 1969; Fujiwara et al. 1977; Shoemaker and Wolfe 1982). The total number striking a satellite is proportional to the cross-sectional area, r_s^2. The probability of disruption of a satellite is thus

$$P \propto r_s^2 N (>m_1) \propto \left(\frac{m_1}{m_s} \right)^{1-q} r_s^{5-3q} , \tag{19}$$

where m_1/m_s is the projectile/target mass ratio required to disrupt a body of mass m_s and radius r_s. If the collision strength E_c is constant with respect to size, then m_1/m_s is constant. However, E_c may decrease with increasing size due to greater material imperfections contained in larger bodies, but eventually will increase again due to gravitational compression. Neglecting for a moment these uncertainties, note that for $q < 5/3$, the probability of disruption (i.e., the collision lifetime) actually increases with decreasing size. Even for $q = 2$, it is proportional to r_s^{-1}, and this proportionality may be largely canceled by E_c increasing with decreasing size. Hence the collision lifetime does not become drastically less with decreasing particle size. Indeed, since a single large particle is reduced into many smaller particles ($n \propto r^{-3}$), then the probability that at least one collision fragment larger than r will survive further disruption becomes proportional to $r^{5-3q} \ln (r_s/r)$. Thus it appears that even if the total meteoritic bombardment was considerably greater than that required to disrupt the original parent body, some large fragments may have survived the bombardment subsequent to the disruption event.

IV. CONCLUSION

During the growth and/or cooling of the giant planets, the dynamical processes which would be expected in a gas/solid circumplanetary disk would remove small particles very efficiently. I therefore conclude that the planetary rings that we see today are the remnants of preexisting satellites, which evolved inward from their sites of origin but were spared infall onto the planet by the conclusion of planetary formation. The tidal forces of the respective planets were too weak to disrupt the already formed bodies, but following a meteoritic disruption of such satellites the planet tides were sufficient to prevent reaccumulation. The subsequent competing processes of viscous spreading, disk truncation by large fragments, erosional griding among colliding particles, and perhaps further catastrophic breakups of shepherd satellites, have led to different configurations in each of the presently known ring systems.

Jupiter's ring is perhaps a case of shepherds without sheep. The presently visible ring particles must be very short-lived, and are undoubtedly replenished by meteoritic erosion of the two satellites orbiting within the ring. Burns et al. (see their chapter; also 1980) suggest that there may be many more smaller satellites which are collisionally isolated from each other, and which provide an even larger collective surface area as a source of erosion of the visible particles. These satellites may be the remnants of the breakup of a single parent satellite. The smaller collisional debris, the sheep, may have been lost through a combination of further meteoritic erosion and Poynting-Robertson drag on the small particles (see Mignard's chapter). If tidal truncation within such a ring occurred, then the inner or imbedded shepherds may have been carried in with the ring debris. Some of the larger fragments which were not quite large enough to truncate the disk would be partially carried along by viscous spreading and/or Poynting-Robertson drag, and thus may have become collisionally isolated from one another, as are the two known satellites in the ring.

Saturn's rings appear to be a case of the opposite extreme: essentially all the mass has been reduced to fragments below the size limit for tidal truncation, so that the ring system is mainly broad, with only a few narrow gaps and confined ringlets caused by imbedded satellites large enough to truncate the disk. The weaker strength of ice in comparison to most rocky materials, with the result that the first generation of fragmentation would end with smaller largest bodies (cf. Eq. 15), may account for the difference in outcomes between Saturn's rings and the others. Also, the greater mass leads to higher viscous torques, so that the satellite mass required to truncate the disk is greater. Finally, the meteoritic erosion flux at Saturn may have disrupted most of the shepherds, even if they once existed.

The Uranian rings can be explained as a system dominated by shepherds (see chapters by Demott and by Elliot and Nicholson). The population of smaller fragments (primarily the ϵ Ring) appears to be overwhelmingly truncated by interstitial satellites, to the extent that the gaps are much wider than the ringlets. The persistence of this configuration may indicate that there has not been much further meteoritic activity following the disruption of the parent body of the rings. The above brief scenarios of origin of each known ring system are not intended to be definitive or unique, but only to indicate that the variety of known ring systems can be plausibly accommodated by differences in parent body mass, composition, and the environment of the various planets in question.

Ever since Saturn was discovered to be encircled by a ring, such structures have been regarded as a key to understanding the formation of the planets and their satellite systems. They have definitely earned this status in the sense that ring systems have motivated critical thinking on dynamical processes relevant to planetary origin: the tendency for an orbiting, collisionally interacting system to flatten into a disk, to spread by viscous shear, and perhaps to be truncated by large imbedded bodies. The dynamical processes of coagulation and fragmentation, density wave generation and propagation, and gravitational instability all have received much study because of their relevance to planetary rings, and all this work has contributed to a greater understanding of the processes of planetary formation.

However, in another sense, planetary rings have failed to provide a key to planetary formation. As I have argued in this chapter, it is unlikely that the present rings are remnants of primordial accretion disks about planets; rather, they are highly evolved structures and their presence cannot be regarded as unambiguous evidence for one mode of planet formation or another. Ring formation appears to be a subsidiary event to satellite formation. Thus, an important constraint on planetary origin is the nature of the various satellite systems of the planets, rather than just the rings that also happen to be present in some cases.

Acknowledgment. The preparation of his chapter was supported at the Jet Propulsion Laboratory of the California Institute of Technology, under contract with the National Aeronautics and Space Administration.

REFERENCES

Aggarwal, H. R., and Oberbeck, V. R. 1974. Roche limit of a solid body. *Astrophys. J.* 191:577–588.

Bodenheimer, P., Grossman, A. S., DeCampli, W. M., March, G., and Pollack, J. B. 1980. Calculations of the evolution of the giant planets. *Icarus* 41:293–308.

Burns, J. A., Showalter, M. R., Cuzzi, J. N., and Pollack, J. B. 1980. Physical processes in Jupiter's ring: Clues to its origin by Jove! *Icarus* 44:339–360.

Cameron, A. G. W. 1978. Physics of the primative solar accretion disk. *Moon Planets* 18: 5–40.

Coradini, A., Federico, C., and Magni, G. 1981. Gravitational instability in satellite disks and formation of regular satellites. *Astron. Astrophys.* 99:255–261.

Dobrovolskis, A. R. 1982. Internal stresses in Phobos and other triaxial bodies. *Icarus* 52:136–148.

Dohnanyi, J. S. 1969. Collisional model of asteroids and their debris. *J. Geophys, Res.* 74:2531–2554.

Fujiwara, A., Kamimoto, G., and Tsukamoto, A. 1977. Destruction of basaltic bodies by high-velocity impact. *Icarus* 31:277–288.

Goldreich, P., and Ward, W. R. 1973. The formation of planetesimals. *Astrophys. J.* 183:1051–1061.

Goldreich, P., and Tremaine, S. 1978. The velocity dispersion in Saturn's rings. *Icarus* 34:227–239.

Greenberg, R., Davis, D. R., Hartmann, W. K., and Chapman, C. R. 1977. Size distribution of particles in planetary rings. *Icarus* 30:769–779.

Greenberg, R., Wacker, J. F., Hartmann, W. K., and Chapman, C. R. 1978. Planetesimals to planets: Numerical simulation of collisional evolution. *Icarus* 35:1–26.

Harris, A. W. 1975. Collisional breakup of particles in a planetary ring. *Icarus* 24:190–192.

Harris, A. W. 1977. An analytical theory of planetary rotation rates. *Icarus* 31:168–174.

Harris, A. W. 1978*a*. The formation of the outer planets. *Lunar Planet. Sci.* IX:459–461.

Harris, A. W. 1978*b*. Satellite formation, II. *Icarus* 34:128–145.

Jeffreys, H. 1947. The relation of cohesion to Roche's limit. *Mon. Not. Roy. Astron. Soc.* 107:260–262.

Lin, D. N. C., and Papaloizou, J. 1979. Tidal torques on accretion discs in binary systems with extreme mass ratios. *Mon. Not. Roy. Astron. Soc.* 186:799–812.

Lin, D. N. C., and Papaloizou, J. 1980. On the structure and evolution of the primordial solar nebula. *Mon. Not. Roy. Astron. Soc.* 191:37–48.

Mizuno, H. 1980. Formation of the giant planets. *Prog. Theor. Phys.* 64:544–557.

Mizuno, H., Nakazawa, K., and Hayashi, C. 1978. Instability of a gaseous envelope surrounding a planetary core and formation of giant planets. *Prog. Theor. Phys.* 60:699–710.

Pollack, J. B., Grossman, A. S., Moore, R., and Gabroske, H. C. 1976. The formation of Saturn's satellites and rings, as influenced by Saturn's contraction history. *Icarus* 29: 35–48.

Pollack, J. B., Grossman, A. S., Moore, R., and Gabroske, H. C. 1977. A calculation of Saturn's gravitational contraction history. *Icarus* 30:111–128.

Safronov, V. S. 1972. *Evolution of the Protoplanetary Cloud and Formation of the Earth and Planets.* (Jerusalem: Israel Prog. for Sci. Trans.) NASA TT F-677.

Shoemaker, E. M., and Wolfe, R. F. 1982. Cratering time scales for the Galilean satellites. In *Satellites of Jupiter,* ed. D. Morrison (Tucson: Univ. of Arizona Press), pp. 277–339.

Smith, B. A., Soderblom, L., Batson, R., Bridges, P., Inge, J., Masursky, H., Shoemaker, E., Beebe, R., Boyce, J., Briggs, G., Bunker, A., Collins, S. A., Hansen, C. J., Johnson, T. V., Mitchell, J. L., Terrile, R. J., Cook, A. F. II, Cuzzi, J., Pollack, J. B., Danielson, G. E., Ingersoll, A. P., Davies, M. E., Hunt, G. E., Morrison, D., Owen, T., Sagan, C., Veverka, J., Strom, R., and Suomi, V. 1982. A new look at the Saturn system: The Voyager 2 images. *Science* 215:504–537.

Stevenson, D. J., Harris, A. W., and Lunine, J. I. 1984. Origins of satellites. In *Satellites of the Solar System,* eds. J. A. Burns, D. Morrison, and M. S. Matthews (Tucson: Univ. of Arizona Press). In preparation.

Weidenschilling, S. J. 1977. Aerodynamics of solid bodies in the solar nebula. *Mon. Not. Roy. Astron. Soc.* 180:57–70.

Weidenschilling, S. J. 1982. Origin of regular satellites. In *The Comparative Study of the Planets,* eds. A. Coradini and M. Fulchignoni (Dordrecht: D. Reidel), pp. 49–59.

Whipple, F. L. 1964. The history of the solar system. *Proc. Nat. Acad. Sci.* 52:565–594.

THE SOLAR NEBULA AND THE PLANETESIMAL DISK

WILLIAM R. WARD
Jet Propulsion Laboratory

Two popular theories of solar system formation are briefly reviewed, then used as background in an examination of several new developments related to planetary ring dynamics that promise to have great impact on future research. Most important are the incorporation of accretion disk and density wave theories into cosmogonic theory. A successful integration of these mechanisms may significantly constrain evolutionary models of the early solar system and also provide new insight into the mechanisms themselves.

This chapter reviews the problem of solar system formation, in particular that of the terrestrial planets, to illustrate several processes believed to be associated with this event which are analogues of ring phenomena discussed elsewhere in this book. Processes dominating particle rings in the current solar system seem most relevant to minimum mass models of the solar nebula which employ impact accretion to accumulate planet-sized objects from smaller ones (see e.g. Safronov 1969; Goldreich and Ward 1973; Greenberg et al. 1978; Wetherill 1980). I primarily focus on these models. However, the other principal approach to this problem, the massive nebula models of Cameron and co-workers, is briefly discussed (see e.g. Cameron 1978; Slattery 1978; Decampli and Cameron 1979). These models rely on gravitational instabilities in the gas phase to initiate planet growth.

Historically, research on galactic dynamics and solar system formation has improved our understanding of ring phenomena. In recent years, however, the virtual explosion of new data from planetary probes and sophisticated groundbased and airborne observing techniques has stimulated theoretical efforts directed at planetary rings themselves. Several concepts, newly applied to planetary problems, have arisen from these studies and promise to reflect significantly on our understanding of early processes in the solar nebula.

I. BASIC MODELS

A. Minimum Mass Nebula with Impact Accretion Among Planetesimals

The minimum-mass model of the solar nebula assumes only enough primordial material to provide for the observed present mass of the planetary system. In the inner solar system, this implies $\sim 10^{28}$ g of condensible material spread over a region on the order of 10 AU^2, i.e. 10^{27} cm^2 in size. This results in an average surface density $\sigma \sim 10$ g cm^{-2}. Since this condensible material is presumed to originate from a solar composition cloud, about two orders of magnitude more material ($\sigma \sim 10^3$ g cm^{-2}) should have constituted the gas phase of the solar nebula. The Sun is assumed to be in place at essentially its present mass (2×10^{33} g) and to be surrounded by the tenuous nebula of primarily hydrogen and helium with total mass (at most) a few percent of the solar mass. The nebula is disk-shaped but not especially thin. Its finite scale height is maintained in hydrostatic equilibrium in which the vertical pressure gradient supports the gas against the vertical component of the solar gravity,

$$(1/\rho) \, dP/dz \sim GM_\odot z/r^3 \; . \tag{1}$$

For an isothermal disk with adiabatic index γ, Eq. (1) can be easily integrated to obtain $\rho = \rho_c \, e^{-(z/h)^2}$, where ρ_c is the central density and $h = (2/\gamma)^{\frac{1}{2}}(c/\Omega)$ is scale height. Other thermal profiles yield similar results, i.e. a disk scale height of order c/Ω where c is the sound speed in the gas phase and Ω is the local orbital frequency (at 1 AU $\Omega = 2 \times 10^{-7}$ s^{-1}). The sound speed for molecular hydrogen is taken to be 7.6×10^3 cm s^{-1} where T is the gas temperature. For temperatures low enough to allow chemical condensations, i.e., $\sim 10^2$ K, the scale height of the nebula at 1 AU is $\sim 10^{12}$ cm.

Fractionation of Solids from Gas Phase. The separation of condensible materials is thought to have resulted from a settling process in the early stages of the nebula. The large surface area of the nebula makes it an efficient radiator, with temperature quickly dropping below the condensation point of refractory material. Grains grow by sweeping up vapor phase material, or by collisions with other grains in which surface forces such as van der Waals attraction or ferromagnetic forces promote grain growth. In the former case, the resulting settling time is a few orbital periods, with typical grains ~ 1 to 10 cm in radius (Lyttleton 1972; Goldreich and Ward 1973). However, if the number of nucleation sites is so large that the vapor phase is exhausted before settling can occur, growth proceeds by coagulation of solid grains. Numerical treatments of this process, including simulations of simultaneous growth and descent, have been performed (by e.g., Weidenschilling 1980). These indicate that the vertical settling process is distinctly nonhomologous; some of the first particles to arrive at the midplane initially condense far from it. Typical descent times are in the range of 10^3 orbital periods. In either case, debris begins to collect in a thin sheet at the midplane of the solar nebula in a relatively short time, producing a structure in some respects reminiscent of a present-day planetary ring.

Gravitational Instability. The next stage of the accumulation process depends in part on the behavior of the gaseous phase of the nebula. If the gas flow is essentially quiescent, the debris sheet can become unstable to self gravity. However, as in planetary rings (see Chapter by Shu) both the rotation of the disk around the primary and the random motions of constituent particles inhibit this process.

Random particle velocities tend to stabilize small regions of the disk. If the characteristic time, $\sim x/v$, for a particle with random velocity v to traverse a distance x is multiplied by the characteristic deceleration imposed by the gravity of the collapsing fragment $\sim GM/x^2$, the total implied velocity change must be $<$ v. Otherwise, the particle's path is reversed and it is brought back into the collapsing fragment. Since the mass M of the collapsing fragment is also dependent on the size of the region, i.e., $M = \pi\sigma x^2$, the condition for stability is equivalent to $x < v^2/\pi G\sigma$.

In contrast, rotation stabilizes large regions of the disk. For simplicity, consider a disk in uniform rotation. A region of size x must be chosen such that the centripetal acceleration required to keep a particle orbiting around the center of a collapsing fragment is $> GM/x^2$, the gravity of the fragment. Since the centripetal acceleration is equal to v^2/x where v $\sim x\Omega$, this condition implies $x > \pi G\sigma/\Omega^2$.

Marginal stability of the disk occurs when the two regimes just overlap, $v^2/\pi G\sigma \simeq \pi G\sigma/\Omega^2$. In reality, the disk is in Keplerian rotation and the stability test is best carried out by perturbation analysis of the fluid equations of motion (Toomre 1964; Goldreich and Lynden-Bell 1965; Goldreich and Ward 1973; Ward 1976a). This procedure yields a dispersion relation

$$\omega^2 = k^2c^2 - 2\pi G\sigma k + \kappa^2 \qquad (2)$$

where k and ω are the wavenumber and frequency of a radial sinusoidal disturbance and κ is the epicycle frequency. For a Keplerian disk, $\kappa = \Omega$. Marginal stability ($\omega = d\omega/dk = 0$) establishes a critical dispersion velocity $v_{cr} = \pi G\sigma/\Omega$ for disk stability. For the values of the parameters used here, $v_{cr} \sim 10$ cm s^{-1}.

The particles routinely suffer inelastic collisions with a frequency of order $\tau\Omega$ where τ is the optical depth of the debris disk. If these collisions damp v below v_{cr} the disk will fragment, with the diameter of the largest region being $\lambda \sim 4\pi^2 G\sigma/\Omega^2 \sim$ (few) $\times$ 10^8 cm (Safronov 1969; Goldreich and Ward 1973). The mass involved, $\sigma\lambda^2 \sim 10^{18}$ g, is equivalent to that of a solid object with a diameter from 1 to 10 km. However, the details of the fragmentation process indicate that these largest regions cannot collapse directly to solid density but tend to undergo a further fragmentation process, producing a hierarchy of scale lengths (Goldreich and Ward 1973). The largest region that can contract directly to solid density is $\sim 10^6$ cm, constituting a mass $\sim 10^{14}$ g and a planetesimal diameter from 0.1 to 1 km. The dynamic collapse time is on the order of an orbital period.

It is instructive at this point to compare these numbers with similar calculations for Saturn's rings. With ring surface density $\sigma \sim 10^2$ g cm^{-2} (Holberg et al. 1982; Esposito et al. 1982) and orbital mean motion $\Omega \sim 1.5 \times 10^{-4}$ s^{-1}, the largest unstable wavelength is $\lambda \sim 1.2 \times 10^4$ cm. This scale is much smaller than many of the features revealed by Voyager cameras and thus is unlikely to account for the complex ringlet structure. Mechanisms proposed for producing various ring features are discussed elsewhere in this book (see e.g. chapters by Stewart et al., by Shu and by Franklin et al.). Nonetheless, portions of Saturn's ring system appear to be very near the stability limit ($Q = v\Omega/\pi G\sigma \sim 1$), and local gravitational instability may have a role in some ring behavior such as the observed azimuthal brightness variation in the A Ring (Colombo et al. 1976). Note that, since the rings lie inside the Roche limit, loose aggregates of the sort likely to result from gravitational clumping will be sheared by tidal stresses and the material returned to general circulation. (Exactly this close proximity to the planet is generally believed to account for the failure of ring material to accumulate into a small number of orbiting objects; see the chapter by Weidenschilling et al. for a discussion of this problem.) This process is very dissipative and could generate an effective viscosity large enough to spread an optically thick ring quickly (Ward and Cameron 1978; Harris and Ward 1983; see also chapters by Stewart et al. and by Harris for discussions of ring evolution via viscous shear stresses). Consequently, if Saturn's rings originated from a narrow source zone (viz., disintegration of a satellite), gravitational instabilities could have been dominant early in the system's evolution.

At this point an important caveat must be mentioned. The fragmentation of the preplanetary disk as described above may only occur if the gas phase does not appreciably excite random particle motions, because if gas turbulence is present (see Sec. II below), it may keep the random velocity above v_{cr}. One possibility is that longer wavelength instabilities caused by gas-particle friction may then operate to increase the local surface density until fragmentation is possible (Ward 1976a). Alternatively, accretion may proceed via individual particle collisions instead of through collective gravitational forces. However, for such accretion the rebound velocity must be $\leq v_e$ (the escape velocity of the colliding particles) in order for their mutual gravitational attraction to hold on to the impact products. Since $v_e \approx 10^{-3}a$ cm s^{-1}, where a is the particle radius, for cm-sized grains $v_e \sim 10^{-4} v_{cr}$. Hence, if the instability is not operating, this also precludes accretion solely due to mutual gravity among particles unless the coefficient of restitution is quite low. In any case, once km-sized objects are predominent, collective gravitational forces cannot be effective at generating further growth and individual particle collisions must indeed become the subject of study.

Impact Accretion. To proceed further, a case must be made that a disk composed of km-sized planetesimals will relax to a state wherein the relative velocities do not greatly exceed v_{cr}. Such a case is made as follows: if particles are assumed to lose a fraction β of their random kinetic energy as a result of collisions, the damping rate of their dispersion velocity is

$$dv^2/dt \sim \beta v^2 (3\sigma\Omega/4\rho_p\, a) \qquad (3)$$

where ρ_p is the body density of the accreting objects. Note that, although the planetesimals are arranged in a disk that is thin in comparison to the solar nebula, it nonetheless has a finite scale height h_d. This height is approximately the distance particles rise with a relative velocity v before the vertical component of the solar gravity returns them to the disk, $h_d \simeq v/\Omega$. The spatial density ρ_d of condensible material is thus simply equal to σ/h_d and the flux $\rho_d v = \sigma\Omega$ is independent of the particles' random velocity v.

On the other hand, near misses among particles tend to excite relative velocities. This scattering process acts like a viscosity in converting directed orbital motion of the whole disk into random motions among the constituent particles (Safronov 1969; Kaula 1979; Stewart and Kaula 1980). A typical increment in velocity produced by a near encounter is given by $v \sim GM/\lambda^2(\lambda/v) \sim GM/v\lambda$ where λ is the impact parameter and v is the relative approach velocity. Summing the contributions from over the range of possible λ's $(< h_d)$ as a random walk process,

$$dv^2/dt \sim (3\sigma\Omega\ G^2M^2/2v^2\rho_p\, a^3)\ \ln\ (h_d/a). \qquad (4)$$

The disk relaxes to the point where its damping and excitation rates are equal (Safronov 1969). This condition turns out to be a stable equilibrium and the dispersion velocity is

$$v \sim v_e \left[\ln(h_d/a)/2\beta\right]^{\frac{1}{4}} \sim (GM/\theta a)^{\frac{1}{2}}\ . \qquad (5)$$

Particle collisions are expected to be reasonably inelastic so that their equilibrium velocity is on the order of their escape velocity, and it is plausible to expect net accretion to result from their mutual collisions. [Note that equilibrium is possible even if the fractional energy dissipated per collision β does not depend on velocity. On the other hand, in a planetary ring it may well be that v $>>$ v_e and the principal source of excitation is (nonaccreting) direct collisions instead of near misses. Random motions increase at a rate $dv^2/dt \sim \nu(rd\Omega/dr)^2$, with the viscosity in turn related to the dispersion velocity as $\nu \sim v^2\Omega^{-1}(\tau + 1/\tau)^{-1}$ (Goldreich and Tremaine 1978). Consequently, equilibrium is only possible if β is dependent on velocity. For real materials, this turns out to be the case (see the chapter by Borderies et al.).] The dimensionless parameter θ, defined by Eq. (5), is often called the Safronov number. Its value may depend on additional assumptions of particle size distribution, gas drag effects, etc., but is usually of order unity (Safronov 1969). In this case the rate of particle growth is simply $da/dt \sim \sigma\Omega/\rho_p \sim 10$ cm yr^{-1} at 1 AU and a 100-km-sized object could be accreted in $\approx 10^6$ yr.

Growth from 1 to 100 km may proceed at a rate considerably faster than that of the above argument through runaway accretion (see e.g. Greenberg et al. 1978; Wetherill 1976; Greenberg 1979, 1980). This process comes about because a very flat particle size distribution with relative velocities on the order of their escape velocities is unstable to the introduction of a somewhat larger object, i.e., the size differential grows rapidly. A larger object of radius R immersed in the swarm is able to bend the trajectory of approaching objects considerably more than a typical target can. As a result, the rate of growth in radius for the large object is enhanced by a factor $1+(R/a)^2$. This rate can become so large that, if valid throughout the accretion process, planet-sized objects would emerge on a very short time scale.

The validity of this argument, however, breaks down before planet-sized objects are achieved because of the system's overall orbital motion about the primary. As the cross section gets large Keplerian shear, not velocity dispersion, determines the approach velocity (see e. g. Ward 1976a,b; Weidenschilling 1977). In effect, the cross section is limited to the Roche lobe of the growing object, $L_2 \sim r(m/3M)^{\frac{1}{3}}$, where r is the distance to the primary and m is the mass of the accreting object. Consequently, rapid accretion will proceed only as long as the amount of material contained in an annulus of width $2L_2$ centered on the object is greater than the mass of the object sweeping it up. This allows the differential rotation of the disk to continue to supply material to the Roche lobe. The limiting size of the rapid growth phase is determined by the condition $2\pi r\sigma(2L_2) \sim m$, which implies a mass limit of order 10^{25} g for the inner solar system in a minimum mass model. The resulting sublunar-sized planetoids have radii $\sim 10^2$ to 10^3km. The growth interval is actually determined by the differential orbital time for particles at the boundaries of the accretion annulus $T \sim \Omega^{-1} (a/L_2)$. Since there are $10^{28}/10^{25} = 10^3$ planetoids at the end of this phase, a/L_2 is $\sim 10^3$ and the growth time should be a few thousand years (see, however, Wetherill and Cox [1982] for other effects that may limit runaway growth).

Terminal Stage. The final stage of the accretion process presents the main bottleneck to the overall scheme outlined here. Since the differential rotation of the disk no longer generates close encounters by simply causing particles to drift to within their Roche lobes, further collisions must rely on substantial radial excursion of the particles. This could be accomplished by orbital eccentricities. The dispersion velocity necessary to guarantee collisions is $\sim ea\Omega$ $\sim \delta a\Omega$ where δa is the differential semimajor axis among the N growing particles of radius R, i.e., $\delta a \sim a/N$. Consequently, the necessary v for impacts is proportional to R^3 while the escape velocity is proportional to R. Thus, the cross section must necessarily shrink toward geometric size. For the final stages of the accretion process, the growth rate for a typical accreting object is again $\sim \sigma\Omega/\rho \sim 10\text{cm yr}^{-1}$, producing terrestrial-sized planets in $\sim 10^7$ to 10^8 yr from the start of accretion, although some sublunar-sized objects may be produced in only a few thousand years.

The situation is even more complicated in that it is not obvious that the high dispersion velocities necessary for impacts can be achieved throughout the growth process. Accretion could stall at a stage with too many planets and with relative velocities too small to promote further growth. This difficult and important problem has been most successfully approached numerically. The first efforts, which followed the evolution of 100 lunar-sized objects in the terrestrial zone with a two-dimensional numerical code, illustrated that stall down of the accretion process was, indeed, a real concern (Cox 1978; Cox and Lewis 1980). Most outcomes resulted in a relatively large number of planets (~ 10) in the inner solar system (see Fig. 1a,b). Extension of the code to three dimensions, however, improves the situation in that there is an enhanced probability of scattering relative to physical collisions (Wetherill 1980; Wetherill and Cox 1982). Results are more promising in that "solar-system-like" outcomes can be obtained as in Fig. 1c. However, not all important physical processes have yet been included in the models and any additional dissipative mechanisms (such as gas drag, eccenticity damping by density waves, collisions with large numbers of small objects perhaps produced by Roche lobe encounters, etc.) may still be capable of stalling the system (Wetherill 1980). Further work on this problem is in progress.

At any rate, even if one is convinced that accretion can overcome such obstacles and go to completion, the time scales obtained by these numerical simulations, $\sim 10^7$ to 10^8 yr, may pose a problem if some of the newer ideas concerning the behavior of the solar nebula are correct. We will return to this point in Sec. II.A, but first we look at the major competitor to the planetesimal theory, the massive nebula theory proposed by Cameron and coworkers.

B. Massive Nebula with Giant Gaseous Protoplanets

Cameron's approach is dictated by the belief that a star of solar mass surrounded by a low-mass nebula is an unlikely outcome for the collapse of an

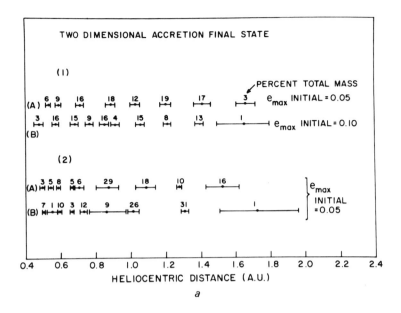

Fig. 1. Numerical calculations of multiplanet accumulation final states. (a) Two-dimensional calculations: (1) Cox (1978) and (2) Wetherill (1980). For the initial eccentricities chosen, an excessive number of terrestrial planets result.

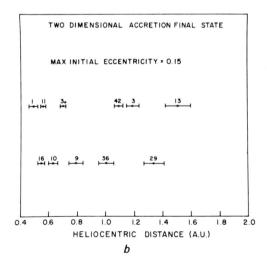

Fig. 1b. Two-dimensional calculations by the same authors showing the smaller number of planets formed if large values of initial eccentricity are used.

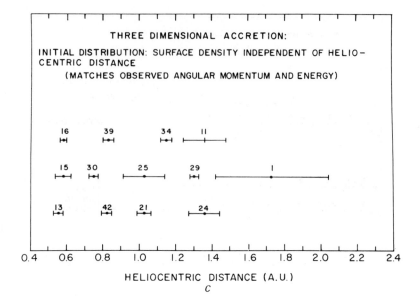

Fig. 1c. Three-dimensional calculations with low initial eccentricity ($e_{max} = 0.05$), leading to a small number of planets (from Wetherill 1980).

interstellar cloud (Cameron 1978). Instead, a typical interstellar cloud fragment would have so much angular momentum that its collapse product would most likely be a large disk with only a small central condensation; the Sun would then be a secondary consequence of the viscous evolution of this disk. Accretion disk studies tell us that since viscous stresses dissipate energy in a gas disk with Keplerian shear, the structure seeks a lower energy configuration wherein most of the material drifts toward the center of the system while a small amount of material drifts farther away carrying most of the angular momentum (Lynden-Bell and Pringle 1974). This is the same process that attempts to spread planetary rings (see chapter by Stewart et al.). If the Sun is to form in the center through this process, the disk must originally have more than a solar mass. Cameron's model disks are also extensive in radius, i.e. $\sim 10^2$ to 10^3 AU; hence, the surface density in these regions is $\sim 10^2$ to 10^3 g cm^{-2}.

Disk Evolution and Stability. The disk is assumed to exhibit vigorous turbulence. A number of mechanisms are proposed for its generation: convective overturn in the vertical direction due to high grain opacity, meridional circulation in the disk, and mismatch between the angular momentum of infalling and already orbiting material. The assumed kinematic viscosity ν is on the order of the sound speed times the scale height and the time scale for viscous evolution of the nebula $T \sim 10^5$ yr. Figure 2a shows numerical calculations of the mass of the disk as a function of time (Cameron 1978). The mass increases during the infall stage and then drops off on a time scale of $\sim 10^5$ yr. The mass loss has two sources: drift towards the center where the Sun is forming, and loss off the surface of the disk itself. In the latter case, mechanical waves produced by the strong turbulence in the nebula propagate into the higher, more tenuous parts of the nebula, depositing their energy, heating the gas, and driving a coronal wind. Figure 2b shows the radius of the disk as a function of time. The growth in radius accompanies the infall stage; the later decrease in radius is a direct result of the coronal wind phenomenon. If mass were lost by its inward drift alone, the outer rim of the nebula would continue to grow, with material convecting the angular momentum content to a larger radius. The same process acts in planetary rings unless they are confined by satellite torques (see chapters by Borderies et al., Dermott, and Stewart et al.).

Since the solar mass is initially distributed throughout a disk, the location in the disk for a given specific orbital angular momentum changes with time as a system evolves. Fig. 2c shows the disk position with time corresponding to the specific orbital angular momenta of the present-day planets. Even material eventually making up the inner planets originated at distances quite remote from the center of the system; hence the gas disk exhibits less stability at great distances. As mentioned earlier, the critical velocity for random motions to stabilize a particle disk is $\sim \pi G\sigma/\Omega$. Broadly speaking, this criterion applies also to gaseous disks. For $\sigma = 10^3$ g cm^{-2} and $R = 100$ AU, $v_{cr} \simeq 10^5$ cm s^{-1}. Since the sound speed is basically a measure of molecular velocities, it is clear that the disk is on the verge of instability at these distances. Cameron suggests that the nebula did indeed break up under self-gravity, producing gaseous protoplanets. Much of his work has been concerned with investigating the evolution of such objects through standard techniques of stellar evolution.

Protoplanetary Objects. Objects with masses on the order of Jupiter's mass contract slowly on the Helmholtz time scale of $\sim 10^5$ yr until their interiors heat to the point where dissociation of hydrogen is possible. This furnishes an energy sink which allows a rapid hydrodynamic collapse. However, in an even shorter time, objects of Jovian mass evolve internally in a manner similar to the first fractionation step of the minimum mass nebula model: grains form and begin to settle together. The interiors of the smaller objects have regions where iron grains may pass through a regime of liquid stability, further promoting efficient grain growth (Slattery 1978; Decampli and Cameron 1979). The settling time through the body of the protoplanet is on the order of hundreds of years. Whereas in the minimum mass nebula

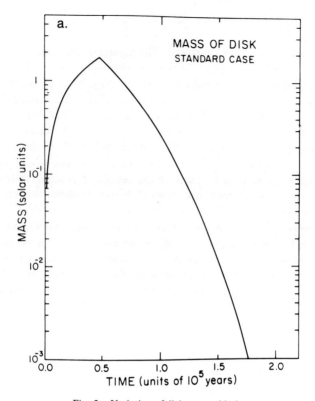

Fig. 2a. Variation of disk mass with time.

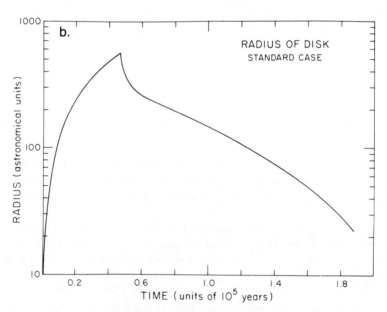

Fig. 2b. Variation of disk radius with time.

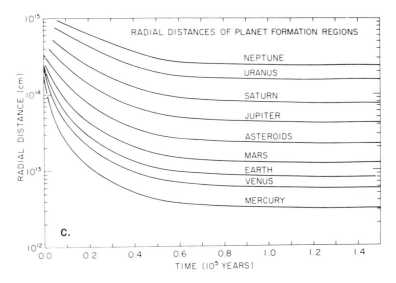

Fig. 2c. Variation with time of the orbital radii of regions of planet formation, centered on specific angular momenta of the present planets (from Cameron 1978).

model this settling process only collected material over one dimension down into a thin disk, Cameron's assumption that the gas phase of the nebula is partitioned into spherically bound structures (protoplanets) allows grains to collect over all three dimensions, forming a planetary core as a direct product. For Jovian-sized protoplanets, such a hypothetical core should have a mass comparable to the mass of terrestrial planets. If the object later undergoes hydrodynamic collapse, the resulting structure resembles an early stage of the giant planets and over $\sim 10^9$ yr does indeed converge toward their structure (Bodenheimer 1976; DeCampli and Cameron 1979).

However, not all protoplanets will follow this route. The gravitational stability of such large distended objects depends in part on the tidal gravitational field exerted by the primary. Since the solar nebula configuration is evolving on a time scale of 10^5 yr, this tidal field is changing. Also the orbital position of these protoplanets is drifting inwards as the nebula concentrates in the center. Basically, the gaseous protoplanets become unstable when their radial dimension exceeds their Roche lobe. Cameron suggests that what historically separated terrestrial and giant planets is just the stability of their precursor protoplanets; the inner solar system constitutes that region where protoplanets became tidally unstable to disruption before they were able to undergo hydrodynamic collapse. The terrestrial planets are the remnant cores of these early structures.

More comprehensive treatments of the material discussed in this section can be found in various review papers including: Wetherill (1980), Cameron

(1978); Ward (1976a), and references therein. The model descriptions provided here are not intended to be complete, but serve only as a background for the discussion of several current problems addressed in Sec. II.

II. CONTEMPORARY ISSUES

This section addresses several new topics which promise to change our thinking in important ways. Some of these concepts, such as the possible role of resonances and density waves in planet formation, have come to us through recent planetary ring research.

A. Nebula Stability and Evolution

Formation of the Nebula. Clearly, an issue of pivotal importance is the mass of the primordial nebula. This is the primary point of connection between theories of solar system origin and of stellar formation. Indeed, it was upon this basis that massive models of the solar nebula were first constructed, i.e., a large ($\sim M_\odot$) disk was thought to be a more probable outcome of the collapse of a typical interstellar cloud (Cameron and Pine 1973; Cameron 1973; Cameron 1978). A critical bifurcation point of nebula models occurs when the disk-to-primary mass ratio is large enough to render the gas structure itself gravitationally unstable. High versus low mass nebulae (with respect to this criterion) follow radically different evolutionary paths.

Recent studies of disk formation have clarified the conditions under which nebulae of various mass are likely to form (Cassen and Moosman 1981; Cassen and Summers 1983). The critical parameter is the total angular momentum J of the cloud fragment M. The collapse phase will occur over a time scale $\tau_M \sim GM/2c^3$ from an initial radius $\sim GM/2c^2$. The resulting nebula radius will, in general, be $\geq R_{CF} = J^2/k^2GM^3$ where k is the gyration constant of the original cloud. However, if the effective viscosity ν of the cloud is large enough to set a diffusion length $R_V = (\nu\tau_M)^{\frac{1}{2}}$, that exceeds R_{CF}, then this property determines the nebula's scale by the end of collapse. Thus, the ratio $P = (R_V/R_{CF})^2 = (\nu/2) (GM/c)^3 (\kappa M/J)^4$ essentially divides high mass ($P \ll 1$) from low mass ($P \gg 1$) models, and the resulting disk-to-primary mass ratio is $\sim kP^{-\frac{1}{4}}$ (Cassen and Summer 1983). Cameron's models use $J \sim 8 \times 10^{53}$ g cm^2 s^{-1}, $M = M_D + M_\odot \sim 2M_\odot$, $c \sim 0.52$ km s^{-1}, $k = 0.4$, and $\nu \sim 10^{17}$–10^{18} cm^2 s^{-1}. Thus $P \sim 0.1$–1.0, yielding comparable disk and primary masses. However, for $J = 5 \times 10^{51}$ g cm^2 s^{-1} (the minimum angular momentum found by augmenting the current solar system to solar abundances), $k = 2/9$, $\nu \simeq 10^{16}$ cm^2 s^{-1}, $c \sim 0.3$ km s^{-1}, and $M \sim M_\odot$, one obtains $P \sim 3 \times 10^4$ and $M_D/M_\odot \sim 10^{-2}$, which is the right order of magnitude for minimum mass models. At present, it does not appear possible to exclude either of these extremes, or intermediate values for that matter, on the strength of current ideas concerning stellar birth. Nevertheless, the motivation for postulating a massive nebula in the first place is weakened by the demonstration that a low-mass nebula can form via interstellar cloud collapse under plausible conditions.

Nebula Fragmentation Scale. Although Cameron makes a case for instability in a massive nebula, this argument is not well connected to the protoplanetary mass required to produce terrestrial-sized objects at protoplanets' cores. It was noted earlier that the typical length scale for instabilities (aside from numerical constants) is $\lambda \sim G\sigma/\Omega^2$, and that the amount of mass involved is $M \sim \sigma\lambda^2 \sim G^2\sigma^3/\Omega^4$. In the case of the minimum mass model, the mass of the debris disk $\sigma R^2 \sim M_d \ll M_\odot$ and consequently $M \ll M_d$. But in Cameron's model the mass of the gas disk $M_g \sim M_\odot$ so that a fragment may constitute a fair fraction of the disk and greatly exceed a Jovian mass (see e.g. Cassen et al. 1981). Even if an argument could be made that a typical instability only has a Jovian mass, there could be up to $\sim 10^3$ such objects. And these in turn would engage in rapid encounters and collisions, similarly to the stage described as following the breakup of the debris disk in the minimum mass model (Lin 1981). It is difficult to determine whether objects would tend to coalesce upon collision or simply to tear each other apart. In fact, since the internal energy dissipation of a fragment may be much less than the inelastic collisions that dominate the breakup of a debris disk, the instability may not collapse but instead assume the form of a spiral density wave (Cassen et al. 1981). At any rate, the sort of quiescent, isolated environment in which the slow evolution of the giant protoplanets have been studied is not at all assured. Cameron (1979) has pointed out that the protoplanets may generate zones of avoidance via tidal action on the disk. However, this endangers the long-term stability of the system by introducing complications concerning the eventual separation of the planets from their precursor disk (see comments on tidal barriers at the end of Sec. II.B). Ironically, a major obstacle to Cameron's model is that, having made the case that a very massive nebula can fragment and bind itself gravitationally, he must then prove essentially the reverse case and account for why 99.9% of the material did not eventually become incorporated into the planetary system.

Turbulence and Accretion Disk Theory. One important contribution of Cameron's work that is potentially germane to any nebula model is his application of accretion disk theory and demonstration that a turbulent nebula evolves very rapidly. Although the mechanism of meridional circulation suggested by Cameron as a driver of turbulence now appears inadequate (Cabot 1982; Cameron 1983), recent models of the minimum-mass nebula nevertheless strongly imply that turbulence is inevitable (Lin and Papaloizou 1980; Lin 1981; Lin and Bodenheimer 1982). In these models the radiative transport properties of grains in the nebula render the structure convectively unstable in the vertical direction. The resulting convective eddies also couple the flow field in the radial direction producing an effective viscosity $\nu \sim v_c^2/\Omega$

where v_c is the convective velocity calculated from mixing length theory. This turns out to be similar in form to the viscosity law adopted by Cameron with the viscosity related to local dissipation and heat transport in a self-consistent manner. The deduced time scale for significant nebula evolution is comparable to that found by Cameron, i.e. $\sim 10^5$ yr.

If unavoidable, this result raises at least two important issues. First, convection may keep the dispersion velocity too high to allow gravitational instability among the small particles that have settled to the midplane. This places new emphasis on finding either a mechanism for growth through the cm to km size range or a mechanism that would cause such convection eventually to die out. One possibility is that turbulence actually promotes growth among very small grains, thereby converting μm-sized particles, which furnish most of the opacity, to cm-sized ones; the resulting drop in disk opacity could conceivably make convection self-limiting (Weidenschilling, personal communication).

A second concern about the 10^5 yr lifetime for the solar nebula is that it is much shorter than the accretion time scales derived in the impact process. Earlier calculations suggested that 100-km-sized objects could form rapidly (in $\ll 10^5$ yr), which would seem to leave the way clear for the nebula to then be dispersed, with sizable objects left behind to accumulate further on a much longer time scale. This might be the case except for the fact that the giant planets, by virtue of their volatile content, must predate nebula loss. In a minimum mass model, giant planets are assumed to have formed by gas accretion on a pre-existing core (see e.g., Mizuno 1980). These cores themselves have accumulated from solid material (including water and ammonia ices) by an impact accretion process similar to that discussed for the formation of terrestrial planets and with a time scale as long and perhaps longer, i.e., $\sim 10^{7-8}$ yr. Hence, there is an apparent dilemma, and one challenge that faces researchers is to find a mechanism which can lead to much more rapid accumulation of planetary-sized objects, both in the inner and outer solar system. At present, only Cameron's model seems capable of forming planets on a time scale as short as 10^5 yr. Note that the terminal stage planetesimal growth rate discussed in Sec. I.A depends only on three parameters: Ω, ρ_p, and σ. Since Ω and ρ_p are clearly not subject to much variation, the key may be in finding ways to enhance the local surface density available for accretion. Increasing the eccentricity and dispersion velocity of the particles does not in general change the average surface density; changing their semimajor axes would be required.

B. Planet-Nebula Tidal Interactions

The application of density wave theory to solar system problems has been one of the major developments of the last five years. These concepts, originally developed to explain the behavior of large stellar disks, have proved to be powerful tools for the investigation of planetary rings, especially in the context of ring perturbations by satellites. These ideas are only beginning to be incorporated into cosmogonic models, and the integration of density-wave theory with accretion-disk theory promises to significantly constrain modeling of early events.

The Torque Density. A planet or planetesimal embedded in a gas disk disturbs the latter's Keplerian velocity field. The perturbations are especially pronounced at Lindblad resonances, i.e., positions where the local epicycle frequency κ of the gas equals the time variation of a particular component of the satellite's disturbing function as seen in the local moving frame, i.e. $\pm \kappa = m(\Omega_p - \Omega)$, where Ω_p is the pattern speed (see the chapter by Shu). The nebula's response involves complex wave phenomena which can redistribute angular momentum. The resulting flow pattern takes on a spiral form that opens up near resonance and couples to the nonspiral potential of the planet, allowing for a mutual torque. Waves launched at the inner (outer) Lindblad resonances carry negative (positive) angular momentum and must be damped before this momentum is permanently transferred to disk material. If the damping length of the waves is long compared to resonance spacing ($x_{\text{damping}} > r\, m^{-2}$), individual gaps do not open and the damping is termed "smooth" (Goldreich and Tremaine 1980). In such a case a torque density can be defined,

$$dT/dr \simeq \text{sgn}(x)\, \text{f}\, G^2 M^2 \sigma r / \Omega^2 x^4 \qquad (6)$$

where x is the distance from a planet of mass M, and f is a constant of order unity. This same basic form was also derived by Lin and Papaloizou (1979) from an impulse point of view (see also Greenberg 1983; chapter by Dermott; chapter by Borderies et al.). For a Keplerian disk, Goldreich and Tremaine (1980) find $f \simeq 2.5$. Equation (6) is only valid for $x \gg c/\Omega$, where c is the gas sound speed. Inside a scale height the torque density departs precipitously from the x^{-4} profile. Figure 3 taken from Goldreich and Tremaine shows the nature of this cut-off. There is some dependence on the stability parameter $Q = c\Omega/\pi\, G\sigma$ of the disk. Particle disks typically have a lower Q than the solar nebula, a point to which we will return below.

Radial Drift. The total torque between disk and planet is obtained by integrating Eq. (6) throughout the disk. Although the sign reversal makes inner and outer torques largely compensatory, the planet may nevertheless experience a net torque as a result of general gradients in disk properties (Goldreich and Tremaine 1980). Because of the steep weakening of torque with distance, disk material in the vicinity of a scale height will be most important. To facilitate computation, we approximate dT/dr inside $x* \simeq 1.5c/\Omega$ with $(dT/dr)_{x=x*}$ where $x*$ is chosen so as to provide the same total disk torque for a constant density disk as the $Q = \infty$ case of Goldreich and Tremaine. Expanding σ to first order in x/r, neglecting

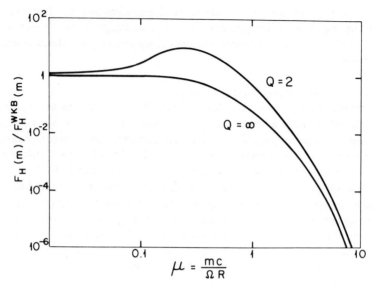

Fig. 3. The ratio of the actual angular momentum flux from a resonance to the flux calculated from a WKB approximation. The azimuthal wavenumber m is assumed $\gg 1$. Q denotes Toomre's stability parameter (from Goldreich and Tremaine 1980).

any slow variation in other quantities, and integrating over the local region yields a net torque which can then be set equal to the rate of change of the planet's angular momentum to crudely estimate an orbital drift rate:

$$\frac{dr}{dt} = -4\left[\frac{d(\ln \sigma)}{d(\ln r)}\right]\left(\frac{M}{M_\odot}\right)\left(\frac{\sigma r^2}{M_\odot}\right)\left(\frac{r\Omega}{c}\right)^2 r\Omega . \qquad (7)$$

Labeling the logarithmic derivative as $S = d(\ln \sigma)/d(\ln r)$, Eq. (7) implies a characteristic drift time $\tau_{\text{drift}} \sim r/\dot{r} \sim 2\mu^{-1} (T_2 a'^{\frac{1}{2}}/\sigma_2 S)$ yr, where $\mu = M/M_\odot, T = T_2 \times 10^2$ K, $\sigma = \sigma_2 \times 10^2$ g cm^{-2} and a' is the heliocentric distance in AU.

By comparison, the orbital drift time due to aerodynamic drag (see e.g. Whipple 1972; Weidenschilling 1977), is of order $\tau_{\text{drag}} \sim 10(C_D\Omega)^{-1}$ $(r\Omega/c)^3(\rho_p R/\sigma) \simeq 10^8(\rho_p R_{\text{km}}/C_D\sigma_2 T_2^{\frac{3}{2}})$ yr. The quantity C_D is the drag coefficient, which for objects large enough to generate a turbulent wake is of order unity. Density wave induced drift is faster than that due to aerodynamic drag for objects with radii $\gtrsim 3 \times 10^2 T_2^{\frac{3}{8}} a'^{\frac{1}{8}}(C_D/S)^{\frac{1}{4}}\rho_p^{-\frac{1}{2}}$ km.

If a drifting object accretes the bulk of the resulting flux of smaller particles across its orbit, its characteristic growth time is $\tau_{\text{growth}} \sim M/\dot{M}$ $\sim M/2\pi r\sigma_d r \sim \mu_d^{-1} T_2 a'^{\frac{1}{2}}/\sigma_2 S$ yr, where σ_d is the surface density of accretable debris and $\mu_d = \pi\sigma_d r^2/M_\odot$. Assuming $\mu_d \sim 6 \times 10^{-6}, T_2 \sim 7,$ $\sigma_2 \sim 10,$ and $a' \sim 1$ for the inner solar system, $\tau_{\text{growth}} \sim 10^5 S^{-1}$ yr; if instead $\mu_d \sim 10^{-4}$ (ice + rock), $T_2 \sim 3$, $\sigma_2 \sim 1$, and $a' \sim 5.2$ are chosen for the Jovian region, a similar time scale is still obtained; finally for $\mu_d \sim 10^{-4}$ (ice + rock), $T_2 \sim 1$, $\sigma_2 \sim 0.1$, and $a' \sim 30$, $\tau_{growth} \sim 10^6$ S^{-1} yr in the outer-most solar system. For planetary-sized objects ($R = R_3 \times 10^3$ km) these times are much shorter than accretion time scales typically ascribed to the process of gravitational relaxation of a planetesimal disk, i.e. $\tau \sim 10^7 R_3 \rho_p a'^{\frac{3}{2}}/\sigma_d$ yr. Hence, density waves have the potential to considerably alter our view of the accumulation process. In particular, Eq. (7), taken at face value appears to compliment aerodynamic drag in providing radial mobility for accreting objects of virtually any size.

Gap Formation. Because Eq. (7) depends most sensitively on disk material within a few scale heights of the planet, radial drift might be aborted if local damping of density waves alters the nebula properties in this region. If the damping is smooth but local (i.e. $h > x_{damping} > r/m^2$) and if the diffusion time scale is very long, a gap of width $x > h$ could be cleared in a time

$$\tau_{gap} \sim (\mu^2 \Omega)^{-1} (x/r)^5 . \tag{8}$$

However, if the planetesimal is of low mass, the time it takes to drift a distance x, $\sim (x/r) \tau_{drift}$, may be less than Eq. (8), thus not allowing enough time to clear a gap. The smallest stable gap width should be of order $c/\Omega \sim h$, implying that a mass $M > M_*$ is necessary for gap development, where

$$M_* \sim C_1 \sigma (c/\Omega)^2 \sim 10^{25} \sigma_2 T_2 a'^3 \text{ g} \tag{9}$$

(Hourigan and Ward 1983). In Eq. (9), C_1 is a constant of order unity. For the values of σ_2, T_2, and a' used above, $M_* \sim 10^{27} - 10^{28}$ g, encouragingly close to planetary scale.

On the other hand, $M > M_*$ does not guarantee a gap. If the nebula has an appreciable viscosity ν, the time necessary to close a gap by diffusion $\tau_{diff} \sim x^2/\nu$, may still be less than Eq. (8). This establishes a second critical mass which must also be exceeded,

$$\mu_\nu \sim (\nu/r^2\Omega)^{\frac{1}{2}} (c/r\Omega)^{\frac{3}{2}} \sim 10^{-4} \alpha (T_2 a')^{\frac{5}{4}} . \tag{10}$$

In Eq. (10), the substitution $\nu = \alpha c^2/\Omega$, $\alpha < 1$, has been made to parameterize an effective turbulent viscosity. Molecular viscosity of the nebula is too small to be dynamically significant. Note that $\mu_\nu > M_*/M_\odot$ for $\alpha > 10^{-4} \sigma_2 a'^{\frac{7}{4}}/T_2^{\frac{1}{4}}$, and then Eq. (10) supersedes Eq. (9) as the gap criterion.

Wave Damping. Neither Eqs. (9) or (10) properly take into account the possibility that wave damping may not be local. Goldreich and Tremaine

1978, 1980) and Shu et al. (1983) consider two damping mechanisms that are effective for particle disks such as Saturn's rings: viscous damping and nonlinear waves (see chapters by Borderies et al. and by Shu). The efficacy of these mechanisms in the solar nebula environment must be examined.

The Lin-Shu dispersion relation for density waves can be written in nondimensional form as

$$2khQ^{-1} - (kh)^2 = (\kappa/\Omega)^2 - m^2(1 - \Omega_s/\Omega)^2 = D/\Omega^2 \tag{11}$$

where $D = 0$ at exact resonance. Figure 4 shows the general form of Eq. (11) for an inner Lindblad resonance. Expanding D to first order in $(x - x_L)/r$ where $x_L \simeq -3/2(r/m)$ locates the mth Lindblad resonance from the planet,

$$D/\Omega^2 \sim 3m(x - x_L)/2r \sim (x - x_L)/(-x_L) . \tag{12}$$

Waves launched at an inner resonance are long trailing waves that propagate toward the planet and carry negative angular momentum at the group velocity

$$c_g \sim \text{sgn}(k)(\pi G\sigma - kc^2)/[m(\Omega - \Omega_s)]. \tag{13}$$

However, the group velocity vanishes when $kh = Q^{-1}$ and $D/\Omega^2 = Q^{-2}$. The location $x_F \sim x_L(1 - Q^{-2})$ where this occurs marks the boundary of a forbidden zone. Inside the zone the wavenumber given by the Lin-Shu dispersion relation is imaginary. If the waves have not damped upon reaching x_F they do not penetrate the forbidden zone, but instead are reflected. For large Q this reflection is nearly complete and the waves become short trailing pressure waves propagating away from the planet. The waves can now leave the vicinity of the resonance eventually to transfer their angular momentum to remote portions of the disk. Such behavior will not contribute to local gap clearing.

Particle disks typically exhibit marginal gravitational stability and thus have $Q \sim 1$. In addition, they are relatively cold and flat, i.e. for Saturn's rings $h/r \sim 10^{-6}$. As a consequence, the forbidden zone lies near the planet, and if waves can propagate that far their radial wavenumber must become of the order of $k \sim h^{-1}$. The scale of oscillation thus shortens considerably. If the angular momentum flux is conserved, this shortening is accompanied by an increase in the amplitude of the density perturbations σ'. The waves become nonlinear and can be expected to shock when $\sigma'/\sigma \sim 1$. This requires a wavenumber

$$k_{NL}h \sim [(3/m)(\sigma h^2/M)(M_0/M)]^{\frac{1}{2}} \tag{14}$$

where M_0 is the mass of the central body. For Saturn's rings, $k_{NL}h \ll 1$ and nonlinearity is predicted well before reflection at x_F.

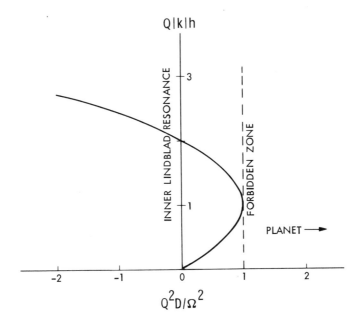

Fig. 4. The nondimensional Lin-Shu dispersion relation as a function of D/Ω^2. The radial wavenumber k is normalized to the inverse scale height $h^{-1} \simeq \Omega/c$, and Q is the Toomre stability parameter.

By contrast, for the solar nebula, $Q \simeq 7 \times 10^2 \, T_2^{\frac{1}{2}}/\sigma_2 \, a'^{\frac{1}{2}}$ and thus the fractional distance to the forbidden zone $(x_F - x_L)/(-x_L) \sim 2 \times 10^{-6} a'^3 \sigma_2^{\,2}/T_2$. From Eq. (14) we expect nonlinear waves when $k_{NL} h \sim \mu^{-1} \, (3\sigma/mM\odot)^{\frac{1}{2}}$ $(c/\Omega) \sim 1.5 \times 10^{-4\mu^{-1}} (\sigma_2 T_2 \, a'^{\,3} \, m^{-1})^{\frac{1}{2}}$. For this to be achieved prior to reflection requires $k_{NL} h < Q^{-1}$ or $\mu > 0.1 \, T_2(m\sigma_2)^{\frac{1}{2}}$. This is unlikely even for a Jovian-size object with $\mu = 10^{-3}$. Subsequent to reflection, Eqs. (11), (12), and (14) can be combined to predict nonlinear waves at a distance

$$(x_{NL} - x_L)/x_L = 2.3 \times 10^{-8} \, (\sigma_2 \, T_2 \, a'^3/\mu^2 m) \, (1 - 18.7\mu\sigma_2^{\frac{1}{2}} \, m^{\frac{1}{2}} \, /T_2). \quad (15)$$

The most important resonances occur near the torque cutoff, i.e., where $m \sim r/h$. For these, $(x_{NL} - x_L)/x_L \sim 6 \times 10^{-10}\mu^{-2}\sigma_2 T_2^{\frac{3}{2}} a'^{\frac{3}{2}}$. For Jupiter this gives $\sim 0.2\sigma_2 T_2^{\frac{3}{2}}$ and local nonlinearity is reasonably likely; for the Earth ($\mu \sim 3 \times 10^{-6}$), yielding $\sim 67\sigma_2 T_2^{\frac{3}{2}}$ and the waves probably do not damp locally.

The second damping mechanism utilizes the same viscosity ν that promotes gap closure by diffusion. Goldreich and Tremaine (1980) give a viscous damping length of

$$x_\nu \sim r(c^3/Q^3 m^2 \nu \, r\Omega^2 \,)^{\frac{1}{3}} \text{ for } Q \sim 1$$

(16)

$$x_\nu \sim r(c^3/m^{\frac{1}{2}}\sigma \, r\Omega^2)^{\frac{2}{3}} \quad \text{for } Q \gg 1.$$

For a particle disk of optical depth τ, $\nu \sim c^2\tau/\Omega(1 + \tau^2)$ and $x_\nu \sim (r/Q)$ $[(1 + \tau^2)/\tau]^{\frac{1}{3}} (h/r)^{\frac{1}{3}} m^{-\frac{2}{3}}$, which for $\tau \sim 1$ and $m \sim r/h$ gives a damping length $\sim h \ll r$. In the case of a turbulent nebula, $\nu \sim \alpha c^2/\Omega$ and

$$x_\nu \sim r\alpha^{-\frac{2}{3}} m^{-\frac{1}{3}}(h/r)^{\frac{2}{3}}$$

(17)

which, for the most important resonances, is $x_\nu \sim \alpha^{-\frac{2}{3}}h$. Since $\alpha < 1$, x_ν exceeds a scale height, which for the nebula is not that small. Comparison of Eqs. (10) and (17) reveals that for an α just small enough for a planet to open a gap, $x_\nu/h \sim 2 \times 10^{-3} (T_2 a')^{\frac{3}{8}} \mu_\nu^{-\frac{2}{3}}$. For Jupiter this value is ~ 2; for the Earth it is ~ 50. The latter in fact violates the assumption of local damping used to derive Eq. (10). Consequently, Eq. (10) probably underestimates the critical mass needed to clear a gap for objects smaller than a Jovian mass, unless other damping mechanisms exist.

Tidal Barriers. The opening of a strong gap and the truncation of the nebula may introduce still another problem. Planet-disk interactions can act as a barrier to the radial flux generated by viscous evolution of the nebula as a whole. The otherwise slowly varying disk gradients can become replaced by a strong density discontinuity across the planet's orbit as nebular material is inhibited from drifting sunward. This may override the stabilizing influence of the gap and initiate orbital decay (Ward 1982). In a sense, the planet becomes linked into the momentum transport of the disk and may ultimately suffer its fate. On the other hand, a disk turbulent enough to prevent a gap for $\mu \sim 10^{-3}$ may be too short-lived to allow accretion of the Jovian core, i.e., the disk's characteristic viscous evolution time is $\tau \approx r^2/\nu \approx \Omega^{-1} \mu^{-2}(h/r)^3 \sim 3T_2^{\frac{3}{2}}a'^3$ yr. Setting $a' \sim 30$ yields $\tau \sim 10^5$ yr.

III. CONCLUSIONS

I have briefly touched on several potentially key issues for future cosmogonical research. Of crucial concern are the large-scale viscous evolution of the nebula and the role of density waves in shaping the concomitant interaction of the disk with the newly formed planetary system. Both these phenomena are so powerful that our view of early solar system events may change markedly as we attempt to incorporate their influence and mutual interplay into model construction. Much progress needs to be made, and I close by mentioning several promising topics of study:

1. The mass of the solar nebula is still not well determined. Although circumstantial evidence has been interpreted as indicating a large mass, the argument is weak (see e.g. Cassen and Summers 1983). The difficulties introduced by the excessive mass of such a system provide an argument at least as persuasive against this model, given our current state of knowledge. Clearly, the discovery and observation of young stars where disk and possibly planet formation are thought to be occurring would help to resolve this question. Barring this, further modeling efforts to explore the divergent ramifications of different mass assumptions may reveal further boundary conditions that shed light on this issue.

2. Was the nebula turbulent? This question is also central to any understanding of early events. Not only does it greatly influence the viability of certain other processes such as gravitational instabilities in the debris disk, but it sets the time scale available for planet formation by controlling the rate of viscous evolution of the nebula. Lin, Papaloizou, and Bodenheimer's argument for turbulence is convincing but not conclusive. The required opacity is furnished by grains which, in turn, remain dispersed vertically by the very convection they are responsible for generating. On the other hand, the grains' size distribution may be modified by accumulation mechanisms, and the long-term ($\sim 10^5$ yr) persistence of turbulence has not yet been established.

3. The density wave mechanism has not been properly adapted to the environment of a hot disk capable of sustaining appreciable pressure gradients. In addition to the dissimilarities between particle and gaseous disks outlined in Sec. II.B, pressure gradients can alter the gas orbital profile $\Omega(r)$. Indeed, it is just this trait that accounts for the systematic differential velocity between gas and particles (which are not so influenced) that brings about aerodynamic drag effects. Pressure gradients can alter both Ω and the epicyclic frequency κ, causing a shift in the locations of the various Lindblad and corotation resonances. Lunine and Stevenson (1982) assert that the resonance locations move away from the planet on both sides if a gap begins to open, weakening the torques and thus inhibiting gap formation. On the other hand, in the absence of a gap resonances are still shifted in position by the pressure gradient associated with the same large-scale gradients in nebula properties used to justify planetesimal drift, Eq. (7). In this case, the sign reversal point of Eq. (6) is not at $x = 0$, but is displaced in the direction of the pressure gradient by $\delta x \sim h^2/r$. This trend may largely compensate for the net torque due to $d\sigma/dr$, and the magnitude and even direction of any planetesimal drift is in reality still problematic. Futhermore, the sign reversal point marks the position of a strong corotation resonance (zero-order in the planet's eccentricity) that could increase the likelihood of nonlinearity and wave shocking, thus promoting gap formation. Clearly, this problem must be addressed before any of these complications can be unraveled.

4. Density waves could be sustained independently by a sub-disk of particles embedded in the nebula. Since the Q of such a structure would be relatively low, the waves would propagate toward corotation, i.e. opposite to waves traveling in the gas medium. In addition to damping mechanism considered in section II.B, gas-particle friction may couple the two components, drawing momentum from the particles. Such an action might interfere with the object's accretion efficiency, possibly opening gaps in the particle disk. On the other hand, the collective behavior required to support wave action should break down when the typical particle spacing becomes greater than the first wave length. For a debris disk M_D composed of N particles of equal mass, this critical condition reads $N_c \sim M_\odot/M_D$; for the inner solar system $N_c \sim 10^5$. This problem has important implications for the planetesimal size distribution and other effects strongly linked to this information such as aerodynamic drag.

5. Models of the viscous evolution (and dispersal) of a nebula should be extended to include post-planetary scenarios, where objects large enough to seriously modify the radial flux may be present. The degree to which survival of the system is in jeopardy is of particular concern. If in fact it proves difficult to extricate the gaseous component from the contingent of newly accumulated planets, this itself could give us valuable insight. A successful account of this situation may provide much needed clues to the sequence and phasing of key early events.

Acknowlegments. The author would like to express appreciation to G. Stewart and S. J. Weidenschilling for several valuable suggestions that have improved this chapter. This work was supported at the Jet Propulsion Laboratory of the California Institute of Technology under a contract sponsored by National Aeronautics and Space Administration. This chapter is based in part on lectures delivered at the Lunar and Planetary Laboratory of the University of Arizona in August 1982, and at the Solar Nebula Workshop held at the Ames Research Center in June 1983.

REFERENCES

Bodenheimer, P. 1976. Contraction models for the evolution of Jupiter. *Icarus* 29:319–325.
Cabot, W. H. 1982. Ph.D. dissertation, University of Rochester. Rochester, N. Y.
Cameron, A. G. W. 1973. Accumulation processes in the primitive solar nebula. *Icarus* 18:407–450.
Cameron, A. G. W. 1978. Physics of the primitive solar accretion disk. *Moon Planets* 18:5–40.
Cameron, A. G. W. 1979. Protoplanets and the primitive solar nebula. *Moon Planets* 21:173–183.
Cameron, A. G. W. 1983. Dissipation of thick accretion disks. Preprint. Center for Astrophysics.
Cameron, A. G. W., and Pine, M. R. 1973. Numerical models of the primitive solar nebula. *Icarus* 18:377–406.

Cassen, P., and Moosman, A. 1981. On the formation of protostellar disks. *Icarus* 48:353–376.

Cassen, P. M., Smith, B. F., Miller, R. H., and Reynolds, R. T. 1981. Numerical experiments on the stability of preplanetary disks. *Icarus* 48:377–392.

Cassen, P., and Summers, A. 1983. Models of the formation of the solar nebula. *Icarus* 53:26–40.

Colombo, G., Goldreich, P., and Harris, A. W. 1976. Spiral structure as an explanation for the asymmetric brightness of Saturn's rings. *Nature* 264:344–345.

Cox, L. P. 1978. Numerical Simulation of the Final Stages of Terrestrial Planet Formation. Ph.D. Dissertation, M.I.T., Cambridge, Mass.

Cox, L. P., and Lewis, J. S. 1980. Numerical simulation of the final stages of terrestrial planet formation. *Icarus* 44:706–721.

DeCampli, W., and Cameron, A. G. W. 1979. Structure and evolution of isolated giant gaseous protoplanets. *Icarus* 38:367–391.

Esposito, L. W., O'Callaghan, M., and West, R. A. 1982. The structure of Saturn's rings: Implications from the Voyager stellar occultation. Submitted to *Icarus*.

Goldreich, P., and Lynden-Bell, D. 1965. I. Gravitational stability of uniformly rotating disks. *Mon. Not. Roy. Astron. Soc.* 130:97–124.

Goldreich, P., and Tremaine, S. 1978. The formation of the Cassini Division in Saturn's rings. *Icarus* 34:240–253.

Goldreich, P., and Tremaine, S. 1980. Disk-Satellite interactions. *Astrophys. J.* 241:425–441.

Goldreich, P., and Ward, W. R. 1973. The formation of planetesimals. *Astrophys. J.* 183:1051–1061.

Greenberg, R. 1979. Growth of large, late-stage planetesimals. *Icarus* 39:141–150.

Greenberg, R. 1980. Collisional growth of planetesimals. *Moon Planets* 22:63–66.

Greenberg, R. 1983. The role of dissipation in shepherding of ring particles. *Icarus* 53-207–218.

Greenberg, R., Wacker, J. F., Hartmann, W. L., and Chapman, C. R. 1978. Planetesimals to planets: Numerical simulation of collisional evolution. *Icarus* 35:1–26.

Harris, A. W., and Ward, W. R. 1982. On the radial structure of planetary rings. Proceedings of I.A.U. Colloquium 75 *Planetary Rings*, ed. A. Brahic, Toulouse, France, August. 1982.

Holberg, J. B., Forrester, W. T., and Lissauer, J. J. 1982. Identification of resonance features within the rings of Saturn. Submitted to *Nature*.

Hourigan, K., and Ward, W. R. 1983. Planetary aggregation and nebula tides. *Lunar Planet. Sci. Conf.* 14:331–332 (abstract).

Kaula, W. M. 1979. Equilibrium velocities of a planetesimal population. *Icarus* 40:262–275.

Lin, D. N. C. 1981. Convective accretion disk model for the primordial solar nebula. *Astrophys. J.* 246:972–984.

Lin, D. N. C., and Bodenheimer, P. 1982. On the evolution of convective-accretion disk models on the primordial solar nebula. *Astrophys. J.* In press.

Lin, D. N. C., and Papaloizou, J. 1980. On the structure and evolution of the primordial solar nebula. *Mon. Not. Roy. Astron. Soc.* 191:37–48.

Lunine, J. I., and Stevenson, D. J. 1982. Formation of the Galilean satellites in a gaseous nebula. *Icarus* 52:14–39.

Lynden-Bell, D., and Pringle, J. E. 1974. The evolution of viscous discs and the origin of the nebular variables. *Mon. Not. Roy. Astron. Soc.* 168:603–637.

Lyttleton, R. A. 1972. *Mon. Not. Roy. Astron. Soc.* 158:463.

Mizuno, H. 1980. Formation of the giant planets. *Prgr. Theor. Phys.* 64:544–557.

Safronov, V. S. 1969. *Evolution of the Protoplanetary Cloud and Formation of the Earth and the Planets*, Moscow. Translation: NASA TT-F-677 (1972).

Shu, F. H., Cuzzi, J. N., and Lissauer, J. J. 1983. Bending waves in Saturn's rings. *Icarus* 53:14–39.

Slattery, W. 1978. Protoplanetary core formation by rainout of iron drops. *Moon Planets* 19:443–456.

Stewart, G., and Kaula, W. M. 1980. A gravitational kinetic theory for planetesimals. *Icarus* 44:154–171.

Toomre, A. 1964. On the gravitational stability of a disk of stars. *Astrophys. J.* 139:1217–1238.

Ward, W. R. 1976a. The formation of the solar system. In *Frontiers of Astrophysics*, ed. E. H. Avrett, (Cambridge, MA: Harvard Univ. Press) pp. 1–40.

Ward, W. R. 1976b. Some remarks on the accretion problem. *Accademia Nazionale dei Lincei*, Rome.

Ward, W. R. 1982. Tidal barriers in the solar nebula, *Lunar Planetary Sci.* xiii:831 (abstract).

Ward, W. R., and Cameron, A. G. W. 1978. Disc Evolution within the Roche limit. *Lunar Planetary Sci.* IX:1205–1207.

Weidenschilling, S. J. 1977. Aerodynamics of solid bodies in the solar nebula. *Mon. Not. Roy. Astron. Soc.* 180:57–70.

Weidenschilling, S. J. 1980. Dust to planetesimals: Settling and coagulation in the solar nebula. *Icarus* 44:172–189.

Wetherill, G. W. 1976. The role of large bodies in the formation of the earth and moon. *Proc. Lunar Planet. Sci. Conf.* 7:3245–3257.

Wetherill, G. 1980. Formation of the terrestrial planets. *Ann. Rev. Astron. Astrophys.* 18:77–113.

Wetherill, G. W., and Cox, L. 1982. Gravitational cross-sections and runaway accretion of planets. Preprint. Carnegie Inst. of Washington.

Whipple, F. 1972. On certain aerodynamic processes for asteroids and comets. In *From Plasma to Planet*, ed. A. Elvius (New York: John Wiley and Sons), pp. 211–232.

PART V

The Future

FUTURE STUDIES OF PLANETARY RINGS BY SPACE PROBES

E. C. STONE
California Institute of Technology

Recent observations by the Pioneer and Voyager missions have been the basis for major advances in our knowledge and understanding of the Jovian and Saturnian rings. Future spaceprobe observations offer further opportunities for studying the ring systems of giant planets.

The first spaceprobe observations of planetary rings were carried out by the Voyager and Pioneer missions to Jupiter and Saturn. Even though limited in time by the brevity of planetary flybys, the missions nevertheless returned a wealth of information about planetary rings which is unobtainable by other means and which in many instances was completely unexpected. As a result, those observations have already had a major impact on our understanding of ring systems as is apparent in many chapters in this book.

This chapter describes further opportunities for ring studies by spaceprobes, including the Voyager 2 flybys of Uranus and Neptune, the Galileo mission to Jupiter, and a possible Saturn Orbiter mission. The results anticipated from these new studies have been inferred from the results obtained from the Voyager and Pioneer encounters with Jupiter and Saturn and are discussed in that context.

I. THE VOYAGER ENCOUNTER WITH URANUS

The Voyager 2 encounter with Uranus in 1986 is a continuation of the exploratory mission to the outer Solar System that began in 1977 with the launch of two essentially identical spacecraft. As shown in Fig. 1, Voyager 1 encountered Jupiter in March 1979, returning numerous new observations

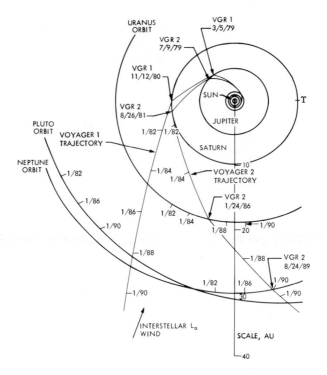

Fig. 1. A view normal to the ecliptic plane of the trajectories of Voyagers 1 and 2.

of that planetary system, including the first image of the Jovian ring as described in *Science* (204:945–1007, 1979). Voyager 2 followed closely behind, providing additional observations of the Jovian system in July 1979 (*Science* 206:925–995, 1979).

Using the gravity assist of the Jovian flybys, both spacecraft continued on to Saturn, Voyager 1 arriving first in November 1980. Among the numerous discoveries about the Saturn system reported in *Science* (212:159–243, 1981) were many new aspects of Saturn's rings, including spokes, satellite-driven spiral density waves, hundreds of features in the B Ring, two small satellites confining the kinked, multistranded F Ring, and the size distribution of particles with radii ranging from several centimeters to 10 meters.

Following nine months later, Voyager 2 returned additional key observations of the Saturnian system (see *Science* 215:499–594, 1982), including a stellar occultation which provided measurements of the optical thickness of Saturn's rings with a radial resolution of ≤300 m. These data indicated the existence of numerous spiral density waves excited by orbital resonances with several of the innermost satellites. It was also found that the outer edge of the B Ring is elliptical as expected from a 2:1 resonant interaction with Mimas.

Voyager 2 used the gravity assist of the Saturn flyby to continue on to Uranus and Neptune. Many of the techniques developed for the study of Saturn's rings will be employed during the Uranus encounter in January 1986, including stellar and radio occultations and high resolution imaging as described below.

At Saturn, studies of the rings and associated phenomena were carried out by many of the eleven scientific investigations on the Voyager spacecraft. The locations of the instruments are shown in Fig. 2, which also depicts several of the major engineering subsystems. A more comprehensive description of the eleven investigations listed in Table I will be found in *Space Science Reviews* (21:75–376, 1977). Although not explicitly noted in Fig. 2, the four remote sensing instruments, ISS, IRIS, PPS, and UVS (see Table I) are mounted on a scan platform with two axes of articulation, permitting the instruments to be pointed at essentially any target.

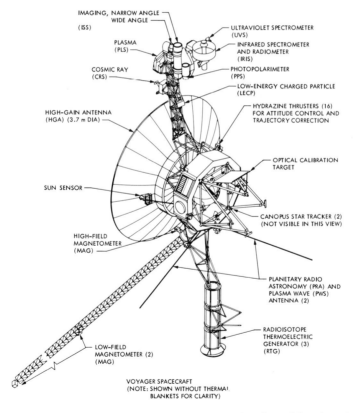

Fig. 2. A drawing of the Voyager spacecraft showing the locations of the science instruments and several spacecraft subsystems. The radio science investigation uses the high gain antenna, the spacecraft X- and S-band transmitters, and an ultrastable oscillator.

TABLE I

Voyager Science Investigations

Investigation Team	Principal Investigator/Institution
Imaging science (ISS)	Smith/Univ. Arizona (team leader)
Infrared spectroscopy and radiometry (IRIS)	Hanel/GSFC
Photopolarimetry (PPS)	Lane/JPL
Ultraviolet spectroscopy (UVS)	Broadfood/Univ. So. California
Radio science (RSS)	Tyler/Stanford Univ. (team leader)
Magnetic fields (MAG)	Ness/GSFC
Plasma (PLS)	Bridge/MIT
Plasma wave (PWS)	Scarf/TRW
Planetary radio astronomy (PRA)	Warwick/Radiophysics, Inc.
Low energy charged particles (LECP)	Krimigis/JHU/APL
Cosmis rays (CRS)	Vogt/Cal. Inst. Tech.

Stellar and Solar Occultation Studies

Since the discovery of the Uranian rings during a stellar occultation observation in 1977, subsequent occultation studies have provided detailed information on the rings' optical depths, eccentricities, precession, and inclination (see, e.g., Elliot et al. 1981, and references therein). Although Voyager 2 observations will add just a few additional stellar occultations to those already observed from Earth, they will provide much higher spatial resolution than Earth-based observations at 2.2 μm that are limited by a Fresnel Zone width of 3.5 km. In contrast, the Voyager 2 spatial resolution will be limited primarily by the instrument sampling times (10 ms for PPS, 320 ms for UVS) and transient response, since the Fresnel Zone width is ≤ 15 m. As a result, the spatial resolution expected at Uranus should be similar to that achieved at Saturn, where the photopolarimeter provided 110 m resolution and the ultraviolet spectrometer 3 km resolution. As at Saturn, comparable resolution of the vertical distribution of ring material is provided by detailed observations of the abruptness of the occultation at the edges of the rings.

Several stars have been identified as candidates for occultation studies during the Uranus encounter. Among these are σ Sagitarii and ϵ Persei. The first of these occultations occurs while the instruments are viewing the illuminated face of the rings (25° phase angle), the second occurs while viewing the other face (140° phase angle). Figures 3 and 4 illustrate the viewing geometries for these two stellar occultations. Since scattered sunlight from the rings will ultimately limit the measurements of regions of low optical depth, observations at several phase angles are of particular value and additional stellar occultations may be included in the sequence of observations.

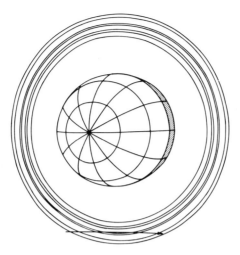

Fig. 3. A view from Voyager 2 of Uranus and its rings during an occultation of σ Sagitarii. The instruments will be viewing the illuminated face of the rings.

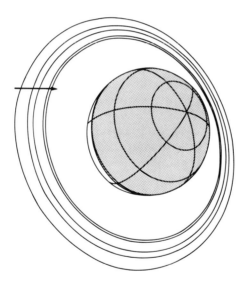

Fig. 4. A view from Voyager 2 of Uranus and its rings during an occultation of ε Persei. The instruments will be viewing the unilluminated face of the rings.

It may also be possible to perform a solar occultation study of the rings using the ultraviolet spectrometer. However, this study will have to take into account both the scattering of sunlight from ring particles in the spectrometer's field of view and the finite size of the Sun which corresponds to a distance of ~ 80 km on the ring plane.

Radio Occultation Studies

Radio occultation studies provide information on the radial variation of the microwave opacity of the rings and on the scattering properties of ring particles. Thus, measurements of the attenuation of the coherent 3.6 and 13 cm radio waves which are transmitted through the rings by the spacecraft yield information on the number of ring particles with radii $\gtrsim 1$ and $\gtrsim 4$ cm, respectively. Simultaneous measurements of the angular distribution of the forward scattering by the ring particles yield information on the number distribution of larger ring particles with radii between 1 and 15 m (see, e.g., Tyler et al. [1983] and Marouf et al. [1983] and references therein for results from the Voyager 1 radio occultation study of Saturn's rings).

The occultation geometry for the Uranus flyby is illustrated in Fig. 5 and summarized in Table II. The spatial resolution of radial opacity variations is related to the size of the Fresnel Zone. Due to the oblique passage of the radio beam through Saturn's rings, the radial component of the Fresnel Zone was ~ 20 km at 3.6 cm and ~ 40 km at 13 cm. However, the radio signals are coherent and the use of an inverse Fresnel transform resulted in a radial resolution of 75 to 300 m (Marouf and Tyler 1983) in the F Ring and C Ring. Because the radio beam is essentially normal to the Uranian ring plane, the radial component of the Fresnel Zone will be only ~ 3 and ~ 6 km at the two wavelengths, so that subkilometer radial resolution should also be achievable.

Another important factor in determining the expected radial resolution is the signal level recorded at Earth, since a weaker signal requires a longer integration time in order to obtain a given signal-to-noise ratio. Although the signal intensity from Uranus' distance will be only $\sim 1/4$ of that from Saturn's, arraying of groundbased antennas will essentially double the receiving aperture so that signal integration time at Uranus will only have to be double that at Saturn. Fortunately, the apparent radial velocity of the radio beam footprint in the Uranian ring plane is only ~ 8 km s^{-1}, about 1/10 that at Saturn where the beam obliquely penetrated the ring plane. As a result, better radial resolution (< 1 km) should be achieved at Uranus than at Saturn.

Imaging Studies

Imaging has been particularly important in both the initial Voyager studies of the Jovian and Saturnian rings (Smith et al. 1979a,b,1981,1982) and in subsequent detailed studies (see, e.g., Jewitt and Danielson 1981; Cuzzi et al. 1981; Lissauer 1982; Esposito et al. 1983; Porco 1983; Shu et al. 1983; and various chapters in this book; see especially Burns et al. and Cuzzi et al.'s chapters). Although the imaging conditions will be difficult at Uranus, both because the ring albedo is low and because of the reduced solar illumination, there are a number of important observations anticipated.

Among the most important observations is a systematic search for small satellites associated with the ring system, such as the shepherding satellites proposed by Goldreich and Tremaine (1979). An inventory of the objects

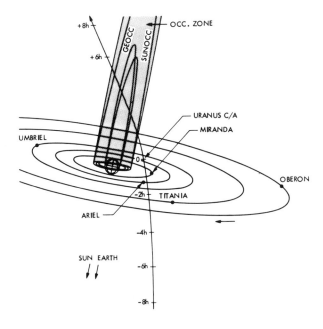

Fig. 5. A view normal to the plane of the trajectory of the Voyager 2 Uranus encounter on 24
January 1986, illustrating the intervals during which Earth (GEOCC) and solar (SUNOCC)
occultations occur.

TABLE II
Voyager 2 Uranus Encounter

Time, closest approach (spacecraft event time, GMT)		24 January 1986 18:00
Radius, closest approach		107,080 km
Radius, ring-plane crossing		115,200 km
Distance (10^3 km) Earth occultation	ϵ Ring	137
	Ring 6	155
	Uranus entrance	186
	Uranus exit	255
	Ring 6	280
	ϵ Ring	298
Distance (10^3 km) Sun occultation	Uranus entrance	177
	Uranus exit	240

orbiting Uranus and an accurate determination of their orbital elements are essential to understanding the dynamics of the ring system. Even with the reduced illumination, objects the size of the F Ring shepherds (1980S26 and 1980S27) with diameters of ~ 100 km should be easily detected, with the smallest detectable objects of unit albedo having diameters < 10 km.

There will also be several sequences of images of the rings, including sequences at low phase angle taken on the approach to Uranus, a sequence in which the ring is imaged in front of the illuminated planet, and sequences taken at high phase angle after closest approach. The expected imaging capability is perhaps best illustrated by considering images taken of the eccentric ringlet in the Maxwell Gap at 1.45 R_S in Saturn's rings. As discussed by Esposito et al. (1983) and Porco (1983), this ring is structurally and dynamically similar to Uranus' ϵ Ring. The two rings have similar optical depths (~ 1 to 2), similar surface mass densities (~ 20 g cm^{-2}), and a similar range of radial widths (~ 20 to ~ 100 km). One significant difference, however, is that the single particle albedo of Uranus' rings (Matthews et al. 1982) is only $\sim 1/8$ of that of the Maxwell Ringlet (Esposito et al. 1983). Taking into account the differences in solar intensity and incidence angles, the ϵ Ring will be $\sim 1/10$ as bright as the Maxwell Ringlet.

Figure 6 is a Voyager 2 image of the C Ring showing the Maxwell Gap and Ringlet. This image was taken with the narrow angle camera, a clear filter, and a 0.72 second shutter. A similar image of the ϵ Ring would require a ten-times longer exposure. Although the spatial resolution of such a long exposure would usually be smear-limited to ~ 30 km/line pair, some of the images may have much better resolution when the residual spacecraft motion is particularly low. The faster wide-angle camera with a geometrical resolution of ~ 25 km/line pair at 3 hr before the closest approach will not be smear-limited.

Although the rings will be relatively dark in backscattered light, they may be considerably brighter when imaged at high phase angles if they contain significant numbers of micron-sized particles. Several wide-angle images can be obtained at 170° phase angle with a spatial resolution of ~ 30 km/line pair while the spacecraft is in the shadow behind Uranus. At Jupiter, narrow-angle images taken at 174° to 176° phase angle displayed not only the ring with an optical depth $\tau \sim 3 \times 10^{-5}$, but also a diffuse halo of particles with $\tau \sim 7 \times 10^{-6}$. Although the illumination at Uranus is only 0.07 of that at Jupiter, the use of the wide-angle camera, which is 9 times faster than the narrow angle, will allow direct imaging of very low optical depth rings of micron-sized particles which are undetectable by stellar or radio occultations.

Trapped Radiation and Other Studies

Rings can also be detected by their absorption of trapped particles, provided that Uranus has a magnetosphere with stably trapped energetic particles. As shown in Fig. 7, detection of absorption features by the low-energy

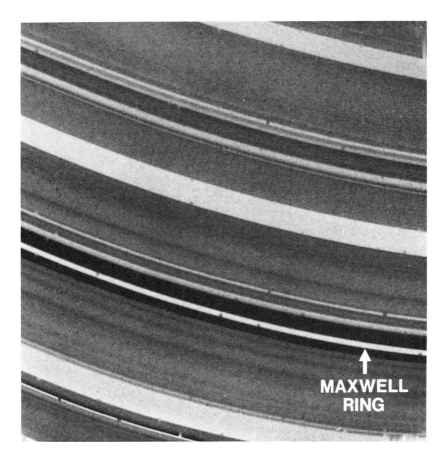

Fig. 6. A Voyager 2 image of the eccentric ringlet in the Maxwell Gap at 1.45 R_S, taken from a range of 542,000 km. There are many similarities between this ringlet and the ϵ Ring at Uranus. However, particles of the ϵ Ring have a much lower albedo.

charged particle experiment and the cosmic ray system will be limited by the spacecraft trajectory to regions outside of ∼4.5 R_U, assuming a magnetic dipole axis that is generally aligned with the planetary spin axis.

Although there are no optically thick rings in the region beyond 4.5 R_U, there might be tenuous rings ($\tau < 0.01$) which are undetectable by Earth-based stellar occultations. Such tenuous rings can produce absorption signatures, as did the Jovian ring ($\tau \sim 3 \times 10^{-5}$) (see, e.g., Acuña and Ness 1976; Ip 1979; Pyle et al. 1983) and the Saturnian G Ring ($\tau \sim 10^{-4}$) (see, e.g., Simpson et al. 1980; Van Allen et al. 1980; Vogt et al. 1982; Van Allen 1982, 1983). As discussed by Thomsen and Van Allen (1979) and Van Allen (1982, 1983), charged particle measurements not only provide information on the existence

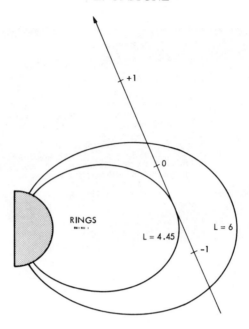

Fig. 7. A representation of the Voyager 2 trajectory through a hypothetical magnetic field with
the dipole axis aligned with the rotation axis of Uranus. In this hypothetical case the
spacecraft would not cross magnetic field lines threading the known rings, but would be
limited to the detection of absorption signatures of matter beyond 4.45 R_U.

of tenuous rings, but also on the optical depth of the ring and on the effective
size of the ring particles. For example, Van Allen found that the particulates
in Saturn's G Ring have an effective radius $r = <r^3>/<r^2>$ in the range
$0.035 \leqslant r \leqslant 0.1$ cm.

The plasma wave and the planetary radio astronomy instruments can also
detect the electrical signals generated by the impact of ring particles as the
spacecraft penetrates the ring plane. Such signals were detected at Saturn as
Voyager 2 penetrated the ring plane just outside of the G Ring (Scarf et al.
1982; Warwick et al. 1982). Gurnett et al. (1983) interpreted the signals as
resulting from the impact of submicron-sized particles with an optical depth of
$\sim 10^{-4}$. The instruments will have a similar sensitivity to any tenuous material
at the location where Voyager 2 penetrates the Uranian ring plane.

II. THE NEPTUNE ENCOUNTER

With the gravity assist of the Uranian flyby, Voyager 2 will continue on
to an encounter with Neptune, arriving there on 24 August 1989. Although
Earth-based stellar occultations indicate an absence of optically thick rings

at Neptune, tenuous rings, such as the Jovian ring or Saturn's G Ring, are not excluded by present observations. Thus, the techniques described above for detecting tenuous rings may provide evidence for previously undetected distributions of particles in orbit about Neptune.

The Neptune flyby trajectory, illustrated in Fig. 8, was chosen to provide a close flyby of Neptune (1.3 R_N) and Triton (~ 44000 km) and opportunities for radio and solar occultation studies of the atmospheres of both bodies. The resulting trajectory limits a solar (or radio) occultation study to regions beyond ~ 1.6 R_N. Although occultation studies are not optimum for detecting tenuous rings, limits on the particle size distribution can be obtained as was done for the Jovian ring (see, e.g., Tyler 1981).

As at Jupiter and Saturn, other techniques can be employed to search for tenuous rings. Specifically, an edge-on view of the ring plane will occur as the spacecraft passes inbound through Neptune's equatorial plane at 3.66 R_N (1 R_N = 24300 km), and a high phase angle view will be available as the spacecraft passes through Neptune's shadow. The Jovian ring, with an optical depth $\sim 3 \times 10^{-5}$ was visible from both such viewing geometries. An even more tenuous halo, with $\sim 7 \times 10^{-6}$, was visible in forward-scattered light. Thus, even with the reduced illumination at Neptune, it should be possible to detect a tenuous ring of micron-sized particles with $\tau \gtrsim 10^{-3}$. A stellar occultation study with a sufficiently bright star may also provide measurements of tenuous rings with $\tau \gtrsim 10^{-2}$.

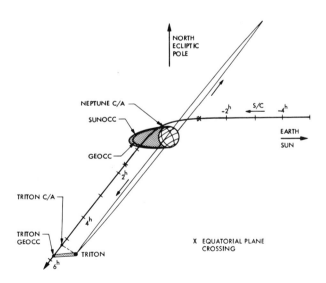

Fig. 8. A view normal to the plane of the trajectory of the Voyager 2 encounter with Neptune on 24 August 1989, illustrating the periods during which Earth (GEOCC) and solar (SUNOCC) occultations occur.

If Neptune has a magnetic field and stable trapped particles, the flyby trajectory will provide the opportunity to observe absorption signatures of ring particles beyond $\sim 2.45\ R_N$, assuming the axis of the magnetic field is aligned with Neptune's rotation axis. The inbound and outbound ring-plane crossings also provide *in situ* detection of small ring particles by the plasma wave and planetary radio astronomy instruments.

With the flyby of Neptune, the Voyager spacecraft will have completed their investigations of the rings of the giant outer planets. The Galileo Mission to Jupiter will continue studies of the Jovian ring, while further spaceprobe studies of Saturn's rings await a new mission to place a properly instrumented spacecraft into orbit about Saturn.

III. THE GALILEO MISSION TO JUPITER

The Galileo Mission to Jupiter is scheduled to be launched in 1986 and to arrive at Jupiter in 1988. The spacecraft shown in Fig. 9 will eject an atmospheric probe prior to being inserted into orbit about Jupiter. Although ring studies are not a primary objective of the mission, both the probe and the orbiter carry instruments that can provide new information on the ring. Table III contains a listing of all the Galileo instruments that are described in more detail in the Galileo *Science Requirements Document* (JPL document 625-50). An overview of the entire mission, its scientific objectives, and the characteristics of the orbiter and probe are provided by Johnson and Yeates (1983).

Most of the ring observations will be made by instruments on the orbiter. After injection at radial distance of 4 R_J and an orbital inclination of $\sim 5°$, subsequent close encounters with the Galilean satellites are used to raise perijove to $\geqslant 9\ R_J$ and to reduce the orbital inclination to a few tenths of a degree. Twelve orbits over a 20-month period will provide opportunities for ring-plane crossing studies and nearly edge-on views of the ring over a complete range of phase angles. Among the instruments on the orbiter that may provide new information on the rings are the solid-state imaging system (SSI), the ultraviolet spectrometer (UVS), and the dust detector (DDS).

The imaging system on the Galileo orbiter uses a spare Voyager optical system with 1500 mm focal length, with the Voyager vidicon sensor replaced by an 800 × 800 pixel charge-coupled device (CCD) which is more than 25 times faster, a particular advantage in imaging the tenuous Jovian ring. The inherent photometric accuracy of the CCD will facilitate a study of the phase angle dependence of the scattering properties of the ring particles for wavelengths between 0.42 and 1.1 μm, potentially providing information on the size distribution of micron-sized particles.

The long observing time will also permit a much more accurate determination of the orbital elements of Metis and Adrastea and a search for other small satellites that may both supply ring material and dynamically control its

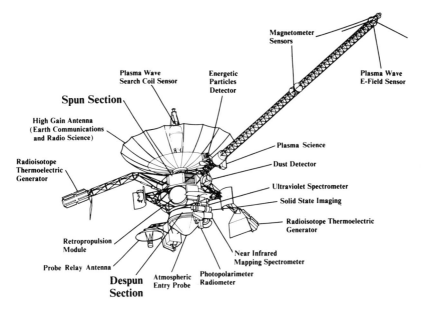

Fig. 9. A drawing of the Galileo spacecraft showing the locations of the science instruments and several spacecraft subsystems. The probe also carries a number of instruments.

TABLE III

Galileo's Scientific Payload

Probe Experiment	Principal Investigator/Institution
Atmospheric structure instrument (ASI)	Seiff/ARC
Neutral mass spectrometer (NMS)	Niemann/GSFC
Helium abundance detector (HAD)	VonZahn/Univ. of Bonn
Nephelometer (NEP)	Ragent/ARC
Net-flux radiometer (NFR)	Boese/ARC
Lightning and energetic particles (LRD/EPI)	Lanzerotti/Bell Labs
Orbiter Experiment	**Principal Investigation/Institution**
Solid-state imaging (SSI)	Belton/KPNO (team leader)
Near-infrared mapping spectrometer (NIMS)	Carlson/JPL
Ultraviolet spectrometer (UVS)	Hord/Univ. of Colorado
Photopolarimeter-radiometer (PPR)	Hansen/Goddard Inst. for Space Studies
Magnetometer (MAG)	Kivelson/UCLA
Energetic-particle detector (EPD)	Williams/APL
Plasma detector (PLS)	Frank/Univ. of Iowa
Plasma-wave spectrometer (PWS)	Gurnett/Univ. of Iowa
Dust detector (DDS)	Grün/Max-Planck Inst.
Radio science (RS): Celestial Mechanics	Anderson/JPL (team leader)
Radio science (RS): Propagation	Howard/Stanford (team leader)

spatial distribution. At closest approach to the ring, the spatial resolution of the camera will be ~ 10 km/line pair.

The ultraviolet spectrometer, operating in a wavelength range between 1150 and 4300 Å with sampling times down to 1 ms, may also provide significant new information on the Jovian ring. The sampling times correspond to a spatial sampling of < 100 m. Although the instrument sensitivity is limited to normal optical depths of > 10^{-3}, the low inclination of the orbit provides long slant paths through the tenuous ring. Thus, stellar occultation studies of the Jovian ring should be possible.

Although the orbiter remains well outside of the ring, the dust detector can make *in situ* measurements of small particles that may be more widely dispersed throughout the Jovian neighborhood. The detector is sensitive to particles with masses between 10^{-6} and 10^{-16} g, corresponding to particles with diameters ranging from ~ 0.02 to 50 μm. Up to 100 particles with speeds from 2 to 50 km s^{-1} can be analyzed each second.

The probe will also return information on the distribution of ring material as indicated by the absorption of trapped radiation. The probe contains a heavily shielded (≥ 0.87 g cm^{-2}) energetic particle detector providing measurements of the fluxes of electrons, protons, α particles, and heavier nuclei such as oxygen and sulfur. The detector will be turned on at 4 R_J, returning flux measurements every 0.02 R_J as the probe plunges toward the Jovian atmosphere, providing some spatial resolution of the absorption signature of the bright ring which is ~ 0.08 R_J in width.

IV. A SATURN ORBITER MISSION

Further spaceprobe studies of Saturn's rings await the placement of a properly instrumented spacecraft into an inclined orbit about Saturn. Such a mission could address questions about the macrostructure of the rings on the scale of $\geq 10^3$ km, the microstructure of the rings on smaller scales, the mass of the rings and the particle size distribution, the particle composition, and the electromagnetic properties of the ring material. The comprehensive longitudinal and temporal coverage provided by an orbiter would make possible a number of important studies which cannot otherwise be adequately addressed.

Among the key instruments for such studies would be a high-speed, multiband photometer that could undertake multiple stellar occultation measurements with high spatial resolution. As has been illustrated by Elliot et al. (1981) for Uranus, multiple stellar occultations provide important information on dynamically significant characteristics such as the density profile, eccentricity, and precession of the edges of gaps in the rings. Comprehensive imaging from an orbiter can also provide a detailed description of dynamically controlled azimuthal variations in various ring structures (see, e.g., Porco 1983) and waves (see, e.g., Shu et al. 1983). Imaging would also permit definitive searches for moonlets within the empty gaps.

An orbiting spacecraft also provides multiple opportunities for studying the physical thickness of the rings. Thus, the thickness of the ring at sharp edges can be estimated from the comparison of edge profiles acquired for different slant paths of the occulted starlight, with multiple radio occultations providing similar information for particles with diameters larger than ~ 2 cm. High-speed photometry of the edge of the ring as the spacecraft passes through the ring plane can provide additional information on the apparent thickness of the rings.

Multiple radio occultations can also yield the size distribution of centimeter and meter-sized particles throughout the ring system, especially if the rings are more open than during the Voyager encounter and therefore offer a smaller slant optical depth through the very dense B Ring. In addition, ring scattering experiments in which the spacecraft radio beam is scattered obliquely off the rings should allow a determination of the size distribution of particles with radii in the centimeter to meter range, complementing X band and S band occultation data that allow a determination of the distribution of particles with radii r in the range $1 \lesssim r \lesssim 4$ cm and $1 \lesssim r \lesssim 10$ m. It would also be of interest to consider approaches yielding information on the distribution of particles with $r > 10$ m. It is possible, for example, that kilometer-sized particles create wakes in the ring that can be directly imaged or statistically sampled by multiple stellar and radio occultations.

Significant improvements in spatial resolution will also be possible with an increased signal-to-noise ratio resulting from greater transmitter power, larger collecting aperture on Earth, and improved occultation geometry. During the Voyager radio occultations, these factors resulted in radial resolutions of 300 m in the inner C Ring, ~ 75 m in the F Ring, and several kilometers in the A Ring and parts of the B Ring (Marouf and Tyler 1983). Since, in principle, the resolution is limited only by the size of the spacecraft antenna (typically ~ 5 m) significant improvements should be readily achievable. In addition, analysis of the near-forward scatter signal can provide a measure of the ring thickness in different regions of the rings (Zebker and Tyler 1983).

Compositional studies with high spatial resolution would also be desirable. Such studies could be undertaken, for example, with an imaging spectrometer operating in the near infrared and a photopolarimeter similar to those on the Galileo orbiter.

With an orbiter it would also be possible to better characterize the electromagnetic properties of the spokes through high time resolution imaging of their radial and azimuthal development and through longer term synoptic studies of their dependence on Saturnian longitude and local time. It would also be desirable to better determine the density of the hydrogen cloud associated with the ring.

V. CONCLUSION

Recent spaceprobe observations of the rings of Jupiter and Saturn have provided an unparalleled increase in our knowledge and understanding of planetary rings. Employing many of the same observational techniques at Uranus and Neptune, Voyager 2 offers further opportunities for studying the nature of ring systems of giant planets. The Galileo mission to Jupiter provides the first opportunity for long-term spaceprobe studies of a planetary ring system, offering the prospect of studies and discoveries beyond those accessible to the earlier Pioneer and Voyager flybys. A properly instrumented Saturn orbiter would not only provide a similar opportunity for studies of Saturn's rings, but is likely the only means by which it will be possible to adequately address the nature of the diverse phenomena displayed by this prototypical planetary ring system.

Acknowledgments. I appreciate helpful comments by S. A. Collins, J. N. Cuzzi, J. B. Holberg, T. V. Johnson, A. L. Lane, and E. D. Miner. This work was partially supported by the National Aeronautics and Space Administration.

REFERENCES

Acuna, M. H., and Ness, N. F. 1976. The main magnetic field of Jupiter. *J. Geophys. Res.* 81:2917–2922.

Cuzzi, J. N., Lissauer, J. J., and Shu, F. H. 1981. Density waves in Saturn's rings. *Nature* 292:703–707.

Elliot, J. L., French, R. G., Frogel, J. A., Elias, J. H., Mink, D. J., and Liller, W. 1981. Orbits of nine Uranian rings. *Astron. J.* 86:444–455.

Esposito, L. W., Borderies, N., Cuzzi, J. N., Goldreich, P., Holberg, J. B., Lane, A. L., Lissauer, J. J., Marouf, E. A., Pomphrey, R. B., Terrile, R. J., Tyler, G. L. 1983. Eccentric ringlet in the Maxwell gap at 1.45 Saturn radii: Multi-instrument Voyager observations. *Science* 222:57–60.

Goldreich, P., and Tremaine, S. 1979. Towards a theory for the Uranian rings. *Nature* 277:97–99.

Gurnett, D. A., Grün, E., Gallagher, D., Kurth, W. S., and Scarf, F. L. 1983. Micron-sized particles detected near Saturn by the Voyager 2 plasma wave instrument. *Icarus* 53:236–254.

Ip, W. H. 1979. On the Pioneer 11 observations of the ring of Jupiter. *Nature* 280:478–479.

Jewitt, D. C., and Danielson, G. E., 1981. The Jovian ring. *J. Geophys. Res.* 86:8691–8698.

Johnson, T. V., and Yeates, C. M. 1983. Return to Jupiter: Project Galileo. *Sky Telescope* 66:99–106.

Lissauer, J. J. 1982. Dynamics of Saturn's rings. Ph.D. dissertation, University of California, Berkeley.

Marouf, E. A., and Tyler, G. L. 1983. Radio occultation of Saturn's rings: Is less than 75 meters resolution achievable? *Bull. Amer. Astron. Soc.* 15:817 (abstract).

Marouf, E. A., Tyler, G. L., Zebker, H. A., Simpson, R. A., and Eshleman, V. R. 1983. Particle size distributions in Saturn's rings from Voyager 1 radio occultation. *Icarus* 54:189–211.

Matthews, K., Neugebauer, G., and Nicholson, P. D. 1982. Maps of the rings of Uranus at a wavelength of 2.2 microns. *Icarus* 52:126–135.

Porco, C. 1983. Voyager observations of Saturn's rings. 1. The eccentric rings at 1.29, 1.45, 1.95, and 2.27 R_S. 2. The periodic variation of spokes. Ph.D. dissertation, California Institute of Technology.

Pyle, K. R., McKibben, R. B., and Simpson, J. A. 1983. Pioneer 11 observations of trapped particle absorption by the Jovian ring and the satellites 1979, J1, J2, J3. *J. Geophys. Res.* 88:45–48.

Scarf, F. L., Gurnett, D. A., Kurth, W. S., and Poynter, R. L. 1982. Voyager 2 plasma wave observations at Saturn. *Science* 215:587–594.

Shu, F. H., Cuzzi, J. N., and Lissauer, J. J. 1983. Bending waves in Saturn's rings. *Icarus* 53:185–206.

Simpson, J. A., Bastian, T. S., Chenette, D. L., Lentz, G. A., McKibben, R. B., Pyle, K. R., and Tuzzolino, A. J. 1980. Saturnian trapped radiation and its absorption by satellites and rings: The first results from Pioneer 11. *Science* 207:411–415.

Smith, B. A., Soderblom, L. A., Beebe, R., Boyce, J., Briggs, G., Carr, M., Collins, S. A., Cook, A. F., Danielson, G. E., Davies, M. E., Hunt, G. E., Ingersoll, A., Johnson, T. V., Masursky, H., McCauley, J., Morrison, D., Owen, T., Sagan, C., Shoemaker, E. M., Strom, R., Suomi, V. E., Veverka, J. 1979a. The Galilean satellites and Jupiter: Voyager 2 imaging science results. *Science* 206:927–950.

Smith, B. A., Soderblom, L. A., Johnson, T. V., Ingersoll, A. P., Collins, S. A., Shoemaker, E. M., Hunt, G. E., Masursky, H., Carr, M. H., Davies, M. E., Cook, A. F., Boyce, J., Danielson, G. E., Owen, T., Sagan, C., Beebe, R. F., Veverka, J., Strom, R. G., McCauley, J. F., Morrison, D., Briggs, G. A., Suomi, V. E., 1979b. The Jovian system through the eyes of Voyager 1. *Science* 204:951–972.

Smith, B. A., Soderblom, L., Beebe, R., Boyce, J., Briggs, G., Bunker, A., Collins, S. A., Hansen, C. J., Johnson, T. V., Mitchell, J. L., Terrile, R. J., Carr, M., Cook, A. F., Cuzzi, J., Pollack, J. B., Danielson, G. E., Ingersoll, A., Davies, M. E., Hunt, G. E., Masursky, H., Shoemaker, E., Morrison, D., Owen, T., Sagan, C., Veverka, J., Strom, R., Suomi, V. E. 1981. Encounter with Saturn: Voyager 1 imaging science results. *Science* 212:163–191.

Smith, B. A., Soderblom, L., Batson, R., Bridges, P., Inge, J., Masursky, H., Shoemaker, E., Beebe, R., Boyce, J., Briggs, G., Bunker, A., Collins, S. A., Hansen, C. J., Johnson, T. V., Mitchell, J. L., Terrile, R. J., Cook, A. F., Cuzzi, J., Pollack, J. B., Danielson, G. E., Ingersoll, A. P., Davies, M. E., Hunt, G. E., Morrison, D., Owen, T., Sagan, C., Veverka, J., Strom, R., Suomi, V. E. 1982. A new look at the Saturn system: The Voyager 2 images. *Science* 215:504–537.

Thomsen, M. F., and Van Allen, J. A. 1979. On the inference of properties of Saturn's Ring E from energetic charged particle observations. *Geophys. Res. Letters* 6:893–896.

Tyler, G. L., Marouf, E. A., Simpson, R. A., Zebker, H. A., and Eshleman, V. R. 1983. The microwave opacity of Saturn's rings at wavelengths of 3.6 and 13 cm from Voyager 1 radio occultation. *Icarus* 54:160–188.

Tyler, G. L., Marouf, E. A., and Wood, G. E. 1981. Radio occultation of Jupiter's ring: Bounds on optical depth and particle size and a comparison with optical and infrared results. *J. Geophys. Res.* 86:8699–8703.

Van Allen, J. A. 1982. Findings on rings and inner satellites of Saturn by Pioneer 11. *Icarus* 51:509–527.

Van Allen, J. A. 1983. Absorption of energetic protons by Saturn's Ring G. *J. Geophys. Res.* 88:6911–6918.

Van Allen, J. A., Thomsen, M. F., Randall, B. A., Rairden, R. L., and Grosskreutz, C. L. 1980. Saturn's magnetosphere, rings, and inner satellites. *Science* 207:415–421.

Vogt, R. E., Chenette, D. L., Cummings, A. C., Garrard, T. L., Stone, E. C., Schardt, A. W., Trainor, J. H., Lal, N. and McDonald, F. B. 1982. Energetic charged particles in Saturn's magnetosphere: Voyager 2 results. *Science* 215:577–582.

Warwick, J. W., Evans, D. R., Romig, J. H., Alexander, J. K., Desch, M. D., Kaiser, M. L., Aubier, M., Leblanc, T., Lecacheux, A., and Pedersen, B. M. 1982. Planetary radio astronomy observations from Voyager 2 near Saturn. *Science* 215:582–587.

Zebker, H. A., and Tyler, G. L. 1983. Thickness of Saturn's rings from Voyager 1 observation of microwave scatter. *Science* 1984. In press.

FUTURE OBSERVATIONS OF PLANETARY RINGS FROM GROUNDBASED OBSERVATORIES AND EARTH-ORBITING SATELLITES

BRADFORD A. SMITH
University of Arizona

Three of the outer planets of our solar system are known to possess ring systems and, among the three known systems, all have one or more components that are detectable from the vicinity of the Earth. Among the more promising methods for continuing study are direct imaging and occultations, obtainable both from the ground and from Earth-orbiting facilities such as Space Telescope. Relevant future observations made from the vicinity of Earth are expected to provide new knowledge of the dynamical and photometric properties of these outer solar system planetary rings.

Among the three known planetary ring systems of the outer solar system, two were discovered from the Earth and the third, we now realize, certainly could have been. The rings of Saturn have been known to exist since the early years following the invention of the telescope, while those of Uranus were discovered only as recently as 1977. The Jupiter ring, although first seen by Voyager 1 in 1979, has since been recorded in both the visual and the infrared regions of the spectrum with groundbased telescopes (Smith and Reitsema 1980; Becklin and Wynn-Williams 1979). Although a Neptune ring had been reported by Guinan et al. (1982), recent stellar occultations have shown conclusively that this planet does not have a ring system, at least not one that resembles that of Uranus or the bright rings of Saturn (Vilas et al. 1983). Neptune rings of low optical thickness, similar to the E and G Rings of Saturn or the Jupiter rings, however, cannot yet be ruled out.

Most of what we now know about the Jupiter and Saturn ring systems is based on observations by planetary spacecraft, particularly Voyager, while our knowledge of the Uranus ring system has evolved solely from ground-based telescopes. Yet, there are large gaps in our understanding of all planetary rings and, in this chapter, we address the question of what we might expect to learn through future observations conducted from the vicinity of the Earth, including the use of both groundbased telescopes and earth-orbiting instruments.

Although the continuing development in our understanding of the physical and dynamical properties of planetary rings will in part be guided by further analysis and possible reinterpretation of existing data, it is clear that substantial progress must depend on the acquisition of new data. There will be a need for new observations to test hypotheses generated in response to earlier data and to explore further the phenomena inadequately covered in the past. The most useful of all future observations will certainly be those obtained by planetary probes, and the prospects for such new data are discussed in the chapter by Stone. But the future of planetary exploration by remote probes is uncertain at best and we are thus compelled to ask ourselves just what might be accomplished from the vicinity of the Earth, including, of course, both the Earth's surface and the nearby space within which Earth-orbiting satellites might be placed. Most groundbased observations, however, will not compete favorably with those which are potentially attainable through the use of powerful Earth-orbiting telescopes such as Space Telescope, to be launched by the United States as early as 1986.

In principle, of course, it is possible to construct very large telescopes in nearby space, with apertures sufficiently large to rival even the spatial resolution which can be reached by remote space probes. However, such very large apertures are not now realistically included in the future plans of NASA, ESA or the Soviet Union; furthermore, from our terrestrial vantage point, anchored permanently within the central core of our solar system, it is impossible to view the outer planets at the higher, and often more desirable, planetocentric latitudes and phase angles reachable by planetary probes. Therefore, for purposes of this discussion, we must be content with what might be accomplished with limited spatial resolution and very limited phase angle coverage.

Radial Structure

The radial structure of both the Uranus rings and the optically thick rings of Saturn has been well determined by stellar occultations, using groundbased telescopes and Voyager photometric instruments, respectively. Occultations of relatively bright stars by planetary rings measure radii, structure and optical depth with spatial resolution limited only by the angular diameter subtended by the stellar disk or, ultimately, by the size of the first Fresnel Zone at the distance of the planet,

$$a = 0.547 \, (r\lambda)^{\frac{1}{2}} \qquad (1)$$

where a is the diameter of the first half-period, annular Fresnel Zone (km), r is the distance to the planetary ring being observed (AU) and λ is the wavelength of light employed in the observation (μm). At the mean distance of Uranus, for example, the diameter of the Fresnel Zone at a wavelength of 0.9 μm is 2.3 km; for comparison, the effective diameter of a typical star, e.g., 0.25 milliarcsec, would be 3.5 km.

The detailed radial structure of the Jupiter ring system and that of the optically thin rings of Saturn, however, is poorly known. Rings having an optical thickness of $\leqslant 0.01$ are difficult, if not impossible, to detect by the occultation method, which is limited by scintillation noise introduced by the terrestrial atmosphere. What little has been learned about these optically thin rings is derived from deconvolution of smeared optical images obtained from both spacecraft and groundbased telescopes.

In general, the mechanisms responsible for the structure of the Uranus and optically-thick Saturn rings is not well understood. While various types of satellite-resonance phenomena have been identified in the A, B and C Rings of Saturn, many of the radial features, and perhaps most of those in the B Ring, remain unexplained. Clues to the responsible mechanisms, however, may be found in possible time variations of the radial structure. Stellar occultations by the Uranus rings will continue to provide new data through the use of groundbased facilities. However, the high speed photometer (HSP) on Space Telescope (Hall 1982), a multispectral instrument with a time resolution of 10 μs, will improve the accuracy of groundbased occultations by eliminating the noise caused by terrestrial atmospheric scintillation. Several occultations of stars by the Uranus rings occur each year.

Because of the high surface brightness of the optically thick Saturn rings, occultations of stars sufficiently bright to reveal fine radial structure occur very infrequently for groundbased telescopes, averaging scarcely more than one per century. The culprit in this case is the Earth's atmosphere, which typically spreads the starlight into a disk of $\geqslant 2$ arcsec, thereby requiring a photometric field of view at least as large. However, a 2 arcsec circular area of the B Ring has a brightness approximately equal to $m_v = 5.5$, and therefore requires an occultation of a star nearly as bright for an acceptable signal-to-noise ratio (SNR). It is here that the Space Telescope HSP can make a significant improvement. Its 0.4 arcsec field of view provides a 3.5 magnitude advantage over the groundbased limit, corresponding to a 50-fold increase in the number of available stars (Allen 1973); this alone provides nearly one opportunity per year. Furthermore, by observing stars that radiate strongly in the ultraviolet, filters can be selected to suppress even more the contribution from the rings, thereby extending the magnitude limit of suitable stars and still further increasing the annual number of observational opportunities. The HSP can record radial structure in the rings of Saturn as narrow as 1.0 km when measured at 300 nm. This resolution exceeds by a factor of 5 the highest resolution recorded by the Voyager cameras and approaches the resolution obtained by the Voyager photopolarimeter and the ultraviolet spectrometer during the occultation of δ Scorpii (see chapter by Cuzzi et al.). Such observations, carried out routinely, would create a baseline for establishing time variability in the optical-depth radial profiles of the Uranus and Saturn systems. In the A, B, C, D and F Rings of Saturn alone, a substantial body of data could be created for investigating ring particle resonances, density waves and discontinuous ringlets. From an analytical point of view, more might be learned from systematic coverage at good resolution than from the excellent but brief snapshot views provided by Voyagers 1 and 2.

The structure of the low optical thickness E and G Rings of Saturn and the brighter component of the Jupiter rings can also be studied from the vicinity of the Earth; only the G Ring of Saturn has thus far gone undetected by groundbased telescopes. Quantitative measurements of the radial structure, optical depth, and any time variations of these optically thin rings will be crucial to an understanding of production and depletion mechanisms for the ring particles; only particles which are meter-size or larger have orbits that are stable over the age of the solar system (see chapter by Burns et al.).

The principal problem with groundbased observations of faint rings is the limited resolution imposed by the terrestrial atmosphere (approximately 5,000 km and 10,000 km, respectively, for Jupiter and Saturn) and the corresponding reduction in SNR. In principle, atmospherically blurred images can be reconstructed if the point-spread function is well known and the SNR is very high; however, it is the poor SNR in groundbased images of these faint Jupiter and Saturn rings that imposes severe constrains on the current studies of radial structure.

Once again, we can look to Earth-orbiting instruments to help solve the resolution problem. The wide field/planetary camera (WF/PC) and the faint object camera (FOC) are two imaging instruments that will be carried aboard Space Telescope (Hall 1982). The WF/PC, by using image restoration techniques, will achieve the same resolution of the Saturn rings (250 km) as was recorded by the Voyager cameras only 10 days before Saturn encounter (see Fig. 1). Resolution of the Jupiter ring will be approximately 125 km. But, there is the possibility that even higher resolution might be achieved; if the optical surfaces of the Space Telescope mirrors are figured as well as we now believe, the FOC with its greater focal-plane scale may exceed the resolution of the WF/PC by nearly a factor of two in the ultraviolet (200 nm) region of the spectrum. If so, resolution at Jupiter, Saturn and Uranus could be as high as 65, 125, and 300 km, respectively (see Table 1).

Spokes

The spokes in the B Ring of Saturn were discovered by Voyager 1 in 1980. Although they were reobserved more systematically by Voyager 2 the following year, we still do not have a good understanding of how they form, how they dissipate, why they are found only in the outer B Ring or, for that

Fig. 1. This view of Saturn and its brighter rings was taken by Voyager 2 on 13 August 1981 at a distance of 12.7 million km. The spatial resolution is 235 km, comparable to the better images to be obtained by Space Telescope. Note the dusky spokes in the middle of the B Ring and the structure in the C Ring.

matter, what they are. Some appear to move at Keplerian rates under the influence of a central gravitational force field (see, e.g., Grün et al. 1982 and the chapter by Grün et al.), while others spend part of their lifetime stationary in a frame of reference that is corotational with the magnetic field of Saturn (Eplee and Smith 1983, 1984). Very few spokes have been studied throughout their entire lifespans because the Voyager camera sequences always lacked either temporal or areal continuity; at no time was continuous coverage of the whole visible ring system recorded at high resolution, i.e., resolution better than 200 km. Furthermore, because the spokes have been recorded only by Voyager, their spectral reflective properties can be determined only over the limited spectral range of the Voyager cameras, essentially 400 to 600 mm.

TABLE I

Resolution Potentially Obtainable with Space Telescope[a]

	Wide Field/ Planetary Camera	Faint Object Camera
Jupiter	125 km	65 km
Saturn	250 km	125 km
Uranus	600 km	300 km
Neptune	900 km	450 km

[a]The numbers are only approximate and depend very much on the final optical performance and pointing stability of the Space Telescope. The WF/PC in its planetary mode and the FOC in its f/288 configuration slightly and significantly oversample, respectively, the point-spread function. Resolution also depends upon the contrast of the scene in such a way that no values given could be any more than a guide to the reader. The values shown are close to full-width at half-maximum, which is itself very close to the Rayleigh criterion.

The relatively poor state of our understanding of the physical and dynamical properties of spokes is such that almost any new information would be welcome. To date, it appears that no one has been successful in recording these features with groundbased telescopes, although it is not clear that it cannot be done; in fact, it is doubtful that the problem is related solely to spatial resolution. Light from Saturn and the rings themselves, scattered and diffracted by the Earth's atmosphere and within the telescope, would tend to suppress the visibility of spokes even in images which have resolution adequate to reveal them. By employing a combination of especially clean telescope optics, a focal-place coronagraph (Larson and Reitsema 1979) and a CCD detector at times of good image quality, it may be possible to obtain groundbased recordings of the Saturn ring spokes. Such observations would be useful in determining their spectral properties and any possible effects due to the changing planetocentic declination of the Sun (seasonal effects); but, it is not likely that observing conditions would remain optimal over intervals long enough to support dynamical and life-cycle studies. The WF/PC and FOC on Space Telescope can provide the necessary resolution (see Fig. 1 and Table I) but, for all celestial objects located near the ecliptic, the maximum interval for continuous observing is only 30–45 min out of each 100 min orbit; thus, we cannot escape those problems imposed by noncontinuous data sets, the same problems encountered with the Voyager images. Statistically, one can partially overcome the problems of a 0.4 observing duty cycle by extending the total observing time. Typical spoke lifetimes are ~5 hr (Grün et al. 1983; Eplee and Smith, 1983, 1984) and thus the accumulation of several tens of hours of observing time by recording the rings for 30–45 min per orbit would likely yield many events of spoke formation, acceleration and dissipation. However, the willingness of the astronomical community to commit this unique facility to such long observing sequences is questionable at best. We must still hope that some compromises are possible, for our experience with the Voyager data have taught us that sporadic observations are of relatively little value for dynamical studies.

Photometric Properties

Among the observations necessary to characterize the physical properties of ring particles are spectral reflectance and polarization and the dependence of these parameters on varying illumination and viewing geometries. Yet, such data are either poorly determine or nonexistent for the Jupiter and Uranus rings and for the faint rings of Saturn. Through the use of clean optics, focal-plane coronagraphs and CCD detectors, two of these rings can be recorded with groundbased telescopes, vis., the brighter component of the Jupiter ring system and the E Ring of Saturn (see Fig. 2). The observations, however, are very difficult and the photometric errors relatively large. Space Telescope should have little difficulty recording photometrically useful images of these two rings and seven of the Uranus rings, but may have difficulty with the inner and outer components of the Jupiter ring and the Saturn D and G Rings; much will depend on the ability of the Space Telescope to reject scattered light from Jupiter, Saturn and the bright rings of Saturn.

Satellite Motions

Voyager observations of Saturn ring gaps, edges, density and bending waves and the multistranded F Ring show abundant evidence for mutual perturbations between satellites and ring particles. It might be reasonable to suppose that similar interactions also take place between the two innermost satellites of Jupiter, Adrastea and Metis, and the brightest component of the Jupiter ring, and between as yet unseen satellites and the Uranus rings. Although the first-order effect in these mutual perturbations is that of the satellite acting on the much less massive ring particles, ring particles can collectively perturb the small, close-in satellites as well. Thus, careful studies of the motions of these small satellites should yield quantitative results on ring mass distribution and viscosity. Most of these perturbing satellites are beyond the reach of groundbased telescopes, and those that can be detected are subject to positional errors of approximately 0.2 arcsec. Space Telescope, however, should detect all of the presently known satellites of Jupiter and Saturn. Voyager 2 will fly past Uranus in January 1986 and will look for ring-associated satellites; it is likely that Space Telescope will eventually detect these as well. The fine guidance system (FGS) of the Space Telescope (Hall 1982) also functions as an astrometric instrument with an accuracy of approximately 2 milliarcsec, and special use of the WF/PC can achieve essentially the same accuracy. In this particular application, Space Telescope, with an appropriate time base of observations, can greatly exceed the astrometric capability of flyby planetary spacecraft such as Voyager.

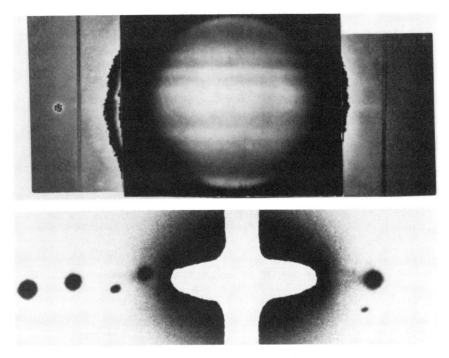

Fig. 2. (top) Composite image of Jupiter photographed with a 1.5-m telescope and a corona-graph in a deep methane absorption band at a wavelength of 0.89 μm. The Jupiter ring is visible on both sides of the planet. The inner satellite, Amalthea, is visible at the left. (bottom) Several satellites and the faint E Ring of Saturn, taken on 1 April 1981 with a coronagraph when the Earth was $5°5$ out of the ring plane. Without a coronagraph, the E Ring can be photographed only when the Earth is in the ring plane. Note that a mask is used to black out the light from Saturn itself and that the image is reproduced here as a negative.

Conclusions

Space Telescope is not, of course, the only astronomical facility to be placed in Earth orbit that could contribute to our understanding of planetary ring systems; others include Spacelab, Shuttle Infrared Telescope Facility (SIRTF), and proposed facilities such as Magellan and Solstice. Rather, Space Telescope has been used as an example because its expected capabilities are presently better understood. Earth-orbiting observatories that will surely follow Space Telescope are expected to employ infrared instruments, which will improve our knowledge of the thermal properties and composition of ring particles, and ultraviolet instruments that can search for ring "atmospheres," molecular or ionic species that may be sputtered into the ring environment.

Although Space Telescope cannot compete with planetary space probes in most areas of investigation, it can and will contribute greatly to new studies of planetary ring systems in the latter part of this decade and, if the current hiatus in outer solar system exploration (beyond Voyager and Galileo) continues, it may become our primary source of new knowledge of planetary rings for many years to come.

REFERENCES

Allen, C. W. 1973. *Astrophysical Quantities,* 3rd ed. (London: Athlone Press, Univ. of London).

Becklin, E. E., and Wynn-Williams, C. G. 1979. Detection of Jupiter's ring at 2.2 μm. *Nature* 279:400–401.

Eplee, R. E., Jr., and Smith, B. A. 1983. Dynamics of spokes in Saturn's B Ring. Proceedings of *I.A.U. Colloquium No. 75 Planetary Rings,* ed. A. Brahic, Toulouse, France, Aug. 1982.

Eplee, R. E., Jr., and Smith, B. A. 1984. Spokes in Saturn's Rings: Dynamical and reflectance properties. Submitted to *Icarus.*

Grün, E., Morfill, G. E., Terrile, R. J., Johnson, T. V., and Schwehm, G. 1983. The evolution of spokes in Saturn's B Ring. *Icarus* 54:227–252.

Guinan, E. F., Harris, C. C., and Maloney, F. P. 1982. Evidence for a ring system of Neptune. *Bull. Amer. Astron. Soc.* 14:658 (abstract).

Hall, D. M. B. ed. 1982. *The Space Telescope Observatory,* NASA Conf. Publ., NASA CP-2244.

Larson, S. M., and Reitsema, H. J. 1979. A Planetary coronagraph. *Bull. Amer. Astron. Soc.* 11:558 (abstract).

Smith, B. A., and Reitsema, H. J. 1980. CCD observations of Jupiter's ring and Amalthea. *Satellites of Jupiter* IAU Colloquium No. 57, Kailua Kona, Hawaii.

Vilas, F., Hubbard, W. B., Frecker, J. E., Hunten, D. M., Gehrels, T., Lebofsky, L. A., Smith, B. A., Tholen, D. J., Zellner, B. H., Wisniewski, W., Gehrels, J.-A., Capron, B., Reitsema, H. J., Waterworth, M. D., Fu, H. H., Wu, H. H., Avey, H. P., and Page, A. A. 1983. Occultations by Pallas (83/5/29) and Neptune (83/6/15): Preliminary results. *Bull. Amer. Astron. Soc.* 15:816 (abstract).

UNSOLVED PROBLEMS IN PLANETARY RING DYNAMICS

NICOLE BORDERIES, PETER GOLDREICH
California Institute of Technology

and

SCOTT TREMAINE
Massachusetts Institute of Technology

We discuss a number of unsolved problems in planetary ring dynamics and offer some thoughts concerning their solution. The discussion is specialized to the rings of Saturn because they are the best studied observationally. The depth of treatment varies from problem to problem and reflects both limitations in our understanding and the relative importance which we attach to different topics. We make no attempt to present material in a tutorial style. We deal with phenomena that are microscopic and macroscopic on the scale set by the resolution of the Voyager cameras. We relate the physics of particle collisions (Sec. I) to the local velocity dispersion (Sec. II) and the particle size distribution (Sec. III). We discuss mechanisms which structure the rings (Sec. IV) and affect the orbital evolution of satellites which interact with them (Sec. V).

I. PARTICLE COLLISIONS

A. Mechanical Properties of the Particles

The mechanical properties are largely characterized by the elastic modulus E and the yield modulus σ_Y. These moduli are properly described by tensors but for our rough calculations their scalar magnitudes suffice.

The elastic modulus is the ratio of the stress to the strain. It has the dimensions of force per unit area or energy per unit volume. The latter reflects its close relation to the molecular bond energy per unit volume. For common materials the molecular bond energy is of order one electron volt and the molecular size is of order one angstrom. It follows that the elastic modulus is $\sim 10^{12}$ dyne cm^{-2}. For ice its value is smaller, $\sim 10^{11}$ dyne cm^{-2} (Hobbs 1974), because the hydrogen bond is weak.

The yield modulus is the limiting stress beyond which ice fractures. Its value decreases with increasing temperature, especially near to the melting temperature. We have been able to find only a single reference to measurements of the yield stress of ice at temperatures comparable to those in Saturn's rings. These experiments obtained values $\sim 3 \times 10^{8}$ dyne cm^{-2} at $T = 77$ K for the compressive yield stress (Parameswaran and Jones 1975).

For future discussion we adopt the values $E = 10^{11}$ dyne cm^{-2} and $\sigma_Y = 10^{8}$ dyne cm^{-2}. We choose a smaller value for σ_Y than quoted above because collisions generate comparable compressive, shear, and tensile stresses, and the compressive yield modulus of ice is larger than either the shear or tensile yield modulus. We also use $\mu = 0.36$ for the Poisson ratio. It must be kept in mind that the numerical values we are adopting refer to solid and chemically pure ice and thus may not directly apply to the properties of the particles in Saturn's rings.

B. Dynamics of Collisions

The imperfectly elastic nature of the impacts is described by the coefficient of restitution ϵ, the factor by which the relative normal velocity is reduced following a collision. We need to relate ϵ to E and σ_Y.

The Hertz (1881) law of contact describes purely elastic, low-velocity, impacts. When applied to the collision of two identical smooth spheres of radius R, density ρ, and relative impact velocity v, it gives the following expressions for the maximum radius of the area of contact a and the maximum stress σ_M:

$$a/R \simeq (\rho v^2/E)^{\frac{1}{5}}, \tag{1}$$

$$\sigma_M/E \simeq 0.7 \, (\rho v^2/E)^{\frac{1}{5}}. \tag{2}$$

From the latter relation we find that the yield stress of ice is reached at an impact velocity of 0.03 cm s^{-1}.

For higher velocity impacts, account must be taken of brittle fracture near the area of contact. The best theory that we have been able to find to apply in this case is due to Andrews (1930). It is designed to treat impacts of soft metals and includes plastic as well as elastic deformation. It is far from rigorous even for its intended application and we shall apply it where failure occurs by brittle fracture rather than by plastic flow. Thus the conclusions we derive from Andrews' theory are to be taken with caution.

Andrews views an impact as taking place in the three stages illustrated in Fig. 1. There is an initial elastic compression which lasts until the critical stress is reached. It is followed by continued compression during which there is an inner circular plastic zone surrounded by an elastic annulus. The stress in the plastic zone is set equal to σ_Y. The restitution which finishes the process is incomplete leaving the sphere permanently flattened. Andrews' theory predicts $\epsilon = 1$ for $v/v^* \leq 1$ and

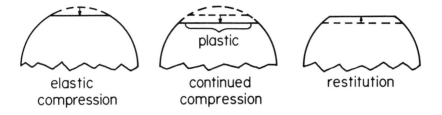

elastic
compression

plastic
continued
compression

restitution

Fig. 1. Three stages of impact in Andrews' theory.

$$\epsilon = \left\{ \left[\frac{-2}{3} \left(\frac{v^*}{v} \right)^2 + \left[\frac{10}{3} \left(\frac{v^*}{v} \right)^2 - \frac{5}{9} \left(\frac{v^*}{v} \right)^4 \right]^{\frac{1}{2}} \right] \right\}^{\frac{1}{2}} \tag{3}$$

for $v/v^* \geq 1$ where

$$v^* = \frac{\pi^2 (1 - \mu^2)^2 \sigma_Y^{\frac{5}{2}}}{10^{\frac{1}{2}} E^2 \rho^{\frac{1}{2}}} . \tag{4}$$

For ice, Eq. (4) gives $v^* \sim 0.03$ cm s^{-1}. Fig. 2 shows ϵ as a function of v/v^*. It will be interesting to compare this theoretical prediction with measurements of ϵ (v), for example in the experiments with ice now in progress by Lin and colleagues (see chapter by Stewart et al.).

Collisions between actual ring particles may differ considerably from those considered here. The particles are not of a single size and their surfaces are probably rough. The radii of curvature on a rough surface are smaller than the particle radius so the stress generated in the collision of rough particles is greater than that generated in a similar impact of smooth particles. Consequently, surface roughness lowers the coefficient of restitution (see chapter by Weidenschilling et al.).

II. VELOCITY DISPERSION

The velocity dispersion in a planetary ring depends upon the mechanical properties of the ring particles as expressed through the variation of the coefficient of restitution with impact velocity.

A. Unperturbed Disks

There is a relation between the equilibrium coefficient of restitution and the optical depth τ in a Keplerian disk (Goldreich and Tremaine 1978a). This relation depends upon the shape and size distribution of the particles. The function ϵ (τ) obtained from the collisional Boltzmann equation, for spherical particles of a single radius, is shown in Fig. 3. It is determined by requiring

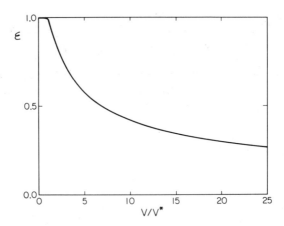

Fig. 2. The coefficient of restitution as a function of impact velocity.

that the rate at which energy is fed into random motions by the viscous stress balance the rate at which the energy of random motions is dissipated in inelastic collisions. The following gedanken experiment may help to clarify the point. Imagine setting up initial conditions such that every particle moves on a circular equatorial orbit. Collisions would occur as a consequence of differential rotation and the finite size of the particles. At first the impacts would be very gentle and almost perfectly elastic and the random velocity would grow. After a few collision times the radial distance traversed between collisions would be larger than the particle radius. The random velocity would begin to increase exponentially on the collision time scale and as it did so the impacts would become less elastic. Eventually the rate at which energy was being dissipated would balance the rate at which it was being converted into random motions by collisional stresses acting on the differential rotation. At this stage the random velocity would have reached equilibrium. The coefficient of restitution plotted in Fig. 3 is that which pertains to this equilibrium.

The monotonic increase of ϵ with increasing τ reflects the monotonic decrease of the radial distance traversed between collisions with increasing τ (at a fixed value of the random velocity). The latter implies that, per collision, the efficiency of conversion of orbital energy into the energy of random motions decreases with increasing optical depth. Thus, at equilibrium, the fraction of the relative kinetic energy dissipated per collision must be a monotonically decreasing function of τ.

We combine the ϵ (v) relation obtained from Andrews' theory with the ϵ (τ) relation described here to derive predictions for the random velocity v and the vertical thickness as functions of τ. The results are displayed in Fig. 4. They provide the theoretical justification for why Saturn's rings are so flat. We have neglected gravitational interactions between particles in the discussion of the random velocity. Pure gravitational scatterings act like physical

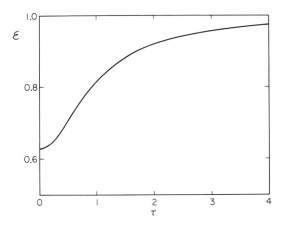

Fig. 3. The equilibrium value of the coefficient of restitution as a function of optical depth.

collisions with $\epsilon = 1$. Gravitational scatterings of small particles by large ones could significantly increase the random velocities of the small particles (Cuzzi et al. 1979).

B. Perturbed Disks

In regions of the rings where satellites exert torques, near sharp edges or resonances, the local rate of energy dissipation is enhanced. An argument we have given elsewhere (Borderies et al. 1982) shows that even in perturbed regions the shear cannot be much larger than the orbital angular velocity Ω unless $\tau \gg 1$. Thus the viscosity, and consequently the velocity dispersion and ring thickness, must be enhanced in perturbed regions. In some places v and h may be increased above their unperturbed values by between one and two orders of magnitude.

III. PARTICLE SIZE

The distribution of particle size must depend upon the mechanical properties of the particles. The detailed physics which determines this distribution is not known.

A. Erosion Time Scale

Is the current size distribution the result of an equilibrium or of initial conditions? The particles are constantly being eroded by collisions and they accrete the resulting debris. The erosion time scale is the reciprocal of the product of the collision rate $\Omega\tau$ and the fractional mass lost per collision f. In a Keplerian disk the impact velocity is typically a few times v^*, the minimum impact velocity at which fracture occurs. We estimate f to be of order $(a^*/R)^3$, where a^* is the radius of the contact area in an impact at relative velocity v^*.

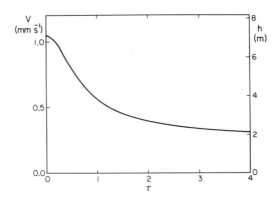

Fig. 4. The random velocity and disk thickness as a function of optical depth.

From Eqs. (1) and (4) and the values of E and σ_Y adopted in Sec. I, we obtain

$$t_{\text{erosion}} \simeq \frac{1}{\Omega\tau} \left(\frac{E}{\sigma_Y} \right)^3 \simeq \frac{10^5}{\tau} \, \text{yr} . \tag{5}$$

Since the estimated erosion time scale is much shorter than the age of the solar system we conclude that the particle size distribution results from an equilibrium between ongoing processes.

Three assumptions implicit in the derivation of the erosion time scale are worth mentioning: (1) we have assumed that a significant portion of the fractured material is lost from the particles; (2) we have neglected surface roughness which could affect the volume of fractured material; (3) we have assumed that v* given by Eq. (4) is a typical value for the impact velocity. This assumption is invalid if v* is smaller than ΩR. In this case the collision velocities would be of order ΩR. This qualification clarifies the seemingly unphysical conclusion that might otherwise be drawn from Eq. (5), namely, that the erosion time scale diverges as σ_Y approaches zero.

B. Range of Particle Sizes

If the processes responsible for establishing the particle size distribution did not single out any characteristic length scales, the distribution would be a power law. In fact, the differential number density per unit radius deduced from the Voyager radio occultation (Marouf et al. 1983) may be crudely fit by the power law

$$n(R) = K R^{-3.3} \tag{6}$$

for particle radii between a few centimeters and a few meters. The distribution is cut off outside this range although precise details of the cutoffs, in particular the lower one, are lacking. The approximate power law is such that the larger particles contain most of the mass and the smaller ones provide most of the surface area. Independent information supports the existence of these two cutoffs. The observation that the optical depth is similar at visible and radio wavelengths proves that subcentimeter particles do not dominate the surface area. An upper cutoff is indicated because the radio determination of the surface mass density due to particles in the centimeter-to-meter-size range is, if anything, somewhat greater than that determined from the wavelengths of density and bending waves which provide a measure of the surface mass density from particles of all sizes. The radio determination of the surface mass density is based on the assumptions that the particles are made of solid ice and that the filling factor is low. In fact, the previous comparison suggests that one or both of these assumptions is invalid, a possibility to which we shall return later.

1. Upper Cutoff. Consider two critical orbital radii related to gravitational binding (see also chapter by Weidenschilling et al.). The inner r_1 is the limit inside which small particles are not gravitationally bound to a synchronously rotating, rigid sphere along the line passing from the center of the planet through the center of the sphere. The outer r_2 is the Roche limit within which no equilibrium exists for a synchronously rotating, incompressible, homogeneous fluid.

$$r_1/R_p = 1.44 \, (\rho_p \, /\rho)^{\frac{1}{3}}, \qquad (7)$$

$$r_2 /R_p = 2.46 \, (\rho_p \, /\rho)^{\frac{1}{3}}, \qquad (8)$$

where ρ_p and ρ are the mean densities of the planet and the sphere and R_p is the planet's radius. The Roche limit r_2 is larger than r_1 because the quadrupole potential of a tidally deformed fluid enhances the disruptive effect of the tidal potential.

The two critical orbital radii are precisely defined but their application to planetary rings is fuzzy. A requirement for the persistence of a ring is that the particles do not form large satellites. Gravity is almost certainly the dominant force involved in satellite formation but it is difficult to see how to deduce a precise criterion for the location of the outermost possible boundary for a planetary ring.

The gravitational binding of small particles on the surface of a larger rigid body would be an appropriate criterion if the small particles eventually became chemically bound to the larger body. The large body would grow by devouring small particles and would remain rigid. On the other hand, if small particles clustered gravitationally about a larger body and chemical bonding did not occur, the assemblage would have low shear strength and thus be mechanically similar to an incompressible, homogeneous, fluid body. Even if we could decide which of these scenarios described more closely the behavior of particles in Saturn's rings, we could not use r_1 or r_2 to make precise predictions because the bodies are not synchronously rotating.

It is of interest to compute the densities of particles for which r_1 or r_2 would coincide with the outer edge of Saturn's A Ring which is located at $2.26\ R_p$. We find $\rho \simeq 0.2$ g cm^{-3} and $\rho \simeq 0.8$ g cm^{-3}, respectively. It is plausible that the particles in Saturn's rings are less dense than solid ice for which $\rho \simeq 0.9$ g cm^{-3}; the Voyager radio occultation results offer some support for this view (Marouf et al. 1983). We conclude that permanent, gravitational binding of the ring particles does not generally occur, but that our understanding of the reason for this is incomplete (see chapter by Weidenschilling et al. for additional discussion of these issues).

Although gravity does not bind ring particles, molecular bonds can. The tidal stress in a particle of radius R is approximately $\sigma_t \simeq \rho\ (\Omega R)^2$. It is equal to the yield stress for $R = 10^3$ km. Obviously, this criterion is irrelevant to the upper cutoff on the size of particles in Saturn's rings.

A plausible argument can be made for identifying the vertical scale height of the rings with the upper cutoff. Suppose that the principal processes which determine the size distribution are collisional erosion and the accretion of the resulting debris. The accretion rate per unit area should be the same for all bodies smaller than the vertical scale height. The erosion of bodies will primarily be due to collisions with particles whose size is near the lower cutoff since they dominate the total area. The erosion rate per unit surface area should be similar for all bodies smaller than the vertical scale height. Typical impact velocities on these bodies are of order the random velocity. For particles large enough to stick out of the main particle layer, i.e. for those with $R > h$, due to differential rotation the typical impact velocity is of order ΩR which is larger than the random velocity. The ratio of the average erosion rate per unit area to the average accretion rate per unit area should be independent of size for $R < h$ and increase with R for $R > h$ since the volume of the material fractured during a collision must be an increasing function of the impact velocity; Andrews' theory predicts that it is proportional to the 3/2 power of v for v $\gg$ v*. It follows that if the net effect of erosion and accretion is to make $dR/dt = 0$ for bodies with $R < h$, then $dR/dt < 0$ for bodies with $R > h$ and the upper cutoff is naturally explained.

A problem with this picture is that the vertical thickness of the rings deduced from the damping lengths of density and bending waves appears to be somewhat larger than the upper cutoff, at least in the A Ring (Cuzzi et al. 1981; Lane et al. 1982; Lissauer et al. 1982; Shu et al. 1983). However, a smaller estimate is derived in Sec. V.B.5.

2. Lower Cutoff. The lower cutoff may be due to the sticking of small particles to bodies of comparable or larger size following collisions. To assess this possibility we compare the elastic energy W_E stored during an impact to the binding energy W_B which would be released if the material bonded across the contact area. The elastic energy is a significant fraction of the relative kinetic energy $M v^2/2$. The surface binding energy is approximately $W_B \simeq E i a^2$, where $i \simeq 10^{-8}$ cm is the bond length and a is the radius of the contact area. We evaluate the ratio of these energies for $v = v^*$, since as we have seen, v is of order a few times v^* in unperturbed regions of the rings. We obtain

$$\frac{W_B}{W_E} \simeq \frac{E i a^{*2}}{\rho R^3 v^{*2}} \simeq \frac{i}{R} \left(\frac{E}{\sigma_Y}\right)^3 \simeq \frac{10 \text{ cm}}{R} . \tag{9}$$

This result suggests that if molecular bonds form across the contact area particles smaller than a few centimeters may stick during collisions.

The sticking hypothesis needs to be critically examined; it is not obvious that molecular bonds form during low-velocity impacts and the fractures which occur in the contact region complicate the dynamics. Also, it is difficult to reconcile the sticking hypothesis with the observation that in certain portions of the rings micron size particles make a detectable contribution to the optical depth at visible wavelengths.

3. Thermal Stresses. Thermal stresses are produced as the particles spin and move in and out of Saturn's shadow. D. Stevenson (personal communication) has pointed out that thermal stresses may be responsible for a size-dependent erosion rate. These stresses are of order

$$\sigma_T \simeq \alpha \Delta T E , \tag{10}$$

where α is the coefficient of linear thermal expansion and ΔT is the temperature variation. The value $\alpha = 5 \times 10^{-6}$ K^{-1} is appropriate for solid ice at the ring temperature (Hobbs 1974) and eclipse cooling measurements indicate that temperature variations of order 10 K penetrate to depths of several millimeters (Froidevaux and Ingersoll 1980; Froidevaux et al. 1981). Thus σ_T is of order 5×10^6 dyne cm^{-2} which, although smaller than the yield stress, is large enough to suggest that fractures may result from thermal stress.

It is probably at least as difficult to evaluate the erosion rate due to thermal stresses as that due to collisional stresses. The thermal and collisional time scales are of the same order of magnitude and the thermal and collisional stresses penetrate to comparable depths. Unlike the collisional stresses which are concentrated around the small area of contact, the thermal stresses are spread over a large fraction of the particle surface.

The spin angular velocity must decrease with increasing particle size as a consequence of the tendency of elastic collisions to promote equipartition of kinetic energy, although due to the imperfectly elastic nature of the collisions equipartition is not achieved. The erosion rate per unit area due to thermal stresses should increase with particle size because the depth of penetration of the thermal wave is proportional to the square root of the spin period.

4. Commentary. Our discussion of the physical processes that determine the size distribution is far from conclusive. Even some of the assumptions made in deducing the distribution from the Voyager radio occultation data are open to question. It is crucial that the particles have shapes that are not too irregular and that they are well separated (Marouf et al. 1982, 1983). Otherwise, some of the scattering attributed to small particles may be due to surface roughness on larger bodies and the gravitational clustering of small particles, such as proposed to explain the azimuthal asymmetry of the A Ring (Colombo et al. 1976), could be partially responsible for the scattering attributed to the larger bodies.

The high abundance of micron-size particles in the F Ring and the outer A Ring suggests that small particles are collisionally produced since the random velocity is enhanced in those regions by satellite perturbations. The surface area of the micron size dust is so large that only a small fraction of it could be produced on the collisional time scale. Most likely the dust grains adhere weakly to the surfaces of larger particles and are shaken off in impacts. The high abundance of micron-sized particles in the B Ring, where perturbations due to external satellites are not especially significant, hints that an enhanced velocity dispersion may be maintained in that ring by small imbedded satellites. These satellites could be in part responsible for the qualitative morphological differences between the A and B Rings.

IV. LARGE-SCALE STRUCTURE OF RINGS

Collisions conserve angular momentum and dissipate energy. As a consequence of differential rotation, angular momentum is transferred outward and the ring spreads. The characteristic spreading time for a ring of radial width Δr is

$$t_{\text{spread}} \simeq (\Delta r)^2 / \nu . \tag{11}$$

The kinematic viscosity ν is given by

$$\nu \simeq 0.5 \frac{v^2}{\Omega} \frac{\tau}{1 + \tau^2} , \tag{12}$$

where v is the random velocity (Goldreich and Tremaine 1978a). We apply Eqs. (11) and (12) to estimate v in Saturn's rings by assuming that the rings have spread from a much narrower initial state to their present width over the age of the solar system. This procedure yields v $\simeq$ 0.2 cm s^{-1} which is between the theoretical estimate given in Sec. II and the value obtained from the damping lengths of density and bending waves.

The overall radial width of the rings could be explained by viscous diffusion but the great amount and variety of structure seen on all smaller scales implies that other mechanisms shape the detailed morphology. Among the outstanding features to explain are: the differences between the classical A, B and C Rings, the multiple ringlets which characterize the B Ring, the sharp outer edges of the A and B Rings and the clear gaps which often contain narrow elliptical ringlets. Two general mechanisms have been proposed to account for some of the structure seen in the rings. They are the viscous instability and satellite perturbations. We discuss them below.

A. Viscous Instability

The torque exerted by the ring material inside radius r on the material outside this radius is given by (Lynden-Bell and Pringle 1974)

$$T_v = 3\pi\nu\Sigma\Omega r^2 . \tag{13}$$

This torque is the rate at which angular momentum flows outward across the circle of radius r; it is also referred to as the viscous angular momentum luminosity L_H. The product $\nu\Sigma$ determines local structure since it is the only part of T_v which can vary on a scale small compared to r. We have computed $\nu\Sigma$ as a function of optical depth from the $v(\tau)$ relation shown in Fig. 4 and the expression for the kinematic viscosity given by Eq. (12). The result is displayed in Fig. 5. Note that this theoretical result, which predicts that $\nu\Sigma$ has a single maximum at $\tau = \tau_M \simeq 0.5$, is based on the assumption that the filling factor is small. For sufficiently large τ, the random velocity would be so low that this assumption would be invalid. We expect that a more complete theory would show that $\nu\Sigma$ increases for sufficiently large τ.

The viscous instability occurs in regions where $\nu\Sigma$ decreases with increasing τ, i.e. for $\tau > \tau_M$. The instability arises as follows (Lin and Bodenheimer 1981; Lukkari 1981; Ward 1981; chapter by Stewart et al.). The net viscous torque on a ringlet is proportional to $\partial(\nu\Sigma)/\partial r = \partial(\nu\Sigma)/\partial\tau \cdot \partial\tau/\partial r$. If $\partial(\nu\Sigma)/\partial\tau < 0$, ringlets move toward regions of enhanced optical depth and away from regions of reduced optical depth; thus a uniform disk with $\tau > \tau_M \simeq 0.5$ is unstable. The instability drives the disk toward a final state in which there are contiguous regions of high and low optical depth with identical values of T_v or $\nu\Sigma$.

The viscous instability is the leading contender for explaining the multi-ringlet structure of the B Ring. However, the current version of the theory predicts a bimodal optical depth distribution which is not observed. Also, it seems far-fetched to imagine that this instability could be responsible for the structure seen in regions of low average optical depth such as the C Ring and the Cassini Division.

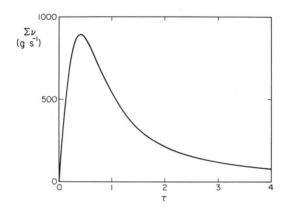

Fig. 5. The product $\nu\Sigma$ as a function of optical depth.

B. Satellite Perturbations

Satellites exert torques on the ring material in the neighborhood of resonances. The density and bending waves seen in Saturn's rings are due to
torques produced by known satellites whose orbits are external to the main
rings. Torques exerted by the shepherd satellites confine the F Ring. The
sharp outer edge of the B Ring is maintained by the torque produced by
Mimas at its 2:1 resonance and the sharp outer edge of the A Ring appears to
be governed by the torque from the coorbital satellites at their 7:6 resonance.
Torques produced by small satellites imbedded in the rings have been proposed to explain much of the small-scale structure. There is good circumstantial evidence that the Encke Division is kept open by small satellites that orbit
within it (see chapter by Cuzzi et al.). However, in spite of a careful search,
the satellites suspected of maintaining the inner and outer clear gaps in the
Cassini Division were not detected by Voyager (Smith et al. 1982). Also, the
absence of clear gaps in the B Ring has been taken as evidence that imbedded
satellites are not responsible for its multi-ringlet structure (Lane et al. 1982).
However, see the final paragraph in Sec. V.C for further discussion of this
point.

V. SATELLITE TORQUES

Satellite torques are clearly associated with a variety of ring features.
Nevertheless, we are far from having achieved a complete understanding of
how they work. Some of the issues that remain to be settled are raised below.
A bewildering variety of satellite torques is discussed in this section. To
reduce confusion we adopt the following conventions: The satellite torques
are exerted on rings and not on satellites. Both satellites and ring particles are

assumed to move on circular orbits. To simplify notation the satellite orbit is taken to be larger than the ring particle orbits. With these conventions, the satellite torques are negative. The discussion can be translated easily to the case where the satellite orbit is smaller than that of the ring particles. We use T to denote a torque; the superscripts L and NL indicate that the perturbation where the torque is applied is either linear or nonlinear; the critical torque which separates the linear and nonlinear torques is indicated by the superscript C; the subscripts m and s denote the torque at an isolated resonance of order m in a ring of uniform surface density and the total torque on a narrow ringlet. The satellite torques will be compared to the viscous torque which is given by Eq. (13).

A. Isolated and Overlapping Resonances

Two formulae for the satellite torque obtained from linear perturbation theory are available. They apply to isolated and overlapping resonances and read (Goldreich and Tremaine 1982)

$$T_m^L \simeq -f_1 \, m^2 \left(\frac{M_s}{M_p}\right)^2 \Sigma \, \Omega^2 \, r^4 \,, \tag{14}$$

$$T_s^L \simeq -f_2 \left(\frac{M_s}{M_p}\right)^2 \frac{\Sigma \, \Omega^2 \, r^7 \Delta r}{x^4} \,, \tag{15}$$

where M_s and M_p are the masses of the satellite and the planet. Equation (14) gives the torque at a Lindblad resonance which is located where

$$\Omega/\Omega_s = (r_s / r)^{\frac{3}{2}} = m/(m-1) \,. \tag{16}$$

Here Ω_s, Ω and r_s, r are the orbital angular velocities and orbital radii of the satellite and ring particle, m is a positive integer and f_1 is a numerical coefficient which is equal to 8.46 for $m \gg 1$. Equation (15) gives the total torque on a narrow ringlet of width Δr which is separated from the satellite orbit by a distance $x = r_s - r \ll r$; $f_2 = 2.51$.

To relate the torque formulae given by Eqs. (14) and (15) we note that for $m \gg 1$, the radial separation between neighboring Lindblad resonances is

$$d \simeq \frac{2 \, r}{3 \, m^2} \,. \tag{17}$$

It is straightforward to derive Eq. (15) from Eqs. (14) and (17). As long as there are several resonances within a narrow ringlet Eq. (15) gives the total satellite torque. It is the correct expression for the coarse grained satellite torque.

It is difficult to decide whether a resonance is truly isolated because the width is not a precise concept. Where we can neglect the collective effects due to self-gravity, it is natural to define the width as the radial distance over which the streamlines of the particle flow are significantly distorted. This definition yields $w \simeq (M_s/M_p)^{\frac{1}{2}} r$. Where self-gravity is important a density wave is launched at the resonance and its first wavelength provides a convenient measure for the resonance width. In this case $w = \{(8\pi^2 G \Sigma r)/[3\Omega^2(m-1)]\}^{\frac{1}{2}}$.

The resonances overlap if $d < w$. With either definition of w the A Ring resonances associated with the coorbital satellites and the shepherd satellites all qualify as isolated resonances and the shepherd satellite resonances in the F Ring overlap.

B. Linear and Nonlinear Satellite Torques

In the derivation of the standard formula for the torque at an isolated resonance, it is implicitly assumed that the ring has uniform surface density and that the disturbance is sufficiently weak so that linear perturbation theory is applicable. We examine these assumptions and the consequences of their violation below.

1. Linear and Nonlinear Density Waves. We consider linear waves first, and for the moment we neglect the effects of dissipation. The perturbation due to a satellite at an inner Lindblad resonance launches a trailing spiral density wave which propagates outward from the resonance. The density wave carries away from the resonance all of the negative angular momentum which the satellite torque adds to the disk. The ring particles undergo coherent oscillations in the wave. At the position of the resonance the oscillation frequency is equal to the epicyclic frequency and it falls steadily below the epicyclic frequency with increasing distance from the resonance. The lowering of the collective oscillation frequency below the epicyclic frequency is accomplished by the self-gravity of the disk material. The direction of propagation of the density wave is determined because self-gravity can lower but not raise the collective oscillation frequency. Since the necessary frequency tuning due to self-gravity increases with distance from the Lindblad resonance, the wavelength shortens in proportion to $(r-r_r)^{-1}$. As a consequence of the decreasing wavelength and the conservation of the angular momentum luminosity carried by the wave, the amplitude of the surface density perturbation increases in proportion to $r-r_r$. Thus, in the absence of damping all density waves would eventually become nonlinear.

We are primarily concerned with estimating the satellite torques at resonances and not with the propagation of density waves. We distinguish linear and nonlinear torques by the fractional surface density perturbation in the first wavelength of the density wave.

2. Linear Torques. For linear torques the important comparison is between the values of $|T_m{}^L|$ and T_v. If $|T_m{}^L| < T_v$, the surface density remains uniform in the vicinity of the resonance and the actual satellite torque is equal to $T_m{}^L$. On the other hand, if $|T_m{}^L| > T_v$, a gap is opened in the ring and the actual satellite torque is reduced to $-T_v$.

3. Nonlinear Torques. Nonlinear satellite torques are of great interest. All of the well-observed density waves in Saturn's rings are nonlinear in their first wavelengths. These include waves excited at the strongest resonances of the shepherd satellites, the coorbital satellites and Mimas.

The theory of nonlinear satellite torques has yet to be worked out. Most likely it will require numerical calculations. However, we can make an educated guess for the value of the torque by an appropriate extension of the linear theory. The basic assumption is that the nonlinear torque excites a density wave whose fractional surface density perturbation is of order unity. As such, the surface density perturbation is independent of the mass of the satellite. Since the satellite torque arises from the interaction between the satellite potential and the perturbed surface density, it must be proportional to the first power of the satellite mass. This is in contrast to the linear torque which is proportional to the second power of the satellite mass.

To obtain an explicit expression for the nonlinear satellite torque we use the linear theory to determine both the critical satellite mass and the critical torque for which the density wave is marginally nonlinear in its first wavelength. The nonlinear torque is obtained by multiplying the critical torque by the ratio of the actual satellite mass to the critical satellite mass. Expressions for the critical mass and the critical torque follow directly from results given in Goldreich and Tremaine (1978*b*). They are

$$M_s{}^C \simeq \frac{f_3 \Sigma r^2}{m} \ , \tag{18}$$

$$T_m{}^C \simeq \frac{-\pi^2 \Sigma^3 \Omega^2 r^8}{6 M_p{}^2} \ . \tag{19}$$

The nonlinear torque then reads

$$T_m{}^{NL} \simeq \frac{-f_4 \, m M_s \, \Sigma^2 \Omega^2 r^6}{M_p{}^2} \ . \tag{20}$$

For $m \gg 1$, $f_3 = 0.44$ and $f_4 = 3.72$.

The density wave associated with a nonlinear satellite torque does not carry away from the resonance all of the angular momentum deposited in the disk. The limitation on its amplitude implies that its angular momentum luminosity decays in proportion to $(r-r_r)^{-2}$ and that it is of order T_m^C within the first wavelength of the resonance. The nonlinear wave enhances the viscous stress which is responsible for its damping.

For nonlinear waves the important comparison is between $|T_m^{NL}|$ and T_v. If $|T_m^{NL}| < T_v$, the actual satellite torque is equal to T_m^{NL}. On the other hand, if $|T_m^{NL}| > T_v$, a gap opens in the ring and the actual torque is reduced to $-T_v$.

4. Satellite Torque at a Sharp Edge. Sharp edges such as the outer edges of Saturn's A and B Rings are maintained by satellite torques. To maintain the edge the satellite torque must be equal to $-T_v$ evaluated in the unperturbed region interior to the edge. In the limit $M_s <$ (or $>$) $M_s^C = f_3 \Sigma \ r^2/m$, the maximum value of the satellite torque on a sharp edge is equal to T_m^L (or T_m^{NL}). The proof of this result follows directly from Eqs. (22)–(25) of Borderies et al. (1982).

5. Ring Viscosity. We apply the results of the preceding subsection to estimate the viscosity in Saturn's rings. The cleanest determination is obtained for the outer A Ring. The strongest resonances in this region are those due to the larger coorbital satellite (Lissauer and Cuzzi 1982). A nonlinear density wave but no gap is found at the position of the 6:5 resonance which implies that the viscous luminosity of angular momentum just inside the resonance exceeds the magnitude of the nonlinear satellite torque. The sharp outer edge coincides with the 7:6 resonance which tells us that the viscous luminosity of angular momentum just inside the edge is smaller than the magnitude of the nonlinear satellite torque. From these deductions and Eqs. (13) and (20) for T_v and T_m^{NL}, we conclude that ν is $> 13(\Sigma/100 \text{ g cm}^{-2})$ cm^2 s^{-1} just inside the 6:5 resonance and $< 16(\Sigma/100 \text{ g cm}^{-2})$ cm^2 s^{-1} just inside the 7:6 resonance. To obtain the corresponding estimates for the velocity dispersion and ring thickness, we adopt $\Sigma = 50$ g cm^{-2} and $\tau = 0.5$ for the outer A Ring and apply Eq. (12). We find $v = 0.07$ cm s^{-1} and $h = 5$ m. These values are not very precise because they were derived using the expression for the non-linear satellite torque which is itself an estimate.

C. Shepherding Torques

There has been considerable discussion of the confinement of narrow ringlets by pairs of small satellites (Goldreich and Tremaine 1979). Recent work has focused on the role of dissipation (Greenberg 1983). The standard explanation of the shepherding mechanism is as follows. Torques due to satellites tend to repel ring material. Therefore, narrow rings can be located between pairs of satellites where the net torque vanishes. The satellites exert a net positive torque on the inner half of the ring and a net negative torque on the outer half. These balance the viscous torque across the midline ring.

The explanation in the previous paragraph leads to a paradox which has thus far escaped attention. The satellites transfer energy as well as angular

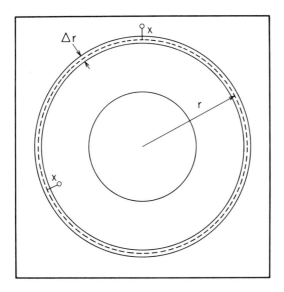

Fig. 6. Illustration of the shepherding geometry.

momentum to the ring. In a steady state the net torque vanishes but the net energy transfer is positive. The energy is dissipated by particle collisions. We now show that the viscous stress associated with the dissipation of energy appears to be so large that the satellites cannot confine the ring. Consider two identical satellites of mass M_s spaced a distance x from a narrow ringlet of width Δr (cf. Fig. 6). We assume that $\Delta r \ll x \ll r$. The ring and the satellite orbits are taken to be coplanar circles. We imagine the ring to be divided into an inner and an outer half. The total torques on the ring from the inner and outer shepherd satellites are denoted by $\mp T_s$. For our purpose we only need to know that the magnitude of T_s decreases with increasing x. It is irrelevant whether the torque is linear or nonlinear. For definiteness we take T_s to be proportional to x^{-q}; $q = 4$ for linear torques and $q = 3$ for nonlinear torques.

The standard treatment of the shepherding mechanism involves equating to zero the net torque on each half of the ring. This procedure yields

$$T_s \simeq \frac{-2x}{q\Delta r} T_v ,$$ (21)

where T_v is evaluated at the midline of the ring.

We extend the standard theory by relating T_s to the energy dissipated in the ring. For simplicity we assume that the ring is in a steady state. The Jacobi constant

$$J = E - \Omega_s H \tag{22}$$

is conserved during an interaction with a satellite whose orbital angular velocity is Ω_s. Here E and H are the total energy and angular momentum of the ring. From this relation we deduce that the satellites transfer energy to the ring at a rate

$$\frac{dE}{dt} \simeq -3 \frac{x}{r} \Omega T_s . \tag{23}$$

In a steady state this energy is dissipated in particle collisions so we have

$$\frac{dE}{dt} \simeq -M_r \frac{v^2}{t_c} \simeq -2\pi \, r\Delta r \, \Sigma v^2 \Omega \tau . \tag{24}$$

In writing Eq. (24) we have set the collision frequency equal to $\Omega\tau$ and assumed that a significant fraction of the relative kinetic energy of colliding particles is dissipated. Next we use Eqs. (12) and (13) to rewrite Eq. (24) as

$$\frac{dE}{dt} \simeq \frac{-4}{3} (1 + \tau^2) \frac{\Delta r}{r} \Omega T_v . \tag{25}$$

Finally, we add the contributions to dE/dt given by Eqs. (23) and (25) which sum to zero to obtain

$$T_s \simeq \frac{-4}{9} (1 + \tau^2) \frac{\Delta r}{x} T_v . \tag{26}$$

Comparison of Eqs. (21) and (26) reveals a contradiction with our beginning assumption that $\Delta r \ll x$. The origin of the problem is clear. The satellites produce torques that tend to confine the ring but they also transfer so much energy to it that confinement is impossible.

Can this paradox be resolved or must the shepherding mechanism be discarded? The answer lies in recognizing the weak link in the chain of arguments which led to the paradox, namely the step relating the rate of energy dissipation to the viscous luminosity of angular momentum. This step would be valid if the satellite perturbations were axisymmetric but they are not. We have shown elsewhere (Borderies et al. 1982, 1983) that in nonaxisymmetric regions, the magnitude and even the sign of the viscous angular momentum luminosity are not simply related to the rate of energy dissipation. We believe that the resolution of the shepherding paradox follows along these lines.

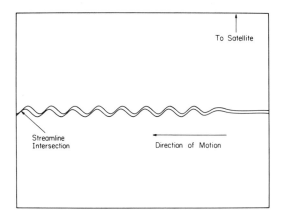

Fig. 7. Collision between neighboring streamlines following perturbation by a shepherding satellite. The horizontal scale is compressed relative to the vertical scale by a factor 75.

Some insight concerning the relation between energy dissipation and angular momentum transport in the shepherding process may be gained from Fig. 7. Shown there are two neighboring streamlines of initially circular test particle orbits perturbed by a shepherd satellite. Since the perturbed streamlines oscillate with slightly different wavelengths ($\lambda = 3\pi x$) they eventually intersect. The intersection occurs at quadrature, i.e., midway between periapse and apoapse. At the point of intersection the orbital angular velocity increases outward. Thus, if we were to assume that all of the energy dissipation and associated angular momentum transport occurred there, we would conclude that the angular momentum flowed inward, rather than outward as it does in unperturbed regions. The foregoing argument illustrates the subtlety of the relation between energy dissipation and angular momentum transport in perturbed regions. A more complete analysis indicates that the balance of the energy dissipation and angular momentum transport occurs well before the hypothetical streamline crossing for $\tau \ll 1$ and close to it for $\tau \gg 1$.

Our assessment of this paradox suggests to us that the accepted description of the shepherding mechanism is seriously flawed. The old picture is correct in so far that the narrow rings are located between pairs of satellites where the net torque vanishes. However, the satellite torques are not responsible for confining the ring material. The confinement is due to the inward transport of angular momentum which accompanies the dissipation of the energy associated with the disturbances created in the ring by the shepherd satellites. This new view removes the problem of understanding the sharp edges of the narrow rings. Also, although less obvious, the minimum sizes predicted for the satellites inferred to shepherd rings are diminished by a factor $\Delta r / x$.

A speculative possibility is that there are regions in Saturn's rings where the optical depth is so high that small imbedded satellites might be unable to clear gaps although they could still produce optical depth variations. Perhaps small satellites may still prove to be the cause of structure in the B Ring.

VI. TIME SCALE FOR SATELLITE ORBIT EVOLUTION

As external satellites extract angular momentum from the rings their orbits expand. Calculations based on the formula for the linear satellite torque predict remarkably short time scales for the recession of close satellites from the rings (Goldreich and Tremaine 1982). Accounting for the effects of non-linearity lengthens the estimated time scales, perhaps by as much as an order of magnitude for 1980S27 and the coorbital satellites, less for 1980S26 and not at all for 1980S28. Thus these short time scales remain perhaps the most intriguing puzzle in planetary ring dynamics.

Since angular momentum may be transferred outward in resonant interactions between satellites (Goldreich 1965; Peale 1976), it is natural to inquire whether the inner satellites in question are involved in any orbital resonances with the more massive outer satellites. If they were, the time scale problem would be alleviated because the angular momentum taken from the rings by the inner satellites would largely go into expanding the orbits of these more massive bodies. We have spent considerable effort fruitlessly searching for appropriate two- and three-body resonances. We identified several close misses and a number of excellent candidates for past or future resonances but failed to find a single active resonance.

The severe nature of the problem is well-illustrated by the system composed of the F Ring and its two shepherd satellites, 1980S26 and 1980S27, hereafter called S26 and S27. Taken by themselves, S26 and S27 could have moved from the outer boundary of the A Ring to their present locations in approximately 2×10^7 yr and 4×10^6 yr, respectively. (These values are based on an assumed $\Sigma = 50$ g cm^{-2}.) The actual situation is more complicated. S27 is trapped between the A Ring and the F Ring and some of the angular momentum it takes from the A Ring is transferred outward to the F Ring. S26 takes angular momentum from the F Ring as well as from the A Ring. Since S27 is located closer to the A Ring and is more massive than S26, and presumably the F Ring as well, in the absence of interactions with additional bodies, the entire system would recede from the A Ring at the rate determined for S27 alone.

The outward movement of the system could be reduced if S26 were involved in a resonance with a more massive outer satellite. When the early results of the orbit determination of S26 showed that its mean motion was close to 3/2 that of Mimas, everything seemed to be falling in place. S26 would transfer angular momentum to Mimas which would then pass most of it along to the more massive Tethys with which it is involved in a resonance. Hope for this solution died when a more accurate orbit determination demonstrated that the suspected resonance between S26 and Mimas was just a close miss.

No resonance has been found linking S26 with one or more outer satellites. What might this imply? Are the F Ring and its shepherd satellites very young? Does the long sought resonance exist awaiting detection? A most interesting possibility is that S26 is transferring angular momentum to Mimas even though the two bodies are not in an exact orbital resonance. This could be accomplished if the motion of S26 were chaotic, i.e., if the value of its mean motion were undergoing a slow random walk. We have proven that if the mean longitude of S26 were subject to a significant random drift, in addition to its dominant secular increase, angular momentum transfer to Mimas would take place by virtue of the near resonance between S26 and Mimas. By significant drift we mean of order one radian on the circulation time scale of the critical argument associated with the near resonance. To check this hypothesis, we must first determine whether the orbital motion of S26 is chaotic. To do so we need to investigate the perturbations of its orbit produced by S27. Solution of this and the other outstanding theoretical problems will await results of ongoing research.

Acknowledgments. N. B. acknowledges support from the National Science Foundation, US-France Exchange Postdoctoral Fellowship and Grant ATP Planétologie. P. G. and S. T. each acknowledge support from both the National Science Foundation and the National Aeronautics and Space Administration.

REFERENCES

Andrews, J. P. 1930. Theory of collision of spheres of soft metal. *Phil. Mag.* 9:593–610.
Borderies, N., Goldreich, P., and Tremaine, S. 1982. Sharp edges of planetary rings. *Nature* 299:209–211.
Borderies, N., Goldreich, P., and Tremaine, S. 1983. Perturbed particle disks. *Icarus* 55:124–132.
Colombo, G., Goldreich, P., and Harris, A. W. 1976. Spiral structure as an explanation for the asymmetric brightness of Saturn's A ring. *Nature* 264:344–345.
Cuzzi, J. N., Durisen, R. H., Burns, J. A., and Hamill, P. 1979. The vertical structure and thickness of Saturn's rings. *Icarus* 38:54–68.
Cuzzi, J. N., Lissauer, J. J., and Shu, F. H. 1981. Density waves in Saturn's rings. *Nature* 292:703–707.
Froidevaux, L., and Ingersoll, A. P. 1980. Temperatures and optical depths of Saturn's rings and a brightness temperature from Titan. *J. Geophys. Res.* 85:5929–5936.
Froidevaux, L., Matthews, K., and Neugebauer, G. 1981. Thermal response of Saturn's ring particles during and after eclipse. *Icarus* 46:18–26.

Goldreich, P. 1965. An explanation of the frequent occurrence of commensurable mean motions in the solar system. *Mon. Not. Roy. Astron. Soc.* 130:159–181.

Goldreich, P., and Tremaine, S. 1978a. The velocity dispersion in Saturn's rings. *Icarus* 34:227–239.

Goldreich, P., and Tremaine, S. 1978b. The formation of the Cassini Division in Saturn's rings. *Icarus* 34:240–253.

Goldreich, P., and Tremaine, S. 1979. Towards a theory for the Uranian rings. *Nature* 277:97–99.

Goldreich, P., and Tremaine, S. 1982. The dynamics of planetary rings. *Ann. Rev. Astron. Astrophys.* 20:249–283.

Greenberg, R. 1983. The role of dissipation in shepherding of ring particles. *Icarus* 53: 207–218.

Hertz, H. 1881. Uber die beruhrung fester elastischer kozper. *J. Reine Angew. Math. (Crelle)* 92:156–171.

Hobbs, P. V. 1974. *Ice Physics* (London: Oxford Univ. Press, Clarendon).

Lane, A. L., Hord, C. W., West, R. A., Esposito, L. W., Coffeen, D. L., Sato, M., Simmons, K. E., Pomphrey, R. B., and Morris, R. B. 1982. Photopolarimetry from Voyager 2: Preliminary results on Saturn, Titan and the rings. *Science* 215:537–543.

Lin, D. N. C., and Bodenheimer, P. 1981. On the stability of Saturn's rings. *Astrophys. J. Lett.* 248:L83–L86.

Lissauer, J. J. and Cuzzi, J. N. 1982. Resonances in Saturn's rings. *Astron. J.* 87:1051–1058.

Lissauer, J. J., Shu, F. H., and Cuzzi, J. N. 1982. Viscosity in Saturn rings. Proceedings of *I.A.U. Colloquium 75. Planetary Rings,* ed. A. Brahic, Toulouse, France, Aug. 1982.

Lukkari, J. 1981. Collisional amplification of density fluctuations in Saturn's rings. *Nature* 292:433–435.

Lynden-Bell, D., and Pringle, J. 1974. The evolution of viscous disks and the origin of the nebular variables. *Mon. Not. Roy. Astron. Soc.* 168:603–637.

Marouf, E. A., Tyler, G. L., and Eshleman, V. R. 1982. Theory of radio occultation by Saturn's rings. *Icarus* 49:161–193.

Marouf, E. A., Tyler, G. L., Zebker, H. A., Simpson, R. A., and Eshleman, V. R. 1983. Particle-size distributions in Saturn's rings from Voyager 1 radio-occultation. *Icarus* 54: 189–211.

Parameswaran, V. R., and Jones, S. J. 1975. Brittle fracture of ice at 77 K. *J. Glaciology* 14:305–315.

Peale, S. J. 1976. Orbital resonances in the solar system. *Ann. Rev. Astron. Astrophys.* 14:215–246.

Shu, F. H., Cuzzi, J. N., and Lissauer, J. J. 1983. Bending waves in Saturn's rings. *Icarus* 53:185–206.

Smith, B. A., Soderblom, L., Batson, R., Bridges, P., Inge, J., Masursky, H., Shoemaker, E., Beebe, R., Boyce, J., Briggs, G., Bunker, A., Collins, S. A., Hansen, C. J., Johnson, T. V., Mitchell, J. L., Terrile, R. J., Cook II, A. F., Cuzzi, J., Pollack, J. B., Danielson, G. E., Ingersoll, A. P., Davies, M. E., Hunt, G. E., Morrison, D., Owen, T., Sagan, C., Veverka, J., Strom, R., and Suomi, V. E. 1982. A new look at the Saturn system: The Voyager 2 images. *Science* 215:504–537.

Ward, W. R. 1981. On the radial structure of Saturn's rings. *Geophys. Res. Letters* 8:641–643.

Appendix

ATLAS OF SATURN'S RINGS

STEWART A. COLLINS, JUDITHANNE DINER,
GLENN W. GARNEAU, ARTHUR L. LANE,
ELLIS D. MINER, S. P. SYNNOTT, RICHARD J. TERRILE
Jet Propulsion Laboratory

JAY B. HOLBERG, BRADFORD A. SMITH
University of Arizona

and

G. LEONARD TYLER
Stanford University

RADIAL PROFILES OF
SATURN RING OPTICAL PROPERTIES

The Voyager spacecraft returned an unprecedented quantity and variety of measurements of the properties of Saturn's rings. An effort is underway to compile and correlate many of these data for publication in an atlas of Voyager Saturn ring data. A preliminary and greatly-abridged version of this compilation is presented here. Specifically, radial profiles of ring opacity (defined here as normal optical thickness) are presented at a resolution of approximately 50 km for wavelengths of 2.6×10^{-5} cm, 3.6 cm and 13 cm. Ring reflectance (at $\lambda = 5 \times 10^{-5}$ cm) is also presented for two geometries of ring viewing (illuminated and unilluminated faces).

Description of Data Sets

The Voyager flybys yielded a wide variety of observations of the rings of Saturn, among which are the following:

1. Radial profiles of optical thickness at 3.6 cm and 13.6 cm using the spacecraft radio signal as the source and Earth-based telemetry antennas as receivers.
2. Radial profiles of optical thickness at about 1300 Å and 2640 Å, using starlight (δ Sco) as the source and the Voyager ultraviolet and photopolarimeter instruments as receivers.
3. Radial profiles of ring albedo for several wavelengths (3000 − 6000 Å) and combinations of lighting and viewing geometry, including both the illuminated and unilluminated surfaces, acquired by the imaging investigation.
4. Azimuthal and time-resolved profiles of ring albedo, derived from imaging data.
5. A record of probable impacts of G Ring particles on the Voyager 2 spacecraft detected by the plasma wave and planetary radio astronomy instruments.
6. Profiles of the number density of magnetospheric charged particles observed by the plasma, low energy charged particle, and cosmic ray instruments.

Figure 1 includes data from observations 1–3 of those listed above. This appendix presents data on microwave opacity, ultraviolet opacity and visible light reflectance.

Microwave Opacity. On 13 November 1980, Voyager 1 flew behind Saturn's rings (as observed from Earth; see Fig. 2) causing the radio signal to be attenuated by the rings (Eshleman et al. 1977; Tyler et al. 1981; and Tyer et al. 1983). The spacecraft-to-ring distance was approximately 250,000 km during this period so that the radio beam illuminated a large area of the rings at any one time. The angle between the optical path and the ring plane was 5°.9. The received signal consisted of two components: (1) the transmitted beam, and (2) forward-scattered radiation from regions of the rings adjacent to the spacecraft-Earth vector. Forward-scattered radiation is shifted in phase and frequency from the transmitted signal and it is possible to determine the region of the rings through which a signal is scattered from these phase and Doppler shifts.

While the results of processing the scattered signal have been reported for selected regions of the rings (see e.g., Marouf et al. 1983), the data in Fig. 1 are derived only from the transmitted beam. The radial resolution and uncertainty of these data vary as a function of opacity with best resolution and lowest uncertainty occurring in regions of minimum opacity. An indication of the uncertainty of the signal measurement is provided in Fig. 3. Wavelength dependent differences in opacity are evident in several regions (e.g., the C Ring). Within such regions, at least some of the ring particles are of a size comparable to the wavelength of the incident radiation.

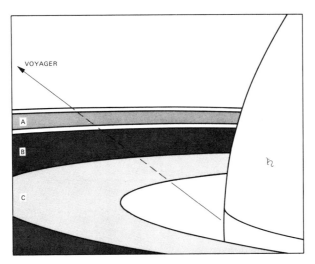

Fig. 2. Voyager 1 radio occultation as viewed from Earth. Spacecraft motion carried Voyager from behind the planet into the eastern ansa and then, successively, behind Rings C, B, A and F. Ring F is indicated by the single outer line in the figure; the Cassini Division is between Rings A and B (figure from Tyler et al. 1983).

Ultraviolet Opacity. On 26 August 1981, the Voyager 2 ultraviolet spectrometer (UVS: Broadfoot et al. 1977) and photopolarimeter (PPS: Lillie et al. 1977) recorded the brightness of the star δ Sco at 1300 Å and 2640 Å, respectively, as it was occulted by the portion of Saturn's rings in Saturn's shadow (Fig. 4). The time resolution (0.32 and 0.01s) of these observations yields a radial resolution of approximately 3.2 and 0.1 km for the UVS and PPS, respectively (Sandel et al. 1982; Lane et al. 1982). The angle between the optical path and the ring plane was 28°7. For Fig. 1, the PPS data, converted to normal optical depth, has been smoothed to a resolution of 50 km.

Visible Light Reflectance. The Voyager trajectories provided the opportunity to photograph Saturn's rings from a variety of perspectives (see Smith et al. 1977). Among these observations were two which provided relatively high spatial resolution (approximately 5 km per sample) throughout all, or nearly all of the ring's radial extent (Fig. 5). Each such observation consisted of a radial scan of approximately 20 narrow-angle images. Voyager 1 returned one of these scans on 12 November 1980 under the following conditions:

Illumination: north surface at an angle of 4°7 above the ring plane;
Viewing: south surface at an angle of approximately 12° below the ring plane;
Phase angle: 45°–48°.

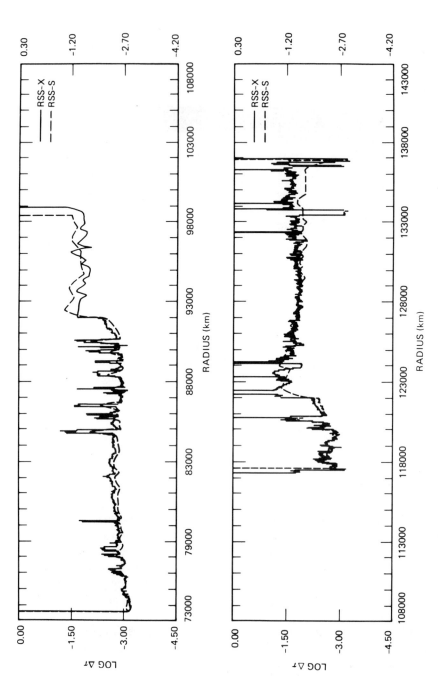

Fig. 3. Uncertainty (1σ) in the microwave opacity measurements plotted in Fig. 1.

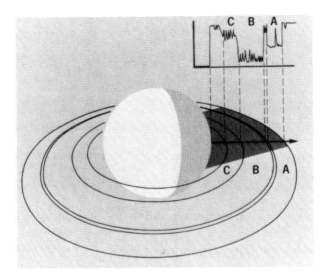

Fig. 4. A diagram showing the apparent path (arrow) of δ Sco as it was occulted by the shadowed portion of Saturn's rings and observed by Voyager 2's PPS and UVS instruments.

Voyager 2 acquired a comparable set of images on 25 August 1981 under significantly different conditions:

Illumination: north surface at an angle of 10° above the ring plane;
Viewing: north surface at an angle of approximately 28° above the ring plane;
Phase angle: 45°−48°.

As reported elsewhere (Simpson et al. 1983) a vector for Saturn's polar axis has been identified which yields excellent registration of the radio and ultraviolet data sets to an accuracy of significantly better than 10 km (1σ). The outer half of the Voyager 2 imaging scan has been geometrically processed using as a benchmark a star that is faintly visible through the A Ring. Other portions of the imaging data have been processed to provide frame-to-frame consistency, and an absolute scale has been adopted which provides accurate registration with the radio and ultraviolet scans for features which are known to be circular.

Acknowledgments. A great many individuals in addition to the authors have contributed to the preparation and analysis of the data sets presented here. Of unique importance to this work are those who have promoted and fostered the consolidation of these previously separate data sets: G. E. Danielson, D. J. Diner, P. Goldreich and E. C. Stone. We appreciate also the

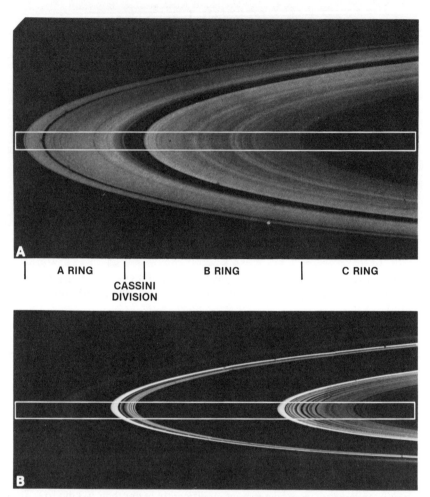

Fig. 5. Voyager 1 photographs showing the rings north (A) and south (B) surfaces. The rectangles indicate the approximate location of the high-resolution imaging mosaics from which were derived the data plotted in Fig. 1. Each imaging mosaic consisted of a linear alignment of approximately 20 narrow-angle images.

generosity of Jet Propulsion Laboratory's Atmospheric Radiation and ATMOS groups, for the use of their data processing equipment. Finally, one of us (JD) has been supported by a grant from the Alfred P. Sloan Foundation during most of this program. This work presents the results of one phase of research carried out at the Jet Propulsion Laboratory under NASA contract.

REFERENCES

Broadfoot, A. L., and the Voyager Ultraviolet Spectrometer Team. 1977. Untraviolet Spectrometer Experiment for the Voyager Mission. *Space Sci. Rev.* 21:183–205.

Eshleman, V. R., and the Voyager Radio Science Team. 1977. Radio Science Investigations with Voyager. *Space Sci. Rev.* 21:207–323.

Lane, A. L., and the Voyager Photopolarimeter Team. 1982. Photopolarimetry from Voyager 2: Preliminary Results on Saturn, Titan, and the Rings. *Science* 215:537–543.

Lillie, Charles F., and the Voyager Photopolarimeter Team. 1977. The Voyager Mission Photopolarimeter Experiment. *Space Sci. Rev.* 21:159–181.

Marouf, E. A., Tyler, G. L., Zebker, H. A., Simpson, R. A., and Eshleman, V. R. 1983. Particle size distributions in Saturn's rings from Voyager I radio occultation. *Icarus* 54:189–211.

Sandel, B. R. and the Voyager Ultraviolet Spectrometer Team. 1982. Extreme ultraviolet observations from the Voyager 2 encounter with Saturn. *Science* 215:548–553.

Simpson, R. A., Tyler, G. L., and Holberg, J. B. 1983. Saturn's pole vector: Geometric corrections based on Voyager UVS and radio occultations. *Astron J.* 88:1531–1536.

Smith, B. A., and the Voyager Imaging Science Team. 1977. Voyager imaging experiment. *Space Sci. Rev.* 21:103–127.

Tyler, G. L., and the Voyager Radio Science Team. 1981. Radio science investigations of the Saturn system with Voyager 1: Preliminary results. *Science* 212:201–206.

Tyler, G. L., Marouf, E. A., Simpson, R. A., Zebker, H. A., and Eshleman, V. R. 1983. The microwave opacity of Saturn's rings at wavelengths of 3.6 and 13 cm from Voyager 1 radio occultation. *Icarus* 54:160–188.

Color Section

LIST OF COLOR ILLUSTRATIONS

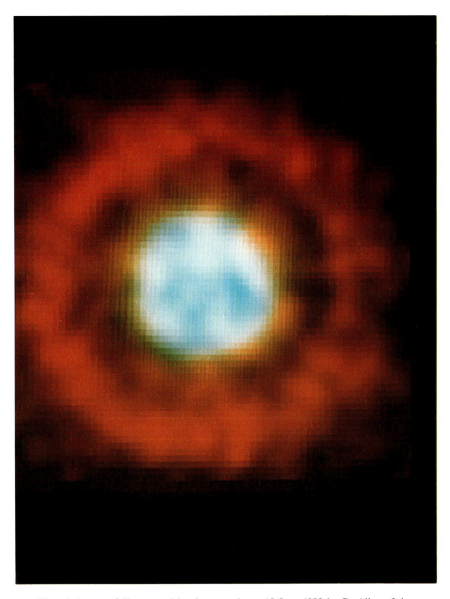

Plate 1. Image of Uranus and its rings, made on 15 June 1983 by D. Allen of the Anglo-Australian Observatory. The brightest part of the rings lies to the southwest. The pixel size is 0.15 arcsec, interpolated from an original 0.30 arcsec. The map was made through a circular aperture of 0.7 arcsec. The red image represents K wavelength (2.2 μm) and the cyan image represents J (1.2 μm) (see chapter by Elliot and Nicholson). Photograph courtesy of the Anglo-Australian Telescope Board.

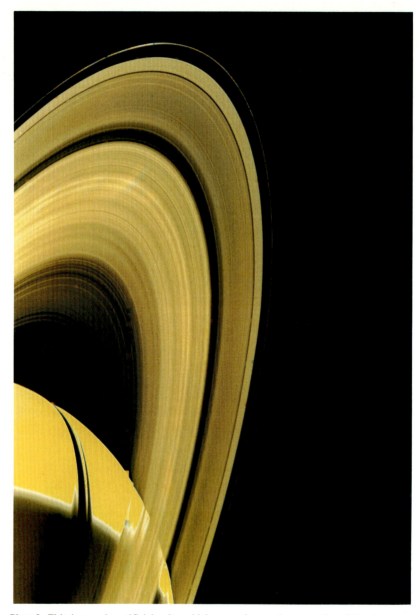

Plate 2. This image, in artificial color which approximates the true color of the rings, was taken by Voyager I after Saturn encounter from about 1.5×10^6 km. The distinctly different appearance of the bright B Ring from the other rings is apparently due to small quantities of dust particles in its many irregularly-spaced features. The innermost C Ring is darker due both to darker particles and fewer of them; its characteristic banded appearance is quite different from the A and B Rings, but similar to the structure in the Cassini Division, the dark area between the bright A and B Rings. The C Ring brightness increases with angle away from the night side of Saturn due to increased illumination from Saturn's lit face. A diffuse, faintly bright spoke is seen crossing the brightest part of the B Ring; such features are probably caused by electrical storms on the planet. The F Ring is faintly visible as a thin strand lying outside of the outer edge of the A Ring. In the outer third of the A Ring lies the Encke Gap, in which several unseen moonlets are probably embedded.

Plate 3. False color view of Saturn's rings. This highly enhanced view was assembled from clear, orange, and ultraviolet frames obtained by Voyager 2 from a range of 8.9×10^6 km (Smith et al. 1982). It shows the relatively blue color of the C Ring and Cassini Division, as well as color differences between the inner and outer B Ring and between these and the A Ring (see chapter by Cuzzi et al.).

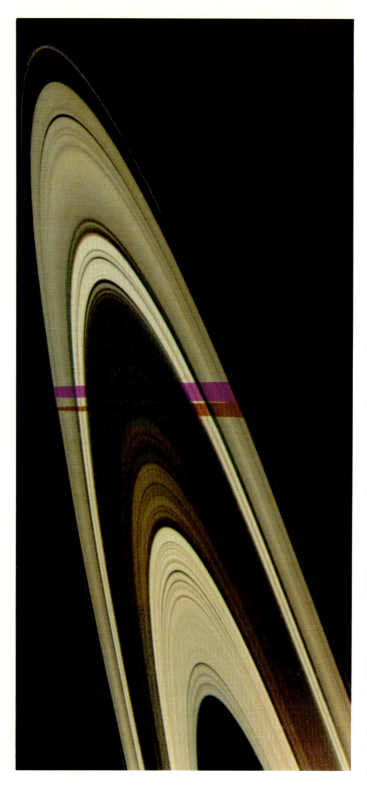

Plate 4. Voyager 1 unlit face mosaic, which illustrates the similarity in structure of the C Ring and Cassini Division, and their relationships with neighboring regions outwards. The optically thick B and inner A Rings are dark because they block most of the light (incident on their opposite surface). The optically thin C Ring and Cassini region are brighter because they contain enough material to scatter a significant amount of light downwards, but not enough material to prevent the scattered light from escaping the ring layer (see chapter by Cuzzi et al. and their appendix).

Plate 6. Saturn's outer C (blue) and inner B (yellow) Rings in false color, as in Color Plate 5 (Smith et al. 1982). The color difference is seen to blur over the optical depth boundary between the rings (outer edge of dark yellow band); see also Fig. 6 in the chapter by Cuzzi et al. The Maxwell Gap and its eccentric ringlet, which appears yellow due to image mis-registration, are in the left center of this frame.

Plate 5. Saturn's A Ring in false color, constructed from Voyager 2 green, violet, and ultraviolet frames (Smith et al. 1982). The Cassini Division (lower right) is distinctly less red than the outer A Ring (cf. Fig. 6 in Cuzzi et al.'s chapter). Also visible, as a sequence of dark features in the outermost A Ring, is a series of orbital resonances with Saturn's near-in ringmoons (cf. Cuzzi et al.'s Figs. 13 and 14). The 325 km wide Encke Division (upper left center) probably contains several embedded moonlets (see Cuzzi et al.'s Fig. 21). However, the cause of the narrower Keeler Gap (closer to the outer edge of the A Ring) and the gaps in the Cassini Division is not known (see chapter by Cuzzi et al., Sec. IV.C).

Plate 7. A computer-simulated view of particles ranging in size from marbles to beach balls populating a 3-meter-square section of Saturn's A Ring. This picture was constructed to illustrate the size distribution of ring particles based on Voyager data, rather than to provide a realistic image of the physical structure of the particles. Nevertheless, its representation of ring particles as perfectly smooth (presumably hard) spheres is typical of the properties often assumed (implicitly or explicitly) in mathematical studies of rings. This idealized image contrasts with the model of particle properties illustrated in Color Plate 8 (see chapter by Weidenschilling et al.). Picture from JPL/NASA.

Plate 8. Dynamic ephemeral bodies (DEB's) in Saturn's rings viewed from just above the plane. House-sized bodies (the larger bodies in the picture) grow in a matter of days by accretion of much smaller particles, shown as a haze among the larger bodies. The large bodies continually break up due to tidal forces, as in the case of the S-shaped disaggregating swarm in the right foreground. Large bodies are irregularly shaped and lie roughly in a monolayer; smaller particles lie in a many-particle-thick layer. This painting represents a thinly populated locale, where optical thickness $\tau < 1$. Zones with $\tau > 1$ would look similar except that the haze of small particles would hide more completely the background. Painting by W. K. Hartmann, based on the model of particle properties and dynamics discussed in the chapter by Weidenschilling et al.

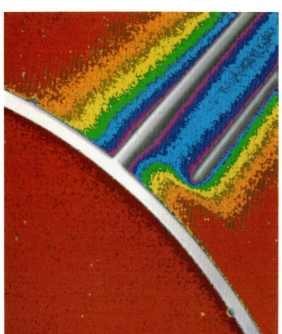

Plate 10. Color isophotes of the planetary shadow on Jupiter's rings (Voyager image FDS 20691.35), indicating that the faint disk is absent and that the halo is toroidal, concentrated just inside the main band (cf. Figs. 6a and 6b in chapter by Burns et al.).

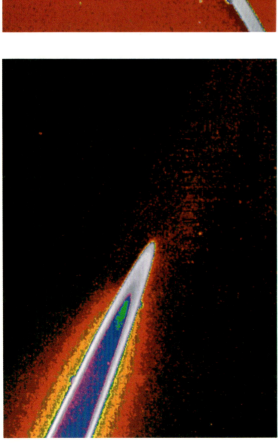

Plate 9. Color isophotes of the ansa region of Jupiter's rings (Voyager image FDS 20693.02) showing that the halo lies inside the main band. The color levels are equally separated in intensity (DN) values. Note the reduced value near the inner ansa and the generally expanding nature of the halo (see chapter by Burns et al.).

Glossary

GLOSSARY

albedo
reflectivity of a body. Bond albedo: ratio of total flux reflected in all directions from a planetary body to total incident flux. Geometric albedo: ratio of observed planetary brightness at a given phase angle to the brightness of a perfectly diffusing disk with the same position and angular size.

angular momentum luminosity
rate at which angular momentum flows outward across a circle of given radius in a planetary ring (see chapter by Borderies et al.).

ansa (anses or ansae, plural)
portion of rings that appears farthest from the disk of a planet. In early Saturn drawings, the "handles" of the ring were so called because they resembled the handles of a vase.

apsidal precession
precession of the points (apsides) at each end of the major axis of an elliptical orbit.

apsidal resonance
resonant orbital perturbations that occur when the free apsidal precession period of an orbiting particle closely matches the sidereal period of a perturbing satellite. Particles in and near the Saturn ringlet at 1.29 R_S have an apsidal resonance with Titan. Apsidal resonance is a type of *eccentricity resonance*, although in some contexts it may be used synonomously with eccentricity resonance. See *resonance*.

bending waves
wave motion in which particles oscillate normal to the ring plane. The wave propagates due to collective gravitational effect of particles on neighboring regions (see chapter by Shu). The wave can be driven by *inclination resonance*.

Boltzmann constant
the constant in the ideal gas law, $k = 1.38 \times 10^{-23}$ Joule/K. It also appears in the related expression for the Maxwell-Boltzmann distribution of momenta in an equilibrium-fluid swarm of particles, derived from the Boltzmann equation (see below).

Boltzmann equation	a fundamental equation of kinetic theory which describes the distribution of velocities and positions of an ensemble of particles. It may be used to obtain the equations of continuity and mass conservation of fluid flow.
coefficient of restitution	a parameter denoting the ratio of rebound to impact velocities for collisions of solid bodies.
coorbital satellites	two of Saturn's satellites, Janus (1980S1) and Epimetheus (1980S3), which have the same mean distance from the planet and which execute *horseshoe orbits* with respect to one another.
corotation resonance	an eccentricity resonance that depends on the eccentricity of the perturbing satellite, rather than on the eccentric motion of the perturbed particle. Contrast this with *Lindblad resonance*. See also *eccentricity resonance*.
cosmogony	study of the origin of the cosmos. It now generally refers to study of the origin and formation of the solar system.
cosmology	study of the origin and formation of the universe— literally, study of the cosmos.
Coulomb collisions	collisions between charged particles, mediated by the electrostatic force between them.
DEB's	dynamic ephemeral bodies, ring particles hypothesized to consist of agglomerations of small particles, with very short lifetimes between their rapid accretion and tidal disruption (see chapter by Weidenschilling et al.).
Debye length	a characteristic distance in a plasma beyond which the electric field of a charged particle is shielded by particles with charges of the opposite sign.
Delaunay variables	a set of orbital elements, which can be expressed in terms of Keplerian elements, and for which the equations of variation due to perturbations take the form of canonical equations of classical mechanics.
density waves	wave motion in a flat disk of particles, often driven by resonant perturbations by a distant satellite. The wave propagates due to collective gravitational effect of particles (i.e., local surface density variations) on particles in neighboring regions. The theory was originally developed to explain spiral structure of galaxies (see chapter by Shu). The wave can be driven by *eccentricity resonance*.

despinning — slowing of a body's rotation rate, e.g., due to tidal dissipation or impacts by many small particles.

dispersion velocity — see *random velocity*.

disturbing function — gravitational potential experienced by a satellite (or ring particle) due to the presence of another satellite or perturbing mass.

eccentricity resonance — oscillatory resonant response, relative to circular motion, of a satellite or ring particle to a periodic gravitational perturbing force in the plane of motion due to another satellite. It is sometimes called *horizontal resonance* or *apsidal resonance*. Eccentricity resonances can drive *density waves* in planetary rings.

elastic modulus — a parameter describing a material's elasticity: the ratio of stress to strain.

epicyclic motion — planar motion of an orbiting body in a noncircular orbit relative to a circular reference orbit.

Epstein drag — a law describing the drag of a medium (e.g., a gas) on an object moving through it. The Epstein drag law applies when the object is smaller than the mean free path of the constituent particles (or molecules) of the medium.

escape velocity — the speed an object must attain to escape from a gravitational field.

Fokker-Planck approximation — a means of describing evolution of an ensemble of particles subject to random forces, by assuming that the random effects are frequent, compared with the rate of evolution of the system.

Fresnel diffraction — the wave interference effects obtained when a light source or an observing plane is at a finite distance from a diffracting aperture.

Fresnel scale (or zone) — the separation distance (or area) between intensity maxima on the observing plane in *Fresnel diffraction*. This distance limits the resolution of occultation measurements of sharp edges, unless the diffraction can be numerically removed.

FWHM — full width at half maximum, the width of a spectral line, of a ring feature, or of any other feature on a quantitative trace at half-maximum intensity.

Gauss's law	a law that states for any closed surface, the total integrated normal gravitational (or electrostatic) force is proportional to the total mass (or charge) within the surface.
glory	phenomenon of wavelength-dependent interference that (as the term is used traditionally) creates circles of color around an observer's shadow in cloud or fog. In the general sense, it refers to the strong enhancement of reflectivity by nearly spherical particles close to the direction of exact backscatter.
gravitational instability	condition in which slight rearrangements or concentrations of a relatively uniform distribution of mass can, by their gravitational effect, initiate substantial further contraction of mass into even more localized concentrations.
Green's function	type of function used in reducing boundary value problems with partial differential equations into integral form.
Helmholtz time scale	time scale for Kelvin-Helmholtz contraction, the contraction of a star (or a planet like Jupiter) as a consequence of losing gravitational potential energy by radiating thermal energy, in the absence of thermonuclear or other nongravitational energy sources.
Hilda asteroids	the group of asteroids with orbits resonant with Jupiter's near the 3:2 commensurability of orbital periods.
horizontal resonance	see *eccentricity resonance* and *resonance*.
horseshoe orbit	motion of an orbiting particle that alternately nearly overtakes another orbiting body and then slows down so as to be nearly overtaken by the other body. In a reference frame rotating with the orbit of the other body, the particle follows a horseshoe shaped path. See also *Jacobi constant* and *coorbital satellites*.
impact strength	the critical energy of an impact, per unit volume of the target, above which the target is fragmented catastrophically.
inclination resonance	oscillatory resonant response of a satellite or ring particle to periodic gravitational perturbations (due to another satellite) out of the plane of reference. Inclination resonances can drive *bending waves* in planetary rings. Also called *vertical resonance*. See *resonance*.

Jacobi constant, or integral	the only known integral of motion in the restricted three-body problem, where a massless particle moves under the influence of two point masses in circular orbit around one another. This constant is a pseudo-energy in a rotating coordinate system in which centrifugal force is represented by a potential field.
Jeans instability	gravitational instability in an idealized, infinite, homogeneous medium. See *gravitational instability*.
Keplerian disk	disk of particles in Keplerian motion. See *Keplerian motion*.
Keplerian motion	motion of freely orbiting bodies in an inverse-square-law force field.
Keplerian shear	shearing motion of an ensemble of particles, each on a nearly circular, Keplerian orbit. Orbital velocity decreases with orbital radius, yielding the shear. Viscous drag on such shear, due to collisions, plays a key role in ring processes.
Kirkwood gaps	gaps in the semimajor-axis distribution of asteroids, located at positions of orbital periods commensurable with Jupiter's period.
Kronecker delta	mathematical symbol δ_{ij}, which equals 1 when indices i and j are equal, or equals zero otherwise.
Lagrange points	equilibrium points in the restricted three-body problem. See *Jacobi constant*. The two stable Lagrange points each form an equilateral triangle with the two massive bodies.
Laplace coefficients	coefficients appearing in the expanded form of the disturbing function, the gravitational potential due to one orbiting perturbing body on another orbiting perturbed body. The coefficients are functions of the ratio of semimajor axes of the two orbiting bodies.
libration	oscillation of an angular quantity about a fixed value. It often describes a stable orbital or rotational configuration.
Lindblad resonance	eccentricity-type resonance of first order in the eccentricity of the perturbed particle. See also *eccentricity resonance* and *resonance*. This terminology comes from galactic dynamics; in the 1920s Lindblad invoked such resonances to explain spiral arms.

Lorentz force	force on a charged particle due to the presence of an electric field and motion across a magnetic field. The latter component is the cross product of the particle's velocity with the magnetic flux density.
Lyman-α	a line in the far-ultraviolet part of the spectrum at 1215.67Å, strongly absorbed and easily emitted by hydrogen atoms and associated with the ground state. It is a part of the Lyman spectral series.
magnetosphere	region surrounding a planet, in which charged particles are controlled by the magnetic field of the planet, rather than by the Sun's magnetic field, which is carried by the solar wind.
mass extinction coefficient	optical thickness divided by surface mass density. In cgs units, mass extinction coefficient is cm^2/g. It is sometimes called specific opacity.
Maxwellian plasma or gas	one whose molecules have a Maxwellian velocity distribution.
Mie scattering	scattering of light by particles with size comparable to the wavelength. Exact albedo and phase functions are highly oscillatory with direction and wavelength; however, slight irregularities and distributions of size smooth out these features, leaving predominantly forward-scattering behavior (see appendix to chapter by Cuzzi et al.).
moom	small satellite in or near a ring that may be the source of ring particles. The word is a combination of moon and mom (see chapter by Burns et al.).
Newcomb operator	notation used in the expansion of the planetary disturbing function. See also *Laplace coefficients*.
nodal precession	precession of the points of intersection between the orbital plane of a satellite and the reference plane.
occultation	the phenomenon of one celestial body apparently passing in front of another. In the case of the Voyager radio occultation, the radio transmitter passed behind Saturn's rings.
opacity	a loosely defined term referring to the ability of a medium to extinguish radiation of any given wavelength. In various applications opacity has been used to mean (a) optical thickness divided by physical

thickness, (b) *optical* or *radio thickness,* or (c) *mass extinction coefficient.* in the latter case it is usually called specific opacity.

opposition effect
abrupt increase in brightness of a body or ring near zero phase angle.

optical depth
the intensity of light passing through a medium consisting of a field of particles decreases exponentially with distance through the medium. The depth of penetration into the medium (e.g., into a planetary atmosphere or ring) can be expressed in terms of the number of factors of *e* of diminishment. That number is the *optical depth.* The number of factors of *e* diminishment through the entire medium (e.g., from one side to the other of a planetary ring) is called the *optical thickness.* Strictly speaking, optical thickness depends on the direction of propagation relative to the normal to the ring surface. In practice the term optical thickness is often used synonymously with *normal optical thickness.* In ring studies, the term *optical depth* is often used to mean optical thickness. These terms generally refer to a particular wavelength.

optical thickness
see *optical depth.*

phase angle
Earth-object-Sun angle.

Poisson ratio
ratio of transverse contraction strain to longitudinal elongation strain for a material under tensile stress.

Poynting-Robertson or (PR) drag
a loss of orbital angular momentum by orbiting particles associated with their absorption and reemission of solar radiation (see Mignard's chapter).

radio thickness or radio depth
equivalent of optical thickness, measured at radio wavelength. See *optical depth.*

random velocity
a ring particle's velocity at any instant can be characterized as the sum of (a) the velocity it would have in a circular (usually Keplerian) orbit at its distance from the center of force (planet), and (b) a random component (related to its orbital eccentricity and inclination in the Keplerian case). A population of ring particles is generally assumed to have a characteristic random velocity value, sometimes called *velocity dispersion, dispersion velocity, sound speed,* or thermal velocity (the last two by analogy with gas dynamics).

random relative velocity — the relative velocities at which ring particles approach one another prior to gravitational or collisional interactions as governed by *random velocity* (see definition), rather than by systematic *Keplerian shear* (see definition). The mean random relative velocity is roughly equal to the mean random velocity.

reflectivity — ratio of reflected intensity of light to incident (usually solar) flux, generally given as a *geometric albedo*.

regolith — surface layer of loose, fragmental debris produced by impacts on the surface of a body.

relative velocity — velocity between ring particles. It is often used to denote *random relative velocity* (see above).

resonance — selective response of any periodic system to an external stimulus of the same frequency as the natural frequency of the system. Ring particles can have a resonant response to the periodic driving force of a satellite's gravity.

Reynolds number — a dimensionless parameter that governs the conditions for the occurrence of turbulence in fluids.

ringmoon — a small satellite in or near planetary rings.

rms — root mean square, the square root of the mean square value of a set of numbers.

Roche limit — the minimum distance at which a zero-strength satellite, influenced by its own gravitation and by that of a primary mass about which it describes a Keplerian orbit, can exist. Such a satellite, with the same mean density as its primary but much smaller in size, will break up at 2.44 times the radius of the primary (see Weidenschilling et al.'s chapter).

Roche lobe — the potential well around each of two massive bodies in circular orbit around one another.

Roche zone — the space inside the Roche limit.

Safronov number — a parameter Θ relating the random velocity in a swarm of particles to the escape velocity of a characteristic size particle (usually the largest in the swarm): $\Theta = $ (escape velocity/random velocity)^2/2.

scale height	height over which atmospheric (or ring) density changes by a factor e. It is often used to connote physical thickness of a ring.
shepherd satellite	a satellite in orbit near the edge of a planetary ring, that exerts a gravitational torque and thus maintains the integrity of the planetary ring by preventing it from spreading.
shielding length in a plasma.	see *Debye length*.
sound speed	in a gas, the speed of sound approximately equal to the rms speed of the molecules. By analogy, for planetary rings, sound speed is sometimes used to denote random velocity of particles. See *random velocity*.
sphere of influence	the region around a body (orbiting a primary) in which the motion of a test particle is dominated by the gravity of the orbiting body, not by the primary.
sputtering	expulsion of atoms from a solid, caused by impact of energetic particles.
Stokes drag	viscous drag law stating that the force which retards a body moving through a fluid is directly proportional to the velocity, the radius of the body, and the viscosity of the fluid. Stokes drag applies for low *Reynolds number* and if the mean free path of fluid molecules is small compared with the body's size. See *Epstein drag*.
surface of section	technique of identifying integrals of motion of a dynamical system by examining the locus of the intersection of trajectories with a two-dimensional surface in phase space.
Taylor series	the expansion of a mathematical function $f(x)$ about a particular value of $x=c$, as a power series in $(x-c)$. The first-order term gives the linear approximation to the function.
Tisserand's relation	the *Jacobi constant* expressed in terms of Keplerian orbital elements. It was originally used as a criterion for comet identification: although individual orbital elements can change under the influence of Jovian perturbation, the Jacobi constant may remain intact.
typical particle size	a loosely defined term which should be carefully defined for any given application.

van der Waals attraction	the relatively weak attraction forces operative between neutral atoms and molecules.
velocity dispersion	see *random velocity*.
vertical resonance	see *inclination resonance* and *resonance*.
viscosity	the resistance that a fluid system offers to flow when it is subjected to a shear stress. It is a measure of the internal friction that arises when there are velocity gradients within the system. See *Keplerian shear*.
WKB (or WKBJ) approximation	the Wentzel-Kramers-Brillouin(-Jeffreys) functional form used as an approximate solution to the Schrodinger wave equation, but also used in density wave theory for galaxies and planetary rings (see chapter by Shu). Jeffrey's work with this mathematical technique preceded that of W, K, and B. Now that wave mechanics are known to be crucial to the behavior of rings, this contribution joins Jeffreys' more direct studies of ring dynamics as a major early contribution to the subject of this book.
yield strength	the stress at which a material exhibits deviation from proportionality of strain to stress.

Acknowledgments

ACKNOWLEDGMENTS

The following people helped to make this book possible, in organizing, writing, refereeing, or otherwise. Members of the Advisory Committee for this book, who also served on the Scientific Committee of the Planetary Rings Colloquium, are indicated with an asterisk (*).

D. A. Allen, Anglo-Australian Observatory, Epping, N.S.W., Australia.
*A. Ammar, Centre Nat. d'Etudes Spatiales, Paris, France.
G. Arrhenius, University of California at San Diego, La Jolla, California.
W. A. Baum, Lowell Observatory, Flagstaff, Arizona.
M. J. S. Belton, Kitt Peak National Observatory, Tucson, Arizona.
*J.-L. Bertaux, Service d'Aéronomie, Verrières-le-Buisson, France.
*J.-P. Bibring, Lab. Rene Bernas, Orsay, France.
*J. Blamont, Service d'Aéronomie, Verrières-le-Buisson, France.
*M. Bobrov, Academy of Sciences, Moscow, USSR.
P. Bodenheimer, Lick Observatory, University of California, Santa Cruz, California.
*A. Boishot, Observatoire de Paris, Meudon, France.
*N. Borderies, G.R.G.S., Centre Nat. d'Etudes Spatiales, Toulouse, France.
R. A. Brown, Space Telescope Science Institute, Baltimore, Maryland.
W. E. Brunk, NASA Headquarters, Washington, D.C.
J. Buitrago, Instituto de Astrofisica de Canarios, Tenerife.
E. Bukoski, Tucson Typographic Service, Tucson, Arizona.
*J. A. Burns, Cornell University, Ithaca, New York.
P. Cassen, NASA Ames Research Center, Moffett Field, California.
A. Cazenave, G.R.G.S., Centre Nat. d'Etudes Spatiales, Toulouse, France.
C. R. Chapman, Planetary Science Institute, Tucson, Arizona.
S. A. Collins, Jet Propulsion Laboratory, Pasadena, California.
G. Consolmagno, Massachusetts Institute of Technology, Cambridge, Massachusetts.
C. Cook, University of Arizona Press, Tucson, Arizona.
S. Cox, University of Arizona Press, Tucson, Arizona.
J. N. Cuzzi, NASA Ames Research Center, Moffett Field, California.
G. E. Danielson, California Institute of Technology, Pasadena, California.
D. R. Davis, Planetary Science Institute, Tucson, Arizona.
K. Denomy, Lunar and Planetary Laboratory, University of Arizona, Tucson, Arizona.
I. de Pater, Lunar and Planetary Laboratory, University of Arizona, Tucson, Arizona.
S. F. Dermott, Cornell University, Ithaca, New York.
J. Diner, Jet Propulsion Laboratory, Pasadena, California.
R. H. Durisen, Indiana University, Bloomington, Indiana.
*J. L. Elliot, Massachusetts Institute of Technology, Cambridge, Massachusetts.
R. E. Eplee, Lunar and Planetary Laboratory, University of Arizona, Tucson, Arizona.
L. W. Esposito, University of Colorado, Boulder, Colorado.
N. Evans, NASA Headquarters, Washington, D.C.
J. Fortuno, Tucson Typographic Service, Tucson, Arizona.
F. A. Franklin, Center for Astrophysics, Cambridge, Massachusetts.
H. Fuhrman, Jet Propulsion Laboratory, Pasadena, California.
G. W. Garneau, Jet Propulsion Laboratory, Pasadena, California.
*D. Gautier, Observatoire de Paris, Meudon, France.
M. Gehrels, Lunar and Planetary Laboratory, University of Arizona, Tucson, Arizona.
*T. Gehrels, Lunar and Planetary Laboratory, University of Arizona, Tucson, Arizona.
L. Geisinger, Tucson Typographic Service, Tucson, Arizona.
O. Gingerich, Center for Astrophysics, Cambridge, Massachusetts.
*P. Goldreich, California Institute of Technology, Pasadena, California.
D. Gordon, Tucson Typographic Service, Tucson, Arizona.
*E. Grün, Max Planck Inst., Heidelberg, West Germany.
M. Guerin, Centre Nat. d'Etudes Spatiales, Toulouse, France.
K. K. Hankey, Planetary Science Institute, Tucson, Arizona.
A. W. Harris, Jet Propulsion Laboratory, Pasadena, California.

W. K. Hartmann, Planetary Science Institute, Tucson, Arizona.
M. Hénon, Observatoire de Nice, France.
J. B. Holberg, Lunar and Planetary Laboratory, University of Arizona, Tucson, Arizona.
W. B. Hubbard, Lunar and Planetary Laboratory, University of Arizona, Tucson, Arizona.
D. M. Hunten, Lunar and Planetary Laboratory, University of Arizona, Tucson, Arizona.
W.-H. Ip, Max Planck Inst., Lindau, West Germany.
W. M. Irvine, University of Massachusetts, Amherst, Massachusetts.
T. V. Johnson, Jet Propulsion Laboratory, Pasadena, California.
J. R. Jokipii, Lunar and Planetary Laboratory, University of Arizona, Tucson, Arizona.
*Y. Kozai, Tokyo Observatory, Japan.
P. L. Lamy, Lab. d'Astronomie Spatiale, Marseilles, France.
A. L. Lane, Jet Propulsion Laboratory, Pasadena, California.
*Y. Langevin, Lab. Rene Bernas, Orsay, France.
*A. Lecacheux, Observatoire de Paris, Meudon, France.
*J. Lecacheux, Observatoire de Paris, Meudon, France.
C. Laverlochere, Centre Nat. d'Etudes Spatiales, Toulouse, France.
M. Lecar, Center for Astrophysics, Cambridge, Massachusetts.
D. Levy, Planetary Science Institute, Tucson, Arizona.
D. N. C. Lin, Lick Observatory, University of California, Santa Cruz, California.
J. J. Lissauer, NASA Ames Research Center, Moffett Field, California.
S. Marinus, Lunar and Planetary Laboratory, University of Arizona, Tucson, Arizona.
E. A. Marouf, Center for Radar Astronomy, Stanford University, Palo Alto, California.
M. A. Matthews, Lunar and Planetary Laboratory, University of Arizona, Tucson, Arizona.
M. S. Matthews, Lunar and Planetary Laboratory, University of Arizona, Tucson, Arizona.
P. K. McBride, Planetary Science Institute, Tucson, Arizona.
G. McLaughlin, Lunar and Planetary Laboratory, University of Arizona, Tucson, Arizona.
D. A. Mendis, University of California at San Diego, La Jolla, California.
F. Mignard, C.E.R.G.A., Grasse, France.
E. D. Miner, Jet Propulsion Laboratory, Pasadena, California.
G. Morfill, Max Planck Inst., Garching, West Germany.
D. Morrison, University of Hawaii, Honolulu, Hawaii.
T. S. Mullin, Lunar and Planetary Laboratory, University of Arizona, Tucson, Arizona.
P. D. Nicholson, Cornell University, Ithaca, New York.
T. Owen, State University of New York, Stony Brook, New York.
S. J. Ostro, Cornell University, Ithaca, New York.
S. J. Peale, University of California at Santa Barbara, California.
B. M. Pederson, Observatoire de Paris, Meudon, France.
R. Pellat, École Polytechnique, Palaiseau, France.
G. H. Pettengill, Massachusetts Institute of Technology, Cambridge, Massachusetts.
B. Phillips, Tucson Typographic Service, Tucson, Arizona.
J. B. Pollack, NASA Ames Research Center, Moffett Field, California.
C. Porco, Lunar and Planetary Laboratory, University of Arizona, Tucson, Arizona.
V. S. Safronov, Schmidt Inst. of Earth Physics, Moscow, USSR.
H. Shaffer, University of Arizona Press, Tucson, Arizona.
M. R. Showalter, Cornell University, Ithaca, N.Y.
F. Shu, University of California, Berkeley, California.
B. Sicardy, Observatoire de Paris, Meudon, France.
B. A. Smith, Lunar and Planetary Laboratory, University of Arizona, Tucson, Arizona.
G. R. Stewart, NASA Ames Research Center, Moffett Field, California.
E. Stone, California Institute of Technology, Pasadena, California.
S. P. Synnott, Jet Propulsion Laboratory, Pasadena, California.
R. J. Terrile, Jet Propulsion Laboratory, Pasadena, California.
A. Toomre, Massachusetts Institute of Technology, Cambridge, Massachusetts.
M. Townsend, University of Arizona Press, Tucson, Arizona.
S. Tremaine, Massachusetts Institute of Technology, Cambridge, Massachusetts.
G. L. Tyler, Center for Radar Astronomy, Stanford University, Palo Alto, California.
A. Van Helden, Rice University, Houston, Texas.
W. R. Ward, Jet Propulsion Laboratory, Pasadena, California.

S. J. Weidenschilling, Planetary Science Institute, Tucson, Arizona.
P. R. Weissman, Jet Propulsion Laboratory, Pasadena, California.
G. W. Wetherill, Dept. of Terrestrial Magnetism, Carnegie Inst., Washington, D.C.
E. A. Whitaker, Lunar and Planetary Laboratory, University of Arizona, Tucson, Arizona.
W. E. Wiesel, Air Force Inst. of Tech., Wright-Patterson AFB, Ohio.
L. L. Wilkening, University of Arizona, Tucson, Arizona.
C. M. Yeates, Jet Propulsion Laboratory, Pasadena, California.
C. F. Yoder, Jet Propulsion Laboratory, Pasadena, California.

In addition to the general support for preparation of this book acknowedged in the Preface, the following authors wish to acknowledge specific funds involved in supporting the preparation of their chapters:

Bodenheimer, P.; NSF Grants AST80 17054 and ASTI81 00163.
Burns, J. A.; NASA Grant NAGW-310 and Intergovernmental Personnel Exchange Act Grant from NASA Ames.
Chapman, C. R.; NASW 3516.
Collins, S. A.; NASA Contract NAS7-918.
Davis, D. R.; NASW 3516.
Dermott, S. F.; NASA Grant NAGW-392.
Diner, J.; NASA Contract NAS7-918 and Alfred P. Sloan Foundation Grant B1982-16.
Durisen, R. H.; NASA-Ames-Indiana Univ. Consortium Agreement NCA2-OR335-101.
Elliot, J.; NASA Grants NSG 2342 and NSG 7526 and NSF Grant AST 8209825.
Esposito, L. W.; NASA Grant NAGW-389.
Garneau, G.; NASA Contract NAS7-918.
Goldreich, P.; NASA Grant NGL-05-002-003 and NSF Grant 80-20005.
Greenberg, R.; NASW 3516.
Harris, A. W.; NASA Contract NAS7-918.
Holberg, J. B.; NASA Grant NAGW-62.
Lin, D. N. C.; NSF Grants AST80 17054 and AST81 00163.
Marouf, E. A.; NASA Grant NAGW-341 and JPL Contract 953618.
Mendis, D. A.; NASA Grants NSG 7102, NSG 7623, and NAGW 399.
Miner, E.; NASA Contract NAS7-918.
Showalter, M. R.; NASA Grant NAGW-310.
Shu, F. H.; NSF Grant PHY79-19884.
Smith, B. A.; NASA Grant NGL 03-002-002.
Stewart, G. R.; NSF Grants AST 80 17054 and AST 81 00163.
Stone, E. C.; NASA Grant NAGW-200 and Contract NAS7-918.
Tremaine, S.; NASA Grants NSG-7643, NGL 22-009-638, and NSF Grant AST 82-10463.
Tyler, G. L.; NASA Grant NAGW-341 and JPL Contract 953618.
Ward, W. R.; NASA Contract NAS7-918.
Weidenschilling, S. J.; NASA Contract NASW 3516.

Index

INDEX

[777]